Topics in Applied Physics Volume 68

Topics in Applied Physics Founded by Helmut K.V. Lotsch

Volumes 1–56 are listed on the back inside cover

Light Scattering in Solids VI

Recent Results, Including High-T_c Superconductivity

Edited by M. Cardona and G. Güntherodt

With Contributions by
M. Cardona   E. Ehrenfreund   G. Güntherodt
J. A. Kash   W. von der Osten   J. B. Page
A. K. Ramdas   S. Rodriguez   C. Thomsen
J. C. Tsang   Y. Yacoby   E. Zirngiebl

With 267 Figures

Springer-Verlag
Berlin Heidelberg GmbH

Prof. Dr. *Manuel Cardona*

Max-Planck-Institut für Festkörperforschung
Heisenbertstraße 1
D-7000 Stuttgart 80

Prof. Dr. *Gernot Güntherodt*

RWTH Aachen – 2. Physik. Institut
Physikzentrum Melaten
Templergraben 55
D-5100 Aachen

ISBN 978-3-662-31094-6   ISBN 978-3-540-46892-9 (eBook)
DOI 10.1007/978-3-540-46892-9

Typesetting: Druckhaus „Thomas Müntzer", Bad Langensalza/Thüringen
54/3140/543210 — Printed on acid-free paper

Preface

This volume is the sixth of a series (Topics in Applied Physics, Vols. 8, 50, 51, 54, 66, 68) devoted to the scattering of light by solids, both as a phenomenon and as a spectroscopic technique. It includes a list of contents for the whole series.

Since the last volume appeared, light scattering has continued to demonstrate its power as a spectroscopic technique for investigating low energy elementary excitations in solids. Because of their technological and also scientific interest semiconductor microstructures still continue to occupy a prominent place in the field, shared more recently by the newly discovered high-T_c superconductors. Beside the use as a spectroscopic technique, considerable progress has been made in the fundamental understanding of the phenomena involved, including line shapes, absolute scattering efficiencies (i.e., cross sections) and resonance profiles.

Translational symmetry is a dominant feature in the theory of light scattering in solids. It breaks down, however, in the case of localized centers whose scattering shows features similar to those found in light scattering by molecules. We have included in this volume a chapter (Chap. 2) concerned with time correlator techniques in the theory of scattering by localized excitations. The other chapters are devoted mainly to experimental questions. Chapter 3 treats light scattering by phonons in conjugate organic polymers, while Chap. 4 discusses scattering by phonons, electrons and magnetic excitations in semimagnetic semiconductors. Chapter 5 is devoted to intermetallic compounds of rare earth and actinide elements comprising intermediate valence and heavy fermion materials. Chapter 6 deals exclusively with high-T_c superconductors, a field which since 1987 has been at the center of attention in solid state physics. Chapter 7 discusses scattering phenomena in the classical photosensitive materials AgCl and AgBr. Finally, Chap. 8 describes time-resolved techniques and phenomena, which have become observable thanks to improvements in pulsed lasers and detection systems.

The editing of a book with eight different chapters involving authors in several countries and three continents is not an easy task. It requires coordination and synchronization of submission deadlines, probably the most taxing of the editorial chores. The editors would like to thank all authors for their patience and cooperation. Inefficient public mail systems, which were a nightmare at the time earlier volumes were edited, have been replaced by express mail organizations and the nowadays nearly omnipresent bitnet and telefax which we have grown accustomed to take for granted.

Sample preparation has again played an essential role in most of the work described in the experimental chapters. Light scattering is not a very demanding technique in this respect, and it is possible to use relatively small samples. Nevertheless good quality crystals are required if one is to exploit the full potential of the technique in determining the symmetry of excitations. We would like to acknowledge, in particular, sample preparation efforts by many groups in the field of high T_c superconductors. Thanks are also due to S. Birtel, I. Dahl and A. Schüren for secretarial help and masterly use of modern word processing systems.

Contents

Contributors

Cardona, Manuel
  Max-Planck-Institut für Festkörperforschung, Heisenbergstr. 1
  D-7000 Stuttgart 80, Federal Republic of Germany
Ehrenfreund, Eitan
  Physics Department, Technion-Israel Institute of Technology
  Haifa, Israel
Güntherodt, Gernot
  2. Physikalisches Institut, RWTH Aachen, Templergraben 55
  D-5100 Aachen, Federal Republic of Germany
Kash, Jeffrey Allen
  IBM, T. J. Watson Research Center, P.O. Box 218
  Yorktown Heights, NY 10598, USA
Page, John B.
  Physics Department, Arizona State University, Tempe
  AZ 85287, USA
Ramdas, Anant
  Department of Physics, Purdue University, West Lafayette
  IN 47907, USA
Rodriguez, Sergio
  Department of Physics, Purdue University, West Lafayette
  IN 47907, USA
Thomsen, C
  Max-Planck-Institut für Festkörperforschung, Heisenbergstr. 1
  D-7000 Stuttgart 80, Federal Republic of Germany
Tsang, James, C.
  IBM, T. J. Watson Research Center, P.O. Box 218
  Yorktown Heights, NY 10598, USA
von der Osten, W.
  Fachbereich Physik, Universität-GH Paderborn, Warburger Str. 100A
  D-4790 Paderborn, Federal Republic of Germany
Yacoby, Yizhak
  Racah Institute of Physics, The Hebrew University
  Jerusalem, Israel
Zirngiebl, Eberhard
  Bayer AG, ZF-TPE 6, Geb. E41
  D-5090 Leverkusen, Federal Republic of Germany

Topics in Applied Physics, Vol. 68
© Springer-Verlag Berlin Heidelberg 1991

1. Introduction

M. Cardona and G. Güntherodt

With 8 Figures

> −Primo, l'abbondanza di specchi. Se c'è uno specchio, è stadio umano, vuoi vederti. E lì non ti vedi. Ti cerchi, cerchi la tua posizione nello spazio in cui lo specchio ti dica "tu sei lì, e sei tu", e molto patisci, e t'affanni, perché gli specchi ... concavi o convessi che siano, ti deludono, ti deridono: arretrando ti trovi, poi ti sposti, e ti perdi ... E non ti senti solo incerto di te ma degli stessi oggetti collocati fra te e un altro specchio. Certo, la fisica sa dirti che cosa e perché avviene: poni uno specchio concavo che raccolga i raggi emanati dall'oggetto ... e lo specchio rinvierà i raggi incidenti in modo che tu non veda l'oggetto, ben delineato, dentro lo specchio, ma lo intuisca fantomatico, evanescente, a mezz'aria, e rovesciato, fuori dallo specchio. Naturalmente basterà che tu ti muova di poco e l'effetto svanisce.
>
> *Umberto Eco:* Il pendolo di Foucault (Bompiani, Milano 1988)

This volume is the sixth in the series *Light Scattering in Solids* (LSS) which appears in the collection *Topics in Applied Physics*. The first volume was published in 1975 and was originally intended to be the only treatise on the subject. A second, corrected and updated edition was issued in 1983 [1.1]. It also includes the tables of contents of Vols. II–IV. Volumes II [1.2] and III [1.3] also appeared in 1982, Volume IV [1.4] was published in 1984 and Volume V [1.5] in 1989.

During the past few years Raman spectroscopy has continued to play a prominent role in condensed matter physics and chemistry. This progress has been due, in part, to the application of Raman spectroscopy to new materials such as the high temperature superconductors. Also, advances in instrumentation, including multichannel detection and time-resolved pulsed laser techniques, have extended the applicability of Raman spectroscopy to new phenomena and different materials. At the same time the investigations of electronic and vibrational excitations in low-dimensional semiconductor microstructures are still being pursued extremely actively. Hence the need for the present new volume of the series.

1.1 Contents of Previous Volumes and Related Recent Developments

Volume I of this series [1.1] was devoted to the foundations of linear and non-linear Raman and Brillouin spectroscopy as applied to solids. At the time when it appeared, the capabilities of Raman spectroscopy to investigate low energy excitations in bulk solids (phonons, excitons, magnons, plasmons, electron–hole pairs, ...) had been well established. Particular attention was being paid to materials opaque to the scattering laser radiation (metals and semiconductors) but the data base available was rather small and the theoretical understanding of the data was limited. Increasing interest was being directed to measurements of the scattering efficiency as a function of laser frequency (resonant scattering), which had become possible through developments in the field of tunable lasers. Most resonant profiles, however, were still being presented in "arbitrary units", a fact which led to loss of information on absolute electron–phonon coupling constants. Today these profiles are presented by the leading groups in absolute scattering efficiency units at least for bulk materials (this is unfortunately still not the case for semiconductor superlattices and other microstructures).

Chapter 1 of [1.1] by M. Cardona contains a historical background, while Chap. 2, by A. Pinczuk and E. Burstein, presents an introduction to the theory of Raman scattering by phonons and electronic excitations. This work is based on the treatment of the linear electron–phonon interaction by perturbation theory as appropriate to extended excitations in solids. Molecular spectroscopists, however, are used to another formalism involving a non-perturbative treatment (Frank–Condon) of this interaction, which requires the use of different vibrational Hamiltonians for ground and excited electronic states. Such treatments may also be required for processes involving localized states in solids. A detailed discussion of the non-perturbative methods and the conditions under which they lead to expressions similar to those obtained by perturbation theory is given in Chap. 2 of the present volume.

Chapter 3 of [1.1] by R. M. Martin and L. M. Falicov dwells on formal aspects of resonant Raman scattering by phonons. First order processes are treated in third order perturbation theory (2 photons + 1 phonon) involving electronic excitations, with either uncorrelated or correlated electron–hole pairs, i.e. excitons. Experimental data available at the time, in arbitrary efficiency units, were not good enough for quantitative comparison with the theory. Such comparison has been made recently for many semiconductors. It has been shown that at the lowest direct edge (E_0) both discrete and continuum exciton intermediate eigenstates are required for quantitative agreement between experiment and theory (Fig. 1.1) for the E_0 and $E_0 + \Delta_0$ resonances of GaP [1.6]. Double resonances, in which both electronic intermediate states are exactly at

Fig. 1.1. Resonant Raman profiles corresponding to scattering by LO phonons ($\hbar\omega_0$) at the lowest direct edges (E_0 and $E_0 + \Delta_0$) of GaP. R_{hh} and R_{so} represent exciton Rydbergs at the E_0 and $E_0 + \Delta_0$ edges. Note that incoming (at $E_0 - R_{hh}$) and outgoing (at $E_0 - R_{hh} + \hbar\omega_0$) resonances are observed. From [1.6]

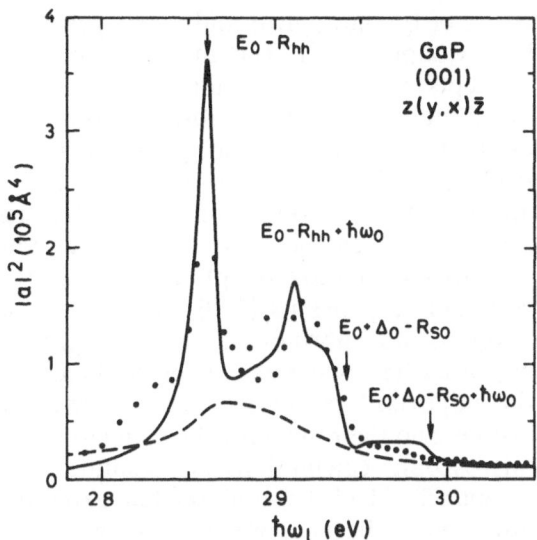

resonance are discussed in Chap. 3 of [1.1]. Together with triple resonances in the scattering by two phonons, they have recently been the object of considerable experimental and theoretical attention [1.7, 8].

Chapter 4 of [1.1] by M. V. Klein reviews scattering by free electrons and electrons bound to impurities in semiconductors. This work was complemented in Chap. 2 of [1.4], where similar scattering in semiconductor microstructures was also treated. These articles are of interest for the understanding of scattering by electronic excitations in high T_c superconductors (Chap. 6 of the present volume).

Chapter 5 of [1.1] by M. H. Brodsky is devoted to scattering by phonons in amorphous semiconductors. Further information on the topic was presented in Chap. 2 of [1.2]. Raman scattering is still being widely used for the investigation and characterization of glasses and amorphous semiconductors.

Chapter 6 of [1.1] by A. S. Pine is devoted to the classical type of Brillouin scattering in semiconductors. At the time it was written, the beautiful and interesting phenomenon of polariton-mediated resonant Brillouin scattering had been predicted but not yet been found experimentally. Chapter 7 of [1.3] is devoted to this polariton-mediated scattering. A discussion of the current methods of multipass and tandem Fabry-Perot Brillouin spectroscopy is given in Chap. 6 of [1.3].

Chapter 7 of [1.1] by Y. R. Shen covers stimulated Raman scattering and some related coherent non-linear phenomena. Further developments in the field of non-linear Raman spectroscopies, in particular of the hyper-Raman variety, are discussed in Chap. 4 of [1.2].

LSS II [1.2] is devoted to basic theoretical concepts and also to multichannel detection and instrumentation for non-linear spectroscopies.

A list of typographical and other corrections to this volume appears in [1.4]. Chapter 2 of [1.1] contains implicitly or explicitly many of the basic concepts of light-scattering spectroscopy used throughout the series, including a brief discussion of non-perturbative Frank–Condon-type calculations of the scattering efficiencies which is taken up in full detail in Chap. 2 of the present volume. It covers quasi-elastic Rayleigh, Brillouin, and Raman scattering. Concerning the first of these, we should mention that it has been the subject of considerable recent interest since it may lead to information on photon localization in disordered systems (e.g. colloidal suspensions [1.9]) Concerning resonant Raman scattering by phonons, an enormous amount of progress has been made since [1.2] appeared. This will be the subject of a chapter in a future volume. Among the important points we mention measurements of absolute resonant profiles and their theoretical treatment including excitonic effects (Fig. 1.1), microscopic calculations of second order Raman scattering and its resonances [1.11, 12] and related experimental results. Investigation of interference effects between Fröhlich and deformation potential scattering mechanism ([1.13–15] and Fig. 1.2), observations of separate incoming and outgoing resonances at low temperatures (Figs. 1.1, 2) and measurements of Faust–Henry coefficients, [1.16]. Resonant Brillouin scattering has also been investigated, both theoretically and experimentally [1.17] and also quasi-elastic (Rayleigh-like) scattering in diamond [1.18]. Attention has also been paid to scattering by polytypes (e.g. of SiC) and the relative intensities of their various optical modes as predicted by a bond polarizability model [1.19]. Moreover, vibrational modes localized at donors and/or acceptors have proven to be very useful for structural investigation and characterization of semiconductors [1.20].

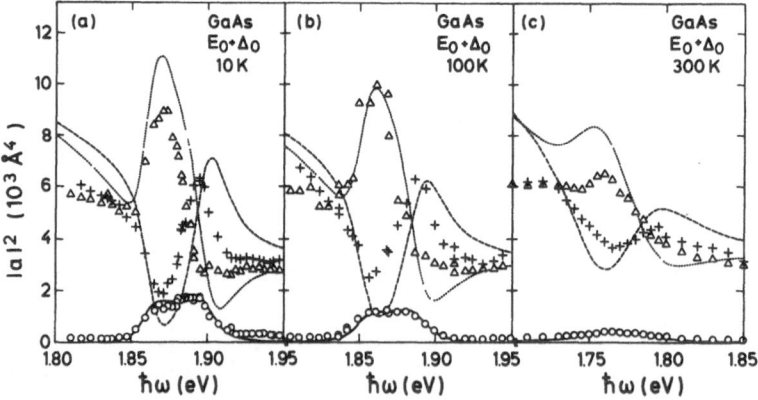

Fig. 1.2. Resonant Raman profiles of scattering by LO phonons in GaAs displaying interferences between Fröhlich and deformation potential mechanism at three different temperatures [1.13, 14]. The dots correspond to polarizations parallel to a crystal axis while the triangles and crosses represent polarizations parallel to [1$\bar{1}$0] and [110], respectively. The curves are theoretical predictions.

Polarion-mediated processes, vital in Brillouin scattering near sharp excitons, have also been shown to be important in Raman scattering [1.21–23]. Light scattering has been playing a growing role in the characterization of semiconductors, for both scientific and technological purposes. We mention as examples the investigation of small crystalline particles [1.25], of coupled plasmon–phonon modes [1.15, 25], effects of ion implantation [1.26] and isotopic fluctuations [1.27] on phonons, and spectra of local modes of donors and acceptors [1.28, 31]. Finally, among recent developments we mention resonant Raman scattering by phonons in a magnetic field (Fig. 1.3). In this technique the spectrometer is set with a fixed laser frequency at the frequency of a phonon or mutliphonon peak and the magnetic field is swept. Resonances related to interband transitions between Landau levels are observed [1.29]. They yield information on the Landau level structure and, correspondingly, on the shape of energy bands in zero field [1.30]. The reader is also referred to the reviews in [1.31, 32].

Chapter 3 of [1.2] by R. K. Chang and M. B. Long was an early report on multichannel detection as applied to light scattering spectroscopy. The technique has been gaining in relevance ever since, especially with the commercial availability of complete multichannel systems. More details can be found in Chap. 6 of [1.4].

Chapter 4 of [1.2] is devoted to non-linear spectroscopies, especially hyper-Raman scattering (two incident photons produce a photon plus an elementary excitation of the solid, usually a phonon). References to more recent work can be found on page 11 of [1.4] (Refs. [1.16, 17]). Very recent publications deal with resonant hyper-Raman [1.33] and hyper-Rayleigh scattering [1.34]. New phonon–polariton coupled modes, which

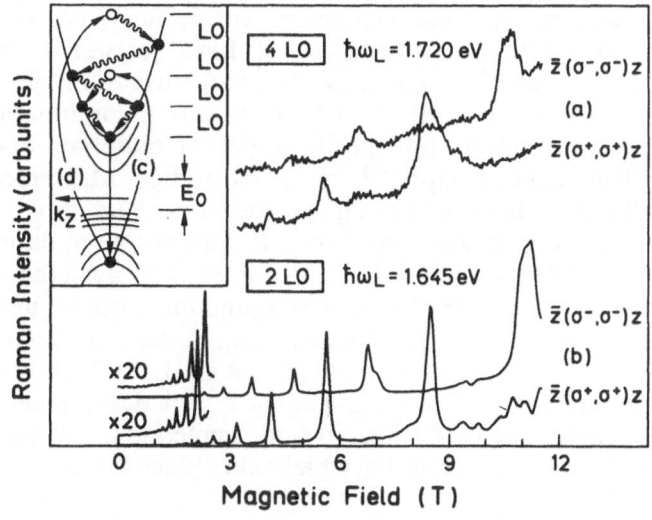

Fig. 1.3. Magneto-Raman profiles of (a) $4\hbar\omega_{LO}$ and (b) $2\hbar\omega_{LO}$ in $\bar{z}(\sigma^+, \sigma^+)z$ configuration. The difference in the exciting laser energies is close to $2\hbar\omega_{LO}$. Inset: (c) Dominant two-phonon and (d) four-phonon Raman processes with electron intermediate states. Straight lines indicate interband optical transitions; wavy lines correspond to LO phonons. From [1.30]

should be detectable by high excitation Raman spectroscopy, have been discussed recently [1.35].

Chapter 2 of [1.3] by M. S. Dresselhaus and G. Dresselhaus, a husband-and-wife team, discusses Raman scattering in graphite intercalation compounds. More recent information on these interesting quasi-two-dimensional materials is contained in [1.36].

Chapter 3 of [1.3] by D. J. Lockwood discusses electronic and magnetic excitations in the metal halides, a large family of materials with rather diverse structures (perovskite, rutile, $CdCl_2$). The magnetic excitations considered here are in part similar to those treated in Chaps. 4 and 5 of [1.4] and Chap. 5 of this volume. A book devoted exclusively to light scattering by magnetic solids has also been published [1.37].

Chapter 4 of [1.3] by W. Hayes, deals with Raman Scattering in superionic conductors. Raman spectroscopy provides information about the dynamics of the mobile ions in these technologically important materials (solid electrolytes, solid state batteries).

Chapter 5 of [1.3] by M. V. Klein discusses phonons and anomalies related to their interaction with electronic excitations in transition metal compounds. This field, pioneered by Prof. Klein, regained interest with the advent of high T_c superconductors and the article will be very useful to the reader as background to Chap. 6 of the present volume, especially in relation to phonon anomalies at T_c, Fano lineshapes and electronic backgrounds.

In Chap. 6 of [1.3] J. R. Sandercock describes the multiple pass and tandem Fabry–Perot interferometer techniques which he developed. Most of the existing Brillouin scattering data for opaque materials (e.g. metals) have been obtained with these spectrometers which are particularly useful in the study of spin-wave excitations (Chap. 7 of [1.5]).

Finally in [1.3] an article by C. Weisbuch and R. G. Ulbrich discusses polariton-mediated Brillouin scattering, first discovered by these authors in 1977 for GaAs. Although interest has now subsided after a period of great activity, these phenomena are among the aesthetically most pleasing ones observed in the field of light scattering. They were predicted theoretically by Brenig, Zeyher, and Birman in 1972.

Volume IV of LSS [1.4] represents a continuation of [1.3] in that it contains results for specific types of materials and specific classes of phenomena. Chapter 2 by G. Abstreiter, M. Cardona, and A. Pinczuk, discusses light scattering by electronic and mixed electronic–vibronic excitations in semiconductors. It contains an updating of Chap. 4 of [1.1] plus a discussion of the scarce data then available for two-dimensional structures (superlattices, heterojunctions, quantum wells, Schottky barriers). The power of light scattering in the investigation of such structures had been recognized by Burstein et al. [1.38]. This work was brought up to date in Chap. 4 of [1.5]. Among recent developments in light scattering by electronic excitations in bulk semiconductors we mention unscreened low frequency scattering which takes place in many-valley electron systems [1.39].

Chapter 3 of [1.4] by S. Geschwind and R. Romestein is devoted to spin-flip Raman scattering and its resonant aspects, as exemplified mainly by their work on CdS.

Chapter 4 of [1.4] by G. Güntherodt and R. Zeyher is devoted to the theory of effects of ordered and disordered magnetic moments on Raman scattering by phonons, and has found its application mainly in measurements for rare earth chalcogenides, cadmium chromium spinels and vanadium dihalides.

In Chap. 5 of [1.4], by G. Güntherodt and R. Merlin, the rich variety of materials aspects and physics phenomena exhibited by the rare earth chalcogenides, ranging from magnetic semiconductors, semiconductor–metal transitions and intermediate valence behavior to magnetic or superconducting metals, has been investigated by Raman scattering.

Chapter 6 of [1.4] by A. Otto treats the subject of surface-enhanced Raman scattering (SERS). Such enhancements, of up to 6 orders of magnitude, are observed for the vibronic scattering by molecules adsorbed on roughened metallic surfaces. The subject is still active, e.g. [1.40, 41]. We should mention that the instrumental sensitivity is nowadays sufficient to observe "surface-unenhanced" Raman scattering produced by physisorbed monolayers [1.42]. The theory of the SERS phenomenon is discussed in Chap. 7 of [1.4] by R. Zeyher and K. Arya. It is generally accepted that the enhancement is due in part to electromagnetic resonances and in part to quantum-mechanical effects related to the details of the chemisorption process.

The final chapter of [1.4], by B. A. Weinstein and R. Zallen, discusses effects of hydrostatic pressure on the phonon spectra of semiconductors. Most of the data have been gathered with the diamond anvil cell which enables one to perform optical measurements routinely at pressures up

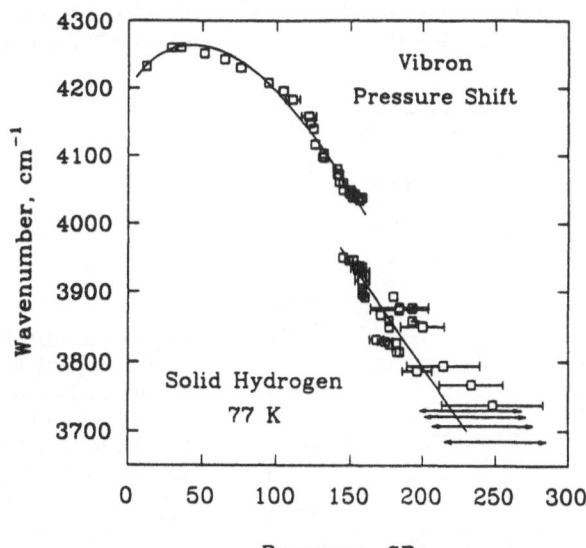

Fig. 1.4. Pressure dependence of the vibron frequency of solid hydrogen displaying at about 160 GPa the phase transition which has been attributed to the formation of metallic hydrogen. From [1.43]

to 500 kbar (50 GPa) with relatively modest equipment. The present state of the art has extended the pressure range up to ~250 GPa [1.43] and helped to determine a phase transition in solid hydrogen (Fig. 1.4) possibly associated with the appearance of a metallic phase [1.43, 44].

The final chapter of [1.4] also contains a list of references to the dependence of phonon frequencies and IR effective changes on uniaxial stress. This work has received considerable attention recently in connection with the appearance of strain in lattice-mismatched semiconductor microstructures (see also Chap. 3 of [1.5]). We mention here recent work on uniaxial stress applied to bulk GaAs [1.45], AlSb [1.46], InP [1.47], and Si [1.48].

Volume V of LSS [1.5] is devoted to low-dimensional metallic and semiconducting structures (superlattices, multiple quantum wells, etc.). For a review of current work, the reader should consult the proceedings of two recent NATO conferences [1.49, 50] and of a conference held in Trieste [1.51]. Chapter 2 by D. L. Mills presents the formal theory of elementary excitations (applicable to phonons, plasmons, magnons) in periodic layer structures on the basis of the appropriate response functions of the consituent bulk materials.

Chapter 3 of [1.5] is concerned with lattice vibrations in superlattices. The concepts of folded acoustic, confined optical and interface modes are discussed in detail. Interface modes, in particular, had been the subject of considerable controversy until about the time that chapter was written. In the meantime it has become clear that the interface modes are intimately related to the IR-active confined optical modes, i.e., for [001] superlattices those with odd index $m(m - 1$ is the number of nodes of the confined vibrational pattern), mainly that with $m = 1$. When the k vector is tilted from the superlattice axis, the confined modes evolve into interface modes even for $k \rightarrow 0$: because of electrostatic interactions these modes disperse even for $k \rightarrow 0$ (i.e., they are singular) when the angle θ between k and the

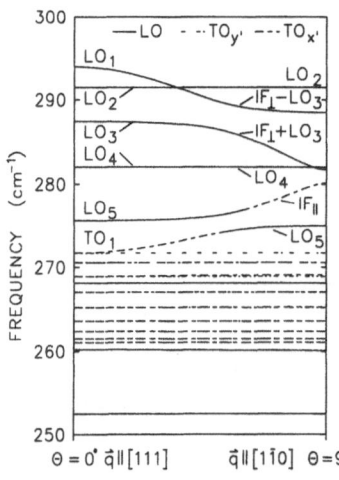

Fig. 1.5. Calculated angular dispersion of the optical modes of a [111] $(GaAs)_9(AlAs)_{10}$ superlattice for an infinitesimally small q with q in the $(11\bar{2})$ plane. The solid lines refer to the LO modes polarized along z' ([111]) while the dashed and the dashed-dotted lines correspond to $TO_{x'}$ and $TO_{y'}$ polarizations. Identical dispersion curves (but different polarizations) are obtained for q in the $(1\bar{1}0)$ plane. From [1.52]

Fig. 1.6. Dispersion relations of interface modes measured for a GaAs/AlAs superlattice at 300 K with the parameters given in the figure. The abscissa represents the in-plane wavevector $k_\parallel$. The solid curve was calculated with [Ref. 1.5, eq. (3.41)] for the experimental conditions. The dashed curve gives the calculated dispersion relations for $k_z = 0$ and $\pi/(d_1 + d_2)$. From [1.58]

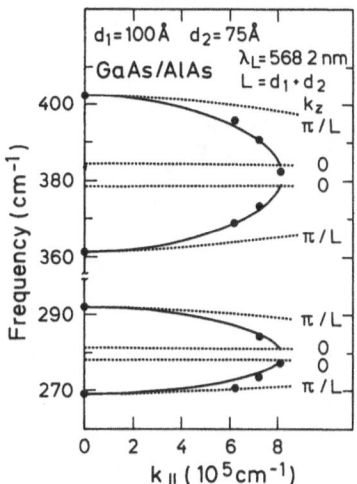

axis is changed from 0 to $\pi/2$, Fig. 1.5 [1.52]. (For a discussion see M. Cardona in [1.49–51, 53] and also [1.54, 57]).

A direct measurement of the dispersion relations of interface modes has recently been performed (Fig. 1.6) [1.58]. The field of vibrations in superlattices, as investigated by Raman scattering, has continued to be very fruitful, especially since it can be used advantageously for sample characterization: both the amount of strain [1.59] and the quality of the interface [1.60] are easily revealed. Superlattices with orientations other than [100] have also been investigated [1.52, 59]. Accurate information on the dispersion relations of the bulk components can be obtained from the Raman spectra of confined optical modes [1.61] as shown in Fig. 1.7. Recently, the existence of well-defined dispersion relations $\omega(k)$ has also been demonstrated in semiconductor alloys $(Al_xGa_{1-x}As)$ [1.62].

Chapter 4 of [1.5] by A. Pinczuk and G. Abstreiter describes light scattering by electronic excitations in superlattices. This work covers a wide variety of phenomena, including intra- and intersubband excitations of electrons with and without applied magnetic fields, excitations of holes in the valence bands, and effects of shallow impurities. It was generally believed at that time that for polarized scattering ($\hat{e}_L \parallel \hat{e}_S$) collective excitations are observed while the depolarized configuration ($\hat{e}_L \perp \hat{e}_S$) enables us to investigate single particle (spin flip) excitations. Pinczuk et al. have recently shown [1.63] that even in the latter case many-body exchange effects (equivalent to electron–hole exciton interaction) play a significant role (Fig. 1.8). Relaxation of electronic excitations in super-lattices, as observed in time-resolved Raman spectroscopy, has recently received considerable attention [1.64]. It is reviewed in Chap. 8 of the present volume.

In Chap. 5 of [1.5], R. Merlin discusses vibronic and electronic states in quasi-periodic (Fibonacci), deterministic aperiodic (Thue-Morse), and

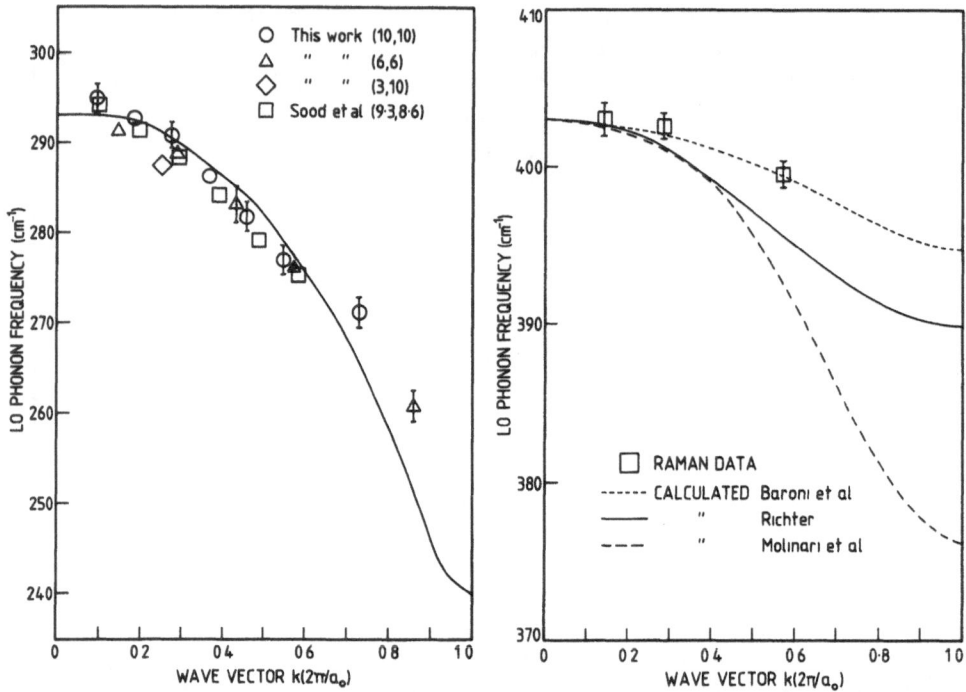

Fig. 1.7. Dispersion relations of GaAs (right) and AlAs (left) for **k** along [Ī00] determined from the confined phonons of GaAs/AlAs superlattices. The GaAs curve represents a fit to neutron spectroscopy data. The three curves for AlAs are the results of three different lattice dynamical calculations, no neutron data are available for AlAs. From [1.61]

random superlattices. Such vibronic excitations have been observed in Raman spectroscopy. Existing data are compared with theoretical predictions.

Chapter 6 of [1.5] by J. C. Tsang represents an updated version of Chap. 3 of [1.2] with regards to multichannel detection. Diode arrays, mepsicrons, and charge coupled devices (CCDs) are reviewed and compared. Examples of applications to the spectroscopy of weakly scattering surface layers and interfaces are given.

Chapter 7 of [1.5] by M. H. Grimsditch is devoted to the investigtation of the elastic properties of metallic superlattices by Brillouin spectroscopy. It has been known for some time that the elastic stiffness constants, C_{ij}, and bulk moduli obtained in this manner are often considerably softer (e.g. smaller) than those calculated from the C_{ij}'s of the bulk constituents using straightforward averaging techniques. The reasons for these anomalies are still unknown. The reader will find a more recent review by M. H. Grimsditch and I. K. Schuler in [1.53]. Finally, Chap. 8 of [1.5] is devoted, after a theoretical introduction, mainly to experimental aspects of light scattering by spin waves in magnetic thin films and layered structures.

Fig. 1.8. Light scattering spectra of intersubband excitations in a 2D electron gas in a GaAs quantum well. The main peak in the polarized spectrum is raised in energy with respect to the single-particle gap due to polarization corrections. The depolarized peak corresponds to spin density excitations and is now known to be downshifted by excitonic effects. The broad SPE structure has been assigned to the bare single-particle excitations. From [1.63]

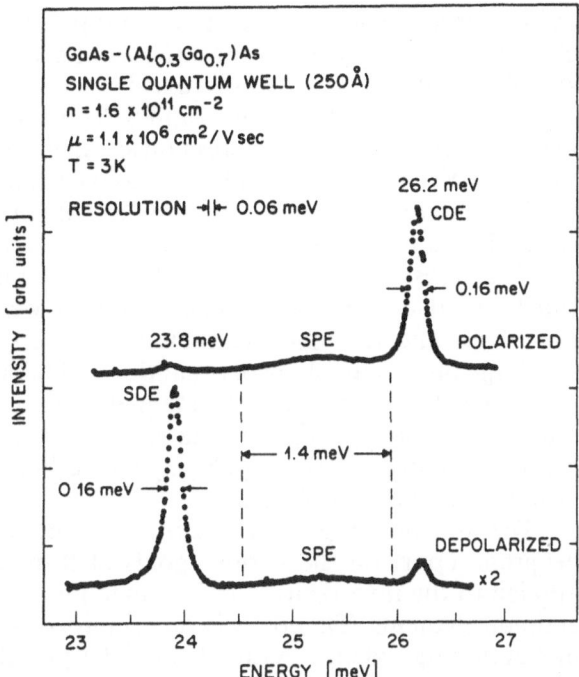

1.2 Contents of this Volume

After this introductory chapter, which follows the usual format of previous volumes, we have chosen to place a contribution by J. B. Page (Chap. 2) devoted to the formal theory of Raman scattering, which is not usually known to solid state practitioners of the technique. This theory, based on finite-temperature many-body techniques recast in the *time-correlator* framework, finds its main application in the interpretation of Raman spectra of large molecules in liquid solution and of complex impurities in solids. Instead of treating the electron–phonon interaction as a weak perturbation on electronic states by perturbation theory, exact adiabatic ground and excited states are used within the Franck–Condon scheme. The perturbation treatment is usually adequate for extended phonons in large periodic solids. For localized states, however, the vibrational Hamiltonian of the ground and the excited states may be rather different. The time-correlator technique offers considerable computational advantages over calculations of multiphonon scattering in the frequency domain. Model calculations are presented and applied to the interpretation of complex Raman spectra of molecules such as azulene and β-carotene.

In Chap. 3, Y. Yacoby and E. Ehrenfreund discuss scattering in conjugated polymers, molecules somewhat similar to those of Chap. 2, but which, because of the translational invariance of their linear chains, can be treated by standard solid state methods. Emphasis is placed on trans- and cis-polyacetylene as model substances but other conjugated polymers such as polythiophenes are also discussed.

Chapter 4 by A. K. Ramdas and S. Rodriguez discusses light scattering in semimagnetic semiconductors (also called dilute magnetic semiconductors, DMS). These materials are conventional III-V, II-VI, or IV-VI semiconductors in which the metallic constituent has been partly substituted by a magnetic transition-metal ion. They are thus to be regarded as mixed crystals with considerable magnetic interactions. Most of them can be used as superlattice and quantum well constituents. Because of these facts and the many variable parameters involved, DMSs give rise to a large phenomenology of excitations which can be observed by Raman scattering. The presence of the magnetic moments of the transition metal ions leads to strong effects in magnetic fields (e.g. Faraday rotation). The canonical family is perhaps $Cd_{1-x}Mn_xTe$ but Fe, Co and other ions can also be used instead of Mn.

The aim of Chap. 5 by E. Zirngiebl and G. Güntherodt is to give a progress report on the various goals of Raman and Brillouin scattering applied to the investigation of intermetallic compounds of rare-earth and actinide elements. The focus is on Kondo-type compounds, intermediate or fluctuating valence materials and the popular class of heavy fermion compounds. Raman scattering has proven to be a most valuable tool, complementary to neutron scattering, in observing crystalline electric-field excitations in metals in which such localized excitations interact strongly with the conduction electrons. From a systematic study of phonon Raman scattering, it was possible to determine the charge relaxation rate of fluctuating valence compounds, a characteristic that is not directly accessible by quasi-elastic light scattering. On the other hand, the spin fluctuation rate of heavy fermion compounds has been observed by quasi-elastic Raman scattering, contrary to predictions of Fermi liquid theory. The comparable sound and Fermi velocities in heavy fermion compounds were found to give rise to strong quasi-elastic Brillouin scattering from electron density fluctuations and thus to an enhanced Landau–Placzek ratio, which has no analogue in ordinary metals. This chapter is an extension of the previously reported work by Güntherodt and Merlin in Chap. 5 of Ref. [1.4], where Raman scattering in metallic and valence fluctuating rare-earth chalcogenides had been reported.

In Chap. 6, C. Thomsen discusses the very topical field of light scattering by high temperature superconductors. The chapter contains a short introduction to the state of the art in high T_c superconductivity, which should be useful to Raman scatterers not familiar with the field. The rest of the article brings up to date previous review work [1.65] that appeared in a series of two volumes on high T_c superconductivity which should be useful to readers who want to become more deeply acquainted with the field. This chapter also contains information on phonons, electronic excitations, superconducting gaps, electron–phonon coupling, and magnons as observed by Raman spectroscopy. Of particular interest is the fact that the scattering profiles for phonons in the visible and near UV can be reproduced rather well, in absolute units, by band structure calculations based on the LMTO (linear muffin tin orbital) LDA (local density approximation for exchange and correlation) method. Softening and hardening of certain phonons occurs below T_c. This provides a reliable determination of the superconducting gap on the basis of the strong coupling theory developed by Zeyher and Zwicknagl [1.66].

Chapter 7 addresses light scattering phenomena in the rock-salt-type silver halides (AgBr, AgCl). The author, W. von der Osten, has pioneered this very fruitful field for several years. These materials, of great interest in the photographic process, are unusual in that they possess an indirect lowest absorption edge with initial valence states at the L-point of the Brillouin zone and final states at Γ. The shift of the initial states from the Γ- to the L-point, which is found in the alkali halides and in zincblende-type Cu and Ag halides, is due to hybridization between halogen p- and Ag$4d$-states, forbidden at Γ because of parity but allowed away from Γ. This feature allows one to study two-phonon scattering (scattering by one phonon is dipole forbidden) resonant at the indirect gaps, i.e. involving phonons at the L-point. The chapter also discusses time-resolved (picosecond) scattering, a technique which, in combination with the cw measurements, has enabled the author's group to extract a remarkable amount of information on electron–phonon coupling constants. The article concludes with a discussion of the quantum beats recently observed in resonant Raman scattering.

Finally, Chap. 8, by J. A. Kash and J. C. Tsang, is concerned with dynamic processes in bulk semiconductors and semiconductor microstructures, involving the relaxation of electrons and phonons. The techniques used for these investigations are light scattering and secondary light emission (luminescence) of the pump-probe variety, developed in part by the authors. Considerable information about electron–phonon coupling constants (i.e. deformation potentials) is obtained and compiled. Recent work has also appeared in [1.67].

References

1.1   M. Cardona (ed.): *Light Scattering in Solids I: Introductory Concepts*, 2nd ed., Topics Appl. Phys. **8** (Springer, Berlin, Heidelberg 1982).

1.2   M. Cardona, G. Güntherodt (eds.): *Light Scattering in Solids II: Basic Concepts and Instrumentation*, Topics Appl. Phys. **50** (Springer, Berlin, Heidelberg 1982)

1.3   M. Cardona, G. Güntherodt (eds.): *Light Scattering in Solids III: Recent Results*, Topics Appl. Phys. **51** (Springer, Berlin, Heidelberg 1982)

1.4   M. Cardona, G. Güntherodt (eds.): *Light Scattering in Solids IV: Electronic Scattering, Spin Effects, SERS, and Morphic Effects*, Topics Appl. Phys. **54** (Springer, Berlin, Heidelberg 1984)

1.5   M. Cardona, G. Güntherodt (eds.): *Light Scattering in Solids V: Superlattices and other Microstructures*, Topics Appl. Phys. **66** (Springer, Berlin, Heidelberg 1989)
1.6   A. Cantarero, C. Trallero-Giner, M. Cardona: Phys. Rev. B **39**, 8388 (1989)
1.7   C. Trallero-Giner, A. Alexandrou, M. Cardona: Phys. Rev. B **38**, 10744 (1988)
1.8   A. Alexandrou: In Proc. 20th Int. Conf. on the Physics of Semiconductors, Tessaloniki (1990) in press
1.9   S. Fraden, G. Maret: Phys. Rev. Lett. **65**, 512 (1990)
1.10   M. Cardona, P.B. Allen, Helv. Phys. Acta **58**, 307 (1985)
1.11   C. Grein: To be published
1.12   R. K. Soni, R. Gupta, K. P. Jain: Phys. Rev. B **33**, 5560 (1986)
1.13   J. Menéndez, M. Cardona: Phys. Rev. Lett. **51**, 1297 (1983)
1.14   A. Cantarero, C. Trallero-Giner, M. Cardona: Phys. Rev. B **40**, 12290 (1989)
1.15   V. Vorlíček, I. Gregora, W. Kauschke, J. Menéndez, M. Cardona: Phys. Rev. B **42**, 5802 (1990)
1.16   S. Zekong, B. Prevot, C. Schwab: Phys. Status Solidi B **150**, 65 (1988)
1.17   B.H. Bairamov, A.V. Gol'tsev, V.V. Toporov, R. Laiho, T. Levola: Phys. Rev. B **33**, 5875 (1986)
1.18   H.E. Jackson, R.T. Harley, S.M. Lindsay, M.W. Anderson: Phys. Rev. Lett. **54**, 459 (1985)
1.19   S. Nakashima, K. Tahara: Phys. Rev. B **40**, 6345 (1989)
1.20   J. Wagner, M. Ramsteiner: IEEE J. Quantum Electronics **25**, 993 (1988); J. Wagner: Proc. of the 4th Int. Conf. on Shallow Impurities in Semiconductors (London, 1990); R. Addinal et al.: Semicond. Sci. Technol. (in press); J. Wagner, M. Ramsteiner, W. Stolz, M. Hauser, K. Ploog: Appl. Phys. Lett. **55**, 978 (1989)
1.21   F. Meseguer, J.C. Merle, M. Cardona: Solid State Commun. **50**, 709 (1984)
1.22   Z. G. Koinov: J. Phys. Conden. Matter **1**, 9853 (1989)
1.23   A.A. Gogolin: To be published
1.24   Y. Sasaki, Y. Nishina, M. Sato, K. Okamura: Phys. Rev. B **40**, 1762 (1989); S. Hayashi, H. Sauda, M. Agata, K. Yamamoto: Phys. Rev. B **40**, 5544 (1989)
1.25   B.H. Bairamov, V.V. Toporov, N.V. Agrinskaya, E.A. Samedov, G. Irmer, J. Monecke: Phys. Status Solidi B **146**, K161 (1988)
1.26   M. Holtz, R. Zallen, O. Brafman: Phys. Rev. B **38**, 6097 (1988)
1.27   H.D. Fuchs, C.H. Grein, C. Thomsen, M. Cardona, W.L. Hansen, E.E. Haller, K. Itoh: Phys. Rev. B **43**, 4835 (1991)
1.28   P. Galtier, G. Martinez: Phys. Rev. B **38**, 10543 (1988)
1.29   G. Ambrazevičius, M. Cardona, R. Merlin: Phys. Rev. Lett. **59**, 700 (1987)
1.30   T. Ruf, M. Cardona: Phys. Rev. B **41**, 10747 (1990)
1.31   M.V. Klein: Vibrational Raman Scattering from Crystals, *Dynamical Properties of Solids*, ed. by G.K. Horton, A.A. Maradudin (North-Holland, Amsterdam 1990)
1.32   M. Cardona, Proc. of SPIE, Vol. 822, Int. Conf. on Raman and Luminescence Spectroscopy in Technology, San Diego, 1987 (SPIE, Bellingham 1987) p. 2
1.33   K. Inoue, K. Watanabe: Phys. Rev. Lett. **39**, 1977 (1989)
1.34   H. Vogt: Phys. Rev. B **41**, 1184 (1990)
1.35   B.S. Wang, J.L. Birman: Solid State Commun. **75**, 867 (1990); A.L. Ivanov, L.K. Keldysh: Sov. Phys. JETP **57**, 234 (1983)
1.36   H. Zabel, S.A. Solin (eds.): *Graphite Intercalation Compounds* (Springer, Berlin Heidelberg, Vol. I, 1990; Vol. II, in press)
1.37   M.G. Cottam, D.J. Lockwook: *Light Scattering by Magnetic Solids* (Wiley, New York, 1986)
1.38   E. Burstein, A. Pinczuk, S. Buchner: *Physics of Semiconductors*, 1978, ed. by B.L.H. Wilson (Institute of Physics, London 1979) p. 123
1.39   G. Contreras, A.K. Sood, M. Cardona: Phys. Rev. B **32**, 924 (1985); ibid, B **32**, 930 (1958)
1.40   S. Hayashi, R. Koh, Y. Ichiyama, K. Yamamoto: Phys. Rev. Lett. **60**, 1085 (1988)
1.41   S. Hayashi, R. Koh, K. Yamamoto, H. Ishida: Jap. J. Appl. Phys. **28**, 1440 (1989)
1.42   D. Kirk Veirs, V.K.F. Chia, G.M. Rosenblatt: Langmuir **5**, 633 (1989)

1.43   R.J. Hemley, H.K. Mao: Phys. Rev. Lett. **63**, 1393 (1989); ibid., **61**, 857 (1988)
1.44   H.K. Mao, R.J. Hemley, M. Hanfland: Phys. Rev. Lett. **65**, 484 (1990); see also A.L. Ruoff and C.A. Vanderborgh: Phys. Rev. Lett. **66**, 754 (1991)
1.45   P. Wickboldt, E. Anastassakis, R. Sauer, M. Cardona: Phys. Rev. B **35**, 1362 (1987)
1.46   E. Anastassakis, M. Cardona: Solid State Commun. **63**, 893 (1987)
1.47   E. Anastassakis, Y.S. Raptis, M. Hünermann, W. Richter, M. Cardona: Phys. Rev. B **38**, 7702 (1988)
1.48   E. Anastassakis, A. Cantarero, M. Cardona: Phys. Rev. B **41**, 7529 (1990)
1.49   A. Fasolino, P. Lugli (eds.): *Spectroscopy of Semiconductor Microstructures* (Plenum, New York 1989)
1.50   Proceedings of the NATO symposium on light scattering in superlattices, Mount Tremblant, Quebec (Plenum, New York 1991)
1.51   Superlattices and Microstructure **5** (1989)
1.52   Z.V. Popović, M. Cardona, E. Richter, D. Strauch, L. Tapfer, K. Ploog: Phys. Rev. B **41**, 5904 (1990)
1.53   M. Cardona: In Proceedings of the SLAFS, Cuzco, Peru, 1990 (Springer, Berlin, Heidelberg 1991)
1.54   J. Menéndez: J. Luminescence **44**, 285 (1989)
1.55   K. Huang, B. Zhu: Phys. Rev. B **38**, 2183 (1988)
1.56   D.L. Lin, R. Chen, T.F. Georg: Solid State Commun. **73**, 799 (1990)
1.57   F. Bechstedt, H. Gerocke: J. Phys. Conden. Matter **2**, 4363 (1990)
1.58   A. Huber, T. Egeler, W. Ettmüller, H. Rothfritz, G. Tränkle, G. Abstreiter: Proceedings of the Int. Conf. on Superl. and Microstr., Berlin 1990, to be published
1.59   E. Friess, H. Brugger, K. Eberl, G. Krötz, G. Abstreiter: Solid State Commun. **69**, 899 (1989)
1.60   S. Wilke: Solid State Commun. **73**, 399 (1990)
1.61   D. Mowbray, M. Cardona, H. Fuchs: Phys. Rev. B **43**, 1591 (1991)
1.62   B. Jusserand, D. Paquet, F. Mollot: Phys. Rev. Lett. **63**, 2397 (1989)
1.63   A. Pinczuk, S. Schmitt-Rink, G. Danan, J.P. Valladares, L.N. Pfeiffer, K.W. West: Phys. Rev. Lett. **63**, 1633 (1989)
1.64   M.C. Tatham, J.F. Ryan, C.T. Foxon: Phys. Rev. Lett. **63**, 1637 (1989)
1.65   C. Thomsen, M. Cardona: In *Physical Properties of High Temperature Superconductors I*, ed. by D.M. Ginsberg (World Scientific, Singapore 1989) p. 409
1.66   R. Zeyher and G. Zwicknagl, Z. Phys. B — Cond. Matter **78**, 175 (1990)
1.67   S.E. Ralph, G.J. Wolga: Phys. Rev. B **15**, 1353 (1990)

2. Many-Body Approach to the Theory of Resonance Raman Scattering by Vibronic Systems

J. B. Page

With 9 Figures

Resonance Raman (RR) scattering is a powerful technique for detailed and selective investigations of the coupling between electronic and vibrational (phonon) states of molecules and solids [2.1]. This is particularly so for polyatomic molecules and solid state defect systems having localized electronic states and intermediate electron-phonon (e-ph) coupling stengths. Such systems have optical absorption bands with resolved vibrational structure arising from transitions involving individual vibronic levels, and they include numerous examples of importance in solid-state physics, organic chemistry and biophysics. Owing to multimode interference effects, RR excitation profiles of these systems (the variation of a mode's RR scattering intensity as the incident light frequency is tuned through an optical absorption band) can be highly structured and vary greatly from mode to mode. Hence the study of these mode-dependent electronic spectra can yield detailed information on the coupling between specific electronic and vibrational states. The room-temperature profile and absorption data in Fig. 2.1 for the azulene molecule ($C_{10}H_8$) in CS_2 typify such spectra. Of central importance in such studies is the information which can be obtained on the vibrational potential energy surface for the resonant electronic *excited* state. Hence RR investigations form a very active area of interdisciplinary research in physical chemistry and solid-state physics. Refs. [2.2–6] and references therein survey aspects of this field up to 1982.

The systems of greatest interest range from polyatomic organic molecules to defect crystals and thus involve many normal modes. Moreover, RR experiments on molecular systems are often carried out at room temperature. These two features render impractical the traditional theoretical approach of direct summation over the individual contributions of the many intermediate vibronic levels. This "sum-over-states" method stems largely from the pioneering work of *Albrecht* [2.7] and is adequate for small molecules, but it rapidly becomes intractable as the number of normal modes increases. This is especially so when temperature effects are included by means of explicit Boltzmann averages over the initial vibrational states, as is discussed here in Sect. 2.1.5.

The theoretical problem, then, is to deal with the large number of vibrational degrees of freedom, including the necessary thermal averages. The usual solid-state approach for RR scattering by perfect crystals is inappropriate, since it uses delocalized electronic states (Bloch states) and

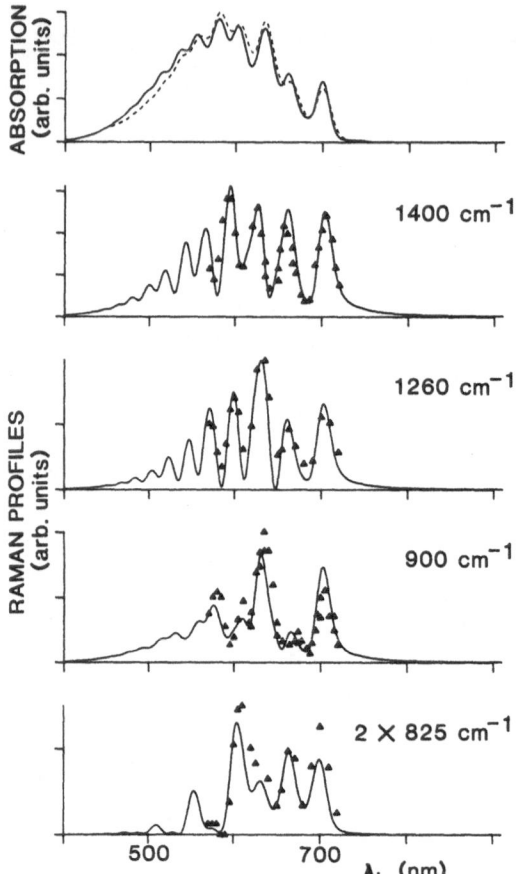

Fig. 2.1. Room temperature absorption and RR profiles for azulene ($C_{10}H_8$) in CS_2, plotted versus the wavelength of the incident light. The dashed curve and the triangles are the experimental data of [2.16]. The solid curves are the results of the explicit seven-mode model calculations summarized here in Sect. 2.4.2.b, and they include the relative intensities of the profiles. The theoretical absorption was scaled to the measured absorption via least squares, and a single scale factor was used for all of the theoretical profiles. The experimental profile relative intensities are given by the tic marks. The profiles are labeled by the observed mode frequencies in the electronic ground state — note that the lower profile is for second harmonic scattering. After [2.17]

treats the e-ph coupling perturbatively. Ref. [2.8] introduces this approach, which has had extensive application to semiconductors. However, for the class of systems of interest here, the electronic states are localized and a full Born–Oppenheimer basis is necessary. In this description, the e-ph interaction arises from the difference between the vibrational potential energy functions in the electronic excited and ground states. Not only is this interaction generally involved in producing the final state vibrational quanta (Raman phonons), it also plays a central role in determining the multimode vibrational structure in the lineshapes of the RR excitation profiles and absorption, necessitating a nonperturbative theory.

Over the past several years our group has developed a unified theoretical approach [2.9–27] to this class of problems, based on generalizations of nonzero temperature phonon many-body techniques within the *time-correlator* framework [2.28, 29]. As a result, non-perturbative and numerically tractable analytic solutions have been achieved for a number of practically important general models, and we have used them to analyze experiments on systems whose vibrationally structured RR profiles and optical absorption spectra provide stringent tests of the theory. This chapter will highlight the most important features of our approach and the principal results.

The results allow one to carry out explicit multimode calculations via *one*-dimensional fast Fourier transforms of convenient analytic functions of the time, the model parameters and the temperature, with computing times that are *orders of magnitude* shorter than those for the conventional sum-over-states method. They allow a rigorous and convenient treatment of multimode homogeneous thermal broadening and inhomogeneous (site) broadening in both the absorption and RR profiles. Moreover, they automatically lead to useful "renormalizations" (rescalings) of the coupling parameters in the various models. Finally, they give a transparent analytic framework for converting model parameters computed from ab initio calculations directly into RR profiles and the absorption for systems too complicated to treat with conventional techniques.

Complementary to the modeling aspects, the results bring out very usefully the close connection between RR profiles and the optical absorption, such that one can often compute profiles directly from the *measured* temperature-dependent absorption, plus model parameters for just a *restricted* subset of modes. Frequently, this subset is only the Raman mode itself, so that one can extract information on the model parameters directly from absorption and profile data *one mode at a time*, without the need for a complete set of profile measurements. Such "transform" calculations (so-called because they involve a Kramers-Kronig transform of the absorption) are easily carried out on typical laboratory computers, and they have become an important adjunct to complete multimode model calculations. Transform calculations of RR profiles from measured absorption spectra were introduced in [2.9], based on a relation given in [2.28] for a restricted but important set of model assumptions. The transform approach has been extensively generalized and applied, both by our group [2.9, 12, 13, 15, 16, 19, 21–25] and others [2.30–50].

The time-correlator framework stems from *Hizhnyakov* and *Tehver's* [2.28] reformulation of the resonant part of the usual frequency-domain Kramers-Heisenberg (KH) formula for light scattering [2.51] as a triple-time Fourier integral, analogous to *Lax's* earlier treatment of the optical absorption [2.52]. *Rebane* and co-workers [2.29] have extensively used this approach in general studies of the nature of resonance secondary radiation by impurities in solids. In [2.28], this framework was used to develop a pair-correlations approximation scheme which is exact for the case of linear e-ph coupling (displaced oscillators), but is inadequate for other, realistic, cases such as linear plus *quadratic* e-ph coupling (general mode mixing, frequency shifts and equilibrium position shifts with electronic excitation). We did not make the pair-correlations approximation of [2.28], but instead exploited the time-correlator framework as a natural basis for the use of suitably adapted many-body techniques, i.e. generalized diagrammatic and/or boson operator algebra methods. Exact results were thereby achieved for a wide variety of realistic models, the most comprehensive of which includes arbitrary linear and quadratic e-ph coupling with *simultaneous* linear and quadratic non-Condon coupling [2.20, 21]. Although various special cases of this model have been considered previously within a variety of restrictive approximations, our work is the first to obtain non-perturbative closed form results for the simultaneous presence of all of these inter-actions.

The power of the results obtained via the correlator/many-body approach is both a reflection of the theoretical techniques employed and the rather extensive analytic effort involved — in this approach most of the work in evaluating RR profiles or absorption spectra is analytic, yielding finite closed expressions and relatively simple numerical algorithms for complex systems. Of central importance, both for explicit multimode modeling and for absorption → RR profile transform calculations, is an especially useful separation of the RR scattering into orders, which corresponds to a particular grouping of the diagrammatic series for the scattering. This follows very naturally from a linked cluster theorem and is discussed in Sect. 2.3.4. In view of the importance of the theoretical approach, an attempt is made here to discuss enough of the theory that its main steps are clear, together with the details of its time-correlator basis, the specific model assumptions, and a survey of our results and applications. The aim is not only to demonstrate the practical aspects of the results, but also to trace their origin in the theoretical techniques.

2.1 Time-Correlator Formulation

As noted above, the time-correlator expressions for the optical absorption and RR scattering offer convenient starting points for a many-body approach. The derivations of these expressions from their more familiar sum-over-states counterparts are given in [Ref. 2.11, Appendix A] and are summarized in [2.22]. For completeness, we will now enumerate the assumptions and discuss some the main points of the derivations. It is illuminating to do this in terms of an electronic polarizability operator [2.11], which allows a unified treatment of both scattering and absorption.

2.1.1 Electronic Polarizability Operator

Within the dipole approximation, linear response theory gives the electronic polarizability for light of frequency ω_L as

$$P_{\alpha\beta}(\omega_L) = \hbar^{-1} Av_a \sum_c \frac{\langle a| M_\alpha |c\rangle \langle c| M_\beta |a\rangle}{\omega_{ca} - (\omega_L + i\gamma_c)} + (NR), \tag{2.1}$$

where M_α is the α component of the system's electronic dipole moment operator $M = \sum_e q_e r_e$; $|a\rangle$ and $|c\rangle$ are the initial and intermediate states; $\omega_{ca} = (E_c - E_a)/\hbar$ is the corresponding transition frequency; γ_c accounts for the lifetime of the intermediate state $|c\rangle$ and Av_a denotes a thermal average over the initial states. The symbol NR stands for a second, non-resonant, term which is obtained from the first term by switching α and β and changing the sign of $(\omega_L + i\gamma_c)$. Because the electronic dipole moment operator is Hermitian, NR is just the complex conjugate of the first term, with ω_L replaced by $-\omega_L$. Several assumptions will now be made which are appropriate for the wide class of systems under consideration here. A full adiabatic (Born-Oppenheimer) basis is assumed, according to which the states of the system may be written as products of electronic and vibrational eigenfunctions: $|n\rangle \rightarrow |n\rangle |nv\rangle \equiv \varphi_n(u, r) \psi_{nv}(u)$. The indices n and v denote the electronic and vibrational quantum numbers, respectively. The electronic eigenfunctions $\varphi_n(u, r)$ depend parametrically on the displacements of the atoms, represented schematically by u, and on the electronic coordinates r. The vibrational potential energy functions in the different electronic states are the corresponding adiabatic electronic energies $E_n(u)$, which include the nuclear-nuclear potential energy. The vibrational Hamiltonian for electronic state n will be written as $H_n(u, p)$, where p represents the atomic momentum operators. The vibrational states thus satisfy $H_n |nv\rangle = E_{nv} |nv\rangle$. If the system is in thermal equilibrium at a temperature $T = (k\beta)^{-1}$ such that only the vibrational levels $|gv\rangle$ for the electronic ground state are initially populated, and if a constant width γ_e is assumed to account for the lifetime of each of the vibrational levels in the excited electronic state $|e\rangle$, equation (2.1) becomes

$$P_{\alpha\beta}(\omega_L) = (\hbar z)^{-1} \sum_v \exp(-\beta E_{gv})$$

$$\times \sum_{e, v'} \frac{\langle gv| M_\alpha^{ge}(u) |ev'\rangle \langle ev'| M_\beta^{eg}(u') |gv\rangle}{\omega_{ev', gv} - \omega_L - i\gamma_e} + (NR). \tag{2.2}$$

Here v' labels the vibrational levels in the electronic excited state $|e\rangle$, $M_\alpha^{eg}(u) \equiv \langle e| M_\alpha |g\rangle$ is the $|g\rangle \rightarrow |e\rangle$ electronic transition dipole matrix element for atomic configuration u, and $z \equiv \sum_v \exp(-\beta E_{gv})$ is the vibrational partition function. Because of the reality of the electronic dipole moment operator, $[M_\alpha^{eg}(u)]^*$ is equal to $M_\alpha^{ge}(u)$.

Equation (2.2) applies to the *entire* system, i.e. the molecule plus host lattice or solvent. In practice, the eigenstates in this equation are approximate, and the widths γ_e arise from all radiative and non-radiative decay processes which contribute to the lifetime of the intermediate vibrational states $|ev'\rangle$. Our focus here is on *multimode* effects in the RR profiles and absorption of complex systems, and when the resonant electronic transition is coupled to a reservoir of low-frequency molecular, host, or solvent modes, multimode homogeneous broadening can easily *dominate* that due to the individual states' intrinsic widths, which are then unimportant. Such broadening arises from closely spaced unresolved vibronic peaks due to the reservoir modes and is temperature dependent. As detailed in [2.13, 14, 17], our analytic solutions give a very convenient treatment of this type of broadening, as well as *inhomogeneous* (site) broadening, and we will return to this point in Sect. 2.4.4.

Equation (2.2) can be rewritten identically as

$$P_{\alpha\beta}(\omega_L) = \langle \hat{P}_{\alpha\beta}(\omega_L) + \hat{P}_{\alpha\beta}^\dagger(-\omega_L)\rangle \,, \tag{2.3}$$

where the bracket $\langle ...\rangle \equiv z^{-1}\, \mathrm{Tr}\,\{\exp\,(-\beta H_g)\,...\}$ is a thermal average over the vibrational states of the electronic ground state, and the resonant electronic polarizability *operator* is defined by

$$\hat{P}_{\alpha\beta}(\omega_L) \equiv \hbar^{-1} \sum_{e,\,v',\,v''} \frac{M_\alpha^{ge}(u)\,|ev'\rangle\,\langle ev'|\,M_\beta^{eg}(u')\,|gv''\rangle\,\langle gv''|}{\omega_{ev',\,gv''} - \omega_L - i\gamma_e}. \tag{2.4}$$

The caret is used here to distinguish this operator from the electronic polarizability $P_{\alpha\beta}(\omega_L)$. As is evident from (2.3), $\hat{P}_{\alpha\beta}^\dagger(-\omega_L)$ is just the operator corresponding to the non-resonant contribution (NR) of (2.2). The vibrational eigenfunctions in (2.4) may be eliminated by using the identity $(a - ib)^{-1} = i \int_0^\infty dt \exp\,[-it(a - ib)]\ (b > 0)$, the operator identity $\exp\,(itH_n/\hbar)\,|nv\rangle = \exp\,(itE_{nv}/\hbar)\,|nv\rangle$, and the completeness relation $\sum_v |nv\rangle\,\langle nv| = 1$ satisfied by the vibrational eigenfunctions. The result is

$$\hat{P}_{\alpha\beta}(\omega_L) \equiv i\hbar^{-1} \sum_e \int_0^\infty dt \exp\,[it(\omega_L + i\gamma_e)]\,M_\alpha^{ge}(u)$$

$$\times \exp\,[-itH_e(u,\,p)/\hbar]\,M_\beta^{eg}(u)\,\exp\,[itH_g(u,\,p)/\hbar]\,. \tag{2.5}$$

Clearly, $\hat{P}_{\alpha\beta}(\omega_L)$ depends upon both the vibrational potential and kinetic energies and is an operator in the vibrational Hilbert space. The dependence of H_g and H_e on the atomic momenta p, which introduces computational complications, is only relevant close to resonance (Sect. 2.1.4).

2.1.2 Optical Absorption

Again within the dipole approximation, the standard Fermi golden rule expression for the optical absorption at frequency ω_L and light polarization α may be written

$$\alpha(\omega_L) = B\pi\hbar^{-1}\omega_L Av_a \sum_c |\langle c| M_\alpha |a\rangle|^2 \, \delta(\omega_{ca} - \omega_L), \tag{2.6}$$

where B is a constant independent of ω_L, and the remaining notation is the same as in (2.1). If lifetime broadening γ_c is included in the excited states, the delta functions are replaced by Lorentzians. Then under the same assumptions used to obtain (2.2), (2.6) becomes

$$\alpha(\omega_L) = B\omega_L \, \mathrm{Im} \left\{ (\hbar z)^{-1} \sum_v \exp(-\beta E_{gv}) \sum_{e,\, v'} \frac{|\langle ev'| M_\alpha^{eg}(u) |gv\rangle|^2}{\omega_{ev',\, gv} - \omega_L - i\gamma_e} \right\}, \tag{2.7}$$

where Im denotes the imaginary part. Comparison with (2.2–4) immediately gives

$$\alpha(\omega_L) = B\omega_L \, \mathrm{Im} \left\{ \langle \hat{P}_{\alpha\alpha}(\omega_L) \rangle \right\}. \tag{2.8}$$

If the time-domain version of $\hat{P}_{\alpha\beta}(\omega_L)$ in (2.5) is now substituted and use is made of the invariance of the trace in the thermal average to cyclic permutations of its arguments, one obtains the time-correlator formulation of the optical absorption:

$$\alpha(\omega_L) = B\hbar^{-1}\omega_L \, \mathrm{Re} \left\{ \sum_e \int_0^\infty dt \, \exp[it(\omega_L + i\gamma_e)] \, \eta_e'(t) \right\}, \tag{2.9}$$

where the absorption correlator $\eta_e'(t)$ is given by

$$\eta_e'(t) = \langle \exp(itH_g/\hbar) \, M_\alpha^{ge}(u) \exp(-itH_e/\hbar) \, M_\alpha^{eg}(u) \rangle. \tag{2.10}$$

Again because of the cyclic invariance of the trace, $\eta_e'(t)$ satisfies the identity $\eta_e'(-t) = [\eta_e'(t)]^*$. Equation (2.9) was obtained by *Lax* [2.52] directly from the golden rule expression (2.6), and its form is useful for the development of a many-body approach for complex multimode systems.

2.1.3 Resonance Raman Scattering

Within the dipole approximation, the KH formula for the differential scattering cross section may be written as

$$\left(\frac{d^2\sigma}{d\Omega \, d\omega} \right)_{\eta\eta'} = \hbar^{-2}c^{-4}\omega_L\omega_S^3 \, Av_a \sum_b$$

$$\times \left| \sum_c \frac{\langle b| M_\eta |c\rangle \langle c| M_{\eta'} |a\rangle}{\omega_{ca} - \omega_L - i\gamma_c} + (NR)' \right|^2 \delta(\omega - \omega_{ab}). \tag{2.11}$$

The notation here is identical with that of (2.1), with the additions that ω_S is the frequency of the scattered light; $\omega = \omega_S - \omega_L$ is the Raman shift; $|b\rangle$ denotes the final states; and η and η' are the polarization directions of the scattered and incident light, respectively. The symbol $(NR)'$ represents the non-resonant part of the Raman amplitude, and it is obtained from the first, resonant, part by taking the complex conjugate, switching a and b in the matrix elements and replacing ω_L by $-\omega_S$. Under the same assumptions used to obtain (2.2), (2.11) becomes

$$
\left(\frac{d^2\sigma}{d\Omega\,d\omega}\right)_{\eta\eta'} = \hbar^{-2}c^{-4}\omega_L\omega_S^3 z^{-1} \sum_{v} \exp\left(-\beta E_{gv}\right) \sum_{v''}
$$

$$
\times \left| \sum_{e,\,v'} \frac{\langle gv''| M_\eta^{ge}(u) |ev'\rangle \langle ev'| M_{\eta'}^{eg}(u') |gv\rangle}{\omega_{ev',\,gv} - \omega_L - i\gamma_e} + (NR)' \right|^2
$$

$$
\times \delta(\omega - \omega_{gv,\,gv''}) . \tag{2.12}
$$

Interference effects in RR profiles arise from the strong variation of the $\sum_{e,\,v'}$ sum in the resonant part of this equation as ω_L is varied through the vibrational levels of the excited electronic states.

Using the eigenstate representation of the electronic polarizability operator (2.4), one can identically rewrite the Raman amplitude appearing in (2.12) as $\hbar \langle gv''| \hat{P}_{\eta\eta'}(\omega_L) + \hat{P}_{\eta\eta'}^\dagger(-\omega_L) |gv\rangle$. In obtaining the second, non-resonant, term of this expression, use is made of the identity $\omega_{ev',\,gv} + \omega_S = \omega_{ev',\,gv''} + \omega_L$, which follows from the energy-conserving delta function. Note also that the dummy summation variable v'' in (2.4) should not be confused with the final vibrational state in (2.12). We thus have

$$
\left(\frac{d^2\sigma}{d\Omega\,d\omega}\right)_{\eta\eta'} = c^{-4}\omega_L\omega_S^3 z^{-1} \sum_{v} \exp\left(-\beta E_{gv}\right) \sum_{v''} |\langle gv''| \hat{P}_{\eta\eta'}(\omega_L)
$$

$$
+ \hat{P}_{\eta\eta'}^\dagger(-\omega_L) |gv\rangle|^2 \, \delta(\omega - \omega_{gv,\,gv''}) . \tag{2.13}
$$

Now we rewrite the delta function in this equation as $\delta(\omega - \omega_{gv,\,gv''})$ $= (2\pi)^{-1} \int_{-\infty}^{\infty} dt \exp\left[-it(\omega - \omega_{gv,\,gv''})\right]$, take the complex square of the Raman amplitude, replace $\exp(it\omega_{gv})$ and $\exp(-it\omega_{gv''})$ by their operator equivalents, and use the completeness relation satisfied by the vibrational eigenfunctions in the ground electronic state, to obtain

$$
\left(\frac{d^2\sigma}{d\Omega\,d\omega}\right)_{\eta\eta'} = c^{-4}(2\pi)^{-1}\, \omega_L\omega_S^3 \int_{-\infty}^{\infty} dt \exp\left(-it\omega\right)
$$

$$
\times \langle[\hat{P}_{\eta\eta'}^\dagger(\omega_L, t) + \hat{P}_{\eta\eta'}(-\omega_L, t)] [\hat{P}_{\eta\eta'}(\omega_L)
$$

$$
+ \hat{P}_{\eta\eta'}^\dagger(-\omega_L)]\rangle . \tag{2.14}
$$

Here $\hat{P}_{\eta\eta'}^\dagger(\omega_L, t) \equiv \exp(itH_g/\hbar)\,\hat{P}_{\eta\eta'}^\dagger(\omega_L)\exp(-itH_g/\hbar)$ is the operator $\hat{P}_{\eta\eta'}^\dagger(\omega_L)$ in the interaction representation at time t. Again, the operators with positive and negative ω_L arguments in (2.14) just correspond to the resonant and off-resonant terms of the Raman amplitude. If we now retain only the resonant terms and define a unit vector $\boldsymbol{\eta}$ along the direction of the scattered polarization, so that $\hat{P}_{\eta\eta'}(\omega_L) = \sum_{\alpha\gamma}\eta_\alpha\eta'_\gamma\hat{P}_{\alpha\gamma}(\omega_L)$, we can rewrite the RR cross section as

$$\left(\frac{d^2\sigma}{d\Omega\,d\omega}\right)_{\eta\eta'} = c^{-4}\omega_L\omega_S^3 \sum_{\alpha\beta\gamma\lambda}\eta_\alpha\eta_\beta\eta'_\gamma\eta'_\lambda\, i_{\alpha\gamma\beta\lambda}^{RR}(\omega),\qquad(2.15)$$

where the RR tensor $i_{\alpha\gamma\beta\lambda}^{RR}(\omega)$ is given by

$$i_{\alpha\gamma\beta\lambda}^{RR}(\omega) = (2\pi)^{-1}\int_{-\infty}^{\infty}dt\exp(-it\omega)\langle\hat{P}_{\alpha\gamma}^\dagger(\omega_L, t)\,\hat{P}_{\beta\lambda}(\omega_L)\rangle.\qquad(2.16)$$

The indices in the Raman tensor refer to a Cartesian coordinate system fixed in the scatterer. Finally, we change the integration variable to $\mu = -t$, use (2.5) for the operators $\hat{P}_{\alpha\gamma}^\dagger(\omega_L)$ and $\hat{P}_{\beta\lambda}(\omega_L)$ in this equation, and use the cyclic invariance of the trace, to obtain the RR tensor in the time-correlator form

$$i_{\alpha\gamma\beta\lambda}^{RR}(\omega) = (2\pi\hbar^2)^{-1}\int_{-\infty}^{\infty}d\mu\exp(i\omega\mu)\int_0^\infty dt'\int_0^\infty dt\exp[-i\omega_L(t'-t)]$$

$$\times\sum_{e',e}\exp(-\gamma_{e'}t' - \gamma_e t)\,A'_{e'e}(t', t, \mu),\qquad(2.17)$$

where the resonance Raman correlator $A'_{e'e}(t', t, \mu)$ is defined as

$$A'_{e'e}(t', t, \mu) \equiv \langle M_\gamma^{ge'}(\boldsymbol{u})\exp(it'H_{e'}/\hbar)\,M_\alpha^{e'g}(\boldsymbol{u})\exp(i\mu H_g/\hbar)$$

$$\times M_\beta^{ge}(\boldsymbol{u})\exp(-itH_e/\hbar)\,M_\lambda^{eg}(\boldsymbol{u})$$

$$\times\exp[-i(\mu + t' - t)H_g/\hbar]\rangle.\qquad(2.18)$$

Note that the $\int d\mu$ integration in (2.17) comes from the δ-function in (2.13); it thus selects out the energy-conserving final vibrational states.

The Raman correlator (2.18) and the absorption correlator (2.10) are thermal averages of products of $\boldsymbol{u}$-dependent electronic transition dipole matrix elements and time-evolution operators for the vibrational system in the electronic excited and ground states. The more complicated form of the Raman correlator, compared with that for the absorption, just reflects the more complex nature of the Raman process, the amplitude for which is of second-order in the matter-light interaction. For the case of a single resonant electronic state and within the Condon approximation, which takes $M_\alpha^{eg}(\boldsymbol{u})$ to be independent of $\boldsymbol{u}$, (2.10) and (2.18) for the correlators are equivalent to those given by *Hizhnyakov* and *Tehver* [2.28].

2.1.4 Off-Resonance Limit

Before proceeding with the RR problem, it is instructive to briefly discuss the off-resonance limit within the above framework. First, we note that (2.14), which gives the Raman scattering for any value of the incident light frequency ω_L within our assumptions, is formally the same as for the case of *off*-resonance Raman scattering, e.g. (Ref. 2.53, eq. (2)], if one replaces the electronic polarizability operator expression $\hat{P}_{\eta\eta'}(\omega_L) + \hat{P}_{\eta\eta'}^\dagger(-\omega_L)$ by $P_{\eta\eta'}(\omega_L, \boldsymbol{u}) \equiv$ the electronic polarizability for a *fixed* atomic configuration $\boldsymbol{u}$. Indeed, it is straightforward to study (2.14) in the off-resonance limit, by using (2.4), the eigenstate representation of $\hat{P}_{\eta\eta'}(\omega_L)$. When ω_L is much less than any of the transition frequencies $\omega_{ev',gv''}$, one can replace these frequencies by a "typical" transition frequency and use closure to eliminate the vibrational eigenstates in (2.4). As a result, $\boldsymbol{u}'$ becomes equal to $\boldsymbol{u}$, and it is then clear that the obvious transition frequency to use in this "closure" approximation is just $\omega_{eg}(\boldsymbol{u}) \equiv [E_e(\boldsymbol{u}) - E_g(\boldsymbol{u})]/\hbar$, the $|g\rangle \rightarrow |e\rangle$ electronic transition frequency for fixed atomic configuration $\boldsymbol{u}$. Hence in the off-resonance limit one gets

$$\hat{P}_{\eta\eta'}(\omega_L) + \hat{P}_{\eta\eta'}^\dagger(-\omega_L) \rightarrow \hbar^{-1} \sum_e \left(\frac{M_\eta^{ge}(\boldsymbol{u}) \, M_{\eta'}^{eg}(\boldsymbol{u})}{\omega_{eg}(\boldsymbol{u}) - \omega_L} + \frac{M_{\eta'}^{ge}(\boldsymbol{u}) \, M_\eta^{eg}(\boldsymbol{u})}{\omega_{eg}(\boldsymbol{u}) + \omega_L} \right)$$
$$= P_{\eta\eta'}(\omega_L, \boldsymbol{u}) , \tag{2.19}$$

the "clamped" electronic polarizability introduced just above. The damping terms are of course neglected in this limit. It is now simple to transform this quantity into the time domain, just as in the derivation of (2.5). The result is simply $\hat{P}_{\eta\eta'}(\omega_L) + \hat{P}_{\eta\eta'}^\dagger(-\omega_L)$, with $\hat{P}_{\eta\eta'}(\omega_L)$ given by (2.5), *except* that the kinetic energy is dropped from the vibrational Hamiltonians H_e and H_g, i.e. $H_e(\boldsymbol{u}, \boldsymbol{p}) \rightarrow E_e(\boldsymbol{u})$ and $H_g(\boldsymbol{u}, \boldsymbol{p}) \rightarrow E_g(\boldsymbol{u})$. We thus see that the off-resonance limit is obtained from the time-correlator formulation of RR scattering if one adds the off-resonance terms to the resonance terms in the correlator expression and then *neglects* the vibrational kinetic energy operators *inside* the thermal average. The result is the standard off-resonance theory in terms of clamped electronic polarizabilities. On the other hand, for the case of *resonance* Raman scattering, the necessary presence of the kinetic energy operators prevents the operators within the thermal average in the correlator from commuting, requiring a more complicated theoretical treatment.[1]

---

[1] Although (2.19) is not valid near resonance, it does provide an heuristic indication as to why the Condon approximation for Raman scattering is often very good near resonance. First-order off-resonance Raman scattering is determined by derivatives of (2.19) with respect to atomic displacements, evaluated at the equilibrium configuration $\boldsymbol{u} = 0$ in the electronic ground state. As ω_L approaches resonance, the most divergent term in each polarizability derivative is seen to have a *second*-order pole at the resonant frequency, and a numerator involving just electronic transition dipole matrix elements evaluated at $\boldsymbol{u} = 0$, i.e. within the Condon approximation. The other divergent terms have only first-order poles. Their numerators involve *derivatives* of the matrix elements, and they are thus non-Condon terms.

2.1.5 The Multimode Problem

It is easy to appreciate the computational difficulties associated with the conventional sum-over-states approach for complex systems. Equations (2.7) and (2.12) are the basis for this approach — within specific models, the individual terms of these equations are evaluated and then summed by brute force. For harmonic systems, the vibrations can be classified into N normal modes, and if n_i vibrational levels are included in each mode in the electronic excited state, the computing time for a $T = 0\,\mathrm{K}$ calculation would increase roughly as $(n_i)^N$. For $T \neq 0\,\mathrm{K}$, if one does the thermal average directly by including m_j vibrational levels in each mode for the ground electronic state, the computing time would increase roughly as $(n_i m_j)^N$ [2.14]. This factor can become truly formidable. For the simplest situation of purely *linear* e-ph coupling (displaced oscillators) and within the Condon approximation, the optical absorption was evaluated for a hypothetical 10 mode system at $T = 0\,\mathrm{K}$, with the sums truncated at 6 vibrational levels in each mode. The resulting computing time on an IBM 3081 was 95 min, for just a *single* value of the frequency ω_L [2.21]. On the other hand, it will be seen later that the computing time for the correlator expressions resulting from the same model increases just *linearly* with N, leading to orders of magnitude improvement. For instance, within the same linear coupling model, but for a 100 mode system, the entire absorption and all 100 first-order RR profiles for $T = 300\,\mathrm{K}$ were obtained in but 1 min on the same computer. Even in our much more sophisticated models including mode frequency shifts, equilibrium position shifts, and mode mixing under electronic excitation, *plus* linear and quadratic non-Condon coupling, the computing time varies roughly as $N^3 \ll (n_i m_j)^N$, for large N [2.14, 21]. Beyond this purely computational advantage, the parallel treatment of the absorption and RR profiles in the correlator approach usefully reveals their close connection, leading to the transform methods to be discussed in Sect. 2.4.3.

2.2 Description of the General Model

Although the preceding formulation applies for the case of multiple electronic excited states, our use of a full Born–Oppenheimer basis precludes the direct inclusion of dynamical Jahn–Teller (JT) effects. These arise from the fact that, in general, the electronic states within a degenerate manifold can be coupled by certain non-totally symmetric vibrations. The case of degenerate levels coupled to *non*-JT-active vibrations *can* be treated within the present formulation—for instance in [2.54] the correlator approach was used to explain the changes in the defect-induced RR scattering by F-centers when the incident light frequency is tuned through two excited electronic levels, each degenerate and coupled to non-JT-active vibrations. Applications to the case of a *single* degenerate excited electronic level coupled to totally symmetric (and hence non-JT-active) vibrations are given in [2.12, 24]. Theoretical discussions of aspects of the use of the correlator approach for degenerate electronic states with arbitrary coupling to non-JT-active and weak coupling to JT-active vibrations are found in [2.55] and additional references therein. Applications of some of those results to experimental profiles for JT-active modes are discussed in [2.56].

For simplicity, we will assume here that there is a single, non-degenerate, resonant excited electronic state $|e\rangle$. Moreover, although we have obtained exact solutions for arbitrary components of the Raman tensor and for vibrations of any symmetry within the general model to be described here [2.21], the following discussion will be restricted to the important case when just totally symmetric modes are included. Accordingly, we will consider a single, diagonal, component of the Raman tensor and will henceforth drop the Cartesian indices from the $M_\alpha^{ge}(u)$'s. [2] Also, since the electronic eigenfunctions $\varphi_n(u, r)$ may be taken to be real without loss of generality, we note that $M^{ge}(u) = M^{eg}(u)$.

As stated in the introduction, our most general model encompasses arbitrary linear and quadratic e-ph coupling, together with arbitrary linear and quadratic non-Condon coupling. The e-ph coupling is given by the *change* $\Delta H \equiv H_e - H_g$ under electronic excitation, so that the model corresponds to arbitrary harmonic vibrational Hamiltonians H_g and H_e for the electronic ground and excited states. Although normal coordinates are well defined for both H_g and H_e, they are different in each case, reflecting the changed atomic equilibrium configuration and force constants upon electronic excitation. The fact that H_g and H_e are not simultaneously diagonalizable provides the major complexity to the present problem. Because the thermal averages are over H_g, the normal coordinates for H_g are the natural variables in terms of which to parameterize the model. Our notation will closely follow that in [2.20, 22].

---

[2] Equation (2.15) for the RR cross section applies to the case of a single scatterer or to a system of scatterers with identical orientations. The Cartesian indices in the Raman tensor of (2.16) refer to axes fixed in the scatterers. For a system of randomly oriented scatterers such as molecules in solution, one should, for fixed incident and scattered polarization directions, average (2.15) over orientations. For the common case of 90° scattering geometry, this leads to the usual expressions [2.5] in terms of Placzek invariants. For the present case of a nondegenerate resonant electronic state and just totally symmetric modes, the resulting depolarization ratio is $I_\perp/I_\parallel = 1/3$, for any value of the incident light frequency ω_L. However, for the general case when both totally symmetric and non-totally symmetric vibrations are included, along with linear and quadratic e-ph *and* linear and quadratic non-Condon coupling, the polarization properties of RR scattering are affected, giving rise to depolarization dispersion for some cases, as discussed in [2.21].

2.2.1 Vibrational Hamiltonians and Electron–Phonon Coupling

The normal coordinates and e-ph coupling in the present model are discussed in detail in [Ref. 2.22, Sect. II. B and Appendix A]. The definitions and some major points will now be reviewed.

In the harmonic approximation, H_g may be written in two familiar forms:

$$H_g = \tfrac{1}{2} \sum_{f=1}^{N} (p_{g,f}^2 + \omega_{g,f}^2 d_{g,f}^2) = \sum_{f=1}^{N} \hbar\omega_{g,f}(a_f^\dagger a_f + \tfrac{1}{2}). \tag{2.20}$$

Here $\omega_{g,f}$ is the frequency of normal mode f, for the system in the electronic ground state, and $d_{g,f}$ and $p_{g,f}$ are the corresponding normal coordinate and canonically conjugate momentum operators. These operators satisfy the commutation relations $[d_{g,f}, p_{g,f'}] = i\hbar\delta_{ff'}$, and the number N of modes is arbitrary. In the second form of H_g above, $a_f^\dagger$ and a_f are the usual creation and annihilation operators for mode f. They are defined by $a_f \equiv (2\hbar\omega_{g,f})^{-1/2}(p_{g,f} - i\omega_{g,f}d_{g,f})$ and satisfy the commutation relations $[a_f, a_{f'}^\dagger] = \delta_{ff'}$.

The most general harmonic form for H_e may be written in terms of the electronic ground state normal modes as

$$H_e = H_g + \hbar\omega_{eg} + \sum_f L_f \mathscr{A}_f + \tfrac{1}{2} \sum_{ff'} V_{ff'} \mathscr{A}_f \mathscr{A}_{f'}, \tag{2.21}$$

where $\mathscr{A}_f \equiv i(a_f - a_f^\dagger) = (2\omega_{g,f}/\hbar)^{1/2} d_{g,f}$ is a convenient rescaling of the normal coordinates, and $\hbar\omega_{eg}$ is the "vertical" electronic transition frequency for the system fixed at its electronic ground state equilibrium configuration. The linear and quadratic e-ph coupling coefficients $\{L_f\}$ and $\{V_{ff'}\}$ are arbitrary, so that this equation embodies the most general linear plus quadratic e-ph coupling $\Delta H = H_e - H_g$. The linear e-ph coupling introduces equilibrium position shifts upon electronic excitation. The quadratic e-ph coupling arises from the changes in the force constants under electronic excitation, and it introduces normal coordinate changes (i.e. mode mixing or "Duschinsky rotation" [2.57]), normal mode frequency shifts and additional equilibrium position shifts. As is detailed in [Ref. 2.22, Sect. II.B], the mode mixing and frequency shifts are related to the $N \times N$ matrix $V \equiv \{V_{ff'}\}$ by the equation

$$V = \frac{\hbar}{2} \omega_g^{-1/2} [\tilde{R}\omega_e^2 R - \omega_g^2] \omega_g^{-1/2}, \tag{2.22}$$

where $\omega_e \equiv \{\omega_{e,f}\}$ and $\omega_g \equiv \{\omega_{g,f}\}$ are $N \times N$ diagonal matrices of the normal mode frequencies in the electronic excited and ground states, respectively. The $N \times N$ mode mixing matrix $R \equiv \{R_{ff'}\}$ relates the normal coordinates in the electronic excited and ground states via the equation $d_{e,f} = \sum_{f'} R_{ff'} d_{g,f'} + \Delta_f$, or in matrix notation,

$$d_e = Rd_g + \Delta. \tag{2.23}$$

Here the $N \times 1$ vectors $d_e \equiv \{d_{e,f}\}$ and $d_g \equiv \{d_{g,f}\}$ give the excited and ground state normal coordinates, respectively, for a given configuration of the atoms. In terms of the $d_{e,f}$'s, the excited state vibrational potential energy is given by $(PE)_e = (1/2)\sum_f \omega_{e,f}^2 d_{e,f}^2 + (PE)_{e,min}$. The vector $\Delta \equiv \{\Delta_f\}$ accounts for the atomic equilibrium position shifts under electronic excitation and is given by $\Delta = (2/\hbar)^{1/2} \omega_e^{-2} R\omega_g^{1/2} L$, where $L \equiv \{L_f\}$ is the $N \times 1$ vector composed of the linear e-ph coupling coefficients.

Thus the normal coordinates in the electronic excited and ground states are formally related by the "rotation" R and the shift Δ, as is shown schematically in Fig. 2.2. The mode mixing matrix R is orthogonal $[\tilde{R}R = I]$, being the matrix of the orthogonal transformation which brings H_e into diagonal form. Note that except in the frequency-shift limit of no mode mixing, there is no correspondence between the e-state and g-state normal modes; i.e., the specific form of R is determined only after the modes in each electronic state are labeled. A general element $R_{ff'}$, is just a scalar product of the normalized mode displacement patterns for modes (e, f) and (g, f'), as shown in [2.22], which should be consulted for additional details.

Section I of [2.22] summarizes some other approaches to the problem of quadratic e-ph coupling effects in RR scattering, within the Condon approximation. Much of that work involves sum-over-states methods and is not practicable for the complex multimode systems of interest here. Refs. [2.58-61] listed here are, however, aimed at multimode systems. In [2.58] and [2.59], closed expressions are obtained for $T = 0$ K RR profiles for arbitrary linear and quadratic e-ph coupling, using different methods than those employed here: a density operator method is used in [2.58], while [2.59] uses semiclassical wavepacket propagation techniques. In both

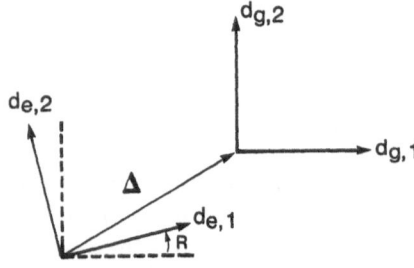

Fig. 2.2. Schematic illustration of the effect of linear and quadratic e-ph coupling, for a two-mode system. The axes for the electronic excited state normal coordinates are rotated and displaced relative to those for the ground state, according to (2.23)

of these references, the $T = 0\,\mathrm{K}$ RR profiles are obtained as *one*-dimensional Fourier transforms of closed expressions. Our work shows that this important feature can be retained for *finite* temperatures. Ref. [2.60] uses matrix methods to develop a numerical scheme for evaluating the $T \neq 0\,\mathrm{K}$ RR scattering for multimode systems. As discussed in [2.22], this approach does not separate the RR scattering into orders, and it involves solving a matrix eigenvalue problem at each time point of the triple-time Fourier transform (2.17) for the scattering. The method of [2.60] is thus more computationally intensive than that to be described here. In [2.61], the statistical density matrix/stochastic line broadening approach [2.62] is used to study RR scattering from molecules in condensed phases and having linear and quadratic e-ph coupling. The molecular degrees of freedom are treated explicitly, while the interaction with the solvent or host is treated via an assumed phenomenological stochastic variation of the resonant energy gap.[3] In [2.61], closed expressions are obtained for the *amplitudes* of the RR scattering for each initial vibrational state of the molecule in terms of one-dimensional Fourier transforms, necessitating an explicit thermal average of the resulting transition probabilities. In our approach, the necessary thermal averages are carried out analytically, resulting in a single one-dimensional Fourier transform for the profile. Moreover, it allows a straightforward treatment of multimode homogeneous thermal broadening due to a reservoir of host or solvent modes, as mentioned in Sect. 2.1.1 and discussed later in Sect. 2.4.4.

2.2.2 Non-Condon Coupling

The Condon approximation takes the electronic transition dipole matrix element $M^{ge}(u)$ to be independent of the atomic configuration u. Specifically, it sets this quantity equal to its value for $u = \mathbf{0}$, the equilibrium configuration in the electronic ground state: $M^{ge}(u) \approx M^{ge}(\mathbf{0})$. Within this approximation, the lineshapes of the optical absorption and RR excitation profiles are determined solely by overlap integrals $\langle ev' \mid gv \rangle$ between the vibrational levels in the electronic excited and ground states, as is evident from (2.7) and (2.12).

Non-Condon coupling arises from the explicit u-dependence of $M^{ge}(u)$. Our most general model includes arbitrary first- *and* second-order

---

[3] The stochastic approach has proven useful for studies of the nature of the broadening of individual spectral lines and the separation of the entire resonance secondary emission into components such as RR scattering, hot luminescence, etc. Both the stochastic approach and the present Kramers–Heisenberg treatment of resonance light scattering are phenomenological, in that neither the intermediate state widths γ_e in the KH approach nor the energy gap modulation in the stochastic approach are treated microscopically. Aspects of both approaches are discussed in [2.63].

non-Condon coupling by expanding $M^{ge}(u)$ about $u = 0$, through quadratic terms in the displacements, or, equivalently, in the rescaled normal coordinates $\{\mathscr{A}_f\}$ appearing in (2.21). This gives

$$M^{ge}(u) = M^{ge}(0) \left[1 + \sum_f m_f \mathscr{A}_f + \tfrac{1}{2} \sum_{ff'} Q_{ff'} \mathscr{A}_f \mathscr{A}_{f'} \right]. \tag{2.24}$$

We are using a full adiabatic electronic basis set. An alternative approach is to use a "crude" adiabatic electronic basis, in which the electronic states are just those with the nuclei fixed in the equilibrium configuration $u = 0$ for the electronic ground state. In this basis, one can in principle obtain the full adiabatic electronic eigenfunctions $\varphi_n(u, r)$ and corresponding eigenvalues $E_n(u)$ by applying perturbation theory, with the perturbation being $\Delta U(u, r) = U(u, r) - U(0, r)$, where $U(u, r)$ is the system's *total* potential energy. In practice, one expands $\Delta U(u, r)$ about $u = 0$ and works consistently to the desired order in u. For the commonly considered case of a single resonant nondegenerate (crude) electronic excited state e coupled in this way to just one other nondegenerate (crude) excited state e', it is readily shown that the linear and quadratic non-Condon coupling coefficients m_f and $Q_{ff'}$ defined by (2.24) arise from vibrational coupling between the (crude) states e and e'. Specifically, these coefficients are, respectively, proportional to matrix elements of $(\partial \Delta U/\partial \mathscr{A}_f)_0$ and $(\partial^2 \Delta U/\partial \mathscr{A}_f \, \partial \mathscr{A}_{f'})_0$ between the (crude) electronic excited states, with the derivatives evaluated at $u = 0$. This is an example of the so-called Herzberg-Teller vibronic coupling scheme [2.64]. It not only provides insight into the origin of non-Condon coupling, but can also be used to show that modes (f, f') which have nonzero linear non-Condon coupling coefficients m_f and $m_{f'}$, *necessarily* have a nonzero quadratic e-ph coupling coefficient $V_{ff'}$. This results in vibrational frequency shifts under electronic excitation, and, when the modes f and f' are different, it also produces mode mixing between them. This connection was shown some twenty years ago [2.65], and it necessitates a theory which treats both quadratic e-ph coupling and non-Condon coupling together. This is done in the present model, which for consistency takes both the e-ph coupling and the non-Condon coupling to second order. It should also be noted that while non-Condon coupling involving two or more modes requires mode mixing between them, the converse is *not* true–the precise admixture of non-Condon to mode mixing depends upon the details of the electronic states. We therefore keep the values of the various coupling constants open, treating an arbitrary admixture.

Reference [2.66] treated $T = 0\,\mathrm{K}$ RR scattering for systems with simultaneous linear non-Condon and frequency shifts, but no mode mixing. The general problem, including quadratic non-Condon terms, was considered in [2.67], but only perturbative results were obtained for the small mode mixing case and $T = 0\,\mathrm{K}$. Moreover, the results of [2.66, 67] are in the sum-over-states form, as are those of the additional representative non-Condon studies in [2.68]. Some previous work for the case of non-Condon and only *linear* e-ph coupling is given in [2.33, 34]. For this case, *Chan* [2.15] used the correlator framework to obtain exact nonzero temperature results well-suited to multimode modeling and absorption → profile transform techniques.

The present model is very general and reduces to many interesting special cases when various of the e-ph and/or non-Condon coupling parameters are set equal to zero. In particular, the basic but important "standard assumptions" model [2.9, 11, 13, 14] to be discussed here in Sect. 2.3.3, results when one neglects non-Condon and quadratic e-ph coupling altogether. Then only the mode equilibrium positions are shifted under electronic excitation. Even in this simplest case, multimode and temperature effects for complex systems are difficult to treat via the traditional sum-over-states approach, as noted in Sect. 2.1.5.

2.3 The Many-Body Approach

In the following, we will highlight only some of the most important aspects of our many-body methods, referring the reader to [2.15, 20–22] for details.

2.3.1 Undetermined Coefficients Method

It is convenient to rewrite (2.24) in a more compact form, namely $M^{ge}(u) \equiv M^{ge}(0)\,[1 + \Delta M(u)]$, where the fractional non-Condon correction $\Delta M(u)$ is defined by

$$\Delta M(u) \equiv \sum_f m_f \mathcal{A}_f + \tfrac{1}{2} \sum_{ff'} Q_{ff'} \mathcal{A}_f \mathcal{A}_{f'} . \qquad (2.25)$$

a) Absorption

For the present model, the optical absorption (2.9) may be rewritten

$$\alpha(\omega_L) = B\hbar^{-1} |M(0)|^2 \, \omega_L \, \mathrm{Re} \int_0^\infty dt \, \exp\left[it(\omega_L + i\gamma)\right] \eta(t), \qquad (2.26)$$

where

$$\eta(t) = \langle \exp\left(itH_g/\hbar\right) [1 + \Delta M(u)] \exp\left(-itH_e/\hbar\right) [1 + \Delta M(u)] \rangle . \qquad (2.27)$$

The absorption correlator $\eta(t)$ is just a rescaling of $\eta'_e(t)$ of (2.10).

Within the Condon approximation ($\{m_f = 0, \ Q_{ff'} = 0\}$), a diagrammatic technique was developed to obtain a useful exact analytic expression for $\eta(t)$ for the case of arbitrary linear plus quadratic e-ph coupling [2.10, 22]. To generalize this approach with non-Condon coupling added, we have adapted the undetermined coefficients techniques used by *Chan* [2.15] for the case of just linear e-ph and linear non-Condon coupling. Consider the function

$$g(\Lambda_1, \Lambda_2; t) \equiv \langle \exp{(itH_g/\hbar)} \exp{[\Lambda_1 \, \Delta M(\boldsymbol{u})]}$$
$$\times \exp{(-itH_e/\hbar)} \exp{[\Lambda_2 \, \Delta M(\boldsymbol{u})]} \rangle , \qquad (2.28)$$

where Λ_1 and Λ_2 are undetermined coefficients. If one expands this function in a Taylor series in terms of Λ_1 and Λ_2, neglects all terms in which either of these coefficients appears more than once, and then sets $\Lambda_1 = \Lambda_2 = 1$, the result will be $\eta(t)$. Hence our strategy is to find the analytic form of $g(\Lambda_1, \Lambda_2; t)$ and then expand it to obtain $\eta(t)$. This function will be called the auxiliary absorption correlator.

For the case of just linear non-Condon coupling and linear e-ph coupling (displaced oscillators), this function was evaluated algebraically in [2.15], by means of a canonical transformation and boson operator identities. For the present much more complicated case with quadratic e-ph and quadratic non-Condon coupling added, our approach is based upon a generalization of the diagrammatic technique detailed in [2.22]. To this end, we insert the identity operator in the form $I = \exp{(-itH_g/\hbar)}$ $\exp{(itH_g/\hbar)}$ ahead of the factor $\exp{(-itH_e/\hbar)}$ in (2.28), giving

$$g(\Lambda_1, \Lambda_2; t) \equiv \langle \exp{[\Lambda_1 \, \Delta M(\boldsymbol{u}, t)]} \exp{(itH_g/\hbar)}$$
$$\times \exp{(-itH_e/\hbar)} \exp{[\Lambda_2 \, \Delta M(\boldsymbol{u})]} \rangle , \qquad (2.29)$$

where $\Delta M(\boldsymbol{u}, t) \equiv \exp{(itH_g/\hbar)} \, \Delta M(\boldsymbol{u}) \exp{(-itH_g/\hbar)}$ is $\Delta M(\boldsymbol{u})$ in the interaction representation at time t. Next we use the familiar operator identity

$$\exp{(itH_g/\hbar)} \exp{(-itH_e/\hbar)} \equiv T_+ \exp{\left[-i\hbar^{-1} \int_0^t ds \, \Delta H(s) \right]}, \qquad (2.30)$$

where $\Delta H(s)$ is ΔH in the interaction representation at time s. The time-ordering operator T_+ arranges products of $\Delta H(s)$ operators arising from the expansion of the exponential so that operators with later time arguments are to the left of those for earlier times. The above identity is valid for nonnegative t, and a derivation is given in [Ref. 2.11, Appendix A]. The substitution of (2.30) into (2.29) gives

$$g(\Lambda_1, \Lambda_2; t) = \exp{(-i\omega_{eg}t)}$$
$$\times \left\langle T_+ \exp{\left(\Lambda_1 \, \Delta M(\boldsymbol{u}, t) + \Lambda_2 \, \Delta M(\boldsymbol{u}) - i\hbar^{-1} \int_0^t ds \, \delta H(s) \right)} \right\rangle , \qquad (2.31)$$

where the $\Delta M(\boldsymbol{u})$ operators without time arguments are to be regarded as being for zero time, and for convenience we have separated out the constant term in ΔH by writing $\Delta H \equiv \hbar\omega_{eg} + \delta H$. Written explicitly in terms of the rescaled normal coordinates $\{\mathscr{A}_f\}$, this equation is

$$
g(\Lambda_1, \Lambda_2; t) = \exp\left(-i\omega_{eg}t\right) \left\langle T_+ \exp\left(\Lambda_1 \left[\sum_f m_f \mathscr{A}_f(t)\right.\right.\right.
$$

$$
\left. + \tfrac{1}{2}\sum_{ff'} Q_{ff'}\mathscr{A}_f(t)\,\mathscr{A}_{f'}(t)\right] + \Lambda_2\left[\sum_f m_f \mathscr{A}_f + \tfrac{1}{2}\sum_{ff'} Q_{ff'}\mathscr{A}_f \mathscr{A}_{f'}\right]
$$

$$
\left.\left. - i\hbar^{-1}\int_0^t ds\left[\sum_f L_f \mathscr{A}_f(s) + \tfrac{1}{2}\sum_{ff'} V_{ff'}\mathscr{A}_f(s)\,\mathscr{A}_{f'}(s)\right]\right]\right\rangle . \tag{2.32}
$$

This form of $g(\Lambda_1, \Lambda_2; t)$ is tailor-made for the development of a diagrammatic approach.

b) RR Scattering

Within the present assumptions, (2.15) and (2.17) give the RR cross section as

$$
\frac{d^2\sigma}{d\omega\, d\Omega} = (2\pi\hbar^2 c^4)^{-1}\,|M(0)|^4\,\omega_L \omega_s^3 \int_{-\infty}^{\infty} d\mu \exp(i\omega\mu) \int_0^{\infty} dt' \int_0^{\infty} dt
$$

$$
\times \exp\left[-i\omega_L(t'-t) - \gamma(t'+t)\right] A(t', t, \mu), \tag{2.33}
$$

where $A(t', t, \mu)$ is a rescaling of $A'_{e'e}(t', t, \mu)$ of (2.18) and is given by

$$
A(t', t, \mu) = \langle[1 + \Delta M(u)] \exp(it'H_e/\hbar)\,[1 + \Delta M(u)] \exp(i\mu H_g/\hbar)
$$

$$
\times [1 + \Delta M(u)] \exp(-itH_e/\hbar)\,[1 + \Delta M(u)] \exp[-i(\mu + t' - t)H_g/\hbar]\rangle . \tag{2.34}
$$

This correlator can be treated by an undetermined coefficients method analogous to that for the simpler absorption case treated above. Thus we introduce a function [2.20, 21].

$$
G(\Lambda'_1, \Lambda'_2, \Lambda_1, \Lambda_2; t', t, \mu) = \langle\exp[\Lambda'_2\, \Delta M(u)] \exp(it'H_e/\hbar)
$$

$$
\times \exp[\Lambda'_1\, \Delta M(u)] \exp(i\mu H_g/\hbar) \exp[\Lambda_1\, \Delta M(u)]
$$

$$
\times \exp(-itH_e/\hbar) \exp[\Lambda_2\, \Delta M(u)] \exp(-i\tau H_g)\rangle, \tag{2.35}
$$

where $\tau \equiv \mu + t' - t$ and $\Lambda'_1, \Lambda'_2, \Lambda_1$ and Λ_2 are the undetermined coefficients. Again, if one expands $G(\Lambda'_1, \Lambda'_2, \Lambda_1, \Lambda_2; t', t, \mu)$ in a Taylor series in the Λ_i, neglects all terms in which any of the coefficients appear more than once, and then sets them all equal to one, the result will be $A(t', t, \mu)$. Thus G will be termed the auxiliary RR correlator. To bring G

into a form suitable for a diagrammatic expansion, one can manipulate it through steps analogous to those leading to (2.31) for the absorption case. The result is

$$G(\Lambda_1', \Lambda_2', \Lambda_1, \Lambda_2; t', t, \mu) = \exp\left[i\omega_{eg}(t' - t)\right] \langle T_{s's} \exp(\phi)\rangle, \qquad (2.36)$$

where ϕ is defined by

$$\phi \equiv \Lambda_1' \, \Delta M(\mathbf{u}, t') + \Lambda_2' \, \Delta M(\mathbf{u}) + \Lambda_1 \, \Delta M(\mathbf{u}, t + \tau)$$
$$+ \Lambda_2 \, \Delta M(\mathbf{u}, \tau) + i\hbar^{-1} \int_0^{t'} ds' \, \delta H(s') - i\hbar^{-1} \int_0^t ds \, \delta H(s + \tau). \quad (2.37)$$

The generalized time-ordering operator $T_{s's}$ in (2.36) is defined to keep all of the "primed" operators (those with primed time variables t', s' or zero times) to the left of the "unprimed" operators (those with unprimed time variables $t + \tau, s + \tau$, or τ). Moreover, it orders the primed operators among themselves with T_- (later times to the right of earlier times), and it orders the unprimed operators with T_+. Equation (2.36) is valid for nonnegative values of t' and t. One can show that if $t' = \mu = \Lambda_1' = \Lambda_2' = 0$, then $G(0, 0, \Lambda_1, \Lambda_2; 0, t, 0) = g(\Lambda_1, \Lambda_2; t)$, linking the RR and the absorption auxiliary correlators. Moreover, notice that in the limit when all of the Λ_i are zero, $G(\Lambda_1', \Lambda_2', \Lambda_1, \Lambda_2; t', t, \mu)$ and $g(\Lambda_1, \Lambda_2; t)$ become equal to $A_c(t't, \mu)$ and $\eta_c(t)$, respectively, where the subscript c denotes the correlators for the Condon approximation $\{m_f = 0, Q_{ff'} = 0\}$. In this case, the preceding identity gives $\eta_c(t) = A_c(0, t, 0)$. More detailed derivations of the time-ordered forms of the Condon approximation correlators are given in [2.11, 22].

2.3.2 Diagrammatic Expansion

The diagram technique for this model closely parallels that detailed in [2.22] for the linear plus quadratic e-ph coupling case within the Condon approximation, with the major difference being the proliferation of the types of diagrams to be dealt with, owing to the presence of non-Condon coupling and the Λ_i coefficients. Here just a few major points will be highlighted; for details, the reader should consult [2.21, 22].

The first step is to expand the exponential in (2.36): $\langle T_{s's} \exp(\phi)\rangle = \sum_k$ $(k!)^{-1} \langle T_{s's}\phi^k\rangle$. When (2.37) for ϕ is used and each ϕ^k is multiplied out, a complicated series results. Each term consists of a thermally averaged $T_{s's}$-ordered product of $\mathcal{A}_f$ operators having various time arguments and is multiplied by e-ph and non-Condon coupling parameters, together with

Λ_i coefficients. Moreover, the e-ph coupling parameters are associated with time integrations. For example, a typical term from ϕ^4 is

$$\int_0^{t'} ds' \int_0^t ds \sum_{f_1 f_2 f_3 f_4 f_5 f_6} (i\hbar^{-1} L_{f_1}) \left(\tfrac{1}{2} \Lambda_2' Q_{f_2 f_3}\right) \left(-\tfrac{1}{2} i\hbar^{-1} V_{f_4 f_5}\right)$$
$$\times (\Lambda_1 m_{f_6}) \langle T_{s's} \mathscr{A}_{f_1}(s') \mathscr{A}_{f_2} \mathscr{A}_{f_3} \mathscr{A}_{f_4}(s + \tau) \mathscr{A}_{f_5}(s + \tau) \mathscr{A}_{f_6}(t + \tau) \rangle .$$
$$(2.38)$$

a) Wick's Theorem

The next step is to apply a nonzero temperature version of Wick's theorem, generalized to $T_{s's}$ ordering in [2.11]. This exact result allows one to write a harmonic thermally averaged time-ordered product of an even number of $\mathscr{A}_f$ operators as the sum of all distinct factorizations into thermally averaged time-ordered *pair* products, i.e. the sum of all distinct "pairings." To illustrate, with four operators the result is just

$$\langle T \mathscr{A}_{f_1}(\sigma_1) \mathscr{A}_{f_2}(\sigma_2) \mathscr{A}_{f_3}(\sigma_3) \mathscr{A}_{f_4}(\sigma_4) \rangle$$
$$\equiv \langle T \mathscr{A}_{f_1}(\sigma_1) \mathscr{A}_{f_2}(\sigma_2) \rangle \langle T \mathscr{A}_{f_3}(\sigma_3) \mathscr{A}_{f_4}(\sigma_4) \rangle$$
$$+ \langle T \mathscr{A}_{f_1}(\sigma_1) \mathscr{A}_{f_3}(\sigma_3) \rangle \langle T \mathscr{A}_{f_2}(\sigma_2) \mathscr{A}_{f_4}(\sigma_4) \rangle$$
$$+ \langle T \mathscr{A}_{f_1}(\sigma_1) \mathscr{A}_{f_4}(\sigma_4) \rangle \langle T \mathscr{A}_{f_2}(\sigma_2) \mathscr{A}_{f_3}(\sigma_3) \rangle ,$$

where T can be T_+, T_- or $T_{s's}$, and the σ_i's can be any of the possible time arguments of the $\mathscr{A}_f$ operators. Products with an odd number of $\mathscr{A}_f$ operators vanish when thermally averaged. When the generalized Wick's theorem is applied, each of the several kinds of resulting pairs may be expressed in terms of the three basic types of electronic ground state phonon propagators occurring in [2.10, 22], namely

$$\langle T_{s's} \mathscr{A}_f(s_1') \mathscr{A}_{f'}(s_2') \rangle \equiv \langle T_- \mathscr{A}_f(s_1') \mathscr{A}_{f'}(s_2') \rangle = \delta_{ff'} g_f^*(s_1' - s_2') ,$$
$$(2.39\,\mathrm{a})$$

$$\langle T_{s's} \mathscr{A}_f(s_1 + \tau) \mathscr{A}_{f'}(s_2 + \tau) \rangle \equiv \langle T_+ \mathscr{A}_f(s_1) \mathscr{A}_{f'}(s_2) \rangle = \delta_{ff'} g_f(s_1 - s_2),$$
$$(2.39\,\mathrm{b})$$

$$\langle T_{s's} \mathscr{A}_f(s_1') \mathscr{A}_{f'}(s_2 + \tau) \rangle$$
$$\equiv \langle \mathscr{A}_f(s_1') \mathscr{A}_{f'}(s_2 + \tau) \rangle = \delta_{ff'} h_f(s_1' - s_2 - \tau), \qquad (2.39\,\mathrm{c})$$

where $h_f(s) = (\langle n_f \rangle + 1) \exp(-is\omega_{g,f}) + \langle n_f \rangle \exp(is\omega_{g,f})$ and $g_f(s) = h_f(|s|)$, where $|s|$ represents the absolute value of s. Here $\langle n_f \rangle = [\exp(\beta\hbar\omega_{g,f}) - 1]^{-1}$ is the thermal average phonon occupation number for normal mode f in the electronic ground state. In using these formulas in the present model, one has to allow for the possibility that s_i' can be equal to t' or zero and s_i can be equal to t or zero.

The number of distinct pairings of $2r$ $\mathscr{A}_f$ operators is $(2r)!/(2^r r!)$; hence the six-operator product in (2.38) generates a sum of fifteen pairings. A typical one of the terms so produced is

$$\int_0^{t'} ds' \int_0^t ds \sum_{f_1 f_2 f_3} (i\hbar^{-1} L_{f_1}) g_{f_1}^*(s') \left(\tfrac{1}{2} \Lambda_2' Q_{f_1 f_2}\right) h_{f_2}(-s-\tau)$$
$$\times (-\tfrac{1}{2} i\hbar^{-1} V_{f_2 f_3}) g_{f_3}(s-t) (\Lambda_1 m_{f_3}), \tag{2.40}$$

where (2.39) has been used.

b) Diagrams

At this point it is straightforward to map the terms of the infinite series for $G(\Lambda_1', \Lambda_2', \Lambda_1, \Lambda_2; t', t, \mu)$ onto diagrams having twelve types of vertices, with the vertices connected by lines corresponding to the phonon propagators in (2.39). The twelve vertex types arise from the two non-Condon coupling constants $\{m_f, Q_{ff'}\}$, each of which is multiplied by one of the four $\{\Lambda_i\}$ coefficients, and from the two e-ph coupling coefficients $\{L_f, V_{ff'}\}$, each of which is associated with either a primed or an unprimed time integration. All of the vertices divide into two classes, "$-$" or "$+$". The $-$ vertices are associated with primed Λ_i's or primed time integrations, while the $+$ vertices are associated with unprimed Λ_i's or unprimed time integrations.[4] Figure 2.3 illustrates those diagrams which

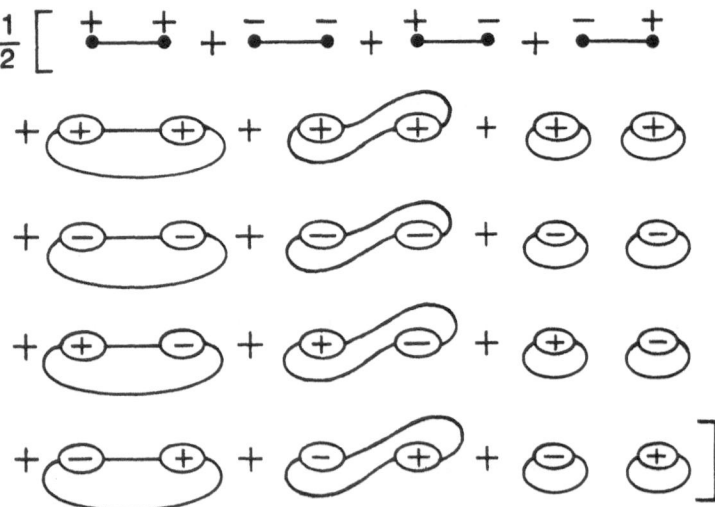

Fig. 2.3. Diagrams arising from the Condon terms of $\langle T_{s's} \phi^2 \rangle / 2$. The lines represent the phonon propagators (2.39), and the dots and ellipses are vertices representing the linear and quadratic e-ph coupling coefficients $\{L_f\}$ and $\{V_{ff'}\}$, respectively. After [2.22]

---

[4] This notation generalizes that used in [2.22], except that for clarity in the following discussion, the $+$ and $-$ labels of that paper have been switched.

arise from ϕ^2 but do not contain any of the eight non-Condon vertices. The remaining diagrams from ϕ^2 are analogous, except for the presence of non-Condon vertices. Details are given in [2.21] and are not needed here.

Note that the phonon propagator $g_f^*(s')$ always connects two minus-type vertices, while $g_f(s)$ always connects two plus-type vertices. These two quantities may be thought of as describing the propagation of ground state phonons between virtual phonon scatterings occurring while the system is in the virtually excited resonant electronic state. The propagator $h_f(s' - s - \tau)$ always connects a $-$ with a $+$ vertex and was thus termed a "neuter" pairing in [2.10, 22]. Since (2.39c) shows that just the neuter pairings depend on the variable $\tau \equiv \mu + t' - t$, it is only these pairings which contribute to the μ dependence of $A(t', t, \mu)$. As noted earlier, the $\int d\mu$ integration in (2.17) [or (2.33)] determines the final vibrational state; hence the neuter pairings play a central role in the separation of the Raman scattering into orders.

c) Linked Cluster Theorem

Every diagram can be classified as either "unlinked" or "linked", the latter being those in which all of the vertices are connected by lines into a single unit. In contrast, an unlinked diagram consists of two or more disconnected portions, as in the 7th, 10th, 13th, and 16th diagrams of Fig. 2.3. A crucial simplification is now obtained by the use of a powerful linked cluster theorem proven by *Abrikosov* et al. [2.69] for the case of T_+ time ordering. It is readily extended to $T_{s's}$ ordering, using the $T_{s's}$ generalization of Wick's theorem proven in [2.11]. The result is

$$\langle T_{s's} \exp (\phi) \rangle = \exp \sum_k \langle T_{s's}\phi^k \rangle_l /k! \,, \tag{2.41}$$

where $\langle T_{s's}\phi^k \rangle_l$ represents only the *linked* diagrams from the diagrammatic expansion of $\langle T_{s's}\phi^k \rangle$.

d) Connection Between Absorption and RR Auxiliary Correlators

The linked cluster theorem directly reveals an important connection between the auxiliary correlators $g(\Lambda_1, \Lambda_2; t)$ and $G(\Lambda_1', \Lambda_2', \Lambda_1, \Lambda_2; t', t, \mu)$. We first note that the linked diagrams can be further classified into three types: "$-$" diagrams,. having all $-$ vertices; "$+$" diagrams having all $+$ vertices; and "neuter" diagrams, having *both* types of vertices (and hence at least one neuter pairing). From the first form of each equation in (2.39), it is seen that all of the lines in the $-$ diagrams involve T_- pairings [$g_f^*(s')$], and all of the lines in the $+$ diagrams involve T_+ pairings [$g_f(s)$]. Hence, using superscripts "$-$", "$+$", and "n" to denote the three types of diagrams, we have $\langle T_{s's}\phi^k\rangle_l = \langle T_-\phi^k\rangle_l^- + \langle T_+\phi^k\rangle_l^+ + \langle T_{s's}\phi^k\rangle_l^n$. Equation (2.41) then factors:

$$\langle T_{s's}\exp(\phi)\rangle = \exp\left[\sum_i \langle T_-\phi^i\rangle_l^- /i!\right]$$

$$\times \exp\left[\sum_j \langle T_+\phi^j\rangle_l^+ /j!\right]\exp\left[\sum_k \langle T_{s's}\phi^k\rangle_l^n/k!\right].$$

(2.42)

It is straightforward to show that the first two factors on the right-hand side of this equation are functions of t' and t, respectively, and that the first factor can be obtained from the second by taking the complex conjugate and replacing t, Λ_1 and Λ_2 by their primed counterparts.

If one now returns to (2.31) for g and makes a diagrammatic expansion following the same steps as done here for G, only the subset of $+$ diagrams occurs in the series, and the result is $g(\Lambda_1, \Lambda_2; t) = \exp(-it\omega_{eg})$ $\times \exp(\sum_j \langle T_+\phi^j\rangle_l^+ /j!)$. Hence g and G are identically related by

$$G(\Lambda'_1, \Lambda'_2, \Lambda_1, \Lambda_2; t', t, \mu) = g^*(\Lambda'_1, \Lambda'_2, t') g(\Lambda_1, \Lambda_2, t)$$

$$\times \exp\left[\sum_k \langle T_{s's}\phi^k\rangle_l^n/k!\right].$$

(2.43)

In the Condon approximation limit $\{\Lambda_i = 0\}$, (2.43) becomes an identity for the correlators themselves:

$$A_c(t', t, \mu) = \eta_c^*(t')\,\eta_c(t)\exp\left[\sum_k \langle T_{s's}\phi^k\rangle_l^n/k!\right].$$

(2.44)

In either the Condon or the non-Condon case, the factorization of the original thermal average in G (or A_c) into a product of two thermal averages involving the corresponding absorption quantities, multiplied by an exponential factor involving just the neuter diagrams, affords a great simplification in the diagram summations needed for separating the RR scattering into orders. Note that (2.43) actually holds for non-Condon and e-ph coupling of *any* order in the normal coordinates, while (2.44) holds for *any* order of e-ph coupling—the only assumption used was that of a harmonic vibrational potential energy in the electronic ground state.

2.3.3 "Standard Assumptions" Case

This commonly used model involves the neglect of non-Condon and quadratic e-ph coupling, and it thus corresponds to just equilibrium position shifts under electronic excitation. As noted in Sect. 2.1.5, this model can still be difficult to handle within the traditional sum-over-states method for multimode systems, even at $T = 0$ K, and the difficulty is compounded when temperature effects are included. As stressed in [2.14], the correlator approach gives very simple and effective exact expressions for this case. Although these may be derived by purely algebraic arguments [2.11], they are much more easily obtained from the diagrammatic expansion at this point.

There are now only linear e-ph coupling vertices, denoted by "dots" in Fig. 2.3. Only one line can attach to each dot, so that the only possible linked diagrams are the four shown in the top row of this figure. The first two are $+$ and $-$ diagrams, and in view of the preceding discussion, they determine $\eta(t)$ and $\eta^*(t')$, respectively. Specifically, the first diagram gives $\eta_{\ell c}(t) = \exp(-it\omega_{eg}) \exp[(1/2) \int_0^t ds_1 \int_0^t ds_2 \sum_f (-i\hbar^{-1}L_f)^2 g_f(s_1 - s_2)]$, where the subscript ℓc (linear, Condon) denotes this case. Using the explicit formula for $g_f(s)$ given just below (2.39), it is then simple to obtain

$$\eta_{\ell c}(t) = \exp(-it\omega_{eg}^0) \exp\left\{-\sum_f \lambda_f \left[(z\langle n_f\rangle + 1)\right.\right.$$

$$\left.\left. \times (1 - \cos \omega_f t) + i \sin \omega_f t\right]\right\}, \tag{2.45}$$

where the Stokes loss parameter λ_f is defined by $\lambda_f \equiv [L_f/(\hbar\omega_f)]^2$, and $\omega_{eg}^0 \equiv \omega_{eg} - \hbar^{-2}\sum_f L_f^2\omega_f^{-1}$ is the frequency of the zero-phonon line in the absorption. Since the normal coordinates and mode frequencies are the same in the ground and excited electronic states in this model, the subscript "g" has been dropped from the frequencies. According to (2.26) the absorption for this case is given by a simple one-dimensional Fourier transform of $\eta_{\ell c}(t)$. Obviously $\eta_{\ell c}(t)$ is easy to compute from the model parameters. Note particularly that the computing time increases just linearly with the number N of normal modes, and that the inclusion of temperature effects ($\langle n_f\rangle \neq 0$) is no more difficult than working at $T = 0$ K.

Having obtained the explicit expressions for the first two factors in (2.44) for the RR correlator, it remains to evaluate the neuter diagrams, which are the third and fourth diagrams in Fig. 2.3. They are equal and give

$$A_{\ell c}(t', t, \mu) = \eta_{\ell c}^*(t') \eta_{\ell c}(t)$$

$$\times \exp\left[\int_0^{t'} ds' \int_0^t ds \sum_f (i\hbar^{-1}L_f)(-i\hbar^{-1}L_f) h_f(s' - s - \tau)\right].$$

Using the explicit formula for $h_f(s)$ given below (2.39), one easily obtains

$$A_{\ell c}(t', t, \mu) = \eta_{\ell c}^*(t') \eta_{\ell c}(t) \exp\left(\sum_f \lambda_f\{(\langle n_f\rangle + 1) \exp(i\omega_f\mu)\right.$$

$$\times [\exp(i\omega_f t') - 1][\exp(-i\omega_f t) - 1] + \langle n_f\rangle$$

$$\left. \times \exp(-i\omega_f\mu)[\exp(-i\omega_f t') - 1][\exp(i\omega_f t) - 1]\}\right). \tag{2.46}$$

Just as for the absorption correlator of (2.45), the computing time for this function is seen to increase only linearly with the number of modes, and non-zero temperatures are as simple to treat as $T = 0$ K. However, unlike the absorption case, one must at this stage perform a *three*-dimensional Fourier transform (2.33) to obtain the scattering. But note that this would give *all* of the scattering within the model; i.e. Stokes and anti-Stokes scattering involving all numbers of scattered phonons. The usual concern in practice is with low-order scattering, in which case the results are even simpler. For clarity, we will now focus on the most commonly occurring situation, namely first-order Stokes scattering.

Recalling that the $\int d\mu \exp(i\omega\mu)$ integration in (2.33) produces the final vibrational states, we see that (2.46) provides a natural means to separate the RR scattering into orders. One simply expands (just) the third factor $[\exp(\sum_f \lambda_f \ldots)]$ in this equation, producing a power series in the quantities $(\langle n_f \rangle + 1) \exp(i\omega_f\mu)$ and $\langle n_f \rangle \exp(-i\omega_f\mu)$. If just the linear terms are kept, the μ-integration in (2.33) produces sums of δ-functions, $\sum_f(\langle n_f \rangle + 1)\delta(\omega + \omega_f)[\ldots]$ and $\sum_f \langle n_f \rangle \delta(\omega - \omega_f)[\ldots]$, corresponding to first-order Stokes and anti-Stokes scattering. Clearly, the constant term in the expansion leads to Rayleigh scattering $[\delta(\omega)]$, while the higher-order terms lead to higher-order sum and difference scattering. The use of (2.46) immediately yields a very simple result for the strength (i.e. profile) of the first-order Stokes scattering by mode f:

$$j_{1,f}(\omega_L) = C(\langle n_f \rangle + 1)\,\omega_L(\omega_L - \omega_f)^3$$
$$\times \left| \int_0^\infty dt\, \exp(i\omega_L t - \gamma t)\, \eta_{\ell c}(t)\, \varrho_{\ell c, f}(t) \right|^2, \qquad (2.47)$$

where $\eta_{\ell c}(t)$ is given by (2.45), and $\varrho_{\ell c, f}(t)$ is given by

$$\varrho_{\ell c, f}(t) = -\lambda_f^{1/2}[1 - \exp(-i\omega_f t)]. \qquad (2.48)$$

The frequency-independent prefactors have been lumped into the constant C. To obtain the profile expression for the case of nth harmonic Stokes scattering at Raman shift $\omega = -n\omega_f$, one simply makes the replacements $(\langle n_f \rangle + 1) \to (\langle n_f \rangle + 1)^n$, $(\omega_L - \omega_f)^3 \to (\omega_L - n\omega_f)^3$ and $\varrho_{\ell c}(t) \to [\varrho_{\ell c}(t)]^n/\sqrt{n!}$ in (2.47). Analogous results hold for *anti*-Stokes scattering.

The form of (2.46) has thus led to a very useful factorization of the entire RR correlator, such that the first-order RR intensity can be obtained as a simple *one*-dimensional Fourier transform of an easily computed function. Moreover, the absorption correlator $\eta_{\ell c}(t)$ occurs as a simple factor in the integrand. These features lead to the powerful absorption → RR profile "transform" technique to be discussed in Sect. 2.4.3. It turns out that even in our most general model with full linear and quadratic coupling in both the e-ph and the non-Condon interactions, one retains the simple structure of (2.47); only the explicit forms of the functions $\eta(t)$ and $\varrho(t)$ in terms of model parameters are different.

Equations (2.26, 33, 45–47) are exact, they contain the e-ph coupling $\{\lambda_f\}$ to *all* orders, and they are much simpler than their sum-over-states counterparts. Sum-over-states frequency-domain analytic expressions could be obtained from these equations by expanding the $\exp\left(\sum_f \lambda_f \ldots\right)$ factors to obtain a harmonic time-series and performing the time integrations in (2.26) and (2.33) analytically. The results would obviously be very complicated infinite series, especially for nonzero temperatures.

Additional ramifications of the time-correlator approach for the standard assumptions case are detailed in [2.11, 13, 14].

2.3.4 Separation of RR Scattering into Orders

While the linked cluster theorem easily led to exact solutions for the preceding standard assumptions case, owing to the fact that the number of linked diagrams was so small, the addition of quadratic e-ph coupling and non-Condon coupling produces an infinity of new diagrams. This renders the subsequent diagram analysis much more complicated. The general analysis is detailed in [2.22], within the Condon approximation. With a good deal of additional analytic effort, it can be extended to our most general model. Briefly, the steps involve developing combinatorial arguments to reduce the diagrams to an infinite set of "standard" diagrams, separating the scattering into orders, and then summing the resulting infinite series of diagrams for each order analytically in terms of one key quantity. This quantity is an infinite series and satisfies a Dyson-like matrix equation, which can be solved in closed form, as shown by *Tonks* [2.70].

The results of this extensive analytic work are exact finite expressions in terms of the model parameters and the temperature, for the absorption and first-order RR scattering within the general model, and for second- and third-order scattering within several special cases of the general model. The solutions are easy to implement numerically, as well as to study analytically, and the results for the RR profiles have exactly the same advantageous structure exhibited by (2.47) above. A crucial step in obtaining these results is the separation of the scattering into orders. A brief discussion of this topic follows, within the context of the general diagrammatic approach [2.21, 22].

The separation into orders is accomplished by expanding the exponential factor in (2.43), resulting in a series of neuter diagrams, multiplied by $g^*(\Lambda'_1, \Lambda'_2; t')\, g(\Lambda_1, \Lambda_2; t)$. The "nth-order" part of the full series for the RR auxiliary correlator G is then defined to be the sum of all of the terms in this series having n neuter pairings. From (2.39c), the definition of h_f immediately following that equation, and the fact that $\tau = \mu + t' - t$, it is seen that each neuter pairing is linear in the factors $(\langle n_f \rangle + 1)$ $\exp(i\omega_{g,f}\mu)$ and $\langle n_f \rangle \exp(-i\omega_{g,f}\mu)$. Hence the nth-order part of G, and therefore of the RR correlator $A(t', t, \mu)$, is of nth order in these factors. For $n = 1$, the $\int d\mu \exp (i\omega\mu)$ integration of the first-order part of the correlator in (2.33) will produce the factors $(\langle n_f \rangle + 1) \delta(\omega + \omega_{g,f})$ [...] and $\langle n_f \rangle \delta(\omega - \omega_{g,f})$ [...], characteristic of first-order Stokes and anti-Stokes scattering by mode f. Clearly, this is just a general version of the explicit calculation done in the preceding subsection for the standard assumptions case.

For general n, one obtains δ-functions at sum and/or difference frequencies involving n vibrational quanta. For instance, the third-order scattering will include terms with the factors $(\langle n_f \rangle + 1) (\langle n_{f'} \rangle + 1) \langle n_{f''} \rangle$ $\delta[\omega + (\omega_{g,f} + \omega_{g,f'} - \omega_{g,f''})]$. For f, f' and f'' all different, this corresponds to a three-phonon scattering term. But note that for $f' = f''$, this term will contribute to the one-phonon Stokes scattering by mode f. Obviously, all such "higher-order" corrections contain factors $(\langle n_{f'} \rangle + 1)$ $\langle n_{f'} \rangle$ requiring thermally populated modes, and hence they disappear at low temperatures. Therefore, *at $T = 0$ K, the first-order Stokes RR spectrum defined above becomes identical with the full one-phonon Stokes spectrum*, where the latter arises from all terms corresponding to the creation of one vibrational quantum in each mode. For $T \neq 0$ K, the full one-phonon Stokes spectrum is an infinite series of temperature-dependent terms, the leading term of which is the first-order spectrum defined here. The higher-order corrections derive from orders 3, 5, ..., and freeze out at $T = 0$ K. Precisely the same features hold for the case of general n-phonon scattering [2.11, 13, 22].

While the "higher-order" corrections only exist for $T \neq 0$ K and may often be computed exactly within specific models, *Chan* and *Page* [2.13] have given detailed analytical and numerical arguments for the standard assumptions case, showing that these corrections are in fact generally negligible for a wide class of practically important multimode systems at room temperature. As a result, the temperature-dependent *nth-order* RR profiles (the theoretically useful quantities) are virtually indistinguishable from the *n-phonon* profiles (the measured quantities). Briefly, a typical system from this class is characterized at room temperature by the following: (1) an optical absorption spectrum with resolved but overlapping vibrational structure due to high frequency modes ($\omega_f \gg kT/\hbar$) and (2) thermal broadening due to a reservoir of weakly Franck-Condon active ($\lambda_{f'}^2 \ll 1$) thermally populated low frequency modes ($\omega_{f'} \lesssim kT/\hbar$), whose frequencies are smaller than the smallest resolvable width of the vibrational structure in the absorption. Additional discussion providing insight into the negligibility of the higher-order corrections is given in Sect. 2.4.5.

The above separation of the RR scattering into orders corresponds to a particular grouping, i.e. resumming, of the full infinite series for the scattering, and it is a key ingredient leading to some of the most useful features of the resulting solutions. It arises very naturally in the many-body approach, being an almost unavoidable consequence of the linked cluster theorem. Our separation scheme is a diagrammatic implementation and generalization of that given algebraically by *Hizhnyakov* and *Tehver* [2.28] in their pair correlation approximation, and we have discussed it extensively in [2.11, 13, 22].

Finally, for the absorption there is no analog of the separation of the RR scattering into orders; our analytic results are exact at all temperatures within the model.

2.4 Results and Applications

2.4.1 Model Expressions

In focusing on a particular order of RR scattering in the diagrammatic approach, one limits the types of diagrams which have to be considered. Moreover, those diagrams in which any of the four A_i coefficients occurs more than once may be excluded. These simplifications notwithstanding, much analytic work must still be carried out to obtain exact solutions, as pointed out in the previous section. After one solves for the first-order part of the auxiliary correlator G, expands it in the $\{A_i\}$ to obtain the first-order part $A_1(t', t, \mu)$ of the RR correlator, and then performs the μ-integration in (2.33), it is found that the resulting expression for the $T \neq 0$ K first-order Stokes profile of mode f has the same simple form as the standard assumptions result (2.47), namely

$$j_{1, f}(\omega_L) = C(\langle n_f \rangle + 1)\, \omega_L (\omega_L - \omega_{g,f})^3$$
$$\times \left| \int_0^\infty dt \exp\left(i\omega_L t - \gamma t\right) \eta(t)\, \varrho_f(t) \right|^2 . \tag{2.49}$$

Thus for $A_1(t', t, \mu)$, the t' and t integrations in (2.33) have factored. With explicit expressions for $\eta(t)$ and $\varrho_f(t)$ available, both the $T \neq 0$ K optical absorption (2.26) and the first-order RR profiles are conveniently computed via *one-dimensional* fast Fourier transform (FFT) techniques. In addition, the presence of the absorption correlator $\eta(t)$ as an explicit factor in the integrand leads to useful absorption $\rightarrow$ profile "transform" techniques to be discussed below.

The exact analytic expressions for $\varrho_f(t)$ and $\eta(t)$ in terms of the model parameters defined in Sects. 2.2.1 and 2.2.2 are given in [2.20]. Even for the full model case of linear and quadratic interactions in *both* the e-ph and the non-Condon couplings, they are convenient to work with, numerically and analytically. Considering the complexity of the full model, they are not particularly lengthy, but for the subsequent discussion it will suffice to give the results in the limit of no quadratic non-Condon coupling $\{Q_{ff'} = 0\}$. This will be done in two steps.

a) Condon Approximation Limit

This case includes full mode mixing, frequency shifts and equilibrium position shifts with electronic excitation. The linear non-Condon coefficients $\{m_f\}$ are all zero, and $\varrho_f(t)$ is given by [2.17, 22][5]

$$\varrho_{c,\,f}(t) = (\langle n_f \rangle + 1)^{-1} (2\omega_{g,\,f})^{1/2} \{\tilde{R} f^+(t)\, [I - \theta(-t)]\, (\omega_e)^{-1} \\ \times (\omega_g)^{3/2}\, L''\}_f , \tag{2.50}$$

with

$$f^{\pm}(t) = \{[\omega^- \pm \theta(-t)\,\omega^+]\, \Gamma^{-1}(t) - [\theta(-t)\,\omega^- \pm \omega^+]\}^{-1} ,$$

$$\theta(t) = \exp(it\omega_e) ,$$

$$\omega^{\pm} = R\omega_g\tilde{R} \pm \omega_e , \qquad L'' = 2^{1/2}(\omega_g)^{-3/2}\, R(\omega_g)^{3/2}\, \xi ,$$

$$\Gamma^{-1}(t) = \exp\left[(-\beta\hbar + it)\, R\omega_g\tilde{R}\right] .$$

All of the boldface quantities are $N \times N$ matrices, except for the $N \times 1$ vectors L'' and $\xi \equiv \{\xi_f = (\lambda_f)^{1/2}\ \text{sgn}\,(L_f) = L_f/(\hbar\omega_{g,\,f})\}$. The diagonal matrices ω_g and ω_e of the normal mode frequencies in the ground and excited electronic states were introduced in Sect. 2.2.1, as were the mode mixing matrix R and the vector L of linear e-ph coupling coefficients. The tilde denotes the transpose and I is the identity matrix. The matrix identity $\exp(R\omega_g\tilde{R}) = R\exp(\omega_g)\,\tilde{R}$, which follows from the fact that R is orthogonal ($R\tilde{R} = I$), is useful in numerically evaluating $\Gamma^{-1}(t)$.

The absorption correlator is given by

$$\eta_c(t) = \exp\left[-it\omega_{eg} + F_1(t) + F_2(t)\right], \tag{2.51}$$

with[6]

$$F_1(t) = \tfrac{1}{2}\,it\tilde{L}''(\omega_g)^3\,(\omega_e)^{-2}\,L'' - \tilde{L}''(\omega_g)^{3/2}\,(\omega_e)^{-2}\,R\omega_g\tilde{R}[\Gamma^{-1}(t) - I] \\ \times f^+(t)\,[I - \theta(-t)]\,(\omega_e)^{-1}\,(\omega_g)^{3/2}\,L'' ,$$

$$F_2(t) = \tfrac{1}{2}\,it\,\text{Tr}\,\{\omega^-\} + N\ln 2 + \tfrac{1}{2}\ln\left(\det\{\omega_g\omega_e[I - \exp(-\beta\hbar\omega_g)]^2\}\right) \\ + \tfrac{1}{2}\ln\{\det[-f^+(t)\,f^-(t)]\} .$$

---

[5] This expression for $\varrho_{c,\,f}(t)$ is written in a more compact form than in [2.17, 22], but they are identical.

[6] The quantities $F_1(t)$ and $F_2(t)$ are identical, respectively, with $F_d(t)$ and $F_e(t)$ of [2.17, 22] — the notation is altered here since we are using the subscript c to denote the results within the Condon approximation.

It is straightforward to show that in the standard assumptions limit $R = I$ and $\omega_e = \omega_g$ of no mode mixing and no frequency shifts under electronic excitation, these formulas reduce to $\eta_{\ell c}(t)$ and $\varrho_{\ell c, f}(t)$ given by (2.45) and (2.48).

b) Linear Non-Condon Included

The results for full linear and quadratic e-ph coupling *plus* linear non-Condon $\{m_f \neq 0\}$ are easily expressed using the quantities defined above. We obtain [2.20, 21]

$$\eta(t) = \eta_c(t)\left[\psi^2(t) + \zeta(t)\right], \tag{2.52a}$$

$$\varrho_f(t) = \varrho_{c,f}(t) + \left[\psi^2(t) + \zeta(t)\right]^{-1}\psi(t)\left\{W(t)\,m\right\}_f, \tag{2.52b}$$

with

$$\psi(t) \equiv 1 + 2\tilde{m}(\omega_g)^{1/2}\,\tilde{R}[\Gamma^{-1}(t) + I]\,f^+(t)\,[I - \theta(-t)]\,(\omega_e)^{-1}\,R(\omega_g)^{3/2}\,\xi,$$

$$\zeta(t) \equiv \tilde{m}(\omega_g)^{1/2}\,\tilde{R}\{[\Gamma^{-1}(t) - I]\,f^-(t)\,[I - \theta(-t)]$$

$$- [\Gamma^{-1}(t) + I]\,f^+(t)\,[I + \theta(-t)]\}\,R(\omega_g)^{1/2}\,m,$$

$$W(t) \equiv -2(\langle n\rangle + I)^{-1}\,(\omega_g)^{1/2}\,\tilde{R}f^+(t)\,[I + \theta(-t)]\,R(\omega_g)^{1/2}.$$

Here $\langle n\rangle$ is a diagonal matrix of the $\langle n_f\rangle$'s, and $m \equiv \{m_f\}$ is the $N \times 1$ vector of the linear non-Condon coefficients.

The exact expressions (2.26, 49–52) for the absorption and first-order RR profiles contain the e-ph interaction to infinite order. Moreover, as required by the basic equations (2.27, 34), the absorption correlator is seen to be of second order in the m_f's, and after $\eta(t)$ and $\varrho_f(t)$ are multiplied together, the first-order profile (2.49) is seen to be of fourth order in the m_f's.

When we add general quadratic non-Condon coupling, nontotally symmetric modes, and multiple (nondegenerate) electronic excited states [2.20, 21], our resulting exact expressions are straightforward extensions of those above. Also, within some special cases of the general model, similar expressions have been obtained for higher-order, summation band scattering [2.12, 21, 22, 26].

These results constitute a general analytic theory for first-order RR scattering by nonzero temperature mulitmode systems having *simultaneous* linear and quadratic e-ph *and* non-Condon coupling. The temperature enters through the quantities β^{-1}, $\langle n_f\rangle$, $\langle n\rangle$ and $\Gamma^{-1}(t)$, and the $T = 0$ K limit results when these are set equal to zero. As noted in [2.20], our earlier correlator theory results for various special cases of the present model are easily recovered in the appropriate limits. The theory contains *interference* terms, i.e. terms requiring the simultaneous presence of non-Condon and quadratic e-ph coupling. Of central importance is that the expressions are not only analytic, but are finite, as opposed to the complicated infinite series typical of the sum-over-states method for much simpler models. This is significant, both for the modeling aspect and for analytic studies, such as those leading to the transform theory or to the useful concept of renormalized model parameters.

2.4.2 Multimode Modeling

a) General Aspects

The structure of the above solutions provides a convenient and efficient modeling procedure. The time-dependent functions are seen to have many common "building blocks", and they are easily computed from the model parameters and the temperature via simple matrix calculations. In the frequency-shift limit ($R = I$) of no mode mixing, all of the matrices become diagonal, and the computing time increases roughly as N, the number of modes, just as for the standard assumptions case (2.45–48). In the worst-case situation of mode mixing involving all N modes, the matrix operations result in the computing time increasing roughly as N^3 for large N. As pointed out in Sect. 2.1.5, this is far superior to the exponential increase occurring in the sum-over-states methods. With $\eta(t)$ and $\varrho_f(t)$ computed, simple FFT calculations immediately give the absorption and first-order profiles. As a simple test, we computed the room-temperature optical absorption plus all ten first-order RR profiles for a hypothetical ten-mode system in the full model, i.e. with nonzero values of *all* of the coupling parameters, including mode mixing and quadratic non-Condon involving all ten modes. Without making any special efforts towards computing efficiency, we found that the cpu time on an IBM 3081 was five minutes. Clearly, this theory greatly extends the possibilities for precise model calculations of RR and absorption spectra, to complex multimode systems and interaction models which were simply inaccessible with conventional methods.

Another important aspect of the time-correlator model expressions is that they allow one to perform direct first-principles calculations of experimentally observed spectra for complex systems, using parameters obtained from ab initio or semiempirical electronic structure calculations [2.71]. Such calculations can produce large numbers of model parameters, with the result that exact computations of absorption and RR spectra are impossible using conventional approaches. The results given here allow the assumptions underlying the ab initio calculations to be carried all the way through to the experimentally measured spectra. As RR experiments are rapidly being extended to probe an ever-widening range of complicated systems and interactions, the present approach should result in these experiments becoming an important tool for the direct refinement of ab initio calculations.

b) Application to Azulene

To illustrate the multimode modeling aspect of the theory, Fig. 2.1 gives the results of our seven-mode model calculations [2.17] of the measured room temperature absorption and four RR profiles of the azulene molecule $(C_{10}H_8)$ in CS_2, within the spectral range 570–720 nm of the 700 nm $S_0 \rightarrow S_1$ electronic transition. One of the profiles is for the second harmonic of a mode, and the calculations include the relative intensities of the profiles. The input data were the position and width of the lowest energy vibronic peak (0-0) in the measured absorption, together with literature values of the electronic ground and excited state mode frequencies observed in $T = 4$ K absorption and emission spectra. We varied the linear e-ph coupling parameters for the seven totally symmetric modes in the model, plus a single mode mixing parameter. There was no non-Condon coupling, so the theory was based on (2.49–51).[7] The results are seen to be in good agreement with the experiments, both for the detailed structure of the individual spectra and for the profile relative intensities. Using a different width and position of the first absorption peak, but otherwise the same values of the model parameters, we obtained similarly good results for the absorption and the first-order profile of another mode of azulene in the more polar solvent methanol [2.17]. Although these calculations led to well-defined values of the parameters within the model, additional calculations were done in [2.17] for a model with no mode mixing but including non-Condon coupling, with good results. Accordingly, we have recently extended the work of [2.17] to predict the second-order *summation* band scattering, in order to discriminate between competing models for this multimode system [2.26].

c) Multimode Calculations with All Couplings Present

Our analytic solutions allow one to study the effects of having several types of complicated interactions present together. Figure 2.4 shows some results from model studies [2.21] of a hypothetical four-mode system containing simultaneous linear non-Condon plus linear and quadratic e-ph coupling. The values of the model parameters are realistic for molecular systems, and the four modes were chosen to have various properties. All four have linear e-ph coupling, given in Table 2.1. Modes 1 and 2 are the only mixed modes, so that the nondiagonal part of R is a 2×2 submatrix, with elements $R_{11} = R_{22} = \cos(\theta)$ and $R_{12} = -R_{21} = \sin(\theta)$, where θ is the angle of normal coordinate rotation under electronic excitation. For these calculations, we took $\sin(\theta) = -0.42$ ($\theta = -25°$). For simplicity these two modes have no frequency shifts under electronic excitation, but they *do* have linear non-Condon coupling.

---

[7] The prefactor $\omega_L \omega_S^3$ in (2.49) has been removed from all of the experimental and calculated profiles in this chapter.

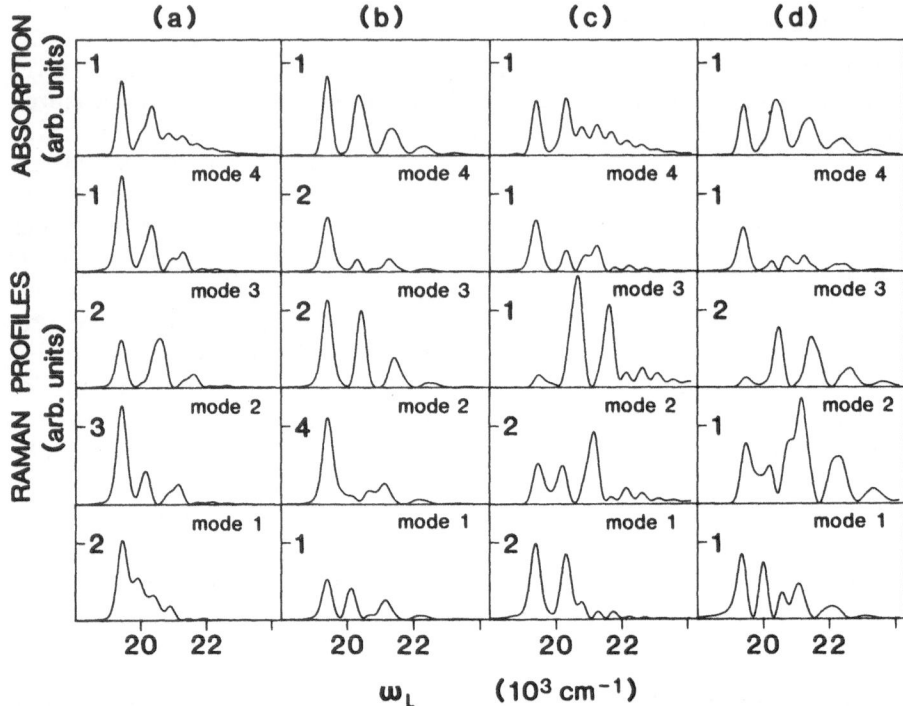

Fig. 2.4. Optical absorption and first-order RR profiles for the four-mode model of Table 2.1 and discussed in the text. (a) Standard assumptions case. (b) Standard assumptions plus quadratic e-ph coupling. (c) Standard assumptions plus linear non-Condon coupling. (d) Full model. After [2.21]

The unmixed mode 3 also has linear non-Condon coupling, together with a frequency shift. The fourth mode obeys the standard assumptions, having no non-Condon and only linear e-ph coupling.

Phenomenological Gaussian broadening of the vibronic peaks was included by introducing a factor $\exp(-\sigma^2 t^2/2)$ into the absorption correlator and setting $\gamma = 0$. Although the precise form of the broadening

Table 2.1. Parameters for the four-mode model of Fig. 2.4. As discussed in the text, modes 1 and 2 are mixed under electronic excitation, with $\sin(\theta) = -0.42$. After [2.21]

Mode index f	Ground state frequency $\omega_{g,f}$ [cm^{-1}]	Excited state frequency $\omega_{e,f}$ [cm^{-1}]	ξ_f	m_f
1	600	600	0.50	−0.15
2	900	900	0.70	0.20
3	1350	1150	0.40	0.20
4	1000	1000	0.40	0.00

is immaterial for these illustrative calculations, it will be seen in Sect. 2.4.4a that a Gaussian form is *required* by our general theory for the case of homogeneous broadening due to a reservoir of low-frequency modes coupled to the resonant transition. For clarity in resolving the vibronic structure, a width (HWHM) of 150 cm^{-1} was used. The zero-phonon (0-0) transition frequency was set at 19400 cm^{-1} and we used $T = 300 \text{ K}$.

Using (2.26, 49, 52) together with standard FFT techniques, we obtained the absorption and first-order profiles shown in Fig. 2.4. The numbers at the ticmarks on the profile vertical axes give the correct relative intensities of all of the profiles. Likewise, the absorptions are plotted at their correct relative intensities.

Columms (a-c) give the results for special cases which do *not* include the full coupling of our model. Column (a) is for the standard assumptions case including only linear e-ph coupling but neglecting quadratic e-ph coupling and non-Condon coupling; column (b) includes linear and quadratic e-ph coupling but no non-Condon coupling; column (c) includes linear e-ph and linear non-Condon coupling but no quadratic e-ph coupling. Column (d) gives the results for the general theory including the simultaneous presence of all of the couplings in the model.

As can be seen, the absorption and all four RR profiles in column (d) differ appreciably from those for any of the special cases shown in the other three columns, both in lineshapes and relative intensities. These differences would easily be detected in experiments. Notice that this holds even for mode 4, which has neither non-Condon nor quadratic e-ph coupling. This is an illustration of multimode interference effects, which in this case arise through the functions $\psi(t)$ and $\zeta(t)$ occurring in $\eta(t)$ and $\varrho_f(t)$. Additional details are given in [2.21].

2.4.3 Transform Theory

Complementary to the explicit modeling, aspects, the many-body approach brings out very usefully the fundamental connection between RR profiles and the optical absorption. As noted earlier, this connection is reflected by the presence of the absorption correlator $\eta(t)$ as a factor in the integrand of the Fourier transform which determines the first-order profiles (2.49). We will see that if the factor $\varrho_f(t)$ in (2.49) is solely a function of the parameters of the Raman mode f, then the mode profile may be computed *directly* from the measured temperature-dependent absorption, plus model parameters for *just* the Raman mode. This is the essence of the "transform" methods, which are proving very fruitful for analyzing RR data for multimode systems [2.9, 12, 13, 15, 16, 19, 21–25, 30–50].

Transform calculations of RR profiles give necessary (but not sufficient) tests of the consistency of absorption and profile data and the underlying model assumptions. More importantly, they provide a method for extracting model parameters from data for complex multimode systems *mode by mode*. The multimode information is automatically carried through the calculation via the use of the measured input absorption. Transform calculations are simple enough to be easily carried out on typical laboratory computers. Hence they provide a powerful adjunct to full-fledged multimode model calculations, which for complex systems require measurements of many mode profiles. Transform techniques have been applied to study a variety of systems, ranging from solid-state impurity systems such as $KClO_4 : MnO_4^-$ [2.12, 24], $CsI : MnO_4^{2-}$ [2.18], etc., to complex organic molecules such as β-carotene [2.9, 13, 19], pyrimidine nucleotides [2.30], azulene [2.16, 39], cytochrome-c [2.35, 36, 45], etc. Moreover, these techniques have been extended to analyze experimental data from resonance CARS (coherent anti-Stokes Raman scattering) [2.37, 43].

Calculations of RR profiles directly from measured absorption spectra [2.9] were originally based on a relation between the absorption and RR profiles given in [2.28] for systems obeying the standard assumptions, i.e. no non-Condon or quadratic e-ph coupling. Temperature effects in transform calculations were studied in detail in [2.13]. Transform techniques have been extended to wider classes of model assumptions by several groups, within various approximations. These assumptions include frequency shifts with electronic excitation [2.12], linear non-Condon coupling [2.15, 33, 34], simultaneous frequency shifts *plus* linear non-Condon coupling [2.23], inhomogenous broadening [2.19, 24], nonadiabatic corrections [2.35] and anharmonicity [2.40]. Recently, we have extended the transform theory so as to encompass simultaneous mode mixing and non-Condon coupling [2.25]. In this case, the results are that one can compute the RR profile of a mode from the measured absorption and model parameters of just a *subset* of modes, namely the Raman mode plus those modes mixed with it. These results are shown in [2.25b] to offer a sensitive means for extracting information about mode mixing in complex systems.

a) Formal Transform Expression

To obtain transform relations, we first introduce the normalized absorption lineshape function $I(\omega_L)$,

$$I(\omega_L) \equiv \left[\int d\omega \, \alpha(\omega)/\omega\right]^{-1} \alpha(\omega_L)/\omega_L \tag{2.53a}$$

$$= [\pi\eta(0)]^{-1} \, \text{Re} \left\{ \int\limits_0^\infty dt \, \exp(i\omega_L t - \gamma t) \, \eta(t) \right\}, \tag{2.53b}$$

where (2.26) has been used to obtain (2.53b). In evaluating the normalization integral, it is clarifying to extend the lower limit of the integration in (2.26) to $t = -\infty$, through the use of the identity $\eta(-t) = \eta^*(t)$. Note that $\eta(0)$ is real. We next define a complex function $\Phi(\omega_L)$,

$$\Phi(\omega_L) \equiv P \int\limits_{-\infty}^\infty d\omega \, I(\omega) \, [\omega - \omega_L]^{-1} + i\pi I(\omega_L), \tag{2.54}$$

where P denotes a Cauchy principle value. Thus the real and imaginary parts of $\Phi(\omega_L)$ are related by a Kramers-Kronig transform, and we see that $\Phi(\omega_L)$ is proportional to the resonant part of the polarizability associated with the $|g\rangle \rightarrow |e\rangle$ electronic transition. Substituting (2.53 b) into (2.54) and carrying out the integration yields

$$\Phi(\omega_L) = [\eta(0)]^{-1} \, \mathrm{i} \int_0^\infty \mathrm{d}t \exp(\mathrm{i}\omega_L t - \gamma t) \, \eta(t), \qquad (2.55)$$

in terms of which the absorption (2.26) may be expressed as

$$\alpha(\omega_L) = B\hbar^{-1} |M(0)|^2 \, \eta(0) \, \omega_L \, \mathrm{Im} \, \Phi(\omega_L). \qquad (2.56)$$

The essential point here is that the function $\Phi(\omega_L)$ may be evaluated directly from the *measured* temperature-dependent absorption, via (2.53 a) and (2.54).

The profiles will now be expressed in terms of the function $\Phi(\omega_L)$. First note that (2.55) gives the identity

$$\mathrm{i} \int_0^\infty \mathrm{d}t \exp(\mathrm{i}\omega_L t - \gamma t) \, \eta(t) \exp(-\mathrm{i}t\omega m) \equiv \eta(0) \, \Phi(\omega_L - m\omega)$$

$$\equiv \eta(0) \, \hat{\sigma}^m(\omega) \, \Phi(\omega_L),$$

where m is an integer and the shift operator $\hat{\sigma}(\omega)$ is defined by $\hat{\sigma}(\omega) \times \Phi(\omega_L) \equiv \Phi(\omega_L - \omega)$. Now if $F[\exp(-\mathrm{i}t\omega)] = \sum_m c_m \exp(-\mathrm{i}t\omega m)$ is any function of $\exp(-\mathrm{i}t\omega)$ which can be expanded in a power series, we have the useful result

$$\mathrm{i} \int_0^\infty \mathrm{d}t \exp(\mathrm{i}\omega_L t - \gamma t) \, \eta(t) \, F[\exp(-\mathrm{i}t\omega)] \equiv \eta(0) \, F[\hat{\sigma}(\omega)] \, \Phi(\omega_L). \qquad (2.57)$$

The integral in (2.49) is of this form, since $\varrho_f(t)$ is a function of the exponential factors $\{\exp(-\mathrm{i}t\omega_{e,f})\}$ and $\{\exp(\mathrm{i}t\omega_{g,f})\}$, which enter via the matrices $\theta(-t)$ and $\Gamma^{-1}(t)$, respectively. The use of (2.57) thus allows one to reexpress (2.49) formally in the frequency domain:

$$j_{1,f}(\omega_L) = C[\eta(0)]^2 \, (\langle n_f \rangle + 1) \, \omega_L (\omega_L - \omega_{g,f})^3 \, |\hat{\varrho}_f \Phi(\omega_L)|^2. \qquad (2.58)$$

The operator $\hat{\varrho}_f$ is just the function $\varrho_f(t)$ of (2.52b), but with the exponential factors $\{\exp(-\mathrm{i}t\omega_{e,f})\}$ and $\{\exp(\mathrm{i}t\omega_{g,f})\}$ replaced by the shift operators $\{\hat{\sigma}(\omega_{e,f})\}$ and $\{\hat{\sigma}(-\omega_{g,f})\}$, respectively. Also, from the model expression (2.52a) for $\eta(t)$, we find $\eta(0) = 1 + \sum_f (2\langle n_f \rangle + 1) \, m_f^2$.

Note that (2.58) holds for *any* model and/or order of RR scattering as long as the profile can be expressed in the generic form (2.49); that is, as the complex square of a one-dimensional Fourier transform in which the absorption correlator $\eta(t)$ appears as an explicit factor. As stated earlier, we have also obtained this form for higher-order summation scattering, within particular models [2.12, 21, 22, 26]. Thus the formal result (2.58) applies to these cases as well, with the specific model dependence entering through the explicit form for the higher-order analogs of the operator $\hat{\varrho}_f$.

b) Condon Approximation Transform Including Frequency Shifts

The formally exact result (2.58) will now be studied for the important case of nonzero vibrational frequency shifts ($\omega_e \neq \omega_g$) but no mode mixing ($R = I$) under electronic excitation, within the Condon approximation {$m_f = 0$}. It is straightforward to show from (2.52b) that $\hat{\varrho}_f$ is then given by [2.12, 22]

$$\hat{\varrho}_f(\omega_L) = -(\lambda'_f)^{1/2} [1 - \hat{\sigma}(\omega_{e,f})] \{1 - \delta_f \hat{\sigma}(\omega_{e,f}) + \langle n_f \rangle$$
$$\times [1 - \hat{\sigma}(\omega_{e,f} - \omega_{g,f})] - \langle n_f \rangle \delta_f [\hat{\sigma}(\omega_{e,f}) - \hat{\sigma}(-\omega_{g,f})]\}^{-1}, \tag{2.59}$$

where $\delta_f \equiv (\omega_{e,f} - \omega_{g,f})/(\omega_{e,f} + \omega_{g,f})$ describes the frequency shifts, and where the *renormalized* Stokes loss parameter λ'_f is defined in terms of the "bare" Stokes loss parameter $\lambda_f = [L_f/(\hbar\omega_{g,f})]^2$, by

$$\lambda'_f \equiv \lambda_f 4\omega^4_{g,f}[\omega_{e,f}(\omega_{g,f} + \omega_{e,f})]^{-2}. \tag{2.60}$$

Keeping in mind that the inverse in the exact operator expression (2.59) is to be understood in the sense of a power series expansion, we see that the use of (2.59) in (2.58) allows the profile to be computed directly from the absorption [used to obtain $\Phi(\omega_L)$] and shift operations involving just the Raman mode f. However, (2.59) takes on a much simpler, approximate, form in the commonly occurring case of small frequency shifts under electronic excitation. Specifically, if $|\delta_f| \ll 1$ and $\langle n_f \rangle |\delta_f| \ll 1$, (2.59) collapses to $\hat{\varrho}_f \simeq -\lambda_f^{1/2}[1 - \hat{\sigma}(\omega_{e,f})]$, where we have used the fact that $1 - \hat{\sigma}(\omega_{e,f} - \omega_{g,f})$ is of first order in δ_f. Then the first-order profile (2.58) reduces to the practically important approximate result [2.12, 22]

$$j_{1,f}(\omega_L) \simeq C(\langle n_f \rangle + 1) \omega_L(\omega_L - \omega_{g,f})^3 \lambda'_f |[1 - \hat{\sigma}(\omega_{e,f})] \Phi(\omega_L)|^2. \tag{2.61}$$

For the case of first-, second-, or third-harmonic Stokes scattering by a mode f, we have shown that (2.61) can be generalized to give [2.12, 22].

$$j_{n,f}(\omega_L) \simeq C(\langle n_f \rangle + 1)^n \omega_L(\omega_L - n\omega_{g,f})^3 (\lambda'_f)^n$$
$$\times |[1 - \hat{\sigma}(\omega_{e,f})]^n \Phi(\omega_L)|^2/n!, \tag{2.62}$$

for $n \leq 3$. For $n = 2$ and 3, the additional condition $|\delta_f/\lambda'_f| \ll 1$ is needed, but again this is not a stringent condition in practice. An expression analogous to (2.62), but in the context of scattering by multiple LO phonons in perfect crystals via the Fröhlich mechanism, is given by eq. 2.252 of [2.1].

Equations (2.62) allow one to compute the RR profile *lineshapes* of a mode f directly from the measured temperature-dependent optical absorption and the mode frequency $\omega_{e,f}$ in the electronic excited state. If the absolute intensities of the measured absorption and RR profiles are known, and if the transformed lineshapes are in good agreement with experiment, one can extract the e-ph coupling parameters λ'_f as simple scaling factors. This can be done one mode at a time and does not require a complete set of profile measurements. *Champion, Albrecht,* and co-workers have very fruitfully exploited this procedure in their studies on cytochrome-c [2.35, 36].

As an illustration of how well the transform method can reproduce the complicated profile lineshapes for multimode systems, Fig. 2.5e gives a comparison of our computed and measured room temperature profile lineshape for the 825 cm^{-1} Raman line of the azulene molecule in methanol [2.16]. The theoretical profile lineshape (solid curve) was computed via (2.53a, 54, 61), using *only* the measured room temperature absorption and the excited state mode frequency $\omega_{e,f} = 857$ cm^{-1}, obtained from $T = 4$ K absorption experiments. To illustrate the simplicity of the calculations, the results of the intermediate steps are shown in panels (a-d). The left-most curve in panel (a) is proportional to $\omega_L^{-1} \times$ the measured room temperature absorption; specifically, it is $\Phi_i(\omega_L) = \pi I(\omega_L)$ of (2.54). The left-most curve in panel (b) is $\Phi_r(\omega_L)$, computed from $\Phi_i(\omega_L)$ via a Kramers-Kronig transform according to (2.54). The right-most curves in panels (a) and (b) are just $\Phi_i(\omega_L - \omega_{e,f})$ and $\Phi_r(\omega_L - \omega_{e,f})$, respectively. Panel (c) shows $\{[1 - \hat{\sigma}(\omega_{e,f})]\,\Phi_i(\omega_L)\}^2 \equiv [\hat{\Delta}(\omega_{e,f})\,\Phi_i(\omega_L)]^2 = [\Phi_i(\omega_L) - \Phi_i(\omega_L - \omega_{e,f})]^2$, and panel (d) gives $[\hat{\Delta}(\omega_{e,f})\,\Phi_r(\omega_L)]^2$. The results of panels (c) and (d) are then added to form $|\hat{\Delta}(\omega_{e,f})\,\Phi(\omega_L)|^2$, which by (2.61) is proportional to the profile lineshape $\times\,(\omega_L\omega_s^3)^{-1}$. Notice how well the strong multimode interference effects in the measured profiles, given by the circles, are reproduced in the calculation.

The power of transform methods is best realized for multimode systems having intermediate e-ph coupling, such that the absorption and RR profiles exhibit well-resolved vibrational structure. Multimode interference then leads to strong variations in the profile lineshapes from mode to mode, as exemplified by the set of measured profiles in Fig. 2.1 for azulene in CS$_2$. In contrast, for systems with broad, structureless, absorption bands whose widths are large compared with the mode frequencies, the profile lineshapes given by (2.61) just become proportional to $|d\Phi(\omega_L)/d\omega_L|^2$, and then *all* modes have the same profile lineshape. F-centers in alkali halides have strongly phonon-broadened absorption bands typifying such behavior, and in earlier work [2.54] we found that this "derivative" approximation works well for studies of the observed variation of the defect-induced first-order RR spectrum (as opposed to profiles) as ω_L is tuned through different *electronic* transitions. However, for the present focus on systems with vibrationally structured absorption bands, it is crucial to use the finite-difference forms (2.61) or (2.62).

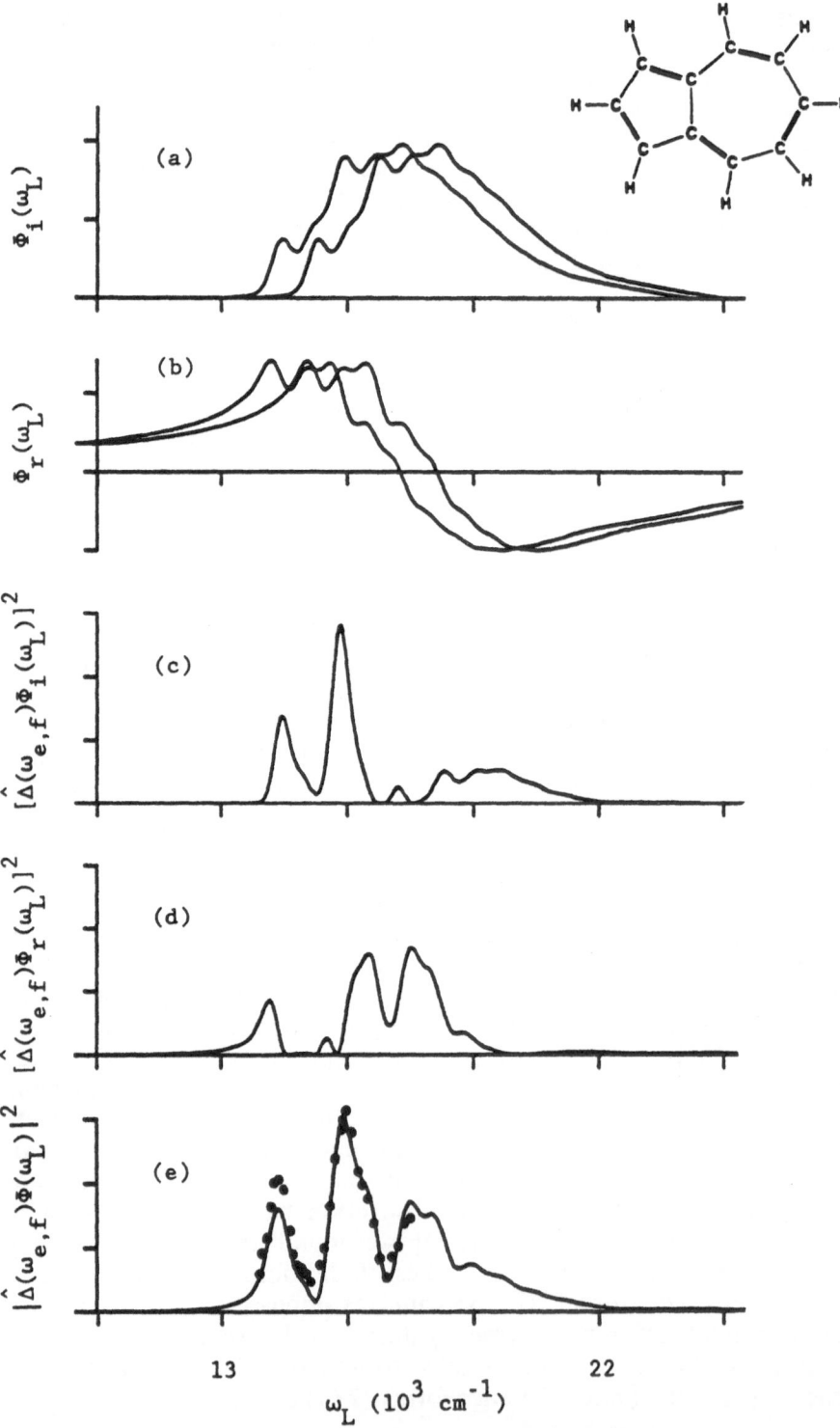

ω_L $(10^3$ $cm^{-1})$

In the $\{\delta_f \to 0\}$ limit *no* vibrational frequency shifts with electronic excitation, $\hat{\sigma}(\omega_{e,f}) \to \hat{\sigma}(\omega_{g,f})$; $\lambda'_f \to \lambda_f$; and (2.62) becomes *exact* for all n. For $n = 1$, this can be seen from the exact expression (2.59) in this limit. This is just the standard-assumptions case, for which the result may be obtained more directly by substituting (2.48) into (2.47) and using (2.57). Since $\hat{\sigma}(\omega_{g,f})\Phi(\omega_L) = \Phi(\omega_S)$, the first-order profiles for this case may be written as

$$j_{1,f}(\omega_L) = C(\langle n_f \rangle + 1)\, \omega_L \omega_S^3 \lambda_f \, |\Phi(\omega_L) - \Phi(\omega_S)|^2 \,. \tag{2.63}$$

This result was given by *Hizhnyakov* and *Tehver* in their pair correlations approach [2.28], and it formed the basis for the transform calculations of RR profiles directly from measured absorption bands, introduced in [2.9] and detailed in [2.11, 13]. In [Ref. 2.13, Appendix A] we generalized (2.63) and obtained an exact expression for the *entire* RR series (i.e. all orders of Stokes and anti-Stokes scattering, for any temperature) in terms of simple shift operations on the function $\Phi(\omega_L)$, within the standard assumptions.

The form (2.63) for the first-order Stokes scattering was also given in [Ref. 2.1, Eq. (2.165)] and [Ref. 2.72, Eq. (2.94)], in the context of the "solid-state" basis mentioned in the introduction. Within the present standard-assumptions context, we see that this form is exact to all orders of the e-ph coupling and for all temperatures. However as we have seen above, when one goes beyond this case, (2.63) is modified such that the first-order profile is no longer proportional to the square magnitude of $\Phi(\omega_L) - \Phi(\omega_S)$. The vibrational structure in the absorption reflects the *excited* state mode frequencies, and for systems with appreciable frequency shifts under electronic excitation, it is important that the excited state frequency be used in the transform, as required by (2.61) or (2.62). This is illustrated in Fig. 2.6, which shows our measured and calculated room temperature first-order profile lineshapes at two different pressures, for the breathing mode in the solid-state impurity system $KClO_4 : MnO_4^-$ [2.24]. The experiments were done with a diamond anvil cell, and the left-hand column is for atmospheric pressure, while the right-hand column is for 10.6 GPa. The upper panel in each column gives the measured absorption, used as input to the transform calculations. The only other input parameter was the mode frequency used in the shift operator in (2.61), and the two lower panels in each column give the results of using the ground state frequency or the excited state frequency at the appropriate

---

◀ **Fig. 2.5.** Transform calculation of the room temperature first-order RR profile lineshape for the 825 cm^{-1} mode of azulene in methanol. The experimental absorption and profile lineshapes of [2.16] are given, respectively, by the left-most curve in panel (a) and by the circles in panel (e). The different panels show the intermediate steps in the transform calculation, as explained in the text. The computed profile lineshape given by the solid curve in panel (e) was multiplied by a simple scaling factor to fit the experimental lineshape. Note that the only input to the profile lineshape calculation is the experimental absorption spectrum and the mode frequency 857 cm^{-1} in the excited electronic state

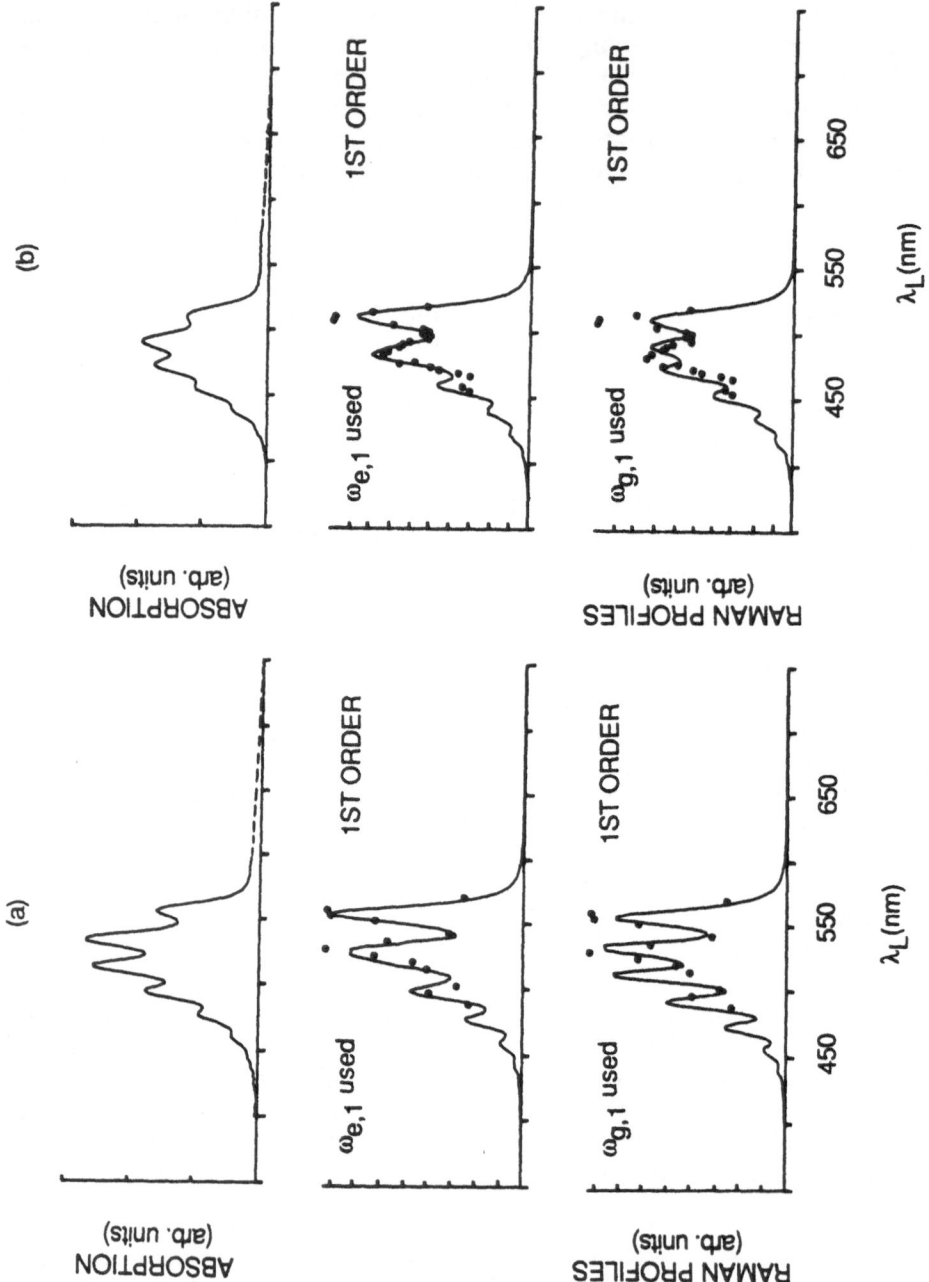

◀ **Fig. 2.6.** Frequency-shift and pressure effects in transform calculations. Shown are the room-temperature absorption and first-order RR profile lineshapes for the breathing mode ($f = 1$) of the solid-state impurity system $KClO_4 : MnO_4^-$, for two different pressures: 1 atm (column a), and 10.6 GPa (column b). The measured absorption spectra used as input to the transform calculations are given in the top panels. The solid lines in the lower four panels are the results of transform calculations using the mode frequency for the excited or ground electronic state in the shift operator in (2.61). For 1 atm, these frequencies are $\omega_{e,1} = 750 \, cm^{-1}$ and $\omega_{g,1} = 846 \, cm^{-1}$. At 10.6 GPa, they are $\omega_{e,1} = 800 \, cm^{-1}$ and $\omega_{g,1} = 900 \, cm^{-1}$. After [2.24]

pressure. The former was obtained from the measured Raman shift, and latter was determined from the spacing of the vibronic peaks in the measured absorption spectra, which are dominated by this mode. Notice that at both pressures the use of $\omega_{g,f}$ gives poor agreement, and even predicts the wrong number of peaks, whereas the use of $\omega_{e,f}$ gives good agreement with the measured profile lineshapes.

Transform calculations offer a useful prelude to bona fide multimode model calculations, since they allow one to combine experimental absorption and profile data to extract information on the parameters of the Raman modes a mode at a time. For instance, if a transform like (2.61) or (2.62) applies, one can compute just profile *lineshapes*, and study the effect of varying $\omega_{e,f}$. To study the renormalized Stokes loss parameters $\{\lambda'_f\}$, one needs to go beyond lineshape analysis and consider intensities. As pointed out earlier, the clearest case is that where *absolute* profile and absorption intensities are measured, so that the $\{\lambda'_f\}$ can be determined by scaling the transform profiles to the data. A very favorable situation occurs if a mode scatters so strongly that its profiles can be measured in different *orders*, i.e. if overtones as well as the fundamental profile can be obtained. In this case, just the easily measured *relative* intensities can be used to extract λ'_f. The straightforward procedure for doing this is to compute the profile lineshapes from the absorption using (2.62) for different orders n, scale the results to the experimental profiles, and take ratios of the resulting scaling parameters. We have used this procedure in [2.12, 18, 24]. Lacking either overtone profiles or absolute intensity measurements, but having measured the *first*-order profile relative intensities for a *set* of modes, one can still obtain the *ratios* $\lambda'_f/\lambda'_{f'}$ for different modes. Such calculations were carried out in [2.16] for azulene, and they yielded very useful information for the subsequent explicit model calculations in [2.17].

Temperature effects are automatically incorporated in transform calculations, by means of the temperature-dependent input optical absorption used to compute $\Phi(\omega_L)$. Figure 2.7 gives the results of our transform calculation of the first-order RR profile lineshapes for three modes of the biologically important molecule β-carotene, in the solvent CS_2, for $T = 172 \, K$ and $T = 300 \, K$ [2.19]. For these calculations, the frequency shifts with electronic excitation were found to be unimportant, so that the standard assumptions transform (2.63) applied. Thus the only input to

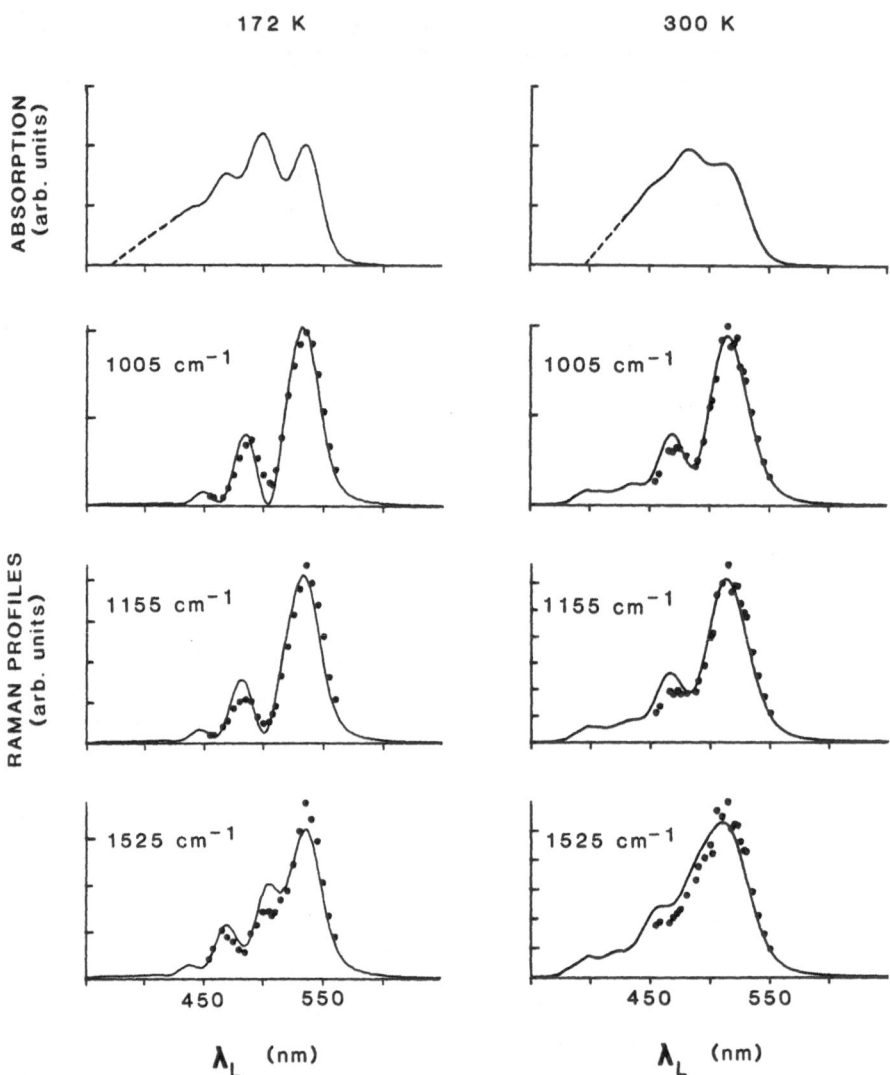

Fig. 2.7. Temperature effects in transform calculations. Shown are transform calculations of the first-order profile lineshapes for three modes of β–carotene in CS_2 at two temperatures, as indicated. The dashed portions of the experimental absorption spectra here and in Fig. 2.6 are extrapolations used for the transform calculations. After [2.19]

these profile lineshape calculations were the measured absorption spectra and the Raman shifts. The agreement between theory and experiment is seen to be good at both temperatures. Thus, although there is considerable thermal broadening in both the absorption and the profiles as the temperature is raised, the absorption and profile lineshapes remain related by the transform relation.

c) Transforms Including Frequency Shifts and Non-Condon Coupling

While the formal transform (2.58) is exact, the preceding results and illustrations pertain to just the Condon approximation. An important generalization is to include non-Condon coupling. For the linear e-ph coupling case of *no* frequency shifts or mode mixing, this was done by *Tehver* [2.33], *Stallard* et al. [2.34], and *Chan* [2.15], with Chan giving the most detailed theoretical discussion. However, we have seen that the inclusion of frequency shifts can be important. Accordingly, generalized transform results have been obtained for the case when there are *simultaneous* frequency shifts and linear non-Condon coupling. These results are detailed and illustrated in [2.23], and the main results will now be briefly sketched.

For this case, we need to obtain the operator $\hat{\varrho}_f$ in (2.58) for the full model of Sect. 2.4.1b, but without mode mixing ($R = I$). From our exact model expression for $\varrho_f(t)$ in (2.52b), we can make a Taylor series expansion of the quantity $[\psi^2(t) + \zeta(t)]^{-1}$ and rewrite $\varrho_f(t)$ as

$$\varrho_f(t) = \varrho_{c,f}(t) + [W(t)\, m]_f + O(m^2)\,, \qquad (2.64)$$

where the last term represents all contributions to $\varrho_f(t)$ of second and higher orders in the linear non-Condon coupling coefficients $\{m_f\}$. This part will henceforth be dropped. As noted in [2.23], this approximation is difficult to study analytically, especially for the $T \neq 0\,\mathrm{K}$ case. We have studied this approximation numerically, and have found that it holds very well for typical parameter values. It is important to note that there is *no* corresponding approximation for the absorption correlator $\eta(t)$, which is *exact* within the model.

The operator $\hat{\varrho}_f$ is obtained by: (1) making the approximation (2.64) for $\varrho_f(t)$; (2) focusing on the profile of an unmixed mode f; (3) assuming as in (2.61) that this mode's frequency shift and thermal population $\langle n_f \rangle$ are small enough that $|\delta_f| \ll 1$ and $|\delta_f| \langle n_f \rangle \ll 1$; and (4) making the shift operator replacements for the time exponentials. Then (2.58) yields a simple approximate transform relation

$$
\begin{aligned}
j_{1,f}(\omega_L) &\simeq C''(\langle n_f \rangle + 1)\, \omega_L \omega_S^3\, |\{\xi'_f[1 - \hat{\sigma}(\omega_{e,f})] \\
&\quad - m'_f[1 + \hat{\sigma}(\omega_{e,f})]\}\, \Phi(\omega_L)|^2\,,
\end{aligned}
\qquad (2.65)
$$

where $\xi'_f \equiv \xi_f 2\omega_{g,f}^2/[\omega_{e,f}(\omega_{g,f} + \omega_{e,f})]$ and $m'_f \equiv m_f 2\omega_{g,f}/(\omega_{g,f} + \omega_{e,f})$ are *renormalized* dimensionless linear e-ph and non-Condon coupling constants, respectively. [Note that $(\xi'_f)^2 = \lambda'_f$.] The frequency-independent constant C'' is defined by $C'' \equiv C[\eta(0)]^2 = C[1 + \sum_f (2\langle n_f \rangle + 1)\, m_f^2]^2$. This new transform still allows one to compute the first-order profile of a mode using the temperature-dependent optical absorption and *just* parameters for the mode. These are the excited state frequency $\omega_{e,f}$, and the renormalized linear e-ph and non-Condon parameters defined above.

For the case of no frequency shifts ($\omega_{e,f} = \omega_{g,f}$), (2.65) reduces to the transform relation obtained via correlator methods in [2.15] for this case. That result has also been discussed and applied to the study of experimental profile data in [2.35, 39]. For the opposite case of no non-Condon coupling, but with frequency shifts present, (2.65) reduces to (2.61) above, as it must. Numerical studies of the transform relation (2.65) for a hypothetical but realistic system having simultaneous frequency shifts and non-Condon coupling are described in [2.23].

It should be pointed out that the assumption of no mode mixing ($R = I$) for the preceding transform relations (2.61–63, 65) is actually too strong. It is easy to see that each of these relations holds, provided only that the Raman mode f *itself* is not involved in mode mixing, in which case $R_{ff'} \propto \delta_{ff'}$. Note that mode mixing involving other modes in the system *will* generally affect the profile of mode f, through the input optical absorption. In this connection, it is interesting to note that the profile of a mode which obeys the standard assumptions is given by the simple result (2.63), even though there could be mode mixing, non-Condon coupling and frequency shifts involving all of the *other* modes in the system.

d) General Transforms Including Mode Mixing and Non-Condon Coupling

Finally, let us turn to the most general transform yet obtained, namely a practicable transform that includes linear non-Condon coupling, mode mixing and frequency shifts [2.25]. The derivation is described in detail in [2.25b], and it follows from the general solutions (2.52) given above, within well-defined approximations. The approximations leading to the new transform relation are: (1) Again neglect the terms of order $\{m_f^2\}$ in (2.64). (2) Assume that the Raman mode f and the modes f' mixed with it are of sufficiently high frequency that they are frozen out ($\langle n_{f'} \rangle \simeq 0$). Thermal broadening due to a reservoir of weakly coupled thermally populated low-frequency modes *is* still allowed, and it enters the transform through the input optical absorption; but for the following transform relation, we are assuming that the reservoir modes are not mixed with the Raman mode f. This is a reasonable assumption for many cases, since the Raman modes are usually high-frequency "internal" molecular modes, whereas the low-frequency modes typically result from weak "external" interactions, i.e. molecule/solvent or impurity/host. (3) Approximate the matrix inverse $[I + X \exp(-it\omega_e)]^{-1}$ as I, where X denotes the product $(\omega_g)^{-1/2} (\omega^-) (\omega^+)^{-1} (\omega_g)^{1/2}$. As discussed in [2.25b], this approximation may be directly checked from the model parameters one obtains in carrying out the resulting transform calculations, and it is found to be valid for a wide range of practically important systems. The resulting transform relation for first-order Stokes scattering is given by

$$j_{1,f}(\omega_L) \simeq C'' \omega_L \omega_S^3 \left| \sum_{f'} (\omega_{g,f'}/\omega_{g,f})^{1/2} R_{f'f}(\xi'_{f'}[1 - \hat{\sigma}(\omega_{e,f'})] \right.$$
$$\left. - (\omega_{g,f}/\omega_{g,f'}) m'_{f'}[1 + \hat{\sigma}(\omega_{e,f'})]) \Phi(\omega_L) \right|^2 , \qquad (2.66)$$

where the renormalized dimensionless linear e-ph and non-Condon coupling constants $\{\xi'_f\}$ and $\{m'_f\}$ are now defined by the matrix equations

$$\xi' \equiv 2\omega_e(\omega_g)^{-1/2} (\omega^+)^{-1} R\omega_g \tilde{R}(\omega_e)^{-2} R(\omega_g)^{3/2} \xi, \qquad (2.67a)$$

$$m' \equiv 2(\omega_g)^{1/2} (\omega^+)^{-1} R(\omega_g)^{1/2} m. \qquad (2.67b)$$

The quantities ξ' and m' are $N \times 1$ vectors, just as are ξ and m. Obviously, in the limit $R = I$ of no mode mixing, the transform relation (2.66) reduces to our earlier result (2.65) for this case [but without the prefactor $(\langle n_f \rangle + 1)$, owing to our assumption here that the Raman mode f is frozen out]. Indeed, all of the preceding transform relations given here are obtainable from (2.66) in the appropriate limits (taking into account the omitted prefactor noted above). The importance of (2.66) for multimode systems is that it allows one to compute the profile of mode f from the optical absorption and the model parameters of just a subset of modes, namely the Raman mode f and those modes f' with which it is mixed. Therefore, this transform will be most useful when the Raman mode is mixed with a relatively small number of modes. At any rate, the new transform always requires fewer model parameters than does an explicit multimode model calculation.

By comparison with (2.65), we see that the mode-mixing transform (2.66) simply mixes the amplitudes of the profiles of the directly mixed modes, with the admixture being determined by the mode mixing matrix $\{R_{ff'}\}$. This very clearly reveals the nature of the intensity borrowing between the profiles of mixed modes.

Figure 2.8 illustrates how the new transform could be used as a sensitive probe of mode mixing. Shown are the results of numerical calculations based on the Condon approximation $\{m_f = 0\}$ version of the transform (2.66), for a hypothetical seven-mode system whose model parameters are the same as those in our multimode model calculations of Fig. 2.1 for azulene in CS_2, except that here we have strongly increased the mixing between the two mixed modes (900 cm^{-1} and 1400 cm^{-1}) in that model. The calculations for Fig. 2.1 were for $\sin(\theta) = -0.25$, whereas those for Fig. 2.8 are for $\sin(\theta) = -0.5$. As is mentioned above and detailed in [2.25b], the approximations leading to (2.66) are readily tested, and they are valid for this model. Results are shown for the absorption and the profiles of the two mixed modes, together with the profile of one of the unmixed modes in the model (1260 cm^{-1}), for comparison. The solid lines give the exact multimode model results for these parameter values, and they are the same in both columns. Notice how the increased mode mixing little affects the model absorption and profiles for the 1260 cm^{-1} and 1400 cm^{-1} modes, compared with the results of Fig. 2.1, whereas the model profile for the 900 cm^{-1} mode is strongly affected by this change. The dashed curves in column (a) are profiles computed from the absorption using the non mode-mixing transform (2.61), and their intensities have

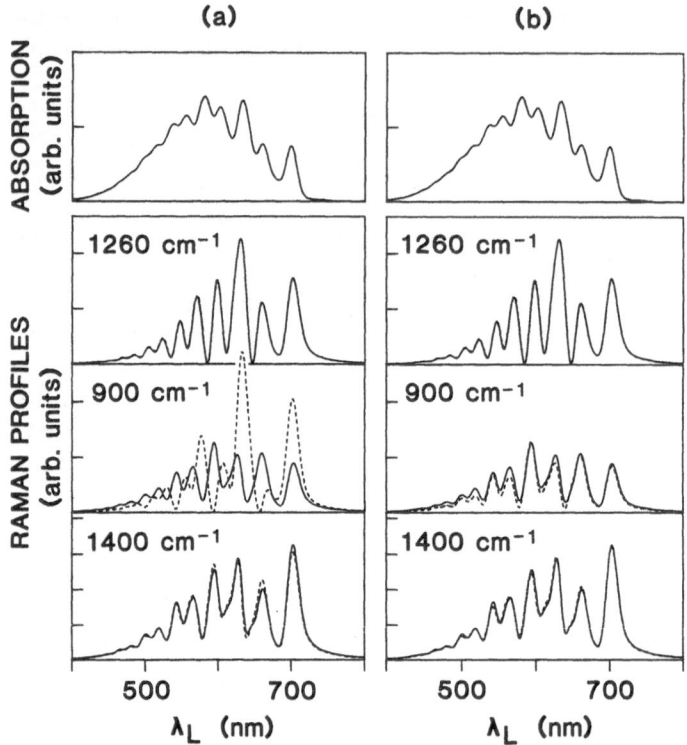

(a)   **(b)**

ABSORPTION (arb. units)

RAMAN PROFILES (arb. units)

1260 cm⁻¹

900 cm⁻¹

1400 cm⁻¹

500   700   500   700

λ_L (nm) λ_L (nm)

Fig. 2.8. Illustration of the use of the general transform (2.66) as a probe of mode mixing for a hypothetical seven-mode system, as described in the text. After [2.25b]

been least-squares fit to the model intensities, as one would do in a typical transform analysis of experimental data. The transform result for the 1260 cm^{-1} profile is perfect, as it must be since this mode is not mixed. The relatively intense profile of the 1400 cm^{-1} mode is also well-reproduced by the non mode-mixing transform, but the weaker 900 cm^{-1} mode profile is seen to be very poorly reproduced. The dashed curves in column (b) were computed with the Condon approximation $\{m_f = 0\}$ version of the mode-mixing transform (2.66), and they are all seen to be in excellent agreement with the exact model calculations. Note that the mode-mixing transform calculations required but *one* additional parameter, namely the mixing angle θ, compared with the non mode-mixing transforms of column (a). As detailed in [2.25b], the intensities of *both* of the transform profiles in column (b) for the two mixed modes were scaled with a *single* least-squares scaling parameter, since the mode-mixing transform requires one to fit the lineshapes and relative intensities of mixed modes *together*. Additional details and calculations are given in [2.25b]. Clearly, this example demonstrates the practical utility of the mode-mixing transform as a sensitive probe of mode mixing, via the study of profile *lineshapes*.

A recent, purely theoretical work on non-Condon/mode-mixing transform relations [2.50] employs different theoretical methods and somewhat different approximations than we have used, but gives results having many features in common with (2.66) above. For instance, in the Condon approximation/no-mode-mixing limit, [2.50] gives $T = 0$ K first- and second-order profile expressions which are identical to our results in [2.12]. An apparent difference in the renormalized Stokes loss parameters between these two references merely reflects different definitions of the "bare" Stokes loss parameters. Numerical studies are not given in [2.50].

2.4.4 Thermal and Inhomogeneous Broadening

a) Thermal Broadening

The time-correlator framework provides a natural setting in which to discuss homogeneous thermal broadening due to a reservoir of low-frequency thermally populated molecular, host-lattice, or solvent modes coupled to the resonant electronic transition. Owing to the large number of modes which could be involved, such broadening is difficult to handle in the traditional sum-over-states methods, and it is therefore common to treat only the high-frequency molecular modes (e.g. the Raman modes) explicitly and assume a phenomenological width $\Gamma(\text{HWHM}) > 100 \text{ cm}^{-1}$, neglecting temperature altogether. Such widths correspond to effective excited state lifetimes $\ll 1$ ps, and are thus usually unrealistically short for the sort of systems under consideration here. But since the "reduced" optical absorption $\alpha(\omega_L)/\omega_L$ is the Fourier transform (2.26) of the absorption correlator $\eta(t)$, the short effective lifetimes imply that $\eta(t)$ is non-negligible only for a short characteristic time $t_c \sim \Gamma^{-1}$. As detailed in [2.13, 14], such behavior results from dephasing in the contributions of the reservoir modes to the correlator. For the simplest, standard-assumptions result (2.45) for $\eta(t)$, it is seen that the reservoir modes contribute the factor

$$\exp\left\{-\sum_{\{f'\}} \lambda_{f'}(2\langle n_{f'}\rangle + 1)\left[1 - \cos(\omega_{f'}t)\right]\right\}\exp\left[-i\sum_{\{f'\}} \lambda_{f'}\sin(\omega_{f'}t)\right],$$

$$(2.68)$$

where the primes denote just the low-frequency reservoir modes. Of course the high-frequency modes also contribute an analogous factor, but the point here is that, provided the reservoir modes satisfy the condition

$$\sum_{\{f'\}} \lambda_{f'}(2\langle n_{f'}\rangle + 1) \gg 1,$$

$$(2.69)$$

dephasing in the first sum in (2.68) allows that expression to be approximated by its short-time expansion $\exp\left(-\sigma_T^2 t^2/2\right)\exp\left(-\mathrm{i}t\sum_{\{f'\}}\lambda_{f'}\omega_{f'}\right)$, where σ_T^2 is given by the temperature-dependent function

$$\sigma_T^2 \equiv \sum_{\{f'\}} \lambda_{f'}(2\langle n_{f'}\rangle + 1)\,\omega_{f'}^2 . \tag{2.70}$$

This means that the entire absorption correlator may be replaced by its "short-time-approximation" (STA) form [2.13, 14]

$$\eta_{\ell c}^{\mathrm{STA}}(t) \equiv \exp\left(-\sigma_T^2 t^2/2\right)\exp\left[-\mathrm{i}t\left(\omega_{\mathrm{eg}}^0 + \sum_{\{f'\}}\lambda_{f'}\omega_{f'}\right)\right]$$

$$\times \exp\left(-\sum_{\{f\}}\lambda_f\{(2\langle n_f\rangle + 1)[1 - \cos(\omega_f t)] + \mathrm{i}\sin(\omega_f t)\}\right),$$

$$\tag{2.71}$$

where the sum $\sum_{\{f\}}$ is now over just the high-frequency modes. Thus the effect of the reservoir modes is to produce Gaussian broadening of each of the high-frequency vibronic peaks, with the temperature-dependent width (HWHM) of $(2\ln 2)^{1/2}\,\sigma_T$, and to produce an "effective" zero-phonon frequency of $\omega_{\mathrm{eg}}^0 + \sum_{\{f'\}}\lambda_{f'}\omega_{f'}$. More detailed discussion and numerical examples of the STA are given in [2.13, 14].[8] For solid-state impurity systems such as F-centers, which have broad, structureless, Gaussian-like absorption bands, the STA applies to *all* of the modes, i.e. there is no set of "high-frequency" modes, and the third exponential factor in (2.71) is missing [2.52].[9]

To illustrate the applicability of the STA, Fig. 2.9 shows results of model calculations [2.13] of the optical absorption lineshape of β-carotene in isopentane at $T = 123$ K and 298 K. The model parameters for the three high frequency Raman modes of this system were obtained from published fits of the low temperature absorption, and we simply used these fixed values and adjusted the Gaussian width to fit the low temperature absorption. The room-temperature lineshape was then predicted with no free parameters, using (2.70) and a simple "Einstein oscillator" model for the reservoir to compute σ_T^2, with quite good results, as shown.

---

[8] It is interesting to note that the same result (2.71) was obtained in [2.73], but in the context of a heuristic discussion of "pure dephasing" effects on absorption and RR profile lineshapes.

[9] In [2.52] a formula equivalent to (2.45), which underlies (2.71) above, was obtained in an approximation valid to order $1/N$, where N is the number of modes. As we have stressed, however, (2.45) is in fact exact, for *any* N.

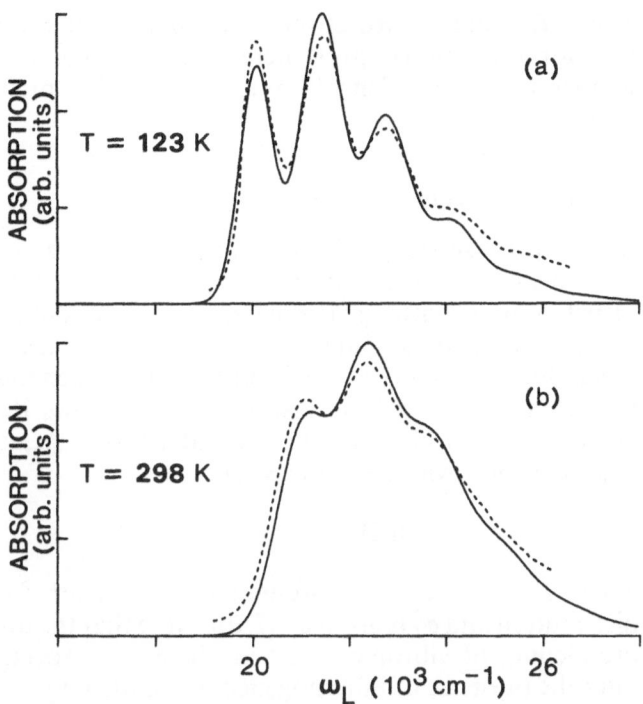

Fig. 2.9. Multimode model calculations of the optical absorption lineshape for β-carotene in isopentane at two temperatures. The width of the vibronic peaks was fit at $T = 123$ K, and the short-time aproximation (STA) was used to predict the $T = 298$ K absorption lineshape with no free parameters, as discussed in the text. The computed absorption lineshapes were scaled via least squares to the experimental data, and the zero-phonon frequency (which does not affect the lineshape) was fit at each temperature. After [2.13]

In [2.13], we studied the temperature broadening of the first-order profiles for β-carotene in isopentane, via transform calculations, with good results similar to those given here in Fig. 2.7 for this molecule in the more polar solvent CS_2. Although at first glance thermal broadening effects for RR profiles would seem more difficult to handle than for the absorption, to which the STA applies, note that the exact standard-assumptions first-order profile expression (2.47) carries *all* of the temperature dependence in the absorption correlator (except for a trivial prefactor). Thus homogeneous thermal broadening of profiles may be treated by focusing on just the absorption. Multimode model calculations based upon (2.47), or transform calculations based on (2.63) then carry the broadening through into the profiles in the proper way, as we have seen.

While the precedng explicit form of the STA was for the standard assumptions case, analogous arguments often hold for our more general models as well, with the specific analytical forms of the temperature-dependent broadening σ_T^2 and effective zero-phonon frequencies being model dependent. Moreover, note that all of the practically important transform relations given here (2.61–63, 65, 66) contain the profile temperature dependence (except for trivial prefactors) solely in the absorption.

b) Inhomogeneous Broadening

As we have stressed [2.19, 24], the STA discussed above also allows a very convenient treatment of inhomogeneous (site) broadening due to a distribution of scatterers in different local environments, for the commonly assumed case of a Gaussian distribution of zero-phonon frequencies. Assuming the validity of the STA for the homogeneous thermal broadening of each scatterer, comparison with (2.71) shows that the *absorption* will again consist of Gaussian broadened vibronic peaks, but with each now having a total width Γ (HWHM) given by

$$\Gamma^2 = \Gamma_I^2 + (2 \ln 2) \, \sigma_T^2, \tag{2.72}$$

where Γ_I is the temperature-independent width for the inhomogeneous distribution and σ_T^2 is given by (2.70). Note that the observation of Gaussian broadening of vibronic peaks in absorption spectra is *not* sufficient to infer the presence of inhomogeneous broadening – one needs temperature dependence measurements in order to separate experimentally the two terms of (2.72). As a practical matter, it is straightforward to assume a value for Γ_I and deconvolve a measured absorption spectrum to obtain the corresponding "homogeneous" absorption. The purely "homogeneous" profiles may then be computed from the homogeneous absorption by means of transform calculations and then superimposed via Γ_I to form the observed, inhomogeneously broadened profiles. Such calculations were employed in [2.19, 24].

2.4.5 Additional Remarks on the "Higher-Order" Profile Corrections

The STA of Sect. 2.4.4a was used in [2.13] as an important component of detailed arguments showing the negligibility of the "higher-order" profile corrections for $T \neq 0$ K, for the broad class of multimode systems characterized here in Sect. 2.3.4. To very briefly sketch how this comes about, we write the exact standard-assumptions first-order profile (2.63) for mode f, together with its lowest-order correction for $T \neq 0$ K [2.13]:

$$j_{1,f}(\omega_L) = C(\langle n_f \rangle + 1) \, \omega_L \omega_S^3 \lambda_f \left(|\hat{A}(\omega_f) \, \Phi(\omega_L)|^2 + \sum_{f'=1}^{N} (1 - \delta_{ff'}/2) \right.$$

$$\left. \times (\langle n_{f'} \rangle + 1) \langle n_{f'} \rangle \, \lambda_{f'}^2 |\hat{A}(\omega_f) \, \hat{A}(\omega_{f'}) \, \hat{A}(-\omega_{f'}) \, \Phi(\omega_L)|^2 + \dots \right).$$

$$\tag{2.73}$$

As in Fig. 2.5, the difference operators in this expression are defined by $\hat{\Delta}(\omega) \equiv 1 - \hat{\sigma}(\omega)$, so that $\hat{\Delta}(\omega)\, \Phi(\omega_L) \equiv \Phi(\omega_L) - \Phi(\omega_L - \omega)$. The profile correction is just the sum $\sum_{f'}$, which extends over all N modes. We note first that because of the factors $(\langle n_{f'} \rangle + 1)\, \langle n_{f'} \rangle$, there is *no* correction at $T = 0$ K, as discussed in Sect. 2.3.4. For nonzero temperatures, the molecular modes in the systems of interest here are typically of frequencies $\gg kT/\hbar$ at room temperature. Such modes cannot contribute to the profile correction-only the thermally populated low frequency modes contribute to the sum over f'. But since these modes have frequencies much lower that those for the high-frequency molecular modes that produce the typical vibronic peak spacings in the absorption, the difference operators involving $\omega_{f'}$ in (2.73) result in the factor $(\langle n_{f'} \rangle + 1)\, \langle n_{f'} \rangle |\hat{\Delta}(\omega_f)\, \hat{\Delta}(\omega_{f'})\, \hat{\Delta}(-\omega_{f'})$ $\times\ \Phi(\omega_L)|^2$ vanishing as $\omega_{f'}^2$ at low frequencies; hence the *individual* higher-order corrections from the thermally-populated low-frequency modes are typically very small [2.13]. Finally, consider a *reservoir* of N_r low-frequency thermally populated modes. As we have seen, the STA shows that they produce Gaussian broadening of the high-frequency vibronic peaks in the absorption, with the width of these peaks being given by (2.70). Accordingly, if the reservoir modes make comparable contributions to the width, the individual $\lambda_{f'}$'s vary roughly as N_r^{-1} for large N_r. But then the sum $\sum_{f'} \lambda_{f'}^2$ over the reservoir modes in (2.73) varies as N_r/N_r^2 and thus *vanishes* in the limit of large N_r. In [2.13] this heuristic argument is replaced by careful analytic and numerical arguments. Recently, *Schomacker* and *Champion* [2.49] carried out detailed experimental studies of temperature effects in RR scattering by heme-proteins, and they verified our STA-based arguments [2.13] concerning the negligibility of the higher-order profile corrections.

Recalling that the e-ph coupling $\{\lambda_f\}$ occurs to infinite order in $\Phi(\omega_L)$, and hence in each term of (2.73), we now have deeper insight into why this particular grouping of the infinite series for the one-phonon RR scattering is so optimal — the leading term (the first-order profile) is exact at $T = 0$ K, and the higher-order corrections are indeed negligible for the wide class of practically important systems characterized in Sect. 2.3.4. Again, this optimal grouping of the RR series emerges almost automatically in our many-body/diagrammatic approach.

2.4.6 Renormalization

In the various models considered here, we have seen several instances where the basic dimensionless linear e-ph coupling parameters $\{\xi_f\}$ [or $\{\lambda_f = \xi_f^2\}$] and linear non-Condon parameters $\{m_f\}$, become "renormalized" as complexity is added to the model. For instance, (2.67a, b) give the renormalizations of these parameters due to both mode mixing and frequency shifts, while the renormalizations due to frequency shifts alone are given just below (2.65), and also by (2.60). One significance of the renormalized parameters is that they are the parameters one would extract by directly analyzing absorption and profile data via transform techniques. Moreover, the mode mixing renormalizations are seen to include those mode mixing effects which formally conform to a linear coupling model. Hence, they are convenient for separating out the mode mixing effects which are intrinsically new, i.e. which cannot be described within any linear coupling model. This aspect was stressed in our illustrative model calculations in [2.10]. The question of how well a formal linear coupling model, but having *renormalized* coupling parameters, can reproduce absorption, emission and RR profiles of systems with mode mixing is discussed in [2.27] and, by means of a different, sum-over-states, methodology in [2.74]. The exact analytic expressions obtained in the correlator/many-body approach are convenient for such studies, and we have found that the useful renormalizations which naturally emerge from the analytic results are *different* for each of these three types of spectra. The renormalized linear coupling parameters can then be used to form "renormalized linear coupling theories", which prove to be very useful for studying the general question of the sensitivity of absorption, emission and RR profiles to mode mixing [2.27].

2.5 Conclusion

We have seen that the application of suitably generalized finite temperature phonon many-body techniques within the time-correlator formulation yields highly practicable theories for analyzing resonance Raman profile and absorption data on complex multimode vibronic systems. The emphasis has been on dealing with the multimode aspect of the problem, and the forms of our solutions are seen to be very convenient for this aspect, both for explicit model calculations and for the complementary absorption → RR profile transform methods which follow naturally from this approach.

Acknowledgements. The theoretical work described here was done in close collaboration with C. K. Chan, H. M. Lu, and D. L. Tonks, each of whom made several key contributions. I would like to thank O. Brafman, R. C. Hanson, B. Khodadoost, S. A. Lee, and C. T. Walker for many helpful discussions on their experiments.

References

2.1   M. Cardona: In *Light Scattering in Solids II*, ed. by M. Cardona, G. Güntherodt (Springer, Berlin, Heidelberg 1982) p. 19
2.2   B. Johnson, W. Peticolas: Ann. Rev. Phys. Chem. **27**, 465 (1976)
2.3   A. Warshel: Ann. Rev. Biophys. Bioeng. **6**, 273 (1977)
2.4   W. Siebrand, M. Z. Zgierski: In *Excited States*, ed. by E. C. Lim (Academic, New York 1979) Vol. 4, p. 1
2.5   O. Sonnich Mortensen, S. Hassing: In *Advances in Infrared and Raman Spectroscopy*, ed. by R. J. H. Clark, R. E. Hester (Heyden, London 1980) Vol. 6, p. 1
2.6   P. M. Champion, A. C. Albrecht: Ann. Rev. Phys. Chem. **33**, 353 (1982)
2.7   A. C. Albrecht: J. Chem. Phys. **34**, 1476 (1961)
2.8   R. M. Martin, L. M. Falicov: In *Light Scattering in Solids* I, 2nd ed., ed. by M. Cardona (Springer, Berlin, Heidelberg 1983), p. 79
2.9   D. L. Tonks, J. B. Page: Chem. Phys. Lett. **66**, 449 (1979)
2.10   D. L. Tonks, J. B. Page: Chem. Phys. Lett. **79**, 247 (1981)
2.11   J. B. Page, D. L. Tonks: J. Chem. Phys. **75**, 5694 (1981)
2.12   D. L. Tonks, J. B. Page: J. Chem. Phys. **76**, 5820 (1982)
2.13   C. K. Chan, J. B. Page: J. Chem. Phys. **79**, 5234 (1983)
2.14   C. K. Chan, J. B. Page: Chem. Phys. Lett. **104**, 609 (1984)
2.15   C. K. Chan: J. Chem. Phys. **81**, 1614 (1984)
2.16   O. Brafman, C. K. Chan, B. Khodadoost, J. B. Page, C. T. Walker: J. Chem. Phys. **80**, 5406 (1984)
2.17   C. K. Chan, J. B. Page, D. L. Tonks, O. Brafman, B. Khodadoost, C. T. Walker: J. Chem. Phys. **82**, 4813 (1985)
2.18   J. B. Page: Cryst. Latt. Defects, Amorph. Mater. **12**, 273 (1985)
2.19   S. A. Lee, C. K. Chan, J. B. Page, C. T. Walker: J. Chem. Phys. **84**, 2497 (1986)
2.20   H. M. Lu, J. B. Page: Chem. Phys. Lett. **131**, 87 (1986)
2.21   H. M. Lu, J. B. Page: To be published; H. M. Lu, Ph. D. Dissertation, Arizona State University, 1988
2.22   D. L. Tonks, J. B. Page: J. Chem. Phys. **88**, 738 (1988)
2.23   H. M. Lu, J. B. Page: J. Chem. Phys. **88**, 3508 (1988)
2.24   B. Khodadoost, S. A. Lee, J. B. Page, R. C. Hanson: Phys. Rev. **B38**, 5288 (1988)
2.25a H. M. Lu, J. B. Page: *Eleventh Intenational Conference on Raman Spectroscopy*, ed. by R. J. H. Clark, D. A. Long (Wiley, Chichester 1988) p. 71
2.25b H. M. Lu, J. B. Page: J. Chem. Phys. **90**, 5315 (1989)
2.26   H. M. Lu, J. B. Page: J. Chem. Phys. **92**, 7038 (1990)
2.27   H. M. Lu, J. B. Page: *Proceedings of the Twelfth International Conference on Raman Spectroscopy*, ed. by J. R. Durig, J. F. Sullivan (Wiley, New York 1990) p. 80; H. M. Lu, J. B. Page: To be published
2.28   V. Hizhnyakov, I. Tehver: Phys. Status Solidi **21**, 755 (1967)
2.29   See, for instance, K. K. Rebane, I. Y. Tehver, V. V. Hizhnyakov: In *Theory of Light Scattering in Condensed Matter*, ed. by B. Bendow, J. L. Birman, V. M. Agranovitch (Plenum, New York 1976) p. 393; K. K. Rebane: In *Luminescence of Inorganic Solids*, ed. by B. DiBartolo (Plenum, New York 1978) p. 495 and references therein
2.30   D. C. Blazej, W. Peticolas: J. Chem. Phys. **72**, 3134 (1980)
2.31   S. Hassing, O. Sonnich Mortensen: J. Chem. Phys. **73**, 1078 (1980)
2.32   P. M. Champion, A. C. Albrecht: Chem. Phys. Lett. **82**, 410 (1981)
2.33   I. J. Tehver: Opt. Commun. **38**, 279 (1981)
2.34   B. R. Stallard, P. M. Champion, P. R. Callis, A. C. Albrecht: J. Chem. Phys. **78**, 712 (1983)
2.35   B. R. Stallard, P. R. Callis, P. M. Champion, A. C. Albrecht: J. Chem. Phys. **80**, 70 (1984)
2.36   K. T. Schomacker, O. Bangcharoenpaurpong, P. M. Champion: J. Chem. Phys. **80**, 4701 (1984)

2.37   M. Pfeiffer, A. Lau, W. Werncke: J. Raman Spectrosc. **15**, 20 (1984)
2.38   D. Lee, B. R. Stallard, P. M. Champion, A. C. Albrecht: J. Phys. Chem. **88**, 6693 (1984)
2.39   J. R. Cable, A. C. Albrecht: J. Chem. Phys. **84**, 1969 (1986)
2.40   T. W. Patapoff, P. Turpin, W. L. Peticolas: J. Phys. Chem. **90**, 2347 (1986)
2.41   V. Srajer, K. T. Schomacker, P. C. Champion: Phys. Rev. Lett. **57**, 1267 (1986)
2.42   P. Li, P. C. Champion: J. Chem. Phys. **88**, 761 (1988)
2.43   N. Watanabe, M. Shimizu, J. Tanaka: J. Raman Spectrosc. **18**, 381 (1987)
2.44   K. T. Schomacker, V. Srajer, P. M. Champion: J. Chem. Phys. **86**, 1796 (1987)
2.45   L. Reinisch, K. T. Schomacker, P. M. Champion: J. Chem. Phys. **87**, 150 (1987)
2.46   O. Bangcharoenpaurpong, P. M. Champion, S. Martinis, S. G. Sligar: J. Chem. Phys. **87**, 4273 (1987)
2.47   R. P. Rava: J. Chem. Phys. **87**, 3758 (1987)
2.48   C. M. Jones, S. A. Asher: J. Chem. Phys. **89**, 2649 (1988)
2.49   K. T. Schomacker, P. M. Champion: J. Chem. Phys. **90**, 5982 (1989)
2.50   V. Hizhnyakov, I. Tehver: J. Raman Spectrosc. **19**, 383 (1988)
2.51   H. A. Kramers, W. Heisenberg: Z. Phys. **31**, 681 (1925); R. Loudon, *The Quantum Theory of Light*, 2nd ed. (Clarendon, Oxford 1983) p. 311
2.52   M. Lax: J. Chem. Phys. **20**, 1752 (1952)
2.53   R. T. Harley, J. B. Page, C. T. Walker: Phys. Rev. B **3**, 1365 (1971)
2.54   D. Robbins, J. B. Page: Phys. Rev. Lett. **38**, 365 (1977); D. Robbins, J. B. Page: In *Lattice Dynamics*, ed. by M. Balkanski (Flammarion, Paris 1978) p. 264
2.55   V. Hizhnyakov: Phys. Rev. B **30**, 3490 (1984); V. Hizhnyakov, I. Tehver, G. Zavt: J. Raman Spectrosc. **21**, 231 (1990)
2.56   G. E. Blumberg, L. A. Rebane: Sov. Phys. JETP **67**, 2293 (1989)
2.57   F. Duschinsky: Acta Physiochim. URSS **7**, 551 (1937)
2.58   V. Hizhnyakov, I. Tehver: Opt. Commun. **32**, 419 (1980)
2.59   D. J. Tannor, E. J. Heller: J. Chem. Phys. **77**, 202 (1982)
2.60   R. Friesner, M. Pettitt, J. M. Jean: J. Chem. Phys. **82**, 2918 (1985)
2.61   J. Sue, Y. J. Yan, S. Mukamel: J. Chem. Phys. **85**, 462 (1986)
2.62   See, for instance, T. Takagahara, E. Hanamura, R. Kubo: J. Phys. Soc. Jpn. **43**, 802, 811, 1522 (1977) and references therein
2.63   A. Kotani, Y. Toyozawa: J. Phys. Soc. Jpn. **41**, 1699 (1976); Y. Toyozawa, A. Kotani, A. Sumi: J. Phys. Soc. Jpn. **42**, 1495 (1977)
2.64   G. Herzberg, E. Teller: Z. Phys. Chem. **B21**, 410 (1933)
2.65   B. Sharf, B. Honig: Chem. Phys. Lett. **7**, 132 (1970); G. J. Small: J. Chem. Phys. **54**, 3300 (1971)
2.66   K. Seo, J. Shiraishi, Y. Fujimura: Chem. Phys. Lett. **92**, 415 (1982)
2.67   S. Hassing, O. Sonnich Mortensen: J. Molec. Spectrosc. **87**, 1 (1981)
2.68   P. M. Champion: Chem. Phys. Lett. **86**, 231 (1982); W. H. Henneker, W. Siebrand, M. Z. Zgierski: J. Chem. Phys. **74**, 6560 (1981); Y. Fujimura, S. H. Lin: Chem. Phys. Lett. **63**, 199 (1979)
2.69   A. A. Abrikosov, L. P. Gorkov, I. E. Dzyaloshinski: *Methods of Quantum Field Theory in Statistical Physics* (Prentice-Hall, Englewood Cliffs 1963) Sect. 15
2.70   D. L. Tonks: Phys. Rev. B **22**, 6420 (1980)
2.71   A. Warshel, M. Karplus: J. Am. Chem. Soc. **94**, 5612 (1972); J. Del Bene, H. H. Jaffé: J. Chem. Phys. **48**, 1807 (1968)
2.72   A. Pinczuk, E. Burstein: In *Light Scattering in Solids* I 2nd ed., ed. by M. Cardona (Springer, Berlin, Heidelberg 1983) p. 23
2.73   R. A. Harris, R. A. Mathies, W. T. Pollard: J. Chem. Phys. **85**, 3744 (1986)
2.74   M. Z. Zgierski: Chem. Phys. **108**, 61 (1986)

3. Resonant Raman Scattering in Conjugated Polymers

Y. Yacoby and E. Ehrenfreund

With 53 Figures

In this chapter we review the resonant Raman scattering (RRS) of polyacetylene (PA) and several other conjugated polymers.

Conjugated polymers are polymers with a continuous line of alternating single and double carbon–carbon bonds. The CC bonds consist of one σ-type bond with a wave function mainly in the carbon–carbon line, and a contribution of a π-type bond, distributed unequally between the so-called single and double bonds. The π orbitals play a very important role in the properties of these polymers as well as in their RRS.

Conjugated polymers, in their undoped state, are semiconductors. The top valence bands (highest occupied molecular orbitals or HOMO) and the lowest conduction bands (lowest unoccupied molecular orbitals or LUMO) are composed mainly of the π orbitals. In contrast to the σ orbitals, π orbitals are not highly localized, namely the alternating single and double bonds have two localized σ orbitals and one partially localized π orbital. Thus, π electrons also have a finite probability of being found on the so-called single bonds. The delocalization of the π orbitals has several important consequences. First, it means that the π bands have a relatively large energy spread of about 10 eV, so electrons and holes in these bands have a relatively small effective mass, accounting in this way for the relatively large conductance of these systems when doped. Second, the bond length depends on the probability of finding π electrons on that bond. The larger the probability, the shorter the bond. This property leads to a large electron–phonon coupling between lattice deformations and the valence electrons. Third, defects at one place along the chain affect the properties of the chain over a relatively long distance. Consequently, such defects may produce a change in the electronic energy states, delocalized or partially delocalized intra-gap electronic states and a significant change in the vibrational properties of the chain. Thus, in conjugated polymers and in the polyenes, the delocalization of the π orbitals is responsible for the chain-length dependence of the electronic energy gap and of the A_g vibrational frequencies.

The quasi-one-dimensional nature of the polymers has several important consequences. The density of electronic states and the density of vibrational states are strongly influenced by the quasi-one-dimensionality. The highest point symmetry is D_{2h} which includes a 180° rotation and two perpendicular reflection planes. The Raman active modes are therefore of the A_g and B_g species.

Resonant Raman scattering conditions are obtained in conjugated polymers when the exciting photon energy is approximately equal to the $\pi - \pi^*$ band gap. The resonant enhancement in some of these systems is very strong, because of the fast change in the density of electronic states with the energy, and because of the large electron–phonon interaction. The electron–phonon coupling is large mainly for the carbon–carbon vibrations, because these vibrations are directly affected by the π electrons, which form the electronic gap.

In this chapter, we shall emphasise the effects of conjugation on the RRS. Most of the work related to these effects has been done on *trans*-PA, with related work on *cis*-PA. The chemical structure of several conjugated polymers is shown in Fig. 3.1. In some of these, the effects of the conjugation on the RRS spectra have been observed in the form of dispersion. These effects will be briefly discussed in this review. A considerable amount of other informations has also been obtained from the RRS spectra. However, this information is, so far, almost always material specific and does not seem ripe yet for summary in a review form.

Various properties of these polymers have been discussed in a number of review articles and books. An excellent classified literature list on almost all aspects of these systems up to 1987 has been provided in the review paper of *Roth* and *Bleier* [3.1]. A more recent review was published by *Heeger* et al. [3.2]. These reviews also provide an excellent description of

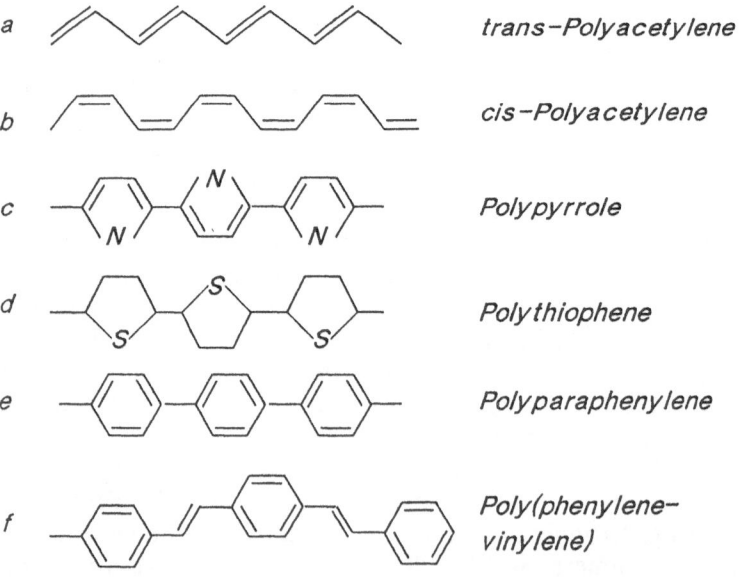

a   trans–Polyacetylene

b   cis–Polyacetylene

c   Polypyrrole

d   Polythiophene

e   Polyparaphenylene

f   Poly(phenylene–vinylene)

Fig. 3.1. Chemical structure of the most important polymers with conjugated double bonds that are frequently discussed within the context of synthetic metals

the important elementary excitations in conjugated polymers namely: solitons, charged solitons, polarons and bipolarons. Some of these concepts are important in understanding the RRS spectra in these systems.

Defects that affect the π bands change both the electronic energy gap and the vibrational frequencies of some of the phonons. Consequently, in certain systems such as the polyenes and some polymers, one finds a well-defined relation between the photon energy which maximizes the resonance enhancement and the phonon frequencies. This relation is the origin of the well known dispersion of the phonon frequencies with the excitation laser photon energy observed in *trans*-polyacetylene and some other systems. The defects that have been of interest are defects which limit the length of the uninterrupted conjugation along a polymer chain. Doping by various atoms and molecules may take electrons, from, or supply extra electrons to, the chain. The charge transfer changes the force constants, producing localized and resonant vibrations.

The morphology of some of the conjugated polymers is quite complicated. For example, polyacetylene is grown in the form of films. The chains are in the film plane forming fibrils which are partially crystalline and partially amorphous [3.3]. This complex morphology affects the details of the RRS, which in turn provides information on the morphology, on the *cis–trans* transition and on the doping process. These features will also be discussed in this chapter.

The experimental results and their phenomenological interpretation will be discussed in the first part, Sect. 3.1. Since most of the in-depth work was done on ployacetylene, this will be discussed first and in detail. Other conjugated polymers showing dispersion effects will then be briefly discussed, and wherever relevant compared to PA. The deeper theoretical understanding of RRS, mainly in PA, will be discussed in Sect. 3.2.

3.1 Resonant Raman Scattering, Experimental Results and Phenomenological Interpretation

3.1.1 General Properties

The simplest conjugated polymer is PA. An important contribution to the investigation of this system was made by *Shirakawa* et al. [3.4], who developed a way to grow PA films. PA has two modifications, *trans*-PA and *cis*-PA, shown in Fig. 3.1. *Trans*-PA has C_{2h} point group symmetry. It has two carbon and two hydrogen atoms per unit cell. Its ground state is doubly degenerate, with two states differing from each other in the order of the single and double bonds. This structure enables the existence of mobile solitons, which consist of a transition region between the two ground states. *cis*-PA has D_{2h} symmetry with four carbon and four hydrogen atoms per unit cell. Its ground state is nondegenerate. Consequently, it has no mobile solitons, but in principle can have charged polarons and bipolarons which may then affect the optical absorption and the RRS of the system.

The morphology of PA is very complicated: *trans*-PA is the thermodynamically stable modification. However, when polymerized it is in the *cis*-rich form, with a relatively small *trans* component which is always present. When grown by the Shirakawa method, it polymerizes in the form of films that consist of disoriented fibrils. The polymer chains are more or less along the fibrils and partly in a three-dimensional microcrystalline structure, partly in an amorphous structure [3.3]. *cis*-PA can be isomerized into the stable *trans*-PA either by heating to 180 °C [3.5] or by doping with various dopants. The morphology has an important effect on the details of the RRS in PA. These effects will be discussed later.

PA can be stretched causing the fibrils to extend and align along the stretching direction. This process facilitated the measurement of polarized spectra which improved the understanding of RRS.

The electronic properties are very important in determining the resonance enhancement. PA is a semiconductor like all other conjugated polymers. The band gap energies of *trans*-PA and *cis*-PA are 1.6–1.7 [3.6]

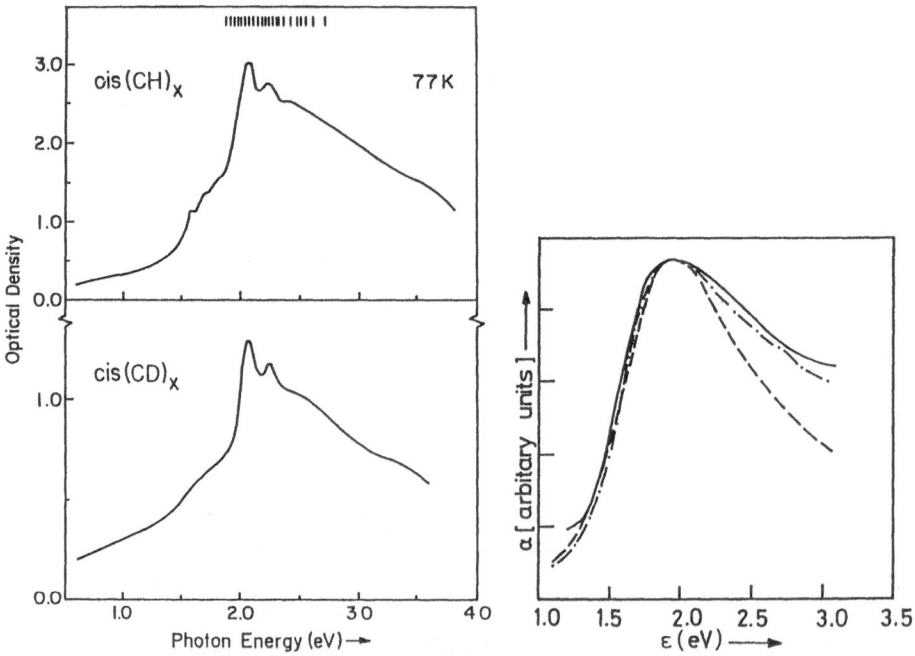

Fig. 3.2. Optical absorption spectra of polyacetylene. (a) *cis*-(CH)$_x$ at 77 K [3.7]. (b) *trans*-(CH)$_x$ as measured (————) and as calculated for good (— — —) and bad (—·—·—) samples [3.6]

and 2.07 eV [3.7], respectively. The absorption spectra of *trans*-PA and of *cis*-(CH)$_x$ and (CD)$_x$ are shown in Fig. 3.2 [3.6, 7]. Notice that the absorption rises rather sharply and falls gradually as would be expected for a system with an approximately one-dimensional density of states:

$$\varrho(\varepsilon) \simeq C(\varepsilon - \varepsilon_g)^{-1/2} . \tag{3.1}$$

However, it is not as sharp as would be expected. The broadened spectrum is a result of several effects: The energy gap of *trans*-PA depends strongly on the chain length as seen in all polyenes. The fact that it is partly amorphous also contributes to the broadening. In addition there is broadening resulting from the interchain interactions [3.8]. In the *cis* spectra, the rise in the absorptions is steeper than in *trans*-PA and vibronic peaks can also be observed [3.7]. PA can be doped by a variety of doping agents both oxidizing and reducing. Doping affects almost all the properties of PA. Most dopants, at large concentrations, reduce the optical absorption at the gap and produce a new band in the near infrared. This is probably related to the creation of charged solitons in *trans*-PA. An example of the effect of doping on the absorption spectra of *trans*-PA [3.9] is shown in Fig. 3.3.

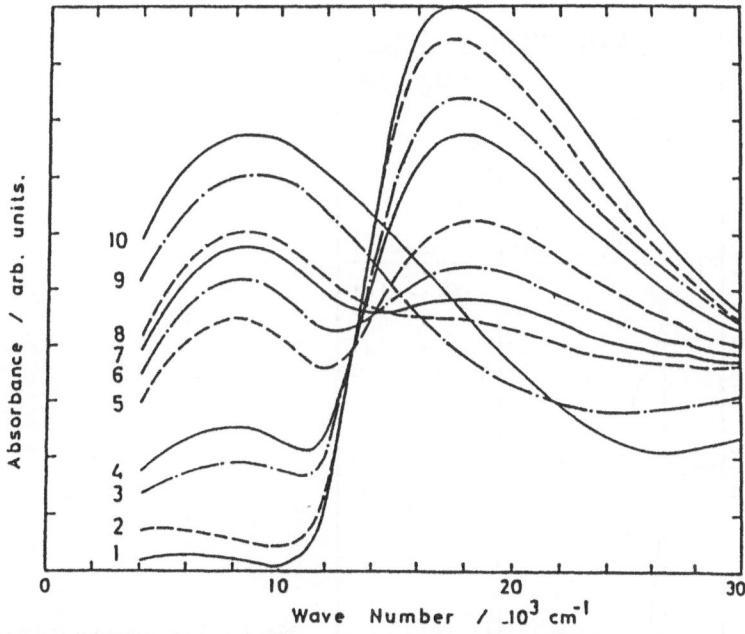

Fig. 3.3. The change of visible and near-IR spectra with doping. (1) undoped, (2) $y = 0.015$, (3) $y = 0.036$, (4) $y = 0.047$, (5) $y = 0.085$, (6) $y = 0.095$, (7) $y = 0.0108$, (8) $y = 0.116$, (9) $y = 0.133$, (10) $y = 0.146$. Note: 10000 cm^{-1} = 1.24 eV. y is the molar fraction of the dopant. From [3.9]

3.1.2 Main Experimental Features of the Resonant Raman Spectra

The experimental unpolarized spectra of *trans-* and *cis-*PA have been measurd by a number of researchers. The first spectra were obtained on totally disordered samples [3.10–12]. Examples of such spectra for several excitation wavelengths for *trans-*PA [3.13] and *cis-*PA are shown in Figs. 3.4, 5 respectively. The *trans* spectrum contains two strong and broad bands at ~ 1070 and ~ 1450 cm^{-1}. The shapes of these two lines are quite unusual. They are very asymmetric, with a sharp rise at the low energy side. The phonon energy of this rise is the same for all spectra independent of the wavelength of the exciting laser line. On the other hand, the band shape at higher phonon energies changes drastically with the excitation laser wavelength. At long wavelengths, the high-energy tail is quite narrow. At shorter wavelengths, this tail increases and in the UV the low and high phonon energy parts even split. No drastic changes in line shape are observed in the weaker lines. However, as the excting laser wavelength becomes shorter, the 1290 cm^{-1} line becomes less symmetric and broader and its relative intensity increases. The line shapes are to some extent sample dependent. The spectrum of a sample subjected to mechanical

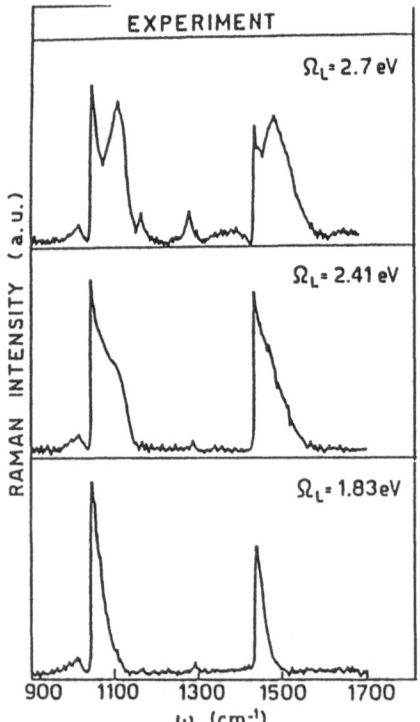

Fig. 3.4. Experimental RRS spectra of *trans-*(CH)$_x$ for different incident light frequencies ω_L [3.13]

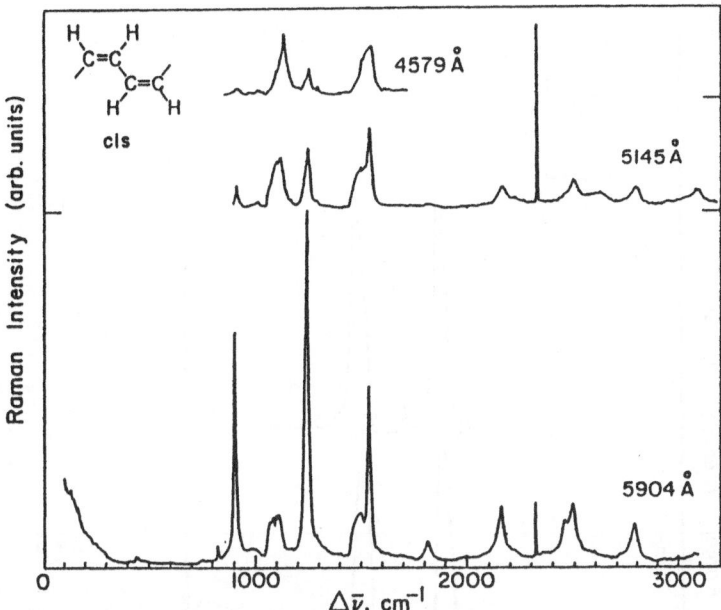

Fig. 3.5. Raman spectra of *cis*-rich $(CH)_x$ at 77 K. The sharp line at 2321 cm^{-1} is from liquid nitrogen [3.14]

treatment [3.15] is shown in Fig. 3.6. Curve (a) belongs to the sample as-grown. Curve (b) was measured after rubbing the sample surface with a 0.5 micron alumina abrasive powder. Measurements at larger Stokes energies reveal three two-phonon peaks which are the combination bands of the two strong fundamental bands at 1070 and 1450 cm^{-1}. An example

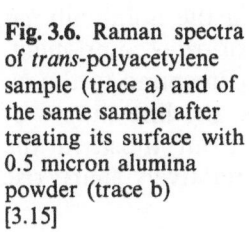

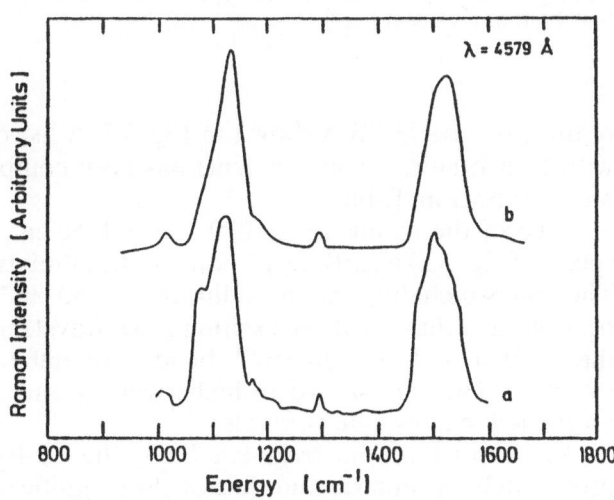

Fig. 3.6. Raman spectra of *trans*-polyacetylene sample (trace a) and of the same sample after treating its surface with 0.5 micron alumina powder (trace b) [3.15]

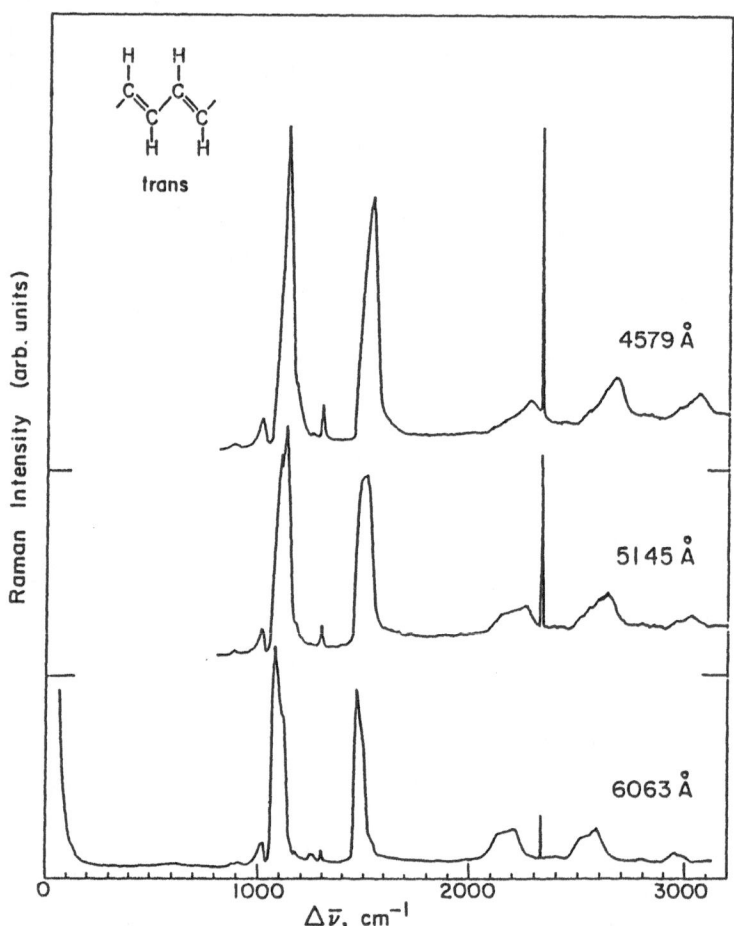

Fig. 3.7. Raman spectra of *trans*-(CH)$_x$ at 77 K. The sharp line at 2321 cm^{-1} is from the liquid nitrogen coolant [3.14]

of these spectra [3.14] is shown in Fig. 3.7. A list of all *trans* Raman lines with their qualitative assignments has been compiled by *Kuzmany* [3.12] and is shown in Table 3.1.

Two of the bands, at ~ 1070 and ~ 1450 cm^{-1}, in the nominally *cis* spectra (Fig. 3.5) clearly correspond to the admixture of *trans* component. The others including the sharp line at ~ 1550 cm^{-1} correspond to the *cis* part. Notice that, at these exciting laser wavelengths, the line shapes of the ~ 1070 and ~ 1450 cm^{-1} bands are different from those in the *trans*-PA. They are shifted to higher energy and the relatively sharp rise on the low energy side is missing.

In contrast to the *trans*-PA lines, the *cis*-PA lines are sharp, approximately symmetric and do not shift significantly with the excitation

Table. 3.1 Raman lines of cis $(CH)_x$ and cis $(CD)_x$

$(CH)_x$ [cm^{-1}]		$(CD)_x$ [cm^{-1}]		v_{CH}/v_{CD}	assignment	
295	vw	265	vw	1.11	B_{3g}	C–C–C deformation, out of plane
445	w	405	vw	1.09	B_{2g}	C–C–C deformation, in plane
820	vw	830	vw			not identified
915	s	870	s	1.06	B_{2g}	C–C stretch plus contribution from trans in the deuterated polymer
985	m	767	w	1.28	B_{3g}	C–H deformation out of plane
1008	m	745	m	1.35	B_{1g}	C–H deformation in plane
1070–						from residual trans $(CH)_x$
1130	s	870	m			
1170	w	930	w	1.26	B_{2g}	C–H deformation in plane
1230–	sh	965–	sh	1.28	A_g	C–C stretch with considerable addition
1252	s	978	s			of C–H deformation in plane
1292	w	1205	s	1.07	A_g	C–H deformation in plane with considerable addition of C–C stretch in the deuterated polymer
1490	sh	1390	sh			from residual trans $(CH)_x$
1544	vs	1460	vs	1.05	A_g	C=C stretch
3030	m	2260	m	1.34	B_{2g}	C–H stretch
3090	m	2315	w	1.34	A_g	C–H stretch

sh – shoulder, vw – very weak, w – weak, m – medium, s – strong, vs – very strong

laser wavelength. A list with all cis-PA lines and their tentative assignment was compiled by *Kuzmany* [3.12] and is shown in Table 3.2.

An example of the Raman spectrum of cis-PA over a broad energy range [3.7] is shown in Fig. 3.8. It displays a large number of overtones extending from the laser line at 2.47 eV to 1.4 eV. Beyond the first overtone series it is difficult to identify the bands, because in most cases more than one combination of phonons agree with the observed energy.

Table 3.2 Raman lines of trans $(CH)_x$ and trans $(CD)_x$

$(CH)_x$ [cm^{-1}]		$(CD)_x$ [cm^{-1}]		v_{CH}/v_{CD}	assigment	
878	w	808	w			not identified
1008	s	745	m	1.35	B_g	C–H deformation out of plane
1090–	s	856–	s	1.28	A_g	stretch with considerable addition
1120	s	870	s			of C–H deformation in plane
1170	w	960	w			not identified
1292	w	1207	w	1.07	A_g	C–C stretch
		1295	w			not identified
1470–	vs	1370–	vs	1.07	A_g	C = C stretch
1520	vs	1450	vs			
2990	w	2230	m	1.34	A_g	C–H stretch

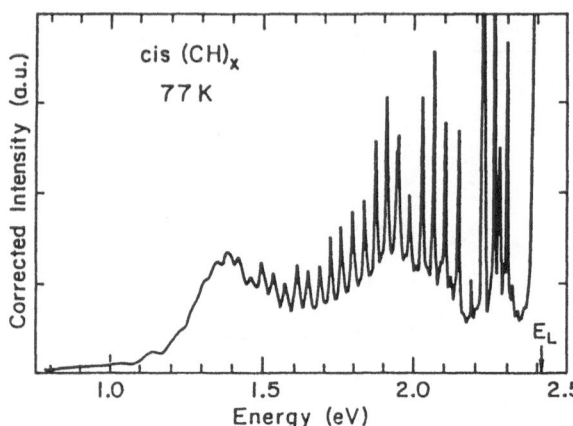

Fig. 3.8. Spectrum of outgoing light measured with extended detector range to show the 1.5 eV emission band. Ordinate is proportional to quantum efficiency $\times 1/\omega_l\omega_s^3$ [3.7]

3.1.3 Phenomenological Interpretation of the Raman Spectra

The three-dimensional crystalline structure seems to have only a minor influence on the Raman spectra. Davidov splittings and low frequency interchain modes would be expected if interchain effects were very strong. These effects have never been observed [3.7]. However, interchain interaction effects have been observed under pressure and will be discussed later. Thus, in the discussion here, we shall be concerned only with the chain structure.

As mentioned previously, *trans*-PA has C_{2h} point group symmetry and four atoms per chain unit cell. Thus, it has eight optical modes and one rotation around the chain axis. The vibrational species of *trans*-PA are [3.12]:

$$\Gamma_{\text{trans}} = 4A_{\text{g}} + B_{\text{g}} + A_{\text{u}} + 2B_{\text{u}} . \tag{3.2}$$

cis-PA has D_{2h} point group symmetry and eight atoms per unit cell. It thus has 20 vibrational species and one rotation. The vibrational species of *cis*-PA are [3.12]:

$$\Gamma_{\text{cis}} = 4A_{\text{g}} + B_{1\text{g}} + 4B_{2\text{g}} + 2B_{3\text{g}} + 2A_{\text{u}} + 3B_{1\text{u}} + B_{2\text{u}} + 3B_{3\text{u}} . \tag{3.3}$$

a) Lattice Dynamics of *trans*-Polyacetylene

Phenomenological lattice dynamics calculations of *trans*-PA have been carried out by several groups [3.16–18]. The calculations use an infinite-chain model with periodic boundary conditions. The force field is expressed in terms of the internal coordinates, which include bond-stretching and bond-bending coordinates. Unfortunately, the number of force constants involved is very large. In the model of *Schügerl* and *Kuzmany* [3.17], there are 24 such constants. In the model of *Jumeau* et at. [3.18], only 12 force constants are considered, but neglecting the others is not clearly justified. Four of the force constants in the *Schügerl-Kuzmany* model [3.17] are indeed very small and may be neglected; the others, however, are not so small. To overcome this difficulty Schügerl and Kuzmany impose three requirements:

i) Both Raman and IR active modes must be reproduced simultaneously for $(CH)_x$ and $(CD)_x$.

ii) The force field must be in reasonable agreement with force constants calculated from MO theories and well-known values from the literture.

iii) The normal coordinates for those modes that are resonance enhanced should exhibit a strong $C=C$ stretch component.

The last requirement stems from the fact that the resonance enhancement involves the $\pi - \pi^*$ transitions. These electronic orbitals are assumed to be mainly, but not exclusively, involved in the double bonds.

The calculations of the three groups just mentioned reproduce the phonon frequencies of both $(CH)_x$ and $(CD)_x$ rather well. This analysis also provides a clear picture of eigenvectors of the vibrational modes. In spite of the fact that the force fields are different, the eigenvectors are quite similar. An example of the eigenvectors [3.17] is given in Fig. 3.9.

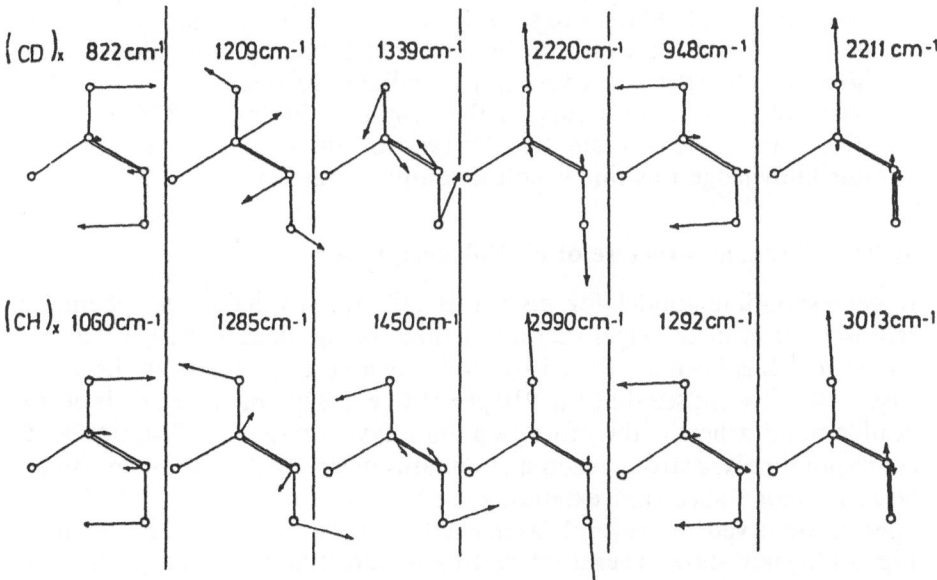

Fig. 3.9. Normal coordinates of the six in-plane modes of *trans*-polyacetylene (infinite chain) [3.17]

On the basis of these models the authors attempted to explain some of the interesting features observed in the experimental Raman spectra.

i) The fact that vibrational modes 1 and 3 (in Fig. 3.9) include an important $C=C$ stretch component is consistent with the large resonance enhancement of these bands in $(CH)_x$. This feature was of course introduced as a requirement. In contrast, the $C=C$ stretch component in mode 2 is very small and indeed its resonance enhancement is very small too. In $(CD)_x$ on the other hand, the $C=C$ bond stretching component of mode 2 is much larger, even larger than that of mode 1. This is qualitatively consistent with the observation that the intensity of mode 2 is larger, relative to mode 1, in $(CD)_x$ than in $(CH)_x$. A quantitative comparison, between the observed intensities and those calculated from the eigenvectors failed, showing that the theory is too crude for such a comparison [3.18].

ii) *Schügerl* and *Kuzmany* calculated the sensitivity of the phonons to changes in the double-bond force constant [3.17]. They indeed find that the frequencies of modes 1 and 3 strongly depend on the values of the double-bond force constant. This is necessarily expected in their model, because it was actually introduced into the calculation by requirement (iii) above. Assuming that *trans*-PA is composed of chains with various lengths of undisturbed conjugated bonds, and that the length affects mainly the gap and the double-bond force constant, the sensitivity of modes 1 and 3 to the double-bond force constant explains qualitatively the width of the first and third bands and the changes in the line shapes produced by different laser excitation frequencies. *Jumeau* et al. [3.18] tried to see if the density of vibrational states could be tied in with the width of the observed Raman bands. They found, however, that the two could not be reliably related.

iii) None of these models suggest the origin of the line at 878 cm^{-1}, in $(CH)_x$, which is consistently observed in all spectra. To the best of our knowledge this line is still a complete mistery.

b) The Overtone Structure of *cis*-Polyacetylene

A corresponding model for *cis*-PA is still not available. A schematic representation of the eigenvectors of three of the lines at 910, 1250 and 1540 cm^{-1} has been suggested by *Lichtmann* et al. [3.7] and is shown in Fig. 3.10. Here we see that the 910 cm^{-1} line has almost no stretch of the double bond, whereas the other two lines have a large $C=C$ stretch and, correspondingly, a strong resonance enhancement. *cis*-PA has an unusually large overtone spectrum extending all the way to 9000 cm^{-1} [3.7, 19]. Spectra observed for several laser excitation wavelengths are shown in Fig. 3.11. They show several interesting features: first, the sharp lines are indeed Raman lines shifting to lower energies with increasing laser line wavelength, in such a way that the shift relative to the laser line is constant. On the other hand, the broad features at 1.9 eV and at ~ 1.4 eV (Fig. 3.8)

Fig. 3.10. Schematic representation of eigenvectors for the three principal Raman-active modes of cis-$(CH)_x$ [3.7]

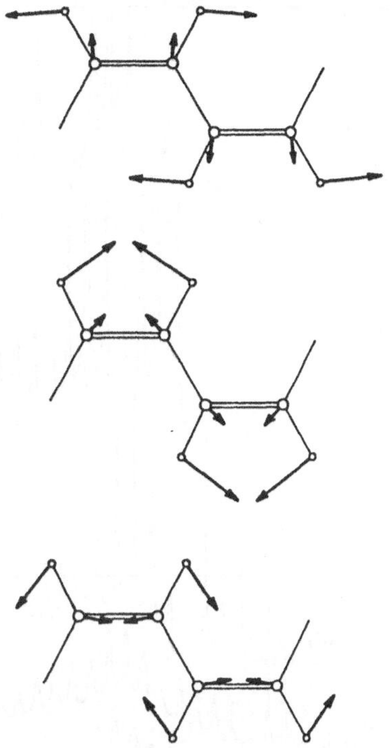

910 cm^{-1}

1250 cm^{-1}

1540 cm^{-1}

are independent of the laser wavelength, indicating that they orginate from fluorescence [3.7]. The intensity of the Raman lines has a broad maximum at ~ 1.9 eV consistent with the Raman excitation spectra. The spectra of the fundamental lines and the first overtones [3.7, 20] are shown in Figs. 3.12, 13. The peak of the resonance enhancement curves is at ~ 2.1 eV. It coincides with the first vibronic peak on the absorption curve shown in Fig. 3.2, and it is just about 0.2 eV, namely about the energy of one phonon, above the energy of 1.9 eV, where the Raman lines have their broad maximum. Siebrand et al. [3.20] developed a theoretical model to explain the Raman excitation spectra. In this model, they assume that in the excited state when an electron has been excited from the π to the π^* band, the mean atomic positions are displaced. It is this displacement that produces the overtones and is responsible for electron-phonon coupling and the resonant Raman scattering. The resonance enhancement curves which they calculate in this way contain more structure than experimentally observed. They reduce this extra structure by introducing inhomogeneous broadening, but the results are not completely satisfactory. An example of their fit to the experimental data is shown in Figs. 3.12, 13. *Tubino* et al. [3.21] proposed a phenomenological model to explain the enhancement when the emitted photon energy is about 1.9 eV: They

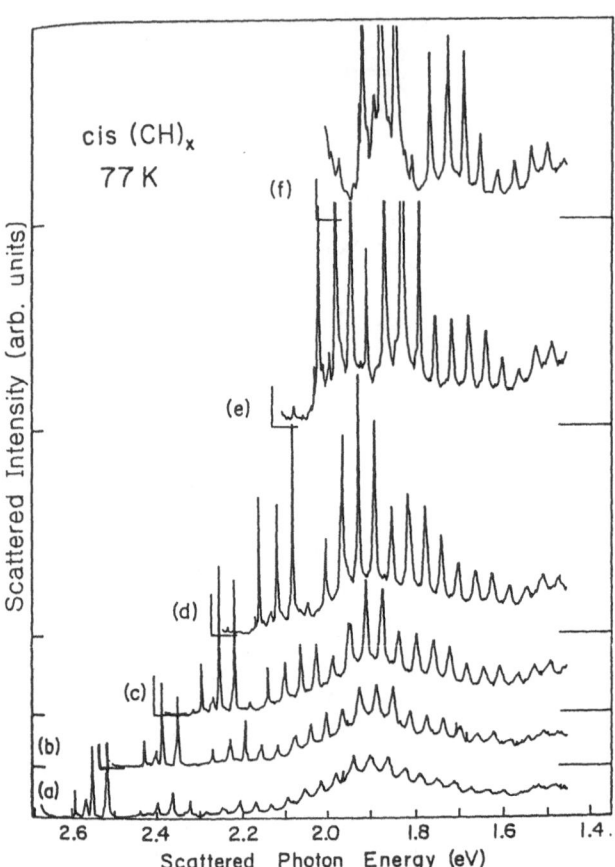

Fig. 3.11. Spectra of outgoing light for *cis*-(CH)$_x$ at 77 K for six different laser wavelengths: (a) 457.9; (b) 488.0; (c) 514.5; (d) 545.0; (e) 580.0; and (f) 610.0 nm [3.7]

propose a localized gap state that becomes weakly allowed due to its interaction with a u (odd parity) phonon. A comparison, between the theoretical and experimental results they obtain, is shown in Fig. 3.14. A discussion of the overtones in *cis*-PA on the basis of the amplitude mode model is presented in Sect. 3.2.

c) Phenomenological Interpretation of the Dispersion Effect

The features that have attracted most of the attention in these Raman spectra are the two large broad *trans*-PA bands and the line in between. The key to their phenomenological interpretation was provided by *Lichtmann* et al. [3.14] who realized that the behavior of these lines is tied to the resonant Raman scattering of the polyenes [3.16]. In the polyene molecules, the frequencies of the two lines corresponding roughly to the single- and double-bond vibrations change in a systematic way with the length of the conjugated backbone of the polymer. Moreover, the fundamental absorption band also changes with the chain length. The

Fig. 3.12. Excitation profiles for the three totally symmetric Raman fundamentals in *cis*-$(CH)_x$. The curves are calculated for a model of three displaced harmonic oscillators and a single resonance state; broken lines include only homogeneous broadening, solid lines both inhomogeneous and homogeneous broadening [3.20]

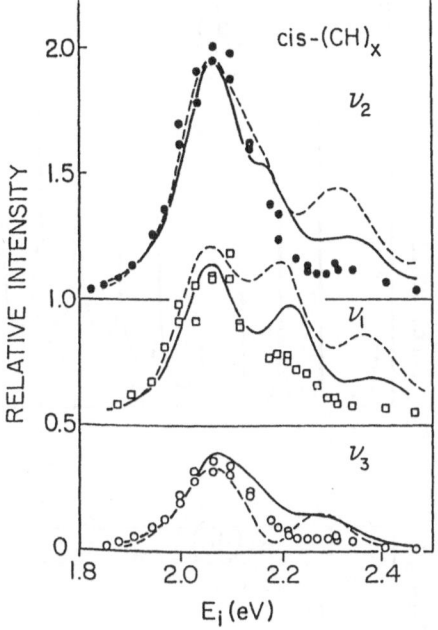

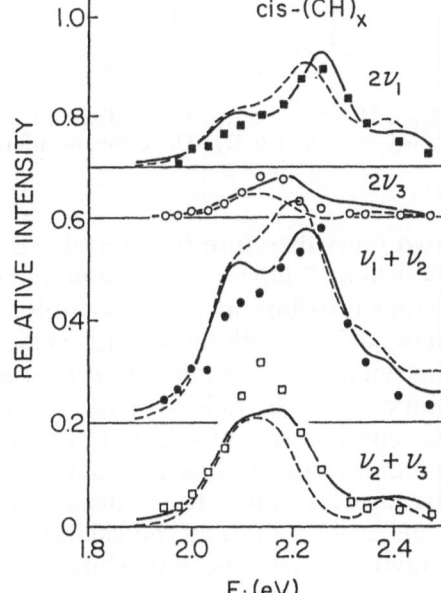

Fig. 3.13. Same as Fig. 3.12 for overtones and combination bands, as indicated. The intensities are fully determined by those of Fig. 3.12 and contain no arbitrary scaling factors [3.20]

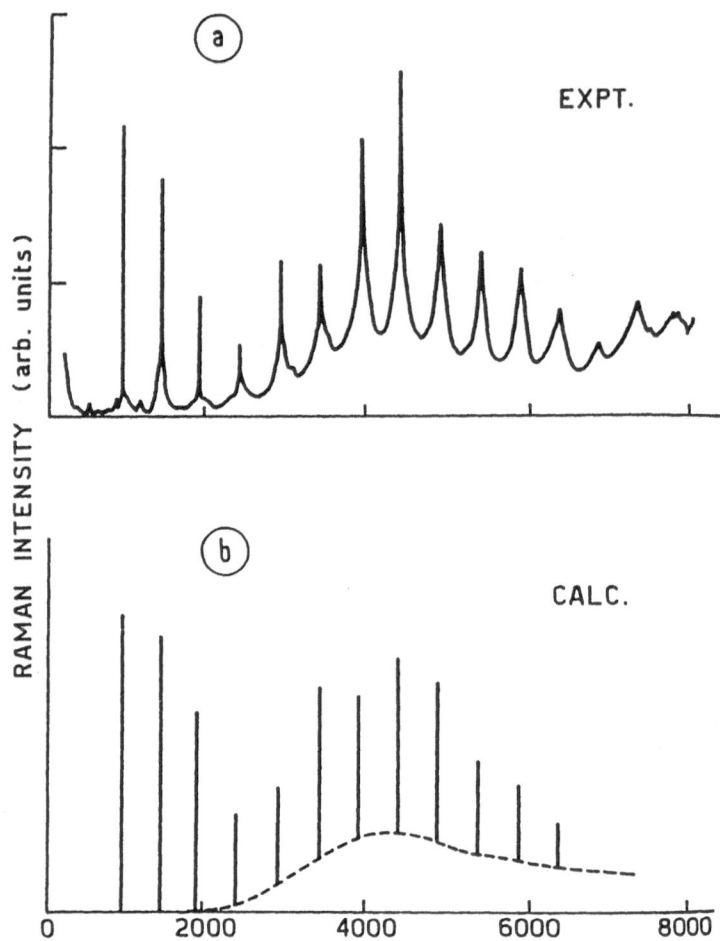

Fig. 3.14. Observed (a) and calculated (b) resonance Raman intensities for the multiphonon processes in *trans*-$(CD)_x$. This comparison is only indicative. Exciting line: 514.5 nm [3.21]

two must therefore be interrelated [3.16]. The qualitative interpretation is then as follows: It is assumed that *trans*-PA samples consist of chains with a distribution of uninterrupted lengths. Chains with different lengths have different phonon frequencies. This, therefore, explains the broad linewidths observed in both *trans*-PA samples and the corresponding *trans* lines in *cis*-PA-rich samples. As the wavelength of the exciting laser becomes shorter, chains with shorter lengths of uninterrupted conjugation become resonant. It is assumed that shorter chains have higher vibrational frequencies similar to the behaviour observed in the polyenes [3.16]. One thus expects that, as the excitation wavelength decreases, the center of gravity of the line will shift upwards, in qualitative agreement with experiment.

A different approach was suggested by *Mele* [3.22], who propsed that the lines could be a result of hot luminescence. However, *Eckhardt* et al. [3.23] showed later that the lines are indeed due to Raman scattering.

Several more quantitative phenomenological models were put forward to explain the results. *Kuzmany* [3.12] and *Kuzmany* et al. [3.6] proposed a Franck-Condon type model and later a particle in a box type model [3.24]. In these models the distribution of the segments is described by an inverse power law: $P(N) = 1/(N + 1)^x$. To obtain a good fit with the experimental results a very large inhomogenous damping constant, about a factor of 5 larger than damping constants encountered in $\pi - \pi^*$ transitions in finite conjugated systems had to be assumed [3.6]. This broadening could also be a manifestation of the three-dimensional nature of the polymer. The exponent of the assumed power law distribution was found to be $x = 2.1$. As pointed out by the authors, this distribution is approximately equivalent to a bimodal log normal distribution [3.24]. A bimodal normal distribution was further developed by *Brivio* and *Mulazzi* [3.13]. They assumed a distribution of the form:

$$P_1(N) = (2\pi\sigma_1^2)^{-1/2} \exp\left[-(N - N_1)^2/2\sigma_1^2\right] G, \tag{3.4}$$

$$P_2(N) = (2\pi\sigma_2^2)^{-1/2} \exp\left[-(N - N_2)/2\sigma_2^2\right](1 - G), \tag{3.5}$$

where G is the fraction of long chains. In addition, they assumed the following forms for the phonon frequencies of the short and long distributions:

For long chains:

$$\omega_1(N) = (1060 + D/N^2)\,\mathrm{cm}^{-1}; \qquad \omega_3(N) = (1450 + D/N^2)\,\mathrm{cm}^{-1}, \tag{3.6}$$

where $D = 1.5 \times 10^4\,\mathrm{cm}^{-1}$. For short chains:

$$\bar{\omega}_1(N) = (1060 + B_1/N)\,\mathrm{cm}^{-1}; \qquad \bar{\omega}_3(N) = (1450 + B_2/N)\,\mathrm{cm}^{-1}, \tag{3.7}$$

where $B_1 = 600\,\mathrm{cm}^{-1}$ and $B_2 = 500\,\mathrm{cm}^{-1}$. A comparison between the experimental and theoretical line shapes thus obtained is shown in Fig. 3.15. The fact that the experimental line shapes depend strongly on the way the samples were prepared and stored indicates that the morphology does indeed play an important role. The bimodal distribution suggests that two types of phase are present. One containig mainly very long chains the other very short ones. This is consistent with the picture provided by neutron scattering [3.3], namely, that the samples are composed of crystalline parts probably with long chains and amorphous parts with mostly short chains.

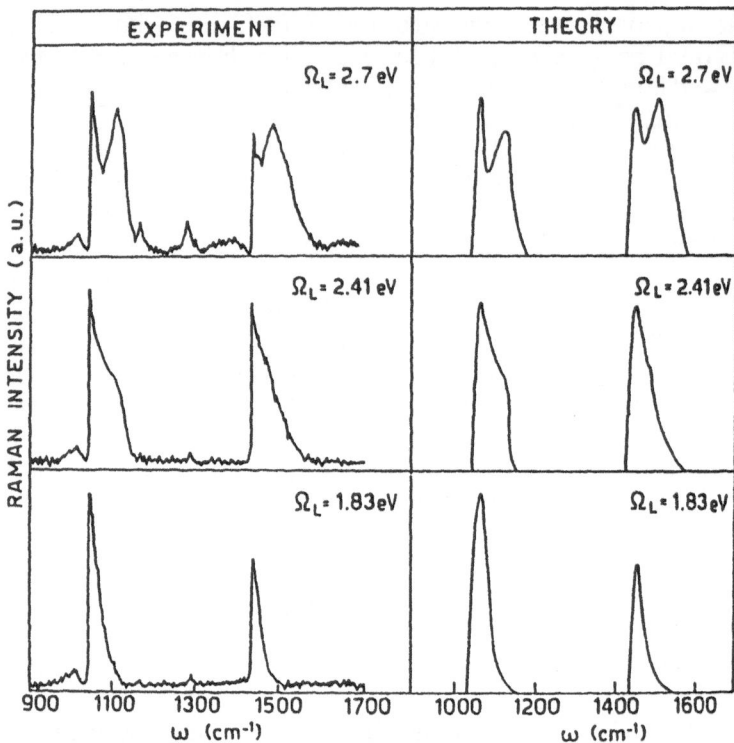

Fig. 3.15. Experimental and theoretical RRS spectra of *trans*-(CH)$_x$ for different incident light frequencies ω_L [3.13]

The main problem with the bimodal distribution suggested by *Kuzmany* et al. [3.6] and *Mulazzi* et al. [3.13] is that there is really no a priori justification for the assumed normal chain length distribution.

A slightly different approach was proposed by *Yacoby* and *Roth* [3.15]. They make two assumptions:

i) The positions of the defects that limit the length of undisturbed conjugated bonds are independent of each other.

ii) The chain length dependences of ω_{3T} and of the electronic band gap are essentially equal, except for a constant shift, to the experimentally measured dependencies in the polyenes. The resultant phonon energy vs. gap energy dependence is shown in Fig. 3.16.

Following assumption (i) the probability that the chain length will be between *l and l + dl* is given by:

$$P(l)\,dl = l_0^{-1} \exp\left(-l/l_0\right) dl\,, \tag{3.8}$$

where l_0 is an average number of double bonds in one chain. Using this distribution function, and the contribution of the long chains part of the

Fig. 3.16. The dependence of phonon energy on electronic gap energy. This curve is the one experimentally measured in the polyenes [3.16] but shifted so as to fit the electronic gap energy and the phonon energy of long uninterrupted segments of *trans*-polyacetylene [3.15]

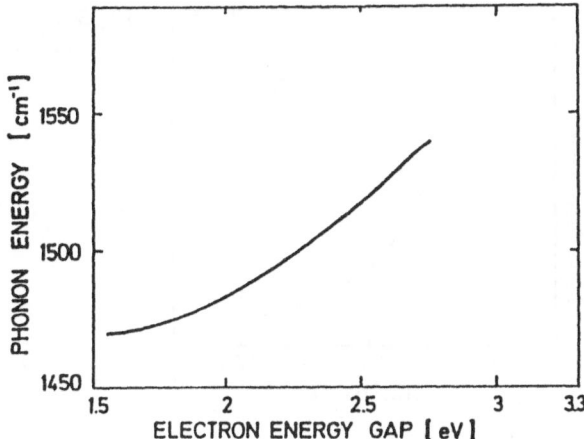

sample, they obtain a very good fit with the experimental results shown in Fig. 3.17. In the process of calculating the resonant Raman cross section, it is necessary to calculate the absorption spectrum and fit it to the experimental curve. To obtain this fit one has to assume an inhomogeneous broadening of ~0.33 eV, which, although large, is only about half the value assumed by *Kuzmany* et al. [3.6].

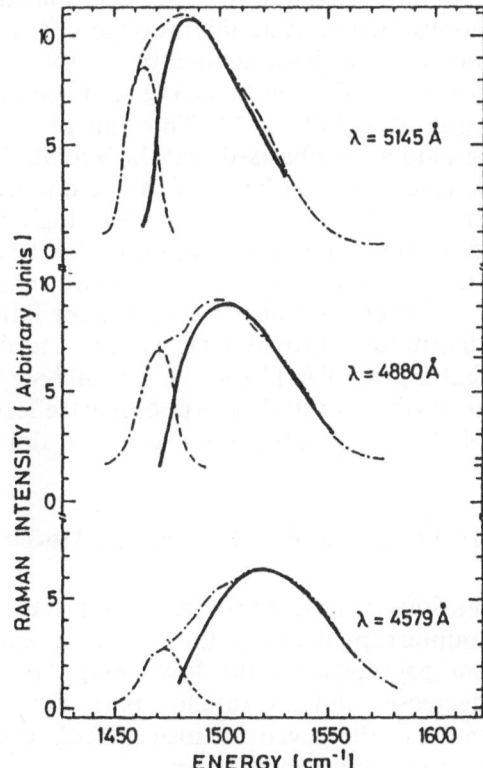

Fig. 3.17. Fit of the model to the experimental Raman band ω_{3T} for the three laser lines. (—·—·—) experimental curve: (————) fitted amorphous contribution; (- - - -) crystalline contribution [3.15]

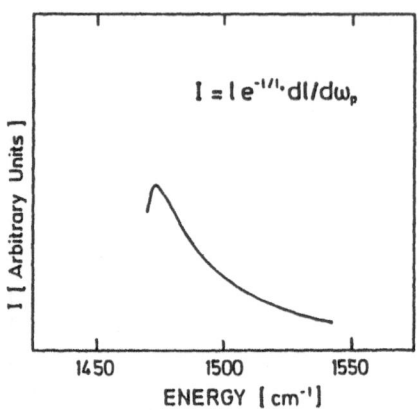

Fig. 3.18. The spectral distribution as a function of phonon energy [3.15]

Calculations based on the amplitude mode model [3.25], which will be discussed in greater detail in Sect. 3.2, led to the conclusion that a single peak distribution $P(\tilde{\lambda})$ as a function of the renormalization parameter, $\tilde{\lambda}$, can explain several key features of the observed experimental results. Since $\tilde{\lambda}$ and the phonon frequency ω_p are linearly dependent, to a first approximation, it appears that the two types of distribution are in contradiction. Actually they are not. To compare the two, both distributions should be presented in the form $P(\omega_p)\, d\omega_p$. When presented in this form both distributions, the short chain and the long chain [3.15], have a peak at $\sim 1475\ \text{cm}^{-1}$. This can be seen in Fig. 3.18 [3.15]. However, it should be emphasised that the bimodal distribution of conjugation lengths is essential in order to obtain a quantitative fit to the experimental line shapes. The reason is that the high frequency tail of the short chain distribution is strongly amplified by the resonance, and in this way controls the high frequency part of the two main Raman bands.

The dependence of the resonant Raman scattering on the chain length distribution provides an excellent tool to study various morphological features of PA [3.13]. This tool has been extensively used by several researchers to study morphological effects during isomerization, conditions of storage, polarization effects and doping.

d) The Effect of Defects on the Morphology

Various defects have been found to affect the effective length of an uninterrupted chain. *Knoll* and *Kuzmany* [3.26] have shown that, upon compactification, the low energy part of the two main *trans* bands decreases and the satellite parts shift to higher energies. These results indicate that even intrinsic defects decrease the effective length of the uninterrupted conjugation.

Schen et al. [3.27] found a clear correlation between the intensity of the low energy peaks of the 1070 and 1450 cm^{-1} bands and the molecular weights of the chains. From this they conclude that a 100,000 Dalton (g/mol) polymer is composed mainly of long *trans* conjugated segments. In contrast, the 10,000 Dalton polymer and below contains a significantly larger portion of short uninterrupted conjugated chains.

A number of different chemical defects have been investigated. In this section, we shall discuss only some of them. The dopants will be discussed in Sect. 3.1.5. *Knoll* and *Kuzmany* [3.26] and *Schen* et al. [3.27] investigated the effect of exposing the samples to oxygen or air. An example of the results is shown in Fig. 3.19. Here again it is clearly seen that exposure

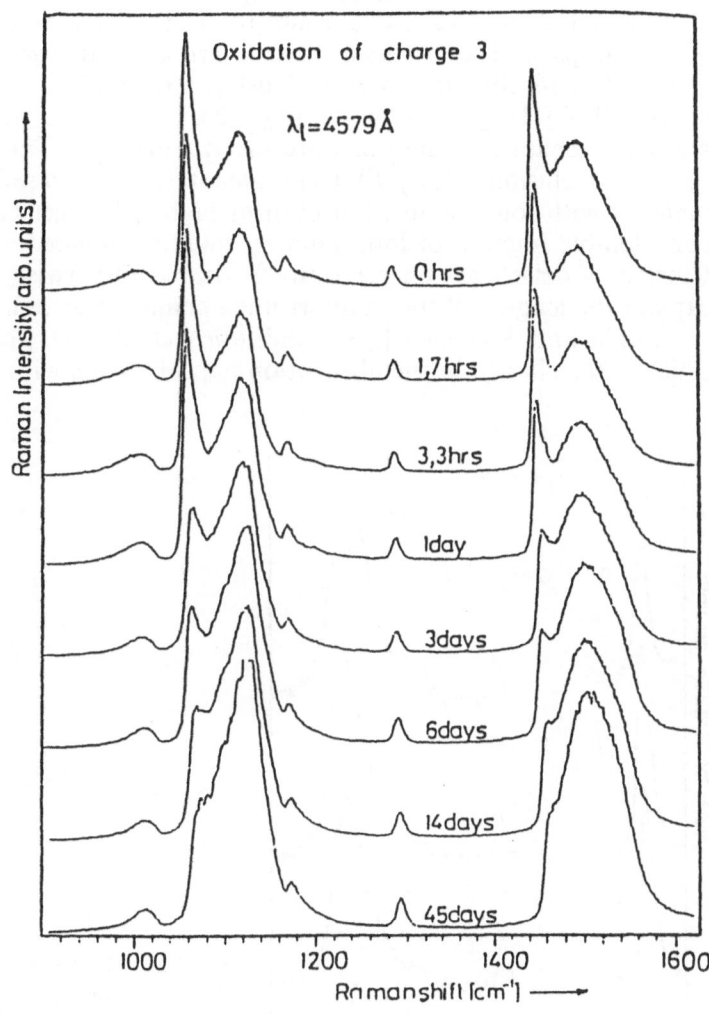

Fig. 3.19. Dependence of line shapes on exposure to air [3.26]

to oxygen decreases the low energy peaks in the 1070 and 1450 cm^{-1} bands and causes an increase and a frequency shift in the satellite peak. These results again indicate that exposure to oxygen reduces the fraction of the material with long uninterrupted chains and shortens the length of the average uninterrupted conjugated chain. Analysing the chain length distributions, and comparing them to the weight uptake of the material while exposed to oxygen, suggests that only a fraction of the oxygen affects the chain length. A large fraction of the oxygen reacts at chain ends and does not affect the length of uninterrupted conjugations.

Another type of chemical segmentation has been effected by doping the material with $AlCl_4$ and letting it interact with water [3.28]. This treatment produces sp^3 defects with a carbonyl group on a neighboring carbon. Another interesting type of segmentation was achieved by *Mulazzi* et al. [3.29]. They added deuterium to $(CH)_x$ samples and hydrogen to $(CD)_x$ samples. This process produces carbons in the sp^3 orbital configuration. The results were observed using RRS with $\lambda_L = 351.1$ nm. The spectra of $(CDH_y)_x$ are shown in Fig. 3.20. In these spectra the long chain and short chain contributions are nicely separated. The results lead to several conclusions [3.29]. (i) The segmentation decreases the fraction of material with long chains, but even at high sp^3 defect concentrations a considerable fraction of long uninterrupted conjugated chains is left (ii) Clusters of defects are energetically favorable. Otherwise, it is difficult to explain the lengths of the uninterrupted chains observed.

Surján and *Kuzmany* [3.30] and *Kürti* et al. [3.31] calculated theoretically the change in the absorption gap of chains with various defects

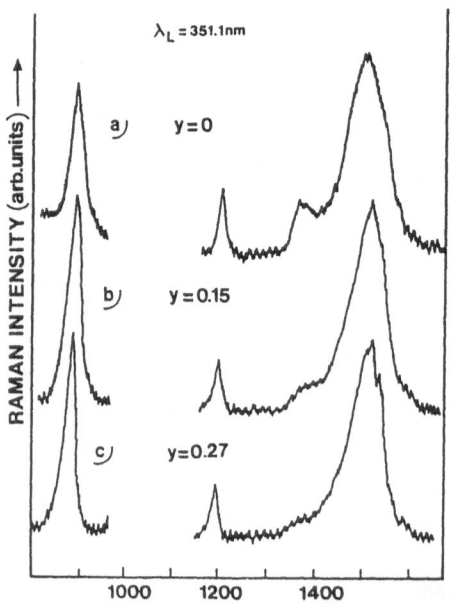

Fig. 3.20. RRS spectra of $(CDH_y)_x$ $T = 20\,°C$ for $\lambda_L = 351.1$ nm (a) $y = 0$; (b) $y = 0.15$ (c) $y = 0.27$ [3.29]

Table 3.3. Interruption of conjugation in polyenes

Defect	Segment	$\Delta\varepsilon$ [eV]	Interruption [%]
twist (60°)	C8	0.8	40
carbonyl	C9	0.85	45
sp^3-carbon	C9	1.33	70
twist (60°)	C30	0.3	16
cis-segment	C30	0.04	1.7
10 twist (9° each)	C60	0.2	18
10 twist (60° each)	C100	0.25 (0.35)	
10 carbonyls	C100	0.40 (0.60)	

accommodated along them. They used both the Hückel and the CNDO/S-CI methods. The first was used with long chains, the second was used to evaluate the effect of sp^3 type defects but can handle only short chain molecules. The defects they treated included rotation around the "double" C=C bonds, the single C–C bonds, local impurities, cis segments, chain bending and sp^3 type defects. The results, summarized by *Kuzmany* and *Kürti* [3.32], are shown in Table 3.3. It is seen that sp^3 carbonyl side groups and large C=C twists produce the largest effects. All the others produce only small disturbances. Even defects that produce a large disturbance are not sufficient, when single, to produce the large changes in gap energy experimentally observed. This leads to the conclusion that defects tend to cluster producing domains with very high defect density and regions which are largely defect free.

Theoretical calculations by *Brédas* et al. [3.33] on segmented PA, $(CHD_y)_x$, using the MNDO method, also lead to the conclusion that the sp^3 defects cluster in groups leaving a relatively large portion of long conjugation lengths.

e) The Effect of Hydrostatic Pressure on the Raman Spectra

The effect of hydrostatic pressure on the RRS of polyacetylene was studied by *Coter* et al. [3.34] up to 8 kbar and by *Yacoby* and *Roth* [3.35] up to 44 kbar. The main effect of the hydrostatic pressure is to shift the main two *trans* bands to higher energies. However, the leading edge of the bands remains essentially unshifted. The results can be seen in Fig. 3.21. It is important to note, that the effect of the hydrostatic pressure is fully reversible. The reversibility means that the effect of the pressure does not involve changes in the structure. The authors interpreted the results in terms of a change in the electronic gap due to changes in the interchain interaction with pressure. As the gap decreases, shorter chains scatter resonantly, shifting the observed Raman bands to higher energies. The change in the electronic gap energy with pressure deduced from these experiments is compared to direct measurements of the absorption edge [3.36] in Fig. 3.22, and the two are in excellent agreement.

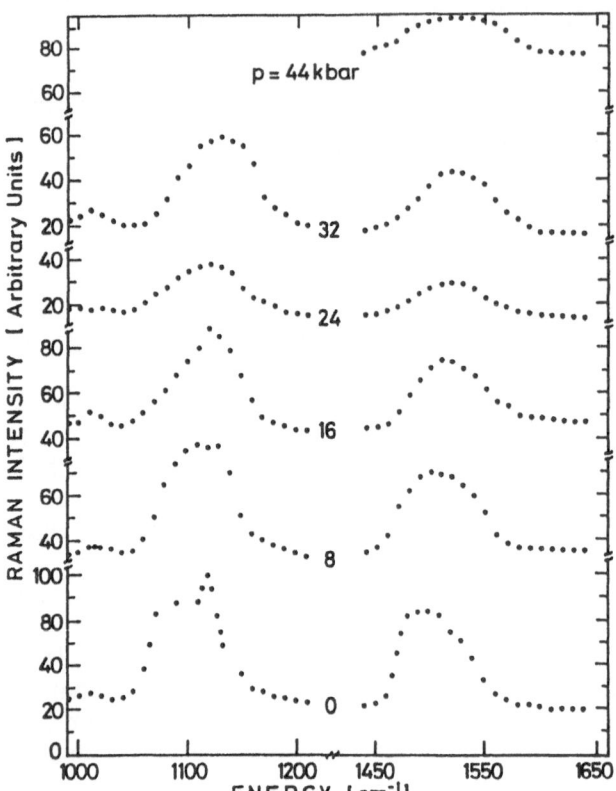

Fig. 3.21. The Raman scattering spectra of *trans*-PA for different hydrostatic pressures

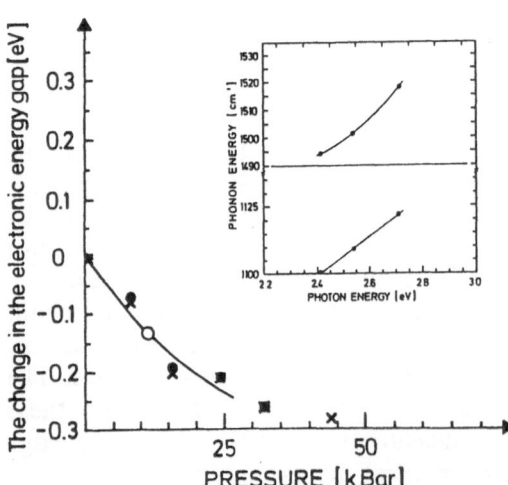

Fig. 3.22. Pressure dependence of the change of the electronic energy gap. ○ 1100 cm⁻¹ band; × 1450 cm⁻¹; ● from Moses et al. [3.36]; Solid curve calculated. Inset: the energy of the midpoints of the two main Raman bands as a function of the laser excitation photon energy. These curves are used in calculating the change of the energy gap with pressure

3.1.4 Polarized Resonant Raman Scattering

The chain directions in as-grown polyacetylene films are totally disordered. Thus, the only additional information that can be obtained from polarized Raman scattering in these films is the depolarization ratio. Several groups have succeeded in obtaining highly oriented films by stretching them by a ratio as large as 7 : 1 [3.37–39]. Using such films fully polarized resonant scattering experiments were carried out by a number of groups leading to a deeper understanding of the spectra and the morphology of the samples [3.40–48].

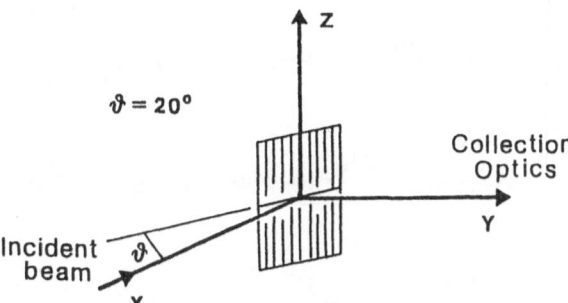

Fig. 3.23. Scattering geometry of Raman experiments. θ defines the angle between the incident laser beam and the plane of the film [3.40]

The scattering geometry used in many cases is shown in Fig. 3.23. The polarized scattering intensities of the totally symmetric vibrational modes can be expressed in the following form [3.49] (the (X, Y, Z) coordinates are as marked in Fig. 3.23, z is parallel to the chains, y is perpendicular to the chain plane and x is perpendicular to both):

$$I_{X(ZZ)Y} \sim f\alpha_{zz}^2 + \tfrac{1}{45}(1 - f)(45\bar{\alpha}^2 + 4\gamma^2 - 5\delta^2), \tag{3.9}$$

$$I_{X(ZX)Y} \sim \tfrac{1}{2} f\alpha_{zy}^2 + \tfrac{1}{15}(1 - f)\gamma^2, \tag{3.10}$$

$$I_{X(YZ)Y} \sim \tfrac{1}{2} f\alpha_{yz}^2 + \tfrac{1}{15}(1 - f)\gamma^2, \tag{3.11}$$

$$I_{X(YX)Y} \sim \tfrac{1}{2} f(\alpha_{xx} - \alpha_{yy})^2 + \tfrac{1}{15}(1 - f)\gamma^2, \tag{3.12}$$

where $\alpha_{\varrho\sigma}$ are the scattering tensor elements f is the molar fraction of chains oriented in the z direction (the rest are randomly oriented), and

$$\bar{\alpha}^2 = \tfrac{1}{9}(\alpha_{xx} + \alpha_{yy} + \alpha_{zz})^2, \tag{3.13}$$

$$\gamma^2 = \tfrac{1}{2}[(\alpha_{xx} - \alpha_{yy})^2 + (\alpha_{yy} - \alpha_{zz})^2 + (\alpha_{zz} - \alpha_{xx})^2$$
$$+ 3(\alpha_{yz}^2 + \alpha_{zy}^2) + \delta^2], \tag{3.14}$$

$$\delta^2 = (\alpha_{yz} - \alpha_{zy})^2. \tag{3.15}$$

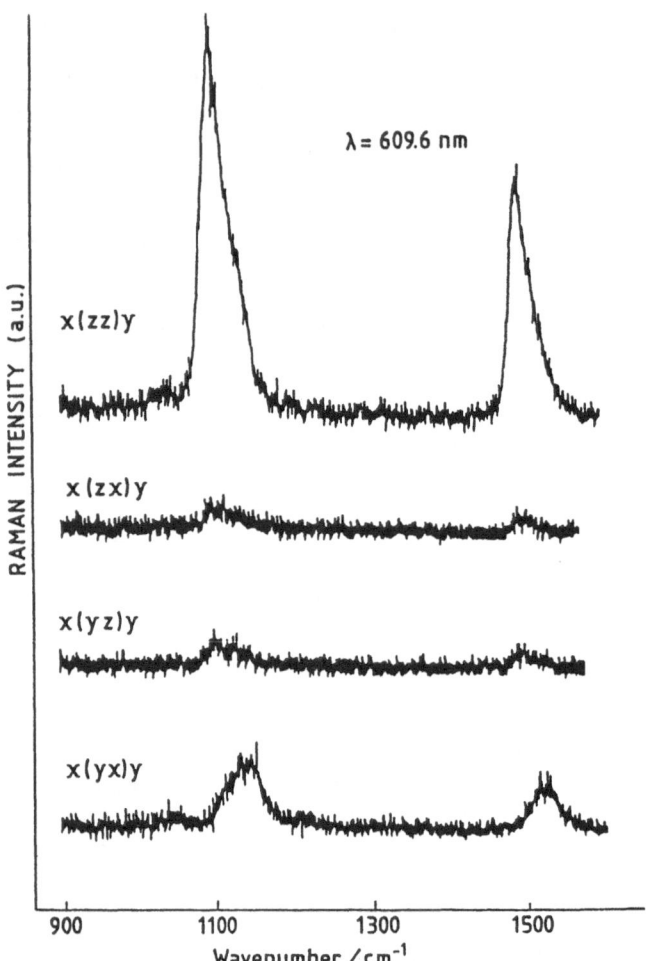

λ = 609.6 nm

Fig. 3.24. Observed Raman intensities of *trans*-polyacetylene for four different scattering configurations. Excitation wavelength: 609.6 nm; power at the sample: 13 mW; spectral slitwidth: 10 cm^{-1}; range: 200 cps: time constant: 2 s; scan rate: 20 cm^{-1}/min. Corrected intensities are obtained multiplying by 1, 0.4, 0.4, 0.04 from top to bottom, respectively [3.41]

a) Polarized Raman Spetra of *trans*-Polyacetylene

An example of the experimental results is shown in Fig. 3.24 [3.41]. By far the strongest spectrum is the $X(ZZ)Y$ spectrum in which both the incident and scattered light beams are polarized parallel to the chains. The $X(YX)Y$ spectrum is weaker and the $X(YZ)Y$, and the $X(ZX)Y$ spectra seem to be the weakest. An important feature to notice is that the depolarization at the high energy side of both main Raman bands is much larger than that of the low energy side. In evaluating the depolarization ratios quantitatively, it is necessary to take the anistropy of the sample into account. *Masetti* et al. [3.41], *Faulques* et al. [3.42] and other authors have calculated the corrections resulting from the anisotropy of the absorption and reflection coefficients and found that in fact the weakest spectrum of

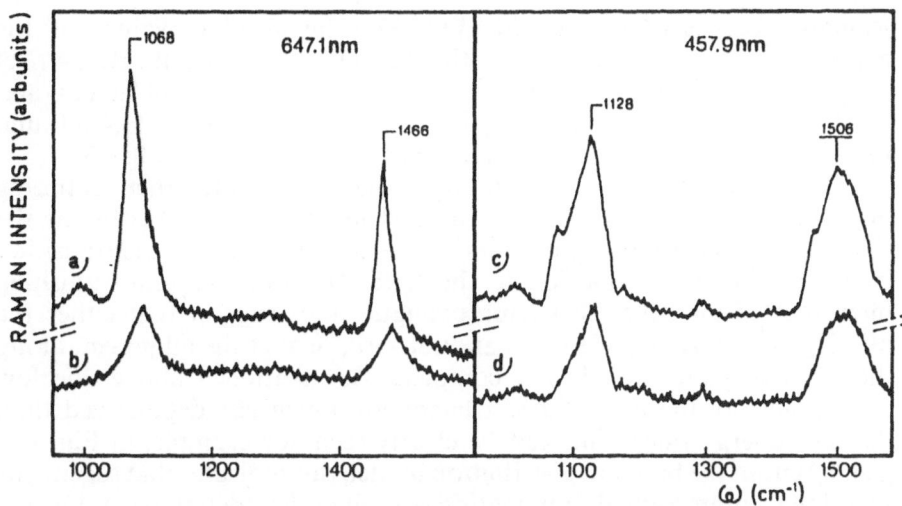

Fig. 3.25. Experimental spectra of stretched undoped *trans*-(CH)$_x$ at 77 K. (a) and (b) ∥ ∥ and ⊥ ⊥ polarized spectra for $\lambda_L = 647.1$ nm, respectively; (c) and (d) ∥ ∥ and ⊥ ⊥ polarized spectra for $\lambda_L = 457.9$ nm respectively [3.45]

Fig. 3.26. Calculated spectra of stretched undoped *trans*-(CH)$_x$. (a) ∥ ∥ polarized spectra for $\lambda_L = 647.1$ nm; (b) and (c) ∥ ∥ and ⊥ ⊥ polarized spectra for $\lambda_L = 457.9$ nm respectively [3.45]

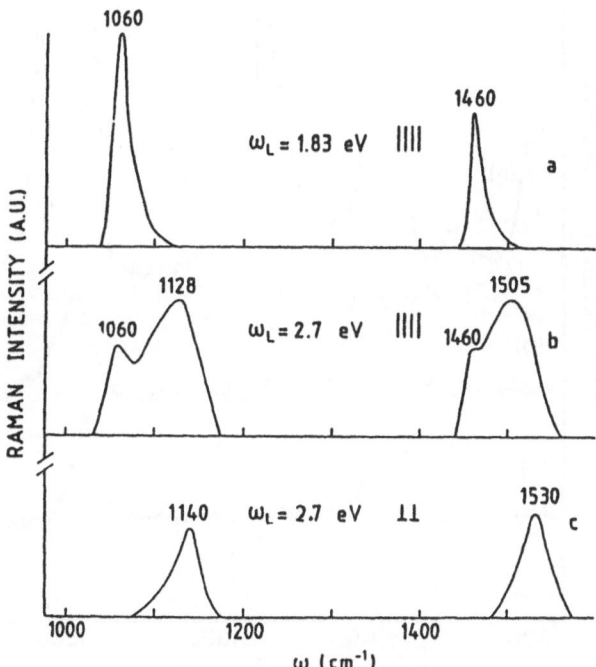

Raman cross sections was the $X(YX)Y$. The absolute values of the depolarization ratios depend on the excitation wavelength. *Masetti* et al. [3.41] found that the depolarization ratios at $\lambda_L = 609.6$ nm are: $I_{YZ}/I_{ZZ} = I_{ZX}/I_{ZZ} \simeq 0.03$ and $I_{YX}/I_{ZZ} = 0.01$. *Faulques* et al. [3.42] found for $\lambda_L = 457.9$ nm, $I_{XZ}/I_{ZZ} \simeq I_{ZY}/I_{ZZ} \simeq 0.02$ and $I_{YX}/I_{ZZ} \simeq 0.013$.

The fact that $I_{X(ZZ)Y}$ is much larger than any of the other scattering intensities indicates that the α_{zz} tensor element in (3.9–15) is by far the strongest. This is to be expected if the $\pi - \pi^*$ virtual electronic transition is the dominant contribution to the RRS. However, the non-vanishing intensities of the other scattering configurations suggest that either the off-diagonal scattering elements are not zero, or that the alignment along the stretching direction is not complete. As mentioned above, the low energy sides of the main Raman bands are much less depolarized than the high energy sides. This can be clearly seen, for example, in Fig. 3.25 [3.45]. Within the bimodal distribution model, this indicates that segments with long uninterrupted conjugation are much less depolarized. On the other hand, segments with only short uninterrupted chains, of the order

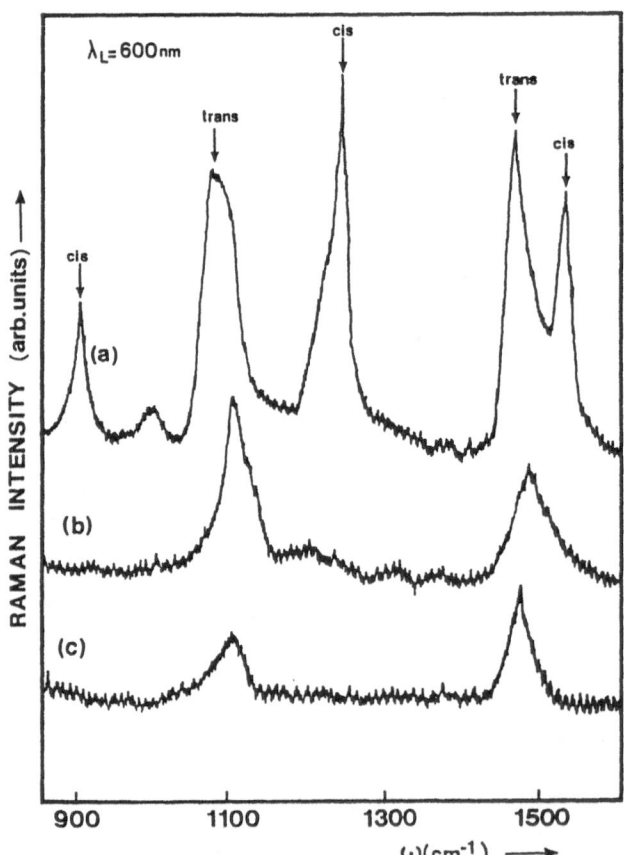

Fig. 3.27. Experimental RRS spectra of the *cis*-rich sample for $\lambda_L = 600$ nm at $T = 77$ K: (a) ∥ ∥ polarized spectrum, (b) ⊥ ⊥ polarized spectrum, (c) ⊥ ∥ polarized spectrum [3.47]

of 10 cells, produce larger depolarization. The result of such a model calculation [3.45] is shown in Fig. 3.26. The depolarization may result either from the misalignment of the short segments and/or from contributions of electronic transitions, which in these short samples produce dipole moments that do not lie along the chain axis.

b) Polarized Raman Scattering of *cis*-Polyacetylene

The polarized Raman spectra of *cis*-PA have been measured only recently [3.47, 48, 50]. Some examples of the specta obtained by *Perrin* et al. [3.47] using excitation wavelengths $\lambda_L = 600$ and $\lambda_L = 676.4$ nm, respectively are shown in Figs. 3.27, 28. In Fig. 3.28, only the *trans* component Raman bands are observed, indicating that at this wavelength the scattering in the *cis* part of the sample is no longer strongly resonant. At the shorter wavelength both parts are observed. However, only the *trans* part of the sample is depolarized. Measurements by *Lanzani* et al. [3.48] show a small depolarization of *cis* lines as well. The discrepancy may be a result

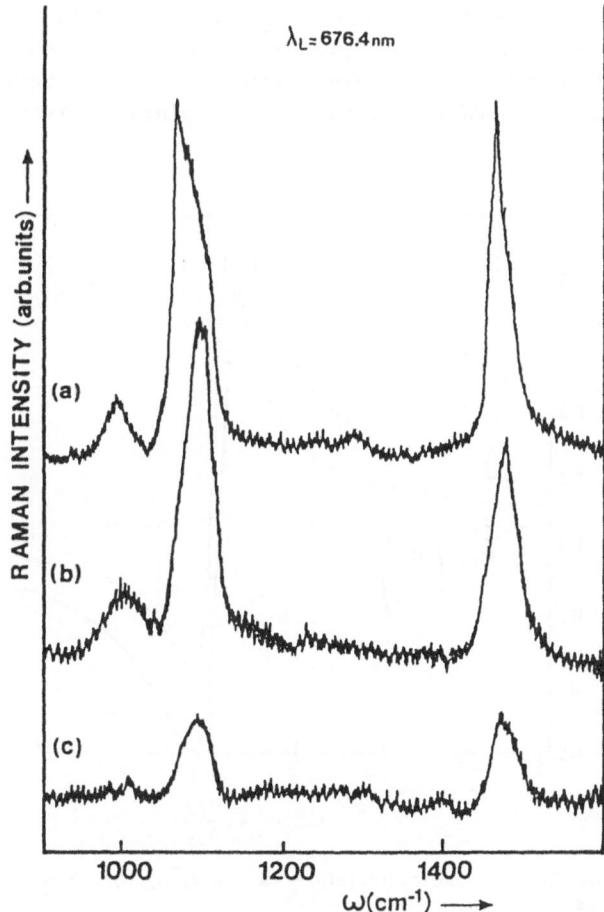

Fig. 3.28. Experimental RRS spectra of the *cis*-rich sample for $\lambda_L = 676.4$ nm at $T = 77$ K: (a) $\parallel \parallel$ polarized spectrum, (b) $\perp \perp$ polarized spectrum, (c) $\perp \parallel$ polarized spectrum [3.47]

of the difference in the samples. These results lead to the conclusions: (i) The *cis* part is indeed composed of long uninterrupted conjugated segments. (ii) These segments are very well aligned along the stretching direction. These results are consistent with the previous contention that the *cis* part of the samples is composed almost entirely of long uninterrupted conjugated segments. The length of the uninterrupted segments decreases in the process of the isomerization.

c) Depolarization in Unstretched Samples

Assuming that, in an unstretched sample, the chain orientations are completely disordered, one expects the depolarization ratio to be $\varrho_\perp / \varrho_\parallel = 0.33$. This is in complete contrast with experiments, which yield values of about 0.6 [3.51] for both *trans* and *cis* samples. Brafman et al. [3.52] find depolarization ratios ranging from 0.54 at the low energy side of the Raman bands to 0.65 at the high energy side. *Lanzani* et al. [3.48] calculated the depolarization ratio assuming that it arises from non-vanishing Raman scattering tensor elements α_{xx} and α_{yy}. The results are shown in Fig. 3.29. The experimentally observed results are consistent with a ratio $k = \alpha_{zz}/\alpha_{xx}$ from -10 to -4. These values seem to be too small to be consistent with the depolarization observed in the stretched samples.

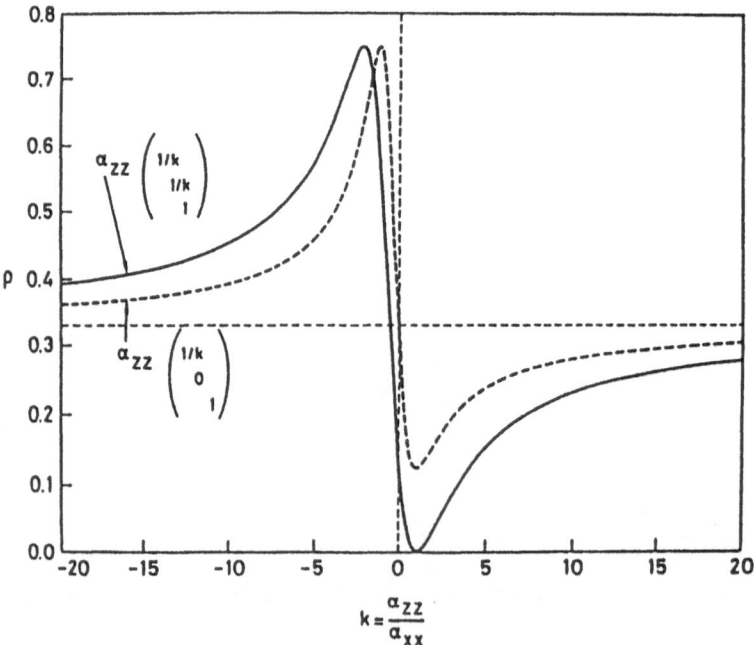

Fig. 3.29. Depolarization ratio ϱ vs. $k = \alpha_{zz}/\alpha_{xx}$ for two different possible scattering tensors [3.48]

It seems to us that the depolarization of the incident and scattered beams cannot be neglected in view of the fact that the distances between the "spaghetti-like" fibers are of the order of the exciting light wavelength and the absorption coefficient is large so that the absorption by one fiber is not at all negligible. This means that the light should feel the spaghetti-like structure of the material and therefore be depolarized by it.

3.1.5 Resonant Raman Scattering of Doped Polyacetylene

Doping PA with various dopants affects both its vibrational and electronic properties. Both effects are important in understanding the RRS of doped PA. The effects of various dopants on the absorption spectra is qualitatively similar. The main absorption band of pristine PA decreases and the absorption in the IR increases forming a broad band. An example of the absorption spectra of an n-type sample are shown in Fig. 3.3. At high doping concentrations, the pristine absorption band decreases drastically and the absorption in the near IR increases forming a well-defined band at ~ 0.8 eV [3.9]. The absorption spectra of an iodine doped *cis* PA sample [3.53] is shown in Fig. 3.30. Qualitatively, the behavior is similar to the

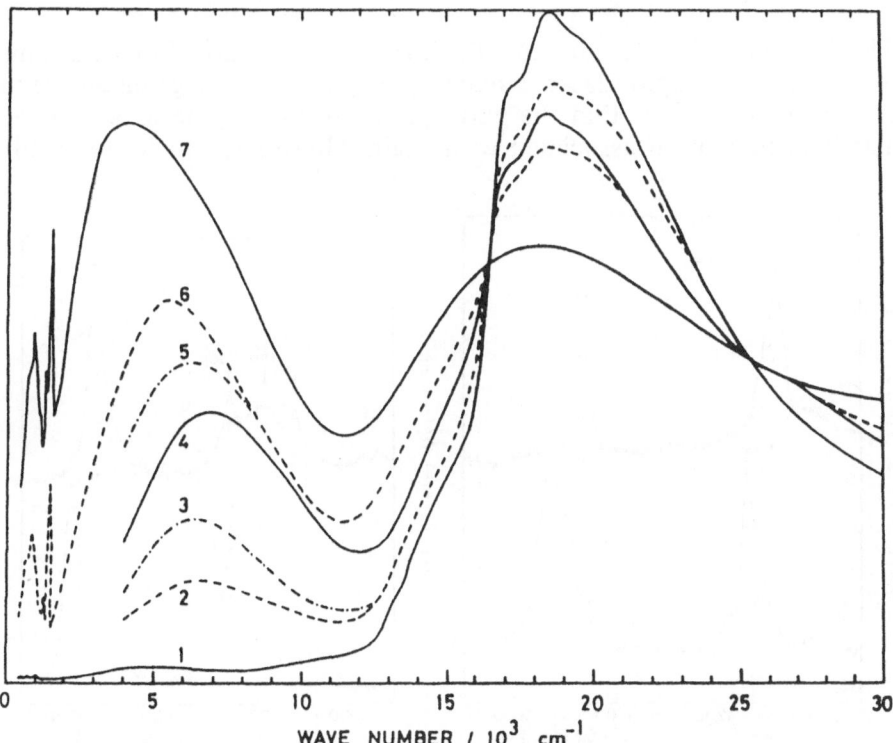

Fig. 3.30. Change of the absorption spectra of *cis*-$(CHI_y)_x$ films: (1) $y = 0$; (2) $y = 0.044$; (3) $y = 0.084$; (4) $y = 0.10$; (5) $y = 0.22$; (6) $y = 0.35$; (7) $y = 0.64$. From [3.53]

n-doped PA, but the decrease in the pristine absorption is smaller and the peak of the IR band is between 0.5 and 0.85 eV. From these results one can expect that, as the doping concentration increases, the peak of the RRS enhancement curve will tend to move to lower frequencies.

a) RRS of *trans*-Polyacetylene Doped with Alkali Metals

A number of authors have studied the RRS of alkali doped PA [3.54–58]. Examples of spectra of K- and Na-doped samples are shown in Figs. 3.31, 32, respectively. Measurements were also done on Li-, Rb- and Cs-doped *trans*-PA [3.58] with qualitatively similar results. The RRS have some important features common to all five dopants:

i) At low doping concentrations, less than 3 mol%, there is very little effect on the *trans* lines with perhaps a slight change towards shorter segments [3.55].

ii) At high doping concentrations two new strong lines are observed, at $1260–1280 \text{ cm}^{-1}$ and at $1545–1595 \text{ cm}^{-1}$. The frequency of the first line is almost independent of the exciting laser frequency, whereas the second shifts considerably towards higher frequencies at higher excitation energies.

The fact that in all four dopants the line frequencies are about the same for the same excitation laser frequency, in spite of the large difference in their masses, suggests that the participation of the doping ions is small. The lines have been associated with chain vibrations, which are locally

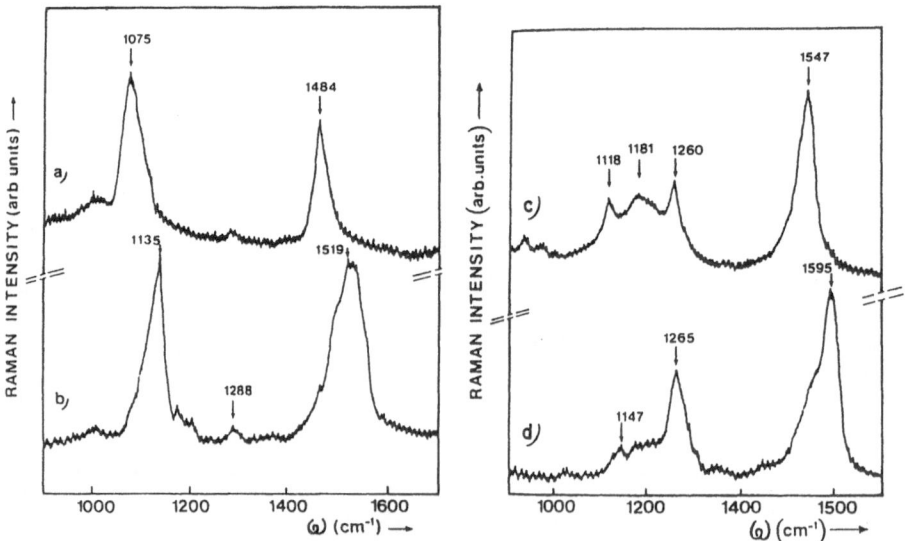

Fig. 3.31. Raman spectra of $[\text{CHK}_y]_x$; $T = 20\,°\text{C}$, (a) $y < 3\%$ $\lambda_L = 676$ nm (b) $y < 3\%$ $\lambda_L = 457.9$ nm (c) $y > 5\%$ $\lambda_L = 676$ nm; (d) $y > 5\%$ $\lambda_L = 459.7$ nm [3.56]

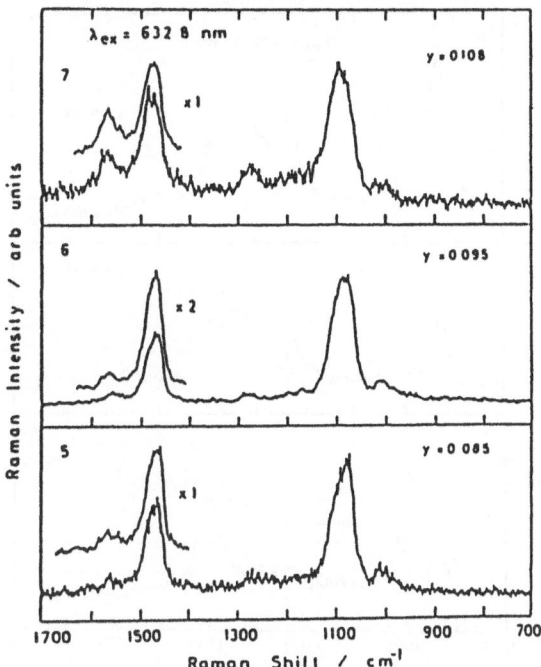

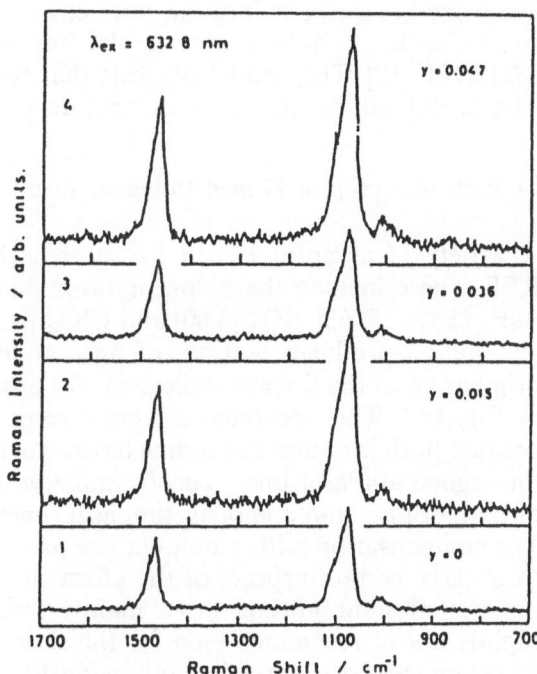

Fig. 3.32

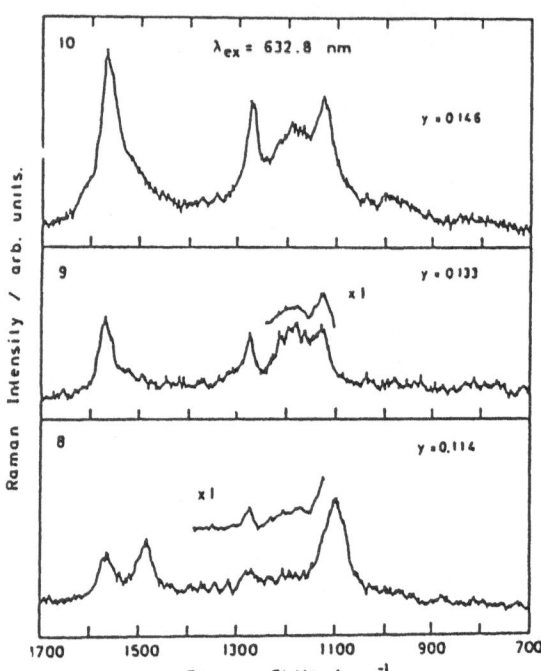

Fig. 3.32. The change of Raman spectra upon doping with Na [3.9]

modified through a change in the intrachain force constant. This change is assumed to be a result of the extra electron transferred from the dopant to the chain. A quantitative model has been proposed by *Brivio* and *Mulazzi* [3.59]. This model suggests that the extra lines are local modes. The model will be discussed in some detail in Sect. 3.2.

b) RRS of Acceptor Doped Polyacetylene

A variety of acceptor doped PA systems have been investigated using RRS. These include the halogens bromine and iodine [3.15, 53, 60–62], AsF_5 [3.12, 15, 63], SO_3 [3.60] and ClO_4 [3.54, 63]. Doping with bromine and iodine produces a series of lines at low frequencies [3.10, 54, 60], originating in the dopant molecules. An example of such spetra is shown in Fig. 3.33. This spectrum is very strong and shows many overtones, because both bromine and iodine have a strong resonance enhancement at the argone ion laser lines. *Yacoby* and *Roth* [3.15] investigated the effects of both doping and compensation as a function of doping concentration. The compensation with ammonia gas provides, in a sense, a clearer way to observe certain aspects of the effect of the dopant on the sample. In these studies the authors used mainly the 457.9 nm lines because it is slightly above resonance even for the short segments, so all parts of the spectrum can be observed simultaneously.

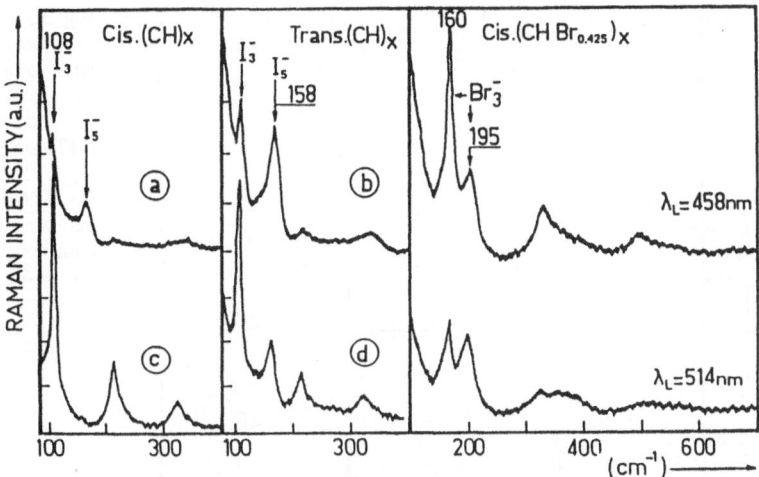

Fig. 3.33. Raman spectra of halogen doped $(CH)_x$ at $T = 78$ K; (a) and (b) after doping with iodine for *cis* and *trans*-$(CH)_x$ respectively; (c) and (d) after 16 h of dynamical pumping. The two spectra in the right side show bromine doped $(CH)_x$ for two different excitation wavelengths [3.54].

The Raman spectra as a function of Br concentration [3.15] are shown in Fig. 3.34. The main effect of the dopant is to reduce the initial rise at the low energy side of the main Raman bands. This clearly shows that the long chain contribution decreases relative to the short chain contribution. The effect of iodine, shown in Fig. 3.35, is somewhat different. At

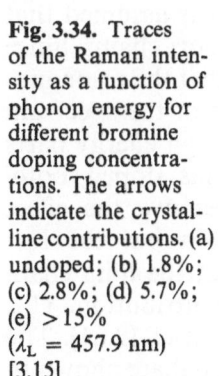

Fig. 3.34. Traces of the Raman intensity as a function of phonon energy for different bromine doping concentrations. The arrows indicate the crystalline contributions. (a) undoped; (b) 1.8%; (c) 2.8%; (d) 5.7%; (e) >15% ($\lambda_L = 457.9$ nm) [3.15]

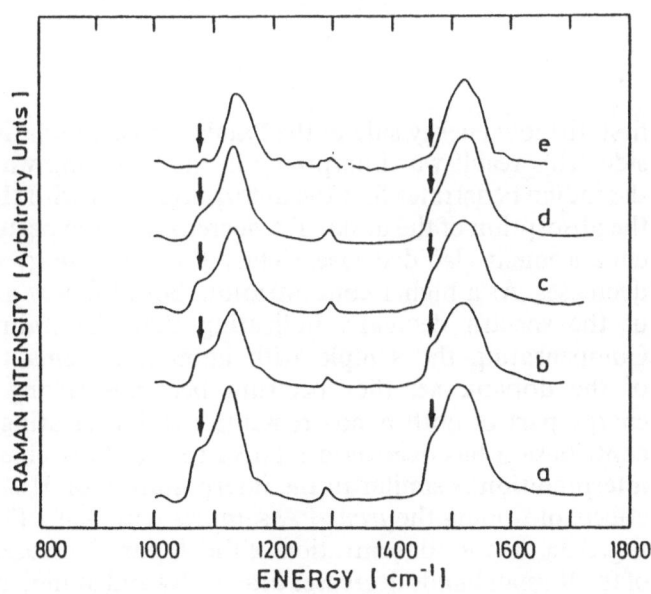

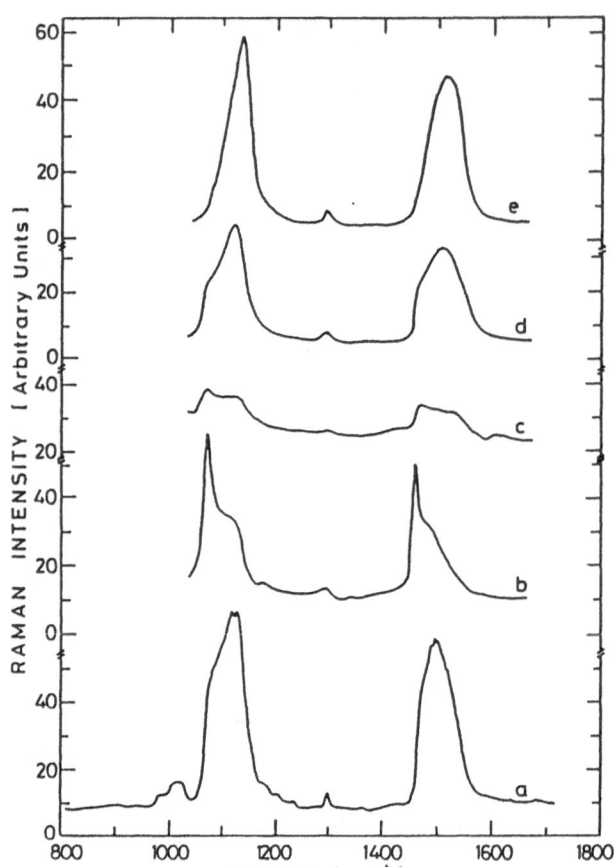

Fig. 3.35. Raman traces of *trans*-PA doped with iodine and compensated by ammonia. (a) undoped; (b) 3.1% uncompensated; (c) 6.1% umcompensated; (d) 3.1% compensated; (e) 6.1% compensated ($\lambda_L = 457.9$ nm [3.15]

first, the low energy side of the bands increases relative to the high energy side. This result was interpreted in the following way. It is assumed that the iodine penetrates first the amorphous part with the short chains. Since the absorption of the doped PA decreases, it is expected that the resonance enhancement also decreases. Thus, the high energy part of the spectrum decreases. At a higher concentration, both the low and high energy parts of the spectra decrease, indicating that the iodine has doped both. Compensating the sample with ammonia seems to remove the effect of the dopant, i.e. the spectrum becomes strong again but the low energy part of both bands remains small, indicating that the long chain contribution has decreased relative to the short chain contribution. This interpretation is similar to the interpretation of *Mulazzi* et al. [3.61]. The effects of doping the *trans*-PA samples with AsF$_5$ [3.15, 63] are shown in Fig. 3.36. As the concentration of the dopant increases the low energy side of the Raman bands decreases drastically and at high concentrations broad

Fig. 3.36. Raman traces of *trans*-polyacetylene doped with AsF$_5$ and compensated by ammonia. (a) undoped; (b) 3.7% uncompensated; (c) 9.9% uncompensated; (d) 11.1% uncompensated; (e) 1.1% compensated; (f) 3.7% compensated; (g) 11.1% compensated. (λ_L = 457.9 nm) [3.15]

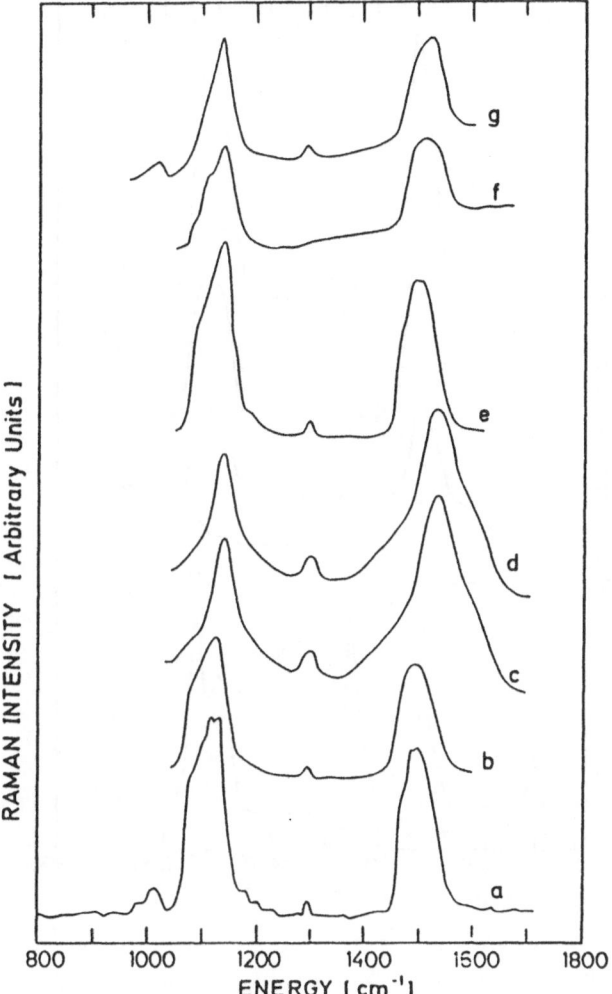

backgrounds and a broad line at 1560 cm^{-1} show up. The origin of the broad backgrounds and the high energy line is not clear. However, upon compensation, they disappear and the only effect left is the decrease in the leading edges of the main Raman bands. This again indicates that in this process the short chain contribtutions have strongly increased relative to the long ones.

c) RRS of Doped *cis*-Polyacetylene

Many authors have investigated the effect of doping *cis*-PA with various dopants [3.12, 15, 54, 55, 60, 62]. In all cases it is shown that a high doping concentration converts the *cis* sample into *trans*. However, it is

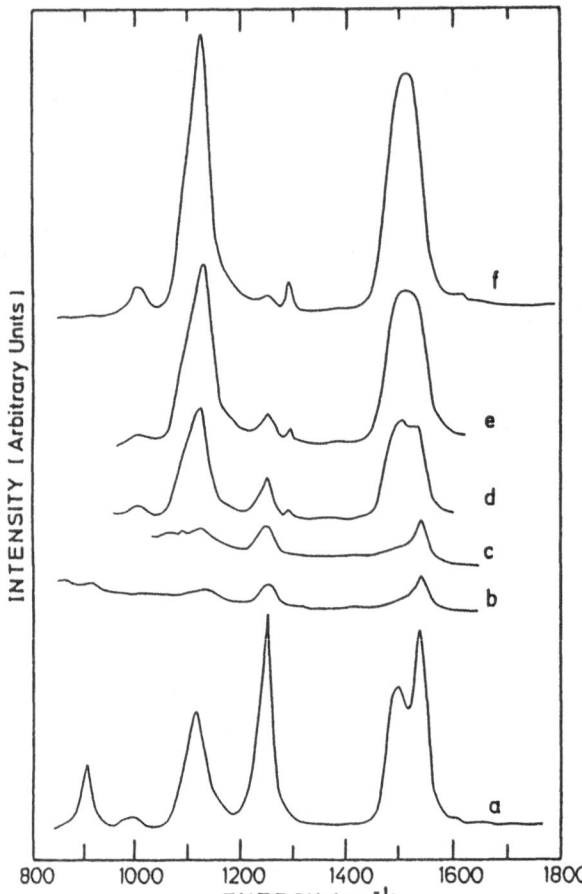

Fig. 3.37. Raman traces of nominally *cis* samples doped with AsF$_5$ and compensated by ammonia. (a) as-grown; (b) 2.0% un-compensated; (c) 4.5% uncompensated; (d) 6.5% uncompensated; (e) 7.95% uncompensated; (f) 9.9% compensated (λ_L = 457.9 nm) [3.15]

not clear whether all parts that are doped convert to *trans*. There is strong evidence that the doping is not uniform. This can be seen for example in Fig. 3.37 which shows the spectra of AsF$_5$-doped *cis*-PA at various doping concentrations with and without compensation. The ratio between the *cis* 1255 cm^{-1} line and the *trans* 1120 cm^{-1} line is shown as a function of concentration in Fig. 3.38. Note, that up to about 5%, there is very little change in the ratio indicating that the *cis* part has not converted. Beyond that, it converts very fast. This is interpreted as indicating that AsF$_5$ dopes first the *trans* part and only then dopes the *cis* part converting it into *trans*. The behavior of the iodine-doped sample is quite different (Fig. 3.39). Doping the sample with 10.7% iodine still leaves the sample in the *cis*-rich form. Compensating it converts it completely into *trans*. These results suggest that the iodine at this concentration penetrates into both *cis* and *trans* parts of the sample, but does not convert *cis* into

Fig. 3.38. The ratio between the *cis* 1255 cm^{-1} and the *trans* peak at 1120 cm^{-1} as a function of AsF$_5$ concentration [3.15]

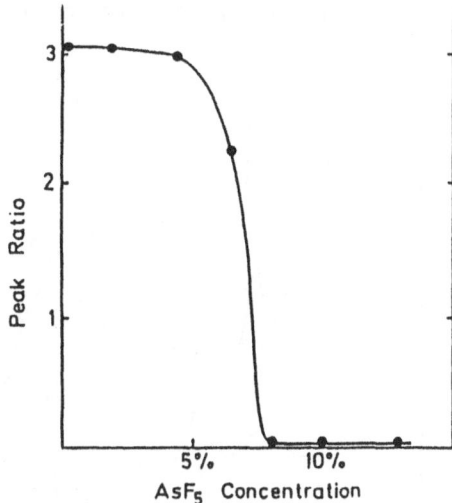

trans. Only compensation converts it into *trans.* It should be emphasised that exposing an as-grown sample to ammonia does not affect the RRS at all.

3.1.6 Dispersion Effects in Other Polymers

Resonant Raman scattering in polyacetylene is particularly interesting for two main reasons:
i) It displays a strong dispersion of the Raman spectra with the laser excitation energy.
ii) The changes in the electronic gap energy, in the phonon frequencies, and in the Raman and IR intensities, are all related to one parameter.

The question is: does this sort of behavior exist in other polymers as well. The dispersion of the RRS with the exciting laser frequency has been a key feature in observing this behavior in PA. The main requirement for the existence of the dipersion is that the sample be composed of a distribution of species with different phonon frequencies and electronic gap energies. In this section we survey the materials and spectra where such a dispersion has been observed.

Dispersion was searched for but not observed in two cases: Poly-paraphenylene-vinylene (PPV) [3.64] and polypyrrole and its 2,5-^{13}C-substituted and C-deuterated analogues [3.65]. Differences in the spectra obtained with different laser excitation lines were observed during the electrochemical oxidation and reduction of polyaniline [3.66]. These differences are not clear enough to be discussed here.

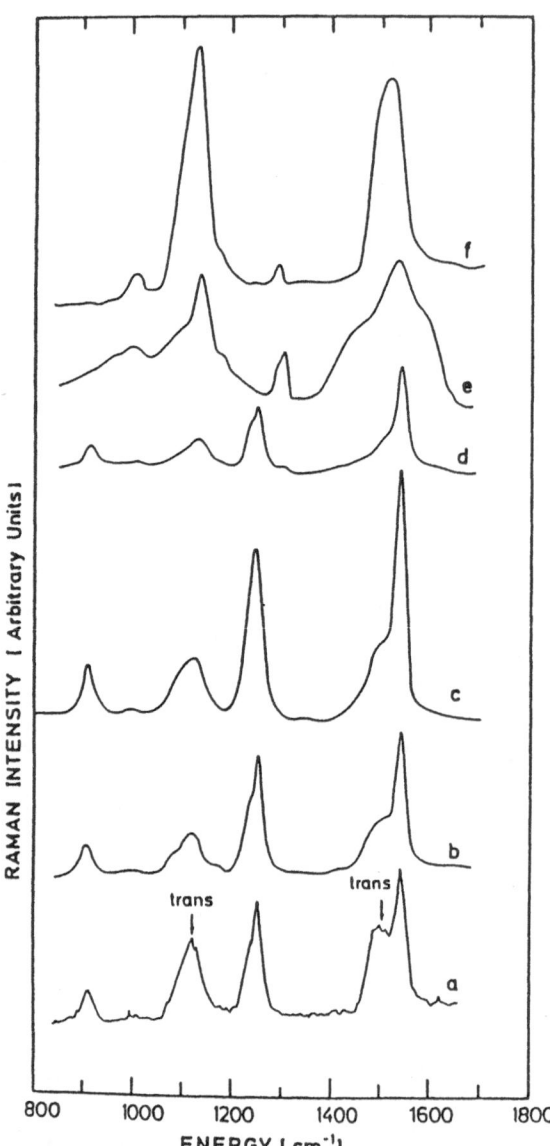

Fig. 3.39. Raman traces of nominally *cis* samples doped with iodine and compensated by ammonia. (a) as-grown exposed to ammonia; (b) 4.4% uncompensated; (c) 7.2% uncompensated; (d) 10.7% uncompensated; (e) 7.2% compensated; (f) 10.7% compensated (λ_L = 457.9 nm) [3.15]

The dispersion effect has been quite clearly observed in electropolymerized polyparaphenylene [3.67]. Clear dispersion has been observed by *Wallhöfer* et al. [3.68] in polyisothia-naphthene. The spectra for various laser excitation energies are shown in Fig. 3.40. The dispersion effect is clearly observed in the 447 and ~ 1500 cm^{-1} lines. The lines behave in a way that is similar but not as pronounced as the CC lines in *trans*-PA. Namely, at higher laser excitation frequencies the lines broaden and their

Fig. 3.40. Raman spectra of neutral PITN, excited with various laser line [3.68]

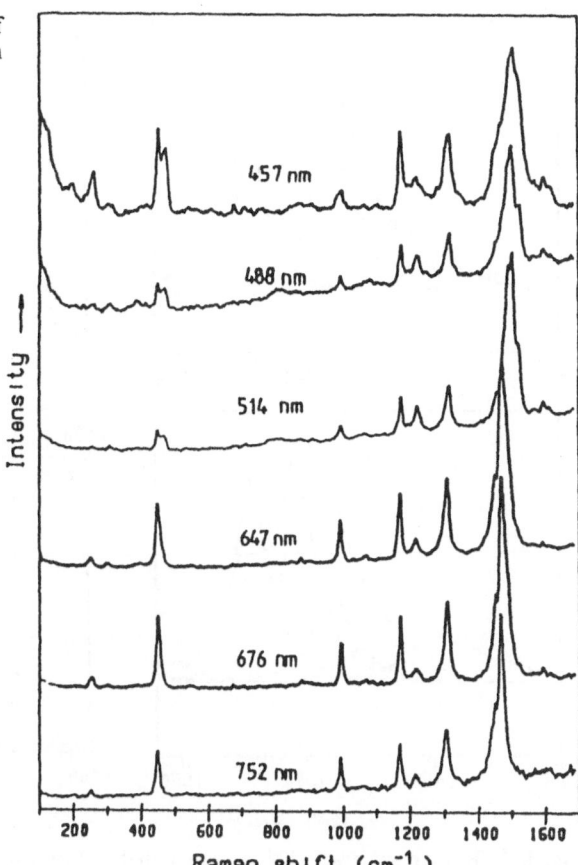

center of gravity shifts to higher frequencies. The high energy line at $\sim 1500\ cm^{-1}$ was decomposed into contributions of a single, double and up to 5 monomer contributions. This is shown in Fig. 3.41 for four polymerization potentials. Another case showing a clear dispersion effect is poly(1.6-heptadiyne) [3.69] shown in Fig. 3.42. The line showing the dispersion is centered at $\sim 1500\ cm^{-1}$.

At present it is not clear why the dispersion effect has been observed in only a few cases. One possibility is that the polymers are prepared in most cases with very long chains, in which case no dispersion is expected as for *cis*-PA. It seems to us that, in order to obtain a unified view of RRS in polymers, the search for dispersion effects in polymers should be continued. In particular, the introduction of defects which limit the conjugation length may prove fruitful.

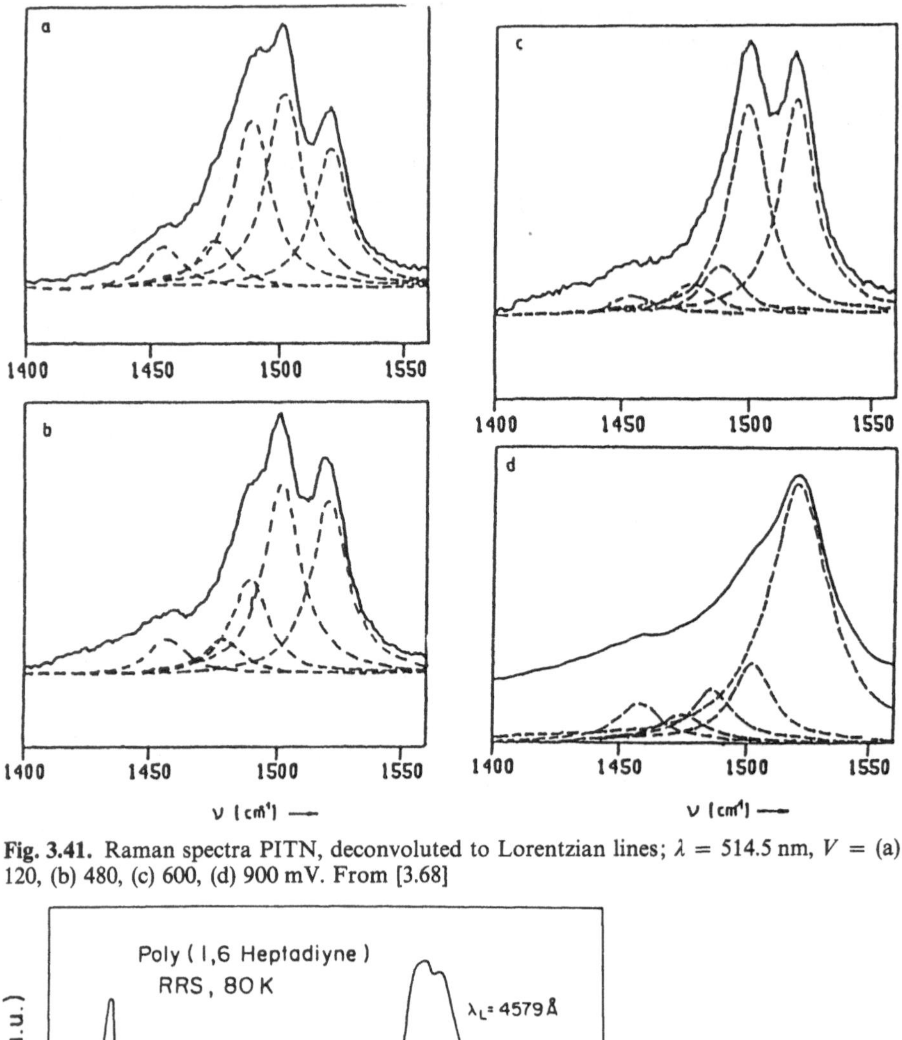

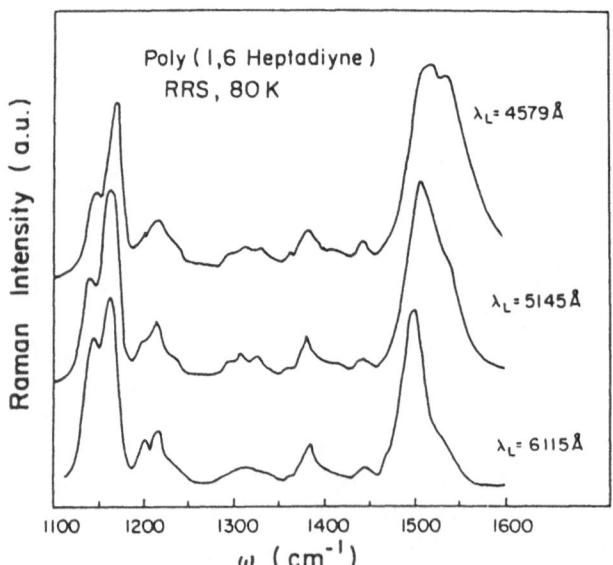

Fig. 3.41. Raman spectra PITN, deconvoluted to Lorentzian lines; $\lambda = 514.5$ nm, $V =$ (a) 120, (b) 480, (c) 600, (d) 900 mV. From [3.68]

Fig. 3.42. RRS spectra of poly(1,6-heptadiyne) at 80 K, excited with different laser frequencies [3.69]

3.2. Theoretical Aspects

The phenomenological interpretation presented in Sect. 3.1 is based on the "conjugation length" approach. This approach has provided a practical tool for the understanding of the RRS phenomena in polyacetylene and in some other conjugated polymers. Within this model it was shown that the length of uninterrupted conjugation plays an important role both in the determination of the vibrational frequencies and in the resonance enhancement of the Raman scattering. In addition, it was shown that various defects affect the length of the uninterrupted conjugation producing changes in the morphology of the samples and their RRS spectra.

A fundamental theory of RRS of conjugated polymers should provide a unifying picture of the various phenomena. In essence it should point out the basic parameters which control the vibrational frequencies, the $\pi - \pi^*$ electronic gap, the relative intensities in both *cis* and *trans*-PA and the frequencies observed in doped polyacetylene. Moreover, it should show in a fundamental way, how the various defects that influence the uninterrupted conjugation length affect the basic parameters leading to the observed RRS spectra.

The presentation in this section will, to a large extent, be based on the "amplitude modes" approach. Central to this approach is the recognition that the atomic vibrations are accompanied by a modulation of the energy gap. Within this model many features, such as relative intensities of the RRS, can be understood. It also yields the relation of the force constant to the energy gap, which is the key to understanding the origin of the dispersion effect. This will be discussed with respect to the conjugation length and other models. The Green's function analogy to the amplitude modes model will also be discussed.

3.2.1 Amplitude Modes

a) General Concepts

The special role of the π-electrons in the electronic properties of conjugated polymers has been recognized for a long time. In his pioneering studies, *Lennard-Jones* [3.70] first introduced the assumption that the total electronic energy is the sum of the adiabatic potential due to the σ-electrons and of the π-electron energies. The force constants for the atomic vibrations, which are obtained by taking the second derivative of the total electronic energy with respect to the normal coordinates, will then be the sum of the "bare" force constant due to the σ-electrons and a (negative) contribution from the π-electrons. In the present discussion we limit ourselves to Raman-active optical modes for which the contribution of the delocalized π-electrons to the force constant is non-negligible. Thus, only the few planar vibrations that affect the carbon−carbon bonding are considered. For these modes, the bare force constant is significantly lowered, or "renormalized", due to the π-electrons.

Let us consider *trans*-PA as a model system. In this system there exist four in-plane Raman active A_g modes [3.17], see (3.2). The highest mode (at $\simeq 2990$ cm^{-1} for *trans*-$(CH)_x$) is mainly a C-H stretch (Fig. 3.9) and we neglect its weak coupling to the π-electrons. In the Hückel approximation, with nearest neighbor transfer integrals (or resonance integrals) only, the π-electron energy, E_π, depends only on the difference, $\Delta_d = t_2 - t_1$, of the transfer integrals of neighboring bonds. If the transfer integral is a function of the C-C distances only, then using the normal coordinate variables u_n (the displacement of the nth coordinate) one can write,

$$\Delta_d = \sum_{n=1}^{N} g_n u_n, \tag{3.16}$$

where $g_n = \partial t / \partial u_n$ [$t = (t_1 + t_2)/2$] is the electron-phonon coupling of the nth mode. The equilibrium value of Δ_d is obtained by minimizing the total energy with respect to the u_n's. For infinitely long *trans*-PA, the "dimerization gap" corresponds to the non-zero equilibrium value of $2\Delta_d$, so that the atomic vibrational modes are amplitude modes of the electronic energy gap. Amplitude modes are not restricted to the Hückel approximation. In general, in any system for which the dependence of E_π on u_n is only via the single parameter Δ_d defined by (3.16), the atomic vibrational modes correspond to amplitude modes of Δ_d.

Amplitude modes provide a unified viewpoint [3.71] in our understanding of both the totally symmetric Raman active modes and local infrared active vibrations (IRAV) induced by charged defects in conjugated polymers.

b) Formal Results

The main assumption [3.71, 72] that enters the amplitude mode formalism is that the π-electron energy, E_π, depends on the individual coordinates u_n only via the total ionic field (or dimerization gap function) Δ_d defined by (3.16). In general, E_π includes such effects as electron–electron interaction, finite conjugation length and disorder [3.73]. The contribution of E_π to the total force constant is proportional to the second derivative with respect to the vibrational amplitudes u_n. Moreover, since these coordinates enter into E_π only via Δ_d, one can write the total classical potential in the harmonic approximation as

$$U^{\text{harm}} = \frac{1}{2}\left(\frac{\partial^2 E_\pi}{\partial \Delta_d^2}\right)_{\text{equil}} \delta^2 + \sum_{n=1}^{N} \frac{1}{2} K_n u_n^2, \tag{3.17}$$

where δ is the small variation in Δ around its equilibrium value and $K_n = \partial^2 E_\sigma / \partial u_n^2$ is the elastic force constant for the nth mode. Using this harmonic potential, one can now write down the equations of motion and solve for the eigenfrequencies, ω_n, of the system. The latter are given by the solutions of the following equation [3.72]:

$$\mathrm{Re}\,\{D_0(\omega)\} = -(1 - 2\tilde{\lambda})^{-1}, \tag{3.18}$$

where

$$2\tilde{\lambda} = 1 + \sum_{n=1}^{N} \frac{|g_n^2|}{K_n} \left(\frac{\partial^2 E_\pi}{\partial \Delta_n^2} \right)_{\mathrm{equil}}, \tag{3.19}$$

and

$$D_0(\omega) = \sum_{n=1}^{N} \tilde{\lambda}_n \frac{(\omega_n^0)^2}{\omega^2 - (\omega_n^0)^2 - i\omega\gamma_n}, \tag{3.20}$$

and ω_n^0, γ_n, are the bare vibration frequency and natural width of the nth mode, respectively. An example for the function $D_0(\omega)$ for a three-oscillator system is shown in Fig. 3.43. In (3.20) the quantity $\tilde{\lambda}_n$ is a measure of the coupling of the nth mode and is given by

$$\tilde{\lambda}_n = \frac{|g_n|^2}{\sum_{n=1}^{N} |g_n|^2}. \tag{3.21}$$

$\tilde{\lambda}_n$ represents the geometry of the nth bare mode, i.e. it is related to the projection of u_n on the bond direction. Note also, that since $\partial^2 E_\pi / \partial \Delta_d^2$ is in general negative, $2\tilde{\lambda}$ will be smaller than unity.

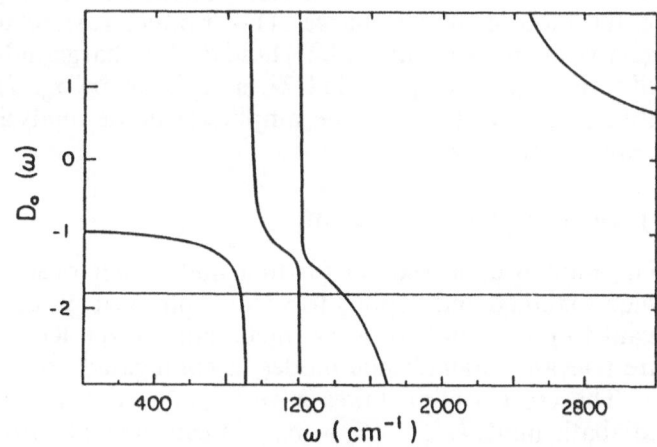

Fig. 3.43. The function $D_0(\omega)$. The horizontal line marks the value of $-(1 - 2\tilde{\lambda})^{-1}$ which determines the RRS frequencies at a given $\tilde{\lambda}$, which corresponds to a given laser excitation energy [3.72]

The most important feature of (3.18) is the unique parameter $\tilde{\lambda}$ which renormalizes all the modes; it is not necessary to treat the different modes separately since they are all coupled by the same field.

Equation (3.18) yields the "product rule", a useful relation for the practical analysis of the RRS data,

$$\prod_{n=1}^{N} (\omega_n/\omega_n^0)^2 = 2\tilde{\lambda}. \tag{3.22}$$

Applying (3.22) to a single mode system we see the meaning of $\tilde{\lambda}$ as a renormalizing factor: the ratio $(\omega_n/\omega_n^0)^2$ gives directly the ratio of the actual force constant to the bare force constant K.

Equation (3.18) determines not only the eigenfrequencies of the system but also the RRS intensities. Since the radiation couples to the vibrations through the π-electrons, it is assumed [3.73] that the RRS intensity of each mode, $I(\omega_n)$, is proportional to $|\delta(\omega_n)|^2$, where $\delta(\omega_n)$ is the amplitude of the oscillations in Δ_d at $\omega = \omega_n$. From the equations of motion it can thus be deduced that

$$I(\omega_n) \propto \frac{1}{(1 - 2\tilde{\lambda})^2 \, \text{Im} \, \{D_0(\omega_n)\}} \rightarrow \left(\frac{\partial}{\partial \omega} \, \text{Re} \, \{D_0(\omega_n)\} \right)^{-1}. \tag{3.23}$$

The second relation in (3.23) is obtained assuming a small natural width for the vibrations.

It is instructive to note here that, when charge is added to a dimerized chain, e.g. trans-PA, a defect in the dimerization pattern results and IR activity due to local modes appears [3.72]. The eigenfrequencies of these modes are given by a formula similar to (3.18), except that the parameter $2\tilde{\lambda}$ is replaced by α_p which describes the renormalized force constant for the local mode. For the translational mode, or the vibrations of the center of gravity coordinates of the defect, α_p is refered to as a "pinning parameter" and is a measure of the friction (caused by the dopants, for instance) in the motion of the charge. The product rule relation (3.22) and the relative intensity formula (3.23) hold for the charge-induced infrared-active vibrations (IRAV), provided $2\tilde{\lambda}$ is replaced by α_p. The relevance of the absorption by IRAV to the amplitude mode analysis will be discussed below (Sect. 3.2.2).

c) The Raman Cross Section

The Raman cross section for first and higher order processes has been widely studied and applied to various physical systems. In this section we want to point out several characteristics of the RRS cross section which are relevant to amplitude modes in conjugated polymers.

The cross section takes a very simple and transparent form in the adiabatic limit: $\hbar\omega_n \ll |E_g - \omega_L|$. It can then be written (for processes of all orders) as the product of two factors [3.25, 73–78]: (i) a resonant enhancement factor which determines the strength of the scattering and is dependent on $\hbar\omega_L/E_g$, and (ii) a phononic factor which gives the Raman shifts and relative intensities of the lines.

The first order cross section can be written as,

$$\sigma_R^{(1)} \propto |\mathscr{A}(\hbar\omega_L/E_g)|^2 \, \mathrm{Im} \left\{ \frac{D_0(\omega)}{1 + (1 - 2\tilde{\lambda}) D_0(\omega)} \right\} \tag{3.24}$$

where $\mathscr{A}$ is the Raman amplitude, which contains the resonance effect. For quasi-one-dimensional systems, it is strongly peaked at the onset of the valence (HOMO) to conduction (LUMO) bands optical transitions. Analytic expressions for $\mathscr{A}$ were obtained for the simple infinite Hückel-type (or Peierls-type) conjugated polymers. In that case, $E_g = 2\Delta$ and $\mathscr{A} = \partial\sigma/\partial\Delta$, where $\sigma(\omega)$ is the dynamic conductivity [3.73], and $|\mathscr{A}|^2$ diverges as $(\hbar\omega_L - 2\Delta)^{-3}$ due to the $(E - 2\Delta)^{-1/2}$ divergence of the one-dimensional density of states. This divergence is smeared out by the introduction of finite life time, $\hbar/\Gamma$, for the electronic levels. For finite Γ, the divergence of $|\mathscr{A}|^2$ at $\hbar\omega_L = 2\Delta$ is suppressed to a peak whose height is proportional to Γ^{-3} [3.77, 79]. Figure 3.44 shows the first order cross section function $|\mathscr{A}|^2$ for a finite value of Γ: the asymmetric shape is evident.

The second factor in (3.24) solely determines the exact positions and relative intensities of the RRS lines. The line positions ω_n are given by the condition $1 + (1 - 2\tilde{\lambda}) \, \mathrm{Re}\,\{D_0(\omega)\} = 0$, which is equivalent to (3.18), while the relative intensities are proportional to $[\mathrm{Im}\,\{D_0(\omega_n)\}]^{-1}$, which is equivalent to (3.23). Experimental confirmation of the relative intensity formula was found in the RRS of the various isotopes of *trans*-polyacetylene [3.25], and will be discussed below in Sect. 3.2.2.

The cross section for second and higher order RRS from amplitude modes, has been calculated by several authors [3.75, 76, 79]. In the adiabatic limit it can be written as [3.79]

$$\sigma_R^{(2)} \propto \left| \frac{\partial^2\sigma}{\partial\Delta^2} \right|^2 \int \mathrm{Im}\,\{D(\omega')\} \, \mathrm{Im}\,\{D(\omega - \omega')\} \, d\omega' \,, \tag{3.25}$$

where

$$D(\omega) = \frac{D_0(\omega)}{1 + (1 - 2\tilde{\lambda}) D_0(\omega)}, \tag{3.26}$$

and $\sigma(\omega)$ is the frequency-dependent conductivity of the polymer. The convolution $\int \mathrm{Im}\,\{D(\omega')\} \, \mathrm{Im}\,\{D(\omega - \omega')\} \, d\omega$, is purely phononic and it gives rise to strong peaks at all frequencies ω which are the sums of two first order Raman frequencies. The resonant enhancement factor $|\partial^2\sigma/\partial\Delta^2|^2$ is strongly peaked at $\hbar\omega_L = 2\Delta$, and is shown in Fig. 3.44 for the damped

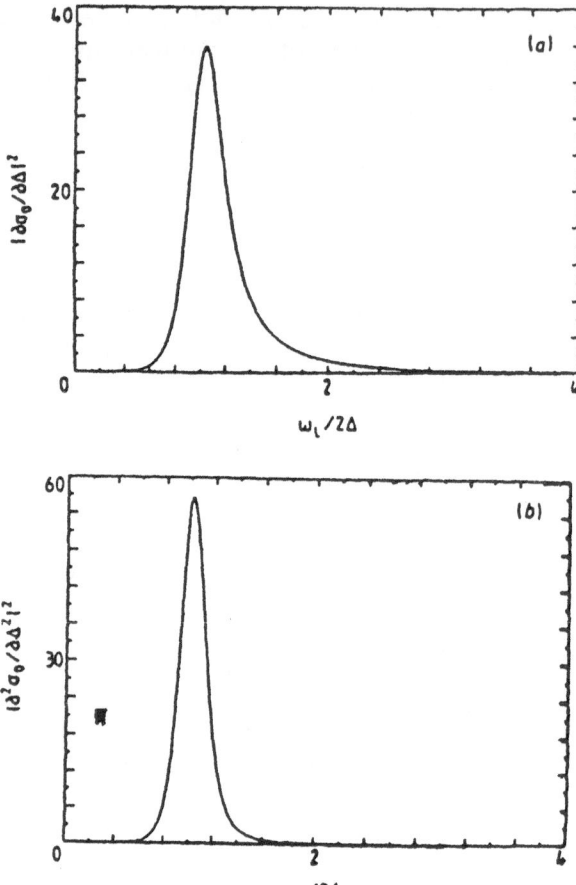

Fig. 3.44. The first and second order Raman cross section, (a) and (b) respectively, as a function of $\omega_L/2\Delta$ for $\Gamma = 0.2$. From [3.79]

Peierls model, with $\Gamma/2\Delta = 0.2$. It was shown that the relative enhancement of the second order with respect to the first order is determined by the damping parameter Γ and for the Peierls model is proportional to Γ^{-2}. An application of this model to the overtone structure in cis-$(CH)_x$ will be given in Sect. 3.2.6b.

A non-adiabatic treatment of the RRS cross section from amplitude modes, in which the finite vibration frequencies ω_n are taken into account has also been given [3.76, 77]. There are no appreciable changes in the cross section, except that the resonance enhancements of the incoming and outgoing beams have to be taken into account separately. This effect changes, of course, the relative intensities among the RRS lines. It has been observed, for instance, in cis-$(CH)_x$ where the lines at resonance with the photoluminescence center are enhanced [3.7].

3.2.2 Experimental Evidence for Amplitude Modes in *trans*-Polyacetylene

trans-PA is the best system in which the above concepts can be tested, since it features a "dispersion" of the RRS lines with the excitation wavelength; the three RRS lines are blue shifted with decreasing excitation wavelength (Sect. 3.1.2). This dispersion is ascribed to spatially inhomogeneous *trans*-PA samples in which resonance conditions in different regions are met at different excitation wavelengths leading to different Raman shifts. The inhomogeneity may arise, for instance, due to the presence of *trans*-PA chains with different conjugation lengths. In general, it is assumed that there is an inhomogeneity in the optical gap, E_g. Since the cross section $\mathscr{A}$ is sharply peaked at $\hbar\omega_L = E_g$ (Sect. 3.2.1c), varying the laser excitation frequency photoselects different parts of the sample with $E_g \simeq \hbar\omega_L$. On the other hand, $D_0(\omega)$ depends only on the bare parameters and these are not sensitive to the π-electrons and are therefore assumed to be uniform throughout the sample. In an ideal *trans*-PA the optical gap is identical with the dimerization gap, $2\Delta_d$, and, since $\Delta_d = \Delta_d(\tilde{\lambda})$, the inhomogeneity in the optical gap leads to a distribution in $\tilde{\lambda}$ and in the RRS frequencies, ω_n. Thus, using the experimental data of the RRS frequencies measured with many laser excitation wavelengths in the range 350–750 nm, the five parameters of $D_0(\omega)$ have been deduced using (3.18) [3.25]. This procedure yielded a unique function $D_0(\omega)$ which can account simultaneously for all the RRS frequencies and relative intensities at different ω_L, by changing $\tilde{\lambda}$ alone in (3.18). Furthermore, while $D_0(\omega)$ is different for the different isotopes (*trans*-$(CH)_x$, *trans*-$(CD)_x$ and *trans*-$(^{13}CH)_x$), $\tilde{\lambda}$ is found to be isotope independent and is a function of ω_L (or E_g) only. Fig. 3.43 displays the function $D_0(\omega)$ for *trans*-$(CD)_x$ showing the way in which ω_n is determined by ω_L. At a given ω_L the resonance conditions are met by a certain $\tilde{\lambda}$, so that the intersection of $D_0(\omega)$ with the horizontal line drawn at $-(1 - 2\tilde{\lambda})^{-1}$ yield the RRS frequencies.

Experimental confirmation of the relative intensity formula (3.23) was found in the RRS of the various isotopes of *trans*-PA [3.25, 73, 78]. As an example for the overall fit of the amplitude modes model, Fig. 3.45 shows the experimental relative intensities for *trans*-$(CH)_x$ and *trans*-$(CD)_x$ and the theoretical fits using (3.23).

An additional remarkable feature is that the same $D_0(\omega)$ which describes the RRS results also determines the frequencies, relative intensities and line shapes of the infrared active vibrations (IRAV) of local charged defects added to the chains, either by doping or by photogeneration [3.25, 72, 73]. The IRAV frequencies are given by (3.18) with $\tilde{\lambda}$ replaced by a pinning parameter α_p. α_p describes the reduction in the bare force constant due to the finite region in which the charged defect can move: $\alpha_p = 0$ corresponds to a free charged defect for which the lowest IRAV frequency is zero (the Goldstone mode), while $\alpha_p > 0$ represents pinned defects for which the lowest frequency is finite (the pinning mode).

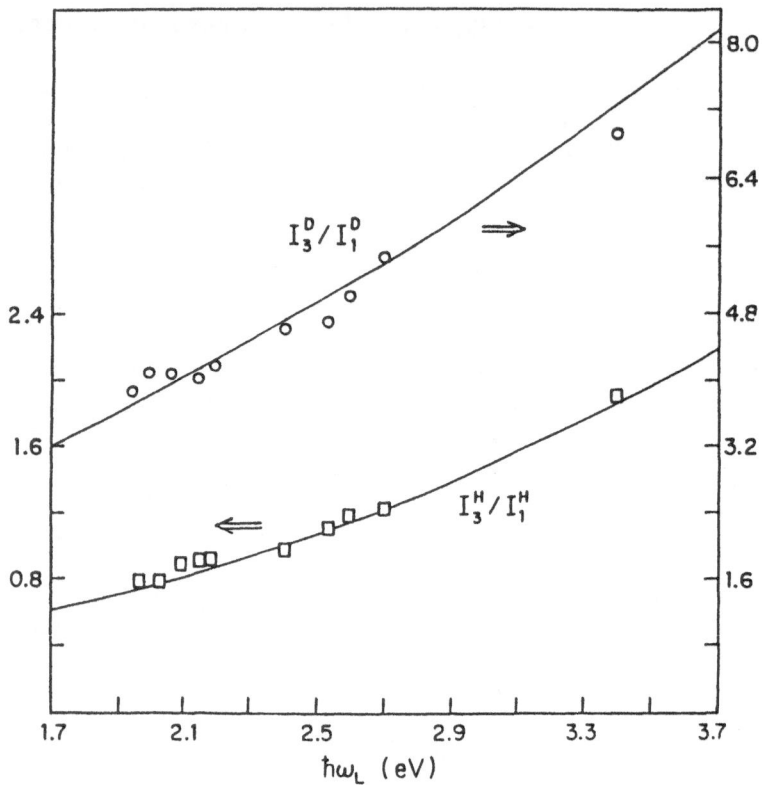

Fig. 3.45. The ratios $I(\omega_3)/I(\omega_1)$ for *trans*-(CH)$_x$ (□) and *trans*-(CD)$_x$ (○) as a function of ω_L. The solid lines are the theoretical fits using (3.23). From [3.73]

Again, similar to the behaviour of $\tilde{\lambda}$, α_p is isotope independent and is dependent only on the type of the defects.

The fact that a single function $D_0(\omega)$ can explain simultaneously the frequencies and relative intensities of both RRS (all laser excitation energies) and IRAV (doping and photogenerated) is by itself strong evidence for the existence of amplitude modes in *trans*-PA. Furthermore, applying the amplitude mode model for the RRS, one can deduce two important characteristics of *trans*-PA samples. They are: (i) the dependence of $\tilde{\lambda}$ on $\hbar\omega_L$ or on the gap, and (ii) the distribution function $P(\tilde{\lambda})$ which is at the origin of the dispersion effect. Figure 3.46 shows the experimentally determined $\tilde{\lambda}$ [using the product rule relation, (3.22)] vs. $\ln(\hbar\omega_L)$ for *trans*-(CH)$_x$ and *trans*-(CD)$_x$; the straight line is apparent, suggesting logarithmic behavior of $\tilde{\lambda}^{-1}$ with the electronic gap. We shall discuss this point further (Sect. 3.2.4) in connection with specific models of *trans*-PA. The experimentally determined distribution $P(\tilde{\lambda})$ is shown in Fig. 3.47. Note that it has only one peak and is relatively narrow; its width

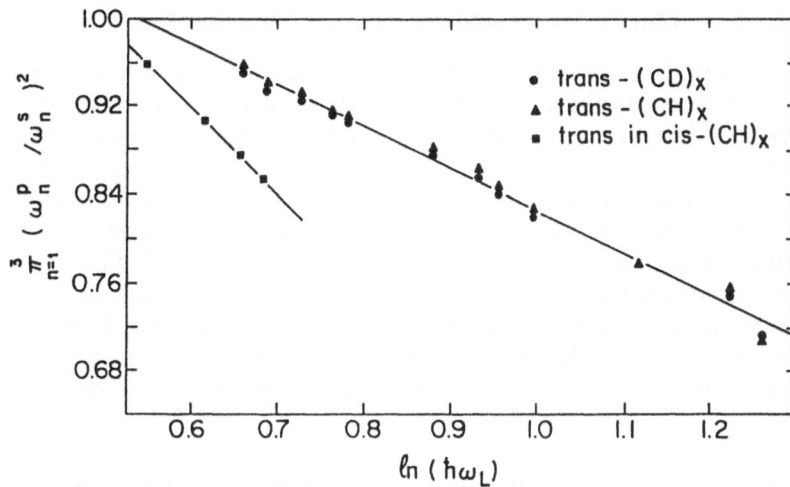

Fig. 3.46. The experimental product relation $\prod_{n=1}^{3} (\omega_n^p/\omega_n^s)^2$ for *trans*-(CH)$_x$ (▲), *trans*-(CD)$_x$ (●) and *trans*-in-*cis*-(CH)$_x$ (■) as a function of ln ω_L. From [3.73]

corresponds to a gap distribution with a width of only 0.2 eV; more homogeneous *trans*-PA samples show even smaller widths [3.80]. The relation of this distribution to the distribution of conjugation lengths was discussed in Sect. 3.1.3 c. In Sect. 3.2.4 we shall briefly discuss some possible origins of this distribution which are compatible with the logarithmic behavior shown in Fig. 3.46.

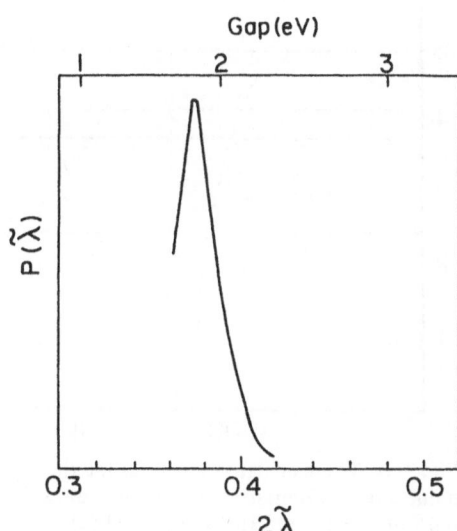

Fig. 3.47. The experimentally deduced distribution function $P(\tilde{\lambda})$ for *trans*-(CH)$_x$. From [3.78]

3.2.3 Amplitude Modes and the Green's Function Formalism

The coupling among the different A_g modes, which is an essential part of the multimode amplitude modes formalism, was treated recently using the Green's function formalism [3.81–83]. This formalism allowed the definition of an effective vibrational coordinate, denoted hereafter as $\mathscr{R}$, which determines the Raman spectrum. For *trans*-PA, for instance, $\mathscr{R} = R_{C-C} - R_{C=C}$ the difference between the lengths of the single and double bonds. This effective coordinate is analogous in some way to the ionic field Δ_d, defined in (3.16), since the corresponding force constant is a function of $\mathscr{R}$ only, similar to the case of the amplitude modes formalism where the force constant is a function of Δ_d only. Writing out the simplified dynamical matrix for the modes of the A_g species, it was possible to write the force constant as a sum

$$F_{\mathscr{R}} = F_{\mathscr{R}}^0 + \Delta F, \tag{3.27}$$

where $F_{\mathscr{R}}^0$ includes only the shortest range interaction. The contribution ΔF arises from the long range interaction which takes into account the full π-electron system of the whole chain. Using this recipe, a "product rule", analogous to (3.22) was obtained [3.81]; namely $\Pi(\omega_n/\bar{\omega}_n^0)^2$ depends only on the π-electron potential ($\bar{\omega}_n^0$ are the eigenfrequencies for $F_{\mathscr{R}}^0$). For a given set of $\bar{\omega}_n^0$, or starting parameters, each value of ΔF, determines uniquely a set of N ($N = 3$ for polyacetylene) eigenfrequencies. In this way, an N-branch plot of $F_{\mathscr{R}}$ vs. ω could be obtained, as shown in Fig. 3.48 for PA. This is analogous to the $D_0(\omega)$ curves shown in Fig. 3.43.

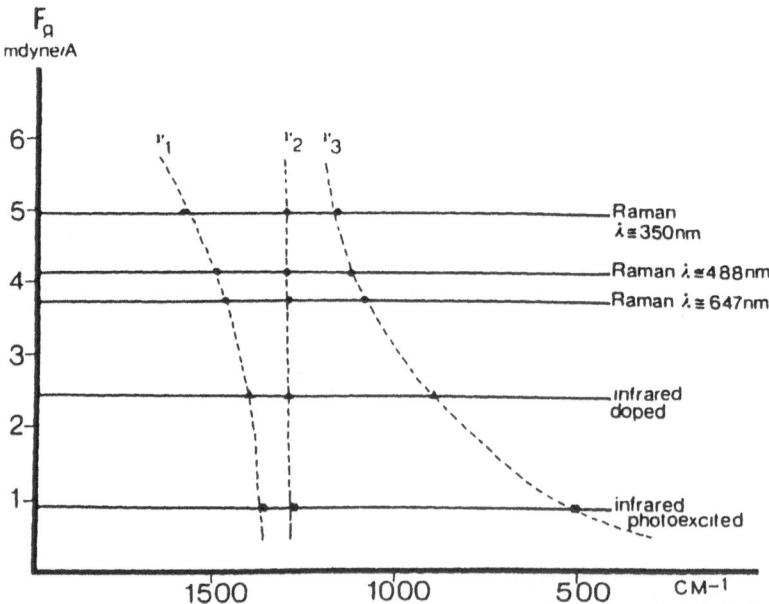

Fig. 3.48. The calculated $F_{\mathscr{R}}$ vs. ω for the A_g normal modes of *trans*-$(CH)_x$. Experimental points are: (●) Raman shifts for different excitation wavelengths; (▲) Doping induced IRAV; (■) Photogenerated IRAV. From [3.85]

Model calculations for $F_\mathcal{R}$ were performed, in which ΔF was taken as the sum of terms, f_n, each representing the contribution to the force constant from a unit cell n double bonds away. It was found [3.84] that $F_\mathcal{R}$ is inversely proportional to the extent of the interaction among the C-C bonds: it decreases when the conjugation length increases and/or the chain is perturbed by a charge added by doping or photogeneration. In this way, the force constant and hence the Raman frequencies could, in principle, be calculated as a function of the conjugation length. Furthermore, the relative intensities of the RRS lines were shown [3.85] to be proportional to the content of the $\mathcal{R}$ coordinate in the particular line, and are thus inversely proportional to the slope of $F_\mathcal{R}$ at ω_n, cf. (3.23).

3.2.4 The relation Between the Force Constant and the Gap

a) *trans*-Polyacetylene

As discussed in detail in the preceding sections, the dispersion of the RRS lines with the laser energy in *trans*-PA is the most prominent characteristic of the spatial inhomogeneity in the samples. Assuming that at a given photon excitation energy one photo-selects that part of the conjugated polymers having an energy gap equal to the photon energy, the dispersion yields a relation between the energy gap and the RRS frequencies, or the force constant. Within the amplitude modes formalism, this yields the relation $\tilde{\lambda}$ vs. Δ (or $\hbar\omega_L$); the equivalent relation within the Green's function formalism (Sect. 3.2.3) is $F_\mathcal{R}$ vs. $\hbar\omega_L$. In *trans*-PA, detailed analysis [3.25, 78] of the RRS frequencies as a function of the laser photon energy yielded a logarithmic behavior of $\tilde{\lambda}^{-1}$ vs. ω_L:

$$\tilde{\lambda}_p/\tilde{\lambda} = 1 - \tilde{c}\ln(\hbar\omega_L/2\Delta_0), \tag{3.28}$$

where $\tilde{\lambda}_p \simeq 0.37$ is the value of $\tilde{\lambda}$ for the laser-wavelength independent primary lines (Sect. 3.1.2) and $2\Delta_0 \simeq 1.7$ eV is the corresponding gap. The value $\tilde{c} = 0.37$ for the slope in (3.28) was found to be isotope independent; i.e. it is the same for *trans*-$(CH)_x$, *trans*-$(CD)_x$ and *trans*-$(^{13}CH)_x$. Likewise, using the analysis of *Zerbi* et al. [3.85], one could find the experimentally determined relation between $F_\mathcal{R}$ and ω_L. It can be shown that the data could approximately be described by a logarithmic behavior:

$$F_\mathcal{R}^{-1} = c - c_\mathcal{R}\ln(\hbar\omega_L/2\Delta_0) \tag{3.29}$$

where $c \simeq 0.35$ Å/mdyne and $c_\mathcal{R} \simeq 0.12$ Å/mdyne. Again, by analyzing the data for the different isotopes [3.84], we find that the slope $c_\mathcal{R}$ is isotope independent; i.e. it is a property of the π-electron system and not of the atomic vibrations. Using the definition of $\tilde{\lambda}$, (3.19), and identifying $F_\mathcal{R}$ with the total force constant, we relate the two slopes, $\tilde{c}$ and $c_\mathcal{R}$, via,

$$c_\mathcal{R}K_\sigma \simeq \tilde{c}/\tilde{\lambda}_p, \tag{3.30}$$

where K_σ is some weighted average σ-bond force constant (in *trans*-PA it is very nearly the C–C stretching force constant). Taking $K_\sigma \simeq 50\,eV/$ $Å^2 \simeq 8\,mdyne/Å$ [3.78, 86], we find $c_{\mathscr{R}} K_\sigma \simeq \tilde{c}/\tilde{\lambda}_p \simeq 1$ in agreement with (3.30).

An important conclusion can thus be drawn: the "renormalized" force constant in *trans*-PA varies logarithmically with the energy gap. Therefore, any theory that explains the dispersion effect (Sect. 3.2.2) in *trans*-PA should account at the same time for the logarithmic behaviour of the force constant.

There are several models which result in a logarithmic behaviour:
i) One-dimensional Peierls systems.
    For this case,

$$\Delta = 2E_c \exp(-1/2\lambda), \tag{3.31}$$

where E_c is an energy parameter which approximately equals the band width, and λ is the dimensionless electron–phonon coupling constant. For a Hückel alternating chain, e.g., with single and double bond transfer integrals t_1 and t_2, respectively, the two parameters E_c and λ can be expressed as follows: $E_c = 4t$ where $t = (t_1 t_2)^{1/2}$, and $\lambda = (\pi t)^{-1} \sum (g_n^2/K_n)$ (see Sect. 3.2.1a, b for the definitions of g_n and K_n). For long chains, it has been shown that $\tilde{\lambda} = \lambda$ [3.72]. Thus the logarithmic dependence of $\tilde{\lambda}^{-1}$ on Δ (or ω_L) can be accounted for by a distribution of the electron-phonon coupling constants. Such a distribution may be the result of finite localization lengths, due, for instance, to finite conjugation lengths.
ii) Peierls systems including electron–electron interactions.
    When the effects of the electron–electron interactions on the Peierls model are not too severe the dependence of $\tilde{\lambda}^{-1}$ on ω_L is still approximately logarithmic [3.87], and the dispersion effect can be understood in terms of a distribution of the electron–electron (and also electron–phonon) coupling constant [3.78]. Since the electron-electron interaction is more important in shorter chains, a distribution in chain lengths may result in a distribution of the electron–electron coupling constant.
iii) A conjugation length dependent force constant.
    Model calculations of $F_{\mathscr{R}}$ (Sect. 3.2.3) for *trans*-PA chains of various conjugation lengths [3.82–84] revealed that the total force constant decreases with the conjugation lenght. Since the gap also decreases with the conjugation length, an increase of the force constant with the gap is expected, in qualitative agreement with the experiment. However, detailed calculations of the energy gap were not reported in this study, thus the observed behaviour of the force constant vs. the gap could not be verified by this model.

Fig. 3.49. (a, c) RRS spectrum of all *trans*-(CH)$_x$ and a 95%-*cis*-5%-*trans* sample. (b, d) The function $D_0(\omega)$ for the *trans* and *cis* isomers and the respective ω_n^0, $\tilde{\lambda}_n$ values [3.88]

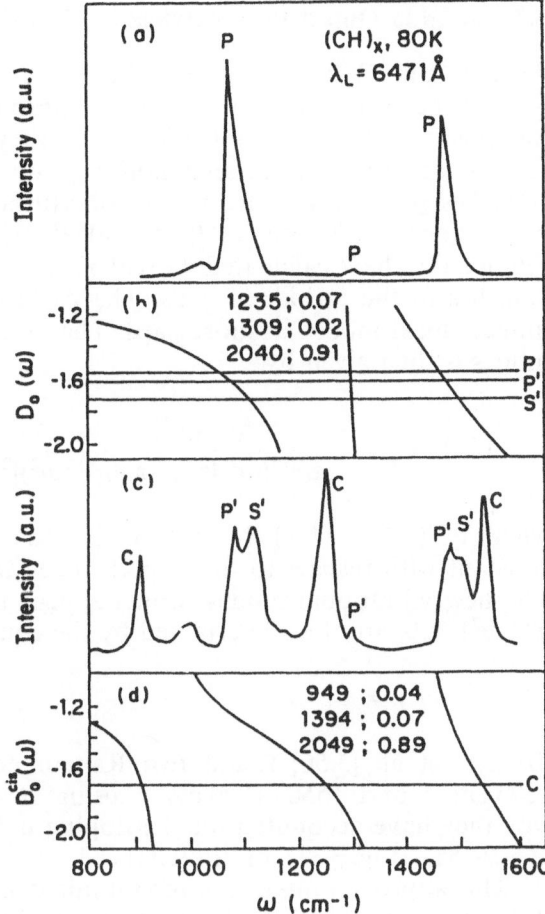

b) *trans-cis* Polyacetylene Mixtures

The dispersion of the RRS frequencies of the *trans* isomer of *trans-cis*-(CH)$_x$ mixtures $t(c)$ is quite different from that of the all *trans* frequencies. This is shown in Fig. 3.49 for a 5% *trans* in *cis* sample, exhibiting the following features:

i) the primary frequencies for each line are higher than those of all *trans* sample.

ii) the satellites shift to substantially higher frequencies and are more sensitive to small changes in ω_L compared with all *trans* samples.

The resulting $\tilde{\lambda}$ vs. ω_L, found by *Vardeny* et al. [3.88], is shown in Fig. 3.46, indicating a linear relation $\tilde{\lambda}/\tilde{\lambda}_p = 1 + \ln(\hbar\omega_L/2\Delta_0)$ (cf. (3.28) for the all *trans* samples where $1/\tilde{\lambda}$ is linear in ln ω_L). It was therefore concluded that the type of disorder in the two types of *trans*-PA are different. It is possible that the effect of the majority of the *cis* chains in such diluted samples is to induce an inhomogeneous "extrinsic gap" into *trans*-PA. The dispersion effect in these samples is thus due to the inhomogeneity induced by the *cis* environment.

3.2.5 RRS in Doped Polyacetylene – Green's Function Approach

Mulazzi et al. [3.56] have studied the RRS modes of *n*-doped *trans*-$(CH)_x$, using a perturbed Green's function approach. Utilizing the calculated dispersion relation for *trans*-$(CH)_x$ [3.89] they have calculated the "unperturbed" density of vibrational modes, $\varrho^{(0)}(\omega^2)$, for the acoustic and the A_g branches. $\varrho^{(0)}(\omega^2)$ is related to the unperturbed Green's function, $G^{(0)}(\omega^2)$, by: $\varrho^{(0)}(\omega^2) = (1/\pi) \, \text{Im} \, \{G^{(0)}(\omega^2 + i0^+)\}$. Applying the Kramers-Kronig relation they have calculated the real part, $\text{Re} \, \{G^{(0)}(\omega^2)\}$. Note that $\varrho^{(0)}(\omega^2)$ vanishes in the forbidden gaps. The perturbation caused by the defect induces local modes in those gaps. The perturbed density of vibrational modes is then given by

$$\varrho(\omega^2) = \frac{\varrho^{(0)}(\omega^2)}{[1 - \Lambda \tilde{\varrho}^{(0)}(\omega^2)]^2 + [\pi \Lambda \varrho^{(0)}(\omega^2)]^2} \, , \tag{3.32}$$

where $\tilde{\varrho}^{(0)}(\omega^2) = (1/\pi) \, \text{Re} \, \{G^{(0)}(\omega^2)\}$, and Λ is the change in the force constant with respect to the unperturbed lattice of the polymeric chain. The new vibrational modes, which appear in the forbidden gaps where $\varrho^{(0)}(\omega^2) \equiv 0$, are thus determined by the equation

$$\Lambda \tilde{\varrho}^{(0)}(\omega^2) = 1 \, . \tag{3.33}$$

Mulazzi et al. [3.56] found two Raman active local modes at 1270–1275 cm^{-1} and 1550–1590 cm^{-1} using $\Lambda = 2$–4.5×10^6 cm^{-2}. In this way they have accounted for the doping induced RRS lines in $[CHK_y]_x$ at $y > 5\%$ (Fig. 3.31) and in $[CHNa_y]_x$ at $y > 11\%$ (Fig. 3.32).

This approach predicts a monotonic dependence of the intensities of the additional lines on the doping level. On the other hand, the experiments (Figs. 3.31, 32) show that these extra lines appear and start growing only above a certain doping level ($\simeq 5\%$ in $[CHK_y]_x$ and $\simeq 11\%$ in $[CHNa_y]_x$). It was suggested [3.2] that at these concentrations a semiconductor-to-metal phase transition takes place. It is quite possible that this phase transition plays an important role in the appearance of these extra lines, so that the approach presented in this section may not be adequate.

3.2.6 Application of the Amplitude Modes to Various Systems

a) $[(C_2H_2)_y \, (C_2D_2)_{1-y}]_x$

The copolymer *trans*-$[(C_2H_{2y} \, (C_2D_2)_{1-y}]_x$ can be thought of as a mixed *trans*-$(CH)_x$ and *trans*-$(CD)_x$ chains. It provides an excellent example for the application of the amplitude mode scheme. The RRS spectra for several concentrations y, taken at $\lambda_L = 671.5$ nm, are shown in Fig. 3.50

Fig. 3.50. RRS spectra of $[(C_2H_2)_y (C_2D_2)_{1-y}]_x$ at different concentrations y [3.91]

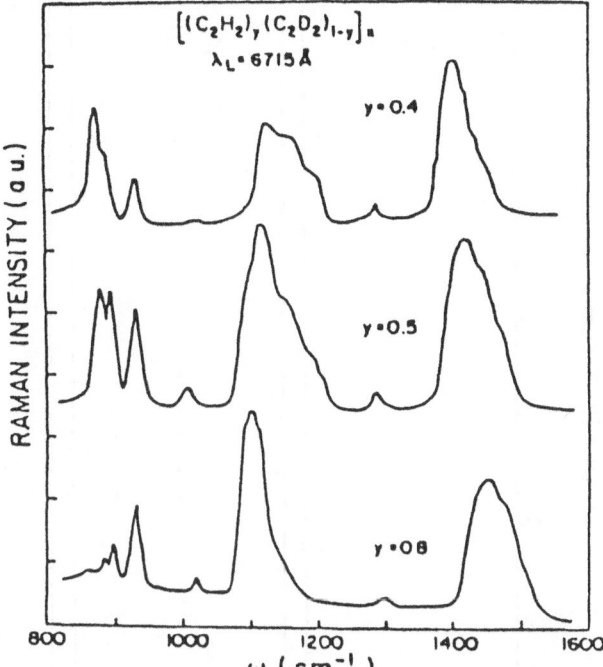

[3.90, 91]. The frequency variation of each of the lines with isotope concentration is summarized in Fig. 3.51 [3.91, 92]. The data show the following features:

i) The highest frequency line shows a one-mode behavior while the others show a two-mode behavior.

ii) There are more lines than the five expected for a mixed crystal having three modes for each of the constituents.

iii) Contrary to the usual expectations, the 1065 cm^{-1} line of *trans*-(CH)$_x$ (at 670 nm) increases in frequency as the D concentration increases, in contrast with the 1460 and 1290 cm^{-1} lines. This behavior reverts to "normal" at 457.9 nm.

Feature (ii) is probably the result of some weak lines which are usually neglected in the *trans*-(CH)$_x$ and *trans*-(CD)$_x$ RRS. Here they become resonantly enhanced, since they interact with one of the six (three for each isotope) "normal" A_g modes. Feature (iii), on the other hand, poses a more fundamental question: why a mode with cosiderable H (or D) character increases in frequency upon replacing H with D?

Analysis by the amplitude modes formalism [3.92] gave elegant answers to these puzzling features. On one hand, it explained the dispersion with the exciting laser energy, including the relative intensities of each of the RRS and IRAV lines. On the other hand, it showed that the bare

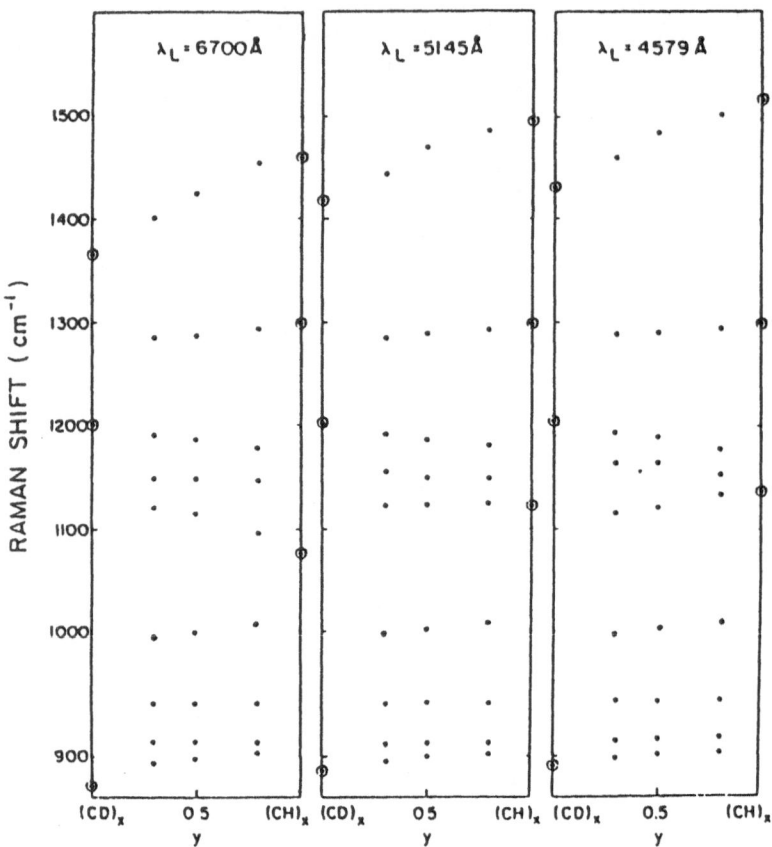

Fig. 3.51. Concentration dependence of the RRS frequencies of $[(C_2H_2)_{0.5} (C_2D_2)_{0.5}]_x$ at various excitation energies: 1.85, 2.41 and 2.71 eV. From [3.92]

frequencies behave "normally" upon increasing the D concentration, see Fig. 3.52. The peculiar behavior of the *trans*-$(CH)_x$ 1065 cm^{-1} line [feature (iii) above] is controlled by the exact form of $D_0(\omega)$. Note also the near isotope independence of the highest bare frequency (at $\simeq 2050$ cm^{-1}), strengthening an earlier conclusion [3.78] that this bare mode is pure C-C stretch of the *trans*-PA skeleton.

b) *cis*-Polyacetylene

The overtone spectra of *cis*-$(CH)_x$ were discussed in Sect. 3.1.3. Using the present amplitude modes approach (Sect. 3.2.1c), it was possible [3.79] to account for the relative intensities among the second order lines and the total intensity of the second order lines relative to the first order ones using a damping parameter Γ (Sect. 3.2.1c) given by $\Gamma/2\Delta = 0.2$. A fit of the theoretical and experimental first and second order RRS in

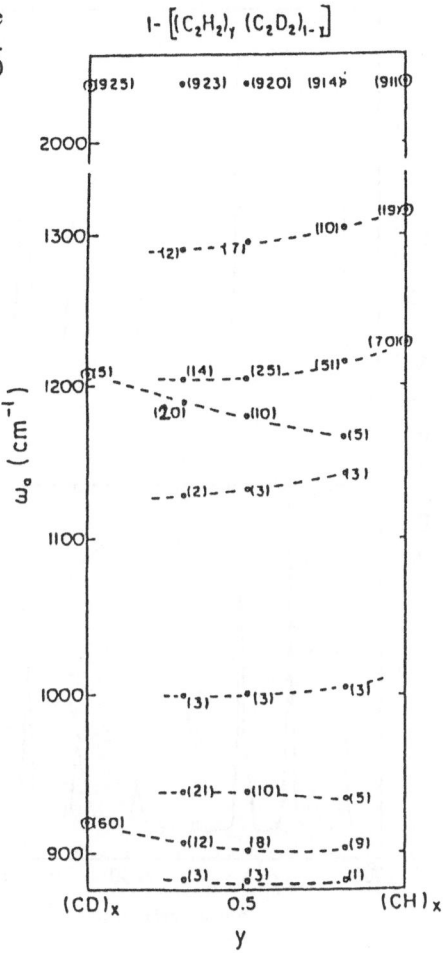

Fig. 3.52. Concentration dependence of the bare phonon frequencies for $[(C_2H_2)_y (C_2D_2)_{1-y}]_x$. The number in parenthesis denotes $\bar{\lambda}_n \times 1000$ (3.21). From [3.92]

cis-(CH)$_x$ is shown in Fig. 3.53. It is thus concluded that the strong second-order RRS in *cis*-(CH)$_x$ can be explained by its quasi-one-dimensional nature.

c) Polythiophenes

The chemical structure of polythiophene is shown in Fig. 3.1. Using symmetry considerations, *Zerbi* et al. have found that it has five A_g modes that affect the C-C separation [3.93]. Experimentally, the RRS consists of a strong line, three weak lines and some other very weak lines. There is hardly any dispersion with the laser frequency. The IRAV (either doped or photogenerated) shows three broad bands and several weaker bands. This situation makes it very difficult to analyze the data using the amplitude modes model. However, assuming the ap-

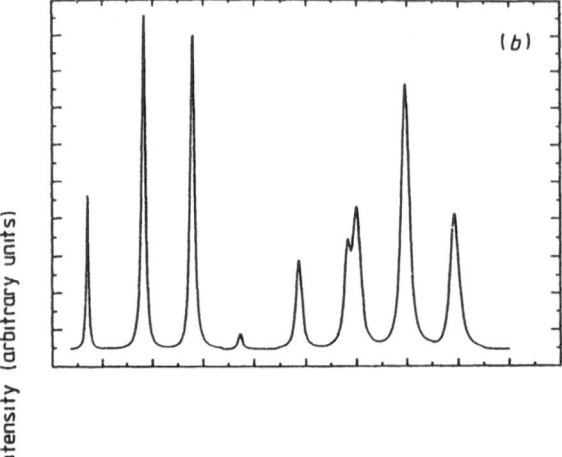

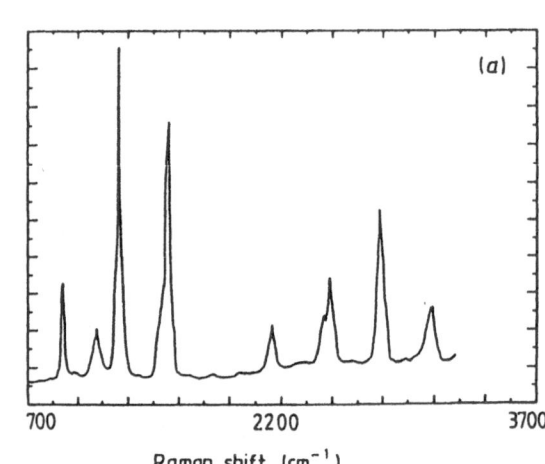

Raman intensity (arbitrary units)

Raman shift (cm^{-1})

700   2200   3700

Fig. 3.53. First and second order RRS spectra for *cis*-(CH)$_x$: (a) measured, and (b) calculated for $\Gamma/2\Delta = 0.2$. From [3.79]

plicability of the model and that the bare C-C stretch frequency is similar to that of *trans*-(CH)$_x$, it was possible to find a function $D_0(\omega)$ that simultaneously accounted for the frequencies and relative intensities of the strong and weak RRS and IRAV bands [3.94]. The most important result of this analysis is the value found for $\tilde{\lambda}$: $2\tilde{\lambda} \simeq 0.45$–$0.47$. Comparing this value to the value found for *trans*-PA, $2\tilde{\lambda} = 0.37$, it was concluded that the non-degenerate ground state of polythiophene (Fig. 3.1) gives rise to the increase in $\tilde{\lambda}$. This increase corresponds to a similar increase observed in the optical gap of polythiophene with respect to *trans*-PA. Similar observations were reported for poly(3-methylthiophene) [3.95].

Acknowledgments. This work was supported by the US-Israel Binational Science Foundation (BSF), Jerusalem Israel, under contract numbers 86-00131 and 86-00073.

References

3.1   S. Roth, H. Bleier: Adv. Phys. **36**, 385 (1987)
3.2   A. J. Heeger, S. Kivelson, J. R. Schrieffer, W.-P. Su: Rev. Mod. Phys. **60**, 781 (1988)
3.3   H. W. Hänslin, C. Riedel, K. Menke, S. Roth: Macromolec. Chem. **185**, 387 (1984)
3.4   H. Shirakawa, S. Ikeda: Polym J. **2**, 231 (1971)
3.5   H. Shirakawa, T. Ito, S. Ikeda: Macromolec. Chem. **179**, 1565 (1978)
3.6   H. Kuzmany, E. A. Imhoff, D. B. Fitchen, A. Sarhangi: Phys. Rev. **B26**, 7109 (1982)
3.7   L. S. Lichtmann, E. A. Imhoff, A. Sarhangi, D. B. Fitchen: J. Chem. Phys. **81**, 168 (1984)
3.8   J. Ashkenazy, E. Ehrenfreund, Z. Vardeny, O. Brafman: Molec. Cryst. Liq. Cryst. **117**, 193 (1985)
3.9   J. Tanaka, Y. Saito, M. Shimizu, M. Tanaka: Synthetic Metals **17**, 307 (1987)
3.10   S. Lefrant, L. S. Lichtmann, H. Temkin, D. B. Fitchen, D. C. Midler, G. E. Whitwell II, J. M. Burlitch: Solid State Commun. **29**, 191 (1979)
3.11   G. B. Street, T. C. Clarke: Adv. Chem. **186**, 177 (1980)
3.12   H. Kuzmany: Phys. Status Solidi b **97**, 521 (1980)
3.13   G. P. Brivio, E. Mulazzi: Phys. Rev. **B30**, 876 (1984)
3.14   L. S. Lichtmann, D. B. Fitchen, H. Temkin: Synthetic Metals **1**, 139 (1979/80)
3.15   Y. Yacoby, S. Roth: Synthetic Metals **13**, 299 (1986)
3.16   L. Rimai, M. E. Heyde, D. Gill: J. Am. Chem. Soc. **95**, 4493 (1973)
3.17   F. B. Schügerl, H. Kuzmany: J. Chem. Phys. **74**, 953 (1981)
3.18   D. Jumeau, S. Lafrant, E. Faulques, J. B. Buisson: J. de Phys. **44**, 819 (1984)
3.19   L. S. Lichtman, A. Sarhangi, D. B. Fitchen: Solid State Commun. **36**, 869 (1980)
3.20   W. Siebrand, M. Z. Sgierski: J. Chem. Phys. **81**, 185 (1984)
3.21   R. Tubino, L. Piseri, G. Dellepiane, J. L. Birman, U. Pedretti: Solid State Commun. **49**, 161 (1984)
3.22   E. J. Mele: Solid State Commun. **44**, 827 (1982)
3.23   H. Eckhardt, S. W. Steinhouser, R. R. Chance, M. Schott, R. Silbey: Solid State Commun. **55**, 1075 (1985)
3.24   H. Kuzmany: Pure Appl. Chem. **57**, 235 (1985)
3.25   Z. Vardeny, E. Ehrenfreund, O. Brafman, B. Horovitz: Phys. Rev. Lett. **51**, 2326 (1983)
3.26   P. Knoll, M. Kuzmany: Molec. Cryst. Liq. Cryst. **106**, 317 (1984)
3.27   M. A. Schen, S. Lefrant, E. Perrin, J. C. W. Chien, E. Mulazzi: Synthetic Metals **28**, D287 (1989)
3.28   C. Budrowski, J. Przyluski, S. Roth, H. Kuzmany: Synthetic Metals **16**, 291 (1986)
3.29   E. Mulazzi, G. P. Brivio, S. Lefrant, E. Perrin, G. A. Arbuckle, A. G. McDiarmid: Synthetic Metals **28**, D309 (1989)
3.30   P. R. Surján, H. Kuzmany: Phys. Rev. **B33**, 2615 (1986)
3.31   J. Kürti, H. Kuzmany, G. Nagele: J. Molec. Elec. **3**, 135 (1987)
3.32   H. Kuzmany, J. Kürti: Synthetic Metals **21**, 95 (1987)
3.33   J. L. Brédas, J. M. Toussaint, G. Hennico, J. Delhalle, J. M. André, A. J. Epstein, A. G. MacDiarmid: In *Electronic Properties of Conjugated Polymers*, ed. by H. Kuzmany, M. Mehring, S. Roth, Springer Ser. Solid Sciences Vol. **76** (Springer, Berlin, Heidelberg 1987) p. 48
3.34   F. Coter, Z. Vardeny, O. Brafman, E. Ehrenfreund, J. Ashkenazi: Molec. Cryst. Liq. Cryst. **177**, 373 (1985)
3.35   Y. Yacoby, S. Roth: Solid State Commun. **56**, 319 (1985)
3.36   D. Moses, A. Feldblum, E. Ehrenfreund, A. J. Heeger, T. C. Chung, A. G. McDiarmid: Phys. Rev. **B26**, 3361 (1982)
3.37   H. Edwards, W. J. Feast: Polym. Commun. **21**, 595 (1980); D. C. Bott, X. K. Chai, J. H. Edwards, W. J. Feast, R. H. Friend, M. E. Horton: J. de Phys. **44**, C3–143 (1983)
3.38   G. Leising: Polym. Bull. **11**, 401 (1984); Polym. Commun. **25**, 201 (1984)

3.39   G. Lugli, V. Pederetti, G. Perego: Molec. Cryst. Liq. Cryst. 117, 43 (1985)
3.40   G. Masetti, E. Campani, G. Gorini, L. Piseri, R. Tubino, P. Piaggio, G. Dellepiane:
       Solid State Commun. 55, 737 (1985)
3.41   G. Masetti, E. Campani, G. Gorini, R. Tubino, P. Piaggio, G. Dellepiane: J. Chem.
       Phys. 108, 141 (1986)
3.42   E. Faulques, E. Rzepka, S. Lefrant, E. Mulazzi, G. P. Brivio, G. Leising: Phys. Rev.
       B33, 8622 (1986)
3.43   P. Knoll, H. Kuzmany, G. Leising: In *Electronic Properties of Conjugated Polymers*,
       ed. by H. Kuzmany, M. Mehring, S. Roth, Springer Ser. in Solid State Sciences Vol.
       76 (Springer, Berlin, Heidelberg 1987) p. 134
3.44   R. Dorsinville, M. Szalkiewicz, R. Tubino, J. L. Birman, R. R. Alfano, G. Delle-
       piane, V. Pedretti: Synthetic Metals 17, 509 (1987)
3.45   E. Mulazzi, G. P. Brivio, S. Lefrant, E. Faulques, E. Perrin: Synthetic Metals 17, 325
       (1987)
3.46   S. Lefrant, E. Perrin, E. Mulazzi: Synthetic Metals 28, D295 (1989)
3.47   E. Perrin, E. Faulques, S. Lefrant, E. Mulazzi, G. Leising: Phys. Rev. B38, 10645
       (1988)
3.48   G. Lanzani, S. Luzzati, R. Tubino, G. Dellepiane: J. Chem. Phys. 91, 732 (1989)
3.49   L. Margulies, M. Stockburger: J. Raman Spectrosc. 8, 26 (1979)
3.50   E. Perrin, E. Faulques, S. Lefrant, E. Mulazzi, G. Leising: Synthetic Metals 28, D317
       (1989)
3.51   H. Kuzmany, P. Knoll: J. Raman Spectrosc. 17, 89 (1986)
3.52   O. Brafman, E. Ehrenfreund, Z. Vardeny: in Proc. Second International Conference
       on Phonon Physics, eds. J. Kollár, N. Kroó, M. Menyhárd and T. Siklos (World Sci.
       Pub. Co., Singapore, 1985) p. 783.
3.53   H. Fujimoto, J. Tanaka, M. Tanaka, T. Kishi: Synthetic Metals 16, 133 (1986)
3.54   E. Faulques, S. Lefrant: J. de Phys. Colloq. 44, C3–337 (1983)
3.55   S. Lefrant, E. Faulques, A. Chentli, F. Rachdi, P. Bernier: Synthetic Metals 17, 313
       (1987)
3.56   E. Mulazzi, S. Lefrant, E. Faulques, E. Perrin: Synthetic Metals 24, 35 (1988)
3.57   E. Mulazzi, S. Lefrant: Synthetic Metals 28, D323 (1989)
3.58   J. Ghanbaja, D. Billaud, E. Perrin, T. Verdon, S. Lefrant: Synthetic Metals 28, D339
       (1989)
3.59   G. P. Brivio, B. Mulazzi: Solid State Commun. 60, 203 (1986)
3.60   I. Harada, Y. Furukawa, M. Tasumi, H. Shirakawa, S. Ikeda: J. Chem. Phys. 73,
       4746 (1980)
3.61   E. Mulazzi, S. Lefrant, E. Perrin, E. Faulques: Phys. Rev. B35, 3028 (1987)
3.62   J. L. Sauvajol, D. Chenouni, M. Rolland, J. L. Ribet: Synthetic Metals 28, D281
       (1989)
3.63   P. Meisterle, H. Kuzmany, G. Nauer: Phys. Rev. B29, 6008 (1984)
3.64   S. Lefrant, E. Perrin, J. P. Buisson, H. Eckhardt, C. C. Han: Synthetic Metals 29, E91
       (1989)
3.65   Y. Furukawa, S. Tazawa, Y. Fuji, I. Harada: Synthetic Metals 24, 329 (1988)
3.66   N. S. Sariciftci, H. Kuzmany: Syynthetic Metals 21, 157 (1987)
3.67   S. Kirchene, S. Lefrent: Y. Pelous, G. Froyer, M. Petit, A. Digua, J. F. Fauvarque:
       Synthetic Metals 17, 607 (1987)
3.68   W. Wallnöfer, E. Faulques, H. Kuzmany, K. Eichinger: Synthetic Metals 28, C533
       (1989)
3.69   R. Zemach, Z. Vardeny, O. Brafman, E. Ehrenfreund, A. J. Epstein, R. J. Weagley,
       H. W. Gibson: Molec. Cryst. Liq. Cryst. 118, 423 (1985)
3.70   J. E. Lennard-Jones: Proc. Roy. Soc. A158, 280 (1937)
3.71   B. Horovitz: Molec. Cryst. Liq. Cryst. 77, 286 (1981)
3.72   B. Horovitz: Solid State Commun. 41, 729 (1982)
3.73   B. Horovitz, Z. Vardeny, E. Ehrenfreund, O. Brafman: J. Phys. C 19, 7291 (1986)
3.74. B. Horovitz, Z. Vardeny, E. Ehrenfreund, O. Brafman: Synthetic Metals 9, 215 (1984)
3.75   D. Schmeltzer, I. Ohana, Y. Yacoby: J. Phys. C 19, 2113 (1986)

3.76   Y. Ono, H. Ito: J. Phys. Soc. Japan **54**, 4828 (1985)
3.77   A. Terai, Y. Ono, Y. Wada, J. Phys. Soc. Japan **55**, 2889 (1986)
3.78   E. Ehrenfreund, Z. Vardeny, O. Brafman, B. Horovitz: Phys. Rev. **B36**, 1535 (1987)
3.79   F. Coter, E. Ehrenfreund, B. Horovitz: J. Phys. Condensed Matter **2**, 239 (1990)
3.80   S. Lefrant: J. de Phys. **44**, C3–247 (1983)
3.81   C. Castiglioni, J. T. Lopez Navarrete, G. Zerbi, M. Gussoni: Solid State Commun. **65**, 625 (1988)
3.82   J. T. Lopez Navarrete, G. Zerbi: Synthetic Metals **28**, C15 (1989)
3.83   G. Zerbi, C. Castiglioni, Bogang Tian, M. Gussoni: In *Electronic Properties of Conjugated Polymers III*, Ed. by H. Kuzmany, M. Mehring, S. Roth, Springer Series in Solid State Sciences, Vol. 91 (Springer, Berlin, Heidelberg 1989) p. 106
3.84   M. Gussoni, C. Castilioni, G. Zerbi: Synthetic Metals **28**, D375 (1989)
3.85   G. Zerbi, C. Castiglioni, J. T. Lopez Navarrete, T. Bogang, M. Gussoni: Synthetic Metals **28**, D359 (1989)
3.86   T. Kakitani: Prog. Theo. Phys. **51**, 656 (1974)
3.87   B. Horovitz, J. Solyom: Phys. Rev. **B32**, 2681 (1985)
3.88   Z. Vardeny, E. Ehrenfreund, O. Brafman, B. Horovitz: Phys. Rev. Lett. **54**, 75 (1985)
3.89   L. Piseri, R. Tubino, L. Paltrinieri, G. Dellepiane: Solid State Commun. **46**, 183 (1983)
3.90   H. Takeuchi, Y. Furukawa, I. Harada, H. Shirakawa: J. Chem. Phys. **80**, 2295 (1984)
3.91   O. Brafman, J. Poplawski, E. Ehrenfreund, Z. Vardeny, J. Tanaka, H. Fujimoto, A. G. MacDiarmid, W. S. Huang: Molec. Cryst. Liq. Cryst. **117**, 363 (1985)
3.92   J. Poplawski, O. Brafman, E. Ehrenfreund, Z. Vardeny: Synthetic Metals **17**, 283 (1987)
3.93   Bogang Tian, G. Zerbi: In *Electronic Properties of Conjugated Polymers III*, Ed. by H. Kuzmany, M. Mehring, S. Roth, Springer Series in Solid State Sciences, Vol. 91 (Springer, Berlin, Heidelberg 1989) p.113
3.94   Z. Vardeny, E. Ehrenfreund, O. Brafman, A. J. Heeger, F. Wudl: Synthetic Metals **18**, 183 (1987)
3.95   J. Poplawski, E. Ehrenfreund, S. Glenis, A. J. Frank: Synthetic Metals **28**, C335 (1989)

4. Raman Scattering in Diluted Magnetic Semiconductors

A. K. Ramdas and S. Rodriguez

With 40 Figures

Diluted magnetic semiconductors (DMSs), also referred to as semi-magnetic semiconductors, are semiconducting alloys in which the host atoms on a sub-lattice are replaced in part by magnetic atoms. DMSs resulting from the random replacement of the group II atoms in the tetrahedrally coordinated II–VI semiconductors by transition metal ions of the iron group (Mn^{2+}, Co^{2+}, Fe^{2+}, ...), e.g., $Cd_{1-x}Mn_xTe$, have been the focus of intense investigations in the past decade. The excitement in these studies has been triggered by the discovery of the extraordinary co-existence of remarkable magnetic and semiconducting phenomena they exhibit [4.1]. In common with their non-magnetic counterparts (e.g., $Hg_{1-x}Cd_xTe$), they exhibit the well-known composition-dependent vibrational spectra [4.2], electronic band structure [4.3] and transport properties (e.g., carrier mobilities). In the context of the unique magnetic properties of DMSs, consider a Mn-based II–VI alloy like $Cd_{1-x}Mn_xTe$. In such a tetrahedrally coordinated semiconducting alloy with the zinc blende structure, the ground state of Mn^{2+} is $^6S_{5/2}$ (ignoring small crystal field effects); these ions thus have an effective magnetic moment of 5.92 Bohr magnetons (μ_B). The exchange interaction between the d-electrons of the Mn^{2+} ions on the one hand, and the s-like conduction band states or the p-like valence band states on the other – the so-called *sp-d exchange interaction* – results in large spin-splittings of the conduction band minimum and the valence band maximum. A huge excitonic Zeeman effect [4.4] and a giant Faraday rotation [4.5] are two of the spectacular phenomena observed in DMSs traced to this *sp-d* interaction. Another aspect of the special magnetic effects observed in DMSs has its origin in the anti-ferromagnetic interaction between neighboring Mn^{2+} ions; at a sufficiently low temperature and a sufficiently large concentration (x), long range magnetic ordering sets in, unmistakable evidence for which manifests itself in magnetic susceptibility [4.6], in Raman scattering [4.7, 4.8], and in magnetic neutron scattering [4.9].

The ease with which Mn^{2+} is incorporated in the II–VI compound semiconductors has naturally led to the successful bulk crystal growth of a wide range of Mn-based DMSs, and in turn, to a concentrated research effort focussed on their properties. The simplicity of the ground state of Mn^{2+} (including a negligible crystal field splitting), has allowed relatively straightforward theoretical analyses of the magnetic phenomena. In recent years, bulk crystal growth of $Hg_{1-x}Fe_xSe$, $Cd_{1-x}Fe_xSe$, and $Cd_{1-x}Co_xSe$

has made possible the observation of special magnetic effects associated with the non-magnetic ground state of Fe^{2+} (the so-called van Vleck paramagnetism) [4.1, 4.10] as well as the magnetic $S = 3/2$ ground state of Co^{2+} in which the orbital angular momentum is quenched in the strong tetrahedral environment [4.10].

It is well-known that closely related to the tetrahedrally coordinated zinc blende III–V compounds, there exist families of II–IV–V_2 chalcopyrite compounds; for example, in the spirit of the virtual crystal approximation, the zinc blende InP and the chalcopyrite $CdGeP_2$ are interrelated in this fashion [4.11]. Once again a random replacement of Cd^{2+} by Mn^{2+} will result in a DMS. Physical studies of such chalcopyrite DMSs are just beginning. Other structures such as stannites as well as "ordered" structures grown by molecular beam epitaxy can be also visualized, offering further opportunities to investigate DMSs based on them.

The tetrahedrally coordinated II–VI compound semiconductors represent an important class of semiconductors in the context of the physics of quantum well structures and their opto-electronic applications. The majority of II–VI compounds are direct band gap semiconductors, with energy gaps spanning an impressive range of values (from zero for the Hg-based compounds to 3.8 eV for ZnS) [4.12]. The plot of "ionicity" vs. "bond-charge" for the tetrahedrally coordinated semiconductors of the zinc blende or the wurtzite structure underscores the dramatic increase in ionicity of the II–VI compounds and their predisposition to assume a rock salt structure [4.13]. Indeed the first order Raman spectrum of CdTe, which at ambient pressure exhibits the characteristic zone center LO-TO pair, disappears at a pressure of ~ 30 kbar; above that pressure, CdTe transforms to a rock salt structure with the characteristic absence of the first order Raman spectrum [4.14]. The increased ionicity translates to an increased Fröhlich interaction (polaron coupling constant). An examination of the lattice parameters and the band gaps characterizing the II–VI and the III–V compounds reveals that a variety of lattice matched, and hence strain-free, II–VI/II–VI as well as II–VI/III–V heterostructures can be visualized. They offer a wide range of total band off-sets. A judicious use of molecular beam epitaxy (MBE) also allows the fabrication of strained layer superlattices or quantum well structures [4.12]. MBE growth techniques have significantly increased the composition range of the II–VI alloys including DMSs; it has enabled the growth of epilayers and quantum well structures not accessible in bulk growth. Even the "magnetic" end members of DMSs like $Cd_{1-x}Mn_xTe$ and $Zn_{1-x}Mn_{1-x}Se$, i.e., MnTe and MnSe in the zinc blende phase [4.12] as well as new structures (cubic $Cd_{1-x}Mn_xSe$ rather than that with the wurtzite structure) have been grown with MBE [4.15]. They illustrate the remarkable expansion of new materials available for physical investigations.

The unique semiconducting and magnetic physical phenomena exhibited by DMSs have been investigated with particular success exploiting Raman spectroscopy. Discovery and delineation of vibrational modes, localized and collective; electronic transitions within the Zeeman multiplet of the ground state of the magnetic ions like Mn^{2+} and Co^{2+}; the spin-flip of electrons bound to donors in large, effective mass orbits; and the collective magnetic excitations in the low temperature, magnetically ordered phases – these are illustrative examples of the successful application of Raman spectroscopy to the physics of DMSs and their heterostructures. These phenomena are addressed in the present chapter.

4.1 Raman Scattering: General Considerations

We recall here the aspects of Raman scattering relevant in the context of DMSs as alloys which can sustain vibrational, electronic, and magnetic excitations [4.16–18]. We note that the collective excitations of a perfect crystal with full translational symmetry are characterized by wave-vectors q_i confined to the fudamental Brillouin zone (BZ). When illuminated with a monochromatic radiation (angular frequency ω_L, wavelength λ_L, and wave-vector k_L), the radiation inelastically scattered at an angle Θ has an angular frequency $\hbar\omega_s = \hbar\omega_L \pm \hbar\omega_i$, where $\hbar\omega_i$ is the energy of the internal excitation of wave-vector q_i. Conservation of wavevector for the typical $\hbar\omega_i$'s yields $|k_L| \sim |k_s|$ and hence

$$|q_i| = 2\,|k_L|\,\sin\left(\frac{\Theta}{2}\right). \qquad (4.1)$$

Thus, in the first order Raman scattering, only the states whose wave-vectors are near the center of the BZ can be excited or de-excited, i.e., $|q_i| \sim 0$. For higher order processes $|\sum q_i| \sim 0$. In the *absence of strict translational symmetry*, as would be the case for alloys such as the DMSs, the conservation of wave-vector is not a strict selection rule and to that extent the first order inelastic scattering may involve excitations spanning the entire BZ.

The selection rule which dictates whether an excitation is Raman-active or not has its physical origin in the *modulation* of the electric susceptibility tensor (χ) by the internal excitation embodied in

$$P = \chi E_L \qquad (4.2)$$

to first order in E_L, the incident electric field, P being the macroscopic polarization, and χ is proportional to α, the polarizability tensor.

In addition to wave vector conservation, Raman transitions are governed by selection rules appropriate to the second rank polarizability tensor α. In this context, symmetry arguments can be used to deduce selection rules including details of the polarization features for different scattering geometries. A transition from a state ν_0 belonging to the irreducible representation Γ_{v_0} of the space group of the crystal to a state v belonging to Γ_v occurs only if $\Gamma_v^* \times \Gamma_\alpha \times \Gamma_{v_0}$ contains the totally symmetric representation. Here Γ_α is the representation generated by the components of α. We also note that, for first order Raman shifts associated with visible exciting radiation, since the wave vectors of the excitations responsible are near the center of the Brillouin zone, one is justified in treating these "zone-center" excitations using the point-group of the crystal [4.19].

The quantum theory of light scattering can be developed in close analogy with the corresponding classical picture described above. Within the framework of the electric-dipole approximation the scattering cross section per unit solid angle Ω resulting in the scattering of a photon of frequency ω_L, wave vector k_L and polarization $\hat{\varepsilon}_L$ into a state characterized by the corresponding quantities ω_S, k_S and $\hat{\varepsilon}_S$ while the scattering system experiences a transition from $|v_0\rangle$ to $|v\rangle$ is

$$\frac{d\sigma}{d\Omega} = (n_S + 1)\, n_L \, \frac{\omega_S^3 \omega_L}{c^4} \, |\hat{\varepsilon}_S \cdot \alpha_{vv_0} \cdot \hat{\varepsilon}_L|^2 \,. \tag{4.3}$$

Here

$$\alpha_{vv_0} = \sum_{v'} \left(\frac{\langle v|\, d\, |v'\rangle \, \langle v'|\, d\, |v_0\rangle}{E_{v'} - E_{v_0} - \hbar\omega_L} + \frac{\langle v'|\, d\, |v_0\rangle \, \langle v|\, d\, |v'\rangle}{E_{v'} - E_v + \hbar\omega_L} \right), \tag{4.4}$$

the states $|v\rangle$ are the energy eigenvectors with eigenvalues E_v and d, the electric dipole moment operator of the system. The quantities n_L and n_S are the photon populations in the incident and scattered beams, respectively. The summation in (4.4) extends over all states $|v'\rangle$ of the scatterer.

Often, it is convenient to describe the motion of the scattering system as consisting of two or more types of excitations coupled by very small interactions. Typical examples are molecules with weakly coupled electronic, vibrational and rotational motions, and crystals where one can, in a first approximation, disregard the electron–phonon interaction. In such cases, the states $|v\rangle$ described above are the stationary states of the system taking due account of the coupling between all modes of motion. To fix the ideas, we think of crystals in which the electron–phonon interaction H_{ep} is sufficiently small to allow a description of the system by states ϕ_v written as products of electronic and vibrational states. Then, the energy eigenstates appearing in (4.4) are

$$|v\rangle = \phi_v + \sum_{v'}{}' \frac{\phi_{v'}\langle\phi_{v'}|\, H_{ep}\, |\phi_v\rangle}{E_{0v} - E_{0v'}} + \dots\,, \tag{4.5a}$$

where the E_{0v} are the energies associated with the ϕ_v. Clearly each E_{0v} is the sum of electronic and phonon energies. The sum omits $v' = v$.

An equivalent procedure is to take H_{ep} and the electron–photon interaction as perturbations. Then, the scattering cross section is expressed as a sum reflecting the use of third order perturbation theory as in [4.20, 21]. For example, a Raman process associated with emission of an optical phonon can be viewed as the virtual creation of an electron–hole pair (step 1), emission by the electron or by the hole of an optical phonon (step 2, H_{ep}) followed by electron–hole recombination with emission of a scattered photon (step 3). The steps involved in this description are implicit in the nature of the states $|v\rangle$ in (4.5a) which explicitly take the electron–phonon interaction into account. The scattering cross-section in such a three-step process can be written as being proportional to

$$
\left| \frac{\langle n_S + 1, 0| H_{eR} |n_S, b\rangle \langle n_q + 1, b| H_{ep} |n_q, a\rangle \langle n_L - 1, a| H_{eR} |n_L, 0\rangle}{(\omega_b + \omega_q - \omega_L)(\omega_a - \omega_L)} \right.
$$

$$
\left. + 5 \text{ additional terms} \right|^2 . \tag{4.5b}
$$

Here

H_{eR} = electron–photon interaction Hamiltonian

H_{ep} = electron–phonon interaction (i.e., Fröhlich interaction with LO-phonons whose population is increased from n_q to n_{q+1} during the process in which an optical phonon is created)

n_L, n_S = occupation of the incident and scattered modes, respectively

$\hbar\omega_a, \hbar\omega_b$ = intermediate excitation energies of the virtual electron–hole pair.

The other five terms in (4.5b) are the remaining permutations of the three steps of the Raman process. It is clear that the first term will dominate when $\omega_a \sim \omega_L$ or $\omega_b \sim (\omega_L - \omega_q) = \omega_S$, producing a *double peak in the frequency dependence of the cross section*. The condition of *"in resonance"* results from the matching of the incident photon energy with that of an electronic excitation whereas *"out resonance"* occurs when the scattered photon energy equals the energy of such a transition [4.21].

Equation (4.4) for α_{vv_0} contains matrix elements between the exact states $|v_0\rangle$, $|v\rangle$ and $|v'\rangle$ as well as the energy eigenvalues E_v. These are related to ϕ_{v_0}, ϕ_v, $\phi_{v'}$ and E_{0v} by standard perturbation theory. It is sometimes useful to view α_{vv_0} as being modulated by the internal motions. The amplitude of the polarization giving rise to a Raman line at $\omega_L - \omega_i$ is of the form

$$
\frac{\partial \alpha_{vv_0}}{\partial Q_i} Q_i = -\sum_{v'} \left[\frac{\langle v_0| \boldsymbol{d} |v'\rangle \langle v'| \boldsymbol{d} |v_0\rangle}{(E_{v'} - E_{v_0} - \hbar\omega_L)^2} + \frac{\langle v'| \boldsymbol{d} |v_0\rangle \langle v_0| \boldsymbol{d} |v'\rangle}{(E_{v'} - E_{v_0} + \hbar\omega_L)^2} \right]
$$

$$
\times \frac{\partial}{\partial Q_i} (E_{v'} - E_{v_0}) Q_i . \tag{4.5c}
$$

Here we consider the matrix elements of $\boldsymbol{d}$ to be slowly varying functions of Q_i.

4.2 Raman Scattering by Vibrational Excitations

In this section we review the vibrational Raman scattering in DMS. We note here that even though inelastic neutron scattering provides, in principle, information about the vibrational spectrum throughout the fundamental BZ, Raman scattering gives more precise frequencies when a given mode appears in the Raman spectrum. It should be emphasized that for the Cd-based DMS crystals, neutron scattering is ruled out due to the large neutron absorption cross section of the naturally occurring Cd. The physical considerations underlying Raman scattering are additional motivations for its study.

4.2.1 Lattice Dynamics

We recall here some of the salient aspects of lattice dynamics. In a perfect crystal with full translational symmetry the atomic vibrations can be described in terms of collective excitations characterized by wave-vectors q_i, confined to the first BZ. Associated with imperfections, chemical or structural in nature, localized vibrations occur.

Consider the motion of the atoms in a crystal with f atoms per primitive cell. Let the displacements from their equilibrium positions be characterized by the coordinates $u_{n\alpha}$ where n designates a particular cell of the crystal and α ($\alpha = 1, 2, 3, \ldots 3f$) corresponds to any one of the $3f$ Cartesian components of the displacements of the f atoms within the cell at n. To first order in the displacements, the equations of motion are

$$M_\alpha \ddot{u}_{n\alpha} = - \sum_{n'\alpha'} C_{\alpha\alpha'}(n - n') u_{n'\alpha'}. \tag{4.6}$$

We take a crystal with lattice translations a_1, a_2, a_3, having dimensions $N_0 a_i$ ($i = 1, 2, 3$), and use periodic boundary conditions with periods $N_0 a_i$. The quantities $C_{\alpha\alpha'}(n - n')$ in (4.6) are real coefficients, symmetric in $n\alpha$ and $n'\alpha'$ and depend only on $n - n'$ in order to satisfy translational symmetry. Assuming solutions of these equations in the form

$$u_{n\alpha} = (NM_\alpha)^{-1/2} e_\alpha \exp[i(q \cdot n - \omega t)], \tag{4.7}$$

where $N = N_0^3$ is the number of primitive cells, we obtain

$$\sum_{\alpha'} [\bar{C}_{\alpha\alpha'}(q) - \omega^2 \delta_{\alpha\alpha'}] e_{\alpha'} = 0, \tag{4.8}$$

where

$$\bar{C}_{\alpha\alpha'}(q) = \sum_{n'} (M_\alpha M_{\alpha'})^{-1/2} C_{\alpha\alpha'}(n - n') \exp[-iq \cdot (n - n')]. \tag{4.9}$$

Here e_α is a component of a $3f$-dimensional vector determined by the eigenvalue problem represented by (4.8); it is usually convenient to select an orthonormal set of eigenvectors. We note that $\bar{C}_{\alpha\alpha'}(\boldsymbol{q})$, which can be regarded as a Fourier transform of the force constants $C_{\alpha\alpha'}(\boldsymbol{n})$, is independent of $\boldsymbol{n}$. Further, the $3f \times 3f$ matrix with components given by (4.4) is Hermitian by virtue of the symmetry of the force constants; it is customary to refer to this matrix as the *dynamical matrix*. The set of linear equations (4.8) has a non-trivial solution only if ω satisfies the secular equation

$$|\bar{C}_{\alpha\alpha'}(\boldsymbol{q}) - \omega^2 \delta_{\alpha\alpha'}| = 0 . \tag{4.10}$$

For each wave vector $\boldsymbol{q}$, (4.10) gives $3f$ solutions for the frequency $\omega(\boldsymbol{q})$ yielding $3f$ branches of the dispersion curves of the vibrational spectrum of the crystal. In order to avoid redundancy, the values of $\boldsymbol{q}$ are restricted to the reduced or the fundamental Brillouin zone. For crystals containing f atoms per primitive cell there are 3 acoustic branches and $3(f-1)$ optical branches. For acoustic modes, ω is linear in $\boldsymbol{q}$ near $\boldsymbol{q} = 0$ and vanishes at the center of the Brillouin zone. However, optical branches have a non-vanishing frequency at the zone center. Except for degenerate polar optical phonons in crystals free from improper symmetry operations [4.23–25], the frequency $\omega(\boldsymbol{q})$ is of the form

$$\omega(\boldsymbol{q}) = \omega(0) + \sum_{ij} a_{ij} q_i q_j \tag{4.11}$$

for small values of $\boldsymbol{q}$.

The phonon excitations of crystals, like other stationary states, can be classified according to the irreducible representations of the group of the wave vector $\boldsymbol{q}$. In the first order Raman effect, only phonons with wave vectors near $\boldsymbol{q} = 0$ need be studied. The eigenvectors of the dynamical matrix of each optical phonon corresponding to $\boldsymbol{q} = 0$ generate one of the irreducible representations of the point group of the crystal. The group theoretical classification of the optical phonons is carried out following the procedure given by Bhagavantam and Venkatarayudu [4.22] or an equivalent procedure [4.26, 27]. One can show that in zincblende crystals at $\boldsymbol{q} = 0$ there are threefold degenerate F_2 optical and F_2 acoustic phonons. Because of the partially ionic character of zinc-blende crystals, the F_2 phonons split (for $\boldsymbol{q}$ small but non-vanishing) into a longitudinal (LO) and two transverse (TO) branches. The F_2 optical modes are both Raman and infrared active. The wurtzite structure, having four atoms per primitive cell, has optical phonons with A_1, E_1, B_1, E_2 symmetries, there being two of each of the last two irreducible representations. Of these A_1 and E_1 are both Raman and infrared active, E_2 is only Raman active, whereas B_1 is active in neither.

For convenience we give in Tables 4.1 and 4.2 the character tables for the irreducible representations of the groups T_d and C_{6v}, the point groups of crystals having the zincblende and wurtzite structures, respectively. The last column of these tables gives functions generating the different irreducible representations, selected in such a manner that these representations are automatically unitary.

In order to explain the polarization features of the Raman spectra of phonons we determine the form of the Raman scattering tensor α, assumed to be symmetric away from resonance, by reducing the representation it generates into its irreducible components. This reduction is $\Gamma(\alpha) = A_1 + E + F_2$ for T_d and $\Gamma(\alpha) = 2A_1 + E_1 + E_2$ for C_{6v}. The forms of α for the transitions from the totally symmetric ground state A_1 to the allowed final states in first order Raman scattering are displayed in Tables 4.3 and 4.4 for T_d and C_{6v}, respectively. The symmetry classification of the Raman lines in an experimental spectrum can be arrived at on the basis of polarization studies for a variety of scattering geometries.

A number of physically interesting aspects of crystal physics have their origin in the presence of imperfections, e.g., foreign atoms, lattice defects, ... Alloys of two or more compounds forming a homogeneous crystalline phase also exhibit novel physical phenomena. These systems lack strict translational symmetry. In the following discussion, the lattice dynamics of such crystals is illustrated with the results in alloys of two compound

Table 4.1. Character table for T_d. X, Y, Z are the components of a polar vector field along the cubic axes x, y, z. R_x, R_y, and R_z are the components of a pseudo-vector along the cubic axes

T_d	E	$8C_3$	$3C_2$	$6\sigma_d$	$6S_4$	Basis functions
A_1	1	1	1	1	1	$X^2 + Y^2 + Z^2$
A_2	1	1	1	-1	-1	
E	2	-1	2	0	0	$2Z^2 - X^2 - Y^2, \sqrt{3}\,(X^2 - Y^2)$
F_1	3	0	-1	-1	1	R_x, R_y, R_z
F_2	3	0	-1	1	-1	$X, Y, Z; YZ, ZX, XY$

Table 4.2. Character table for C_{6v}. $X, Y,$ and Z are the components of a polar vector field with respect to a coordinate system x, y, z in which $\hat{z}$ is along C_6 and the xz plane is a σ_d. R_x, R_y, and R_z are the components of a pseudo-vector along the coordinate axes

C_{6v}	E	C_2	$2C_3$	$2C_6$	$3\sigma_d$	$3\sigma_v$	Basis functions
A_1	1	1	1	1	1	1	$Z; X^2 + Y^2; Z^2$
A_2	1	1	1	1	-1	-1	R_z
B_1	1	-1	1	-1	-1	1	$Y(Y^2 - 3X^2)$
B_2	1	-1	1	-1	1	-1	$X(X^2 - 3Y^2)$
E_1	2	-2	-1	1	0	0	$X, Y; ZX, ZY; R_x, R_y$
E_2	2	2	-1	-1	0	0	$X^2 - Y^2, -2XY$

Table 4.3. Raman tensors for T_d. Note that the tensors associated with A_1, E and F_2 are symmetric whereas those with F_1 are antisymmetric

A_1	E		F_2		F_1	
$\begin{bmatrix} a & 0 & 0 \\ 0 & a & 0 \\ 0 & 0 & a \end{bmatrix}$	$\begin{bmatrix} -b & 0 & 0 \\ 0 & -b & 0 \\ 0 & 0 & 2b \end{bmatrix}$	$,2Z^2 - X^2 - Y^2$	$\begin{bmatrix} 0 & 0 & 0 \\ 0 & 0 & d \\ 0 & d & 0 \end{bmatrix}$	$,X$	$\begin{bmatrix} 0 & 0 & 0 \\ 0 & 0 & c \\ 0 & -c & 0 \end{bmatrix}$	$,R_x$
	$\begin{bmatrix} \sqrt{3}\,b & 0 & 0 \\ 0 & -\sqrt{3}\,b & 0 \\ 0 & 0 & 0 \end{bmatrix}$	$,\sqrt{3}\,(X^2 - Y^2)$	$\begin{bmatrix} 0 & 0 & d \\ 0 & 0 & 0 \\ d & 0 & 0 \end{bmatrix}$	$,Y$	$\begin{bmatrix} 0 & 0 & -c \\ 0 & 0 & 0 \\ c & 0 & 0 \end{bmatrix}$	$,R_y$
			$\begin{bmatrix} 0 & d & 0 \\ d & 0 & 0 \\ 0 & 0 & 0 \end{bmatrix}$	$,Z$	$\begin{bmatrix} 0 & c & 0 \\ -c & 0 & 0 \\ 0 & 0 & 0 \end{bmatrix}$	$,R_z$

semiconductors like CdTe and MnTe in atomic proportions of $(1 - x)/x$, i.e., in $Cd_{1-x}Mn_xTe$. It has been found that for $x \leq 0.75$ it has the zinc-blende structure with Mn atoms replacing at random sites Cd atoms.

One can show that the presence of a Mn atom replacing the heavier Cd in CdTe will result in a high-frequency local mode [4.27]. On the other hand, Cd in the hypothetical MnTe of zinc-blende structure can produce only a gap mode. The variation of the frequency of the local mode of Mn in CdTe as a function of x can be calculated, at least for small x, by considering the mutual interactions of the local modes centered around the different Mn ions. This type of problem has been discussed by *Maradudin* and *Oitmaa* [4.28]. We give first a simple approach similar to that used by *Fröhlich* [4.29] in the context of the long wavelength polar modes in an ionic crystal. When the lattice is excited at frequencies near ω_0, the angular frequency of the local mode, the optical modes of the host are quiescent. If u is the displacement of the Mn atom in a long-wavelength excitation, we have

$$M' \frac{d^2u}{dt^2} = -M'\omega_0^2 u + zeE_1, \tag{4.12}$$

where E_1 is the local electric field at the position of a Mn ion and ze is the effective charge of the Mn ion upon displacement. We have assumed here that the local mode has an inertial mass equal to that of the free Mn atom; in any case (z/M') can be viewed as a phenomenological constant. The local electric field E_1 with the system having uniform polarization P is

$$E_1 = E + \frac{4\pi}{3} P, \tag{4.13}$$

where E is the macroscopic electric field inside the material.

Table 4.4. Raman tensors for C_{6v}

	Symmetric			Antisymmetric	
	A_1	E_1	E_2	A_2	E_1

A_1:
$$\begin{bmatrix} p & 0 & 0 \\ 0 & p & 0 \\ 0 & 0 & q \end{bmatrix}, Z$$

E_1 (Symmetric):
$$\begin{bmatrix} 0 & 0 & r \\ 0 & 0 & 0 \\ r & 0 & 0 \end{bmatrix}, X \qquad \begin{bmatrix} 0 & 0 & 0 \\ 0 & 0 & r \\ 0 & r & 0 \end{bmatrix}, Y$$

E_2:
$$\begin{bmatrix} s & 0 & 0 \\ 0 & -s & 0 \\ 0 & 0 & 0 \end{bmatrix}, X^2 - Y^2 \qquad \begin{bmatrix} 0 & -s & 0 \\ -s & 0 & 0 \\ 0 & 0 & 0 \end{bmatrix}, -2XY$$

A_2:
$$\begin{bmatrix} 0 & a & 0 \\ -a & 0 & 0 \\ 0 & 0 & 0 \end{bmatrix}, R_z$$

E_1 (Antisymmetric):
$$\begin{bmatrix} 0 & 0 & -b \\ 0 & 0 & 0 \\ b & 0 & 0 \end{bmatrix}, X, R_x \qquad \begin{bmatrix} 0 & 0 & 0 \\ 0 & 0 & -b \\ 0 & b & 0 \end{bmatrix}, Y, R_y$$

Each Mn ion gives rise to a dipole moment zeu. Since the probability that a cell be occupied by a Mn atom is x, the contribution to the polarization is $(xzeu/v_0)$, where v_0 is the volume of the primitive cell. However, the solid experiences an electronic polarization, which for frequencies below the optical absorption edge gives rise to a screening of the dipole moments associated with the local modes. This results in a polarization of the medium given by

$$P = \frac{xze}{n^2 v_0} u, \tag{4.14}$$

where n is the index of refraction corresponding to $\varepsilon_\infty = n^2$, the high-frequency dielectric constant. Neglecting the effect of retardation, we note that the fields E, P, and $D = E + 4\pi P$ must obey the equations of electrostatics, i.e., $\nabla \cdot D = 0$ and $\nabla \cdot E = 0$. In a wave with wave vector q these quantities vary as $\exp (iq \cdot r - i\omega t)$. Thus, $q \cdot D = 0$ and, for a longitudinal wave, this yields $D = 0$, and hence

$$E = -4\pi P. \tag{4.15}$$

For a transverse wave, on the other hand, $\nabla \cdot E = -4\pi iq \cdot P = 0$ which, combined with $\nabla \times E = 0$, results in $E = 0$. Hence, for transverse waves,

$$E_1 = \frac{4\pi}{3} P. \tag{4.16}$$

and for longitudinal waves

$$E_1 = -\frac{8\pi}{3} P. \tag{4.17}$$

Thus, from (4.12, 14, 16, 17) we find that for transverse (TO) waves

$$\omega_{TO}^2 = \omega_0^2 - \frac{x}{3} \frac{4\pi z^2 e^2}{3M' v_0 n^2}, \tag{4.18}$$

while for longitudinal (LO) waves

$$\omega_{LO}^2 = \omega_0^2 + \frac{2x}{3} \frac{4\pi z^2 e^2}{3M' v_0 n^2}. \tag{4.19}$$

For

$$\frac{4\pi z^2 e^2}{M' v_0 n^2 \omega_0^2} \ll 1,$$

Equations (4.18, 19) yield

$$\omega_{TO} = \omega_0 - \frac{x}{6} \frac{4\pi z^2 e^2}{M' v_0 n^2 \omega_0}, \tag{4.20}$$

$$\omega_{LO} = \omega_0 + \frac{x}{3} \frac{4\pi z^2 e^2}{M' v_0 n^2 \omega_0}, \tag{4.21}$$

respectively.

The theory outlined above has the advantage of simplicity and provides a satisfactory account of the two-mode behavior of $Cd_{1-x}Mn_xTe$. It neglects the polarization due to ionic displacements of the Cd and Te atoms and all electronic polarizabilities except for the use of the index of refraction n. We outline below the theory proposed by *Genzel* et al., [4.30], which also considers the displacements for $(q = 0)$ of all three types of atoms and the corresponding restoring forces.

The model is based on the following equations of motion for an $AB_{1-x}C_x$ mixed crystal of zinc-blende symmetry, taking into account nearest-neighbor and second-neighbor interactions:

$$m_A \ddot{u}_A = -(1-x) f_b(u_A - u_B) - x f_c(u_A - u_C) + [(1-x) e_b + e_c] E_1, \tag{4.22}$$

$$m_B \ddot{u}_B = -f_b(u_B - u_A) - x f_s(u_B - u_C) - e_b E_1, \tag{4.23}$$

$$m_C \ddot{u}_C = -f_c(u_C - u_A) - (1-x) f_s(u_C - u_B) - e_c E_1, \tag{4.24}$$

and

$$P = (1-x) \left[\frac{e_b}{v_0} (u_A - u_B) + \frac{\alpha_A + \alpha_B}{v_0} E_1 \right] \\ + x \left[\frac{e_c}{v_0} (u_A - u_C) + \frac{\alpha_A + \alpha_C}{v_0} E_1 \right]. \tag{4.25}$$

In these equations, m, u and α are the masses, displacements, and electronic polarizabilities of the ions A, B, and C, respectively. The volume of the primitive cell, v_0, will vary with composition and is given by $v_0 = a^3/4$, where a is the lattice parameter of the zinc-blende crystals. The subscripts b and c on the force constants f and the Szigeti-effective charges, e, refer to the compounds AB and AC, respectively. Equations (4.22–25) are those of *Genzel* et al. [4.30], with additional terms involving the second-neighbor force constant between the B and C ions, f_s. Based on the experimental evidence that the extrapolation of the frequencies of the phonon modes in $Cd_{1-x}Mn_xTe$ and $Zn_{1-x}Mn_xTe$ to $x = 1$ yields the same values, the force constants between the Mn and Te ions, f_c, must be the same in both alloy systems as $x \to 1$. Hence, it is necessary to let the force constants exhibit a dependence on the lattice parameter. This dependence is approximated by a linear function

$$f(x) = F \left(1 + \Theta \, \frac{a_c - a(x)}{a_c} \right), \tag{4.26}$$

where a_c is the lattice parameter of the AC compound (cubic MnTe in this case), $a(x)$ is the lattice constant of the alloy with Mn concentration x and f and Θ are constants. The alloy systems, $Cd_{1-x}Mn_xTe$ and $Zn_{1-x}Mn_xTe$, have similar chemical bonding, and, hence, Θ is assumed to be the same for both. Since the local electric field is given by (4.13), the relationship between E_1 and the electric polarization P is

$$E_1 = P/\xi, \tag{4.27}$$

where $\xi = 3/4\pi$ and $-3/8\pi$ for transverse and longitudinal modes, respectively. As demonstrated by *Genzel* et al. [4.30], the microscopic parameters in (4.22–25) are related to macroscopic parameters according to the usual *Born-Huang* procedure [4.31].

For $x = 0$:
$$\frac{4\pi}{3} \, \frac{\alpha_A + \alpha_B}{v_b} = \frac{\varepsilon_\infty - 1}{\varepsilon_\infty + 2}, \tag{4.28}$$

$$F_b \left(1 + \Theta \, \frac{a_c - a_b}{a_c} \right) = \mu_b \, \frac{\varepsilon_{0b} + 2}{\varepsilon_\infty + 2} \, \omega_{T_b}^2, \tag{4.29}$$

$$\frac{4\pi}{3} \, \frac{e_b^2}{\mu_b v_b} = 3 \, \frac{\varepsilon_{0b} - \varepsilon_{\infty b}}{(\varepsilon_{\infty b} + 2)^2} \, \omega_{T_b}^2. \tag{4.30}$$

For $x = 1$:
$$\frac{4\pi}{3} \, \frac{\alpha_A + \alpha_C}{v_c} = \frac{\varepsilon_\infty - 1}{\varepsilon_{\infty c} + 2}, \tag{4.31}$$

$$F_c = \mu_c \, \frac{\varepsilon_{0c} + 2}{\varepsilon_{\infty c} + 2} \, \omega_{T_c}^2, \tag{4.32}$$

$$\frac{4\pi}{3} \, \frac{e_c^2}{\mu_c v_c} = 3 \, \frac{\varepsilon_{0c} - \varepsilon_{\infty c}}{(\varepsilon_{\infty c} + 2)^2} \, \omega_{T_c}^2. \tag{4.33}$$

Here μ_b and μ_c denote the reduced masses of ions A and B, and of ions A and C, respectively; v_b and v_c are the corresponding volumes of the primitive cells. The static and high frequency dielectric constants of the crystals AB and AC are given by $\varepsilon_{0b}, \varepsilon_{\infty b}, \varepsilon_{0c}$ and $\varepsilon_{\infty c}$, and the corresponding TO phonon frequencies at zero wave vector are ω_{T_b} and ω_{T_c}. In addition to the force constants, the polarizabilities and effective charges should also exhibit an x-dependence, since the nearest-neighbor distance changes with x. To a first approximation, these effects will be neglected and the force constants are assumed to have a linear dependence on the lattice parameter. The eigenfrequencies of (4.22–4.24) have the form

$$\omega^2 = N \pm \sqrt{N^2 - L^2},\tag{4.34}$$

where

$$N = \frac{1}{2m_c}[f_c + (1-x)f_s - xq_c^2]$$

$$+ \frac{1}{2m_B}[f_b + xf_s - (1-x)q_b^2] + \frac{1}{2m_A}[(1-x)f_b + xf_c - q^2],\tag{4.35}$$

$$L^2 = \frac{m_A + (1-x)m_B + xm_C}{m_A m_B m_C}$$

$$\times [f_b f_c + (1-x)f_b f_s + xf_c f_s - xf_b q_c^2 - (1-x)f_c q_b^2 - f_s q^2],\tag{4.36}$$

$$q = (1-x)q_b + xq_c, \qquad q_b = \frac{e_b}{\sqrt{\eta}}, \qquad q_c = \frac{e_c}{\sqrt{\eta}},\tag{4.37}$$

and

$$\eta = v\xi - \alpha_A - (1-x)\alpha_B - x\alpha_C.\tag{4.38}$$

In the limits of $x = 0$ and $x = 1$, the frequencies of the impurity modes will be given by the following expressions.

$$x = 0: \quad \omega_{Ic} = \left[\frac{F_c + F_s}{m_C}\left(1 + \Theta\frac{a_c - a_b}{a_c}\right)\right]^{1/2},\tag{4.39}$$

$$x = 1: \quad \omega_{Ib} = \left(\frac{F_b + F_s}{m_B}\right)^{1/2}.\tag{4.40}$$

The microscopic parameters of (4.34–4.38) are determined from the Born-Huang relationships of (4.28–33) and the boundary conditions expressed in (4.39, 40).

4.2.2 Experimental Results and Discussion

Figure 4.1 shows the room temperature Raman spectrum of $Cd_{0.6}Mn_{0.4}Te$ [4.8]. The scattering configuration is $y'(z'z')\,x'$ where x', y' and z' are along [100], [011] and [0$\bar{1}$1], respectively; it allows the observation of the Raman active phonon features having A_1, E, and F_2 symmetries. The first order Raman spectrum of pure CdTe consists of a pair of LO-TO lines of F_2 symmetry occurring at 140 and 171 cm^{-1}, respectively [4.32, 33]. We note that the alloy exhibits a rather intense and quasi-continuous spectrum below 130 cm^{-1}, in addition to the *two* pairs of relatively sharp lines, one at 143 and 158 cm^{-1} and the other at 189 and 203 cm^{-1}. This clearly shows that the mixed crystal has a spectrum richer and more complex than that of CdTe.

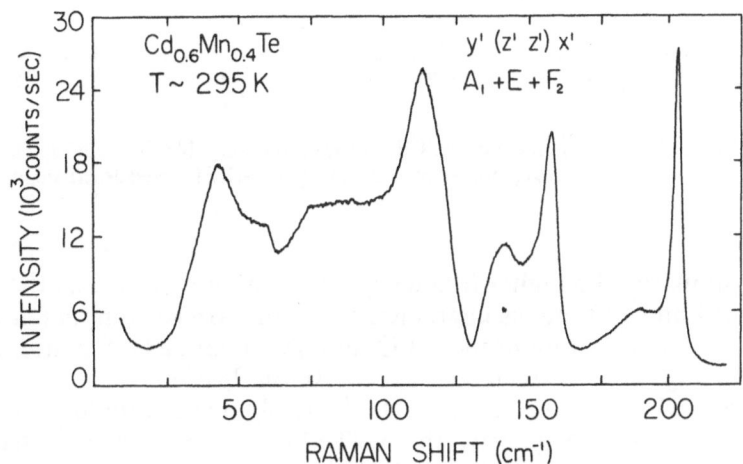

Fig. 4.1. Raman spectrum of $Cd_{0.6}Mn_{0.4}Te$ at 295 K, excited with 6764 Å Kr$^+$ laser line. x', y', and z' are along [100], [011], and [0$\bar{1}$1], respectively. The polarization geometry and the allowed phonon symmetries are also indicated [4.8]

4.2.3 Zone Center Optical Phonons

Polarization features as well as concentration dependence of the spectra provide further insight into their origin. To the extent that the mixed crystal as a whole can be considered to have T_d point group symmetry, the spectrum in Fig. 4.1 can be analyzed in terms of the different Raman active species viz., A_1, E and F_2. Such studies show that the 143–158 cm^{-1} and the 189–203 cm^{-1} pairs are LO-TO split F_2 modes. In other words, $Cd_{1-x}Mn_xTe$ exhibits a "two-mode" behaviour [4.2]. Based on the zone center LO-TO F_2 modes of pure CdTe, one can identify the lower frequency pair as the respective TO and LO modes characteristic of the CdTe component of the mixed crystal and hence are labeled "CdTe like".

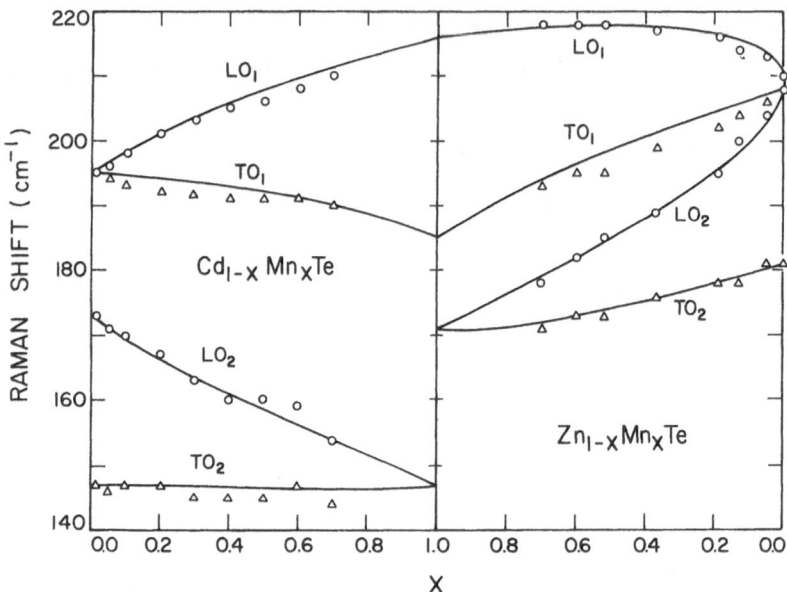

Fig. 4.2. The frequencies of the $Cd_{1-x}Mn_xTe$ and $Zn_{1-x}Mn_xTe$ zone-center optical phonons at $T = 80$ K. The curves were generated using the MREI model described in the text [4.34]

Similarly, the higher frequency pair can be regarded as the "MnTe-like" TO and LO modes characteristic of the MnTe component of the alloy.

The variation of these LO and TO modes as a function of x is shown in Fig. 4.2. It is seen that as $x \to 0$, the MnTe-like modes merge toward one which can be regarded as the triply degenerate local mode of Mn in pure CdTe. Similarly, the CdTe-like modes are expected to become degenerate for $x = 1$, at which point they must correspond to the gap mode of Cd in the "hypothetical" MnTe crystal with point group T_d. The extrapolations shown in Fig. 4.2 beyond $x = 0.7$ reflect these assumptions. We note here that $Cd_{1-x}Mn_xTe$ occurs in a homogeneous, zinc-blende phase over $0 \leq x \leq 0.75$; pure MnTe crystallizes in the hexagonal nickel arsenide structure with C_{6v}^4 symmetry and in this context the zinc-blende structure for MnTe as the end member of $Cd_{1-x}Mn_xTe$ is "hypothetical". We note here that the "hypothetical" MnTe and MnSe have been produced in MBE grown quantum well structures [4.12].

Experimental results for the vibrational Raman spectra for $Zn_{1-x}Mn_xTe$, another zinc-blende DMS, are presented in Figs. 4.3 and 4.4 [4.34]. We first consider the Raman lines which can be traced to the zone-center LO and TO phonons of ZnTe as $x \to 0$ and those that evolve from the impurity mode of Mn in ZnTe as x increases. As can be seen in Fig. 4.3, an inflection labeled I appears on the low frequency side of the LO mode in $Zn_{1-x}Mn_xTe$, $x = 0.05$; we identify it with the band mode of Mn in ZnTe, since it lies between the TO and LO frequencies. As the

Fig. 4.3. Room temperature Raman spectrum of $Zn_{1-x}Mn_xTe$ for (a) $x = 0.003$ and (b) $x = 0.05$; $\lambda_L = 6471$ Å. The scattering geometries are (a) $z'(y'y')x'$ and (b) $x'(y'y')z'$, where x', y', and z' are along [100], [011], and [01$\bar{1}$], respectively. The labeling of the phonon features is discussed in the text [4.34]

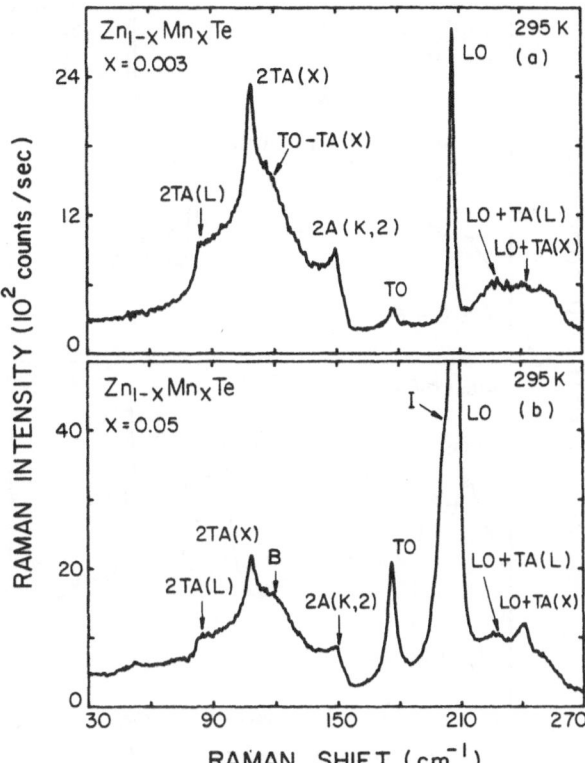

Mn concentration is increased, the impurity mode splits into a transverse (TO_1) and a longitudinal component (LO_2), as illustrated in Fig. 4.4 for $x = 0.37$. The frequencies of the TO and LO modes as a function of x are shown in Fig. 4.2. In contrast to the behavior of the zone center optical phonons in $Cd_{1-x}Mn_xTe$, those in $Zn_{1-x}Mn_xTe$ exhibit a mixed mode behavior intermediate to the one- and two-mode situations. For $x \to 0$, one observes the TO and LO modes of ZnTe and the band mode ZnTe:Mn. As x increases, the band mode splits into the longitudinal mode LO_2 and the transverse mode TO_1. The vibrational modes that evolve from the LO and TO phonons of ZnTe are designated as LO_1 and TO_2. As $x \to 1$, the LO_1 mode, which is the LO phonon of ZnTe, evolves into the LO mode of MnTe. The TO_1 phonon, one of the components evolving from the impurity mode ZnTe:Mn, becomes the TO vibrational mode of MnTe. The two remaining phonons, TO_2 and LO_2, merge to become the gap mode of Zn in the zinc-blende MnTe (MnTe:Zn). As can be seen in Fig. 4.4, the extrapolation of the $Cd_{1-x}Mn_xTe$ and $Zn_{1-x}Mn_xTe$ phonons to $x = 1$ results in the same values for the frequencies of the TO and LO modes for the hypothetical zinc-blende MnTe. This effect has also been observed for other physical characteristics of these alloys. The

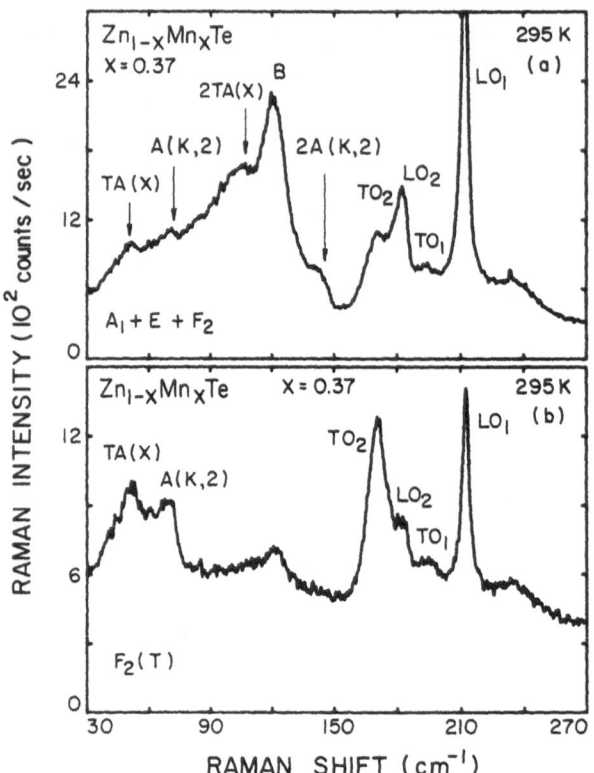

Fig. 4.4. Room temperature Raman spectrum of $Zn_{1-x}Mn_xTe$, $x = 0.37$; $\lambda_L = 6471\,\text{Å}$. The scattering geometries are (a) $x'(y'y')z'$ and (b) $x'(y'x')z'$, where x', y' and z' are along [100], [011], and [01$\bar{1}$], respectively [4.34]

lattice parameters [4.35] and the energy band gaps [4.36] of $Cd_{1-x}Mn_xTe$ and $Zn_{1-x}Mn_xTe$ extrapolate to the same values for $x = 1$.

The curves in Fig. 4.2 were determined from the modified random element isodisplacement (MREI) model. The fundamental assumptions of the random element isodisplacement model are that in the long wavelength limit ($q \sim 0$) the anion and cation of like species vibrate with the same phase and amplitude and that the force each ion experiences is provided by a statistical average of the interaction with its neighbors. This MREI model is a modification of that developed by *Genzel* et al. [4.30] which emphasizes the use of the local field and is completely defined by the macroscopic parameters of the pure end members. The modified theory is described in the theoretical discussion in Sect. 4.2.1 and embodied in equations (4.22–40). The additional constraint that the frequencies of the LO and TO modes for $Cd_{1-x}Mn_xTe$ be equal to the corresponding modes of $Zn_{1-x}Mn_xTe$ when $x = 1$ allows one to incorporate second-neighbor force constants and a linear dependence of the force constants on the lattice parameter into the model without resorting to microscopic fitting parameters. The only "fitting parameters" necessary for this model are the frequencies of the TO and LO modes of MnTe and of the gap modes MnTe:Cd and MnTe:Zn. The frequencies of the vibrational modes in

Table 4.5. Parameters in the MREI model [4.34]

$Cd_{1-x}Mn_xTe$ (experimental)[a]	$Zn_{1-x}Mn_xTe$ (experimental)[a]	"Fitting parameters"
$\omega_{TO}(CdTe) = 147\,cm^{-1}$	$\omega_{TO}(ZnTe) = 181\,cm^{-1}$	$\omega_{TO}(MnTe) = 185\,cm^{-1}$
$\omega_{LO}(CdTe) = 173\,cm^{-1}$	$\omega_{LO}(ZnTe) = 210\,cm^{-1}$	$\omega_{LO}(MnTe) = 216\,cm^{-1}$
$\omega_1(CdTe:Mn) = 195\,cm^{-1}$	$\omega_1(ZnTe:Mn) = 108\,cm^{-1}$	$\omega_1(MnTe:Cd) = 147\,cm^{-1}$
$\varepsilon_0(CdTe) = 9.6$	$\varepsilon_0(ZnTe) = 10.1$	$\omega_1(MnTe:Zn) = 171\,cm^{-1}$
$a = 6.486 - 0.145x\,Å^{b}$	$a = 6.103 - 0.238x\,Å^{b}$	

Resultant force constants [10^6 amu $(cm^{-1})^2$]:
$F_{Mn-Te} = 1.72$, $F_{Cd-Te} = 1.85$, $F_{Zn-Te} = 1.57$, $F_{Cd-Mn} = 0.58$, $F_{Zn-Mn} = 0.34$
[a] $T = 80\,K$, [b] From [4.35]

$Cd_{1-x}Mn_xTe$ and $Zn_{1-x}Mn_xTe$ as a function of x (Fig. 4.2) were determined from the macroscopic parameters of Table 4.5. The resulting force constants are also given in the table. The magnitude of the second-neighbor force constants is about one third that of the nearest-neighbor constants. As illustrated in Fig. 4.2, the curves generated from this MREI model follow the experimental results quite well, except for the TO_1 modes for which there is a significant discrepancy in the curvature of the theory and the experimental results. However, considering the simplifying assumptions of this model, the theory provides an adequate description of the phonon frequencies in $Cd_{1-x}Mn_xTe$ and $Zn_{1-x}Mn_xTe$ over the entire composition range.

4.2.4 Low Frequency Phonon Features

The room temperature Raman spectrum of $Zn_{1-x}Mn_xTe$, $x = 0.003$, shown in Fig. 4.3a is essentially that of ZnTe, which has been studied in detail [4.37]. The prominent features of the low frequency two phonon density of states spectrum have been attributed to two transverse acoustic phonons at point L of the Brillouin zone [2TA(L)], two TA modes at X[2TA(X)], the difference mode of the transverse optical and acoustic phonons at X[TO-TA(X)], and two acoustic phonons of the second type at approximately point K[2A(K,2)]. In addition to the transverse optical (TO) and the longitudinal optical (LO) phonons, there are two higher frequency features that have been identified as the combination of the LO phonons with transverse acoustic phonons at points L and X of the Brillouin zone, LO + TA(L) and LO + TA(X), respectively. The Raman shifts of the low frequency features are given in Table 4.6.

The room temperature vibrational Raman spectrum for $Zn_{1-x}Mn_xTe$, $x = 0.05$ is similar to that for $x = 0.003$, except that the 2TA(L) mode has decreased in frequency and there is an increase in the relative Raman intensity in the region of the difference mode (TO-TA(X)). The label 'B' reflects the emergence of a new feature very close to the TO-TA(X) mode and the evidence for this is apparent in the spectra of the higher

Table 4.6. Low frequency phonon features in $Zn_{1-x}Mn_xTe$ [cm^{-1}]

x	TA(X)	A(K,2)	2TA(L)	2TA(X)	TO-TA(X); B	2A(K,2)
0.003			84	109	119	150
0.05			82	109	119	150
0.13			82	108	119	148
0.19	54	73		108	120	146
0.37	53	72		107	120	143
0.60				101	120	140
0.70	50	69		99	120	128

composition samples. In addition, weak disordered-induced one-phonon features are observable having Raman shifts < 75 cm^{-1}.

The trends continue as the Mn concentration is increased to $x = 0.13$ and $x = 0.19$ with the TA(L), TA(X), and A(K,2) phonons decreasing in frequency. The intensity of the vibrational mode B continues to increase. The low frequency disorder-induced features are attributed to the one-phonon density of states with the peaks being identified as TA(X) and A(K,2); these features also increase in intensity as x increases.

When the heavier Cd atom is replaced by Mn in $Cd_{1-x}Mn_xTe$, one expects new features corresponding to band modes below the TO mode frequency of CdTe. This is borne out by the pronounced low frequency modes seen in Figs. 4.1 and 4.5 in the 20–130 cm^{-1} range. These features persist over the composition range $x = 0.4$–0.75, but are considerably weaker for $x \leqq 0.3$. Also, for any given x, the higher the sample temperature the greater is their intensity. The presence of Mn at random sites within the lattice effectively destroys the exact translational periodicity in the mixed crystals. Thus, although only zone-center ($q \sim 0$) optical phonons are allowed in the first-order Raman spectrum of a perfect crystal, the $q \sim 0$ selection rule is relaxed in the alloys. Consequently, they may exhibit one-phonon Raman scattering due to optic and acoustic phonons with all possible q vectors spanning the BZ. As a substitutional impurity, Mn retains the site symmetry T_d in the mixed crystal. Hence, the new impurity-induced features of the spectrum should again belong to the Raman-active representations of T_d, viz., A_1, E and F_2.

Figure 4.5 displays the striking polarization behavior of these low frequency modes. The asymmetric peak centered at 42 cm^{-1} in Fig 4.5a exhibits a predominant polarization characteristic of transverse F_2 modes. While investigating the side-band absorption associated with the local mode of Be in CdTe, *Sennet* et al. [4.38] performed a shell model calculation of the weighted density of one-phonon states, $S(\omega)/\omega^2$, for CdTe. We note that in the 20–130 cm^{-1} range the main features of Fig. 4.5a bear a striking similarity to their calculated function $S(\omega)/\omega^2$, shown as an inset in the same figure. The one-phonon density of states in CdTe has also been determined experimentally by *Rowe* et al. [4.39] using neutron inelastic

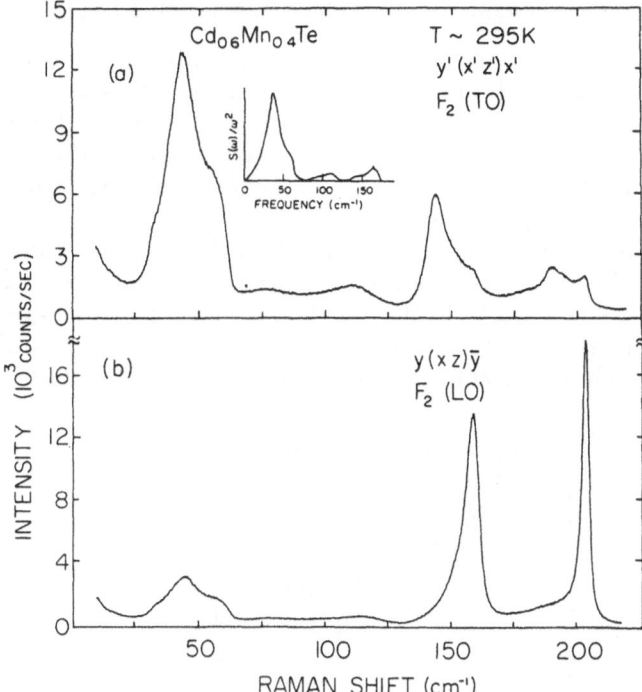

Fig. 4.5. Polarized spectra of $Cd_{0.6}Mn_{0.4}Te$. (a) $y'(x'z')\,x'$; (b) $y(xz)\,\bar{y}$. The significance of the inset in (a) is discussed in the text. The crystallographic axes are the same as in Fig. 4.1 for (a). For (b), x, y, z are the cubic axes [4.8]

scattering. Their results reveal a prominent peak at $\sim 40\ \mathrm{cm}^{-1}$ due to transverse acoustic (TA) phonons and a less intense peak at $\sim 110\ \mathrm{cm}^{-1}$ due to longitudinal acoustic (LA) phonons. Furthermore, their comparison of the frequency distributions of CdTe and InSb shows that as compared to InSb, the optic and LA modes of CdTe show a clear "softening", but the TA modes lie closely within the same range of frequencies for both compounds. We therefore conclude that the peak in Fig. 4.5a at $42\ \mathrm{cm}^{-1}$ has its origins in the disorder-activated, first-order Raman scattering by TA phonons and that it reflects the corresponding TA phonon density of states in the mixed crystal.

The Raman spectra of Figs. 4.3 and 4.4 demonstrate that the low frequency disorder-induced features become more intense in $Zn_{1-x}Mn_xTe$ as the disorder in the lattice increases. The lack of strict translational symmetry in the alloy allows the observation of Raman features associated with nonzero wave vector excitations. However, since the masses of Zn and Mn are more closely matched, the deviation from strict translational symmetry is not as severe as in $Cd_{1-x}Mn_xTe$, for which there is a large difference in the masses of Cd and Mn. This conclusion is based on the observation that the one-phonon density of states features dominate the low frequency Raman spectra in $Cd_{1-x}Mn_xTe$, in contrast to the situation which prevails for $Zn_{1-x}Mn_xTe$.

4.2.5 Resonant and Forbidden Scattering Effects

As we noted in Sect. 4.2, a resonant enhancement of the Raman scattering intensities should occur whenever the energies of the incident ($\hbar\omega_L$) or scattered photons ($\hbar\omega_S$) lie close to the energy of an electronic excitation of the medium. In the case of polar semiconductors, the dependence of the Raman intensities on $\hbar\omega_L$ exhibits several additional features in the vicinity of the fundamental energy gap (E_g) of the crystal; striking effects due to resonant cancellation, multiple LO-phonon scattering as well as forbidden one-LO phonon scattering have been observed, for example, in CdS and GaAs [4.40–44]. These effects provide valuable insight into the different mechanisms of the scattering process; in addition, they are helpful in characterizing the intermediate electronic states which participate in

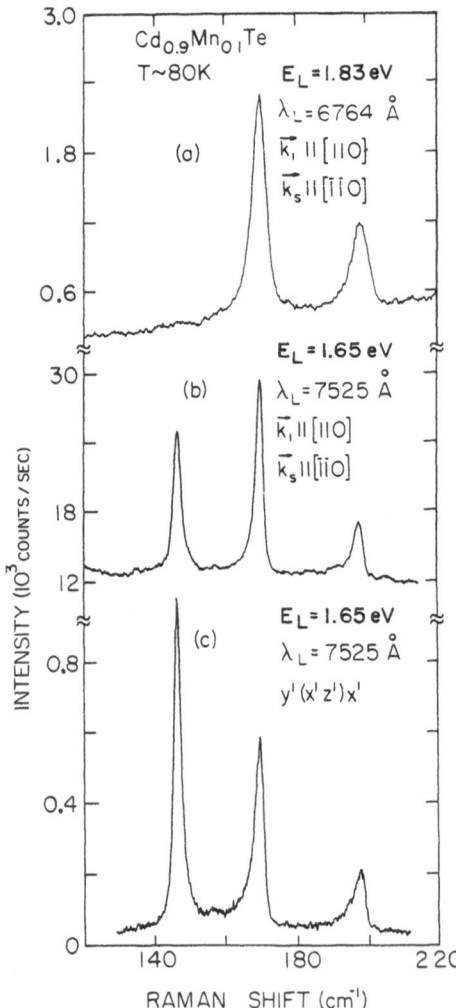

Fig. 4.6. Forbidden LO and allowed TO phonon scattering in $Cd_{0.9}Mn_{0.1}Te$. $T \sim 80$ K. (a) Backscattering along [110] with $\lambda_L = 6764$ Å. (b) Backscattering along [110] with $\lambda_L = 7525$ Å (c) Right angle scattering in the $y'(x'z')x'$ geometry, x', y', and z' being defined in Fig. 4.1; $\lambda_L = 7525$ Å, k_i and k_s being the wave vectors of the incident and scattered photons, respectively [4.8]

the scattering process. Figure 4.6 shows the results for $Cd_{0.9}Mn_{0.1}Te$ at 80 K in the two-mode region. In all the three spectra, the TO phonons are allowed, whereas the LO phonons are forbidden according to the usual polarization selection rules. In contrast to these predictions, the LO phonons are very clearly observed in both scattering geometries and for both λ_L's used in the experiment. The CdTe-like TO phonon is extremely weak for $\hbar\omega_L > E_g$ (1.72 eV) as seen in Fig. 4.6a but appears quite strongly in traces (b) and (c) where $\hbar\omega_L < E_g$, although quite close to the band gap.

Forbidden and resonant one-LO-phonon scattering have been reported earlier in other semiconductors for exciting radiation energies above as well as below E_g. It was shown [4.40, 41] that (i) q-dependent, intraband Fröhlich interactions, (ii) q-independent, impurity-induced effects, and (iii) surface electric fields at the depletion layer are the three main processes, in general, that can cause forbidden LO scattering. The resonance of the allowed TO modes, on the other hand, can be readily understood on the basis of the deformation-potential interaction [4.40]. It appears that the forbidden LO scattering seen in Fig. 4.6 is dominated by the possible presence of impurities in the sample [4.8].

Figure 4.7 shows the results obtained on a p-type CdTe at 80 K. With $\hbar\omega_L > E_g$, the direct band gap ($\sim$1.58 eV), and in the backscattering

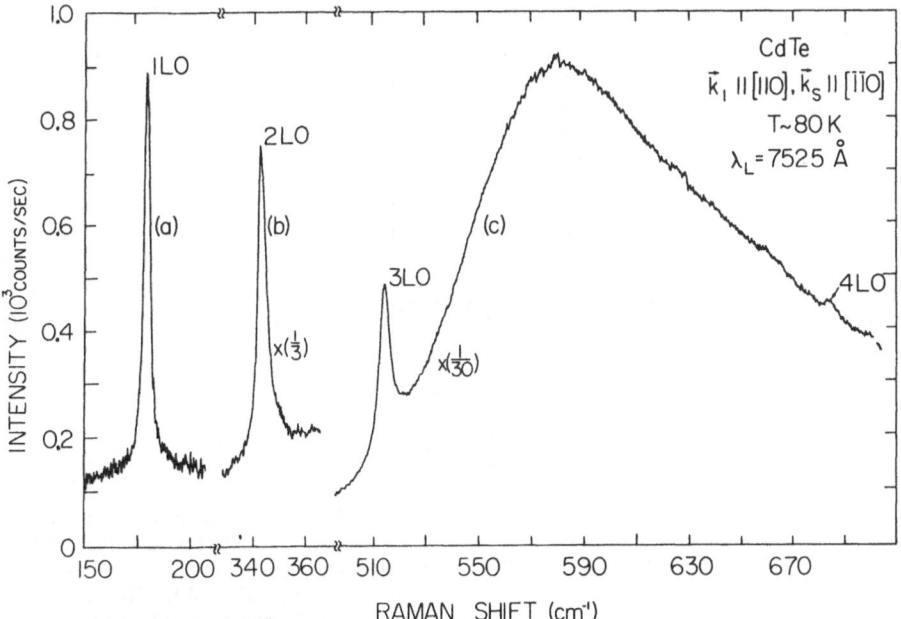

Fig. 4.7. Forbidden one-LO and allowed multiple LO scattering in CdTe at 80 K. $\lambda_L = 7525$ Å. Note that for traces (b) and (c) the respective intensity scale factors are 3 and 30 times greater as compared to that of (a); k_i [110] and $k_s \parallel [\bar{1}\bar{1}0]$ [4.8]

geometry with $q \parallel [110]$, forbidden 1 LO phonon scattering is observed, as well as allowed multiple-LO phonon scattering up to 4 LO. However, the allowed TO phonon is not observed; this is consistent with earlier observations by *Mooradian* and *Harman* [4.45] when they employed exciting photon energies greater than E_g. The 1 LO peak is more intense with $E_i \parallel E_s$ than for $E_i \perp E_s$, where E_i and E_s are the electric vectors of the incident and scattered photons, respectively. The 2 LO and 3 LO scattering lie increasingly closer to E_g and the associated exciton levels; accordingly, the resonant enhancement is greater for the 3 LO than for the 2 LO scattering. The broad, intense peak in trace (c) occurs in the region of the exciton luminescence peaks [4.3].

In [4.8] the authors report a resonant increase of the intensity of overtones and sum of the LO modes, i.e., 2 LO_1, $(LO_1 + LO_2)$, and 2 LO_2 as $\hbar\omega_L$ approaches E_g. Typical spectra are shown in Fig. 4.8 for $x = 0.1$ and $x = 0.3$. For $x = 0.1$, the 2 LO_2 mode is much weaker than 2 LO_1. Understandably, this is due to the low Mn concentration of this alloy.

We conclude this section by drawing attention to the 'outgoing multiphonon resonant Raman scattering' observed by *Menéndez* et al. in $Cd_xHg_{1-x}Te$, by *Feng* et al. in epitaxial CdTe films, and by *Perkowitz* et al. [4.44] in $Cd_{1-x}Mn_xTe$ thin films grown by metal organic vapor deposition (MOCVD); the electronic transition mediating in the three step process envisioned in (4.5b) is identified with a $E_0 + \Delta_0$ transition, where E_0 is the direct band gap and Δ_0 is the spin-orbit splitting of the valence band.

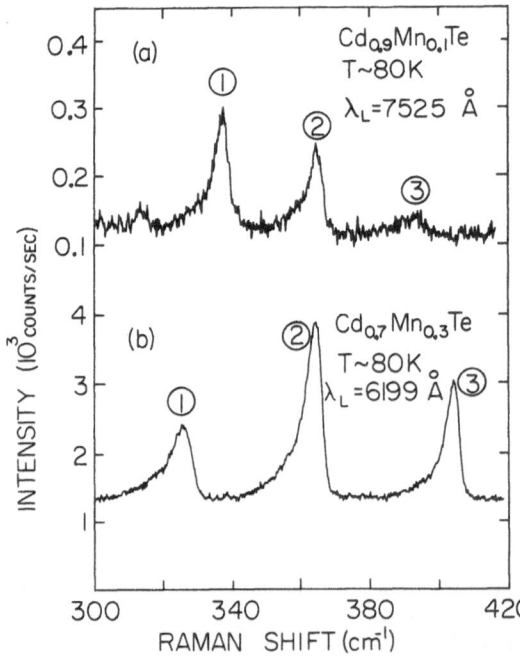

Fig. 4.8. Overtones and combinations of the LO modes observed for $Cd_{1-x}Mn_xTe$ (a) $x = 0.1$ and (b) $x = 0.3$. $T \sim 80$ K. The labels 1, 2, and 3 refer to 2 LO_1, $LO_1 + LO_2$, and 2 LO_2, respectively [4.8]

4.3. Magnetic Excitations

4.3.1 General Considerations

The incomplete d-shell of the magnetic ions in a DMS gives rise to a variety of properties in which their localized magnetic moments play important roles, either individually or collectively through their mutual interactions. In this section the phenomena are in the main illustrated with examples based on Mn-based DMSs in view of the simplicity of the level structure of Mn^{2+} in the tetrahedral environment and the large body of experimental work based on them.

For sufficiently low concentrations, x, of the magnetic ions, or at temperatures above a critical temperature T_N, the material is in a paramagnetic phase [4.6]. The temperature T_N is, of course, a function of x. A transition to an ordered phase occurs below T_N. In these materials, the interaction between neighboring magnetic ions is antiferromagnetic, giving rise to a spin glass or to an antiferromagnetic phase. *Galazka* et al. [4.6], using specific-heat and magnetic-susceptibility measurements, found that $Cd_{1-x}Mn_xTe$ crystals are paramagnetic at all temperatures for $x < 0.17$. For compositions in the range $0.17 < x < 0.75$, the crystals are paramagnetic at high temperatures, and, as the temperature is lowered, a spin glass is obtained for $0.17 < x < 0.60$, while for $0.60 < x < 0.75$, a paramagnetic-to-antiferromagnetic phase transition occurs. It appears that the distinction between spin glass and the antiferromagnetic phases is not clear-cut and the antiferromagnetic interaction effects can be observed even for small x, at sufficiently low temperatures [4.1].

The theory of Raman scattering by magnetic excitations is similar to that for phonons. However, the selection rules for Raman scattering differ from those associated with symmetric polarizability tenors because of the axial nature of the magnetic field and of the magnetization M.

In a magnetic system, the electric susceptibility, χ, is a functional of the magnetization as well as of the other variables describing internal modes of motion. Thus, we write

$$P = \chi(M) \cdot E_L . \tag{4.41}$$

Several microscopic mechanisms for the depedence of χ on M can be envisioned. Magnetostriction can conceivably be one but it is likely to yield extremely small scattering cross sections. Exchange interactions with itinerant or localized electrons, having energies comparable to electrostatic interactions, are expected to be important. This suggests that the Raman features associated with magnetic excitations should exhibit strong resonance enhancement when the energy of the quantum $\hbar\omega_L$ of the exciting radiation is near the energy of an electronic transition, e.g., the direct energy gap.

The modulation of P resulting from magnetic excitations is obtained from a Taylor series expansion of the functional $\chi(M)$. We write

$$P = P^{(0)} + P^{(1)} + P^{(2)} \dots , \tag{4.42}$$

where the successive terms are independent of M, linear in M, second order in M, ... The restrictions imposed by the Onsager reciprocity relations, the lack of absorption in the frequency region of interest and the cubic symmetry require that (see, for example, [4.46])

$$P^{(1)} = iGM \times E_L , \tag{4.43}$$

where G is a constant. The scattering cross-section for a Raman process is proportional to

$$|\hat{\varepsilon}_S \cdot \ddot{P}|^2$$

where, as before, $\hat{\varepsilon}_S$ is the direction of polarization of the scattered radiation. Thus, the scattering cross-section involving a magnetic excitation in first order is of the form

$$\sigma = C \, |(\hat{\varepsilon}_S \times \hat{\varepsilon}_L) \cdot M|^2 , \tag{4.44}$$

where C is an appropriate function of ω_L and ω_S. Thus, Raman scattering does not occur when the polarizations of the incident and scattered radiation are parallel. In the presence of a magnetic field H, M varies according to the Bloch equation

$$\frac{dM}{dt} = \gamma M \times H , \tag{4.45}$$

γ being the gyromagnetic ratio, i.e., the ratio of the magnetic-moment and the angular-momentum densities. For electrons $\gamma = -ge/2mc$ is negative. Considering only inelastic scattering, we need only keep the time-dependent components of M. The solutions of (4.45), taking H parallel to the z-axis, are such that

$$M_x \pm iM_y$$

vary as $\exp[\mp i\gamma Ht]$, respectively. Since M_z and M^2 are constants of the motions, we can write

$$M = (M \sin \Theta \cos \Omega t, M \sin \Theta \sin \Omega t, M \cos \Theta) , \tag{4.46}$$

where

$$\Omega = -\gamma H \tag{4.47}$$

is the Larmor frequency and Θ the angle between M and H.

We consider incident radiation propagating parallel to the z-axis selected along H. For circularly polarized radiation with positive and negative helicity denoted by $\hat{\sigma}_+$ and $\hat{\sigma}_-$, respectively, we write

$$E_L = (\hat{x} \pm i\hat{y}) E_0 \exp[-i\omega_L t]. \tag{4.48}$$

Using (4.43), we obtain

$$P^{(1)} = \mp \hat{z} G M E_0 \sin \Theta \exp[-i(\omega_L \mp \Omega) t]. \tag{4.49}$$

This shows that in this geometry there is a Stokes line with polarization $(\hat{\sigma}_+, \hat{z})$ and an anti-Stokes line with $(\hat{\sigma}_-, \hat{z})$.

In a similar way, if the incident wave propagates at right angles to H but is polarized along H,

$$E_L = E_0 \hat{z} \exp[-i\omega_L t] \tag{4.50}$$

and

$$P^{(1)} = \tfrac{1}{2} G M E_0 \sin \Theta \left\{ [(\hat{x} - i\hat{y}) \exp[-i(\omega_L - \Omega) t] \right.$$
$$\left. - (\hat{x} + i\hat{y}) \exp[-i(\omega + \Omega) t] \right\}. \tag{4.51}$$

Thus, Stokes and anti-Stokes lines occur in the geometries $(\hat{z}, \hat{\sigma}_-)$ and $(\hat{z}, \hat{\sigma}_+)$, respectively. The two cases described above are, of course, related to each other by time reversal symmetry.

In order to describe the magnetic excitations observed in Raman scattering in terms of microscopic models, it is useful to consider the Hamiltonian of the Mn^{2+} ions interacting with one another and with either band electrons or electrons bound to donors. We designate the spin of a Mn^{2+} ion at the site R_i by S_i and the spin of the electron by s. In the presence of a magnetic field H, the Zeeman energies of the Mn^{2+} ion and of the electron are $g\mu_B H \cdot S_i$ and $g^* \mu_B H \cdot s$ where μ_B is the Bohr magneton and g and g^* are the Lande g-factors of Mn^{2+} and the electron, respectively. In addition there are exchange interactions between Mn^{2+} ions at R_i and R_j of the form $-2J_{ij} S_i \cdot S_j$ and between an electron and the Mn^{2+} ions. The Hamiltonian of an electron in mutual interaction with Mn^{2+} ions is

$$H = -\alpha \sum_i S_i \cdot s \, |\psi(R_i)|^2 + g^* \mu_B H \cdot s + g\mu_B H \cdot \sum_i S_i - \sum_{i<j} 2J_{ij} S_i \cdot S_j. \tag{4.52}$$

Here $\psi(R_i)$ is the electronic wavefunction normalized over the primitive cell and evaluated at R_i; αN_0 is the s-d exchange integral, N_0 being the number of primitive cells per unit volume.

4.3.2 Paramagnetic Phase

We now consider Raman transitions between Zeeman sublevels of the individual Mn^{2+} ions in an external magnetic field, the sample being in its paramagnetic phase. In this phase the exchange interaction between Mn^{2+} ions is smaller than the thermal energy $k_B T$ and the ions can be considered as being independent of one another. The $^6S_{5/2}$ ground state of the Mn^{2+} ion has a total spin S = 5/2, orbital angular momentum L = 0 and total angular momentum J = 5/2. In this subsection we will discuss $Cd_{1-x}Mn_x Te$ as an illustrative example. The cubic crystalline field (site symmetry T_d) splits the sixfold degenerate state into a Γ_8 quadruplet state at a, and a Γ_7 doublet at $-2a$, where $3a$ is the crystal field splitting. From electron paramagnetic resonance (EPR) experiments, *Lambe* and *Kikuchi* [4.47] obtained $3a = 0.0084$ cm^{-1} for Mn^{2+} in CdTe. This crystal field splitting is too small to be observed with the resolution of a standard Raman spectrometer and we treat the ground state of Mn^{2+} in $Cd_{1-x}Mn_x Te$ as an atomic $^6S_{5/2}$ level. The application of an external magnetic field, H, results in the removal of the sixfold degeneracy of the ground state, the energy levels being $E(m_S) = g\mu_B H M_S$. Here m_S, the projection of S along H, has the values $-5/2, -3/2, \ldots, +5/2$. These energy levels form the Zeeman multiplet of the ground state of Mn^{2+}.

In the paramagnetic phase, Raman scattering associated with spin-flip transitions between adjacent sublevels of this multiplet has been observed by *Petrou* et al. [4.48]. The results in $Cd_{1-x}Mn_x Te$ are shown in Fig. 4.9 for x = 0.40. As can be seen, a strong Stokes/anti-Stokes pair is observed with a Raman shift of $\omega_{PM} = 5.62 \pm 0.02$ cm^{-1} at room temperature and H = 60 kG. Taking H and incident light parallel to $\hat{z}$, the Stokes line is observed in the $(\hat{\sigma}_+, \hat{z})$ configuration, whereas the anti-Stokes line is seen in $(\hat{\sigma}_-, \hat{z})$. When the incident light propagates at right angles to $H(\hat{z})$, the Stokes component appears in the polarization $(\hat{z}, \hat{\sigma}_-)$, while the anti-Stokes is observed in $(\hat{z}, \hat{\sigma}_+)$. Within experimental error the frequency shift is linear in H. With the energy separation between adjacent sublevels of the Zeeman multiplet given by $\Delta E = g\mu_B H = \hbar\omega_{PM}$, it is found that $g = 2.01 \pm 0.02$. The Raman line at ω_{PM} in $Cd_{1-x}Mn_x Te$ has been observed for a variety of compositions ranging from x = 0.02 to x = 0.70.

Following the arguments given by *Fleury* and *Loudon* [4.49] one considers, as a possible mechanism for the ω_{PM} Raman line, a two-step process having as the intermediate state one of the excited states of the Mn^{2+} ion (L = 1, S = 5/2). Figure 4.10 shows such mechanisms for the Stokes and anti-Stokes components of the ω_{PM} line. For the Stokes component, an incident photon of energy $\hbar\omega_L$ and polarization $\hat{\sigma}_+$ induces a virtual electric dipole transition between an initial and an intermediate state which differ by $\Delta m_J = +1$; it is followed by a second electric dipole transition between the intermediate and final states with $\Delta m_J = 0$. This

Fig. 4.9. Stokes (S) and anti-Stokes (AS) Raman lines at ω_{PM} resulting from $\Delta m_S = \pm 1$ spin-flip transitions within the Zeeman multiplet of Mn^{2+} in $Cd_{0.6}Mn_{0.4}Te$: $-\hat{\sigma}_+, \hat{z}$; $-\cdot-$ $\hat{\sigma}_-, \hat{z}$. The wavelength of the exciting laser line $\lambda_L = 6764$ Å; the applied magnetic field $H = 60$ kG; the temperature $T = 300$ K; x, y, and z are along [001], [1$\bar{1}$1], and [110], respectively [4.48]

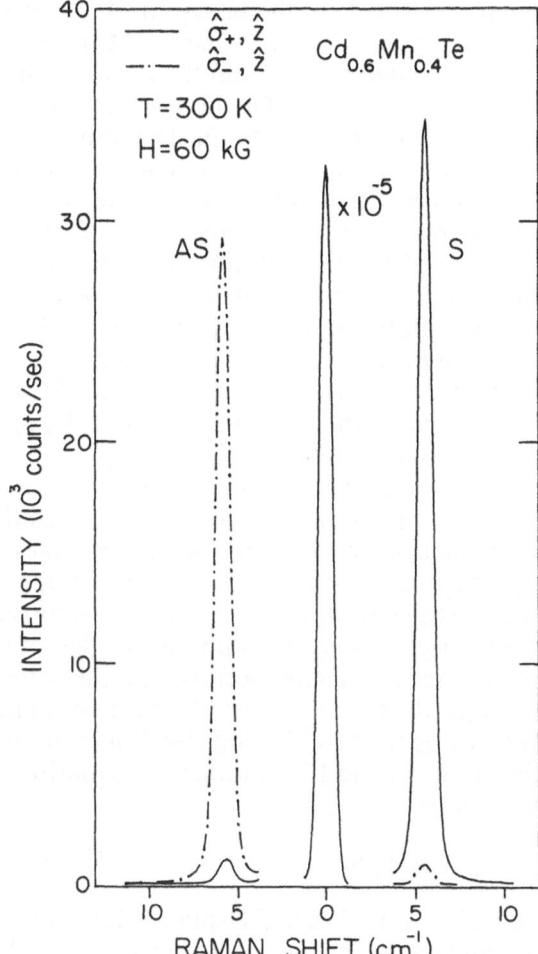

Fig. 4.10. Raman mechanism for the ω_{PM} line involving the internal transitions of the Mn^{2+} ion. The arrows indicate virtual electric-dipole transitions. The energy level scheme is not to scale and the energy difference between the excited and ground states E_0 is much greater than $\hbar\omega_{PM}$ [4.48]

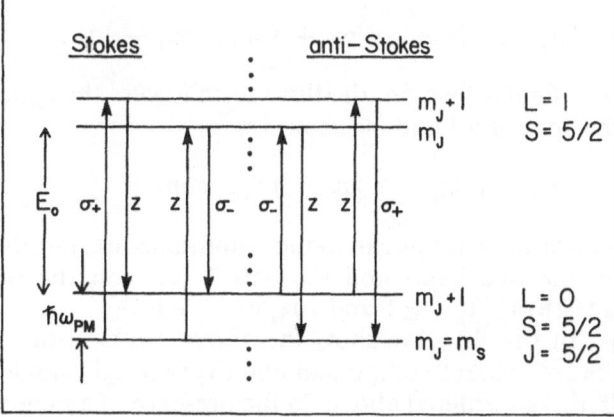

is accompanied by the emission of a scattered photon of polarization $\hat{z}$ and energy $\hbar\omega_S = \hbar\omega_L - \hbar\omega_{PM}$. At the end of this process the Mn^{2+} ion is in an excited state within the Zeeman multiplet differing from the initial state by $\Delta m_S = +1$. The appearance of the Stokes component in the $(\hat{z}, \hat{\sigma}_-)$ configuration is also illustrated in Fig. 4.10. Similar processes can be visualized for the anti-Stokes component having the $(\hat{\sigma}_-, \hat{z})$ or the $(\hat{z}, \hat{\sigma}_+)$ polarization. This mechanism correctly predicts the experimentally observed polarization characteristics of the Stokes and anti-Stokes components of the ω_{PM} Raman line. We note that the selection rules are immediate consequences of conservation of angular momentum of the system comprised of the Mn^{2+} ions and the photon field. It should also be pointed out that the Stokes scattering process in the $(\hat{\sigma}_+, \hat{z})$ configuration is the time reversed conjugate of the $(\hat{z}, \hat{\sigma}_+)$ anti-Stokes process; in the same manner, the $(\hat{z}, \hat{\sigma}_-)$ Stokes and the $(\hat{\sigma}_-, \hat{z})$ anti-Stokes processes are related by time reversal [4.50, 4.51].

Exploiting the variation of the band gap with manganese concentration and/or temperature, it is possible to match the band gap of several samples with the energy of one of the discrete lines of a Kr^+ laser. In addition, a dye laser can also be used to achieve resonant conditions. It is found that the intensity of the ω_{PM} line increases by several orders of magnitude as the laser photon energy approaches that of the band gap. The observation of this resonant enhancement in the intensity of the ω_{PM} Raman line prompts the consideration of a mechanism involving interband transitions. It involves the Mn^{2+}-band electron exchange interaction described by the first term in the Hamiltonian in equation (4.52). The term $S_i \cdot s$ can be written as

$$S_i \cdot s = S_i^{(z)} s^{(z)} + \tfrac{1}{2} S_i^{(+)} s^{(-)} + \tfrac{1}{2} S_i^{(-)} s^{(+)} ; \qquad (4.53)$$

here $S_i^{(\pm)}$ and $s^{(\pm)}$ are the spin raising and lowering operators for a Mn^{2+} ion and band electron, respectively, and $S_i^{(z)}$ and $s^{(z)}$ are the corresponding projections of spin along $\hat{z}$. The second term of (4.53) raises the spin of a Mn^{2+} ion while simultaneously lowering the spin of a band electron, i.e.,

$$|m_S\rangle_{Mn^{2+}} |m_J\rangle_e \rightarrow |m_S + 1\rangle_{Mn^{2+}} |m_J - 1\rangle_e . \qquad (4.54)$$

In a similar fashion, the third term lowers the spin of an ion while raising the spin of a band electron, i.e.,

$$|m_S\rangle_{Mn^{2+}} |m_J\rangle_e \rightarrow |m_S - 1\rangle_{Mn^{2+}} |m_J + 1\rangle_e . \qquad (4.55)$$

Hence, these terms can induce simultaneous spin-flips of the band electrons on the one hand and the Mn^{2+} ions on the other, corresponding to $\Delta M_S(Mn^{2+}) = \pm 1$ and $\Delta m_J(e) = \mp 1$.

In Fig. 4.11, we show the above mechanism for both the Stokes and the anti-Stokes component and the two right-angle geometries $(\hat{\sigma}_\pm, \hat{z})$ and $(\hat{z}, \hat{\sigma}_\mp)$ considered above. In the presence of a magnetic field the Γ_8 valence

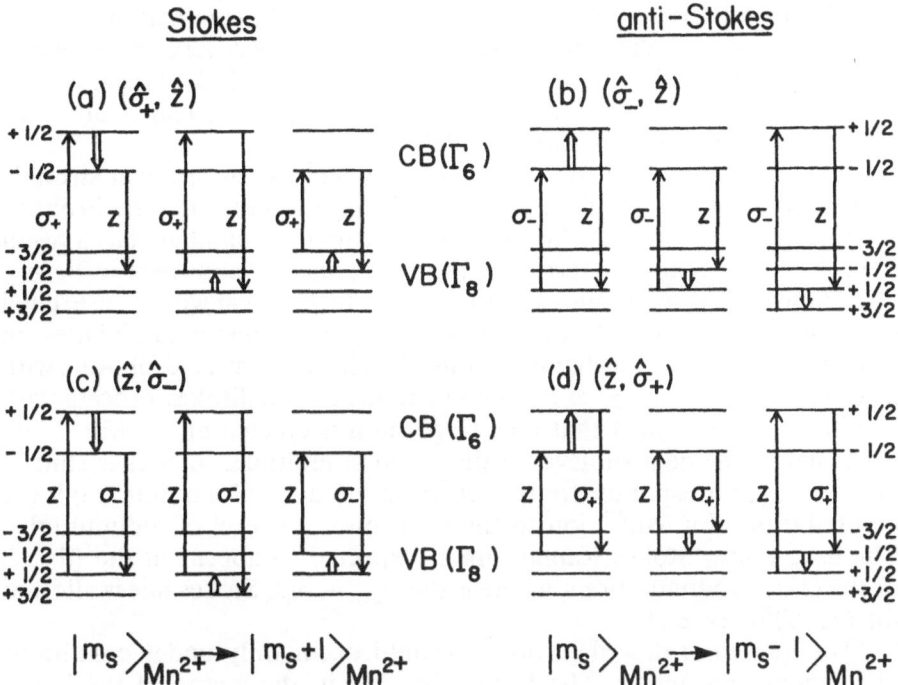

Fig. 4.11. Raman mechanism for the ω_{PM} line involving the band electrons: CB and VB refer to the conduction and valence bands, respectively, which are labeled by the electronic quantum number m_J. The single arrows indicate virtual electric-dipole transitions, while the double arrows refer to transitions induced by the electron-Mn^{2+} exchange interaction. The allowed polarizations for the Stokes component are $(\hat{\sigma}_+, \hat{z})$ (a) and $(\hat{z}, \hat{\sigma}_-)$ (c), while those for the anti-Stokes are $(\hat{\sigma}_-, \hat{z})$ (b) and $(\hat{z}, \hat{\sigma}_+)$ (d) [4.48]

band splits into four subbands with $m_J = -3/2, -1/2, +1/2$, and $+3/2$, and the Γ_6 conduction band splits into $m_J = +1/2$ and $-1/2$ subbands. The possible processes for the Stokes component appearing in the $(\hat{\sigma}_+, \hat{z})$ configuration are shown in Fig. 4.11a. In the first process an incident photon of polarization $\hat{\sigma}_+$ is absorbed, raising an electron to the conduction band with $\Delta m_J = +1$ and creating a hole in the valence band. In the second step the excited electron interacts with a Mn^{2+} ion via the second term of (4.52), resulting in $\Delta m_S(Mn^{2+}) = +1$ and $\Delta m_J(e) = -1$. Finally, the electron and hole recombine emitting a photon of energy $\hbar\omega_S = \hbar\omega_L - \hbar\omega_{PM}$ of polarization $\hat{z}$; the band electrons have thus returned to their ground state, but leaving the Mn^{2+} ion excited to the next sublevel of the Zeeman multiplet. In the other two processes shown in Fig. 4.11a the hole, rather than the excited electron, interacts with the Mn^{2+} ion, resulting however, in identical polarization selection rules. In the same manner the Stokes component in the $(\hat{z}, \hat{\sigma}_-)$ configuration follows from Fig. 4.11c. The anti-Stokes processes for the $(\hat{\sigma}_-, \hat{z})$ and the $(\hat{z}, \hat{\sigma}_+)$ geometries are shown in Figs. 4.11b and d, respectively.

All the observations discussed above and the predictions of the microscopic models considered are in accord with the general phenomenological selection rules embodied in equations (4.48–4.51). We also note that this entire phenomenon is electron paramagnetic resonance observed as Raman shifts, i.e., it is *Raman-EPR*.

It is known that electrons and holes in polar crystals interact strongly with zone center longitudinal optical (LO) phonons through the Fröhlich interaction [4.18]. An LO phonon can be created or annihilated as a result of such an interaction. Referring to the mechanism responsible for the ω_{PM} line, shown in Fig. 4.11, one can visualize a fourth step in which the excited electron or hole interacts with the lattice and creates or annihilates an LO phonon. Such a mechanism would result in a scattered photon with a Raman shift of $\omega_{LO} \pm \omega_{PM}$. The net result for the Stokes process with a shift of $\omega_{LO} + \omega_{PM}$ is that an LO phonon is created and a Mn^{2+} ion is excited to the next sublevel of the Zeeman multiplet. A Stokes shift of $\omega_{LO} - \omega_{PM}$ corresponds to the creation of an LO phonon and the de-excitation of a Mn^{2+} ion to the next lower sublevel of the multiplet. The $\omega_{LO} + \omega_{PM}$ Stokes Raman line is expected to appear in the $(\hat{\sigma}_+, \hat{z})$ or the $(\hat{z}, \hat{\sigma}_-)$ configurations, whereas the $\omega_{LO} - \omega_{PM}$ Stokes line is allowed for $(\hat{\sigma}_-, \hat{z})$ or $(\hat{z}, \hat{\sigma}_+)$.

The new lines described above should occur only under conditions of band-gap resonance. The Raman spectra in the region of the longitudinal and transverse optical (TO) vibrational modes are shown in Fig. 4.12 for $Cd_{1-x}Mn_xTe$ with $x = 0.10$. The 7525 Å Kr^+ laser line was used to excite the spectra. The zero magnetic field LO and TO phonon spectrum is shown in Fig. 4.12b. Here the "CdTe-like" TO and LO and the "MnTe-like" LO modes are quite distinct, while the "MnTe-like" TO appears as a shoulder to the LO. The corresponding Raman spectra, recorded in the presence of a magnetic field of 60 kG and in the $(\hat{z}, \hat{\sigma}_+)$ and $(\hat{z}, \hat{\sigma}_-)$ configurations, are presented in Figs. 4.12a and c, respectively. The additional Raman lines with Stokes shifts $\omega_{LO} \pm \omega_{PM}$ are clearly present with the proper polarization characteristics. The sample temperature was 120 K; at this temperature, the sublevel occupation probability ratio of adjacent levels in the Zeeman multiplet of Mn^{2+} is $\exp(-\hbar\omega_{PM}/k_BT) = 0.94$. Thus, the intensities of the $\omega_{LO} + \omega_{PM}$ and the $\omega_{LO} - \omega_{PM}$ lines are expected to be approximately equal. The $\omega_{LO} \pm \omega_{PM}$ lines can be observed only when the exciting photon energy is strongly resonant with the band gap. There is no evidence of corresponding Raman lines associated with the TO phonons, which would have shifts of $\omega_{TO} \pm \omega_{PM}$. This supports the assumption that the Fröhlich interaction is responsible for the appearance of the new features.

The scattering amplitude for the Raman mechanism discussed above is proportional to the magnitude of the exchange coupling between the ions and the band electrons which is especially strong in these alloys. Adapting Loudon's theory for optical phonons, the scattering cross section

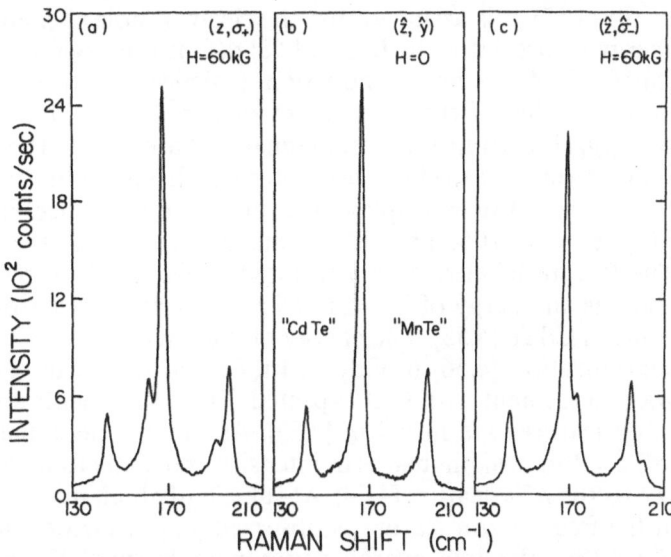

Fig. 4.12. Raman spectra of $Cd_{1-x}Mn_xTe$, $x = 0.10$, showing the combination lines ω_{LO_1} $\pm \omega_{PM}$ and $\omega_{LO_2} \pm \omega_{PM}$, where LO_1 and LO_2 are the "CdTe-like" and "MnTe-like" LO phonons. The sample temperature $T = 120$ K, $H = 60$ kG, and $\lambda_L = 7525$ Å; $\hat{x}$, $\hat{y}$ and $\hat{z}$ are along [110], [001], and [1$\bar{1}$0], respectively [4.48]

can be written as being proportional to

$$\left| \frac{\langle n_S + 1, 0| H_{eR} |n_S, b\rangle \langle m_S + 1, b| H_{ex} |m_S, a\rangle \langle n_L - 1, a| H_{eR} |n_L, 0\rangle}{(\omega_b + \omega_{PM} - \omega_L)(\omega_a - \omega_L)} \right.$$
$$\left. + \ 5 \text{ additional terms} \right|^2 . \tag{4.56}$$

Here H_{ex} is the exchange Hamiltonian describing the exchange interaction between $3d$-localized states of Mn^{2+} and conduction (valence) electrons, the other terms having the same significance as in equation (4.5b).

Once again, the first term will dominate when $\omega_a \sim \omega_L$ or $\omega_b \sim (\omega_L - \omega_{PM}) = \omega_S$, producing a double peak in the frequency dependence of the cross section, corresponding to "in resonance" and "out resonance", respectively.

In (4.56) $\hbar\omega_a$ and $\hbar\omega_b$ are the energies of excitation of two states of the exciton with angular momenta differing by one unit. For example, following *Twardowski* et al. [4.52], for a $(\hat{z}, \hat{\sigma}_+)$ Stokes line

$$\hbar\omega_a = E_x - 3A + B, \qquad \hbar\omega_b = E_x - 3A + 3B, \tag{4.57}$$

where E_x is the exciton energy and

$$A = \tfrac{1}{6} x N_0 \alpha \langle S_z^{Mn} \rangle, \qquad B = \tfrac{1}{6} x N_0 \beta \langle S_z^{Mn} \rangle. \tag{4.58}$$

In (4.57), it is assumed that the electron-hole exchange and correlation energy is insensitive to H. In (4.58) $\alpha(\beta)$ is the exchange integral of the $3d$ states of Mn^{2+} and conduction (valence) electrons and $\langle S_z^{Mn} \rangle$ is the average value of the component of Mn^{2+} spin along H given by $(5/2)$ $B_{5/2}(g\mu_B H/k_B T)$ in the paramagnetic phase. Thus, the energies $\hbar\omega_{a,b}$ are characterized by an effective g factor which is a nonlinear function of (H/T).

Figure 4.13a, b displays the photoluminescence spectrum of Cd_{1-x} $Mn_x Te$, $x = 0.05$, at 5 K in the absence of a magnetic field [4.53]. The feature labeled X is attributed to free exciton recombination [4.54] and has an energy of $E_x = 1.665$ eV, and an increase of 70 meV from the value in CdTe [4.55]. The energy of this exciton varies linearly with Mn^{2+} concentration [4.56] in $Cd_{1-x}Mn_x Te$ and the value measured here is in good agreement with that expected. The assignments and the labels of the other features are noted in [4.53]. As the magnetic field is increased the free exciton peak increases in intensity and shifts to lower energy sweeping across the $A°X$ feature [4.54]. At sufficiently high magnetic fields a splitting in the free exciton feature is observed [4.57]. Two of these are present in Fig. 4.13c; the low energy component labeled X_+ appears in the $\hat{\sigma}_+$ polarization, while the other, X_z, is polarized along $\hat{z}$. The feature

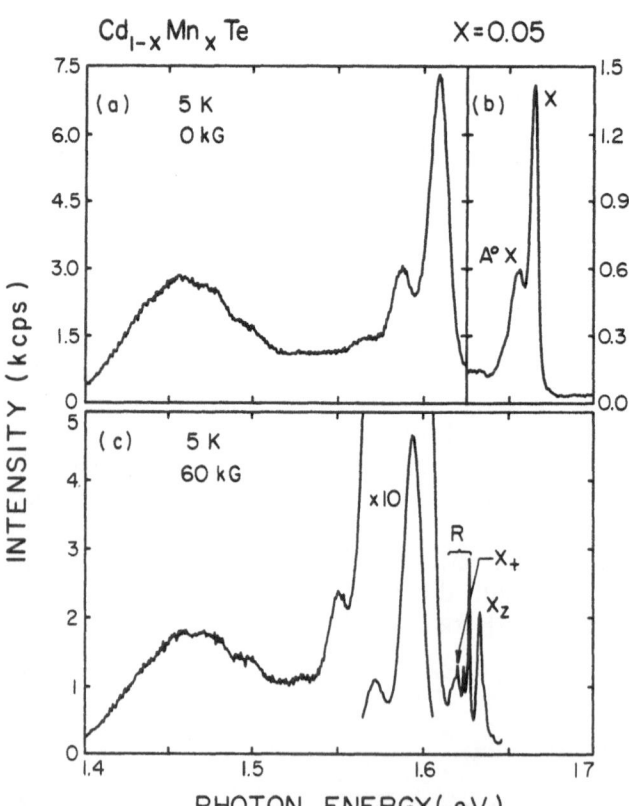

Fig. 4.13. The photoluminescence spectra of $Cd_{0.95}Mn_{0.05}Te$ at $T = 5$ K with $H = 0$ for (a) and (b), $H = 60$ kG for (c), $\lambda_L = 7525$ Å for (a) and (c), $\lambda_L = 6764$ Å for (b), and the laser power $P_L = 25$ mW for all cases. Raman features are denoted by 'R'. In (c) the features between 1.55 and 1.6 eV are also displayed on a scale reduced by 10 [4.53]

associated with the free electron to acceptor transition and its LO phonon replicas shift by 14 meV in a magnetic field of 60 kG.

On the basis of the luminescence spectra, appropriate choices of laser wavelength, sample temperature and magnetic field can be made to achieve conditions of "in resonance" or "out resonance" which can selectively enhance specific features in the Raman spectrum. In the experiments discussed here, the $\omega_{LO} - \omega_{PM}$ line, i.e., the Raman shift with the creation of an LO phonon and the de-excitation of the Mn^{2+} by $\Delta m_S = -1$ involves a virtual transition at an energy $\hbar\omega_a$ for the incident radiation polarized along $\hat{z}$ and another at $\hbar\omega_b$ for scattered radiation having $\hat{\sigma}_+$ polarization. With the 7525 Å ($\hbar\omega_L = 1.648$ eV) laser line, the "in resonance" condition is nearly fulfilled; with A and B in equation (4.57) for $x = 0.05$, $T = 5$ K, and $H = 60$ kG, one can estimate the energies of X_z and X_+, the components which move to lower energies with increasing H, to be 1.636 eV and 1.603 eV, respectively (here $N_0\alpha = 220$ meV and $N_0\beta = -880$ meV given by *Gaj* et al. [4.58]). Thus, it is clear that the "in resonance" condition is satisfied even more closely whereas the "out resonance" for $\omega_{LO} - \omega_{PM}$ can now be realized. The results in Fig. 4.14 show the resonantly enhanced lines at ω_{PM}, $2\omega_{PM}$, $3\omega_{PM}$ and $4\omega_{PM}$. We discuss some of the underlying physical considerations later. Here we

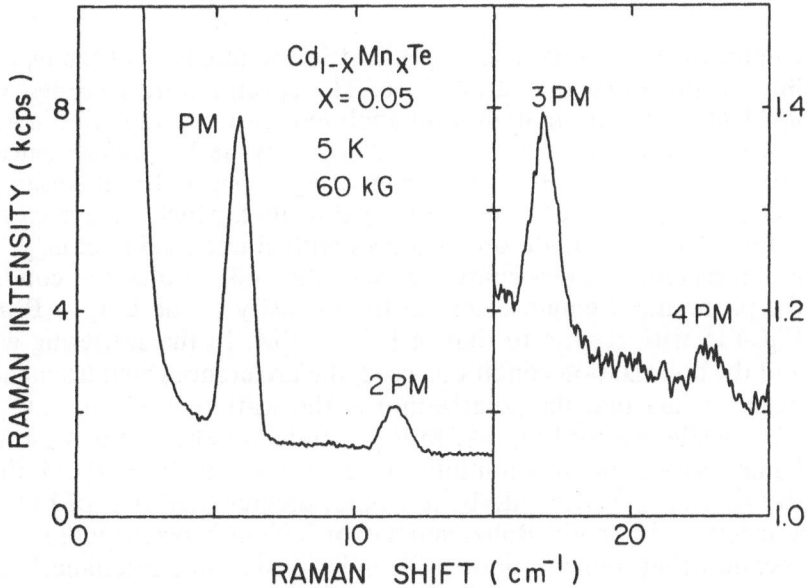

Fig. 4.14. The stokes Raman lines at ω_{PM} ($n = 1 \ldots 4$) from $\Delta m_S = +n$ spin-flip transitions in the Zeeman multiplet of the ground state of the $3d$ shell of Mn^{2+} in $Cd_{0.95}Mn_{0.05}Te$. Exciting wavelength $\lambda_L = 7525$ Å, the laser power $P_L = 90$ mW; applied magnetic field $H = 60$ kG and temperature $T = 5$ K. The spectrum for shifts greather than 15 cm^{-1} is the average of ten scans [4.53]

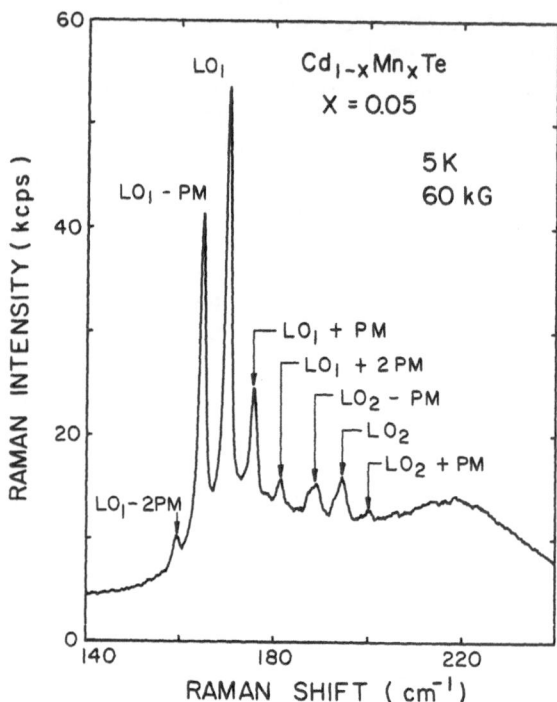

Fig. 4.15. The Raman spectrum of $Cd_{0.95}Mn_{0.05}Te$ in the region of the LO phonons with $T = 5\,K$, $H = 60\,kG$, $\lambda_L = 7525\,\text{Å}$, and $P_L = 30\,mW$. Incident light polarized along $z \parallel H \parallel \hat{k}_S$, and scattered light unanalyzed [4.53]

emphasize the dramatic enhancement in the intensity of the $\omega_{LO_1} - \omega_{PM}$ line as illustrated in Figs. 4.15, 16. The spectra were recorded with the incident polarization along $\hat{z}$, unanalyzed scattered radiation and $\hat{k}_s \parallel H$. The relatively broad feature at $\sim 220\,cm^{-1}$ is the X_+ luminescence feature attributed to the exciton component at $E_x - 3A + 3B$. In addition to the LO_1, $LO_1 \pm PM$, LO_2 and $LO_2 \pm PM$ lines, which have been reported before [4.48], two additional features with Raman shifts of $\omega_{LO_1} \pm 2\omega_{PM}$ are observed. A direct consequence of the "out resonance" conditions is the pronounced enhancement in the intensity of the $LO_1 - PM$ line in Fig. 4.15 with respect to that of $LO_1 + PM$. In the scattering geometry and the polarization conditions used, the preferential enhancement results from the fact that the polarization of the scattered light for $LO_1 + PM$ is $\hat{\sigma}_-$, while that for $LO_1 - PM$ is $\hat{\sigma}_+$, matching that of the X_+ transition. Under non-resonant conditions, for $T = 5\,K$ and $H = 60\,kG$, the intensity of $LO_1 + PM$ would be five times greater than that of $LO_1 - PM$, as calculated from the Boltzmann factor. This enhancement of $LO_1 - PM$ becomes even more pronounced under exact "out resonance" achieved by decreasing the magnetic field to 35 kG and moving the X_+ luminescence feature under the LO_1 and the $LO_1 - PM$ Raman lines; this is illustrated in Fig. 4.16 where the $LO_1 - PM$ Raman line is more intense than the LO_1 line. The $LO_2 - PM$, LO_2 and $LO_2 + PM$ Raman lines show similar effects as can be seen from a comparison of Figs. 4.15 and 4.16. In Fig. 4.17,

Fig. 4.16. The Raman spectrum of $Cd_{0.95}Mn_{0.05}Te$ in the region of the LO phonons with $H = 35\,kG$ and other conditions as in Fig. 4.15 [4.53]

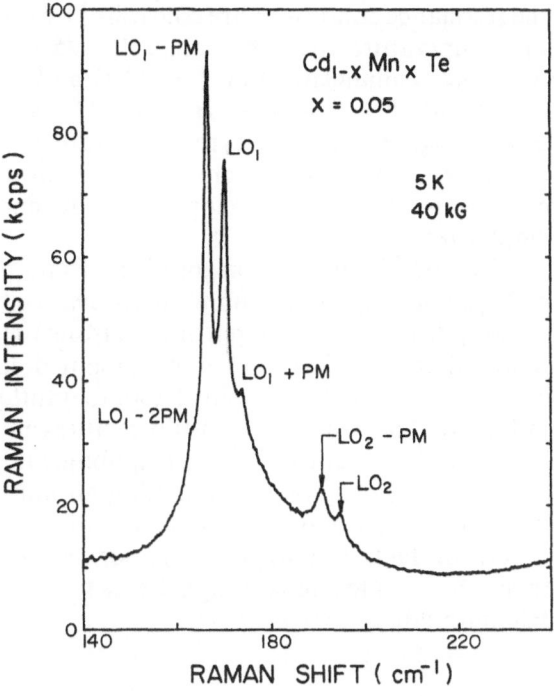

Fig. 4.17. Temperature variation of the resonance Raman scattering in $Cd_{0.95}Mn_{0.05}Te$ in the region of the LO phonons with $H = 60\,kG$, and other conditions as in Fig. 4.15 [4.53]

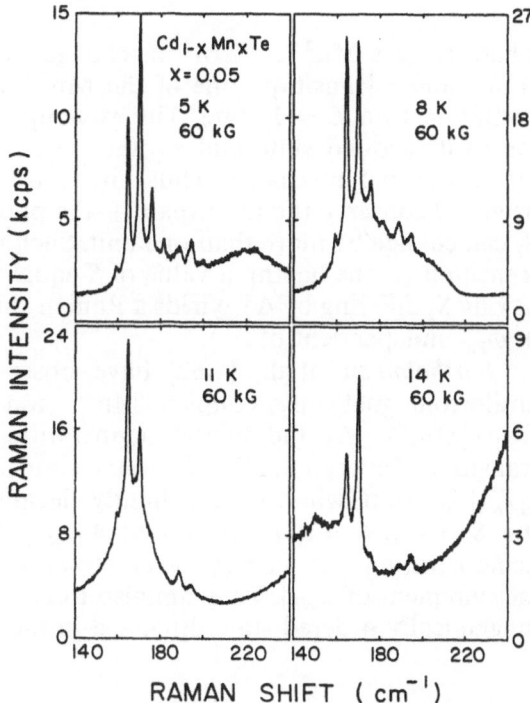

the resonance conditions are controlled by keeping $H = 60$ kG but varying the temperature over the range 4.5–25 K. As can be clearly seen, the resonance enhancement at $T = 11$ K and $H = 60$ kG is almost identical to that at $T = 5$ K and $H = 35$ kG. We note that the Raman shift $\omega_{LO} \pm \omega_{PM}$ is insensitive to temperature variation while the position of the Zeeman component of the exciton and, hence, the resonance conditions are strongly temperature dependent for the reasons already emphasized.

We now discuss the origin of the Raman lines in Fig. 4.14 with shifts of $3\omega_{PM}$ and $4\omega_{PM}$, corresponding to $\Delta m_S = 3$ and 4, respectively. Multiple spin-flip Raman scattering from electrons bound to donors [4.59] has been reported in the literature and explained invoking either an exchange coupling among donor spins [4.60] or mutliple scattering (*Wolff* quoted in [4.61]). The multiple spin-flip features in DMS can be accounted for in terms of excitations within neighbouring pairs of Mn^{2+} ions coupled antiferromagnetically and assuming an anisotropic exchange interaction between the ground state multiplet of one and an excited state of the other. In fact, a pair of neighboring Mn^{2+} ions, say 1 and 2, in their $^6S_{5/2}$ states in a magnetic field $\hat{H} \parallel \hat{z}$ can be described by the Hamiltonian

$$H = g\mu_B H S_z - J_{NN}(S^2 - \tfrac{35}{2}), \tag{4.59}$$

where J_{NN} is a $Mn^{2+} - Mn^{2+}$ exchange interaction and $S = S^{(1)} + S^{(2)}$. In a Raman transition, one of the Mn^{2+} ions can experience a virtual excitation to a $L = 1$ state. The exchange interaction between a Mn^{2+} ion in its ground state and another in an excited state will, in general, lack rotational invariance. Thus, virtual transitions to intermediate states need not conserve the total (pair + the photon) angular momentum and S_z can change by more than one unit. Such transitions followed by a final transition to one having a value of S equal to that of the initial state but having S_z differing by ΔS_z yields a Raman shift of $g\mu_B H \Delta S_z$, i.e., a multiple of ω_{PM}, independent of J.

Bartholomew et al., [4.62] have observed Raman scattering from antiferromagnetically coupled Mn^{2+} ion pairs in $Cd_{1-x}Mn_xS$ and $Cd_{1-x}Mn_xSe$. At the lowest temperature, they observed the Raman transitions between the $S = 0$ ground state to the $S = 1$ excited state lying $2|J_{NN}|$ above it, whereas at a slightly elevated temperature they observed the $S = 1 \rightarrow S = 2$ transition at $4|J_{NN}|$. Such measurements, besides demonstrating the energy level structure of pairs, provide a direct measurement of $|J_{NN}|$. They are also to be viewed as the precursor to the magnetically ordered state discussed in the next section.

4.4 Magnetically Ordered Phase

As mentioned earlier in this section, $Cd_{1-x}Mn_xTe$ exhibits a magnetically ordered low temperature phase for $x > 0.17$. The transition from the paramagnetic to the magnetically ordered phase is accompanied by the appearance of a new Raman feature at low temperatures as shown in Fig. 4.18 [4.8]. Since this excitation is associated with magnetic order, it is attributed to a magnon. A distinct magnon feature is observed in $Cd_{1-x}Mn_xTe$ for the composition range $0.40 \leq x \leq 0.70$. The magnon feature is absent when the incident and the scattered polarizations are parallel and appears when they are crossed in agreement with (4.44) and shown in Fig. 4.19. This was found to be the case for several crystallographic orientations as well as for polycrystalline samples. Such a behavior, irrespective of the crystallographic orientation, is exhibited only by an excitation whose Raman tensor is *antisymmetric*. As the temperature is increased, the Raman shift of the magnon, ω_M, decreases, and above a characteristic Néel temperature $T_N(x)$ the feature is no longer observable. For $x = 0.70$, the temperature dependence of ω_M follows a Brillouin function (see Fig. 4.20).

The coordinates decribing the magnons, the elementary excitations of a system of interacting magnetic dipole moments can be regarded as the Fourier components of the magnetization $M(r, t)$, i.e., the coefficients M_q in the expansion

$$M(r, t) = \sum_q M_q \exp(iqr - i\omega_q t) . \tag{4.60}$$

Fig. 4.18. Raman spectra of $Cd_{0.3}Mn_{0.7}Te$ in the paramagnetic and antiferromagnetic phase; $\lambda_L = 6764$ Å. The phase transition occurs at $T_N \sim 40$ K. The scattering geometry corresponds to VH + VV. Here VH and VV denote incident light vertically polarized (V) and scattered light analyzed horizontally (H) and vertically (V), respectively, the scattering plane being horizontal. (a) $T = 295$ K and (b) $T = 5$ K. M denotes the peak due to magnon scattering [4.8]

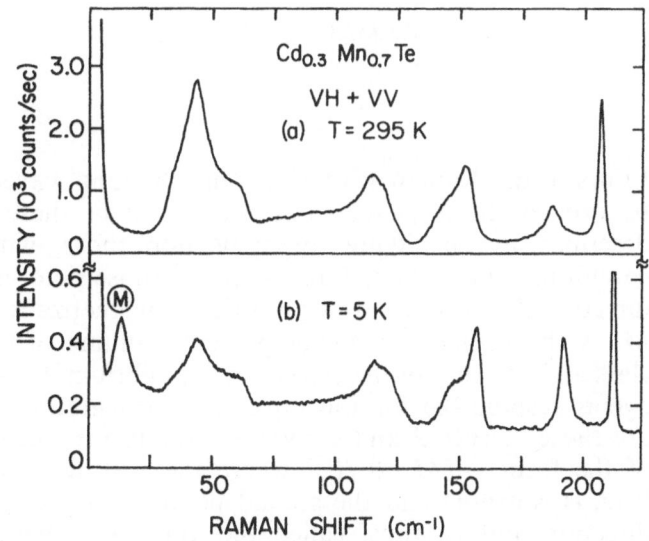

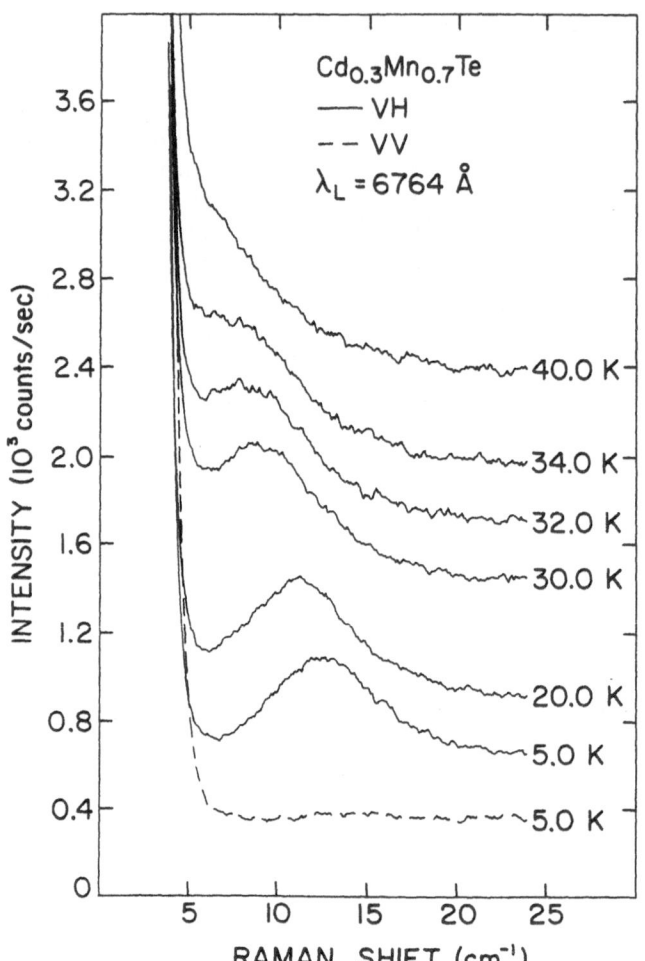

Fig. 4.19. Temperature dependence of the low-frequency Raman spectrum of $Cd_{0.3}Mn_{0.7}Te$. The intensity scale refers to the VV spectrum at 5 K and successive spectra have been displaced vertically upwards for clarity. The sample temperature is indicated for each trace. The broken curve shows the VV spectrum. All the other spectra correspond to VH polarization. $\lambda_L = 6764$ Å, the exciting laser wavelength [4.8]

In first order Raman scattering, only the long wavelength magnons can be excited. In an antiferromagnetic system these excitations can be described by classifying the spins into those, which in the state of equilibrium, point in one direction and those pointing in an anti-parallel direction. This classification gives rise to magnetizations M_1 and M_2 where M_1 is the magnetic moment per unit volume of the spins of the first class and M_2 that of the second. In equilibrium $M_1 + M_2 = 0$. Now the agents responsible for the preferential orientation of a spin of type 1 are those of type 2 and conversely, the former are in a molecular field of the form $-\lambda M_2 + H_A^{(1)}$ where the first term, called the exchange field, is isotropic and the second points in a preferred crystallographic direction and is, thus, called the anisotropy field. The equations of

Fig. 4.20. Temperature dependence of the magnon peak frequency in the magnetically ordered phases of $Cd_{1-x}Mn_xTe$. Triangles: $x = 0.7$, antiferromagnetic phase; circles: $x = 0.4$, spin-glass phase. The solid curve passing through the points for $x = 0.7$ is calculated for $T_N = 40$ K solving equation (4.67) for M_S as a function of $T \leqq T_N$ and assuming ω_M is proportional to M_S and equal to 12.5 cm^{-1} at $T = 0$ K [4.8]

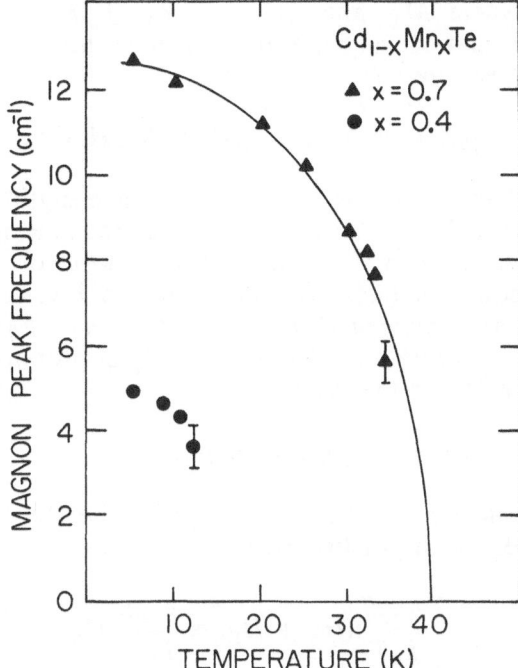

motion of M_1 and M_2 are

$$\frac{dM_1}{dt} = \gamma M_1 \times (H - \lambda M_2 + H_A^{(1)}) \tag{4.61}$$

and

$$\frac{dM_2}{dt} = \gamma M_2 \times (H - \lambda M_1 + H_A^{(2)}), \tag{4.62}$$

where H is an externally applied magnetic field. Supposing, for simplicity, that H and $H_A^{(1)} = -H_A^{(2)}$ are parallel to a direction which we take as the z-axis, taking the exchange fields $-\lambda M_{2,1}$ as approximately $\pm \lambda M_S = \pm H_E$ where M_S is the saturation value of M_1 we find

$$\frac{d}{dt} M_1^{\pm} = \mp i\gamma (H + H_A + H_E) M_1^{\pm} \mp i\gamma H_E M_2^{\pm}, \tag{4.63}$$

$$\frac{d}{dt} M_2^{\pm} = \mp i\gamma (H - H_A - H_E) M_2^{\pm} \pm i\gamma H_E M_1^{\pm}, \tag{4.64}$$

where $M_1^{\pm}$ and $M_2^{\pm}$ are $M_{1x} \pm iM_{1y}$ and $M_{2x} \pm iM_{2y}$, respectively. Equations (4.63) and (4.64) have non-trivial solutions of frequency $\omega_M^{\pm}$ given by

$$\omega_M^{\pm} = \pm\gamma H + |\gamma| (H_A^2 + 2H_AH_E)^{1/2} . \tag{4.65}$$

Thus, when $H = 0$, a long wavelength magnon of frequency $|\gamma| (H_A^2 + 2H_AH_E)^{1/2}$ occurs, manifests itself in Raman scattering and is the M line in Fig. 4.18. The polarization features of this line are those predicted by equation (4.44) and illustrated in Fig. 4.19. We have here the *Raman-antiferromagnetic resonance (Raman-AFMR)*.

We assume that H_A like H_E is proportional to the saturation values of M_1 or M_2 given by

$$M_S = (2xg\mu_BS/a^3) B_S(y) , \tag{4.66}$$

where μ_B is the Bohr magneton, a the lattice constant, and $B_S(y)$ the Brillouin function with

$$y = \frac{g\mu_BS}{k_BT} (H_A + H_E) \equiv \frac{g\mu_BS}{k_BT} M_S\varkappa . \tag{4.67}$$

The Néel temperature, T_N, is then given by

$$k_BT_N = (2x/3a^3) g^2\mu_B^2S(S + 1) \varkappa . \tag{4.68}$$

A numerical solution of equation (4.66) yields M_S as a function of temperature, T, and in turn, the variation of ω_M with T. The best fit to the data shown in Fig. 4.20 for $x = 0.7$ is given by $T_N = 40$ K.

The magnon feature in the presence of an external magnetic field of 60 kG is shown in Fig. 4.21a, c for $Cd_{1-x}Mn_xTe$, $x = 0.70$. The spectrum shown in Fig. 4.21a was recorded at $T = 5$ K with $\lambda_L = 5682$ Å in the $(\hat{\sigma}_-, \hat{z})$ polarization while the spectrum in Fig. 4.21c was observed in the $(\hat{\sigma}_+, \hat{z})$ configuration. In the following we discuss the Stokes components of these Raman features. The Raman shifts of the peaks of the features in Fig. 4.21 are $\omega_{M_-} = 8.5$ cm^{-1}, $\omega_M = 12$ cm^{-1}, and $\omega_{M_+} = 15.5$ cm^{-1}. It can be shown that a one magnon Raman line in an antiferromagnet should split into two components of equal intensity separated by $2g\mu_BH$ at $T = 0$ K, if H is along the anisotropy field, H_A. The observed spacing between M_+ and M_- of 7 cm^{-1} for 60 kG is significantly smaller than $2g\mu_BH = 11$ cm^{-1}. The polarization characteristics of ω_{M_+} and ω_{M_-} are those expected. These results are independent of the crystal orientations with respect to the applied field suggesting that H_A is small compared to the applied field. In [4.8], H_E for $x = 0.7$ was calculated from the observed transition temperature $T_N = 40$ K to be 208 kG, while H_A was determined to be ~ 36 kG as deduced from $\hbar\omega_M = g\mu_B(H_A^2 + 2H_AH_E)^{1/2}$, strictly

Fig. 4.21. Effect of the magnetic field on the magnon feature of $Cd_{1-x}Mn_xTe$, $x = 0.70$, at $T = 5$ K with $\lambda_L = 5682$ Å. x, y, and z are along [110], [$\bar{1}$10], and [001], respectively; (a) ($\hat{\sigma}_-$, $\hat{z}$), $H = 60$ kG; (b) ($\hat{\sigma}_+$, $\hat{z}$), $H = 0$; (c) ($\hat{\sigma}_+$, $\hat{z}$) $H = 60$ kG. Owing to imperfect polarization results, leakage of the fairly strong feature M_+ appears as a small shoulder in the ($\hat{\sigma}_-$, $\hat{z}$) configuration [4.48]

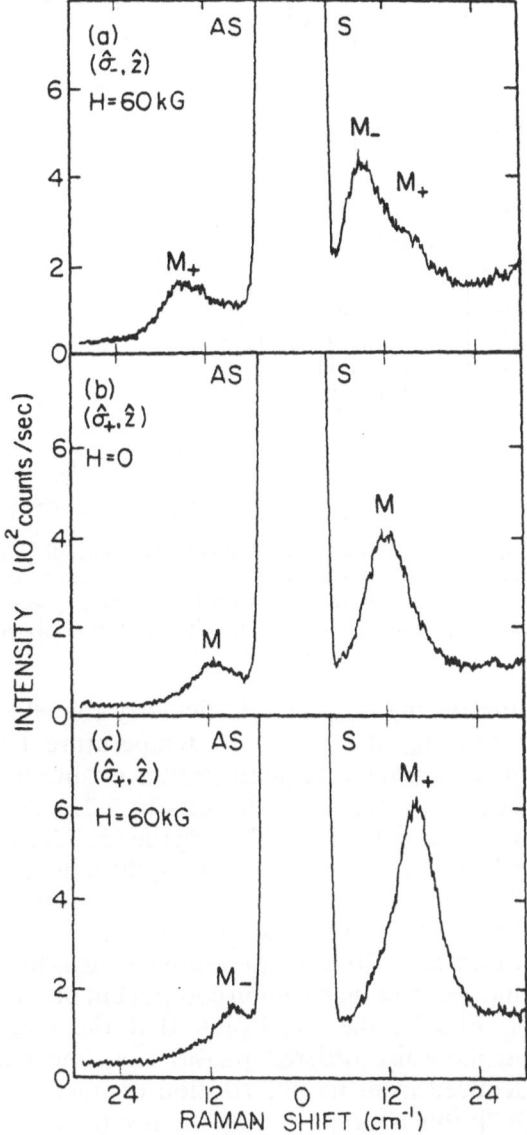

applicable to an antiferromagnet exhibiting long range order. While the value for H_E is reasonable, that for H_A must be viewed as too large in the context of the experimental results [4.48].

We have described the Raman features associated with magnetic excitations appearing in the paramagnetic state as well as in the magnetically ordered phase. The exchange interaction between Mn^{2+} ions is negligible compared to k_BT at high temperatures becoming more important as the temperature is lowered. It is of interest to investigate the effects of tempera-

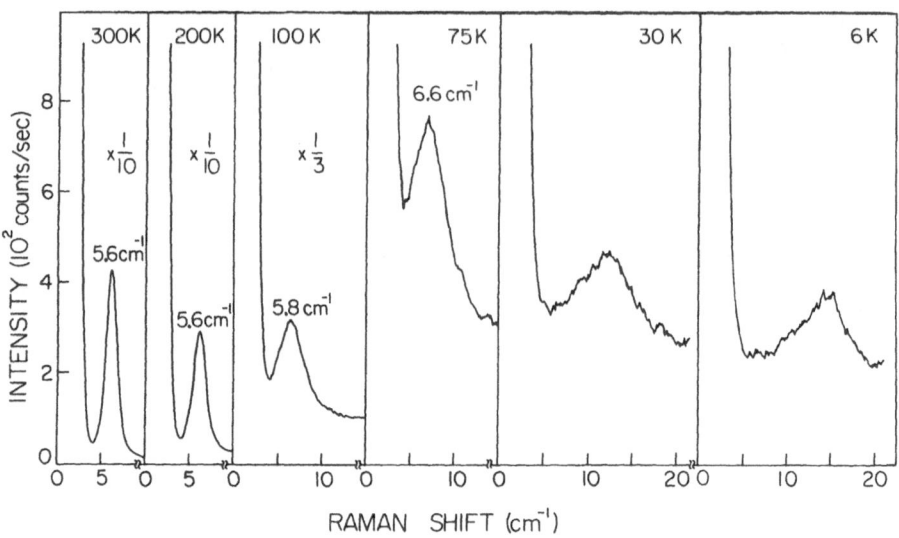

Fig. 4.22. Evolution of the Raman line at ω_{PM} of $Cd_{1-x}Mn_xTe$, $x = 0.70$, into the magnon feature as the temperature is lowered from room temperature to below the Néel temperature. The spectra were recorded with $H = 60$ kG, $\lambda_L = 6764$ Å in the $(\hat{\sigma}_+, \hat{z})$ polarization; x, y, and z are along [110], [$\bar{1}$10], and [001], respectively [4.48]

ture on the ω_{PM} line, particularly the effect of lowering the temperature below T_N, the transition temperature characterizing the magnetically ordered phase. Te temperature evolution of the ω_{PM} line is shown for $Cd_{1-x}Mn_xTe$, $x = 0.70$, in Fig. 4.22. The spectra were recorded for a magnetic field of 60 kG using the $(\hat{\sigma}_+, \hat{z})$ configuration. As the temperature is lowered, the ω_{PM} line initially broadens and then moves towards higher Raman shifts as a consequence of the increased importance of the exchange interaction. An increase in Raman shift is observed at temperatures well above $T_N \sim 40$ K. As the temperature is lowered through and below T_N, the line becomes the magnon component observed in the $(\hat{\sigma}_+, \hat{z})$ configuration.

Finally, the conclusion that the magnetic feature observed in the magnetically ordered phases is a one magnon excitation was initially deduced from its polarization characteristics and temperature behavior [4.7]; this is supported by the results of *Ching* and *Huber* [4.63]. The fact that this feature shows a splitting in the presence of a magnetic field and that the ω_{PM} line of the paramagnetic phase, clearly associated with a single ion excitation, evolves smoothly into the higher energy component of the magnon provides a strong confirmation of this interpretation. In the same spirit, one might expect a two magnon feature associated with the $2\omega_{PM}$ line; however, given the intensity of the $2\omega_{PM}$ line compared to that of the ω_{PM} line, the intensity of such a feature would preclude its observation. A two magnon feature similar to that seen in MnF_2 by *Fleury* and *Loudon* [4.49] would have symmetric polarization characteristics; such a feature has also not been observed in $Cd_{1-x}Mn_xTe$.

4.5 Spin-Flip Raman Scattering

When detectable, spin-flip Raman scattering from free electrons bound in larger 'effective mass' orbits to donors provides a practical means of probing the electronic structure of semiconductors, as dramatically illustrated in DMS. The large Raman shifts associated with spin-flip scattering from electrons in DMS were first observed in $Hg_{1-x}Mn_xTe$ by *Geyer* and *Fan* [4.64]. The first evidence of a finite spin-splitting of the electronic level in the absence of a magnetic field was reported by *Nawrocki* et al. [4.65, 4.66] in the wide gap diluted magnetic semiconductor $Cd_{1-x}Mn_xSe$. The effects of Mn concentration and the antiferromagnetic coupling among the ions on the spin-flip Raman shifts were first observed [4.67] in $Cd_{1-x}Mn_xTe$. These studies established the nature of Raman scattering associated with the spin-flip transitions of electrons bound to donors in DMS. The large Raman shifts depend not only on the applied magnetic field, but also on temperature and manganese concentrations. The far-infrared absorption spectra [4.68] of $Cd_{1-x}Mn_xSe$ also provided

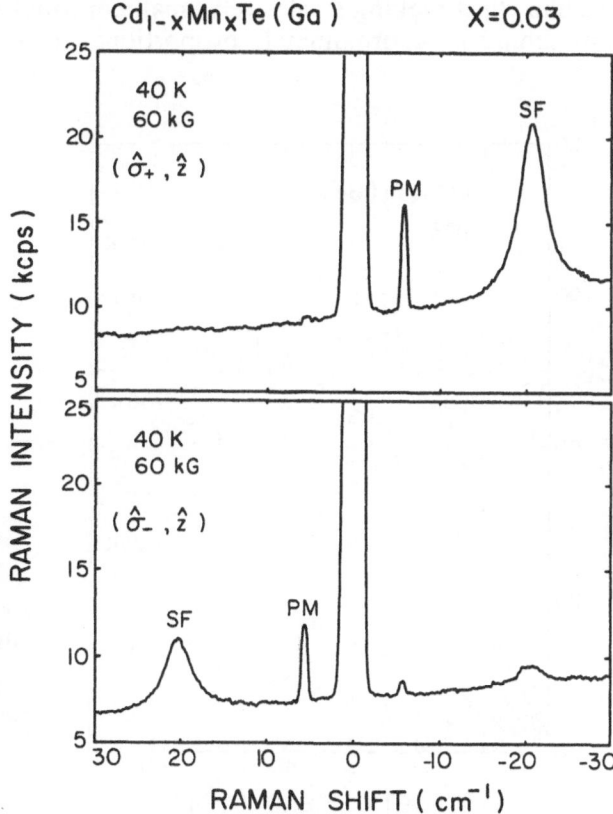

Fig. 4.23. Raman spectra of $Cd_{1-x}Mn_xTe(Ga)$, $x = 0.03$, showing the $\Delta m_s = \pm 1$ transitions within the Zeeman multiplet of Mn^{2+} (PM) and the spin-flip of electrons bound to Ga donors (SF). kcps $\equiv 10^3$ counts/s [4.76]

evidence for these spin-flip transitions. Following these initial reports, there have been several investigations [4.69–76] of spin-flip Raman scattering in DMS. The extensive results currently available allow a detailed comparison with the theory of *Dietl* and *Spałek* [4.77, 78].

The Raman spectra of $Cd_{1-x}Mn_xTe(Ga)$, $x = 0.03$, are shown in Fig. 4.23 for the $(\hat{\sigma}_+, \hat{z})$ and $(\hat{\sigma}_-, \hat{z})$ polarization configurations with $T = 40$ K and $H = 60$ kG. The two Stokes features labeled 'PM' and 'SF' are present only in $(\hat{\sigma}_+, \hat{z})$, while the corresponding anti-Stokes features appear only in the $(\hat{\sigma}_-, \hat{z})$ configuration. The observed width of the 'PM' line is instrument limited, while that of the spin-flip transitions is ~ 3 cm^{-1}. The PM feature is associated with the spin-flip transitions within the Zeeman multiplet of the ground state of Mn^{2+} as discussed in Sect. 4.3.2.

The 'SF' feature of Fig. 4.23 is attributed to spin-flip Raman scattering from electrons bound to gallium donors. It has the same polarization characteristics as those of the 'PM' line appearing [4.67] in the $(\hat{\sigma}_+, \hat{z})$ or $(\hat{z}, \hat{\sigma}_-)$ polarizations for Stokes scattering and in $(\hat{\sigma}_-, \hat{z})$ or $(\hat{z}, \hat{\sigma}_+)$ for anti-Stokes. As illustrated in Fig. 4.24, the peak Raman shift of this spin-flip feature exhibits a strong dependence on both temperature and magnetic field. The primary source of the spin splitting of the electronic level is the exchange coupling with the Mn^{2+} ions [first term of (4.52)] with the Zeeman effect making a relatively small contribution. Hence, the Raman shift should be approximately proportional to the magnetization of the

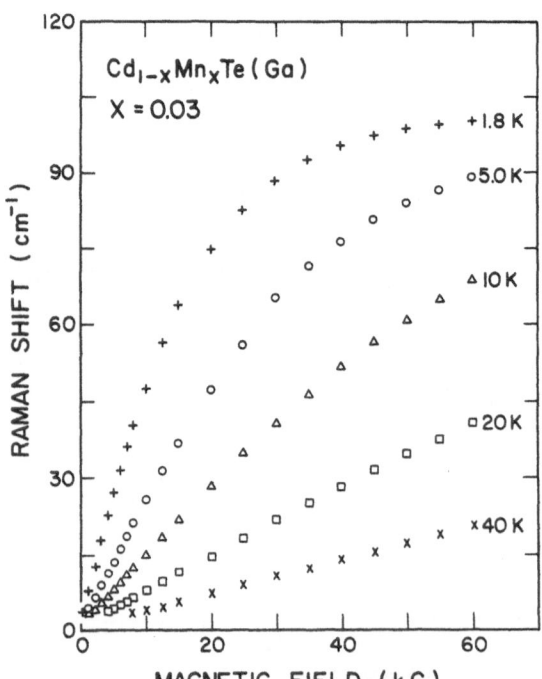

Fig. 4.24. Magnetic field and temperature dependence of the Raman shift associated with the spin-flip of electrons bound to donors in $Cd_{1-x}Mn_xTe(Ga)$, $x = 0.03$ [4.76]

Mn^{2+} ion system, which amplifies the effect of the magnetic field on the electron. As can be seen in Fig. 4.24, a finite Raman shift is observed for zero magnetic field. This effect is attributed by *Dietl* and *Spałek* [4.77] to the "bound magnetic polaron (BMP)": The electron localized on a donor in a diluted magnetic crystal polarizes the magnetic ions within its orbit, creating a spin cloud that exhibits a net magnetic moment. An additional effect on the binding energy of the electron bound to the donor originates from thermodynamic fluctuations of the magnetization and the resulting spin alignment of the magnetic ions around the donor.

According to (4.52), the spin splitting of the donor energy levels in DMS arises from the combined effect of the magnetization of the Mn^{2+} ions and the external field H. Due to the strong s-d coupling, the effect due to the magnetization dominates. The Raman shift associated with spin-flip scattering from the donor states has the form

$$\hbar\omega_0 = \Delta_0 = \frac{\alpha}{g\mu_B} M_0(H) + g^*\mu_B H, \tag{4.69}$$

where M_0 is the macroscopic magnetization. The magnetization is proportional to the thermal average of the Mn^{2+} spin projection along H multiplied by the density of Mn^{2+} ions contributing to the magnetization, yielding

$$\Delta_0 = \bar{x}\alpha N_0 \langle S_z^{Mn} \rangle + g^*\mu_B H. \tag{4.70}$$

Here $\bar{x}$ is the concentration of Mn^{2+} ions that contribute to the magnetization. For small x, the crystal is paramagnetic and the thermal average of the Mn^{2+} spins is

$$\langle S_z^{Mn} \rangle = \frac{5}{2} B_{5/2} \left(\frac{g\mu_B H}{k_B T} \right), \tag{4.71}$$

where $B_{5/2}$ is the Brillouin function B_J for $J = 5/2$.

The compositional dependence of spin-flip Raman scattering has two sources. Within a DMS system, such as $Cd_{1-x}Mn_xTe$, the properties of the spin splitting should show a strong dependence on the Mn concentration. For a given x, these properties should also vary from one DMS system to another.

The spin-flip Raman shifts for $Cd_{1-x}Mn_xTe$ at $T = 1.8$ K are shown in Fig. 4.25 as a function of magnetic field and composition. The results for $x = 0.01$ show the saturation behavior characteristic of the paramagnetic phase. As the Mn concentration is increased to $x = 0.03$ and $x = 0.05$, the Raman shifts increase and the effects of saturation are still clearly evident, but less pronounced. For $x = 0.10$, the deviation from the paramagnetic behavior is quite evident. For $H = 60$ kG, the Raman shift for $x = 0.10$ is only four times that for $x = 0.01$. As x exceeds 0.10, the Raman shifts for a given field actually, decrease; note that the shifts for

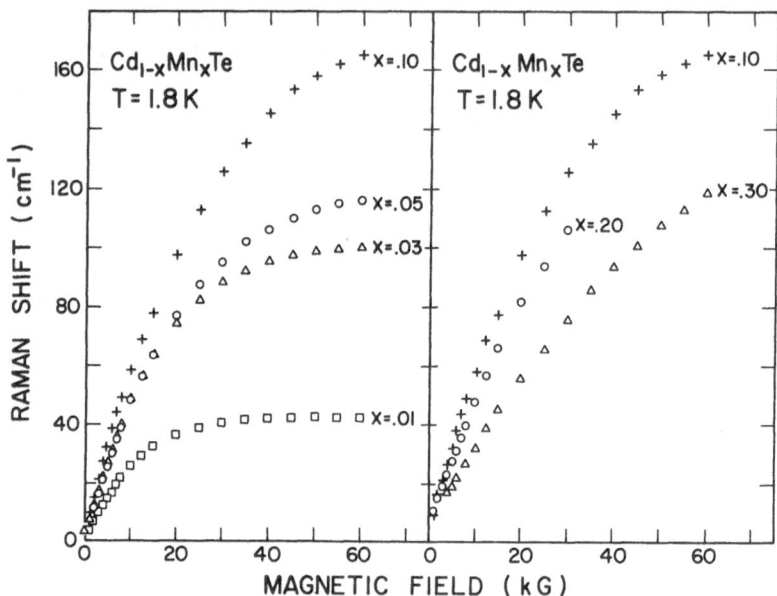

Fig. 4.25. Magnetic field and composition dependence of the peak spin-flip Raman shift in the $Cd_{1-x}Mn_xTe$ samples at $T = 1.8$ K [4.76]

the $x = 0.20$ sample lie below those for the $x = 0.10$ sample. And the Raman shifts for the $Cd_{1-x}Mn_xTe(Ga)$, $x = 0.30$, sample are significantly smaller than those for the $x = 0.10$ and $x = 0.20$ samples. These trends have their origin in the decrease of the mean magnetic field due to the increasing antiferromagentic pairing of Mn^{2+} neighbors.

The low field data and the associated theoretical curves are plotted in Fig. 4.26 for $Cd_{1-x}Mn_xTe(Ga)$, $x = 0.03$, 0.05, and 0.10. As can

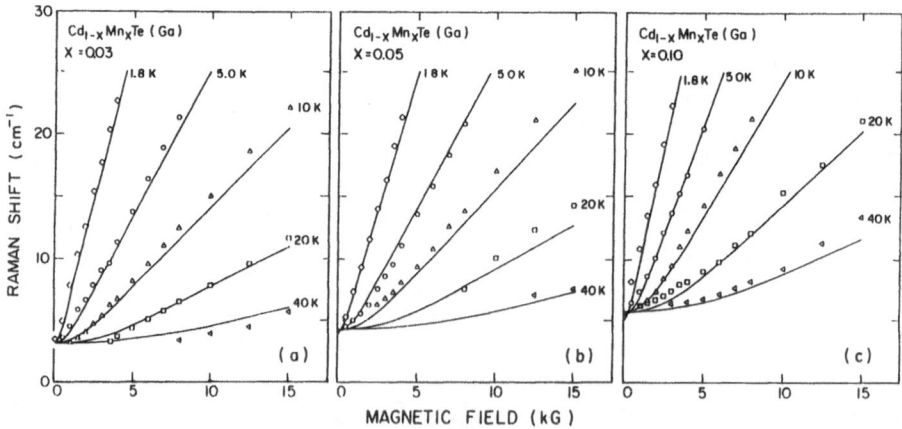

Fig. 4.26. Magnetic field and temperature dependence of the peak spin-flip Raman shift in the $Cd_{1-x}Mn_xTe(Ga)$, $x = 0.05$, 0.10, and 0.20, samples at low fields [4.76]

be seen in Fig. 4.26a, a zero-field shift of 3.5 cm^{-1} was observed in $Cd_{1-x}Mn_xTe(Ga)$, $x = 0.03$, at $T = 1.8$ K. The data for the other two samples, $x = 0.05$ and $x = 0.10$, give evidence of zero-field shifts of ~ 4 and ~ 6 cm^{-1}. The magnetic field and temperature dependence of the data, particularly for $x = 0.10$, are well described by the curves generated from the theory in [4.71, 77, 78] and employed in [4.76]. The Raman line corresponding to the bound magnetic polaron, i.e., the zero-field spin-flip Raman line in $Cd_{1-x}Mn_xSe$, $x = 0.10$, can be clearly seen in Fig. 4.27 for $T = 1.8, 5, 10,$ and 20 K.

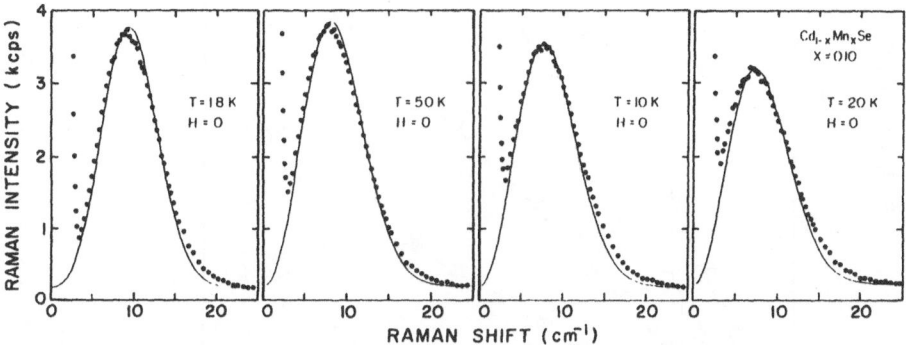

Fig. 4.27. Zero-field spin-flip Raman spectra for the $Cd_{1-x}Mn_xSe$, $x = 0.10$, sample at $T = 1.8, 5, 10$ and 20 K. The scans were recorded in the $(\hat{\sigma}_+, \hat{z})$ polarization [4.76]

Spin-flip Raman scattering from donors in $Cd_{1-x}Co_xSe$ and $Cd_{1-x}Fe_xSe$ has also been observed [4.79]. The contrast between Mn^{2+}, Co^{2+}, and Fe^{2+} with respect to the nature of their ground state is worth noting. The atomic ground state of Mn^{2+}, $^6S_{5/2}$, in a tetrahedral environment characteristic of, say, a zinc blende structure undergoes a crystal field splitting into a Γ_7 doublet and a Γ_8 quadruplet. As we have noted, experimentally this crystal field splitting is very small and can be ignored; as expected, the $g_{Mn^{2+}} = 2$ is consistent with atomic $^6S_{5/2}$ like ground state. Co^{2+} has a $3d^7$ electron configuration. The atomic ground state of Co^{2+}, $^4F_{9/2}$, splits into an orbital singlet (Γ_2) and two higher lying triplets (Γ_5, Γ_4). The crystal field splitting separating the lowest Γ_2 ground state from the Γ_5 and Γ_4 states is so large that all the magnetic phenomena are controlled by Γ_2 along with its total spin of 3/2. Thus, the magnetic behavior of Co^{2+} is qualitatively analogous to that of Mn^{2+}. It is very different from that of Fe^{2+} which has complex level structure and a non-magnetic ground state; the Fe-based DMS's thus exhibit van Vleck paramagnetism [4.10, 80].

Raman scattering associated with the spin-flip of electrons bound to donors in $Cd_{1-x}Co_xSe$ and $Cd_{1-x}Fe_xSe$ has been observed by *Bar-*

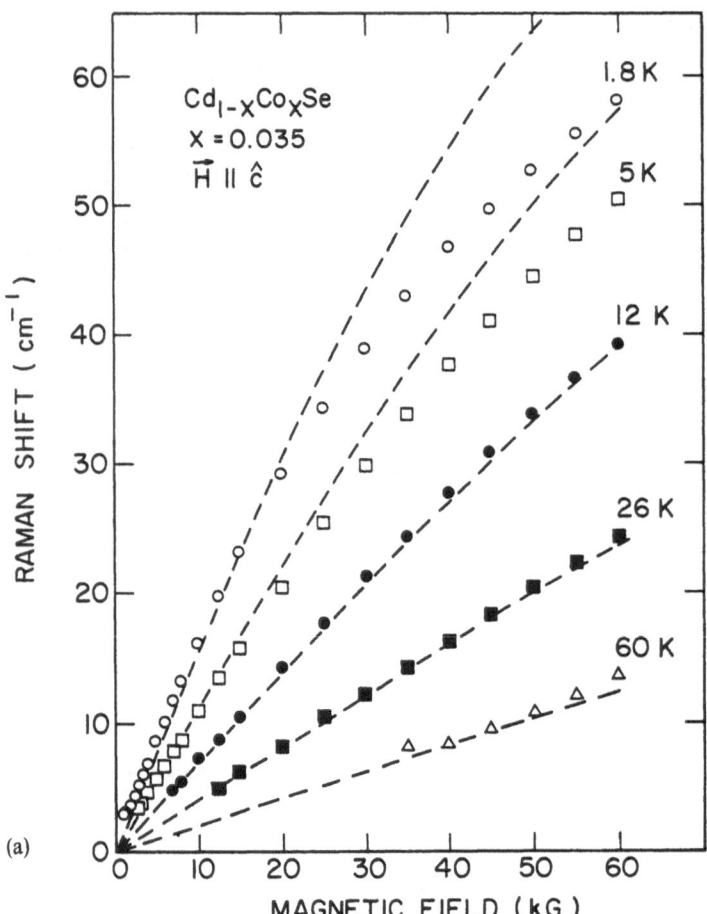

(a)

tholomew et al. and *Suh* et al. [4.79]. The magnetic field dependence of ω_{SF} in $Cd_{1-x}Co_xSe$ at various temperatures shown in Fig. 4.28a indicates a large *s-d* exchange interaction ($\alpha N_0 = 320$ meV) and a clear evidence of a bound magnetic polaron. The data for the spin-flip of donor electrons in $Cd_{1-x}Fe_xSe$ are displayed in Fig. 4.28b. The magnetic field and temperature dependences of ω_{SF} in $Cd_{1-x}Fe_xSe$ are qualitatively different from that in $Cd_{1-x}Mn_xSe$ and $Cd_{1-x}Co_xSe$. The non-magnetic nature of Fe^{2+} is clearly indicated by the zero Raman shift at zero field, i.e., by the absence of the bound magnetic polaron [4.79]. *Heiman* et al. [4.81] have reported and discussed this lack of bound magnetic polaron in $Cd_{1-x}Fe_xSe$ whereas *Scalbert* et al. [4.82] have provided further insights into the bound magnetic polaron problem in $Cd_{1-x}Fe_xSe$ by drawing attention to the anti-crossing of the donor spin-flip Raman line with that of an intra-Fe^{2+}-ion excitation.

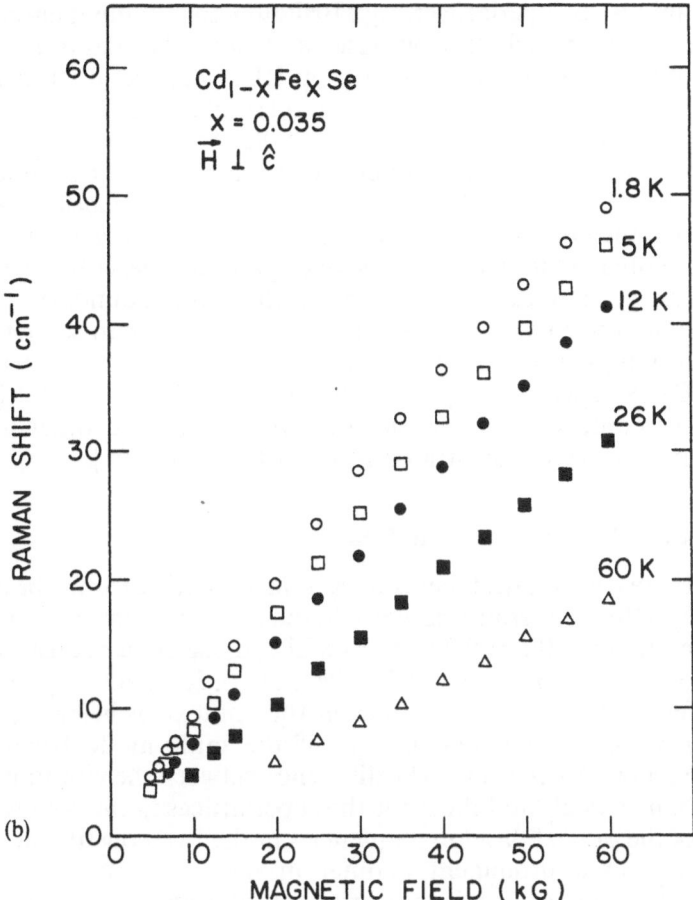

Fig. 4.28. Spin-flip Raman shift from donor-bound electrons in (a) $Cd_{1-x}Co_xSe$ and (b) $Cd_{1-x}Fe_xSe$ as a function of applied magnetic field [4.79]

4.6 Vibrational, Electronic, and Magnetic Excitations in Superlattices

The fabrication of heterostructure of semiconductors by techniques such as molecular beam epitaxy (MBE), or metallo-organic chemical vapor deposition (MOCVD) has been a major technological breakthrough. Single quantum wells, multiple quantum wells, superlattices and many other heterostructures possess properties and exhibit phenomena not encountered in the bulk. Such "synthetic", "modulated" or "engineered" structures are fascinating in the context of fundamental physics just as much as for their technological importance. Electronic and optical properties of multilayer structures have brought out several phenomena unique to superlattices, e.g., Brillouin zone folding effects, plasma dispersion in layered electron gases and multiple quantum well effects. Much of the focus to date has been on III–V semiconductors and their ternary alloys [4.83–85]. Recently, heterostructures of DMS–superlattices–have been successfully fabricated [4.12, 86–88].

DMS superlattices offer the exciting prospect of tuning the electronic potential within the individual layers, *after fabrication*, using external parameters such as temperature and magnetic field. It has also been pointed out [4.89] that by "spin doping" a superlattice, a tunable electronic energy gap can be created at the zone boundary. Here we present illustrative results that demonstrate how Raman scattering is a powerful technique for obtaining special insights into various structural aspects of DMS superlattices [4.90]. The power of Raman scattering techniques in the context of the physics of superlattices and other microstructures is extensively demonstrated in the articles in [4.91].

4.6.1 Vibrational Excitations

In this section the focus is on Raman scattering from vibrational excitations in DMS superlattices with frequencies comparable to those of optical phonons in the bulk DMSs as well as those of the acoustic branch accessible through special superlattice effects. In order to identify some of the relevant features of lattice dynamics in the context of heterostructures involving DMSs, the different aspects of the multi-mode behavior discussed in Sect. 4.2 are relevant. The difference between the vibrational excitations for bulk crystals and those for the superlattices in the context of the matching of the optical phonon frequencies in the two constituent layers comes into play in a prominent manner in the phneomena revealed by Raman scattering. As a relevant example, consider a superlattice consisting of alternating layers of CdTe and $Cd_{1-x}Mn_xTe$. The dispersion curves for LO phonons in bulk CdTe and $Cd_{1-x}Mn_xTe$ ($x = 0.25$) calculated using the linear chain model with nearest neighbor interaction only and assuming that Mn^{2+} ions randomly replace Cd^{2+} ions in the mixed crystal, $Cd_{1-x}Mn_xTe$, are shown in Fig. 4.29. This plot indicates that in $CdTe/Cd_{1-x}Mn_xTe$ superlattices, the "MnTe-like" phonon mode of the $Cd_{1-x}Mn_xTe$ layers cannot propagate into the CdTe layers because of the large attenuation of the vibration in that frequency region (region I). Similarly, in the frequency region between the CdTe zone center LO phonon and the CdTe-like zone center LO phonon of the $Cd_{1-x}Mn_xTe$ layer (region II), the CdTe LO mode cannot be sustained in the $Cd_{1-x}Mn_xTe$ layers. Therefore, these phonon modes can be considered to be confined to their respective layers, resulting in quantized optical phonons which are equivalent to vibrations in the bulk material whose wavevectors are given by $q = (m\pi)/(d + a/2)$, where $d = d_1$ or d_2, the thickness of CdTe (1) or the $Cd_{1-x}Mn_xTe$ (2) layer, and m is an integer [4.85, 92–95]. On the other hand, in the frequency region of the "CdTe-like"

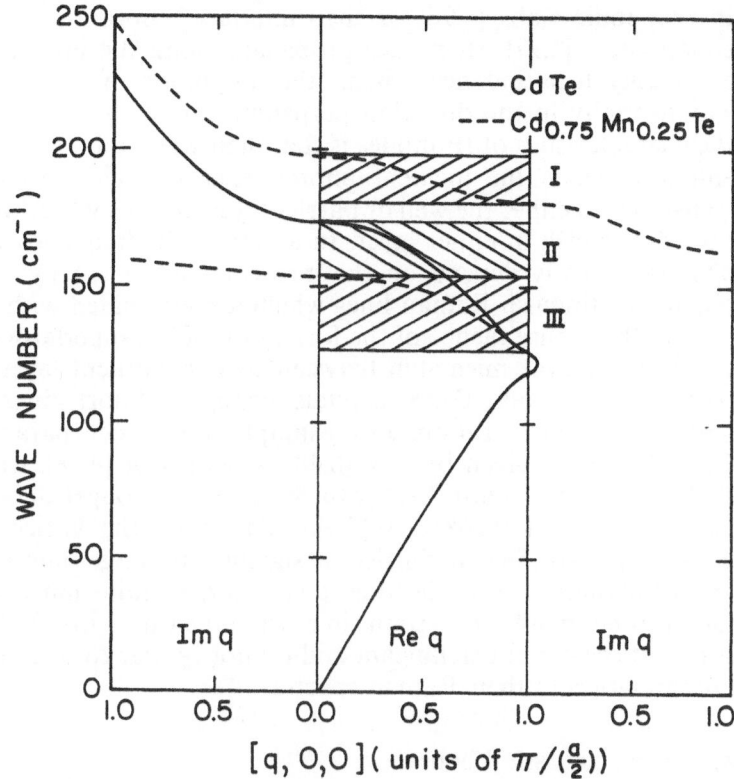

Fig. 4.29. Dispersion curves of LO phonons in bulk CdTe and in $Cd_{0.75}Mn_{0.25}Te$ calculated using the linear chain model. Solid curve, CdTe; dashed curves, $Cd_{0.75}Mn_{0.25}Te$. I and II are frequency ranges of confined optical phonons whereas III corresponds to propagating optical phonons [4.90]

phonon modes in the $Cd_{1-x}Mn_xTe$ layers (region III), the vibrations of $Cd_{1-x}Mn_xTe$ layers and CdTe layers can propagate into both layers: the coupling between excitations originating in different layers may result in an average collective excitation as in the case of the acoustical phonons discussed in Sect. 4.6.1a.

We note that, as in the bulk crystal, in a superlatice one can obtain a selective enhancement of the intensities of the Raman lines associated with optical phonons from the well or the barrier layers by matching either the incident or the scattered photon energy with electronic transitions [4.21, 53]; in superlattices the relevant electronic transitions can be associated with either the well or the barrier.

For finite crystals like thin films, or layers in superlattices and heterostructures, the existence of surfaces or interfaces results in new vibrational excitations in addition to the "bulk" vibrational excitations which we discussed earlier. These vibrational modes are the surface modes in thin ionic slabs [4.96] or the "interface" vibrational (IF) modes in superlattices [4.97]. IF modes propagate along the interface planes and are highly localized near them; the amplitude of an IF mode decays exponentially in the direction perpendicular to the layer plane. One of the characteristics of IF modes is that their Raman intensity is resonantly enhanced when the incident photon energy is close to the electronic transitions of either the well or barrier layer since its vibrational amplitude does not vanish in either layer. In addition, the frequencies of IF modes observed greatly depend on the incident photon energy. These characteristics distinguish Raman lines which are associated with IF modes.

Another consequence of the formation of the superlattice is the strain due to the lattice mismatch between two constituent layers. The lattice parameter in bulk DMS crystals changes almost linearly with the manganese concentration. For example, the lattice parameter of bulk $Cd_{1-x}Mn_xTe$ is given by $a = (6.487 - 0.149x)$ Å [4.98], whereas that of bulk (or epitaxial film) $Zn_{1-x}Mn_xSe$ shows a stronger dependence on x, viz. $a = (5.666 + 0.268 x)$ Å [4.98]. Therefore, the lattice mismatch in $ZnSe/Zn_{1-x}Mn_xSe$ superlattices is significantly larger and the shift of the optical phonon frequencies due to such large strains must be considered. On the other hand, such strains in most of the $Cd_{1-x}Mn_xTe/Cd_{1-y}Mn_yTe$ superlattices are not significant and do not appear to be important in the interpretation of their Raman spectra.

a) Folded Acoustic Phonons

Since the acoustic dispersion curves of the two constituent materials of the superlattice overlap over a wide frequency range, acoustic phonons can propagate through both layers. For long wavelength acoustic phonons one can apply a model in which the superlattice is considered as an elastic continuum composed of two alternating layers characterized by densities ϱ_1 and ϱ_2, and by bulk longitudinal acoustic velocities v_1 and v_2 along the superlattice axis [4.85, 99]. The dispersion relation for the acoustic phonons propagating along the superlattice axis with wavevector q_z is given by [4.100]

$$\cos (q_z D) = \cos \left(\frac{\omega d_1}{v_1}\right) \cos \left(\frac{\omega d_2}{v_2}\right) - (1 + \delta) \sin \left(\frac{\omega d_1}{v_1}\right) \sin \left(\frac{\omega d_2}{v_2}\right),$$

$$(4.72)$$

where d_1 and d_2 are the respective thicknesses of the well and barrier layers, $D = d_1 + d_2$, and $\delta = \frac{1}{2} (\varrho_1 v_1 - \varrho_2 v_2)^2/(\varrho_1 v_1 \varrho_2 v_2)$. From (4.72) one can deduce that the dispersion curve of the average bulk material is folded into the new Brillouin zone; in addition, small gaps open up at the zone center and the boundary when δ is different from zero. As a result of this zone folding, the "zone-center" modes of the new Brillouin zone which can interact with the electromagnetic radiation can be observed in Raman scattering. (For a short derivation of (4.72) see [4.90]).

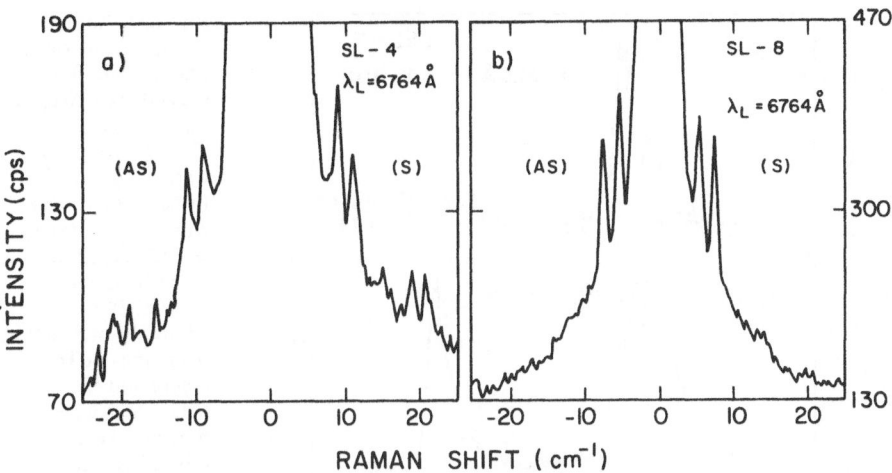

Fig. 4.30. Stokes (S) and anti-Stokes (AS) components of the folded longitudinal acoustic phonons in $Cd_{1-x}Mn_xTe/Cd_{1-y}Mn_yTe$ superlattices with $\hat{z}' \parallel [111]$; (a) $x = 0.11$, $y = 0.50$ (b) $x = 0$, $y = 0.24$ [4.90]

Figure 4.30 shows the low frequency Raman spectra for a [111] $Cd_{1-x}Mn_xTe/Cd_{1-y}Mn_yTe$ superlattice at room temperature. All the spectra were recorded in the $z'(x'x')\bar{z}'$ scattering configuration: here x' and y' are in the plane of the layers with z' along the superlattice axis. The incident laser radiation has an energy $\hbar\omega_L$ intermediate between the band gap of the well and that of the barrier layers. Bulk crystals corresponding to the constituent materials of the superlattices show no Raman lines due to phonons in this spectral region. We attribute these lines to the longitudinal acoustic (LA) phonons folded into the new Brillouin zone, which arises from the additional periodicity of the superlattice.

Using the interpolated values of the densities and elastic moduli for various compositions of bulk $Cd_{1-x}Mn_xTe$, one can calculate the frequencies of LA phonons as a function of the wavevector q. In the backscattering geometry, the wavevector of the scattered radiation is $(4\pi n/\lambda_L)$ where n is the refractive index of the sample. The solutions of (4.72) with the relevant parameters are 8.8 and 10.8 cm^{-1} for the first doublet and 18.6 and 20.5 cm^{-1} for the second doublet, in excellent agreement with the experimental values of 9.2, 11.0, 19.0, and 20.7 cm^{-1}, respectively. The strain associated with the lattice mismatch between the alternate layers estimated to be less than ± 0.007 in both layers. With the Grüneisen constant estimated from the pressure derivative of the bulk elastic constants, ranging from 1.0 to 1.03 for $Cd_{1-x}Mn_xTe$ ($x = 0$ to $x = 0.5$), the change in the frequencies of the LA phonons due to strain are expected to be insignificant in the context of the present measurements. Folded acoustic phonons have been also observed in $ZnSe/Zn_{1-x}Mn_xSe$ [4.90] as well as in $ZnSe/Zn_{1-x}Cd_xSe$ superlattice structures [4.101].

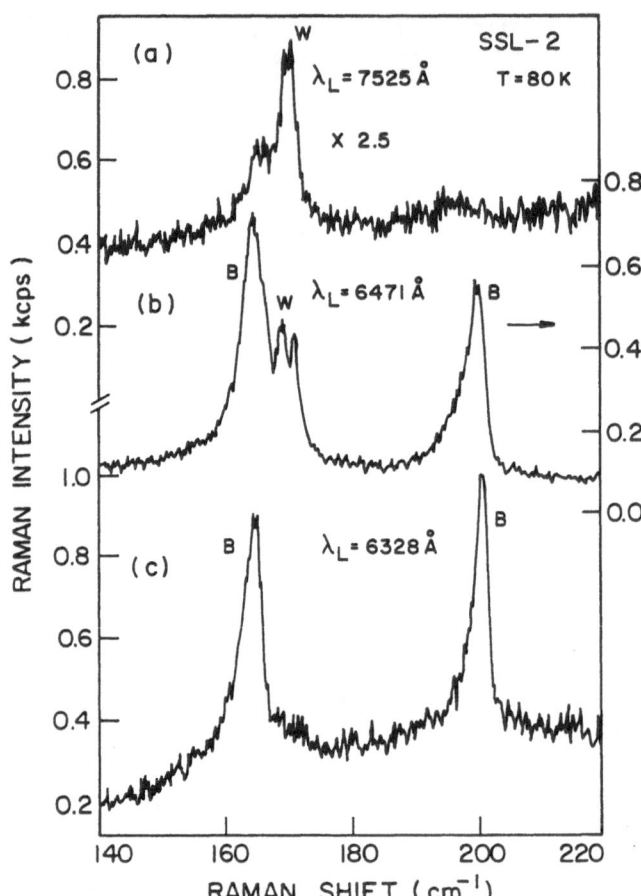

Fig. 4.31. Raman spectra from optical phonons in [001] $CdTe/Cd_{0.75}Mn_{0.25}$ Te obtained at $T = 80$ K. Spectra were excited with three different laser wavelengths. (a) $\lambda_L = 7525$ Å; (b) $\lambda_L = 6471$ Å, and (c) $\lambda_L = 6328$ Å. Phonons originating from the CdTe well layers and the $Cd_{0.75}Mn_{0.25}$ Te barrier layers are labeled W and B, respectively [4.90]

b) Optical Phonons

Figure 4.31 shows the Raman spectra of a [001] $CdTe/Cd_{0.75}Mn_{0.25}$ Te superlattice (SSL-2) at $T = 80$ K in the frequency region of optical phonons with various wavelengths of laser excitation (λ_L). In Fig. 4.31a, obtained with $\lambda_L = 7525$ Å, i.e., with photon energy ($\hbar\omega_L$) close to the lowest electronic transition of the CdTe well, the LO phonon of CdTe appears resonantly enhanced whereas those of the $Cd_{0.75}Mn_{0.25}$ Te barrier are weak. As the incident photon energy is increased to a value lying between the energy gaps of the two layers, LO phonons from the CdTe well as well as those from the $Cd_{0.75}Mn_{0.25}$ Te barrier are observed as shown in Fig. 4.31b. When the incident photon energy approaches the bandgap of $Cd_{0.75}Mn_{0.25}$ Te, only the phonons from the barrier layer are observed as shown in Fig. 4.31c. These spectra illustrate the resonance enhancement of the intensities of phonons associated with the different layers as discussed earlier. By choosing the appropriate incident photon

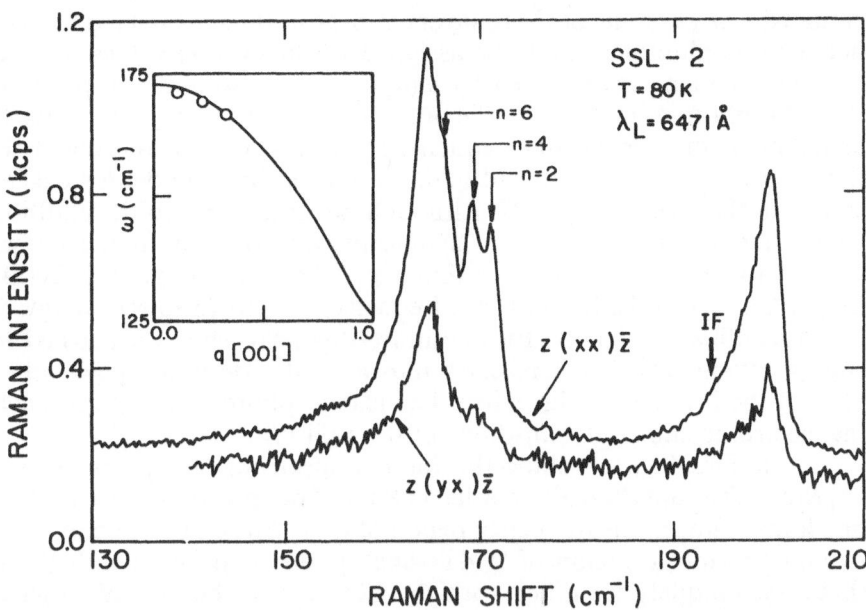

Fig. 4.32. Raman spectra from optical phonons in a [001] $CdTe/Cd_{0.75}Mn_{0.25}Te$ superlattice for different polarizations. Confined optical phonons are labeled with $n = 2, 4,$ and 6. Inset shows the observed frequencies of the confined LO phonons plotted on the bulk CdTe dispersion curve calculated with a linear chain model [4.90]

energy, one can thus obtain a selective resonance enhancement of the LO phonons in either superlattice layer.

In Fig. 4.32, we show once more the Raman spectrum of SSL-2, recorded at 80 K. The peak at ~ 200 cm^{-1} is the "MnTe-like" longitudinal optical (LO) phonon from the barrier layers. The peaks between 160 and 173 cm^{-1} are due to the "CdTe-like" and CdTe LO phonons of the barrier and well layers, respectively. The frequencies and shapes of these Raman lines are different from the LO phonons seen in the bulk crystals. The most prominent peak at 165 cm^{-1} is the "CdTe-like" LO phonon of $Cd_{0.75}Mn_{0.25}Te$, which propagates through both layers. However, the Raman line does not resolve into the multiple peaks expected due to this effect since their frequencies do not change substantially with the variation of the wavevector. The peaks labeled $n = 2, 4,$ and 6 are attributed to the LO phonons confined to the CdTe layers.

In the [001] superlattice, belonging to the point group D_{2d}, the phonons observed in Raman scattering have symmetry A_1 for $z(xx)\bar{z}$ scattering and B_2 for $z(xy)\bar{z}$ scattering [4.90]. Here x, y, z are along the cubic axes, z being the superlattice axis. A_1 phonons, which dominate the Raman spectra of the well under resonance condition, have wavevectors characterized by even m values, while B_2 phonons, which can be observed far from resonance, have wavevectors with odd m values. In Fig. 4.32, the confined LO phonons of CdTe layers appear in the (xx) polarization and hence these phonons should be assigned to the even m values, 2, 4, and 6. Their frequencies are given by $\omega(q_m)$, where $\omega(q)$ is the dispersion of the LO phonon in bulk CdTe and $q_m = (m\pi)/(d_1 + a/2)$. With this assignment, their measured frequencies are in excellent agreement with the dispersion curve calculated with the linear chain model for CdTe as shown in the inset in Fig. 4.32. Although not resolved into the additional peaks due to the absence of significant variation of frequency with wavevector and the large layer thickness, the peak at the "MnTe-like" LO of $Cd_{1-x}Mn_xTe$ layers has a frequency shifted to lower energy due to the confinement effect. Furthermore, this peak shows a marked low energy asymmetry which resolves into a weak additional peak labeled "IF" at the low energy side when the incident photon energy approaches the electronic subband transitions of the well layer.

In Fig. 4.33, we show the Raman spectrum of a [111] CdTe/Cd$_{0.76}$Mn$_{0.24}$Te superlattice, recorded at 80 K. The spectrum consists of only one Raman line in the frequency region of the "MnTe-like" phonon (peak A) and one in the region of "CdTe-like" phonon (peak B) with Raman shifts not uniquely identified with the LO or TO phonons of either the well or the barrier layer; in contrast to the [001] superlattices, this is the case for all λ_L's. Peaks A and B appear in $z'(x'x')\bar{z}'$ (parallel polarization) as well as in $z'(y'x')\bar{z}'$ (crossed polarization), with a larger intensity in the

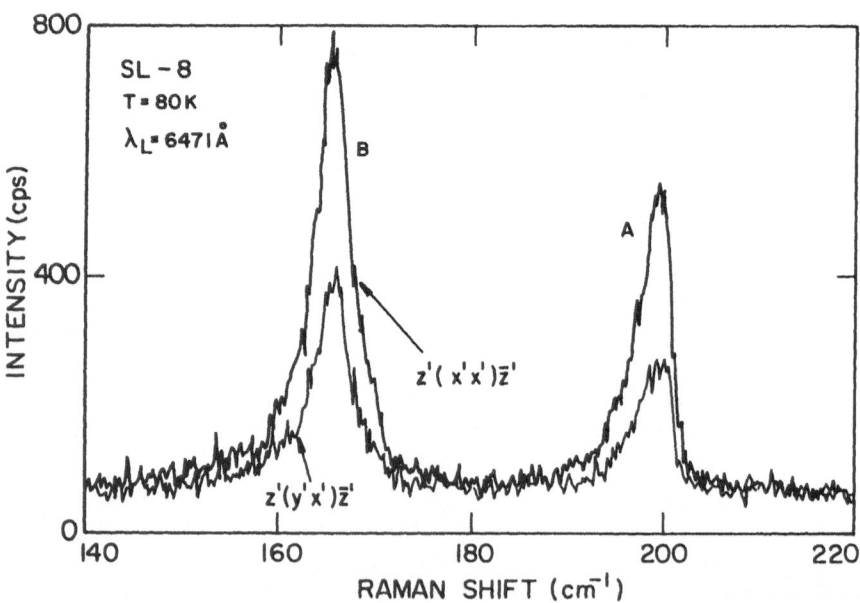

Fig. 4.33. Raman spectra in the frequency region of optical phonons in a [111] CdTe/Cd$_{0.76}$Mn$_{0.24}$Te superlattice recorded at $T = 80$ K, for different polarizations. A and B are the Raman lines in the MnTe-like and CdTe-like phonon regions, respectively [4.90]

former. Here x', y' and z' are along [01$\bar{1}$], [2$\bar{1}\bar{1}$], and [111], respectively. A particularly noteworthy feature of these lines is their marked frequency dependence on λ_L.

For [111] $Cd_{1-x}Mn_xTe/Cd_{1-y}Mn_yTe$ superlattices, the point group symmetry is reduced to C_{3v} from the higher symmetry T_d of the bulk crystal. LO phonons with A_1 symmetry can be observed in $z'(x'x')\bar{z}'$ whereas TO phonons with E symmetry are allowed in both $z'(x'x')\bar{z}'$ and $z'(x'y')\bar{z}'$. LO phonons referred to the new Brillouin zone can show a small variation in its frequency as the energy of the incident laser radiation is changed. However, the observed range of the frequency variation is too large for it to be explained on this basis. Furthermore, the behavior of peak A is even more difficult to understand in this manner because its frequency increases with the incident photon energy, a behavior opposite to that expected for the LO phonons of the new Brillouin zone. Thus, there is a clear difference between Raman scattering from LO phonons in [111] and [001] superlattices.

In the light of the above observations, it is useful to review other experimental observations distinguishing these two types of superlattices. In Fig. 4.34, we show the photoluminescence peak at 5 K and magnetic

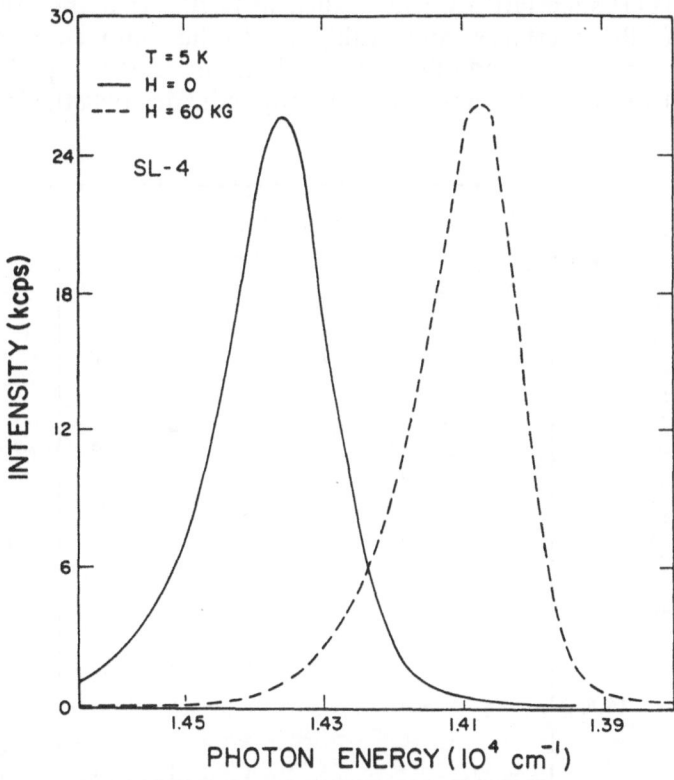

Fig. 4.34. Magnetic-field-induced shift in the position of the luminescence peak in $Cd_{0.89}Mn_{0.11}Te/Cd_{0.50}Mn_{0.50}Te$ superlattice, $\lambda_L = 5682$ Å [4.90]

fields of 0 and 60 kG in a [111] $Cd_{0.89}Mn_{0.11}Te/Cd_{0.50}Mn_{0.50}Te$ superlattice, SL-4. The position of the luminescence peak shifts with magnetic
field towards lower energy with an effective g factor of ~ 100. This
demonstrates that the s–d and p–d exchange-enhanced g factor characterizes the DMS superlattices just as they do the bulk DMSs [4.90]. This
enhancement is both magnetic-field- and temperature-dependent. Furthermore, there is a marked difference between the magnetic field dependence of the photoluminescence spectrum in [111] and [001] superlattices,
as was first observed by *Zhang* et al. and *Chang* et al. [4.102]. The
luminescence spectrum of [111] superlattices exhibit one broad peak arising
from the excitons localized at the heterointerfaces, while for [001]
superlattices, a sharp peak due to intrinsic excitons as well as a weaker
one arising from the localized excitons were observed. It appears that a
significantly larger number of excitons are localized at the heterointerfaces
in the [111] than in the [001] superlattice. A result of this localization is
the observed shift of the exciton luminescence peak position as the
magnetic field is changed from being perpendicular to parallel with respect
to the superlattice axis. In Fig. 4.35, we show a similar result for SL-4.
From the large spectral red shift of the luminescence peak seen in other
[111] superlattices, it is argued in [4.102] that the excitons are localized
at the interfaces, conceivably due to the compositional fluctuation at the
interface. In addition, an evaluation of the superlattice structure by
transmission electron microscopy indicates that the [111] samples have a

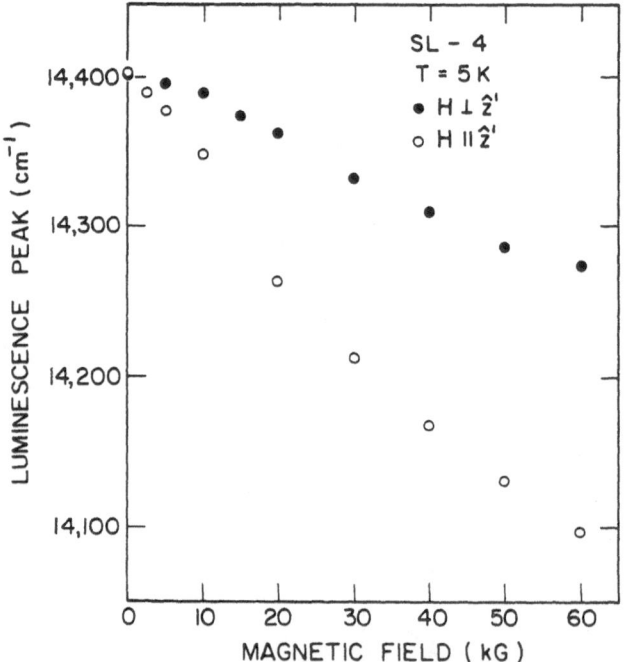

Fig. 4.35. Magnetic-
field-induced shift in the
position of the lumi-
nescence peak in
$Cd_{0.89}Mn_{0.11}Te/$
$Cd_{0.50}Mn_{0.50}Te$ with
the magnetic field direc-
tions parallel or per-
pendicular to the
superlattice axis [4.90]

surface dislocation density approximately an order of magnitude higher than in the [001] samples [4.103]. It also appears that the [001] interfaces are more abrupt than the [111] interfaces. We suggest that the quality of the interface results in the absence of confined or propagating optical phonons in the [111] superlattices and the Raman lines A and B should be ascribed to vibrations associated with the interface. The imperfection of the interfaces suggests the possibility of the excitation of IF modes even in the backscattering experiments in [111] superlattices. Furthermore, the observation of the exciton localization at the interfaces suggests that the IF mode may be strongly enhanced, thereby dominating the Raman spectra in the [111] superlattices.

We can obtain a resonance enhancement of optical phonons or of IF modes by matching the incident or the scattered photon energy with the electronic transitions in the superlattice. Exploiting the magnetic field dependence of the photoluminescence peak with the large effective g factor as demonstrated in Fig. 4.34, we can achieve an "out resonance" either by magnetic field or by temperature tuning. This is illustrated in Fig. 4.36 where we show a striking resonance effect observed for the Raman lines with a magnetic field of 60 kG which shifts the onset of the photoluminescence peak to the region beyond 600 cm^{-1}. The clear observation of overtones and combinations of the two fundamental modes up to fourth

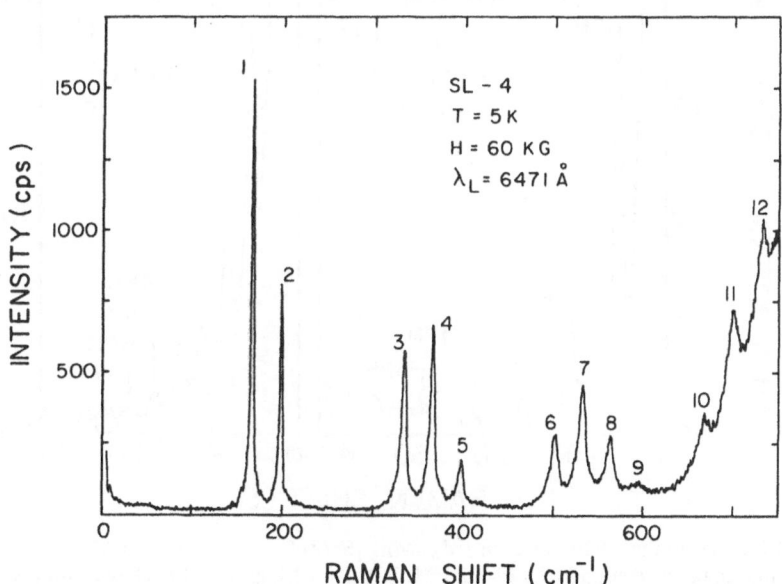

Fig. 4.36. Resonant Raman scattering from interface optical phonons, their overtones and combinations. The line labels denote the following assignments: (1) IF$_1$, (2) IF$_2$, (3) 2 IF$_1$, (4) IF$_1$ + IF$_2$, (5) 2 IF$_2$, (6) 3IF$_1$, (7) 2 IF$_1$ + IF$_2$, (8) IF$_1$ + 2 IF$_2$, (9) 3 IF$_2$, (10) 4 IF$_1$, (11) 3 IF$_1$ + IF$_2$, (12) 2 IF$_1$ + 2 IF$_2$, where IF$_1$ and IF$_2$ are the Raman lines B and A in [111] superlattices [4.90]

order in these superlattices attests to the resonance enhancement. The striking "out resonance" seen for the IF phonons in these [111] superlattices suggests the coupling of these excitations with excitons localized at the interfaces via the Fröhlich electron-phonon interaction. The enhancement of the interface modes once more underscores the exciton localization at the interfaces.

4.6.2 Electronic and Magnetic Excitations

The combination of magnetic and semiconducting properties is a particularly unique feature of DMSs and has emerged in a variety of phenomena in bulk crystals addressed with Raman spectroscopy. The consequences of reduced dimensionality as in a superlattice consisting of constituents with sub-micron dimensions are obviously of great interest. We have already encountered resonance enhancement of Raman cross-sections associated with quantum confined states strongly influenced by the *sp-d* exchange interaction in an external magnetic field. By exploiting modulation doping in which the barrier is "magnetic" and the well is "nonmagnetic", one could investigate spin-flip Raman scattering of electrons and holes; the *sp-d* exchange interaction with the magnetic constituents

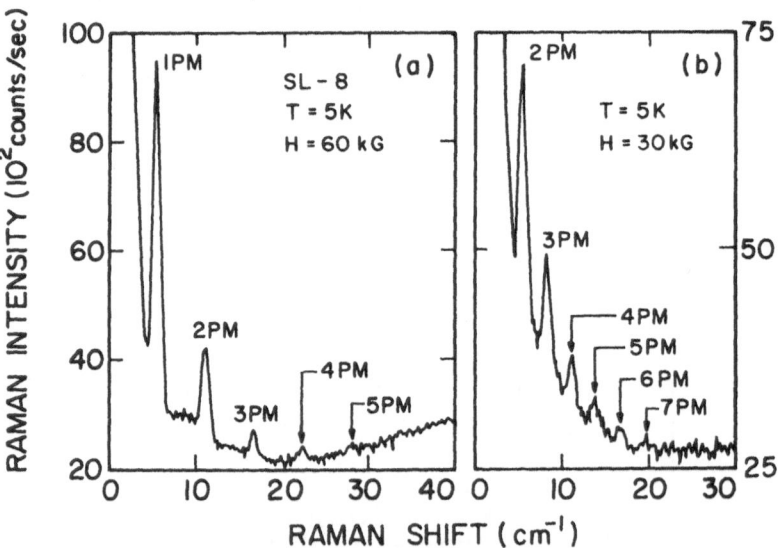

Fig. 4.37 Raman-EPR lines in $Cd_{0.9}Mn_{0.1}Se/ZnSe$ superlattice at $T = 5$ K. The Raman lines in (a) correspond to 1 PM, 2 PM, 3 PM, 4 PM, and 5 PM. They result from transitions within the Zeeman multiplet of Mn^{2+}, with $\Delta m_s = 1, 2, 3, 4,$ and 5, respectively. This spectrum is obtained in the crossed polarization $z(yx)\bar{z}$ with magnetic field $H = 60$ kG along x and incident wavelength of 6764 Å with 57 mW power. The 5 PM line appearing superposed on the photoluminescence shows how closely the resonance condition is fulfilled. The spectrum in (b) shows the additional 6 PM and 7 PM lines observed at a lower magnetic field. In this spectrum the 1 PM line is obscured by the parasitic laser light [4.109]

would then provide insights into the penetration of the quantum confined wavefunction into the barrier. The effect of reduced dimension on magnetic ordering is clearly of fundamental importance. Magnetic superlattices with alternate layers of differing magnetization has been theoretically analyzed by *Mills* [4.104] and by *Villeret* et al. [4.105] and explored experimentally by *Sandercock* [4.106], *Grimsditch* [4.107], and *Grünberg* [4.108]. These issues are of great interest in the context of DMS superlattices and are discussed here.

a) Raman Electron Paramagnetic Resonance (Raman-EPR) of Mn^{2+}.

Figure 4.37 shows the remarkable Raman-EPR spectrum observed in a $Cd_{0.9}Mn_{0.1}Se/ZnSe$ superlattice at 5 K, where $Cd_{0.9}Mn_{0.1}Se$ is the well [4.109]. The observed Raman lines labeled 1 PM, ... and 5 PM have their origin in the transitions with $\Delta m_S = 1, 2, 3, 4$, and 5 within the $J = \frac{5}{2}$ Zeeman multiplet of Mn^{2+}. Figure 4.38 shows the linear dependence of the Raman-EPR shift as a function of magnetic field where the solid lines correspond to $g_{Mn^{2+}} = 2$. In Fig. 4.37b we also observe peaks corresponding to 6 PM and 7 PM (see also Fig. 4.38). The spin-flip features higher than $\Delta m_S = 2$ in a DMS can be accounted for in terms of excitations

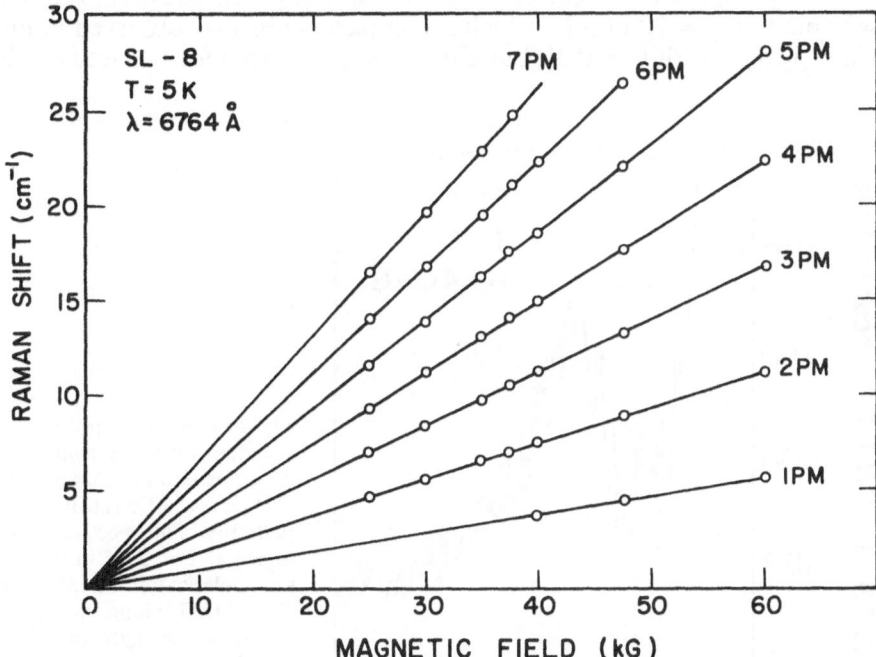

Fig. 4.38 Raman-EPR (PM) shift as a function of magnetic field in the $Cd_{0.9}Mn_{0.1}Se/ZnSe$ superlattice at $T = 5$ K. The solid lines correspond to the Raman-EPR shift given by $n g_{Mn^{2+}} \mu_B H$ with $g_{Mn^{2+}} = 2$ and $n = 1, 2, ..., 7$ [4.109]

within pairs of neighboring Mn^{2+} ions coupled antiferromagnetically and assuming an anisotropic exchange interaction between the ground-state multiplet of one and the excited state of the other. In fact, a pair of neighboring Mn^{2+} in their $^6S_{5/2}$ states in a magnetic field $\boldsymbol{H} \parallel \hat{z}$ can be described by (4.59).

In bulk Mn-based DMSs, at sufficiently high x and low temperatures, the spin flip of isolated Mn^{2+} evolves into the high frequency component of the antiferromagnetic magnon split into a doublet in the presence of a magnetic field. It appears that submicron heterostructures inhibit the formation of this long range order [4.90, 110, 111].

b) Spin Flip from Electrons Bound to Donors

Figure 4.39 shows the Raman spectrum of the $Cd_{0.9}Zn_{0.1}Se/Cd_{0.9}Mn_{0.1}Se$ superlattice, where a Raman line consistent with spin flip is observed. The spectrum is obtained at 5 K in the backscattering configuration $z(xy)\bar{z}$, with a magnetic field of 50 kG along x. As in a bulk DMS, the Raman shift of the donor spin-flip line exhibits a Brillouin-function-like behavior, as can be seen in Fig. 4.40. Since the spin-flip Raman mechanism also involves an interband electronic transition, the observed resonant enhancement for incident frequencies close to excitonic excitations is to be expected. From the slope of the linear portion of the spin-flip data shown in Fig. 4.40, we obtain $g_{eff} = 22$ at 5 K, which is comparable to that observed in bulk $Cd_{1-y}Mn_ySe$ with $y < 0.01$, but much lower than that for the actual value

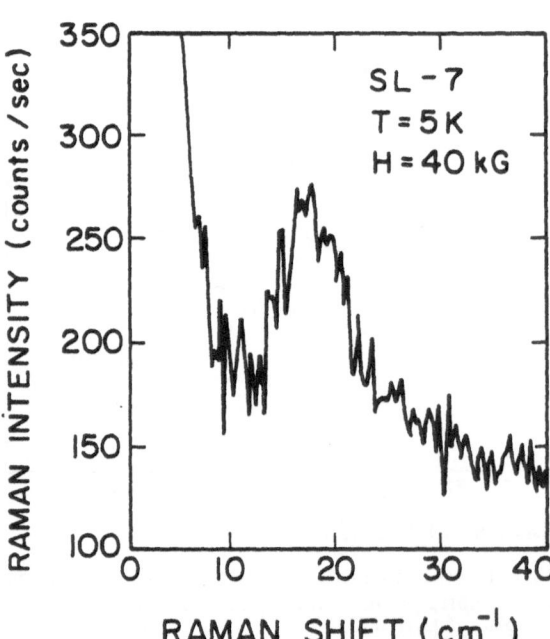

Fig. 4.39. Raman spectrum associated with the spin-flip of electrons bound to donors in a $Cd_{0.9}Zn_{0.1}Se/Cd_{0.9}Mn_{0.1}Se$ superlattice. The spectrum is obtained at $T = 20$ K in the crossed polarization $z(yx)\bar{z}$ with $H = 40$ kG along x and incident wavelength of 6471 Å [4.109]

Fig. 4.40. Magnetic field and temperature dependence of the spin-flip Raman shift in the $Cd_{0.9}Zn_{0.1}Se/$ $Cd_{0.9}Mn_{0.1}Se$ super-lattice with external magnetic field in the plane of the (001) layers. The spectra were obtained in the cross polarization $z(xy)\bar{z}$ with incident laser wavelength $\lambda_L = 6471$ Å [4.109]

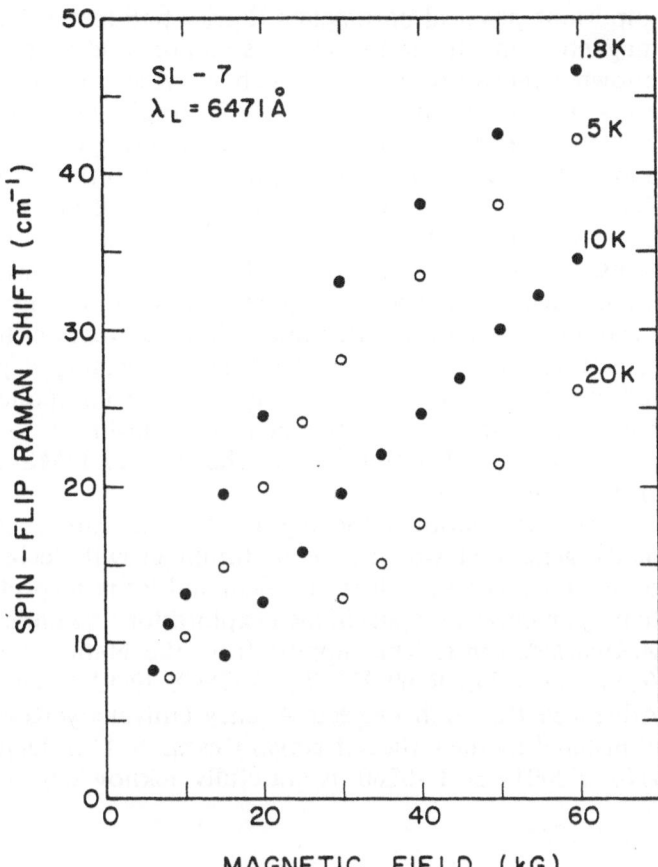

of y in this superlattice. The reason for the smaller Raman shift observed has to be sought in terms of the magnitude of the band offset which could result in electrons being in the quantum well levels while the donors are in the barrier, the well or the interface, the Mn^{2+} ions being in the barrier [4.109].

Finally we draw attention to the spin-flip Raman scattering from $Cd_{1-x}Mn_xTe:In$ epilayers and modulation-doped $Cd_{1-x}Mn_xTe:In/CdTe$ superlattices grown by photoassisted molecular beam epitaxy reported by *Suh* et al. [4.112].

4.7 Concluding Remarks

Diluted magnetic semiconductors is a field where issues significant in the context of magnetism come into sharp focus. The controlled variation of important physical parameters (lattice parameter, band gap, the type of magnetic constituent, free carriers, donors and acceptors incorporated in known concentration, ...), which is typical in semiconductor physics, allows one to establish the microscopic origins of the magnetic interactions. The power and versatility of Raman scattering in addressing these aspects and in discovering unexpected phenomena in the II–VI DMSs have been documented in this chapter. Incorporation of magnetic constituents like Ti^{2+}, V^{2+}, Cr^{2+}, Fe^{2+}, Co^{2+}, Ni^{2+}, and Cu^{2+} in significant concentrations, especially by exploiting MBE [4.113]; systematic doping with donors/acceptors in a predictable fashion [4.114]; novel DMSs based on the tetrahedral coordination but with structures like chalcopyrite, stannite, ...; atomic layer epitaxy permitting, for example, growth of magnetic monolayers separated from non-magnetic components one or several monolayers thick in a predetermined fashion; – progress in these directions is well within reach. The field of DMS will be significantly enriched as a result.

The combination of the physics of DMSs and of the reduced dimensionality generated with non-equilibrium growth techniques like MBE is occupying a central role in this field and Raman spectroscopy is expected to play an important part in their exploration and understanding [4.115].

Acknowledgments: The support from the National Science Foundation (Grant Nos. DMR-89-21717 and DMR-89-13706) and from the Defense Advanced Research Projects Agency-University Research Initiative, administered by the Office of Naval Research, U.S. Department of Defense (No. N00014-86-K-0760) is gratefully acknowledged.

References

4.1   J. K. Furdyna: J. Appl. Phys. **64**, R29 (1988)
4.2   A. S. Barker, A. J. Sievers: Rev. Mod. Phys. **47**, Suppl. 2pp. S1–S179 (1975)
4.3   K. Zanio: *Semiconductors and Semimetals*, Vol. 13, *Cadmium Telluride*, ed. by R. K. Willardson, A. C. Beer (Academic, New York 1978)
4.4   J. A. Gaj, J. Ginter, R. R. Galazka: Phys. Status Solidi B **89**, 655 (1978)
4.5   J. A. Gaj, R. R. Galazka, M. Nawrocki: Solid State Commun. **25**, 193 (1978)
4.6   R. R. Galazka, S. Nagata, P. H. Keesom: Phys. Rev. B **22**, 3344 (1980)
4.7   S. Venugopalan, A. Petrou, R. R. Galazka, A. K. Ramdas: Solid State Commun. **38**, 365 (1981)
4.8   S. Venugopalan, A. Petrou, R. R. Galazka, A. K. Ramdas, S. Rodriguez: Phys. Rev. B **25**, 2681 (1982)
4.9   G. Dolling, T. M. Holden, V. F. Sears, J. K. Furdyna, W. Giriat: J. Appl. Phys. **53**, 7644 (1982)
4.10   See M. Villeret, S. Rodriguez, E. Kartheuser: J. Appl. Phys. **67**, 4221 (1990) for a recent discussion on the energy level spectra of Co^{2+} and Fe^{2+} in DMSs. Figures 2 and 4 in this reference display the effects of the tetrahedral and the trigonal distortions in Co^{2+} and Fe^{2+}, respectively
4.11   See, for example, A. S. Borshchevskii, N. A. Goryunova, F. P. Kesamanly, D. N. Nasledov: Phys. Status Solidi **21**, 9 (1967)

4.12   See, for example, R. L. Gunshor, L. A. Kolodziejski, A. V. Nurmikko, N. Otsuka: *Molecular Beam Epitaxy of II–VI Semiconductor Microstructures*, in *Strained-layer Superlattices*, a special volume edited by T. P. Pearsall in the series *Semiconductors and Semimetals* ed. by R. K. Willardson, A. C. Beer (Academic, New York 1991) vol. 33, pp. 337–409.

4.13   See, M. L. Cohen, J. R. Chelikowsky: *Electronic Structure and Optical Properties of Semiconductors* Springer Ser. Solid-State Sci. Vol. 75 (Springer, Berlin, Heidelberg 1988) Fig. 8.66, p. 139

4.14   A. K. Arora, D. U. Bartholomew, D. L. Peterson, A. K. Ramdas: Phys. Rev. **35**, 7966 (1987)

4.15   N. Samarth, H. Luo, J. K. Furdyna, S. B. Qadri, Y. R. Lee, A. K. Ramdas, N. Otsuka: Appl. Phys. Lett. **54**, 2680 (1989)

4.16   S. Rodriguez, A. K. Ramdas: *Inelastic Light Scattering in Crystals*, in *Highlights of Condensed Matter Theory*, Proceedings of the International School of Physics, "Enrico Fermi" (Course LXXXIX), eds., F. Bassani, F. Fumi, M. P. Tosi (North Holland, Amsterdam 1985) pp. 369–420

4.17   A. K. Ramdas, S. Rodriguez: In *Diluted Magnetic Semiconductors*, Vol. 25 of Semiconductors and Semiconductors, Volume eds., J. K. Furdyna, J. Kossut, Series eds., R. K. Willardson, A. C. Beer (Academic, New York, 1988) pp. 345–412

4.18   W. Hayes, R. Loudon: *Scattering of Light by Crystals* (Wiley, New York 1978)

4.19   S. Bhagavantam, T. Venkatarayudu: *Theory of Groups and Its Application to Physical Problems* (Academic, New York 1969)

4.20   R. Loudon: Proc. Roy. Soc (London) **A275**, 218 (1963)

4.21   R. Loudon: Adv. Physics **13**, 423 (1964); A. S. Barker, R. Loudon: Rev. Mod. Phys. **44**, 18 (1972)

4.22   See Ref. [4.19], p. 140

4.23   A. S. Pine, G. Dresselhaus: Phys. Rev. **188**, 1489 (1969); Phys. Rev. **B4**, 356 (1971)

4.24   M. H. Grimsditch, A. K. Ramdas, S. Rodriguez, V. J. Tekippe: Phys. Rev. B **15**, 5869 (1977)

4.25   W. Imaino, A. K. Ramdas, S. Rodriguez: Phys. Rev. B **22**, 5679 (1980)

4.26   See, for example, L. Couture, J. P. Mathieu: J. Phys. Radium **10**, 145 (1949); M. Cardona: Resonance Phenomena, *Light Scattering in Solids II*, ed by M. Cardona, G. Güntherodt, Topics Appl. Phys. **50**, (Springer, Berlin, Heidelberg 1982) pp. 19–178

4.27   See, for example, I. M. Lifshitz, A. M. Kosevich: Rep. Prog. Phys. **29**, Part I, 217 (1966). See also [4.16]

4.28   A. A. Maradudin, J. Oitmaa: Solid State Commun. **7**, 1143 (1969)

4.29   H. Fröhlich: *Theory of Dielectrics* (Oxford University Press, London 1958)

4.30   L. Genzel, T. P. Martin, C. H. Perry: Phys. Status Solidi **B62**, 83 (1974)

4.31   M. Born, K. Huang: *Dynamical Theory of Crystal Lattices* (Oxford University Press, London 1968)

4.32   A. Mooradian, G. B. Wright: Proceedings of the Ninth International Conference on the Physics of Semiconductors, Moscow, ed. by S. M. Gyvkin (Nauka, Leningrad, 1968) p. 1020

4.33   M. Selders, E. Y. Chen, R. K. Chang: Solid State Commun. **12**, 1057 (1973)

4.34   D. L. Peterson, A. Petrou, W. Giriat, A. K. Ramdas, S. Rodriguez: Phys. Rev. B **33**, 1160 (1986)

4.35   J. K. Furdyna, W. Giriat, D. F. Mitchell, G. I. Sproule: J. Solid State Chem. **46**, 349 (1983)

4.36   Y. R. Lee, A. K. Ramdas, R. L. Aggarwal: Phys. Rev. B **38**, 10600 (1988)

4.37   B. A. Weinstein: Proceedings of the XIII International Conference on the Physics of Semiconductors, Rome, 1976, ed. by F. Fumi (Tipografia, Marves, Rome, 1976) p. 326; R. L. Schmidt, K. Kunc, M. Cardona, H. Bilz: Phys. Rev. B **20**, 3345 (1979)

4.38   C. T. Sennet, D. R. Bosomworth, W. Hayes, A. R. L. Spray: J. Phys. **C2**, 1137 (1969)

4.39   J. M. Rowe, R. M. Nicklow, D. L. Price, K. Zanio: Phys. Rev. B **10**, 671 (1974)

4.40   R. M. Martin, L. M. Falicov: Resonant Raman Scattering in *Light Scattering in Solids*, ed. by M. Cardona, Topics Appl. Phys. **8** (Springer, Berlin, Heidelberg 1975) pp. 79–145

4.41   W. Richter: Resonant Raman Scattering in Semiconductors, *Solid State Physics* Springer Tracts in Modern Physics Vol. 78 (Springer, Berlin, Heidelberg 1976) p. 121

4.42   R. Trommer, G. Arbstreiter, M. Cardona: In *Lattice Dynamics*, ed. by M. Balkanski (Flammarion Sciences, Paris, 1978) p. 189; A. Cantarero, C. Trallero-Giner, M. Cardona: Phys. Rev. B **40**, 12290 (1989)

4.43   A. C. Petrou: Ph.D. Thesis (Purdue University, 1983)

4.44   J. Menéndez, M. Cardona, L. K. Vodopayanov: Phys. Rev. B **31**, 3705 (1985); Z. C. Feng, S. Perkowitz, J. M. Wrobel, J. J. Dubowski: Phys. Rev. B **39**, 12997 (1989); S. Perkowitz, Z. C. Feng, A. Erbil, R. Sudharsanan, K. T. Pollard, A. Rohatgi: In *Raman Scattering, Luminescence and Spectroscopic Instrumentation in Technology*, ed. by F. Adar, J. E. Griffiths, J. M. Lerner, SPIE vol. 1055, p. 76 (1989)

4.45   A. Mooradian, T. C. Harman: In *The Physics of Semimetals and Narrow Gap Semiconductors*, ed. by L. Carter, R. T. Bates (Pergamon, Oxford 1971) p. 297

4.46   L. D. Landau, E. M. Lifshitz: *Electrodynamics of Continuous Media* (Pergamon, Oxford, 1960) p. 331

4.47   J. Lambe, C. Kikuchi: Phys. Rev. **119**, 1256 (1960)

4.48   A. Petrou, D. L. Peterson, S. Venugopalan, R. R. Galazka, A. K. Ramdas, S. Rodriguez: Phys. Rev. B **27**, 3471 (1983)

4.49   P. A. Fleury, R. Loudon: Phys. Rev. **166**, 514 (1968)

4.50   R. Loudon: J. Raman Spectrosc. **7**, 10 (1978)

4.51   M. G. Cottam, D. J. Lockwood: *Light Scattering in Magnetic Solids*, (Wiley, New York 1986)

4.52   A. Twardowski, M. Nawrocki, J. Ginter: Phys. Status Solidi B **96**, 497 (1979)

4.53   D. L. Peterson, D. U. Bartholomew, A. K. Ramdas, S. Rodriguez: Phys. Rev. B **31**, 7932 (1985)

4.54   R. Planel, J. Gaj, C. Benoit a la Guillaume: J. de Phys. Colloq. **41**, C5–39 (1980)

4.55   See Ref. [4.3] p. 100

4.56   Y. R. Lee, A. K. Ramdas: Solid State Commun. **51**, 861 (1984)

4.57   S. M. Rybachenko, O. V. Terletskii, I. B. Miezetskaya, G. S. Oleinik: Fiz. Tekh. Poluprovodn. **15**, 2314 (1981) [English translation: Sov. Phys. Semicond. **15**, 1345 (1981)]

4.58   J. A. Gaj, R. Planel, G. Fishman: Solid State Commun. **29**, 435 (1979)

4.59   Y. Oka, M. Cardona: Phys. Rev. B **23**, 4129 (1981); J. de Phys. Colloq. **42**, C6-459 (1981)

4.60   E. N. Economou, J. Ruvalds, K. L. Ngai: Phys. Rev. Lett. **29**, 110 (1972)

4.61   S. Geschwind, R. Romestain: In *Light Scattering in Solids IV*, ed. by M. Cardona, G. Güntherodt Topics Appl. Phys. **54** (Springer, Berlin, Heidelberg 1984) pp. 151–201

4.62   D. U. Bartholomew, E.-K. Suh, S. Rodriguez, A. K. Ramdas, R. L. Aggarwal: Solid State Commun. **62**, 235 (1987)

4.63   W. Y. Ching, D. L. Huber: Phys. Rev. B **25**, 5761 (1982); Phys. Rev. B **26**, 6164 (1982)

4.64   F. F. Geyer, H. Y. Fan: IEEE J Quantum Electron. **QE-16**, 1365 (1980)

4.65   M. Nawrocki, R. Planel, G. Fishman, R. Galazka: Proceedings of the XV International Conference on the Physics of Semiconductors, ed. by S. Tanaka, Y. Toyozawa, J. Phys. Soc. Jpn. **49**, Suppl. A, 823 (1980)

4.66   M. Nawrocki, R. Planel, G. Fishman, R. Galazka: Phys. Rev. Lett. **46**, 735 (1981)

4.67   D. L. Peterson, A. Petrou, M. Dutta, A. K. Ramdas, S. Rodriguez: Solid State Commun. **43**, 667 (1982)

4.68   M. Dobrowolska, H. D. Drew, J. K. Furdyna, T. Ichiguchi, A. Witkowski, P. A. Wolff: Phys. Rev. Lett. **49**, 845 (1982)

4.69   D. D. Alov, S. I. Gubarev, V. B. Timofeev, B. N. Shepel: Pis'ma Zh. Eksp. Teor. Fiz. **34**, 76 (1981) [English Translation: JETP Lett. **34**, 71 (1981)]

4.70   D. L. Alov, S. I. Gubarev, V. B. Timofeev: Zh. Eksp. Teor. Fiz. **84**, 1806 (1983) [English Translation: Sov. Phys. JETP **57**, 1052 (1983)]

4.71   D. Heiman, P. A. Wolff, J. Warnock: Phys. Rev. B **27**, 4848 (1983)

4.72   D. Heiman, Y. Shapira, S. Foner: Solid State Commun. **45**, 899 (1983); D. Heiman,
         Y. Shapira, S. Foner: Solid State Commun. **51**, 603 (1984)
4.73   D. Heiman, Y. Shapira, S. Foner, B. Khazai, R. Kershaw, R. Dwight, A. Wold:
         Phys. Rev. B **29**, 5634 (1984)
4.74   E. D. Issacs, D. Heiman, P. Becla, Y. Shapira, R. Kershaw, A. Wold: Phys. Rev.
         B **38**, 8412 (1988); E. D. Issacs, D. Heiman, M. J. Graf, B. B. Goldberg, R. Kershaw,
         D. Ridgley, K. Dwight, A. Wold, J. K. Furdyna, J. S. Brooks: Phys. Rev. B **37**, 7108
         (1988)
4.75   K. Douglas, S. Nakashima, J. F. Scott: Phys. Rev. B **29**, 5602 (1984)
4.76   D. L. Peterson, D. U. Bartholomew, U. Debska, A. K. Ramdas, S. Rodriguez: Phys.
         Rev. B **32**, 323 (1985)
4.77   T. Dietl, J. Spałek: Phys. Rev. Lett. **48**, 355 (1982)
4.78   T. Dietl, J. Spałek: Phys. Rev. B **28**, 1548 (1983)
4.79   D. U. Bartholomew, E.-K. Suh, A. K. Ramdas, S. Rodriguez, U. Debska, J. K.
         Furdyna: Phys. Rev. B **39**, 5865 (1989). See also: E.-K. Suh, D. U. Bartholomew,
         J. K. Furdyna, U. Debska, A. K. Ramdas, S. Rodriguez: Bull. Am. Phys. Soc. **32**,
         802 (1987)
4.80   H. Serre, G. Bastard, C. Rigaux, J. Mycielski, J. K. Furdyna: In 4th International
         Conference on the Physics of Narrow-gap Semiconductors, Linz, 1982 ed. by
         E. Gornik, H. Heinrich, L. Palmetshofer (Springer, New York) p. 321
4.81   D. Heiman, A. Petrou, S. H. Bloom, Y. Shapira, E. D. Issacs, W. Giriat: Phys. Rev.
         Lett. **60**, 1876 (1988)
4.82   D. Scalbert, J. A. Gaj, A. Mauger, J. Cernogora, C. Benoit á la Guillaume: Phys.
         Rev. Lett. **62**, 2865 (1989)
4.83   K. Ploog, G. H. Döhler: Adv. Phys. **32**, 285 (1983)
4.84   R. Dingle: *Advances in Solid State Physics*, Vol. 15, ed. by H. J. Queisser (Pergamon
         Vieweg, Braunschweig 1975) p. 21
4.85   M. V. Klein: IEEE J. Quantum Electron. **22**, 1760 (1986)
4.86   L. A. Kolodziejski, T. Sakamoto, R. L. Gunshor, S. Datta: Appl. Phys. Lett. **44**,
         799 (1984)
4.87   R. N. Bicknell, N. C. Giles-Taylor, D. K. Blanks, R. W. Yanka, E. L. Buckland, J.
         F. Schetzina: J. Vac. Sci. Technol. B**3**, 709 (1985)
4.88   S. Datta, J. K. Furdyna, R. L. Gunshor: Superlattices and Microstructures **1**, 327
         (1985)
4.89   M. v. Ortenberg: Phys. Rev. Lett **49**, 1041 (1982)
4.90   E.-K. Suh, D. U. Bartholomew, A. K. Ramdas, S. Rodriguez, S. Venugopalan, L.
         A. Kolodziejski, R. L. Gunshor: Phys. Rev. B **36**, 4316 (1987)
4.91   M. Cardona, G. Güntherodt (eds.): *Light Scattering in Solids V*, Topics Appl. Phys.
         66 (Springer, Berlin, Heidelberg 1989)
4.92   See, for example, C. Colvard, T. A. Gant, M. V. Klein, R. Merlin, R. Fischer, H.
         Morkoc, A. C. Gossard: Phys. Rev. B **31**, 2080 (1985)
4.93   B. Jusserand, D. Paquet: Phys. Rev. Lett., **56**, 1752 (1986); A. K. Sood, J. Menéndez,
         M. Cardona, K. Ploog: Phys. Rev. Lett. **56**, 1753 (1986)
4.94   A. K. Sood, J. Menéndez, M. Cardona, K. Ploog: Phys. Rev. Lett. **54**, 2111 (1985)
4.95   B. Jusserand, M. Cardona: In Ref. [4.91] pp. 49–152
4.96   R. Fuchs, K. L. Kliewer: Phys. Rev. **140**, A2076 (1965)
4.97   A. K. Sood, J. Menéndez, M. Cardona, K. Ploog: Phys. Rev. Lett. **54**, 2115 (1985)
4.98   See N. Bottka, J. Stankiewicz, W. Giriat: J. Appl. Phys. **52**, 4189 (1981)
         for $Cd_{1-x}Mn_xTe$ and L. A. Kolodziejski, R. L. Gunshor, R. Venkatasubramanian,
         T. C. Bonsett, R. Frohne, S. Datta, N. Otsuka, R. B. Bylsma, W. M. Becker, A. V.
         Nurmikko: J. Vac. Sci. Technol. B **4**, 583 (1986) for $Zn_{1-x}Mn_xSe$
4.99   C. Colvard, R. Merlin, M. V. Klein, A. C. Gossard: Phys. Rev. Lett. **45**, 298 (1980)
4.100   S. M. Rytov: Akust. Zh. **2**, 71 (1956). [Sov. Phys. Acoust. **2**, 68 (1956)]
4.101   R. G. Alonso, Eunsoon Oh, A. K. Ramdas, N. Samarth, H. Luo, J. K. Furdyna, L.
         R. Ram Mohan: Unpublished

4.102 X.-C. Zhang, S.-K. Chang, A. V. Nurmikko, L. A. Kolodziejski, R. L. Gunshor, S. Datta: Phys. Rev. B **31**, 4056 (1985); S.-K. Chang, A. V. Nurmikko, L. A. Kolodziejski, R. L. Gunshor: Phys. Rev. B **33**, 2589 (1986)

4.103 R. L. Gunshor, L. A. Kolodziejski, N. Otsuka, S. K. Chang, A. V. Nurmikko: J. Vac. Sci. Technol. A **4**, 2117 (1986)

4.104 D. L. Mills: Collective Excitations in Superlattice Structures, *Light Scattering in Solids V*, ed. by M. Cardona, G. Güntherodt Topics Appl. Phys. **66** (Springer, Berlin, Heidelberg 1989) pp. 13–48

4.105 M. Villeret, S. Rodriguez, E. Kartheuser: Phys. Rev. **39**, 2583 (1989)

4.106 J. R. Sandercock: Trends in Brillouin Scattering: Studies of Opaque Materials, Supported Films and Central Modes, *Light Scattering in Solids III*, ed. by M. Cardona, G. Güntherodt, Topics Appl. Phys. **51** (Springer, Berlin, Heidelberg 1982) pp. 173–206

4.107 M. H. Grimsditch: Brillouin Scattering from Metallic Superlattices, *Light Scattering in Solids V*, ed by M. Cardona, G. Güntherodt Topics Appl. Phys. **66** (Springer, Berlin, Heidelberg 1989) pp. 285–303

4.108 P. Grünberg: Light Scattering from Spin Waves in Thin Films and Layered Magnetic Structures, *Light Scattering in Solids V*, ed. by M. Cardona, G. Güntherodt Topics Appl. Phys. **66** (Springer, Berlin, Heidelberg 1989) pp. 303–335

4.109 R. G. Alonso, E.-K. Suh, A. K. Ramdas, N. Samarth, H. Luo, J. K. Furdyna: Phys. Rev. B **40**, 3720 (1989)

4.110 R. L. Gunshor, L. A. Kolodziejski, N. Otsuka, B. P. Gu, D. Lee, Y. Hefetz, A. V. Nurmikko: Superlattices and Microstructures, **3**, 5 (1987)

4.111 D. D. Awschalom, J. M. Hong, L. L. Chang, G. Grinstein: Phys. Rev. Lett. **59**, 1733 (1987)

4.112 E.-K. Suh, D. U. Bartholomew, A. K. Ramdas, R. N. Bicknell, R. L. Harper, N. C. Giles, J. F. Schetzina: Phys. Rev. B **36**, 9358 (1987)

4.113 B. T. Jonker, J. J. Krebs, G. A. Prinz: Appl. Phys. Lett. **50**, 848 (1987); B. T. Jonker, J. J. Krebs, G. A. Prinz: Appl. Phys. Lett. **53**, 450 (1988)

4.114 R. N. Bicknell, N. C. Giles, J. F. Schetzina: Appl. Phys. Lett. **49**, 1095 (1986); R. N. Bicknell, N. C. Giles, J. F. Schetzina: Appl. Phys. Lett. **49**, 1735 (1986)

4.115 A. V. Nurmikko, Q. Fu, D. Lee, R. L. Gunshor, L. A. Kolodziejski: Proceedings of SPIE Conference on Raman Scattering, Liminescence and Spectroscopic Instrumentation in Technology, ed. by F. Adar, J. E. Griffiths, J. M. Lerner, Vol. **1055**, 47 (1989); Q. Fu, D. Lee, A. V. Nurmikko, L. A. Kolodziejski, R. L. Gunshor: Phys. Rev. B **39**, 3173 (1989); N. Pelekanos, Q. Fu, J. Ding, W. Wałecki, A. V. Nurmikko, S. M. Durbin, J. Han, M. Kobayashi, R. L. Gunshor, Phys. Rev. B **41**, 9966 (1990)

Relevant Publications which have appeared since the review was written.

4.116 S. I. Gubarev, T. Ruf, and M. Cardona, Phys. Rev. B **43**, 1551 (1991). Doubly resonant Raman scattering in the semimagnetic semiconductor $Cd_{0.95}Mn_{0.05}Te$

4.117 W. Limmer, H. Leiderer, K. Jakob, and W. Gebhardt, Phys. Rev. B **42**, 11325 (1990). Resonant Raman scattering by longitudinal-optical phonons in $Zn_{1-x}Mn_xSe$ ($x = 0$, 0.03, 0.1) near the E_0 gap.

4.118 W. J. Keeler, H. Huang, and J. J. Dubowski, Phys. Rev. B **42**, 11355 (1990). Multiphonon resonant Raman scattering in high-manganese-concentration $Cd_{1-x}Mn_xTe$ films.

5. Light Scattering in Rare Earth and Actinide Intermetallic Compounds

Eberhard Zirngiebl and Gernot Güntherodt

With 54 Figures

The advent of light scattering applied to the spectroscopic investigations of metals occurred about a decade ago. Case studies have focused on layered charge-density-wave materials, superconductors, graphite intercalation compounds, thin metallic layers and metallic superlattices. Reviews of these topics can be found in the books on "Light Scattering in Solids" edited by Cardona and Güntherodt [5.1, 2].

5.1 Overview

In this Chapter we give a status report on the progress that has been made in applying light scattering to various rare earth (R) and actinide (A) intermetallic compounds with the emphasis on Kondo- and intermediate-valence(IV)-type materials and on heavy fermion (HF) compounds. In this field of research light scattering can now provide detailed information about localized electronic excitations [such as crystalline electric field (CEF) excitations, spin–orbit split (J) multiplet excitations], phononic and elastic properties as well as about spin and electron density fluctuations. This has not been an easy task, since for the materials under investigation the penetration depth of the light is of the order of about 100 Å, yielding a rather small scattering volume. In particular we want to show how inelastic and quasielastic light scattering has become a valuable complementary tool compared to neutron scattering, with the advantages of high resolution (Raman $\lesssim 0.5$ meV, Brillouin $\lesssim 10$ μeV), strict symmetry selection rules, high ($\lesssim 1000$ meV) as well as extremely small ($\gtrsim 0.01$ meV) energy losses observable, small usable sample size ($\lesssim 1$ mm^3) and independence of rare earth isotopes. The disadvantage that light scattering is limited to small wave vectors of the excitations, is irrelevant in the case of localized electronic excitations. Difficulties due to the small penetration depth of the light can be overcome by careful surface preparation techniques, such as cleaving or fracturing under inert gas atmosphere. Moreover, recent advances in optical multichannel detector technology, as reviewed by *Tsang* in [Ref. 5.2c, Chap. 6], have yielded tremendous improvements in detecting phenomena previously inaccessible to Raman spectroscopy. Consequently, one can expect an additional impact in the near future on the field of research summarized here.

Such an impact has already occurred in applying the novel technique of the tandem Fabry-Perot interferometer, described by *Sandercock* in [Ref. 5.2a, Chap. 6] to the determination of the elastic properties of rare-earth and actinide intermetallic compounds, which are often available only as rather small (< 1 mm^3) or irregularly shaped samples not accessible to ultrasonic measurements. The scattering from bulk acoustic phonons is related to the elasto-optic effect in the case of large penetration depth of the light, for instance for laser excitation above the plasma reflection edge. In most cases, however, the scattering occurs from surface wave excitations due to the surface ripple mechanism. The method then takes advantage of the scattering cross section being proportional to the reflectivity of the sample. The experimental situation is sketched in Fig. 5.1 for the commonly used back-scattering configuration, where k_I, k_S and k_R are the wave vectors of the incident, scattered and reflected photon, respectively. Due to momentum conservation, the wave-vector component of the surface-wave excitation $q_\parallel$ parallel to the surface is the relevant quantity and can be varied by changing the angle of incidence Θ_I with respect to the surface normal.

Another major goal of the tandem Fabry-Perot interferometry has been the possibility of observing quasielastic light scattering on a frequency scale of ± 100 GHz. Hence, besides inelastic excitations due to bulk or surface wave excitations at frequency shifts $\Delta\omega = \pm vq$, where v is the

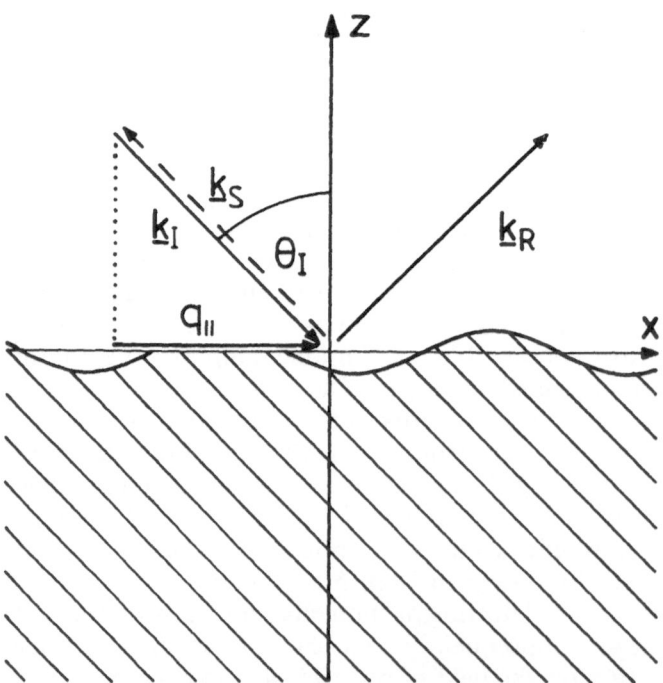

Fig. 5.1. 180° back-scattering geometry ($\Theta_S = \Theta_I$) used for Brillouin scattering experiments. The incident light with wavevector k_I is scattered from surface ripple excitations with wavevector $q_\parallel$. $k_{S(R)}$ denotes the wavevector of the scattered (reflected) beam

Fig. 5.2. Schematic light scattering spectrum showing quasielastic scattering intensity (I_{qe}) due to diffusive or relaxational modes and inelastic Stokes ($+\Delta\omega$) and anti-Stokes ($-\Delta\omega$) scattering intensity (I_{in}) due to propagating modes of frequency $\omega = vq$, where v is the velocity and q the wave vector. The full width at half maximum of the quasielastic intensity is given by $\Gamma = 2Dq^2$, with D the diffusion constant. The Landau-Placzek ratio R is defined as $R = I_{qe}/2I_{in}$

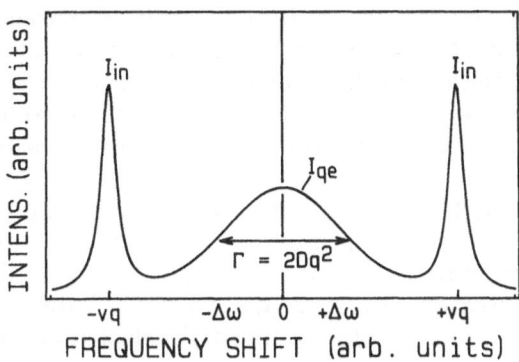

FREQUENCY SHIFT (arb. units)

sound velocity, it is now feasible to observe diffusive, relaxational or fluctuation processes giving rise to quasielastic light scattering of full width at half maximum (FWHM) Γ as sketched in Fig. 5.2. This has enabled the first observation of electron density fluctuations in a heavy fermion (HF) compound and an enhanced Landau-Placzek ratio (Sect. 5.3.2), which has no analog in an ordinary metal.

This chapter is organized according to the types of excitations in rare earth and actinide intermetallics that have been investigated successfully by light scattering. Starting with intraconfigurational and interconfigurational localized electronic excitations in Sects. 5.2.3 and 5.2.4, we turn to CEF excitations in Sects. 5.2.5 and 5.2.6. Spin fluctuations and electron density fluctuations in HF compounds are discussed in Sects. 5.3.1 and 5.3.2, respectively. Section 5.4 is devoted to lattice dynamics in rare earth and actinide intermetallics, i.e. the investigation of optical phonon modes by Raman scattering and of elastic constants by Brillouin scattering. Particular emphasis is placed on the identification of charge fluctuation rates in intermediate valence (IV) compounds by a systematic analysis of optical and acoustic phonon anomalies. Finally in Sect. 5.5 we will deal with light scattering from modes of electronic as well as phononic character resulting from strong electron–phonon coupling. The electronic part of the coupled modes is due to either intraconfigurational J-multiplet or CEF excitations.

5.2 Light Scattering
from 4f and 5f Localized Electronic Excitations

5.2.1 Intermediate Valence Compounds

Among the various classes of rare-earth intermetallic compounds those exhibiting intermediate valence or Kondo behavior have received much attention over the last decade. The occurrence of two different energetically degenerate configurations of the rare earth atoms ($4f^n5d^16s^2$ and $4f^{n+1}6s^2$), together with the strongly localized character of the 4*f* shell are responsible for this seemingly irregular pattern of properties exhibited by these compounds. Despite the near degeneracy of 4*f*, 5*d* and 6*s* binding energies allowing for 4*f* interconfigurational excitations ($4f^{n+1}6s^2 \rightarrow 4f^n5d^16s^2$), the spatial localization of the 4*f* electrons inside the filled 5*s* and 5*p* orbitals of the Xe cores, as discussed for example by *Goldschmidt* [5.3], is the reason why the 4f states largely retain their highly correlated, atomic nature after compound formation. These particular properties of the 4*f* shell, apart from being the cause of some of the most interesting aspects of rare earth research, are also the source of the problems entailed in giving a quantitative description of the electronic structure of these materials. These problems arise because of the inadequacy of the conventional methods of band structure calculation to handle correlated many-electron states. On the other hand, descriptions of the intermediate valent 4*f* shell on the basis of an ionic picture, characterized by single-ion intraconfigurational J-multiplet-level and crystal-field excitations, taking into account hybridization perturbatively, is the appropriate way to describe Kondo- and intermediate valence (IV) behavior [5.4–6]. The ionic picture was first promoted in the interconfigurational fluctuation (ICF) model by *Hirst* [5.7].

In this model the interaction between two 4*f* configurations, both described by their J-multiplet level structure, is parameterized first by the interconfigurational excitation energy (E_x), denoting the energy difference between the ground states of the two J-multiplet systems, and second by the fluctuation temperature (T_f) or interconfigurational mixing width. This model has been applied to the analysis and interpretation of experimental data [5.8–10]. A schematic picture of this model is given in Fig. 5.3. In any case, these modified low lying many-particle electronic excitations (J-multiplet level excitations, crystal-field excitations, interconfigurational

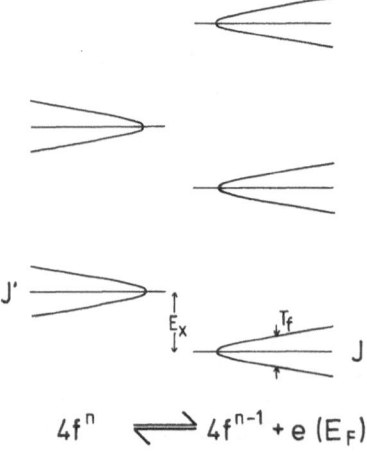

5.3. Schematic energy level diagram of the interconfigurational fluctuation (ICF) model describing valence fluctuations between two 4f configurations ($4f^n$, $4f^{n-1}$), described by their J-multiplet level structure. The basic parameters of the ICF model, E_x and T_f, denote the interconfigurational excitation energy and interconfigurational mixing width, respectively

excitations) profoundly influence the thermodynamic and transport properties of IV and Kondo compounds. Their experimental determination serves not only as spectroscopic test of microscopic theories but is an indispensible prerequisite for any understanding of the fascinating and numerous macroscopic properties of these compounds. A more comprehensive presentation of the many experimental and theoretical aspects of IV and Kondo compounds as well as a compilation of original references can be found in a number of conference proceedings and review articles, for example [5.4–6, 11].

The application of Raman scattering to the rare earth Kondo and IV compounds has resulted in a fruitful area of research as judged from its contributions to the understanding of these materials and from the large number of publications. Later in this section we review Raman scattering experiments that deal with the determination of intraconfigurational excitations (J-multiplet and crystal-field level excitations) as well as interconfigurational excitation energies of valence unstable rare earth compounds.

5.2.2 Heavy Fermion Compounds

Since the discovery of the heavy fermion (HF) superconductor $CeCu_2Si_2$ by *Steglich* et al. in 1979 [5.12], much attention in solid state physics has been focused on heavy fermion compounds like UPt_3, UBe_{13} and URu_2Si_2. The discovery of bulk superconductivity in UPt_3, together with a $T^3 \ln (T/T_{sf})$ term in the low temperature specific heat, where T_{sf} is a spin fluctuation temperature, led to speculations about the important role of spin fluctuations for mediating a non-BCS electron pairing in the heavy fermion superconductors.

All HF compounds are characterized by a similarly high value of the electronic specific heat γ, which has become the standard criterion for the classification of HF systems [5.13]. In a Fermi liquid model, values for γ of about 100 times that of an ordinary metal indicate a very high density of states at the Fermi energy E_F. This can be related to narrow f bands due to a hybridization of d and f states. One then describes these highly correlated electrons and the corresponding many-body effects by attributing a rather high effective mass to the f electrons. As a consequence of the larger radial extent of the $5f$ wave function compared to $4f$ electrons and a possible direct $f - f$ overlap, the tendency towards delocalization, i.e. band formation, is much more pronounced in $5f$ than in $4f$ compounds. For example, one finds well defined crystalline-electric-field (CEF) levels in Kondo-type $4f$ compounds, such as $CeCu_2Si_2$ [5.14] and CeB_6 [5.15]. Neutron scattering and Raman spectroscopy have been versatile and complementary tools in the investigation of the localized $4f$ ground state of CeB_6 and other $4f$-compounds [5.15–17]. On the other hand, not much information is available about CEF levels in metallic actinide compounds.

In this section we review Raman scattering results of the HF compounds $CeCu_2Si_2$, UPt_3, UBe_{13} and URu_2Si_2. We present evidence of at least partially localized $5f$-electron character in the HF compounds UBe_{13} and UPt_3 [5.18a, b, 19]. We discuss investigations of spin fluctuations in the HF compounds UPt_3 [5.18c] and UBe_{13} [5.20a, b] by means of quasielastic Raman scattering, with the emphasis on the spin relaxation rate Γ_s at $q = 0$. This allows us to test the prediction of the Fermi liquid theory that $\Gamma_s = v_F q$, where v_F is the Fermi velocity of the heavy particles. For the heavy fermion compound URu_2Si_2 exhibiting coexisting magnetic order ($T_N = 17.5$ K) and superconductivity ($T_c = 1.5$ K) [5.21–23] quasielastic Raman scattering as a function of temperature [5.24] will be reviewed. The most recent achievements concern the experimental finding of a diffusive mode due to electron density fluctuations in UPt_3 [5.25], theoretically predicted for heavy fermion compounds [5.26a, b].

5.2.3 Electronic Raman scattering in $EuPd_2Si_2$ and $EuCu_2Si_2$

Eu-based IV systems have been studied quite extensively as model compounds, since their fluctuation temperature is quite often in the vicinity of 100 K. Thus, on the one hand large enough to induce quite dramatic modifications in the thermodynamic as well as transport properties of the compound under investigation and, on the other hand, small enough to retain the basic multiplet structure of the $Eu^{2+}4f^6(^7F_7)$ ion. Moreover, Eu-based compounds serve as model systems because of the rather simple ground states of the two $4f$ configurations participating in the valence fluctuation process: $Eu^{2+}(4f^7)$ has only the $J = S = 7/2$ pure spin configuration without any CEF splitting and $Eu^{3+}(4f^6)$ has a $J = 0$ ground state. The $J = 1$ state of Eu^{3+} lies about 550 K [5.27] above the $J = 0$ ground state and can be thermally populated, thus offering the possibility of a strong modification of the valence fluctuation process by temperature variation. Moreover, for Eu-based IV compounds the investigation of the valence and its temperature dependence is easily feasible by Mössbauer- and L_{III}-edge absorption spectroscopies. In particular, IV $EuPd_2Si_2$ [5.28, 29] as well as $EuCu_2Si_2$ [5.9, 30] are sufficiently well characterized to establish helpful relationships between thermodynamic data and spectroscopic results.

Raman scattering measurements were carried out on freshly fractured polycrystals of $EuPd_2Si_2$ between 300 and 4.2 K [5.31]. In Fig. 5.4 the Raman spectra of $EuPd_2Si_2$ obtained with 5309-Å laser excitation are shown for selected temperatures. At the top of the spectra are shown the J-multiplet levels observed by electronic Raman scattering in stable valent isoelectronic SmSe [5.32]. The maxima observed in $EuPd_2Si_2$ are close in energy to the J-multiplet levels of SmSe. This is to be expected, since the J levels of Eu^{3+} should be the same as those of the $Sm^{2+}4f^6(^7F_J)$

Fig. 5.4. Raman spectra of $EuPd_2Si_2$ at different temperatures under 5309-Å laser excitation. Vertical dashed guide lines mark the intraconfigurational excitations ($E_{L,intra}$) of the $Eu^{3+}4f^6(^7F_J)$ configuration; solid lines through spectra are guide lines to the eye. Top of the figure: Electronic Raman scattering due to the $Sm^{2+}4f^6(^7F_J)$ configuration of SmSe

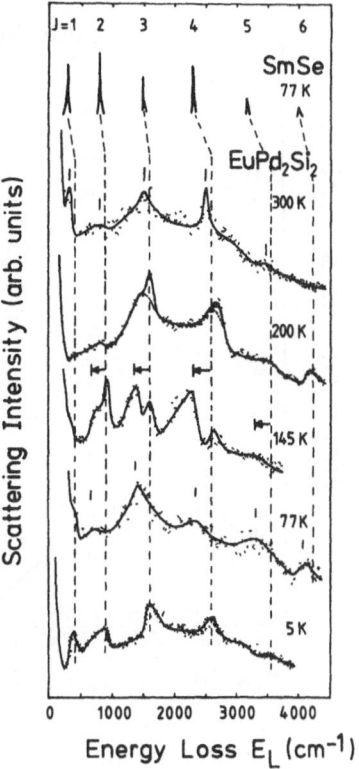

configuration. The J-multiplet levels of IV $EuPd_2Si_2$ at 300 K are quite broad compared to the J levels of divalent SmSe. Upon cooling to 200 K, the J levels of $EuPd_2Si_2$ undergo an asymmetric broadening and this is most clearly seen for the J = 3 and J = 4 levels. This broadening develops into a well-pronounced splitting at 145 K, which is best seen in the case of the J = 2, 3 and 4 levels. At 77 K the splitting of individual J levels has disappeared, yielding a single broadened peak for each one of the J levels. The spectrum at 4.2 K reveals well defined J-multiplet levels similar to the spectrum at 300 K. The reproducibility of the spectral features in Fig. 5.4 is presumably slightly affected by local strains giving rise to frequency shifts of the levels of up to 3%. The overall background in Fig. 5.4 arises mainly from sample imperfections causing luminescence emission.

Figure 5.5 shows the Raman spectra of freshly fractured $EuCu_2Si_2$ at selected temperatures (300, 77 and 4 K) under 5309-Å laser excitation [5.33]. At room temperature one observes a broad background of luminescence emission, which is also present in the low temperature spectra. The double peak structure near 2700 cm^{-1} is close to the energy of the J = 0 → 4 transition of the free Eu^{3+} ion [5.34]. By cooling down to 77 K broad structures emerge near 800 and 1800 cm^{-1}, and the double peak

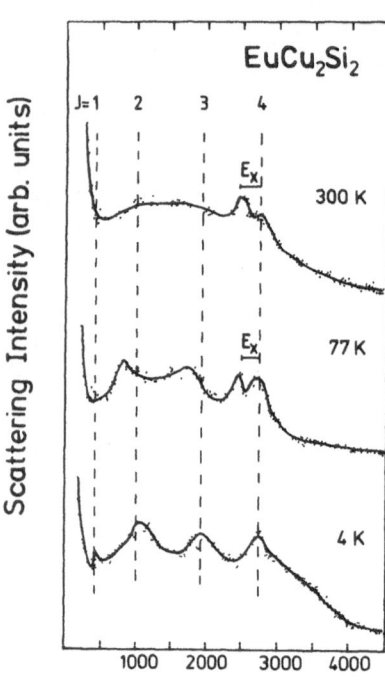

Fig. 5.5. Raman spectra of $EuCu_2Si_2$ at different temperatures under 5309-Å laser excitation. Vertical dashed lines mark the intraconfigurational excitations ($E_{L, intra}$) of the free Eu^{3+} ion having the $4f^6(^7F_{J=0...6})$ configuration. Solid lines through spectra are guides to the eyes

structure near 2700 cm^{-1} becomes even more pronounced. For the latter the splitting vanishes at 4 K, but the other peaks become sharper. Moreover, at 4 K a fourth peak near 400 cm^{-1} can be resolved from the background. For comparison the J-level excitations of the free Eu^{3+}-ion for J = 1 to J = 4 [5.34] are indicated by the dashed vertical lines in Fig. 5.5.

In Fig. 5.6 we show a schematic energy level diagram explaining the electronic Raman scattering from the J-multiplet levels in an IV Eu compound by applying the interconfigurational fluctuation model of intermediate valence [Ref. 5.7, Fig. 5.3] to Raman scattering from IV Eu compounds. The (Eu^{2+}) $4f^7$ and (Eu^{3+}) $4f^6 + e^-$ configurations are nearly degenerate and separated by E_x. When $E_x \neq 0$, two sets of inelastic energy losses $E_{L, intra}$ and $E_{L, inter} = E_{L, intra} - E_x$ are observable in Raman spectroscopy for the J-multiplet levels of the $4f^6$ configuration. These are the intraconfigurational excitations ($E_{L, intra}$) of the J = 1, ..., 6 levels with respect to the $(4f^6)$ J = 0 initial state and the interconfigurational excitations ($E_{L, inter}$) with respect to the $(4f^7)$ J = 7/2 initial state. The interconfigurational excitation energy E_x itself should also show up with respect to the J = 0 initial state. The inelastic scattering intensity of the interconfigurational excitations $(4f^7 \rightarrow 4f^6 + e^-)$ is obtained in second-order perturbation theory and should be stronger than for the magnetic-dipole-allowed intraconfigurational excitations (J = 0 $\rightarrow$ J $\neq$ 0). The latter con-

Fig. 5.6. Schematic representation of the electronic Raman scattering process in an intermediate-valence Eu compound showing an interconfigurational excitation energy $E_x \neq 0$. Indicated are the intraconfigurational energy losses ($E_{L,\text{intra}}$) and the interconfigurational energy losses ($E_{L,\text{inter}} = E_{L,\text{intra}} - E_x$) of the incident photon $\hbar\omega_i$, with respect to the $J = 0$ and $J = 7/2$ initial state, respectively

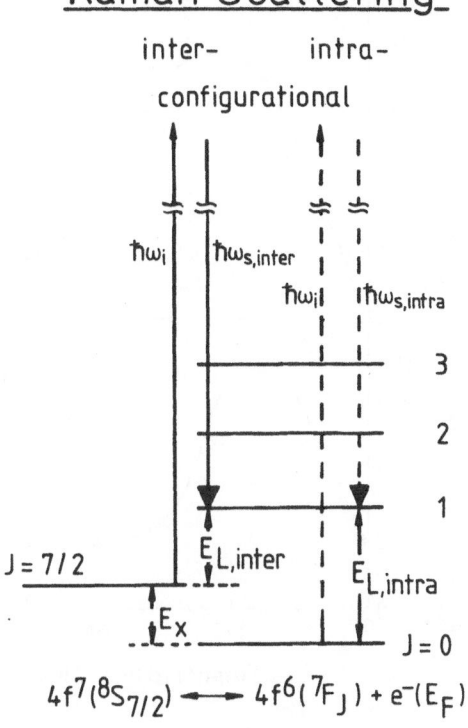

tribute to the scattering cross section in third-order perturbation theory involving spin–orbit coupling.

We now turn to the interconfigurational excitation energy E_x and its temperature dependence, which is given by the common, rigid shift of all interconfigurational excitations ($E_{L,\text{inter}}$) with respect to the intraconfigurational J-multiplet excitations ($E_{L,\text{intra}}$). The latter have been indicated in Figs. 5.4, 5 by the dashed vertical lines. For the case of $EuPd_2Si_2$ at 300 K the more intense interconfigurational excitations $E_{L,\text{inter}}$ (indicated by tick marks) are slightly shifted to lower energy with respect to the dashed lines, implying $E_x \cong 50\ \text{cm}^{-1}$ (70 K). At 200 K a further increase of E_x ($E_x \cong 150$ K) is indicated by the asymmetric broadening of the peaks at their low-energy side. The clearest evidence for a pronounced increase in E_x ($E_x \cong 350$ K) is found in the spectrum at 145 K, in which the $E_{L,\text{intra}}$ and $E_{L,\text{inter}}$ peaks are clearly split. At 77 K this splitting has disappeared, although the energy positions of the dominant interconfigurational excitations $E_{L,\text{inter}}$ (see tick marks) imply that E_x is still about the same as at 145 K. The spectrum at 5 K shows again rather narrow peaks near the positions of the dashed lines, implying that E_x is of the order of the room temperature value. The temperature dependence of E_x for $EuPd_2Si_2$ directly revealed in our Raman spectra is plotted in Fig. 5.7a.

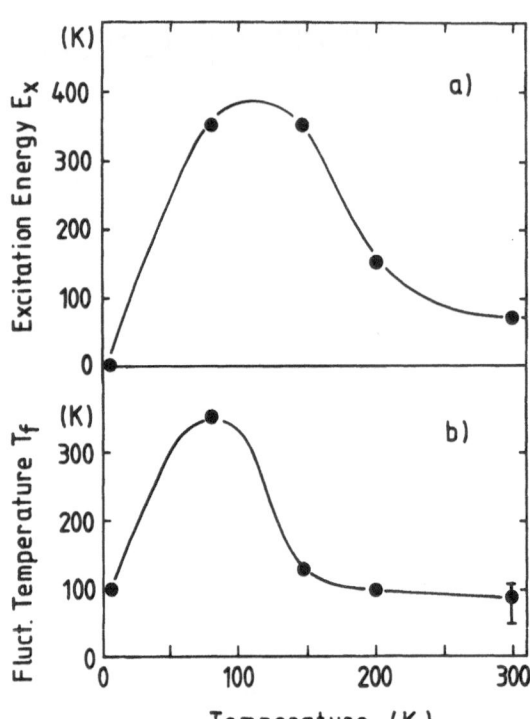

In contrast to $EuPd_2Si_2$, in the case of $EuCu_2Si_2$ $E_{L,inter}$ and $E_{L,intra}$ can be distinguished at 300 and 77 K only for the J = 4 level. For J = 1, 2 and 3 they cannot be separated due to the large inherent linewidths. From the $E_{L,intra}$ and $E_{L,inter}$ excitations of the J = 4 level one deduces an E_x of about 400 K at $T = 300$ K and $E_x \cong 320$ K at 77 K. Only one set of excitations is observable at 4 K, yielding an E_x smaller than the linewidth of the narrowest peak (J = 1). Hence we obtain $E_x < 80$ K.

We will now consider the temperature dependence of T_f, characterizing the quantum-mechanical mixing of the two $4f$ configurations (bottom of Fig. 5.3). This information is mainly contained in the spectral width of the Raman peaks. Although for $EuPd_2Si_2$ the experimentally observed widths of the different J levels differ significantly at 300 K, their additional variations upon cooling below 300 K are very similar. Since we observe the joint mixing width of initial and final state in the scattering process, we ascribe the temperature-dependent common contribution to the width of all energy-loss peaks primarily to T_f of the initial state (ground state). The upper limit of T_f of the ground state of $EuPd_2Si_2$ is set by the narrowest loss peak observable, i.e. J = 1 at 300 and 5 K, and J = 3 at all other temperatures. The values of T_f of $EuPd_2Si_2$ inferred from the spectral features are presented in Fig. 5.7b as a function of temperature. It has to be pointed out that these spectroscopically deduced values of E_x and T_f

show good qualitative agreement with those obtained by *Schmiester* et al. [5.29] for $EuPd_2Si_2$ and by *Röhler* et al. [5.9] for $EuCu_2Si_2$ (see also [5.35]) on very different experimental grounds. By use of the ionic ICF model, *Schmiester* et al. [5.29] have deduced the temperature dependence of E_x and T_f for $EuPd_2Si_2$ from the Mössbauer isomer shift and the magnetic susceptibility.

Moreover, for $EuPd_2Si_2$ there is at least basic agreement between Raman scattering and magnetic neutron scattering [5.36] concerning the temperature dependence of T_f and concerning the existence of a $J = 0 \rightarrow J = 1$ transition of the Eu^{3+}-configuration. Differences in the detailed transition energy of the $J = 0 \rightarrow 1$ excitation and the low temperature behavior are unresolved and await further clarification on more theoretical grounds taking into account the different mechanisms of neutron and Raman scattering.

5.2.4 Electronic Raman scattering in CeS_{1+x} and $CePd_3$

J-multiplet level excitations have also been observed in intermetallic Ce compounds [5.37, 38]. The J-multiplet structure of Ce^{3+} ions is very simple compared to that of Eu^{3+}, since it consists only of the $J = 5/2$ to $J = 7/2$ transition at an energy of about $2150 \, cm^{-1}$ [5.34]. Both J-multiplet levels can be split by crystalline electric fields (CEF), which act more strongly on the Ce $4f^1$-electron than on those of Eu due to the more extended $4f$ radius in Ce compared to Eu. Nevertheless, only the transitions out of the ground state (CEF split ground state of the $J = 5/2$ configuration) should be observable, especially at low temperatures, with the lowest transition energy being roughly $2150 \, cm^{-1}$.

We will first discuss results of *Mörke* et al. [5.37] obtained on a series of stable valent CeS_{1+x} compounds. CeS has the NaCl structure and the first order Raman effect by phonons is symmetry forbidden at the zone center. However, defects can break the translational symmetry of the lattice leading to a relaxation of the q-selection rule. The result is a weighted one-phonon density of states as found, e.g., in LaS [5.39].

The example of CeS is shown at the bottom of Fig. 5.8. The Th_3P_4 crystal structure, within which Ce_2S_3 and Ce_3S_4 crystallize, can accommodate cations (M) and anions (S) in various stoichiometries with the nominal compositions M_2S_3 and M_3S_4. In the M_3S_4 compounds both sublattices are filled, whereas in the M_2S_3 compounds the M sublattice contains vacancies. Therefore it is better to denote the compound $M_{2.67}S_4$. For trivalent rare earth ions the R_3S_4 and R-S samples are metallic, whereas the R_2S_3 compounds are semiconductors. The three investigated Ce_yS_4 samples had the following compositions according to wet chemical analysis:

$$Ce_4S_4 = Ce_{1.00}S_{1.00} \qquad a_0 = 5.7772 \, Å \, ,$$
$$Ce_{2.996}S_4 \qquad a_0 = 8.622 \, Å \, ,$$
$$Ce_{2.717}S_4 \qquad a_0 = 8.625 \, Å \, .$$

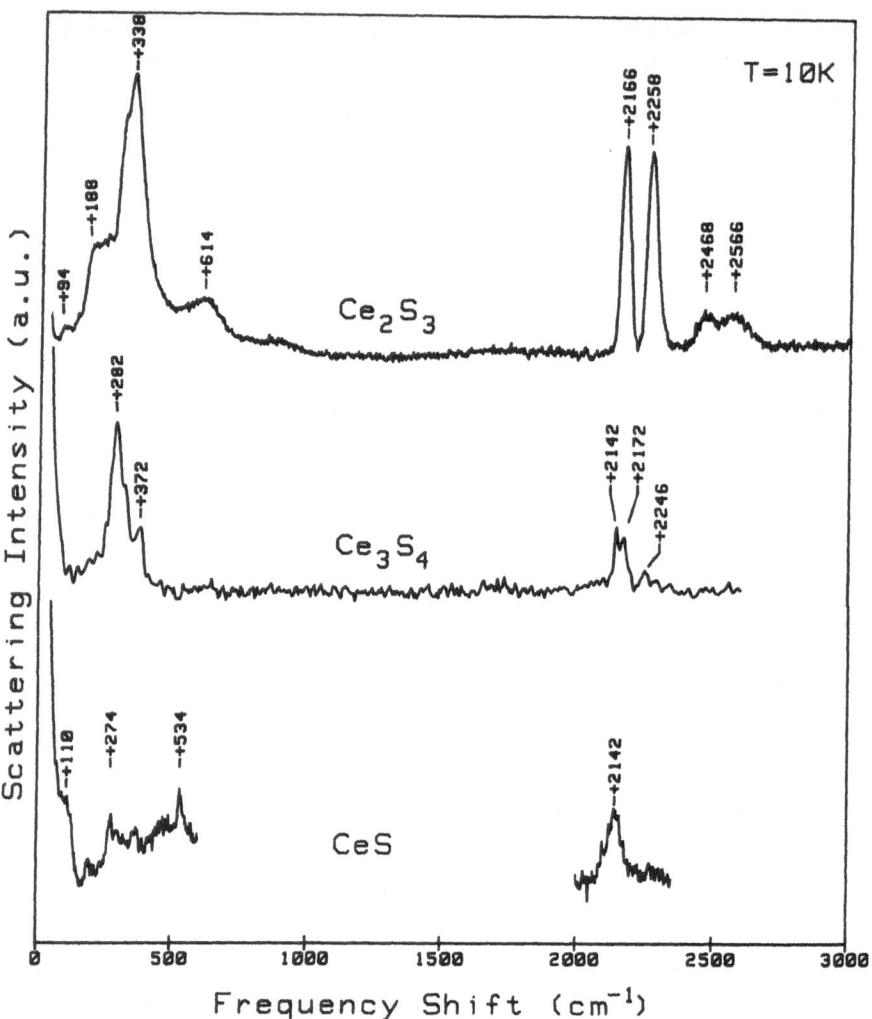

Fig. 5.8. Raman spectra of the series Ce_2S_3, Ce_3S_4 and CeS at 10 K. Especially noteworthy are the CEF-split J-multiplet excitations above 2000 cm^{-1}. From [5.37]

The second compound will be quoted as Ce_3S_4 and the last one as Ce_2S_3. Figure 5.8 shows the Raman spectra of all three compounds [5.37]. In the low frequency range from 90 cm^{-1} to 650 cm^{-1} one observes light scattering from Raman-active phonons, defect-induced phonon density of states scattering and electronic Raman scattering from excitations within the J = 5/2 multiplet. Additional features are seen near 2200 cm^{-1}, with their scattering intensity decreasing by going from the semiconducting Ce_2S_3 to metallic Ce_3S_4 and CeS. These excitations are interpreted as electronic transitions from the ground state of the J = 5/2 configuration to the 4 doublets of the $J = 7/2$ configuration. With increasing metallic behavior not only does the Raman scattering intensity go down, but also the overall crystal field splitting decreases due to the increasing shielding of the sulfur charges by conduction electrons and by a change in site symmetry. The site symmetry of Ce in CeS is cubic ($m3m$), while in the other two compounds the Ce site symmetry is tetragonal ($\bar{4}$). The decrease of the overall crystal field splitting is confirmed by magnetic susceptibility measurements, which reveal a drastic decrease of the crystal field splitting of the $J = 5/2$ ground state [5.37].

The investigation of J-multiplet excitations of Ce has been extended to intermediate valence $CePd_3$ by *Zirngiebl* et al. [5.38]. Figure 5.9 shows the Raman spectrum of $CePd_3$ at 5 K for 5309-Å (2.3-eV) laser excitation. A broad maximum is found near 2600 cm^{-1} (325 meV) as indicated by the dashed line as a guide to the eye. The full width at half maximum (FWHM) amounts to about 2000 cm^{-1} (250 meV). Superimposed on this broad maximum is a relatively sharp peak near 2050 cm^{-1} (256 meV) with FWHM 48 meV. The two maxima in Fig. 5.9 appear in the Raman spectra for different laser excitation energies always at the same energy loss ΔE_R (Stokes shift) with respect to the incident photon energy $\hbar\omega_i$. Thus they are identified as electronic excitations (ΔE_R) with respect to the initial state in Raman scattering (inset of Fig. 5.9). Within the nomenclature introduced for the interpretation of Raman scattering results of IV $EuPd_2Si_2$ the broad maximum near 2600 cm^{-1} (325 meV) in Fig. 5.9 is attributed to the $4f^1 \rightarrow 4f^0$ interconfigurational excitation energy E_x of about 325 meV (2600 cm^{-1}). The one-electron occupied $4f^1$ state above E_F has been observed in BIS (Bremsstrahlen Isochromate Spectroscopy) near 450 meV [5.40] in good agreement with our findings. Both results

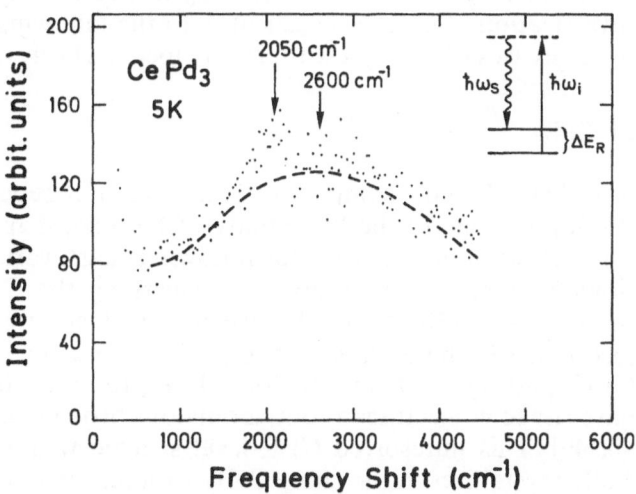

Fig. 5.9. Raman spectrum of $CePd_3$ at 5 K, showing the $4f^1 \rightarrow 4f^0$ excitation near 2600 cm^{-1} (dashed guide line) and the $J = 5/2 \rightarrow 7/2$ excitation near 2050 cm^{-1}. Inset: The energy difference between incident ($\hbar\omega_i$) and scattered ($\hbar\omega_s$) photons yields the energy ΔE_R of the excitation in Raman scattering

indicate that the $4f^1$ level is well below the Fermi level, placing $CePd_3$ into the class of Kondo-like compounds and not into the class of strong IV compounds like $EuPd_2Si_2$, for which E_x is more than an order of magnitude smaller. The FWHM ≈ 250 meV of the interconfigurational excitation obtained in Raman scattering is by about a factor of three smaller compared to that in BIS, most likely due to the much better spectral resolution of Raman scattering. This value is of the order of magnitude of the estimated $f - d$ hybridization width of $CePd_3$ with $\varDelta \approx 100$ meV from optical data at 4.2 K and $\varDelta \approx 150$ meV from X-ray photoemission (XPS) data [5.41].

The peak near 2050 cm^{-1} in Fig. 5.9 coincides with the $^2F_{5/2} \rightarrow {}^2F_{7/2}$ intraionic excitation energy of the Ce^{3+} $4f^1$ configuration. This confirms the previous tentative assignment in infrared spectroscopy by *Hillebrands* et al. [5.42] and proves that ionic configurations may still be preserved in IV compounds. The FWHM ≈ 48 meV is of the same order of magnitude as $2T_f \approx 2\Gamma_s/2 \approx 38$ meV as obtained from quasielastic magnetic neutron scattering [5.43].

5.2.5 Light Scattering from CEF Excitations in R-B$_6$

The investigation of CEF levels in rare earth intermetallic compounds is a classic domain of magnetic neutron scattering [5.44]. The typical energy range of CEF level splitting is of the order of 100 K and best suited for inelastic magnetic scattering experiments. From transition energies and intensities one can identify quite unambiguously CEF level schemes, especially for cubic compounds within the systematics given by *Lea* et al. [5.45]. There are, however, a number of examples where due either to the limited energy range available (the production rate of high energy neutrons ($E > 50$ meV) is quite low at reactors) or to the limited energy resolution of neutron scattering, light scattering can be a complementary tool in the investigation of CEF level schemes. In the following we will discuss some of these cases where light scattering helped to clarify unsolved problems.

a) CeB$_6$

The dense Kondo compound CeB_6 was first reported by *Fisk* in 1969 [5.46a]. In spite of the large body of thermal, magnetic and elastic data for CeB_6 accumulated over the following years, various diverging speculations appeared in the literature concerning the CEF excitations (for a review see [5.15]). Their absence in direct spectroscopic measurements [5.46b] has been puzzling ever since. This led to a wide variety of proposed CEF splittings ranging from 10 K [5.47] to more than 400 K [5.46b] and gave rise to conjectures about anomalous broadenings or splittings [5.46b, 48, 49]. This unresolved CEF level scheme was even more puzzling as CeB_6 crystallizes in the cubic CeB_6 structure (space group O_h^1), where the sixfold degenerate ground state $4f^1$ (J = 5/2) of Ce^{3+} is expected to split only into two levels (a Γ_7 doublet and a Γ_8 quartet), giving rise to only one CEF excitation.

The first spectroscopic insight into the CEF level scheme of CeB_6 finally came from inelastic magnetic neutron scattering using high-energy incident neutrons of up to 185 meV [5.50]. An inelastic magnetic excitation was found near 530 K, still not yet allowing a final conclusion about the CEF level scheme. This inelastic magnetic excitation has been further investigated by Raman spectroscopy [5.15], taking advantage of the high resolution as compared to neutron spectroscopy. The Raman measurements have been carried out on (100) faces of CeB_6 between 300 and 4.2 K using 5309-Å Kr^+ laser excitation. An inelastic excitation at 372 cm^{-1} (530 K) was found at room temperature as shown in Fig. 5.10. Below 20 K the peak shifts by about 10 cm^{-1} to higher energies. At 4.2 K it is much narrower and has an energy of 382 cm^{-1} (Fig. 5.10). The intensity of the peak does not change between 4.2 and 300 K because of the high excitation energy (530 K) compared to the sample temperature. Therefore, one cannot distinguish between a CEF excitation and a phonon by the intensity variation alone. However, the unusually large peak shift and its onset below 20 K rules out a phononic excitation since no lattice anomaly has been observed in this temperature range [5.48]. In addition, the nonmagnetic reference compound LaB_6 shows no excitation in that energy

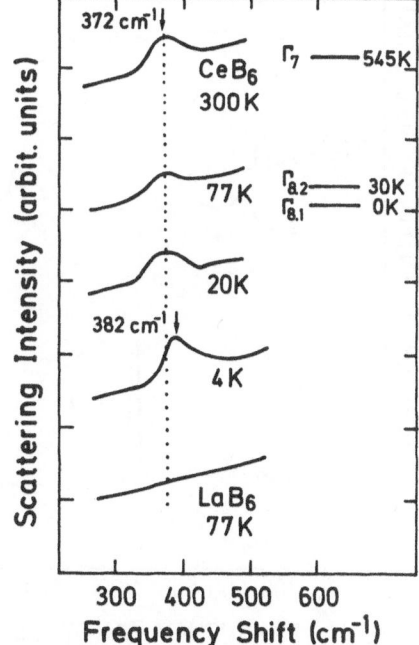

Fig. 5.10. Raman scattering intensities of CeB_6 and LaB_6 at different temperatures showing the Ce^{3+} $\Gamma_8 - \Gamma_7$ CEF transition. Inset: CEF level scheme derived for CeB_6

range at any temperature (Fig. 5.10 for, e.g., 77 K). Hence, *Zirngiebl* et al. concluded that the excitation near 372 cm^{-1} in CeB$_6$ corresponds to the $\Gamma_8 - \Gamma_7$ CEF transition within the $4f^1$ configuration.

In order to further corroborate their CEF level assignment a symmetry analysis by polarized Raman measurements was performed as shown in Fig. 5.11. For comparison the well-known three Raman-active phonons appearing above 600 cm^{-1} have been included. The 372 cm^{-1} peak appears in the $\Gamma_3^+(E_g)$ and $\Gamma_5^+(T_{2g})$ symmetry components. This is consistent with a $\Gamma_8 - \Gamma_7$ transition because the direct symmetry product of initial and final states $|\Gamma_8\rangle \otimes \langle\Gamma_7| = \Gamma_3^+ \oplus \Gamma_4^+ \oplus \Gamma_5^+$ contains the experimentally observed symmetries.

From the symmetry analysis it cannot be decided which is the electronic ground state. However, *Zirngiebl* et al. deduced this information from the anomalous shift of the 372 cm^{-1} peak in Fig. 5.10 for temperatures below 20 K. This behavior is explained by a non-Kramers ground state of Γ_8 symmetry, which is split into two doublets $\Gamma_{8,1}$ and $\Gamma_{8,2}$ with a separation of about 30 K. At temperatures well above 30 K, transitions from both

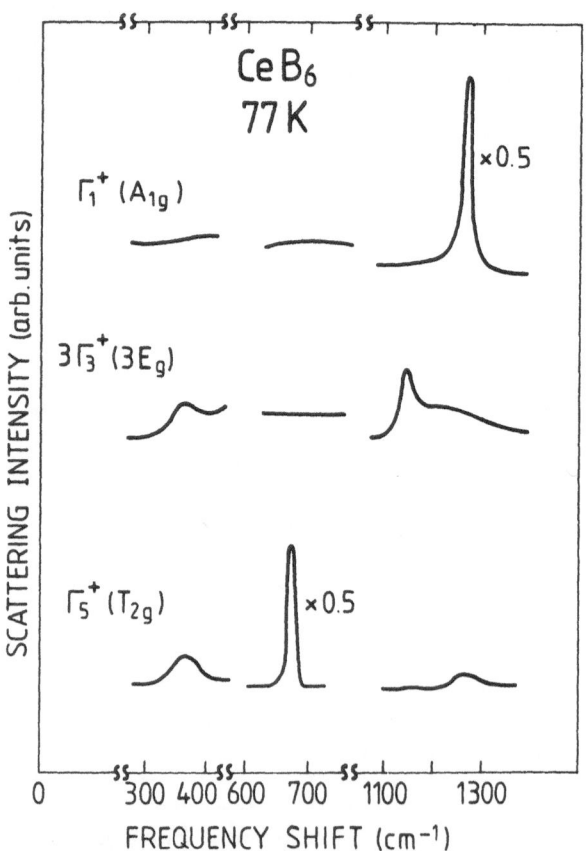

Fig. 5.11. Symmetry analysis of the $\Gamma_8 - \Gamma_7$ CEF transition and the Raman-active phonons of CeB$_6$ at 77 K

$\Gamma'_{8,i}$ doublets to the high-lying Γ_7 level are possible, yielding the measured mean transition energy of $372 \, \text{cm}^{-1}$ (530 K). A double peak structure could not be resolved even in Raman scattering (experimental resolution $3 \, \text{cm}^{-1}$) because of the large inherent width of the peaks. Upon cooling well below 30 K the upper doublet $\Gamma_{8,2}$ becomes thermally depopulated and the scattering takes place between the lower $\Gamma_{8,1}$ doublet and the Γ_7 level. This yields a shift of the observed transition energy upon cooling by $(E_{8,2} - E_{8,1})/2 = 15 \, \text{K} = 10 \, \text{cm}^{-1}$ towards higher energy. The new CEF level scheme derived from the Raman measurements is shown in the inset of Fig. 5.10, providing the simplest interpretation so far of the measured thermal, elastic and magnetic data [5.15].

b) NdB$_6$

The Hund's rule $4f$ ground state of Nd^{3+} $(^4I_{9/2})$ is tenfold degenerate and splits in cubic surroundings into two Γ_8 quartets and a Γ_6 doublet. The corresponding three-level CEF scheme of NdB$_6$ has been investigated by several experimental methods, yielding very different results [5.51, 52]. A CEF level scheme has also been proposed on the basis of neutron scattering [5.53]. Within the examined energy range of 50 meV $(= 400 \, \text{cm}^{-1})$ one broad magnetic excitation was detected at 12 meV $(= 95 \, \text{cm}^{-1})$. From the temperature dependence of the scattering intensity of the excitation and by comparison with the Lea-Leask-Wolf parameters of the other R-B$_6$ the following level scheme was deduced: The ground state $\Gamma_8^{(2)}$ has Γ_8 symmetry, the first excited state $\Gamma_8^{(1)}$ at 12 meV has Γ_8 symmetry and the Γ_6 state lies 24 meV above the ground state. The excitation at 24 meV was not observed in neutron scattering because of the small matrix element for transitions $\Gamma_8^{(2)} \leftrightarrow \Gamma_6$. Therefore three possible transitions can give rise to only one peak in the neutron scattering spectrum. This determination of the CEF level scheme is plausible. It would, however, be more desirable to separate the peak at 12 meV into the two transitions $\Gamma_8^{(2)} \rightarrow \Gamma_8^{(1)}$ and $\Gamma_8^{(1)} \rightarrow \Gamma_6$.

Raman scattering in NdB$_6$ has been reported by *Pofahl* et al. [5.54]. The Raman measurements were carried out on the (110) face of a NdB$_6$ single crystal, which was prepared using the arc floating zone technique [5.55]. The measurements were performed as a function of temperature between 300 K and 7 K, with the sample under vacuum. Several different laser wavelengths (5145 Å, 5130 Å, 5682 Å) were used.

In Fig. 5.12 the Raman spectra of the magnetic CeB$_6$ and NdB$_6$ and of the nonmagnetic LaB$_6$ are shown. The well-known Raman-active T_{2g} phonon of the rare earth hexaborides is seen in all three spectra near $680 \, \text{cm}^{-1}$ together with a phonon density of states near $180 \, \text{cm}^{-1}$. The additional peak at $372 \, \text{cm}^{-1}$ in the spectrum of CeB$_6$ was attributed to the CEF excitation $\Gamma_8(0 \, \text{K}) \rightarrow \Gamma_7(545 \, \text{K})$ as discussed in 5.2.5a. The extra peak at $95 \, \text{cm}^{-1}$ in the spectrum of NdB$_6$ coincides with that observed

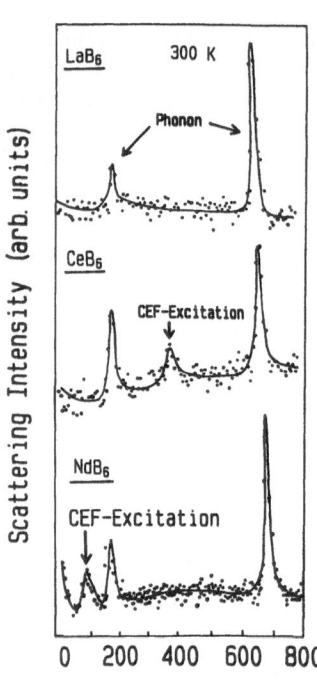

Fig. 5.12. Unpolarized Raman spectra of different R-B$_6$ (R = La, Ce, Nd). Magnetic excitations are observed in CeB$_6$ near 372 cm^{-1} and in NdB$_6$ near 95 cm^{-1}. In all three spectra one observes the Raman-active T_{2g} phonon near 680 cm^{-1} and a phonon density of states near 180 cm^{-1}

in neutron scattering and was the subject of more detailed examinations, such as symmetry analysis and temperature dependence. As the peak does not show up in the Raman spectra of either LaB$_6$ or CeB$_6$ it is reasonable to associate it with excitations within the $4f^3$ configuration of the Nd^{3+} ion.

Polarized Raman measurements and special crystal orientations have been used [5.54] to analyze the mode symmetries as shown in Fig. 5.13. Group theory yields for the symmetries of the different CEF transitions: $\Gamma_8 \otimes \Gamma_8 = \Gamma_1 \oplus \Gamma_2 \oplus \Gamma_3 \oplus 2\Gamma_4 \oplus 2\Gamma_5; \Gamma_6 \otimes \Gamma_8 = \Gamma_3 \oplus \Gamma_4 \oplus \Gamma_5$. In the case of cubic crystal symmetry the Γ_1, Γ_3 and Γ_5 components are Raman-allowed. In Fig. 5.13 the well-known vibrations of the B$_6$ octahedra have been included to demonstrate the corresponding symmetry. The peak near 95 cm^{-1} appears only in Γ_5 symmetry with the Γ_1 and Γ_3 components being zero. The symmetry analysis is consistent with the identification of the 95 cm^{-1} line as due to a crystal field excitation, but does not allow a separation of the two transitions.

Figure 5.14 shows the temperature dependence of the magnetic excitation of NdB$_6$ near 95 cm^{-1} together with the phonon density of states near 170 cm^{-1} [5.54]. At 300 K the peak at 95 cm^{-1} is the center of two transitions $\Gamma_8^{(2)} \rightarrow \Gamma_8^{(1)}$ and the $\Gamma_8^{(1)} \rightarrow \Gamma_6$. The excitation $\Gamma_8^{(2)} \rightarrow \Gamma_6$ (24 meV = 190 cm^{-1}) is neither observable in Raman scattering nor in

Fig. 5.13. Symmetry analysis of the CEF transitions and the Raman-active phonons of NdB_6 for the (110)-face at 77 K. The unpolarized spectrum is shown at the top

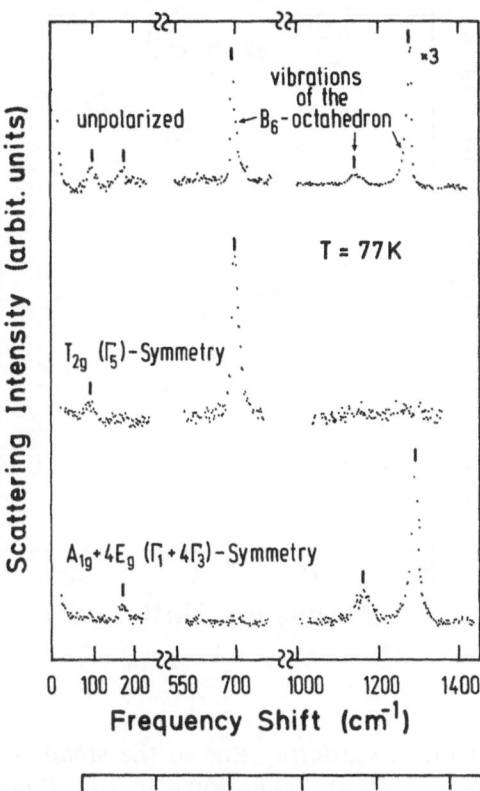

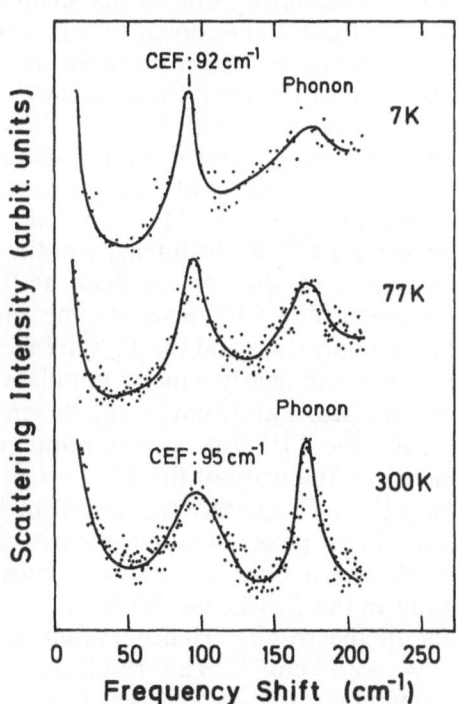

Fig. 5.14. Raman scattering intensities of NdB_6 at different temperatures. The peak at $170\,cm^{-1}$ corresponds to the phonon density of states and decreases upon cooling due to the Bose factor. By cooling down from 300 K to 7 K the center of the CEF transitions shifts from $95\,cm^{-1}$ to $92\,cm^{-1}$

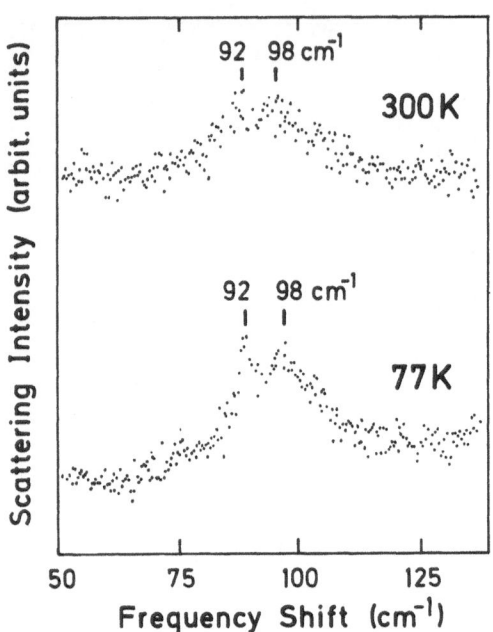

Fig. 5.15. Raman spectra of the CEF transitions of NdB_6 at 300 K and 77 K measured with a high instrumental resolution of 2 cm^{-1} (= 0.25 meV)

neutron scattering due to the small transition matrix element [5.53]. By cooling from 300 K down to 7 K, *Pofahl* et al. [5.54] found a shift of the center of the excitation from 95 cm^{-1} to 92 cm^{-1} and a decrease in the linewidth of the peak from 36 cm^{-1} to 16 cm^{-1} (Fig. 5.14).

By measuring with high resolution of 2 cm^{-1} and by extending the measurement time by a factor of about ten *Pofahl* et al. [5.54] could resolve at 300 K as well as at 77 K two peaks at 92 cm^{-1} and at 98 cm^{-1} as shown in Fig. 5.15. At room temperature both peaks have the same intensity. At 77 K the intensity of the excitation at 98 cm^{-1} has decreased compared to that of the peak at 92 cm^{-1}. These results can only be explained by a CEF level scheme with the $\Gamma_8^{(1)}$ state 92 cm^{-1} above the $\Gamma_8^{(2)}$ ground state and the Γ_6 state 98 cm^{-1} above the $\Gamma_8^{(1)}$ state. At 300 K Γ_8 states are nearly equally populated, yielding comparable intensities of the two peaks at 92 cm^{-1} and 98 cm^{-1} in Fig. 5.15. At low temperatures (77 K), the $\Gamma_8^{(1)}$ state is less populated than the $\Gamma_8^{(2)}$ state, yielding an increased intensity of the $\Gamma_8^{(2)} \rightarrow \Gamma_8^{(1)}$ excitation at 92 cm^{-1} compared to the $\Gamma_8^{(1)} \rightarrow \Gamma_6$ excitation near 98 cm^{-1}. Taking into account the observation of two peaks of equal intensity at high temperatures and the failure to observe a third one due to its weak intensity, one can find only one point in the Lea-Leask-Wolf scheme of Nd^{3+} for cubic crystal fields that fits all the results: Hence *Pofahl* et al. [5.54] obtained $x = -0.82$ and $W = -2.76$ cm^{-1}. The resulting levels, together with the transition probabilities between the levels are indicated in Fig. 5.16.

Fig. 5.16. CEF level scheme of NdB_6 with the values of the corresponding transition matrix elements indicated

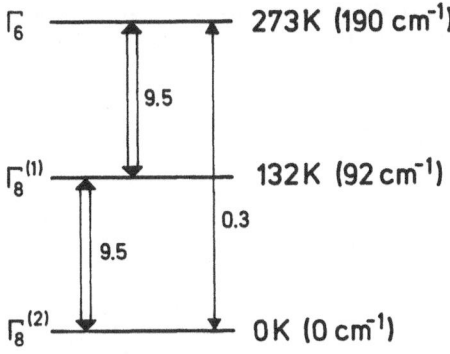

5.2.6 Light Scattering from CEF Excitations in $CeCu_2Si_2$

A great deal of attention has recently been devoted to "heavy fermion" systems, a group of intermetallic compounds, which behave like localized f-moment systems at high temperatures, but yet display many features of a simple metal at low temperatures (for a review of studies of these materials see [5.13, 56]). Of special interest are three such materials, $CeCu_2Si_2$ [5.57], UBe_{13} [5.58] and UPt_3 [5.59], which have been shown to possess a superconducting ground state in which the heavy f-electrons are thought to participate in spite of their room temperature predisposition towards localized magnetism. Given its evident importance, the $4f$-electronic excitation spectrum of $CeCu_2Si_2$ has been widely investigated by neutron scattering to characterize the magnetic fluctuations of the $4f$ electrons and the effects of the CEF on the Ce multiplet. CEF excitations were first observed with levels reported by *Horn* et al. [5.14] at 140 K and 364 K ($100\ cm^{-1}$ and $260\ cm^{-1}$, respectively), whereas subsequent neutron scattering studies, while clearly observing the peak at higher energy, have been unable to confirm the lower energy transition [5.60, 61].

Electronic Raman scattering experiments on oriented single-crystal samples of $CeCu_2Si_2$ were performed by *Cooper* et al. [5.62] using a polarized 4880 or 5145 Å line of an argon laser as an excitation source.

In tetragonal surroundings the Ce^{3+} ($J = 5/2$) multiplet is expected to split into three doublets, two of Γ_7 symmetry and one of Γ_6 symmetry. Electronic transitions between these levels should manifest the symmetries allowed by the direct products of these states:

$$\Gamma_7 \otimes \Gamma_7 = \Gamma_1^+ \oplus \Gamma_2^+ \oplus \Gamma_5^+ (A_{1g} \oplus A_{2g} \oplus E_g),$$
$$\Gamma_7 \otimes \Gamma_6 = \Gamma_3^+ \oplus \Gamma_4^+ \oplus \Gamma_5^+ (B_{1g} \oplus B_{2g} \oplus E_g).$$

Cooper et al. [5.62] reported the observation of crystal-field excitations of $CeCu_2Si_2$ as a broad hump centered roughly at $290\ cm^{-1}$ in the Raman spectra of $CeCu_2Si_2$ as shown in Fig. 5.17. For comparison the spectrum

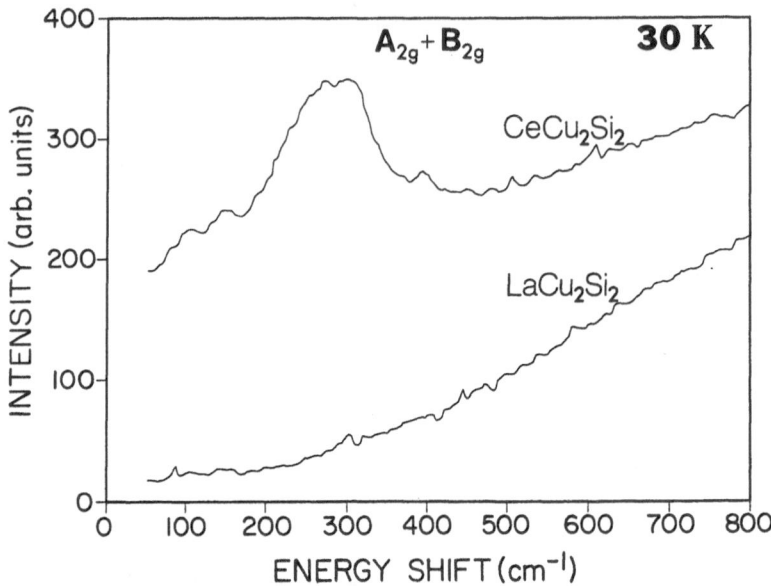

Fig. 5.17. Comparison of the $A_{2g} + B_{2g}$ spectrum of CeCu$_2$Si$_2$ (upper) with that of LaCu$_2$Si$_2$ (lower) at 30 K. Resolution: 10 cm^{-1}. From [5.62]

of the iso-structural d-band metal LaCu$_2$Si$_2$ is also shown, exhibiting no Raman signal around 290 cm^{-1}, thus strongly confirming the interpretation of the 290-cm^{-1} excitation in CeCu$_2$Si$_2$ as being due to $4f$ electrons. This identification is further supported by the temperature dependence of the $A_{2g} + B_{1g}$ spectrum [5.62]. As expected for electronic transitions, the crystal-field peak at 290 cm^{-1} narrows and becomes stronger as the temperature is lowered, mimicking the sharpening Fermi factor. The appearance of the CEF peak in both the ($A_{2g} + B_{1g}$)- and the ($A_{2g} + B_{2g}$)-symmetry type Raman spectra confirms that it has the symmetry of the purely antisymmetric representation of the CeCu$_2$Si$_2$ space group, A_{2g}, characteristic of a $\Gamma_7 - \Gamma_7$ transition. No evidence of a $\Gamma_7 - \Gamma_6$ transition is seen according to *Cooper* et al. [5.62].

The temperature dependence of the CEF-level linewidth [full width at half maximum (FWHM)] is shown in Fig. 5.18 [5.62]. This temperature dependence has been calculated only for cubic Kondo systems [5.63, 64], but the results adequately describe the general features of the observed linewidth in anisotropic CeCu$_2$Si$_2$. In these models, the dominant damping mechanism at high temperatures ($\Delta \ll T$, with Δ the crystal-field splitting) results from elastic scattering (i.e., creation of electron–hole pairs) within each of the crystal-field levels, giving a linear dependence of linewidth on temperature [5.64]:

$$\Gamma = 4\pi [n(E_F)]^2 \, (|J_{77}|^2 + 2\,|J_{88}|^2)\, T \qquad (\Delta \ll T). \tag{5.1}$$

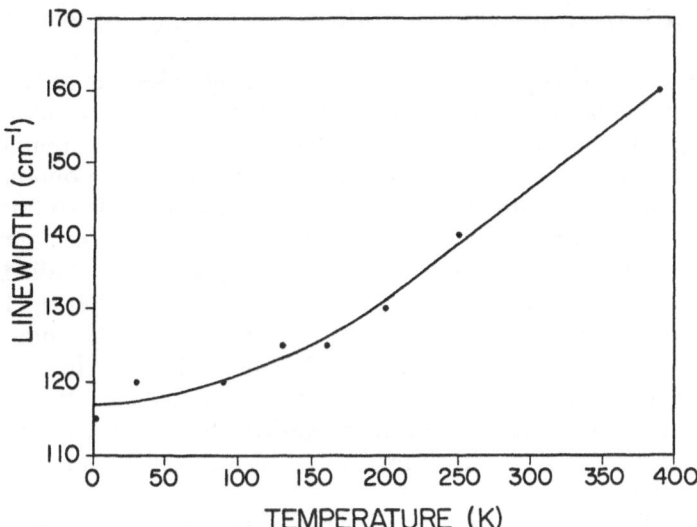

Fig. 5.18. Observed CEF linewidth (FWHM) vs. temperature of the A_{2g} CEF peak of CeCu$_2$Si$_2$. Resolution: 10 cm^{-1}. Line drawn is a guide to the eye. From [5.62]

Here, J_{77} and J_{88} are exchange integrals between electrons within the Γ_7 and Γ_8 crystal-field levels, respectively, while $n(E_F)$ is the conduction-band density of states at the Fermi energy.

At low temperatures ($T \ll \Delta$), damping chiefly results from transitions between the crystal-field levels, promoted by the exchange interaction between the conduction electrons and the Ce^{3+} $4f$ electron. This leads to a saturation of the linewidth at sufficiently low temperatures, as described by [5.64]

$$\Gamma = 4\pi[n(E_F)]^2 \left[|J_{78}|^2 \, \Delta \, \frac{1 + 2 \, e^{-\Delta/T}}{1 - e^{-\Delta/T}} \right] \qquad (T \ll \Delta). \qquad (5.2)$$

The crystal-field splitting is described by Δ in this relation, while J_{78} is the exchange term between the Γ_7 and Γ_8 levels.

An informal application of these results to CeCu$_2$Si$_2$ ($\Delta = 406$ K), presuming equal exchange terms and a weak splitting of the Γ_8 level, indicates that these two mechanisms should be of roughly equal importance down to about 160 K. Below this temperature, the inelastic damping term begins to predominate quickly. This behavior is reflected in the linewidth, observed by *Cooper* et al. [5.62], wherein one notes a linear dependence above 200 K, with saturation occurring at lower temperatures (Fig. 5.18).

5.2.7 Electronic Raman Scattering in $Sm_{1-x}Y_xSe$ and $Sm_{1-x}Y_xS$

Raman scattering in rare earth chalcogenides has already been reviewed by *Güntherodt* and *Merlin* [5.65]. The Sm chalcogenides are mentioned in this section on intermetallic rare earth compounds for the sake of comparison with stable valence reference compounds and for demonstrating the configurational crossover from stable to fluctuating valence.

Spin-orbit and crystal-field split levels of $4f$ states of rare earth ions in insulating hosts have been extensively studied by *Koningstein* et al. [5.66, 67] by means of electronic Raman scattering. In this subsection we discuss electronic Raman scattering from $4f$ spin-orbit split levels in $Sm_{1-x}R_xSe$ and $Sm_{1-x}R_xS$ solid solutions in the vicinity of the $4f$ configuration crossover from stable valence into the intermediate valence (IV) state. Such investigations have a direct bearing not only on the valence fluctuation problem, but also on the fundamentals of Raman scattering.

The ground state of Sm^{2+} in the Sm monochalcogenides is $4f^6(^7F_{J=0,1,\ldots 6})$, with a 0.6 eV wide spin-orbit split ^7F_J multiplet. Electronic Raman scattering from the different J multiplet levels of cleaved (100) faces of semiconducting SmSe has been observed by *Güntherodt* et al. [5.32]. The odd J levels show up for perpendicular incident (E_i) and scattered (E_s) polarization vectors, whereas the opposite is true for the even J levels. It has been shown for SmS that the J = 1 peak (and in principle the J = 3, 5 peaks) appears only in the antisymmetric Γ_{15}^+ component [5.68]. The scattering intensity from either odd or even J levels decreases monotonically with increasing J values. This has been attributed to the higher J levels approaching the $4f - 5d$ excitation gap (the excitation to the J = 6 level near 4010 cm^{-1} coincides with the 0.5 eV gap of SmSe). This conclusion is further supported by the fact that the J $\geq$ 3 levels are not observed in SmS, which shows a $4f - 5d$ gap of about 1200 cm^{-1} (0.15 eV).

Electronic Raman scattering has been observed in SmS, SmSe and SmTe by *Nathan* et al. [5.69] and *Smith* et al. [5.70], with particular emphasis on the temperature dependence of the singlet-triplet (J = 0 $\rightarrow$ 1) excitation. In all three compounds the J = 0 $\rightarrow$ 1 transition has been found to shift to a lower frequency upon cooling below room temperature. This temperature dependence of the singlet–triplet excitation could be fitted by results obtained in the random phase approximation [5.71, 72] using the free-ion spin-orbit coupling constant ($\lambda = 193.5$ cm^{-1}) and by introducing an exchange interaction energy Θ. The latter was found to decrease from 44 cm^{-1} in SmS to 8 cm^{-1} in SmTe. This seems to parallel a decreasing $5d$ admixture into the $4f^6$ ground state with increasing $f - d$ excitation gap from SmS to SmTe.

Substitution of the cation in the solid solution system $Sm_{1-x}R_xSe$ by, e.g., Y or La reduces the lattice parameter and the $4f - 5d$ excitation gap, without yielding the transition into the metallic intermediate valence phase [5.73]. Polarized Raman spectra of $Sm_{1-x}Y_xSe$ for $x = 0$, 0.25, 0.50, 0.75 and 1.0 have been investigated by *Güntherodt* et al. [5.32], together with $Sm_{0.95}La_{0.05}Se$. For the latter sample one observes below $200\ cm^{-1}$ first-order defect-induced Raman scattering from acoustical and optical phonons which is absent in pure SmSe. The $J = 1$ peak of $Sm_{0.95}La_{0.05}Se$ has drastically broadened compared to that of pure SmSe at $275\ cm^{-1}$ and has shifted to $266\ cm^{-1}$. The $J = 1$ peak is broadened further with increasing x and merges with the optical phonon density of states for $x > 0.50$. For $x = 0.75$, the $J = 1$ peak has shifted to about $210\ cm^{-1}$ and is barely seen. The $J = 3$ peak of $Sm_{1-x}Y_xSe$ does not shift with increasing x, but becomes strongly broadened and finally can no longer be resolved for $x > 0.50$. On the other hand, the persistence of the peak related to the $J = 2$ level up to $x = 0.75$ indicates that the $4f - 5d$ gap is still finite, i.e., of the order of 0.1 eV $(800\ cm^{-1})$. This is consistent with the fact that for all values of x $Sm_{1-x}Y_xSe$ does not undergo a transition into the homogeneously intermediate valence phase [5.73]. However, beyond this fact the intensity changes and splittings of the $J = 2$ level with increasing x are unexplained.

The solid solution system $Sm_{1-x}Y_xS$ was first investigated using Raman scattering by *Smith* et al. [5.70] and *Tsang* [5.74] for concentrations near $(x = \lesssim 0.15)$ and beyond $(x > 0.15)$ configuration crossover (CC). Polarized Raman spectra by *Güntherodt* et al. [5.32] have shown a clear separation into phonon and electronic ("magnetic") Raman scattering and the evolution with CC. For SmS $(x = 0)$ phonon scattering is observed below $300\ cm^{-1}$, whereas electronic scattering from the $J = 1$ level is seen near $275\ cm^{-1}$ and from the $J = 2$ level near $780\ cm^{-1}$. The $J \geq 3$ levels are not observed. With increasing x (≤ 0.15) the $J = 1$ level is reduced in intensity, strongly broadened and shifts to $250\ cm^{-1}$ for $x = 0.15$. In the same sequence the $J = 2$ level is subject to strong broadening. Beyond CC $(x \geq 0.15)$ no contributions from the $J = 0 \rightarrow 1$ and $J = 0 \rightarrow 2$ excitations could be identified, contrary to previous unpolarized Raman measurements [5.74]. In the latter no distinction could be made between electronic and phonon contributions.

From the Raman scattering data it can be concluded that the broadening of the $J = 1$ peak must have exceeded at least $200\ cm^{-1}$ (25 meV) FWHM in order to be no longer resolved. On the other hand, a peak near 31 meV $(250\ cm^{-1})$ found in neutron scattering on powdered $Sm_{0.75}Y_{0.25}S$ has been attributed to scattering from the $J = 0 \rightarrow 1$ excitation [5.75]. More recent neutron scattering experiments on polycrystalline $Sm_{0.75}Y_{0.25}S$ also find an inelastic magnetic excitation near 38 meV attributed to a modified $J = 0 \rightarrow J = 1$ transition [5.76]. At present there is no answer to this apparent discrepancy, perhaps suggesting a drastic reduction of the electronic Raman scattering cross section of $Sm_{1-x}Y_xS$ induced by the valence mixing near configuration crossover.

5.2.8 Electronic Raman Scattering in UPt₃

For 5*f* electrons the degree of localization is intermediate between that of 3*d* and 4*f* electrons. In insulating actinide compounds, such as UO_2 [5.18b], 5*f* electrons are generally more localized than in metallic ones. This dependence of the localization on the chemical bonding of the ion in the lattice is in contrast to the situation for the 4*f* electrons.

In actinide metals, especially in compounds like UPt₃ with narrow *f* bands and correlation effects the situation is different to, e.g., UO_2 [5.13]. Figure 5.19 shows the Raman spectrum of UPt₃ at 5 K under 5309 Å laser excitation up to 5000 cm⁻¹ frequency shift [5.18a, b]. We observe strong inelastic scattering between about 1000 and 3000 cm⁻¹ which is also observed for 4762 Å excitation and is thus not due to luminescence. This is in agreement with recent observations of inelastic scattering intensities between 1000 cm⁻¹ and 3000 cm⁻¹ in a UPt₃ single crystal under 5145 Å laser excitation [5.19]. We attribute the inelastic scattering to 5*f* excitations similar to those observed in UO_2 [5.18b]. However, in UPt₃ these inelastic excitations are very broad compared to UO_2, in agreement with the expectation of a stronger tendency towards delocalization in metallic actinide compouds. The origin of these excitations can be

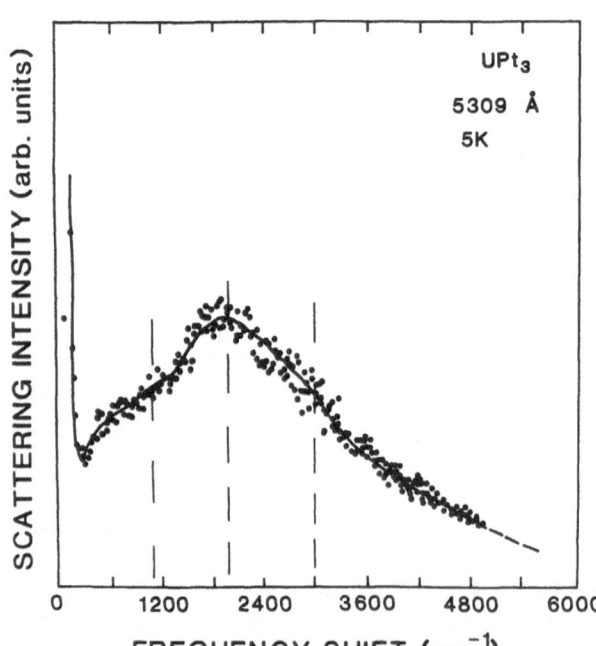

Fig. 5.19. Electronic Raman scattering in UPt₃ that is attributed to intraionic multiplet or CEF excitations

either due to intraionic multiplet levels or to CEF splittings. The splittings of the electronic ground state of this order of magnitude can explain [5.18 b] the temperature dependence of the magnetic susceptibility of UPt_3 between 10 and 1050 K [5.77]. A splitting of roughly 2000 cm^{-1} or 3000 K could describe the bending of the experimental $1/\chi$ vs. T curve at about 750 K. This value of 3000 K should be understood as a rough estimate of an average $5f$ splitting.

5.3 Quasielastic Light Scattering

5.3.1 Spin Fluctuations in UBe_{13}, UPt_3 and URu_2Si_2

Spin fluctuations in the HF compounds UPt_3 and UBe_{13} have been observed by neutron scattering [5.78–80] for momentum transfers $q > 1$ Å. As Raman scattering is a true $q \cong 0$ method, the aim was to test the linear q dependence of the spin relaxation rate Γ_s, predicted by noninteracting Fermi liquid theory. Figure 5.20 shows the quasielastic scattering intensity of UBe_{13} under 5145 Å laser excitation at 350 K and 40 K [5.20a]. In view of the perpendicular orientation of the incident and scattered electric field

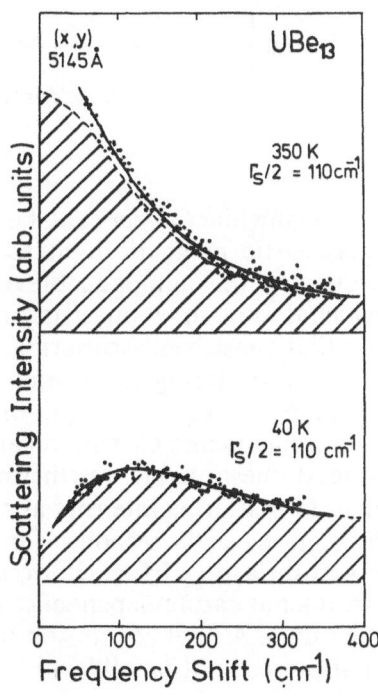

Fig. 5.20. Raman spectra of UBe_{13} [5.20a] obtained with perpendicular polarizations of incident (5145 Å) and scattered light. The hatched area shows the contribution due to spin fluctuations with halfwidth at half maximum, $\Gamma/2$, determined from the fit using (5.3). After [5.20a]

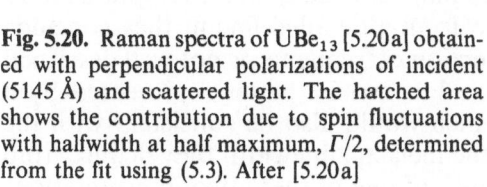

vectors $E_i \perp E_s$ for which it is observed this scattering is identified as magnetic in origin. The scattering intensity $I(\omega)$ has been fitted by

$$I(\omega) \propto [1 + n(\omega)]\, \hbar\omega\, \frac{\Gamma_s/2}{(\Gamma_s/2)^2 + (\hbar\omega)^2}, \qquad (5.3)$$

where $n(\omega)$ is the Bose factor; the Lorentzian is the Fourier transform of the spin-spin correlation function proportional to $\exp(-\Gamma_s t)$ used to describe fluctuating uncorrelated $5f$ spins with a spin relaxation rate Γ_s. The fit based on (5.3) is shown by the hatched area in Fig. 5.20. One obtains an approximately temperature idependent spin relaxation rate $\Gamma_s(q \cong 0) = (110 \pm 10)\,\mathrm{cm}^{-1}\,(= 13.6\,\mathrm{meV})$. This result together with that from neutron scattering for $q = 2\,\text{Å}^{-1}$ [5.80] is shown in Fig. 5.21. The q independence of Γ_s is evidence for the localized nature of the spin fluctuations.

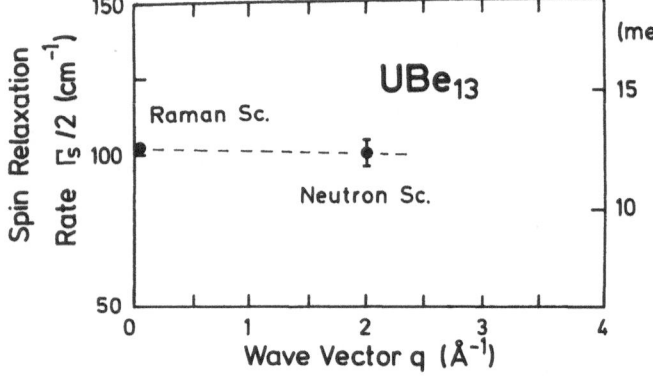

Fig. 5.21. Spin relaxation rate $\Gamma_s/2$ as a function of momentum transfer q for UBe_{13} from Raman scattering [5.20a] and neutron scattering [5.80]

Meanwhile *Cooper* et al. have reported more detailed work concerning quasielastic magnetic light scattering in UBe_{13} [5.20b]. They have investigated the nonmagnetic reference compounds $ThBe_{13}$ and $LaBe_{13}$, which do not show quasielastic light scattering. For $CeBe_{13}$ they also do not find quasielastic scattering, although it has been observed in magnetic neutron scattering [5.43] and should be observable by light scattering assuming comparable spin-orbit coupling for Ce and U. However, comparing U and Ce with regard to magnetic data, this assumption seems at least questionable. Furthermore, from a detailed line fit they conclude that the scattering intensity as a function of frequency is not well described by a quasielastic response function but can be better represented by a broad inelastic CEF excitation, which would also be consistent with the small temperature dependence of the inelastic linewidth they deduce from their data. A final conclusion in favor of one of these two interpretations is at present not possible.

Fig. 5.22. Temperature dependence of the quasielastic Raman spectra of UPt_3 for perpendicular polarizations of incident (5309 Å) and scattered light. The hatched areas mark the Lorentzian lineshape fit (half width at half maximum $\Gamma_s/2$) of the magnetic scattering contributions due to spin fluctuations of relaxation rate Γ_s

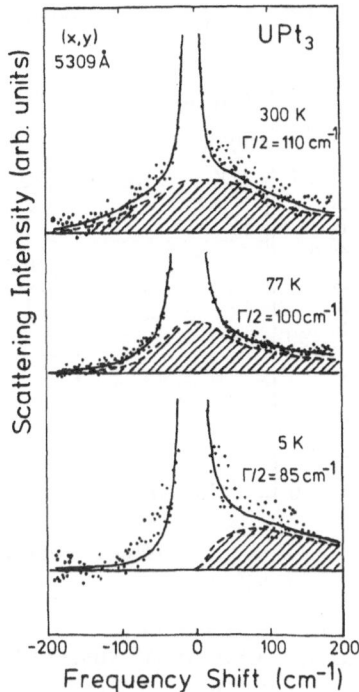

The quasielastic Raman scattering of polycrystalline UPt_3 at 300 K, 77 K and 5 K [5.18c] can be shown to be of magnetic origin because of the polarization selection rules. The pronounced asymmetry at 5 K between the Stokes (energy loss) and the anti-Stokes (energy gain) side of the spectrum is due to the Bose factor. The scattering cross section has been fitted by (5.3). The hatched areas in Fig. 5.22 represent the magnetic scattering contributions of half width at half maximum $\Gamma_s/2 = 13.6$ meV at 300 K and 10.5 meV at 5 K.

In Fig. 5.23 we show the spin relaxation rate $\Gamma_s/2$ as a function of momentum transfer q for UPt_3 [5.81]. The Raman data for different temperatures are shown at $q \approx 0$. Neutron scattering data of polycrystalline UPt_3 for $q > 1$ Å^{-1} at 1.2 K are represented by the solid squares [5.80]. In addition we show also neutron measurements of single crystals at 4.2 K and 1.2 K [5.78] by the crossed circles and the cross, respectively. The Raman results together with the neutron data establish the q independence of Γ_s, which proves the localized nature of the spin fluctuations.

A slight q dependence of the spin relaxation rate has been observed [5.82] in $CeCu_6$ and theoretically explained in [5.83]. All the above results for UBe_{13}, UPt_3 and $CeCu_6$ contradict the expectations from noninteracting Fermi liquid theory that $\Gamma_s \cong v_F q$ for $q \to 0$, where v_F is the Fermi velocity. The observation of a finite-frequency zone-center contribution

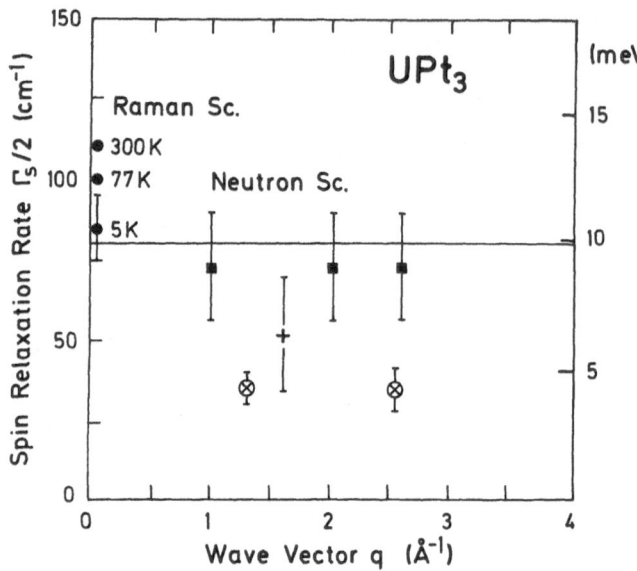

Fig. 5.23. Spin relaxation rate $\Gamma_s/2$ as a function of momentum transfer q for UPt_3. Polycrystalline samples: circles (Raman data), full squares (neutron data [5.80]). Single crystalline samples: cross (1.2 K) and circles (4.2 K) (neutron data [5.78])

to the dynamical susceptibility reflects the fact that the spin or the magnetization is not conserved due to the strong spin-orbit coupling [5.83]. The calculated mean-field zone-center dynamical susceptibility at $T = 0$ K is found to exhibit a maximum at $\omega \approx 4T_K$ [5.83], where T_K or the characteristic temperature T^* sets the energy scale of the excitations involved. From the maximum of the magnetic scattering at 5 K in Fig. 5.22 we deduce $T^* \approx 20$ cm^{-1} (2.5 meV = 30 K) in good agreement with other experimental evidence [5.84].

Heavy fermion URu_2Si_2 with its coexisting magnetic and super-conducting order [5.21–23] is to be distinguished from the other heavy fermion superconductors [5.85] insofar as the same type of "heavy" electrons are thought to be responsible for both transitions in the former material. Studies of URu_2Si_2, including specific-heat [5.21–23] and neutron scattering measurements [5.86] have further suggested that the magnetic transition involves the opening up of an energy gap over a portion of the Fermi surface, possibly driven by a charge- or spin-density wave.

Raman scattering studies were conducted on oriented, single-crystal facets in polycrystalline samples of URu_2Si_2 [5.24]. Figure 5.24 shows an example of the spin-fluctuation scattering for various temperatures, again indicating a steadily decreasing linewidth with decreasing temperature. As expected, the quasielastic scattering (hatched region) is well described by a simple relaxational model, which is illustrated in Fig. 5.24 by fitting to the power spectrum according to (5.3). Also observed in the spectral response is a small inelastic linear term, proportional to $[1 + n(\omega)]\,\omega$, which is illustrated in Fig. 5.24 by a dotted line. The fit to the full spectra (solid line) is given as the sum of this term and the

Fig. 5.24. Spin-fluctuation scattering in URu_2Si_2 at various temperatures [5.24]. The hatched area shows the quasielastic contribution, the dotted line represents the linear (inelastic) contribution, and the solid line displays the sum of these contributions with a small offset. The small feature near 150 cm^{-1} is slight leakage of the B_{1g} phonon into these spectra. All spectra have been offset. From [5.24]

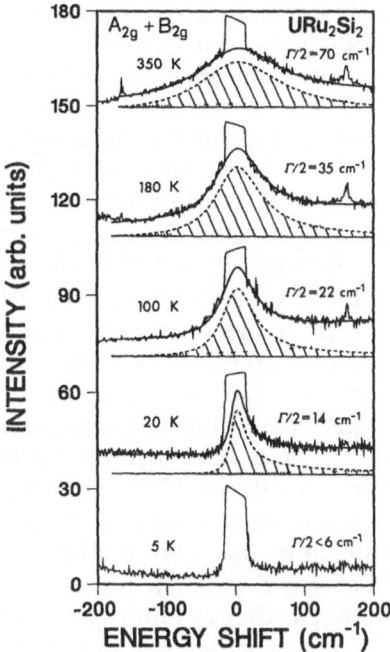

quasielastic contributions, plus a small offset. The small inelastic contribution is presumed due to crystal-field excitations centered at much higher energies (> 1000 cm^{-1}) [5.24].

The temperature dependence of the observed quasielastic linewidths (HWHM) is shown in Fig. 5.25 [5.24]. The linewidths and temperatures for all quasielastic spectra were extracted from a least-squares fit to the full spectra (-200 to $+200$ cm^{-1}). For all cases, the best fits were obtained

Fig. 5.25. Temperature dependence of the quasielastic halfwidth (HWHM) of URu_2Si_2 [5.24]. The estimated errors for all points reflect the range of linewidths for which the best fits were obtained (± 3 cm^{-1}). The solid line is a guide to the eye, while the dashed lines are extrapolations of the linear-in-T and the temperature-independent regimes, as discussed in the text. From [5.24]

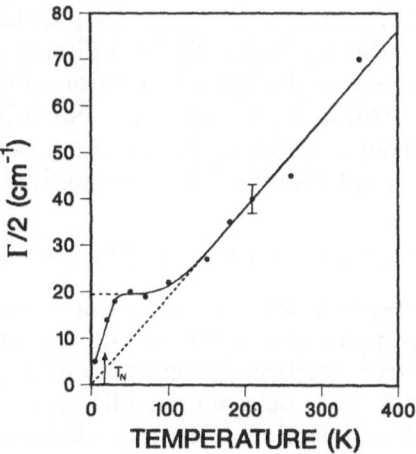

only within a narrow range of linewidths (± 3 cm^{-1}) centered at the points displayed in Fig. 5.25.

As illustrated in Fig. 5.25 several distinct regimes are observed in $\Gamma(T)/2$ vs T. Above roughly 70 K, for example, the quasielastic linewidth exhibits a linear dependence on temperature, $\Gamma/2 = ak_B T$, suggesting that above 70 K the 5f electrons in URu$_2$Si$_2$ relax via a Korringa process, wherein damping of the 5f electrons occurs through the creation of electron–hole pairs within the conducting band.

Between 70 and 30 K, the linewidth is observed to approach a residual value of roughly 19 cm^{-1} ($\cong 27$ K). A temperature-independent linewidth is expected as further thermal (Korringa-type) decreases in the linewidth are interrupted at low temperatures by the U moment-conduction-electron exchange channel. Due to the importance of both the Kondo and Ruderman-Kittel-Kasuya-Yosida (RKKY) interactions in URu$_2$Si$_2$, each is expected to contribute to the residual linewidth, and indeed, it has been suggested [5.87] that in such a case, $h\Gamma(q) = |k_B T_K - \text{J}_{\text{RKKY}}(q)|$ is the relevant energy scale (where T_K is the single-ion Kondo temperature, and $\text{J}_{\text{RKKY}}(q)$ is the q-dependent RKKY exchange parameter). Such a q-dependent linewidth would account for the discrepancy noted between the residual linewidth [$\Gamma_0(q = 0) \approx 27$ K] observed in the Raman experiment, and that reported at higher q by neutron scattering [5.88].

Below 30 K (Fig. 5.25), the spin fluctuation linewidth appears to exhibit an abrupt decrease, and by 5 K there is no observable spin-fluctuation scattering within experimental resolution ($\Gamma/2 < 6$ cm^{-1}). It should be noted, however, that this rapid change in spin-fluctuation scattering may also result from a loss of quasielastic intensity. Indeed, due to the presence of the elastic line, it is difficult to precisely distinguish intensity and linewidth changes below 30 K. This disappearance of quasielastic scattering, within experimental resolution, is believed to corroborate recent evidence [5.21–23, 86] for the formation of an energy gap near $T_N = 17$ K.

As mentioned earlier, the $q = 0$ spin fluctuation linewidths observed in Raman scattering are roughly half those reported at higher q by neutron scattering [5.86, 88]. This discrepancy is much too large to be attributed solely to the lower resolution of neutron scattering, and instead suggests a real q dependence in Γ. Such large dispersion noted in Γ could arise from a strongly q-dependent exchange coupling $J(q)$, indicating some spatial coherence of the spin fluctuations above T_N.

5.3.2 Electron Density Fluctuations in UPt$_3$

Heavy fermion metals exhibit the unusual property of a Fermi velocity comparable to the sound velocity. This has as a consequence a very large electron–phonon coupling which may lead to an appreciable difference between the adiabatic (v_s) and isothermal (v_{th}) sound velocities. The deviation from the adiabatic conditions ($v_s = v_{th}$) should manifest itself in the Landau-Placzek ratio R, which can be described as $R = (v_s - v_{th})/v_{th}$. Hence R is expected to be enhanced with respect to normal metals. From the experimental point of view the Landau-Placzek ratio is defined by the ratio $R = I_{qe}/2I_{in}$ between the intensities of the quasielastic line (I_{qe}) and those of the two inelastic Stokes/anti-Stokes lines ($2I_{in}$). This is schematically illustrated in Fig. 5.2. The enhancement of R is thus expected to show up in the light scattering spectrum as a quasielastic line with an enhanced intensity.

Brillouin scattering measurements have been performed on freshly fractured polycrystalline samples of UPt$_3$ under ^{4}He atmosphere using a triple-pass tandem Fabry-Perot interferometer with a high contrast of 10^{12} [5.25]. The spectra in Fig. 5.26 were taken at three different temperatures for $E_i \parallel E_s$ and $E_i \perp E_s$, where $E_i(E_s)$ is the electric field vector of the incident (scattered) photons. the spectrum at 300 K in Fig. 5.26 shows for $E_i \parallel E_s$ a quasielastic line over a frequency range of about ± 1000 GHz. Since no quasielastic scattering is observed for $E_i \perp E_s$, as

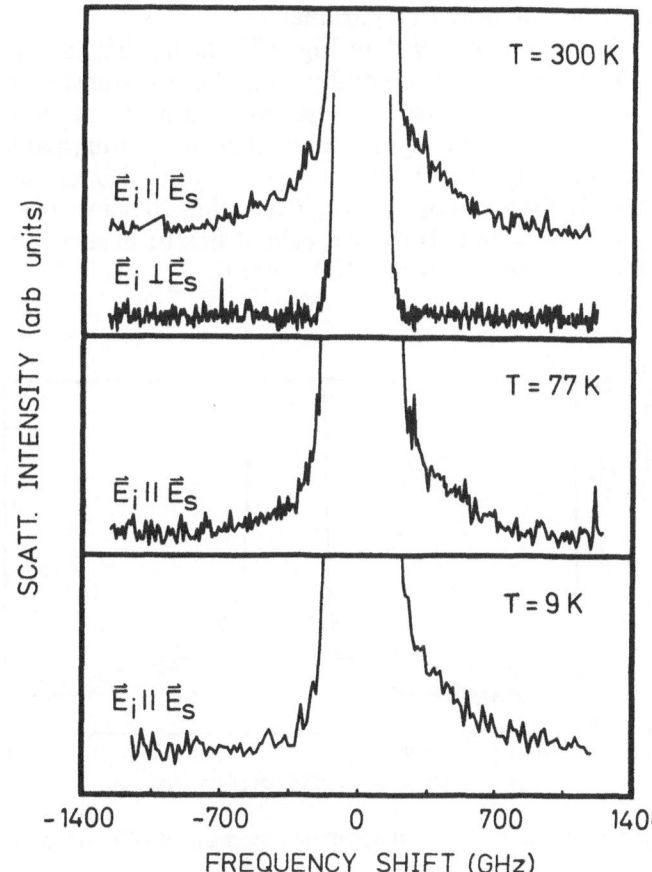

Fig. 5.26. Brillouin scattering spectra of polycrystalline UPt$_3$ at 300 K, 77 K and 9 K for parallel electric field vectors of incident (E_i) and scattered (E_s) photons($E_i \parallel E_s$). No quasielastic scattering is observed at 300 K for $E_i \perp E_s$ and at 9 K in the anti-Stokes (left-hand side) spectrum

shown for 300 K in Fig. 5.26, the scattering must be nonmagnetic. When cooling the sample from 300 K to 9 K the quasielastic line shape becomes increasingly asymmetric. This behavior points to excitations of diffusive or relaxational character which follow a Bose-Einstein distribution function. The line shape of all spectra for $E_i \parallel E_s$ in Fig. 5.26 could be fitted by the spectral function

$$S(\omega) = A(T) \frac{\hbar\omega}{1 - \exp\left(-\hbar\omega/k_B T\right)} \exp\left(-\omega^2/\Gamma^2\right) + L^6(\omega), \qquad (5.4)$$

where $A(T)$ is a temperature- but not frequency-dependent intensity factor and denotes the full width at half maximum of the quasielastic line.

The laser light which is scattered elastically by the surface roughness, was taken into account by means of the transmission function $L^6(\omega)$ of the spectrometer, where L denotes a Lorentzian line shape. (Because of the triple-pass tandem configuration, the Lorentzian-type transmission function of a single Fabry-Perot interferometer had to be taken to the sixth power). The least-squares fit of the data was performed with $A(T)$ and Γ as the only free parameters.

The example at 9 K in Fig. 5.27 shows that the fit matches excellently the measured spectrum and proves that the quasielastic line indeed obeys Bose-Einstein statistics. The linewidth Γ at 9 K is found to be $\Gamma = 20 \pm 2$ K. No significant change of Γ is found at higher temperatures. The intensity factor $A(T)\hbar\omega/[1 - \exp\left(-\hbar\omega/k_B T\right)]$ in (5.4), which for $\hbar\omega \ll k_B T$ is given by $A(T) k_B T$, is within 20% independent of temperature. This implies that $A(T)$ (see below) has to decrease by about a factor of 30 as T increases from 9 K to 300 K.

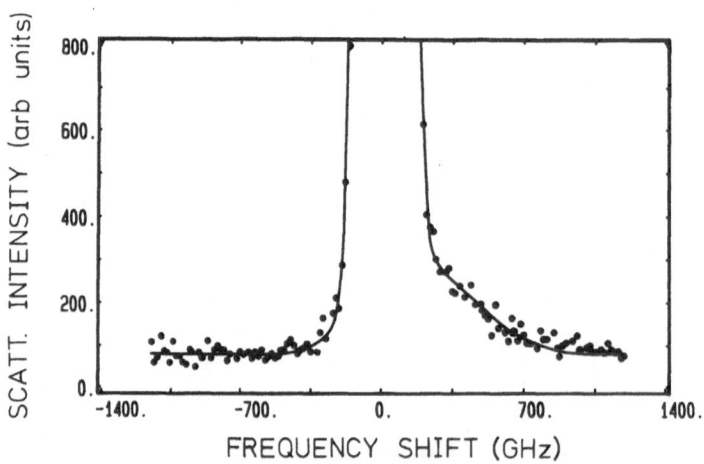

Fig. 5.27. Fit (solid line) of the Brillouin spectrum of UPt$_3$ at 9 K (points see Fig. 5.26) using (5.4)

Quasielastic light scattering due to charge density fluctuations (CDF) can be observed in doped semiconductors [5.89, 90]. Experiments on doped GaAs show that for electron concentrations n_e above 10^{16} cm^{-3} the scattering cross section decreases rapidly as n_e increases [5.89]. However, in n-type Si quasielastic light scattering has been observed for $n = 1.5 \times 10^{20}$ cm^{-3} due to multivalley electronic excitations; in heavily doped p-type Si a similar low-frequency scattering tail is attributed to intravalley excitations within the strongly anisotropic heavy-hole band (for a review see *Abstreiter* et al. in Ref. [5.2b] p. 122). For metals with carrier concentration larger than 10^{22} cm^{-3} the cross section for quasi-elastic light scattering from CDF is expected to decrease by more than eight orders of magnitude compared to GaAs with $n_e = 10^{16}$ cm^{-3} [5.89, 91, 92]. Hence it should become impossible to detect CDF in metals by quasielastic light scattering.

The situation is completely different in UPt$_3$, where the electrons are strongly coupled to the phonon system. It was shown that single-particle excitations or thermal energy density fluctuations (heat diffusion) [5.26a] could not account for the quasielastic scattering of width $\Gamma = 20$ K [5.25]. But rather the experimental value of Γ for $T < T^*$ was found to agree within a factor of two with the diffusive mode due to electron density fluctuations (heavy quasiparticle diffusion) as calculated within a model of hydrodynamic fluctuations in HF systems [5.26b].

The interaction between the heavy fermions and the phonons is taken into account by including, besides the phonon displacement ϕ, the phonon momentum and the thermal energy density, also the charge density. The frequency- and wave-vector dependent displacement correlation function $\chi_{\phi\phi}(q, \omega)$ is obtained by solving the coupled equations of motion using the Mori formalism. The Brillouin scattering intensity is given by

$$I(q, \omega) = \text{Im} \{\omega^{-1}\chi_{\phi\phi}(q, \omega)\} . \tag{5.5}$$

$\chi_{\phi\phi}$ has four poles, two of which describe the inelastic Stokes/anti-Stokes Brillouin intensities I_{in} of the longitudinal acoustic phonon of velocity v_p. The latter is given by $v_p^2 = v_{\text{th}}^2 + \hat{v}^2$, where $\hat{v}$ depends on the strength of the electron-phonon coupling ($\hat{v} = 0$ in normal metals). The other two poles are of relaxational character and give rise to quasi-elastic scattering intensities I_{qe}. They stem besides the heat conduction pole from charge density fluctuations (diffusion of heavy fermions) with full width at half maximum $\Gamma_2 = 2D_2q^2$. In the limit of small wave vectors q, D_2 is given by

$$D_2 \approx \frac{\sigma}{e^2\chi_N} . \tag{5.6}$$

Here σ denotes the specific electric conductivity of the heavy electrons and χ_N the electron density susceptibility in the normal state. The latter equals up to a factor n^2/V^2 the electronic compressibility $\varkappa_e$, where n is the electron number per unit cell and V is the unit cell volume. Assuming that the electron gas and the lattice each contribute equally to the bulk modulus c_B of a metal, it was estimated that $\varkappa \approx 2/c_B$. An estimate of the line width at $T = 9$ K gave $\Gamma_2 = 248$ K [5.25]. The value $\Gamma_2/2 = 40$ K has to be compared with the experimental one $\Gamma = 20$ K. Under the rather crude approximations made, particularly that σ of the heavy electrons is not directly accessible, the good order-of-magnitude agreement proves that the unusually strong nonmagnetic quasielastic scattering intensity I_{qe}, which has no analog in ordinary metals, is due to electron density fluctuations.

Because the theoretical model applies only to the temperature range $T < T^* \approx 20$ K, the behavior of Γ for $T > T^*$ is not yet fully understood. Besides the unknown contribution of the heavy electrons to σ, and thus to Γ_2, there is the unknown temperature dependence of σ which may be strongly affected by interband scattering processes for $T > T^*$. A strong temperature dependence of $A(T)$ in (5.4), however, should nevertheless not be overlooked, as discussed below.

Despite the fact that the spectrum at 9 K in Fig. 5.27 could only be fitted by a Gaussian lineshape, an identification of the prefactor $A(T)$ of the Gaussian in (5.4) with physical quantities is at present not possible, since in the theoretical model [Ref. 5.26b, eq. 45] a Lorentzian is obtained. However, under the more realistic assumption of a frequency-dependent self-energy part of the electrical conductivity, which enters D_2 and $I(q, \omega)$, the lineshape is no longer a Lorentzian. Thus the finding for $T < T^*$ was considered as evidence that improvement of the theoretical model was necessary.

In a first, crude attempt $A(T)$ in (5.4) was identified by the prefactor $D_2 \approx R_2/v^2$ of the Lorentzian in [Ref. 5.26b, eq. 45], where R_2 is the Landau-Placzek ratio, which relates the quasi-elastic charge density fluctuations of width Γ_2 to the inelastic Brillouin lines. The absolute value of R_2 could not be determined, since the inelastic Brillouin scattering from bulk phonons in metals is strongly suppressed in the case of strong absorption of the light. However, since the temperature dependence of the Brillouin scattering intensities in ordinary metals is given by the Bose factor, any deviations of the scattering intensities from this behavior as evidenced by $A(T)$ in (5.4) are related, at least in part, to the Landau-Placzek ratio R_2 independent of whether or not scattering intensity has been normalized to the inelastic Brillouin intensities (following the Bose factor).

The anomalous decrease of $A(T)$ by a factor of 30 from 9 K to 300 K was found to be mainly related to R_2, since D_2 is found either experimentally to be practically temperature independent or theoretically to decrease by a factor of 2 to 3 following (5.6). The qualitatively deduced decrease of R_2 for temperatures well above the characteristic temperature $T^* \approx 20$ K, which characterizes the transition of the heavy fermion system from coherent behavior $(T < T^*)$ to single-ion behavior $(T > T^*)$ indicates that the electron–phonon coupling and thus the diffusion of heavy fermions become weaker with increasing temperature. Consequently the adiabatic conditions ($v_p = v_{th} = v_s$) are restored with increasing temperature.

The anomalously large nonmagnetic quasielastic scattering observed for the first time in a heavy fermion material and, moreover, in any metal (excluding the case of heavily doped semicondutors mentioned above), is due to an enhanced Landau-Placzek ratio. An enhancement by a factor 10^3 to 10^4 was theoretically predicted due to the large electronic Grüneisen parameter (factor 10 to 100) and the large effective mass of the heavy fermions ($m^* \approx 100$) [5.26a, b]. The large Grüneisen parameter characterizes the strong electron–phonon coupling in a HF compound [5.93].

5.4 Phonon Raman Scattering in Rare Earth and Actinide Intermetallics

The primary interest in investigating mixed valence materials using Raman scattering arose from the estimate [5.94] that the inverse time scale of the 4f charge fluctuations, inducing volume changes of up to 15%, may be on the order of phonon frequencies. Hence, early Raman scattering experiments in intermediate valence materials by *Güntherodt* et al. [5.68, 95–97], *Treindl* and *Wachter* [5.98, 99] and *Stüsser* et al. [5.100, 101] have so far been concerned with the investigation of phonon anomalies and their relation to the electron–phonon interaction. In particular, polarized Raman scattering [5.96, 97, 102] has provided an experimental test of the relative importance of the different charge deformabilities introduced in the lattice dynamical model calculations [5.32, 102, 103].

The interaction of phonons with the 4f electrons in IV compounds has been studied theoretically in different and complementary terms by a large body of authors [5.104–110]. On the other hand, the experimental evidence for phonon frequency renormalizations due to valence fluctuations is on the whole rather limited. Phonon anomalies have been observed only for some IV compounds of Sm [5.68, 111–113], Tm [5.98, 114] Yb [5.115, 116] and only in two cases of Ce [5.115, 117, 118]. Moreover, no consistent interpretation of these observed anomalies has been given so far based on microscopic concepts. Therefore, we will review the accumulated data only in the framework of some phenomenological ideas first introduced by *Zirngiebl* et al. [5.119, 120]. We will discuss the behavior of acoustic and optical phonon mode frequencies $\hbar\omega$ in IV compounds according to the following classification regimes: $\hbar\omega \gg \Gamma_c$ and $\hbar\omega \lesssim \Gamma_c$, where Γ_c denotes the charge fluctuation rate. For $\hbar\omega \gg \Gamma_c$ the phonon "sees" a static mixture of divalent and trivalent ions. The mode frequency behavior in, e.g., alloys of stable di- and trivalent R ions as found for

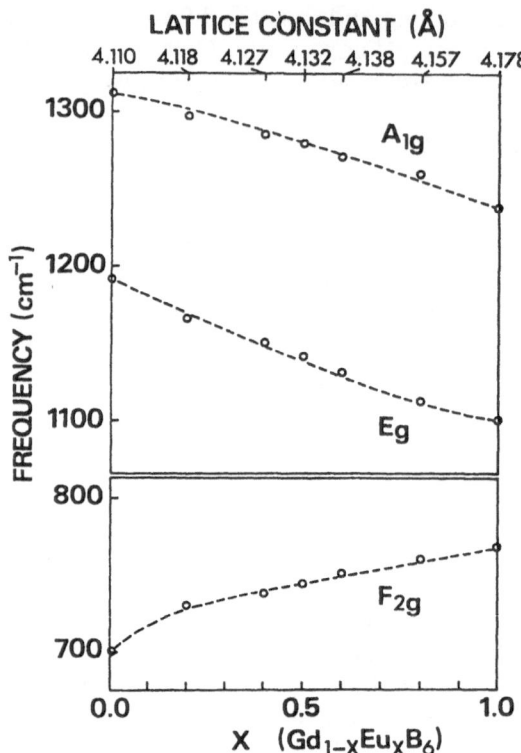

Fig. 5.28. Observed frequencies of three Raman-active modes of $Gd_{1-x}Eu_xB_6$ plotted as a function of x [5.121]. The lattice constants of $Gd_{1-x}Eu_xB_6$ are indicated at the top of the figure. From [5.121]

example in $Gd_{1-x}Eu_xB_6$ by *Ishii* et al. [5.121] is shown in Fig. 5.28. On the basis of this result one expects the phonon mode frequencies of the IV compound for $\hbar\omega \gg T_c$ to be intermediate between those of the di- and trivalent reference compounds according to the valence mixing ratio. In the phonon frequency range characterized by $\hbar\omega \lesssim \Gamma_c$, however, the phonon should soften compared to the stable valent reference compounds, because the charge fluctuation rate can easily follow the movement of the ions and thereby soften the lattice.

It has been emphasized [5.120] that besides a systematic understanding of the occurrence or absence of elastic and phonon anomalies in different IV compounds the concept introduced above allows for a first experimental estimate of charge fluctuation rates. Their direct experimental investigation has not been feasible, in contrast to magnetic relaxation rates, which have been investigated intensively by quasielastic neutron scattering [5.122]. In the following we review Raman and Brillouin scattering from phonons in different IV compounds and their stable valence reference compounds, emphasizing the resulting estimate of the charge fluctuation rate in the IV compound under investigation. A compilation of the different crystal structures and corresponding vibrational mode symmetries of various types of valence fluctuating materials is given in Table 5.1.

Table 5.1. Vibrational mode symmetries of phonon modes in valence fluctuating materials

Structure	Space group	Fluctuating valence materials	Acoustic	Infrared	Raman	Other optical modes
$ThCr_2Si_2$	I4/mmm (D_{4h}^{17}) tetragonal	$CeCu_2Si_2$ $EuCu_2Si_2$ $YbCu_2Si_2$	$\Gamma_3^- + \Gamma_5^-$	$2\Gamma_3^- + 2\Gamma_5^-$	$\Gamma_1^+ + \Gamma_2^+ + 2\Gamma_5^+$	
Cu_3Au	Pm3m (O_h^1) cubic	$CePd_3$, $CeSn_3$	Γ_4^-	$2\Gamma_4^-$		Γ_5^-
CaB_6	Pm3m (O_h^1) cubic	SmB_6, CeB_6	Γ_4^-	$2\Gamma_4^-$	$\Gamma_1^+ + \Gamma_3^+ + \Gamma_5^+$	$\Gamma_4^+ + \Gamma_5^-$
NaCl	Fm3m (O_h^5) cubic	SmS (met.) SmSe, SmTe, TmSe, TmTe, YbS, YnSe, YbTe	Γ_4^-	Γ_4^-		
$NaZn_{13}$	Fm3c (O_h^6) cubic	$CeBe_{13}$	Γ_4^-	$6\Gamma_4^-$	$2\Gamma_1^+ + 4\Gamma_3^+ + 4\Gamma_5^+$	$2\Gamma_2^+ + 5\Gamma_4^+ + \Gamma_1^- + \Gamma_2^- + 2\Gamma_3^- + 6\Gamma_5^-$
$MgCu_2$	Fd3m (O_h^7) cubic	$CeAl_2$ $YbAl_2$	Γ_4^-	$2\Gamma_4^-$	Γ_5^+	$\Gamma_2^- + \Gamma_3^- + \Gamma_5^-$

5.4.1 R-B₆ (R = Y, La, Ce, Nd, Sm, Eu, Gd)

The class of the $R\text{-}B_6$, including the IV compound SmB_6 [5.123] and the Kondo compound CeB_6 [5.124], crystallizes in the simple cubic space group O_h^1 [5.125] as shown in Fig. 5.29a, together with three Raman-active vibrations of the boron octahedron in Fig. 5.29b. Typical Raman spectra of LaB_6 and CeB_6 are shown in Fig. 5.30. One observes the three Raman-active modes near 1300 cm^{-1} (A_{1g}), 1150 cm^{-1} (E_g) and 690 cm^{-1} (T_{2g}) [5.126, 127] as well as first-order symmetry-forbidden scattering intensity near 200 cm^{-1} [5.128]. From a comparison with the measured phonon dispersion of LaB_6 [5.129] the structure at 200 cm^{-1} has been assigned to a T_{1u} optical mode [5.119]. The other low frequency phonons are due to the acoustic branches whose long wavelength limit is described

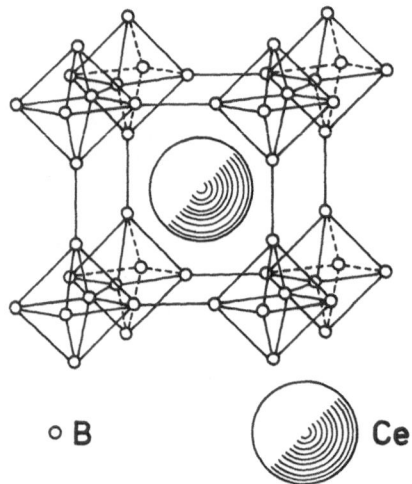

○ B   Ce

Fig. 5.29a. Structure of simple cubic CeB_6 belonging to the space group O_1^h

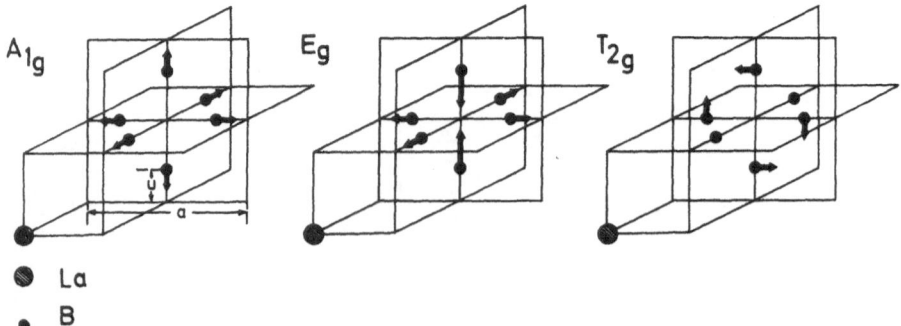

Fig. 5.29b. Eigenvectors of the three Raman-active phonon modes. From [5.127]

Fig. 5.30. Room temperature Raman spectrum of CeB_6 and LaB_6

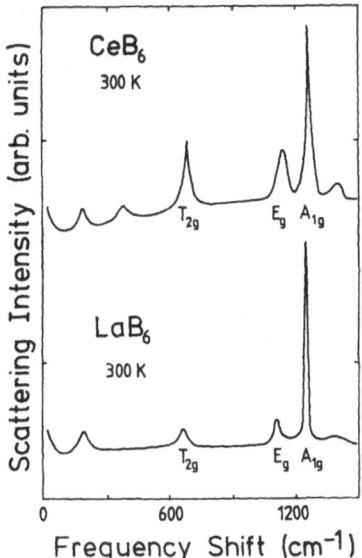

by the elastic constants and, in particular, by the bulk modulus c_B. The elastic properties are accessible by Brillouin scattering or ultrasonic measurements.

In Fig. 5.31 the frequencies of the different symmetry modes of the RB_6 at room temperature are plotted as a function of the lattice parameter a_0 [5.119]. The stable trivalent RB_6 show a linear variation of the mode frequencies with a_0 as indicated by the solid lines. The reference lines for the stable divalent RB_6 (dashed lines) are fixed by EuB_6 and have been drawn parallel to the solid lines. The frequencies of the A_{1g}, E_g and T_{2g} modes of IV SmB_6 lie between the stable di- and trivalent reference lines according to its valence mixing ratio. On the other hand, the T_{1u} mode of SmB_6 shows a softening with respect to the coinciding stable valence reference lines. This softening has already been discussed in the literature [5.128], but not in the general framework presented here.

Fig. 5.32 shows the bulk modulus of RB_6 [5.130] at room temperature as a function of Q/V, where Q denotes the valence and V the unit cell volume. SmB_6 exhibits a significant softening of the bulk modulus compared to the stable valence reference line.

These spectroscopic results can be explained by the above-introduced concept of comparing the phonon frequencies with the charge fluctuation rate [5.119]. The three highest lying phonon modes (A_{1g}, E_g and T_{2g}) of SmB_6 show their frequencies (> 650 cm^{-1}) between the stable di- and trivalent reference lines, exhibiting the behavior of an alloy of stable di- and trivalent R ions. On the other hand, the low lying T_{1u} phonon mode and the bulk modulus of SmB_6 show a softening compared to the reference lines. Therefore it has to be concluded that the charge fluctuation rate is

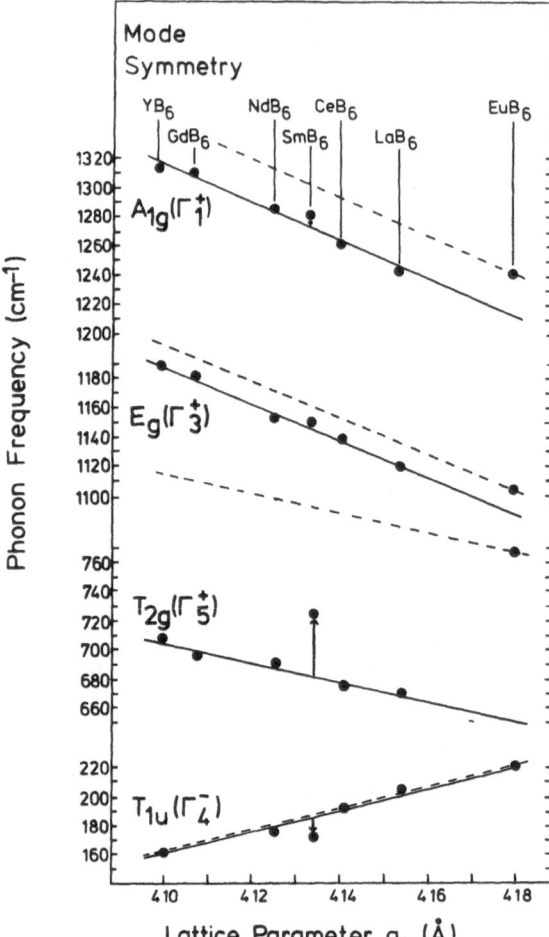

Fig. 5.31. Various phonon mode frequencies of R-B$_6$ as a function of the lattice parameter

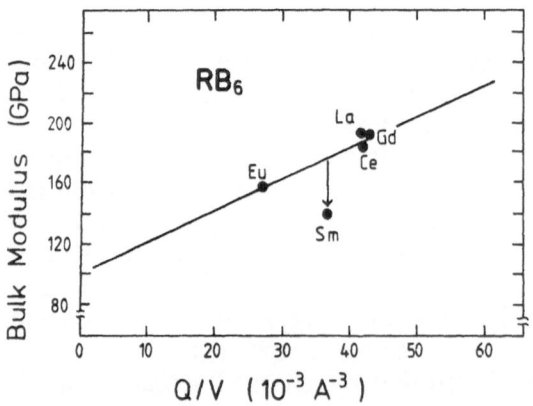

Fig. 5.32. Bulk modulus of various R-B$_6$ plotted as a function of Q/V (Q is the valence of the R-ion and V is the unit cell volume). From [5.130]

above $200 \, cm^{-1}$ (280 K) and well below $650 \, cm^{-1}$ (920 K). This is the same order of magnitude one arrives at by scaling the calculated charge fluctuation rate [5.131] by the spin fluctuation rate $\Gamma_s \approx 100$ K, which has been determined from the magnetic susceptibility [5.132]. Hence one obtains a value of $\Gamma_c = 350$ K.

From the behavior of the T_{2g} mode frequency of SmB_6 compared to the di- and trivalent reference lines a value of the valence $v = 2.65$ can be deduced which is in good agreement with $v = 2.7$ from other measurements [5.132, 133].

5.4.2 R-Al$_2$ (R = Y, La, Ce, Eu, Gd, Dy, Yb)

The lattice dynamics of R-Al$_2$ has been examined experimentally by Raman spectroscopy [5.115, 117] and, in the case of CeAl$_2$, LaAl$_2$ and YAl$_2$, by inelastic neutron scattering [5.134–136]. Moreover, the bulk moduli have been determined from lattice parameter measurements under pressure

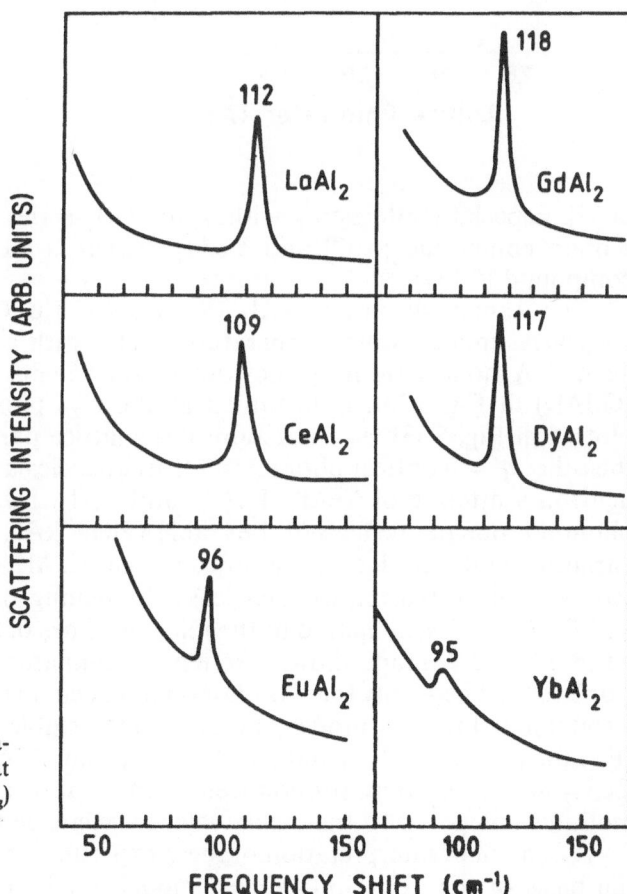

Fig. 5.33a. Unpolarized Raman spectra of R-Al$_2$ at 300 K showing the $\Gamma_{25}'(T_{2g})$ phonon mode

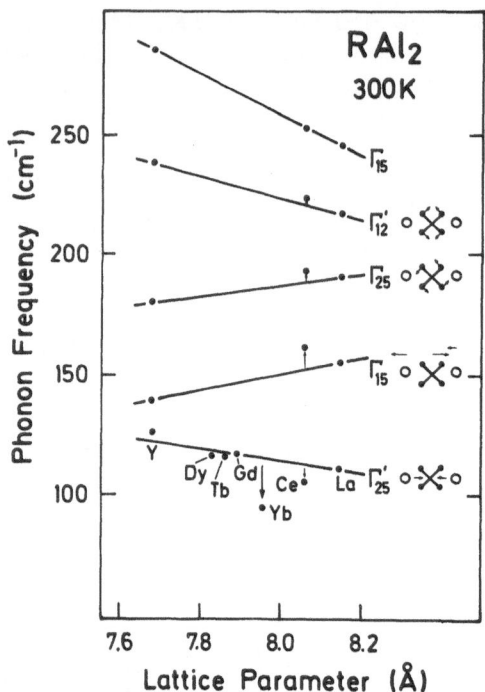

Fig. 5.33b. Phonon frequencies (in decreasing order) of the $\Gamma_{15}(t)$, Γ'_{12}, Γ_{25}, $\Gamma_{15}(l)$ and Γ'_{25} symmetry modes of various R-Al$_2$ taken from Raman scattering [5.115] or neutron scattering [5.134–136]; $\Gamma_{15}(t)$: transverse configuration in the (110) plane, $\Gamma_{15}(l)$: longitudinal configuration

[5.37]. Especially interesting within the RAl$_2$ series are CeAl$_2$, which is a Kondo compound [5.57] and YbAl$_2$, which is an intermediate valence compound [5.137].

The Raman spectra of RAl$_2$ (R=La, Ce, Eu, Gd, Tb, Dy, Yb) have been measured at room temperature [5.115] with a spectral resolution of 4 cm^{-1}. A mode in the frequency range from 95 cm^{-1} (YbAl$_2$) to 118 cm^{-1} (GdAl$_2$) in Fig. 5.33a is identified as the Γ'_{25} phonon, which has been plotted in Fig. 5.33b as a function of the lattice parameter, together with the other $q = 0$ optical phonon mode frequencies available from inelastic neutron scattering of CeAl$_2$, LaAl$_2$, and YAl$_2$. For the Γ'_{25} (Γ^+_5 or T_{2g}) phonon mode frequencies of the stable valent RAl$_2$ one observes a linear variation with the lattice parameter. For CeAl$_2$ one observes (in the sequence of decreasing frequencies) a hardening of the $\Gamma_{15}(t)$, Γ'_{12}, Γ_{25} and $\Gamma_{15}(l)$ modes compared to the reference lines of stable trivalent R-Al$_2$, whereas the Γ'_{25} mode shows a softening compared to the reference lines. For YbAl$_2$, where only the Γ'_{25} phonon has been investigated, one observes a softening of the Γ'_{25} mode compared to the stable valence reference line. (For the notation $\Gamma_{15}(t)$ and $\Gamma_{15}(l)$ see caption of Fig. 5.33b). In addition, CeAl$_2$ and, even more pronounced, YbAl$_2$ show a softening of the bulk modulus compared to the stable valence reference line.

A consistent interpretation of these experimental data can be given on the basis of the above-introduced model assuming a charge relaxation rate Γ_c of roughly 120 cm^{-1} for CeAl$_2$. Phonons with $\hbar\omega \gg \Gamma_c$ show a hardening due to a "static" mixture of tri- and tetravalent Ce ions, whereas phonons with $\hbar\omega \lesssim \Gamma_c$ exhibit a softening compared to the stable valence reference lines. By the same reasoning one can give at least a lower boundary for the charge relaxation rate of YbAl$_2$, i.e. $\Gamma_c \gtrsim 100$ cm^{-1}.

5.4.3 RCu$_2$Si$_2$ (R = Y, La, Ce, Tb, Tm) and URu$_2$Si$_2$

Within the RCu$_2$Si$_2$ series CeCu$_2$Si$_2$ has received a great deal of attention due to its heavy fermion superconductivity [5.12, 13]. The origin of the pairing mechanism is still quite controversial; it has been suggested that it is purely electronic in nature (spin fluctuations) as well as classically phononic [5.138]. In any case, both the dependence of T_c on atomic volume [5.139] and the evidence for elastic anomalies [5.140] found in CeCu$_2$Si$_2$ illustrate that the lattice is also important to the novel properties of this system and makes the investigation of the lattice dynamics by light scattering especially worthwhile. Raman scattering experiments on RCu$_2$Si$_2$ have been reported by *Cooper* et. al. [5.62]. All of their light scattering experiments have been conducted in a near-backscattering geometry using a polarized 4880 or 5145 Å line of an argon laser and a triple stage spectrograph (filter as well as dispersive stage) in conjunction with both multichannel and single-channel detection systems.

The RCu$_2$Si$_2$ compounds crystallize in the tetragonal ThCr$_2$Si$_2$ structure (space group D_{4h}^{17}–I4/mmm) with a unit cell depicted in Fig. 5.34 [5.141]. The Raman-active phonons associated with RCu$_2$Si$_2$ are $A_{1g} + B_{1g} + 2E_g$ ($\Gamma_1^+ + \Gamma_3^+ + 2\Gamma_5^+$). From the site symmetries of Si and

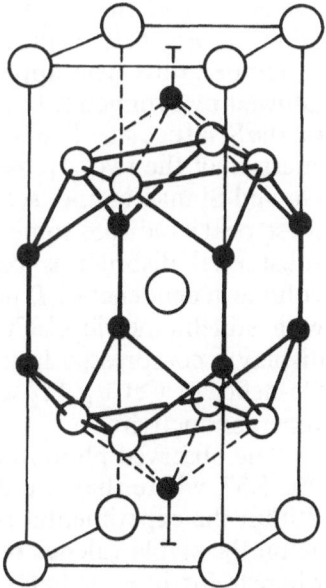

Fig. 5.34. Unit cell of CeCu$_2$Si$_2$ (space group D_{4h}^{17}–I4/mmm)

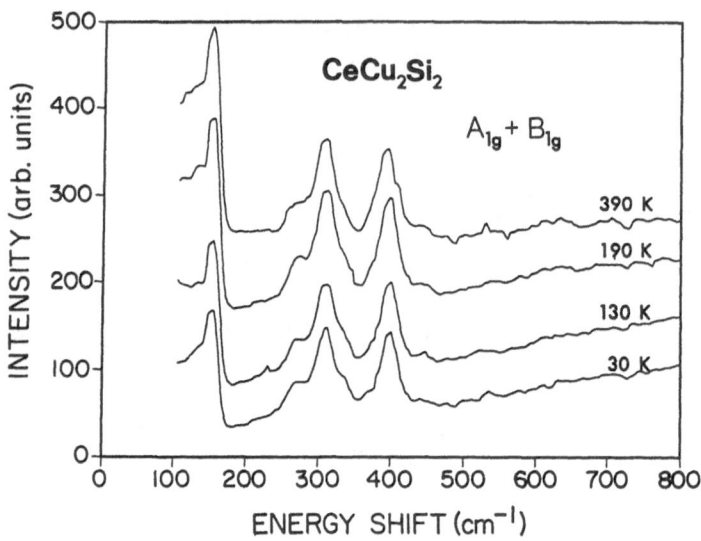

Fig. 5.35. $A_{1g} + B_{1g}$ spectra of CeCu$_2$Si$_2$ at 390 K [5.62]. Resolution: 10 cm^{-1}. The spectra have been shifted upwards by 70 a. u. From [5.62]

Cu in CeCu$_2$Si$_2$, it is known that the A_{1g} phonon involves only Si atoms, while the B_{1g} phonon is purely a Cu mode. Both Si and Cu, however, participate in the two E_g phonons. The Ce atom, sitting at a site of inversion symmetry, is not involved in any Raman-active modes.

As an example, the two A_{1g} modes of CeCu$_2$Si$_2$ near 304 cm^{-1} and 398 cm^{-1} and the B_{1g} mode near 151 cm^{-1} are shown in Fig. 5.35. In addition, Fig. 5.36 displays the two E_g modes of CeCu$_2$Si$_2$ as a function of temperature. The appearance of two A_{1g} modes rather than one, as allowed by symmetry, has been attributed to a static disorder of Cu ions on the Si lattice sites [5.62]. This seems reasonable since, firstly, the observed energies of the two A_{1g} peaks are consistent with a simple scaling of the Cu and Si masses and, secondly, the large widths associated with both of these peaks indicate some degree of disorder in the material. Evidence for substantial disorder is particularly interesting in view of the peculiar volume dependence of T_c noted in CeCu$_2$Si$_2$ [5.139]. Specifically, compared with stoichiometric CeCu$_2$Si$_2$, Cu-rich CeCu$_{2.2}$Si$_2$ has been shown to display a compressed lattice and a high value of T_c (T_c = 0.68 K), while Cu-deficient CeCu$_{1.9}$Si$_2$ exhibits an expanded lattice and an absence of superconductivity.

The observed phonon frequency data for RCu$_2$Si$_2$ are summarized in Fig. 5.37, where they are displayed as a function of the unit cell volume. Within the experimental resolution the phonon frequencies of CeCu$_2$Si$_2$ lie on the stable valence reference line given by LaCu$_2$Si$_2$ and GdCu$_2$Si$_2$, whereas for intermediate valence YbCu$_2$Si$_2$ [5.122] a small softening of

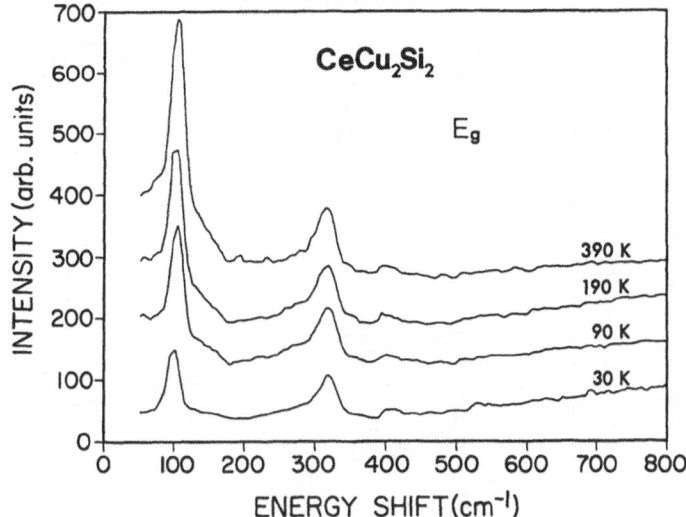

Fig. 5.36. E_g spectra of CeCu$_2$Si$_2$ at various temperatures. Resolution: 10 cm^{-1}. The spectra have been offset. From [5.62]

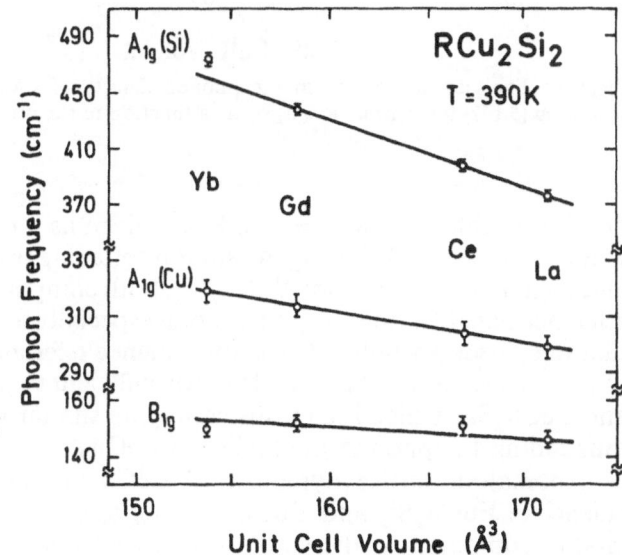

Fig. 5.37. Phonon frequencies of various R-Cu$_2$Si$_2$ as a function of the unit cell volume (see text)

the B_{1g} mode and a small hardening of the A_{1g} (Si) mode compared to the stable valence reference line is observed. For the RCu$_2$Si$_2$ compounds, in adddition to the data of the $q = 0$ Raman-active modes [5.62], there are data available on the low energy part ($E < 300$ K) of the phonon density of states [5.142]. In Fig. 5.38 the peak positions in the phonon

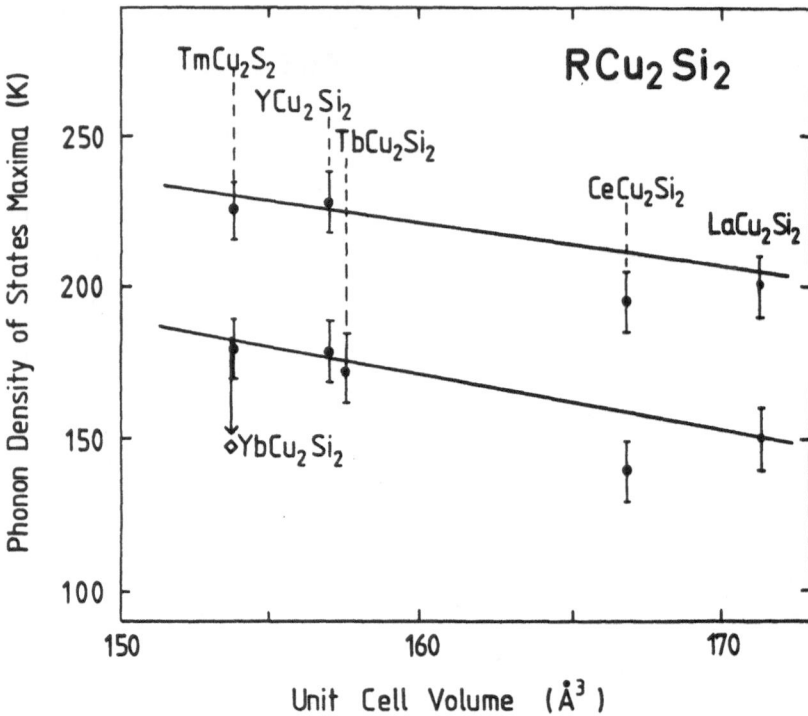

Fig. 5.38. Energies of maxima in the phonon density of states as observed by neutron scattering [5.142] for various R-Cu$_2$Si$_2$ as a function of the unit cell volume

density of states below $E = 300$ K are given as a function of the unit cell volume for various RCu$_2$Si$_2$. A phonon softening is observed for CeCu$_2$Si$_2$ and even more dramatic for YbCu$_2$Si$_2$, both compared to the stable valence reference line. It should be pointed out, especially for the case of CeCu$_2$Si$_2$, that one needs a whole set of stable valence reference compounds to infer the "stable valence behavior". It is not sufficient to just compare LaCu$_2$Si$_2$ and CeCu$_2$Si$_2$, which are in this case quite similar in phonon frequencies, thus hiding the phonon anomalies of CeCu$_2$Si$_2$.

Anomalous lattice dynamics of CeCu$_2$Si$_2$, as well as intermediate valence of EuCu$_2$Si$_2$ and YbCu$_2$Si$_2$, can also be inferred from their bulk moduli compared to the stable valence reference compounds. The bulk moduli of RCu$_2$Si$_2$ have been deduced from Brillouin scattering data, which are given in Fig. 5.39 for CaCu$_2$Si$_2$, YbCu$_2$Si$_2$ and LaCu$_2$Si$_2$ as examples. Fig. 5.40 displays the bulk modulus of various RCu$_2$Si$_2$ compounds as a function of the rare earth charge density Q/V, Q denoting the rare earth valence and V the unit cell volume [5.143]. A drastic softening of the bulk modulus of the intermediate valence RCu$_2$Si$_2$ compounds compared to the stable valence reference line is inferred. Within the above introduced model of electron–phonon coupling in rare earth intermediate

Fig. 5.39. Brillouin spectra at 300 K of polycrystals of (a) $CaCu_2Si_2$ ($\Theta_1 = 72°$), (b) $YbCu_2Si_2$ ($\Theta_1 = 75°$), and (c) $LaCu_2Si_2$ ($\Theta_1 = 72°$). $LaCu_2Si_2$ shows a typical surface ripple spectrum with a cut-off at ω_L. The surface LA peak of $CaCu_2Si_2$ and $YbCu_2Si_2$ at ω_L is indicated

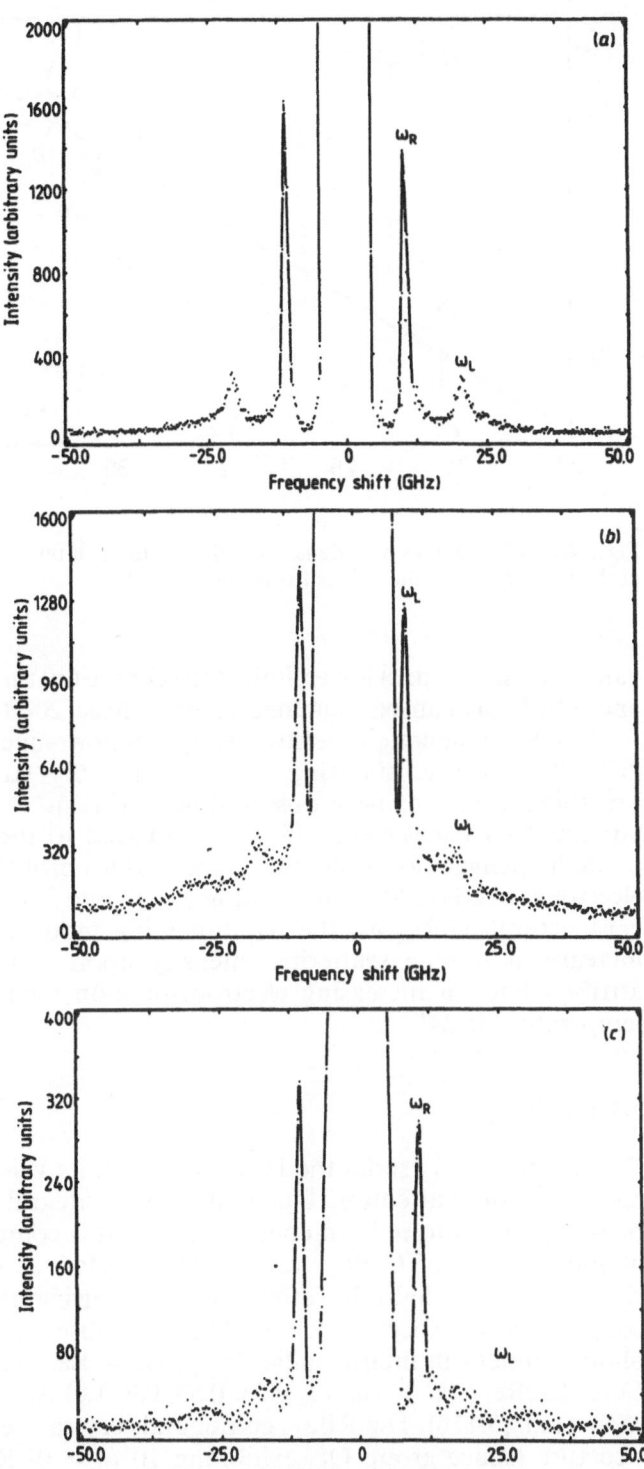

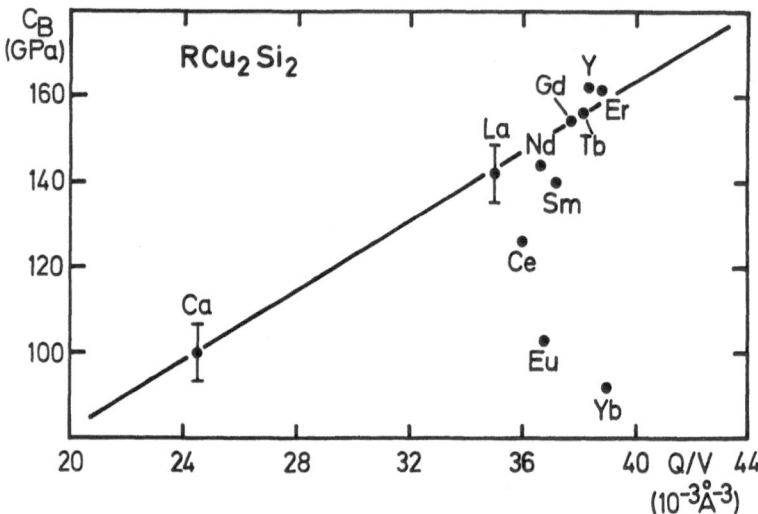

Fig. 5.40. Bulk modulus of various R-Cu$_2$Si$_2$ plotted as a function of Q/V (Q is the valence of the R-ion and V is the unit cell volume)

valence compounds, a lower limit of the charge fluctuation rate of CeCu$_2$Si$_2$ and YbCu$_2$Si$_2$ can be estimated to be at least 200 K.

For the magnetic-ordering heavy fermion superconductor URu$_2$Si$_2$ [5.21–23], also crystallizing in the ThCr$_2$Si$_2$ structure, an A_{1g}-type breathing mode has been observed near 431 cm^{-1} and a B_{1g} mode near 163 cm^{-1} by *Cooper* et al. [5.24]. Compared to the R-Cu$_2$Si$_2$ series the mode frequencies are shifted to higher values probably due to a stronger electronic overlap. Most interesting [5.24], and in contrast to the case of isostructural ThRu$_2$Si$_2$, the A_{1g} breathing mode of URu$_2$Si$_2$ shows an increase in Raman scattering intensity upon cooling, which has been attributed to an increasing electron–phonon coupling with decreasing temperature [5.24].

5.4.4 R-Be$_{13}$

Within the R-Be$_{13}$ series the IV CeBe$_{13}$ with its valence of 3.04 [5.44] has received a lot of attention. Due to its nearly integral valence, large charge relaxation rates have been conjectured for this compound [5.131]. As the phonon modes of CeBe$_{13}$ are to be expected at rather high frequencies due to the light Be atoms, a still considerable electron–phonon coupling was suspected [5.118]. The first observation of Raman-active phonon modes in intermetallic RBe$_{13}$ (R = La, Ce, Gd, Tb, Yb, La) and Ce$_{1-x}$La$_x$Be$_{13}$ (x = 0.0, 0.1, 0.23, 0.55, 0.8, 1.0) was reported by *Blumenröder* et al. [5.118]. The RBe$_{13}$ compounds crystallize in the cubic NaZn$_{13}$ structure (space group O$_h^6$, exhibiting 10 ($q = 0$) Raman-active modes:

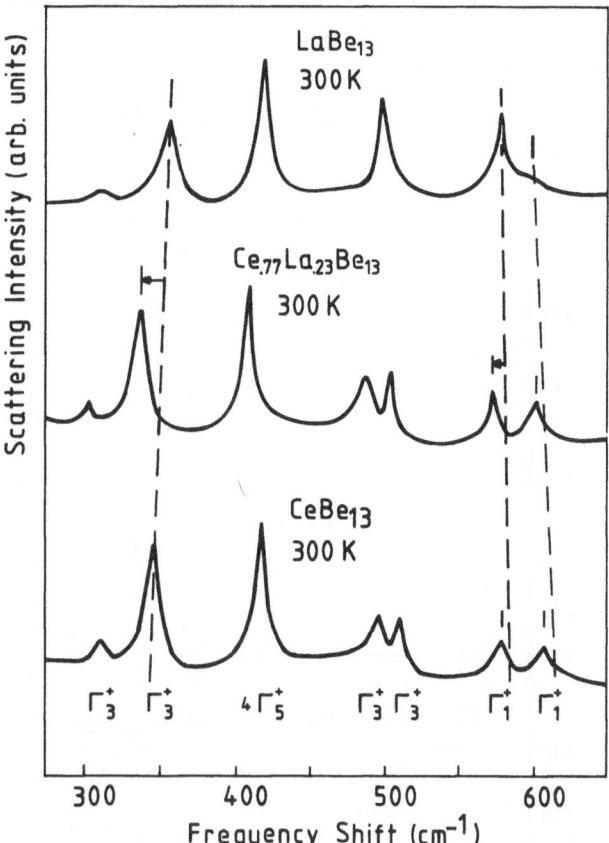

Fig. 5.41. Raman spectra of $LaBe_{13}$, $Ce_{0.77}La_{0.23}Be_{13}$ and $CeBe_{13}$ at room temperature; the behavior of the stable valence reference compounds is indicated for three modes by the dashed lines. The mode symmetry assignment for $CeBe_{13}$ is given at the bottom

$2\Gamma_1^+ + 4\Gamma_3^+ + 4\Gamma_5^+$. Figure 5.41 shows, as an example, the Raman spectra of $LaBe_{13}$, $CeBe_{13}$ and $Ce_{0.77}La_{0.23}Be_{13}$ at 300 K. One observes seven modes, except in $LaBe_{13}$, where the two modes near 500 cm^{-1} are degenerate. The mode assignment [5.118] is also given in Fig. 5.41. Figure 5.42 summarizes the Raman spectroscopic results of RBe_{13}. As a function of the lattice parameter the frequencies of the different phonon modes observed are plotted together with the Debye temperature deduced from specific heat measurements [5.145]. The stable valence reference compounds $LaBe_{13}$, $GdBe_{13}$, $TbBe_{13}$ and $LuBe_{13}$ show a linear variation with lattice constant as indicated in Fig. 5.42 by the solid lines. No anomaly is observed for $YbBe_{13}$, which has been assumed to be intermediate valent on the basis of Mössbauer spectroscopy [5.146].

On the other hand, the two Γ_1^+ modes of $CeBe_{13}$ show a softening of about 2% with respect to the reference line. No other symmetry modes of $CeBe_{13}$ show any anomaly. This applies also to the bulk modulus of $CeBe_{13}$, a fact which is partly indicative of the behavior of the long wavelength acoustic phonons. The bulk modulus of the RBe_{13} series has

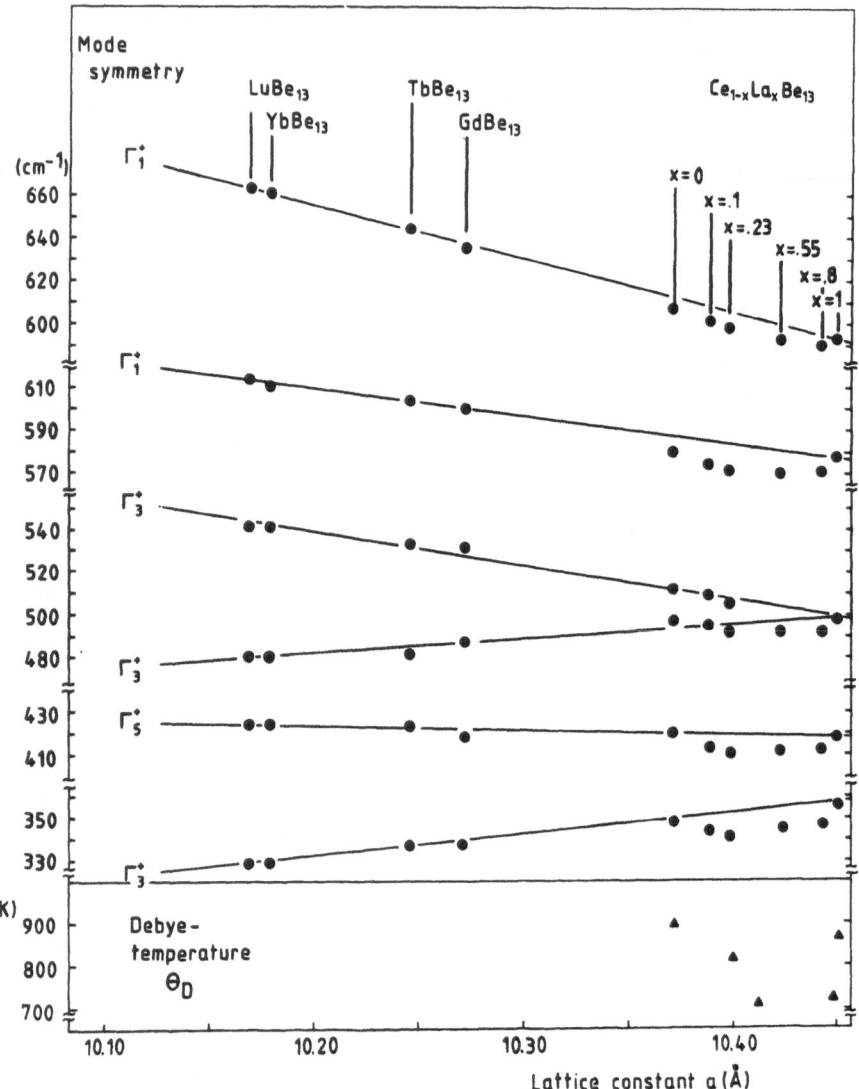

Fig. 5.42. Frequencies of the Raman-active phonon modes and the Debye temperature Θ_D [5.145] of various R-Be$_{13}$ and Ce$_{1-x}$La$_x$Be$_{13}$ compounds as a function of the lattice parameter

been deduced from Brillouin scattering experiments, as shown for LaBe$_{13}$ in Fig. 5.43. The bulk moduli of the various RBe$_{13}$ and ABe$_{13}$ (A = Th, U) are depicted in Fig. 5.44. As a function of Q/V the bulk modulus of the RBe$_{13}$ compounds follows a straight line with CeBe$_{13}$ right on it [5.143]. However, an even stronger softening of the two Γ_1^+ modes is found upon Ce dilution in Ce$_{1-x}$La$_x$Be$_{13}$ as seen for all measured compositions $0.1 \leqq x \leqq 0.8$ in Fig. 5.42. However, unlike CeBe$_{13}$, this mode softening in

Fig. 5.43. Brillouin spectra of polished, polycrystalline LaBe$_{13}$ ($\Theta_1 = 66.3°$), typical for all RBe$_{13}$. LaBe$_{13}$ shows a surface ripple spectrum with a minimum of the scattering intensity at the frequency of the transverse acoustic phonon

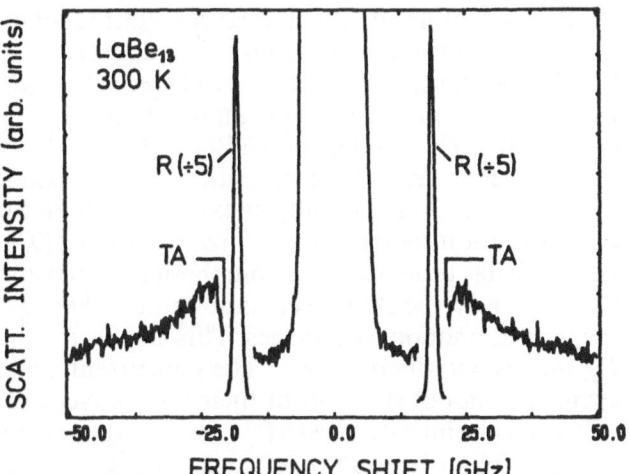

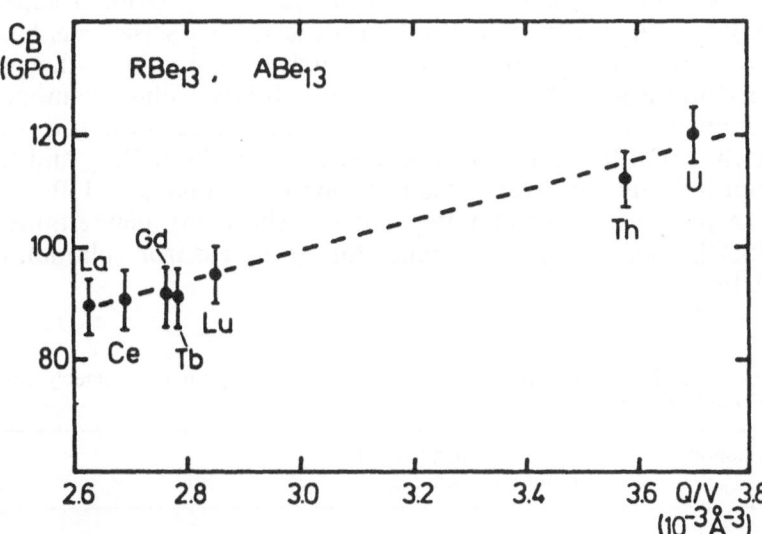

Fig. 5.44. Bulk modulus of various R-Be$_{13}$ and A-Be$_{13}$ (A = actinide) plotted as a function of Q/V (Q is the valence of the R-ion and V is the unit cell volume)

Ce$_{1-x}$La$_x$Be$_{13}$ for $0.1 \leqq x \leqq 0.8$ is also observed for all other symmetry modes with respect to the average behavior of the reference materials. The phonon softening in Ce$_{1-x}$La$_x$Be$_{13}$, for $0.1 \leqq x \leqq 0.8$, independent of the mode symmetry, is also reflected by the behavior of the Debye temperature Θ_D [5.145], which is displayed at the bottom of Fig. 5.42. No temperature dependent phonon anomaly has been observed for the optical phonons of CeBe$_{13}$, contrary to the anomalous softening of the bulk modulus which takes place upon cooling down below 350 K [5.147].

By calibrating the theoretical ratio [5.131] of charge to spin relaxation rates by the experimental spin relaxation rate $\Gamma_s/2 = 20\,\text{meV}$ of CeBe$_{13}$ [5.122], the charge relaxation rate exceeds the highest phonon frequencies by about a factor of ten. This manifests itself in the small softening (2%) of the two Γ_1^+ modes of CeBe$_{13}$ (Fig. 5.42). The dilution of Ce in Ce$_{1-x}$La$_x$Be$_{13}$ leads to a reduction of all relaxation rates, indicated by the decreasing susceptibility maximum [5.145]. Similarly, a reduction of the fluctuation temperature T_f from $\approx 150\,\text{K}$ $(0.0 < x < 0.7)$ to $\approx 50\,\text{K}$ $(x \gtrsim 0.8)$ has been deduced from thermal expansion measurements [5.148]. Consequently, the charge relaxation rate will be lowered, thus approaching the optical phonon frequencies. This is reflected in further softening of the Γ_1^+ modes with increasing x. The concurrent softening of all other $q \approx 0$ symmetry modes stems from their long-wavelength phase-coherent averaging over primarily local (Γ_1^+) breathing-type charge fluctuations. This effect is obviously more pronounced for short-wavelength zone boundary phonons, which contribute most to Θ_D (Fig. 5.42) due to their high density of states.

Phonon Raman scattering from the heavy fermion superconductor UBe$_{13}$, which is isostructural with the R-Be$_{13}$ series, has been reported by *Cooper* et al. [5.20 a, b]. Using single crystal samples they were able to perform a symmetry analysis of the observed phonon modes, which is summarized in Table 5.2. The mode frequencies observed for UBe$_{13}$ lie well within the range of those observed for the R-Be$_{13}$ and therefore do not indicate any strong electron–phonon coupling in UBe$_{13}$. Moreover, the phonon modes of UBe$_{13}$ do not show any temperature anomalies [5.23], again giving no evidence for strong electron–phonon coupling in UBe$_{13}$.

Table 5.2. Mode frequencies of several Raman-active phonon symmetry modes of UBe$_{13}$. From [5.20a, b]

Phonon	Mode frequency [cm^{-1}] (resolution: 2 cm^{-1})
T_{2g}	320
E_g	347
T_{2g}	424
E_g	432
T_{2g}	482
A_{1g}	522
E_g	539
A_{1g}	593
T_{2g}	595
E_g	637

5.4.5 R-S (R = Y, La, Pr, Sm, Eu, Gd, Yb)

Among the various classes of rare earth compounds the fcc (NaCl-type) R-monosulphides are the simplest from a structural and magnetic-exchange point of view. Although closely related in their chemical aspects, the R-S show strongly differing electronic, magnetic and optical properties along with the filling of the 4f shell; R-S can be ferromagentic semiconductors (EuS), metallic antiferromagnets (GdS), superconductors (LaS) or exhibit valence fluctuations after semiconductor–metal transformation induced by external pressure (SmS) or by lattice pressure ($Sm_{1-x}Y_xS$). The interest in Raman studies of this class of compounds arose mainly from the intriguing possibility of investigating widely different types of electron–phonon couplings within one chemical structure. As first order Raman scattering is symmetry forbidden in these fcc compounds, even its observation in these compounds is unusual and has been attributed to strong exciton–LO-phonon coupling through Fröhlich interaction in YbS [5.149a] or to defect-induced Raman scattering in GdS [5.149b], which is modified due to strong intraionic charge deformabilities in superconducting YS and in valence fluctuating SmS and $Sm_{0.75}Y_{0.25}S$ [5.65, 102]. For EuS a strong interaction between the phonon and spin systems has been found to be responsible for the observed spin-disorder-induced Raman scattering as well as the coupled magnon-phonon Raman scattering [5.150–153]. All these results have been extensively reviewed by *Güntherodt* and *Merlin* [5.65] and *Güntherodt* and *Zeyher* [5.153], and will not be discussed in further detail here. In the following we will focus on the phononic properties of intermediate valence, metallic SmS and $Sm_{0.75}Y_{0.25}S$ with respect to the stable valence reference compounds. We will view these results in the more general framework of electron–phonon coupling in IV compounds as discussed at the beginning of Sect. 5.4.

The intermediate valence phase of the solid solution system $Sm_{1-x}Y_xS$ with $x \gtrsim 0.15$ was the first striking example of anomalous electron–lattice interactions associated with valence fluctuations. The bulk modulus is soft for $x \gtrsim 0.15$ due to the soft elastic constant $c_{12} < 0$ [5.154]. Strong phonon anomalies have been identified in the [111]-direction from the Γ to the L point of $Sm_{0.75}Y_{0.25}S$ [5.103, 107, 111, 112, 114]. Besides the LA phonon anomaly for $0 < k < 3/4k_L$ associated with the soft bulk modulus, a corresponding anomaly was found for the LO phonon breathing mode near the L point. In addition, there is an anomaly near the Γ point of similar magnitude as that near the L point [5.103].

The polarized Raman spectrum of a cleaved single crystal of $Sm_{0.75}Y_{0.25}S$ [5.65] is dominated by the Γ_1^+ scattering intensity near 250 cm^{-1} and the weaker scattering near 85 cm^{-1}, superimposed on the rising background. For several reasons, extensively discussed in [5.65], one can conclude that the dominant Γ_1^+ scattering intensities of $Sm_{0.75}Y_{0.25}S$ near 250 cm^{-1} and 85 cm^{-1}, respectively, arise mainly from the LO and LA phonons in the [111] direction, emphasizing scattering from L-point

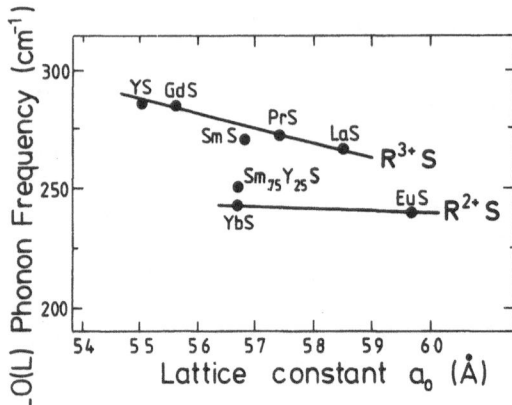

Fig. 5.45. LO(L) phonon frequencies as a function of the lattice parameter of various R-S compounds

phonons and thus allow one to extract a reasonably well-defined value of the LO(L)-phonon frequency from $q = 0$ Raman scattering experiments. The available data on LO(L)-phonon frequencies are depicted in Fig. 5.45 as a function of the lattice parameter. The LO(L) phonons of intermediate valence metallic SmS and $Sm_{0.75}Y_{0.25}S$ lie between the divalent reference line, given by YbS and EuS, and the trivalent reference line spanned by YS, GdS, PrS and LaS, thus exhibiting the behavior of an alloy of di- and trivalent Sm-ions. In addition, metallic SmS as well as $Sm_{0.75}Y_{0.25}S$ exhibit a significant softening of the bulk modulus compared to the stable valence reference line [5.155].

By interpreting these data within the framework described at the beginning of Sect. 5.4, we can estimate a charge fluctuation rate of roughly 100 cm^{-1}, well below the optical phonon frequencies, but within the range of the acoustic phonons. From the position of the LO(L) phonon frequencies of metallic SmS as well as $Sm_{0.75}Y_{0.25}S$ between the di- and trivalent reference lines, one can estimate valences of 2.8 and 2.3, respectively, which agree quite well with the valences deduced from L_{III}-spectroscopy as given by *Allen* et al. [5.156] and by *Weber* et al. [5.157], respectively.

This approach has been tested further by analysis of the temperature dependence of the longitudinal acoustic phonon dispersion of $Sm_{0.75}Y_{0.25}S$ in the [111] direction as deduced by *Mook* et al. [5.112, 114] from neutron scattering experiments and by analysis of the temperature dependence of the bulk modulus inferred from Brillouin scattering experiments. Ultrasonic techniques had failed to determine the elastic constants because of the large lattice expansion upon cooling. Figure 5.46 depicts as an example the Brillouin spectrum of a single crystal of $Sm_{0.75}Y_{0.25}S$ at 300 K. From the various modes observed it was possible to deduce the different elastic constants; they are shown in Fig. 5.47 as a function of temperature. To discuss the concept described at the beginning of Sect. 5.4 for the case of the acoustic phonons of $Sm_{0.75}Y_{0.25}S$ we turn

Fig. 5.46. Brillouin
spectra of acoustic surface
excitations in the [110] −
direction of $Sm_{0.75}Y_{0.25}S$;
R − Rayleigh mode,
LA − longitudinal
acoustic phonon, TA′ −
transverse acoustic
phonon; angle of incidence
of the incoming light: 45°

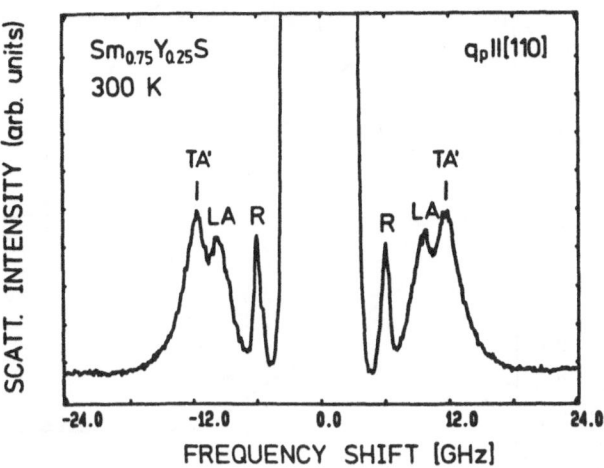

Fig. 5.47. Elastic constants C_{11}, C_{12}, C_{44}
and bulk modulus c_B of $Sm_{0.75}Y_{0.25}S$ as a
function of temperature

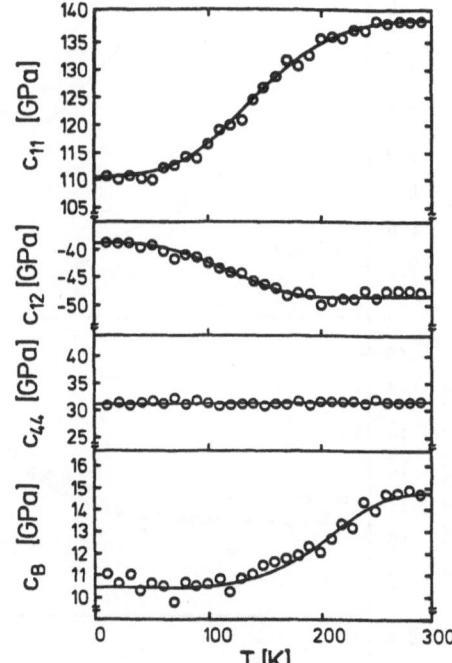

to Fig. 5.48. The longitudinal acoustic phonon branch in [111] direction
shows as a function of temperature the strongest anomaly near 2.0 THz
($q = 0.2$ in Fig. 5.48a). Upon cooling from 300 to 200 K the frequencies
below 2 THz at $q = 0.1$ and $q = 0.2$ (Fig. 5.48a) show a softening, and
from 200 K down to 50 K a hardening. On the other hand, the frequencies
above 2 THz at $q = 0.3$ and $q = 0.35$ (Fig. 5.48a) show only a hardening

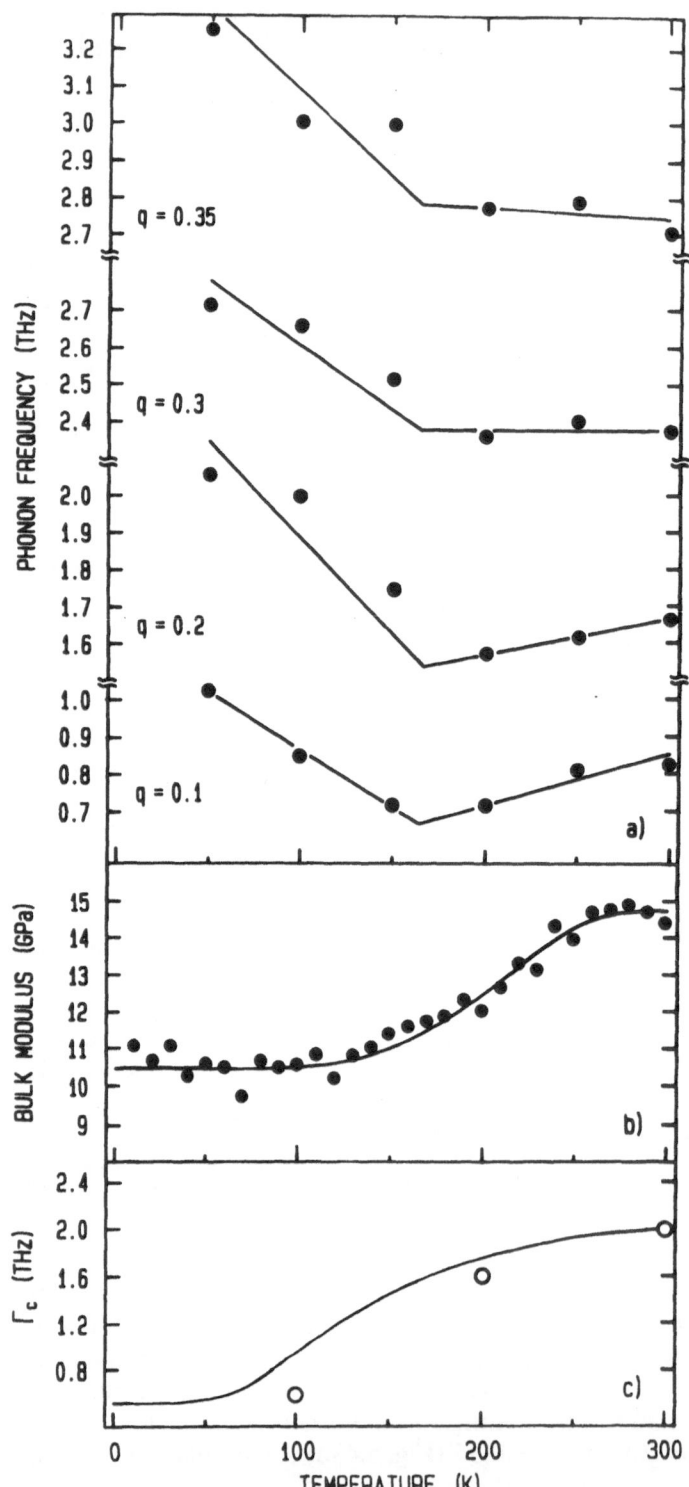

Fig. 5.48. (a) Temperature dependence of the longitudinal-acoustic phonon frequencies of $Sm_{0.75}Y_{0.25}S$ in the [111] − direction for four different values of the wave vector q [5.114]. (b) Temperature dependence of the bulk modulus c_B of $Sm_{0.75}^{\mathsf{i}}Y_{0.25}S$ measured by Brillouin scattering. c_B continues to soften upon cooling below 200 K, unlike the behavior of the phonon-mode frequencies for $q \geqq 0.1$ in part (a). (c) Temperature dependence of the charge relaxation rate Γ_c derived from the experimental data in parts (a) and (b) (open circles) and calculated from theory [5.158] (solid line). The theoretical curve has been matched at 300 K to the experimental value

from 300 to 50 K. For the elastic constants it was found that upon decreasing the temperature from 300 to 4.2 K, c_{11} decreases by 20%, c_{12} increases by 24%, but c_{44} is temperature independent. This results in an overall softening of the bulk modulus c_B with $c_B = c_L$ ([111]) $- 4/3c_{44}$ by 30% upon cooling from 300 to 4.2 K, with the strongest variation near 200 K as shown in Fig. 5.48b.

These results can be understood by assuming a Γ_c which decreases with decreasing temperature. The different temperature dependence of the phonon frequencies in Fig. 5.48a between 300 and 200 K indicates a Γ_c which lies above 1.6 THz and below 2.4 THz (cf. Table 5.4, $\Gamma_c \approx 2.0 \pm 0.5$ THz). Upon cooling from 300 to 200 K, Γ_c is assumed to decrease, resulting in a softening of the low-frequency modes at $q = 0.1$ and 0.2 because of an increasing resonant coupling. In contrast, the higher-frequency modes at $q = 0.3$ and 0.35 are either independent of temperature or harden slightly between 300 and 200 K. The hardening of all phonon modes in Fig. 5.48a upon cooling below 200 K can be explained by Γ_c decreasing below 0.7 THz, thus crossing the phonon modes at $q = 0.1$ and $q = 0.2$. This temperature dependence of Γ_c is corroborated by the maximum in the phonon linewidths near 150 K which is most pronounced for $q = 0.2$ [5.114]. Moreover, the continuous softening of the bulk modulus, i.e., of the longitudinal-acoustic mode in the [111] direction in the $q \approx 0$ limit, from 300 to 4.2 K (Fig. 5.48b) is consistent with the decreasing Γ_c. In Fig. 5.48c we compare the experimentally deduced Γ_c (open circles) with the values predicted by theory (solid line) as given by *Schmidt* and *Müller-Hartmann* [5.158]. Scaling the calculated $\Gamma_c(T)$ by $\Gamma_c(T = 300 K) \approx 2.0$ THz yields good agreement between experiment and theory.

To summarize the results discussed above one should emphasize that from a comparison of the LO(L) phonon frequencies of IV metallic SmS and $Sm_{0.75}Y_{0.25}S$ with those of their stable valence reference compounds in the R-S series a reasonable estimate of charge relaxation rates can be given. Moreover, in the case of intermediate valence Sm ($Sm^{2+} \leftrightarrow Sm^{3+} + e^-$) with the J-multiplet splitting of the Sm^{2+}-configuration well within the range of thermal energies, temperature-dependent changes of valence and charge fluctuation rate are to be expected [5.158]. The valence changes have been observed for, e.g., $Sm_{0.75}Y_{0.25}S$ by *Weber* et al. (Ref. [5.157] and references cited therein). The temperature dependence of the charge fluctuation rate has been deduced from the temperature dependence of phonon anomalies [5.120] in quite good agreement with theoretical predictions [5.158]. To complete the survey on phonon Raman data of R-S compounds we should finally mention that semiconducting, divalent SmS shows a very pronounced softening of the LO(L) phonon mode by 17% below the divalent reference line (at a_0 of EuS) in Fig. 5.45 [5.68, 159], whereas the bulk modulus shows only a slight anomaly. Within the picture of phonon anomalies in rare earth IV compounds discussed at the outset of Sect. 5.4, one can conclude that semiconducting

SmS already exhibits charge fluctuations on the order of $200-300$ cm^{-1}, although it is assumed to be nearly divalent.

In another model calculation given by *Baba* et al. [5.160] the renormalization of the phonon frequencies of SmS has been expressed as a function of the energy gap between the $4f^6$ level and the bottom of the conduction band. A larger phonon softening has been obtained for the semiconducting phase because of the smaller energy gap, as compared to the metallic phase which has a larger (although negative) gap. The microscopic origin of the renomalization has been attributed to the phonon-induced on-site f-d hybridization interaction which is enhanced for the smaller energy gap of the semiconducting phase of SmS.

5.4.6 R-Se

Within the fcc R-Se series TmSe exhibits rather unique properties, since it shows intermediate valence behavior [5.161–163] in combination with low-temperature magnetic order [5.161, 163a, 164, 165]. Another interesting feature is that one can adjust the degree of valence mixing by varying the composition of Tm$_x$Se [5.166]. It has thus been observed that the valence of Tm can be varied between nearly $3+$ for Tm$_{0.87}$Se and $2.7+$ for Tm$_{1.05}$Se.

Intermediate valence behavior of Tm is particularly interesting because, due to the two magnetic $4f$ configurations involved ($4f^{12}$ and $4f^{13}$), two different magnetic relaxation rates have been predicted [5.131]. Using magnetic neutron scattering only one magnetic relaxation rate has been observed so far [5.167, 168], a fact which makes an experimental estimate of the charge relaxation rate particularly worthwhile for comparison with the theoretical predictions.

Raman scattering in TmSe was first investigated by *Treindl* and *Wachter* [5.98, 99] and also by *Stüßer* et al. [5.100]. The connection between strong Raman intensities and phonon anomalies due to strong electron-phonon coupling (see also [5.169]) is clearly illustrated by comparing the Raman spectra of cleaved (100) faces of the superconductor YSe with intermediate valence TmSe and semiconducting SmSe [5.100]. The similarity to the spectrum of Sm$_{0.75}$Y$_{0.25}$S becomes evident by comparing the maxima of the latter near 85 cm^{-1} and 245 cm^{-1}, respectively, with those of TmSe near 70 cm^{-1} and 175 cm^{-1}. The latter two frequencies coincide with the LA and LO phonon anomalies of TmSe in the [111] direction [5.114]. Hence it has been concluded (for a review see [5.65]) that in superconducting or intermediate-valence NaCl-type compounds, the Raman intensity consists of a one-phonon density of states weighted by specific matrix elements of the electron–phonon coupling which are enhanced near phonon anomalies.

Inspecting the LO(L) phonon frequencies of TmSe more closely, a softening of the LO(L) phonon has been found in going from Tm$_{0.87}^{3+}$Se to Tm$_{1.05}^{2.7+}$Se [5.98]. This softening increases linearly with increasing valence

Fig. 5.49. LO(L) phonon frequencies of various R-Se compounds as a function of the lattice parameter

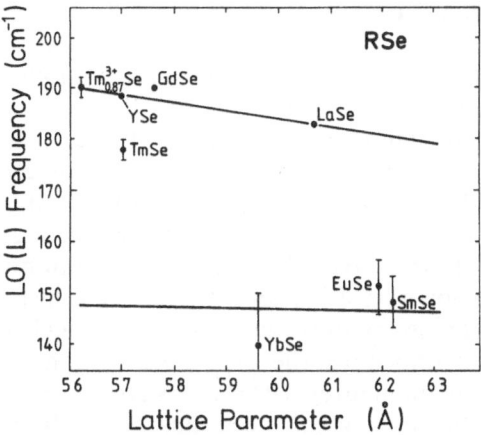

mixing. These data have been summarized in Fig. 5.49 where the LO(L) phonon frequencies of Tm_xSe together with those of various stable valence R-Se have been plotted as a function of the lattice parameter. The LO(L) phonon frequencies have been taken from the literature [5.65, 100, 170]. The stable divalent reference line is given by YbSe, EuSe and SmSe and the trivalent reference line is given by LaSe, GdSe, YSe and $Tm_{0.87}Se$. The LO(L) phonon frequency of intermediate valence $Tm_{1.0}Se$ lies between the di- and trivalent reference lines according to its valence mixing ratio, indicating a charge relaxation rate well below the optical phonon frequencies. On the other hand, the bulk modulus of $Tm_{1.0}Se$ shows a distinct softening compared to the stable valence references. The bulk modulus of SmSe, EuSe and YbSe has been measured by *Jayaraman* et al. [5.155] and that of $Tm_{1.0}Se$ and $Tm_{0.87}Se$ by *Boppart* et al. [5.170]. From the optical phonon anomaly of $Tm_{1.0}Se$ and its soft bulk modulus one can estimate a charge relaxation rate of the order of $50-100$ cm^{-1}. This is even more interesting in view of the fact that *Treindl* and *Wachter* [5.98] find a phonon Raman peak near 60 cm^{-1} which increases in intensity with increasing valence mixing from $x = 0.97$ to $x = 1.05$ and which is absent in $Tm_{0.87}Se$. The 60-cm^{-1} peak, which appears to be connected with the intermediate valence state, has been attributed either to an anomaly in the LA phonon branch or, even more likely, to a localized low-energy electronic excitation near 60 cm^{-1} which may interact with the LA phonon branch. Although no such distinct localized electronic excitation was found in neutron scattering experiment of TmSe for temperatures above 100 K [5.114, 171], a charge relaxation rate of the order of 60 cm^{-1} might well induce strong anomalies within this phonon frequency range resulting in a flattening of the LA-phonon dispersion around 60 cm^{-1}, as assumed by *Boppart* et al. [5.170], as well as an enhancement of the Raman scattering cross section of these LA phonons (increase of the phonon-induced charge deformability) as has been emphasized by *Stüßer* et al. [5.100].

5.4.7 UPt₃

UPt₃ crystallizes in the hexagonal Ni₃Sn structure with space group D_{6h}^4 (P6₃/mmc), which has five Raman-active phonons. These phonons have the symmetries $A_{1g} \oplus E_{1g} \oplus 3E_{2g}$. From the site symmetries of UPt₃, one can determine that the A_{1g} and E_{1g} phonons involve Pt atoms, while the E_{2g} phonons involve both Pt and U.

Cooper et al. [5.19] have been able to observe and identify four of the five Raman-allowed phonons in UPt₃, all of which are shown in Fig. 5.50. The most conspicuous feature observed is a very intense phonon at 150 cm⁻¹, which from the geometry used is clearly identified as the A_{1g} breathing mode of Pt atoms (top spectrum of main figure). (For the sake of completeness it should be mentioned that an optical mode has been observed at 20 meV in a recent optical reflectivity study [5.172]). The A_{1g} phonon is particularly noteworthy in that it exhibits a large anisotropy in its polarizability tensor, given by

$$a(A_{1g}) = \begin{bmatrix} a & 0 & 0 \\ 0 & a & 0 \\ 0 & 0 & b \end{bmatrix}$$

The weakness of the A_{1g} mode in the $(a^2)\,A_{1g} \oplus E_{2g}$ spectrum (second spectrum in main part of Fig. 5.50) compared to that in the $(b^2)\,A_{1g}$ spectrum (top spectrum) indicates a large polarizability of this phonon

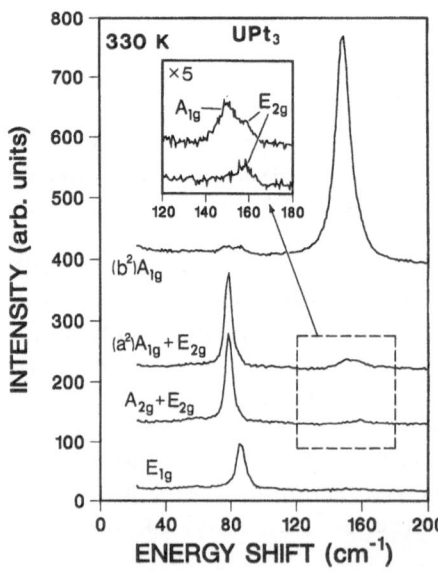

Fig. 5.50. Phonon spectra of UPt₃ at 330 K with symmetries as shown [5.19]. The inset is a 5× enlargement of the $(a^2)\,A_{1g} + E_{2g}$ and $A_{2g} + E_{2g}$ spectra between 120 and 180 cm⁻¹. The small feature near 80 cm⁻¹ in the $(b^2)\,A_{1g}$ spectrum is E_{1g} and E_{2g} leakage due to imperfect alignment of propagation direction k_i and polarization direction E_i along the crystalline axes. All of the spectra have been offset for clarity. Resolution: 3 cm⁻¹. From [5.19]

Table 5.3. Observed Raman-active phonon frequencies in UPt_3 at 330 K. From [5.19]

Phonon	Frequency [cm^{-1}] (Resolution: 3 cm^{-1})
E_{2g}	78
E_{1g}	86
A_{1g}	150
E_{2g}	158

along the c axis compared to that in the basal plane (i.e., $b \gg a$). The further symmetry assignment of the Raman-active modes observed by *Cooper* et al. [5.19] is summarized in Table 5.3. Moreover, *Cooper* et al. [5.19] do not find any indication of strong electron–phonon coupling in UPt_3 from their Raman spectra; in particular, they find no anomalous broadening of the Raman mode near 78 cm^{-1}, which had been reported earlier by *Brenten* et al. [5.18c].

5.4.8 Conclusions

The above analysis of all IV compounds investigated so far with respect to elastic and phononic properties is summarized schematically in Fig. 5.51. The behavior of the long-wavelength ($q \approx 0$) acoustic phonons is represented by the bulk modulus c_B, which is a measure of the isotropic compression of the material and thus has the same (breathing) symmetry as the charge fluctuations. For IV compounds c_B probes anomalies of the longitudinal acoustic phonons. For the optical phonons, two mode frequencies (ω_1, ω_2) are shown schematically for the n^+ and $(n + 1)^+$ valence reference compounds. The order of magnitude of Γ_c has been indicated. The different compounds listed at the bottom of Fig. 5.51 can be classified according to the following four distinct cases.

(i) Intermediate-valent $CeSn_3$ and $CePd_3$ show neither elastic nor optical-phonon anomalies with respect to their trivalent reference compounds [5.173–175]. This is explained by a Γ_c which is far above the highest optical-phonon branches, so that no coupling occurs between Γ_c and the phonons.

(ii) In the case of $CeBe_{13}$ only the two highest-lying optical ($q \approx 0$) phonons of Γ_1^+ symmetry ($\hbar\omega = 610$ cm^{-1} and $\hbar\omega = 580$ cm^{-1}, both represented by $\hbar\omega_1$ in Fig. 5.51) exhibit a softening, whereas the other four optical ($q \approx 0$) phonons (represented by $\hbar\omega_2$ in Fig. 5.51) show no anomalies with respect to their reference lines; see discussion in Sect. 5.4.4. Also, the bulk modulus does not exhibit any anomaly [5.143]. Hence, Γ_c lies above but close to $\hbar\omega_1$.

(iii) The high-frequency optical ($q \approx 0$) phonons ($\hbar\omega > 650$ cm^{-1}) of SmB_6 (represented schematically by $\hbar\omega_1$ in Fig. 5.51) are intermediate

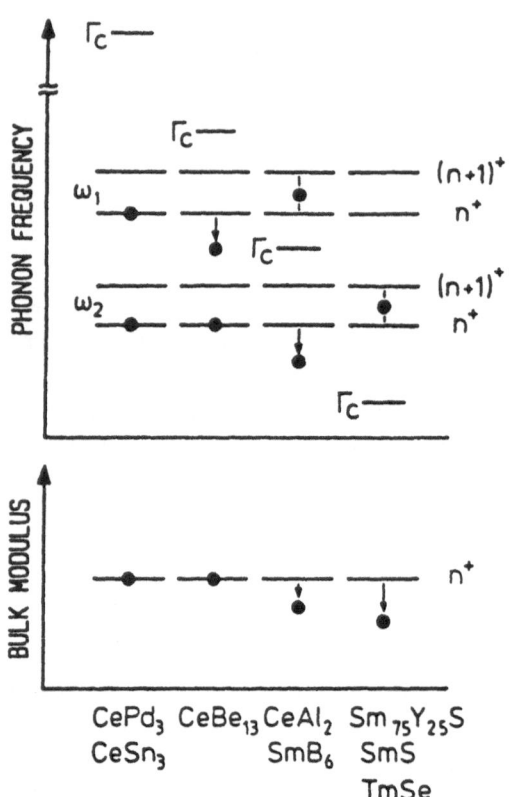

Fig. 5.51. Schematic presentation of the influence of different charge relaxation rates Γ_C of IV rare earth ions on the frequencies of the optical phonon modes (e.g., ω_1 and ω_2) and on the $q \approx 0$ longitudinal acoustic phonon modes, represented by the bulk modulus c_B (see text). For the stable n^+ and $(n + 1)^+$ valent rare earth compounds we show generalized reference lines. Four typical cases are shown with the representative samples given at the bottom

between the n^+ and $(n + 1)^+$ reference lines, whereas both the low-energy optical phonons ($\hbar\omega < 200$ cm^{-1}, represented by $\hbar\omega_2$ in Fig. 5.51) and the bulk modulus become soft as discussed in Section 5.4.1.

From this it follows that $\hbar\omega_2 < \Gamma_c < \hbar\omega_1$. The same refers to CeAl$_2$ with the exception that the 4^+-valent reference line is not known quantitatively.

(iv) The fcc structures SmS, Sm$_{0.75}$Y$_{0.25}$S, and TmSe show longitudinal-optical phonon frequencies near the L point of the Brillouin zone which interpolate between the n^+ and $(n + 1)^+$ reference lines according to the valence mixing ratio. On the other hand, these compounds exhibit a soft bulk modulus. Thus Γ_c lies between the acoustic and optical phonon branches as discussed in Sects. 5.4.5, 6. The analysis of the phonon anomalies in the different IV compounds investigated up to now gives the numerical values of Γ_c listed in Table 5.4. We have also listed theoretical values of the charge relaxation rates obtained by scaling the theoretically predicted [5.131] ratio Γ_c/Γ_s by the experimentally determined spin relaxation rate Γ_s from magnetic susceptibility or neutron scattering. There is reasonable agreement between theoretical and experimental values of Γ_c. Major discrepancies arise only for the "Kondo" systems CeAl$_2$ and

Table 5.4 Experimentally determined charge relaxation rates Γ_c at room temperature compared with theoretical values [5.131], which have been scaled by the measured spin relaxation rates Γ_s. References for the values of Γ_s and the valence are summarized in [5.179]. ($\cong$ means "corresponds to")

Compound	Valence	Γ_s [K]	Γ_c (theory) [K]	Γ_c (expt.) [K]
SmB$_6$	2.7	100	300	300–600
CeAl$_2$	3.02	58	1800	200
YbAl$_2$	2.4	400	600	>180
CeCu$_2$Si$_2$	3.05	70	2100	>200
YbCu$_2$Si$_2$	2.9	40	360	>200
CeBe$_{13}$	3.04	230	6000	≫900
CeSn$_3$	3.02	250	10000	≫300
CePd$_3$	3.26	200	1000	≫300
CeNi$_2$Ge$_2$	3.07	120	1500	≫400
YbNi$_2$Ge$_2$	2.89	90	800	≫400
SmS (metal)	2.8	100	110	120
TmSe	2.58	10	60	120
Sm$_{1-x}$Y$_x$S	2.4	50	80	100
($x = 0.25$)			($\cong$ 1.6 THz)	($\cong$ 2.0 THz)

CeCu$_2$Si$_2$. These discrepancies may be due to the neglect of crystalline-electric-field effects in the theoretical model.

5.5 Light Scattering from "Phonon Bound States"

As described in Sect. 5.4, phonon Raman scattering in IV rare earth compounds has been concerned quite intensively with phonon frequency renormalizations due to strong electron–phonon coupling. However, a coupling of phonons to charge relaxation rates in IV compounds may not only cause phonon softening or a resonance enhancement of the Raman scattering cross section but may also give rise to a split-off phonon-type satellite due to strong nonlinear interactions. The existence of such a phonon bound state due to the strong coupling of localized electronic excitations (or fluctuations) to phonons would indicate a breakdown of the Born-Oppenheimer approximation, which decouples the electronic and phononic systems. The following sections will be devoted to a systematic discussion of the various "phonon bound states" so far observed in Raman scattering from metallic rare earth compounds. Emphasis will be placed on a comparison of Raman scattering results with neutron scattering results from the nominally same phonon mode. Due to the different time scales probed by the two methods (time scale of Raman scattering: 10^{-15} s, neutron scattering: 10^{-13} s), "phonon bound states" may show up rather differently in the two methods. In the following we will discuss the three systems in which phonon bound states are claimed to have been observed, namely $Sm_{0.75}Y_{0.25}S$ and related compounds [5.101], $Tm_{1.0}Se$ and related compounds [5.101] and $CeAl_2$ [5.117]. For a more general discussion of these compounds we refer to Sects. 5.4.5,6,2, respectively.

5.5.1 $Sm_{0.75}Y_{0.25}S$ and Related Compounds

Raman scattering has been performed by *Stüßer* et al. [5.101] for the intermediate-valence phases of the solid solution system $Sm_{1-x}R_xS$ (R = Y, La, Pr, Gd, Tb, Dy, Tm; $0.15 < x < 1.0$). A cation-mass-independent mode has been observed within the gap of acoustic and optical phonon branches for all Sm-rich ($x < 0.5$) intermediate-valence phases. At 300 K the Raman spectra of $Sm_{1-x}R_xS$ with R = Y, Pr, Gd, Dy show a maximum of the scattering intensity near 200 cm^{-1}. This "gap mode" for $x < 0.50$ is due to first-order scattering as demonstrated by its temperature dependence. For $x > 0.5$ the intensity observed in, e.g., $Sm_{0.25}Dy_{0.75}S$ at 300 K near 180 cm^{-1} is due to second-order scattering by phonons as confirmed by its quenching at 80 K. (Difference modes can be safely excluded). The absence of a significant cation-mass effect in $Sm_{1-x}R_xS$ with R = Pr, Gd, Dy rules out an interpretation in which the "gap mode" is viewed as a local vibrational mode of the substituted cations.

The symmetry analysis of the "gap mode" intensity of, e.g., $Sm_{0.78}Dy_{0.22}S$ shows the dominance of the $\Gamma_1^+(A_{1g})$ component, which is consistent with the behavior of the LO(L) phonon near 260 cm^{-1}. Moreover, the "gap mode" frequency appears to follow shifts of the LO(L) phonon frequency upon varying the cation masses. Hence the "gap mode" observed by Raman scattering in $Sm_{1-x}R_xS$ can be considered as a phonon "bound state" of the LO(L)-phonon contributing to the scattering cross section due to the coupling of low frequency charge fluctuations to the lattice vibrations. Similarly, in neutron scattering experiments on IV $Sm_{0.75}Y_{0.25}S$ at room temperature a dispersionless "gap mode" at about $\hbar\omega = 175$ cm^{-1} (21.9 meV) is found, and has been assigned to a localized vibrational mode of the lighter Y ions [5.112]. The frequency position of this "localized" mode agrees well with a simple mass scaling argument

$$\omega = (m_{Sm}/m_Y)^{1/2} \, \omega_{LA(L)} , \tag{7}$$

where $\omega_{LA(L)}$ denotes the frequency of the longitudinal acoustic phonons at the zone boundary L. However, since defect-induced Raman scattering should be rather weak and since the frequency of the "gap mode" in neutron and Raman scattering differ by 25 cm^{-1} (outside of the experimental resolution), we conclude that they are not the same entity and have to be attributed to different origins.

Finally, it is interesting to compare the LO(L) frequencies as measured by Raman and neutron scattering (Table 5.5). The neutron scattering value

Table 5.5. Frequencies of some phonons as measured by neutron scattering, compared to the results of Raman scattering; frequencies of phonon "bound states" observed in Raman scattering together with Raman-scattering intensity-weighted average of Raman LO(L)-phonon frequency and Raman "bound state"

Phonon	Neutron scattering [cm^{-1}]	Raman scattering [cm^{-1}]	Raman "bound state" [cm^{-1}]	Average [cm^{-1}]
LO(L) phonon of $Sm_{0.75}Dy_{0.25}S$	234	253	200	228
LO(L) phonon of TmSe	159	178	142	160
Γ_5^+-phonon of $CeAl_2$ (300 K)	108	109	71	108
Γ_5^+-phonon of $CeAl_2$ (77 K)	97	109	71	99
Γ_5^+-phonon of $CeAl_2$ (25 K)		109	71	82
Γ_5^+-phonon of $CeAl_2$ (4 K)	80	109	71	80

of the LO(L) phonon frequency is soft compared to that measured by Raman scattering. The neutron LO(L)-phonon frequency, on the other hand, compares quite well with the scattering-intensity-weighted average of Raman LO(L)-phonon frequency and "bound state" frequency.

5.5.2 TmSe and Related Compounds

For TmSe similar results have been obtained as for the "bound state" of $Sm_{1-x}R_xS$. In particular, the peak near 145 cm^{-1} in the Raman spectrum of TmSe in Fig. 5.52 has also been identified in Ref. [5.101] as a "gap mode" between the acoustic and optical phonon dispersions. Moreover, from its first-order scattering nature [5.98, 100], its dominant A_{1g} symmetry [5.100] (ruling out any involvement of TO phonons [5.96, 103]) as well as from its frequency shift in parallel to that of the LO(L) phonons in $TmSe_{0.85}Te_{0.15}$ (Fig. 5.52) the "gap mode" can be identified as a bound state of the LO(L) phonon, analogously to the case of $Sm_{1-x}R_xS$. In the case of TmSe no localized phonon mode in the gap between acoustic and optical phonon branches has been observed by neutron scattering [5.111]. Due to the absence of disorder this was to be expected. However, comparing LO(L)-phonon frequencies as measured by Raman and neutron scattering we find again the following discrepancy: The LO(L)-phonon measured by neutron scattering is soft compared to the Raman value but compares quite well with the intensity-weighted average of Raman LO(L)-phonon frequency and Raman "bound state" (Table 5.5).

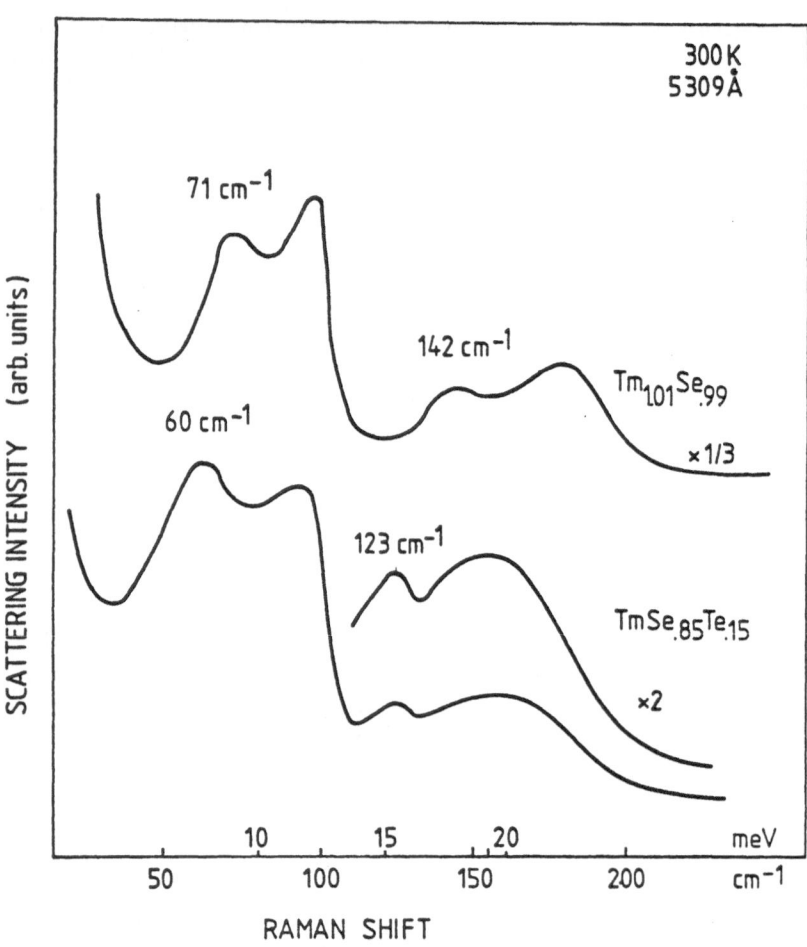

Fig. 5.52. Unpolarized Raman spectra of a (100) cleaved $Tm_{1.01}Se_{0.99}$ and $TmSe_{0.85}Te_{0.15}$ single crystal at 300 K

5.5.3 CeAl$_2$

The Kondo system CeAl$_2$ has attracted continuous interest over the past decade due to its anomalously strong coupling between elastic and magnetic (electronic) subsystems. This results in a wide variety of unusual phenomena, for example, the pronounced softening of the c_{44} elastic mode as observed by *Lüthi* and *Lingner* [5.174] and the splitting of the Γ_8 crystalline-electric-field (CEF) level [5.175]. These features have been interpreted in a model proposed by *Thalmeier* and *Fulde* [5.176] taking into account the large magnetoelastic coupling between the energetically degenerate Γ_8 CEF level and the $\Gamma_5^+(T_{2g})$ optical phonon. Raman scattering in CeAl$_2$ has been studied by *Güntherodt* et al. [5.115, 117] as a function of temperature down to 5 K in order to characterize the Γ_5^+

Fig. 5.53. Raman scattering cross section of $CeAl_2$ as a function of temperature; inset: magnetoelastically coupled phonon–CEF system [5.176]

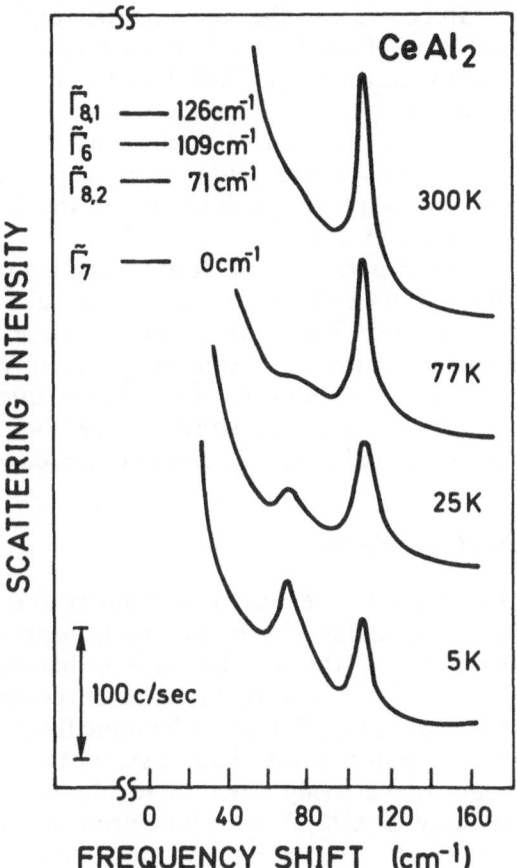

phonon. The results are shown in Fig. 5.53. The Raman peak near 109 cm^{-1} at 300 K does not significantly shift upon cooling to 5 K. Below 77 K a new peak emerges near 71 cm^{-1} (8.9 meV) with a full width at half maximum of 15 cm^{-1}. This additional peak is not observed in the nonmagnetic reference compound $LaAl_2$. The peak position coincides with the lower one of the inelastic transitions observed in neutron scattering [5.175].

The most obvious assignment of this new Raman peak ("phonon bound state") to the transition indicated as $\tilde{\Gamma}_7 - \tilde{\Gamma}_{8(2)}$ by *Thalmeier* and *Fulde* (inset of Fig. 5.53) leads to a severe problem: Since in the theory both $\tilde{\Gamma}_{8(i)}$ levels are symmetrically composed of phonon and CEF states, there is no reason why one should not also observe the high lying $\Gamma_{8(1)}$ CEF level. Another inconsistent feature in this context of the new Raman peak is the fairly large intensity, pointing to a phononic rather than electronic scattering, and making an interpretation of this new mode as a "bound state" of the Γ_5^+ phonon quite straightforward.

In contrast to the Raman scattering results, a temperature dependent softening of the Γ_5^+ mode is observed between 300 and 5 K in phononic neutron scattering [5.134]. This has been attributed by *Thalmeier* [5.177] to the temperature dependence of vertex corrections, which have to be taken into account in higher order perturbation theory. The vertex corrections yield a temperature dependent renormalization of the Γ_5^+ phonon frequency proportional to the difference in occupation of the Γ_7 and Γ_8 CEF levels.

A comparison of the Raman and neutron scattering results for the Γ_5^+ phonon of $CeAl_2$ can be summarized as follows: As soon as one observes the "phonon bound state" in Raman scattering the Γ_5^+ as observed by neutron scattering is soft compared to its Raman counterpart. For all temperatures, however, the Raman intensity-weighted average of the Raman Γ_5^+ phonon energy and the "bound state" energy compare quite well with the Γ_5^+ phonon energy obtained by neutron scattering (Table 5.5).

5.5.4 Conclusions

To reconcile the disparate Raman and neutron scattering results the following point of view has been adopted by *Güntherodt* et al. [5.117]: Similarly to valence fluctuations between two integral valence states $(4f^n \leftrightarrow 4f^{n-1} + e^-)$ in IV rare earth compounds the systems $CeAl_2$, TmSe and $Sm_{0.75}Y_{0.25}S$ show "phononic fluctuations" between a bare phonon and its renormalized bound state. The differences between Raman and neutron scattering from these phononic fluctuations are explained in analogy to XPS (X-ray photoelectron spectroscopy) data vs. Mössbauer isomer shifts from valence fluctuations. The fast probing mechanism "Raman scattering" gives, in analogy to XPS, an instantaneous picture of the dynamical mixture of the two states, while the slow probing mechanism neutron scattering, in analogy to Mössbauer spectroscopy, yields an average over the two fluctuating states; see also Fig. 5.54.

5.6 Final Concluding Remarks

In this chapter we have presented an overview of the rather diverse features observed by Raman scattering in metallic rare earth and actinide compounds, ranging from purely electronic excitations to phonons and strongly coupled electron–phonon excitations. From the materials point of view, Raman scattering has proved to be a useful tool for exploring rare earth and actinide materials at a stage where their purity, perfection, size or isotope composition do not yet allow investigations by neutron scattering. Thus Raman scattering investigations of rare earth and actinide intermetallic compounds have served to advance our understanding of

Fig. 5.54. Comparison of the different experimental methods probing the time scale of "phononic" as well as electronic fluctuations (see text)

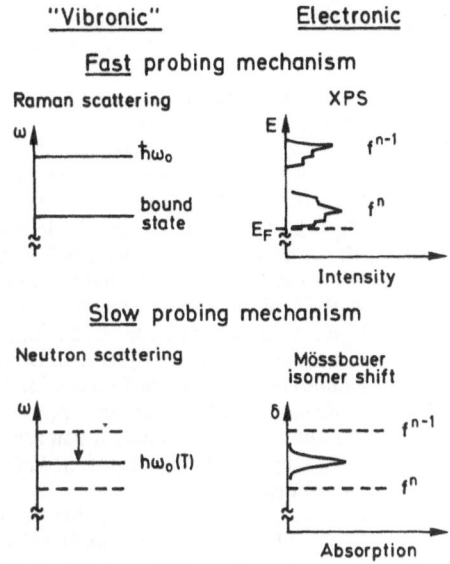

new physical phenomena such as valence fluctuations of rare earth ions in solids as well as heavy fermion behavior.

The field of Raman scattering in rare earth intermediate valence compounds as well as heavy fermion compounds has reached a certain degree of maturity and understanding so that much activity has now shifted to Raman and Brillouin scattering investigations of high-temperature superconductors. The exciting results of Raman scattering from this class of materials will be reviewed in the next chapter.

Acknowledgements. The authors would like to express their gratitude to M. Barth, B. Batlogg, S. Blumenröder, H. Brenten, M. Croft, B. Hillebrands, A. Jayaraman, R. Mock, G. Pofahl, N. Stüsser and J. D. Thompson for their cooperation and participation in various stages of the experimental work. We would like to thank Z. Fisk, E. Holland-Moritz, M. Loewenhaupt, B. Lüthi, E. Müller-Hartmann, P. Thalmeier, and D. Wohlleben for many stimulating discussions.

The preparation, characterization and supply of samples by W. Assmus, M. Beyss, E. Bucher, Z. Fisk, F. Holtzberg, E. Melczer, A. Meyer, E. V. Sampathkumaran, J. L. Smith and K. Winzer are gratefully acknowledged. We are obliged to A. Schüren, M. Jürss-Nysten, I. Kürten, M. Löhrer and H. Mattke for help in preparing the manuscript.

References

5.1   M. Cardona, G. Güntherodt (eds.): *Light Scattering in Solids II* Topics Appl. Phys. **50** (Springer, Berlin, Heidelberg 1982)

5.2a   M. Cardona, G. Güntherodt (eds.): *Light Scattering in Solids III* Topics Appl. Phys. **51** (Springer, Berlin, Heidelberg 1982)

5.2b   M. Cardona, G. Güntherodt (eds.): *Light Scattering in Solids IV* Topics Appl. Phys. **54** (Springer, Berlin, Heidelberg 1984)

5.2c   M. Cardona, G. Güntherodt (eds.): *Light Scattering in Solids V* Topics Appl. Phys. **66** (Springer, Berlin, Heidelberg 1989)

5.3   Z. B. Goldschmidt: In *Handbook on the Physics and Chemistry of Rare Earths* ed. by K. A. Gschneidner, Jr., L. Eyring, Vol. 1 (North-Holland, Amsterdam 1978) p. 1

5.4   E. Müller-Hartmann, B. Roden, D. Wohlleben (eds.): Proceedings of the Fifth Int. Conference on Valence Fluctuations, J. Magn. Magn. Mater. **47 & 48** (1984)

5.5   T. Kasuya (ed.): *Crystalline Field and Anomalous Mixing Effects in f-Electron Systems* (North-Holland, Amsterdam 1985)

5.6   L. C. Gupta, S. K. Malik (eds.): *Valence Fluctuations and Heavy Fermions* (Plenum, New York 1987)

5.7   L. L. Hirst: Phys. Kondens. Mater. **11**, 255 (1970)

5.8   D. Wohlleben: In: *Physics and Chemistry of Electrons and Ions in Condensed Matter*, ed. by J. V. Acrivos, N. F. Mott, A. D. Yoffe, NATO ASI Series, Vol. 130C (Reidel, Dordrecht 1984) p. 85

5.9   J. Röhler, D. Wohlleben, G. Kaindl, H. Balster: Phys. Rev. Lett. **49**, 65 (1982)

5.10   B. Wittershagen, D. Wohlleben: J. Magn. Magn. Mater. **47 & 48**, 79 (1985)

5.11   P. Wachter, H. Boppart (edss.): *Valence Instabilities* (North-Holland, Amsterdam 1982)

5.12   F. Steglich, J. Aarts, C. D. Bredl, W. Lieke, D. Meschede, W. Franz, J. Schäfer: Phys. Rev. Lett. **43**, 1892 (1979)

5.13   G. R. Stewart: Rev. Mod. Phys. **56**, 755 (1984)

5.14   S. Horn, E. Holland-Moritz, M. Loewenhaupt, F. Steglich, H. Scheuer, A. Benoit, J. Flouquet: Phys. Rev. B **23**, 3171 (1981)

5.15   E. Zirngiebl, B. Hillebrands, S. Blumenröder, G. Güntherodt, M. Loewenhaupt, J. M. Carpenter, K. Winzer, Z. Fisk: Phys. Rev. B **30**, 4052 (1984)

5.16   J. T. Hougen, S. Singh: Phys. Rev. Lett. **10**, 406 (1963)

5.17   J. H. Koningstein, O. S. Mortensen: In *The Raman Effect*, ed. by A. Anderson (Marcel Dekker, New York 1973) p. 519

5.18a   S. Blumenröder, H. Brenten, G. Güntherodt, E. Zirngiebl, Z. Fisk: Bull. Am. Phys. Soc. **32**, 719 (1987)

5.18b   S. Blumenröder, E. Zirngiebl, R. Mock, H. Brenten, G. Güntherodt, J. D. Thompson, Z. Fisk, J. Naegele: J. Magn. Magn. Mater. **76–77**, 331 (1988)

5.18c   H. Brenten, E. Zirngiebl, M. S. Wire, S. Blumenröder, G. Pofahl, G. Güntherodt, Z. Fisk: Solid State Commun. **62**, 387 (1987)

5.19   S. L. Cooper, M. V. Klein, Z. Fisk, J. L. Smith: Phys. Rev. B **37**, 2251 (1988)

5.20a   S. L. Cooper, R. T. Demers, M. V. Klein, Z. Fisk, J. L. Smith: Physica B **135**, 49 (1985)

5.20b   S. L. Cooper, M. V. Klein, Z. Fisk, J. L. Smith, H. R. Ott: Phys. Rev. B **35**, 2615 (1987)

5.21   W. Schlabitz, J. Baumann, B. Pollit, U. Rauchschwalbe, H. M. Mayer, U. Ahlheim, C. D. Bredl: Z. Phys. B **62**, 171 (1986)

5.22   T. T. M. Palstra, A. A. Menovsky, J. van den Berg, A. J. Dirkmaat, P. H. Kes, G. J. Nieuwenhuys, J. A. Mydosh: Phys. Rev. Lett. **55**, 2727 (1985)

5.23   M. B. Maple, J. W. Chen, Y. Dalichaouch, T. Kohara, C. Rossel, M. S. Torikachvili, M. W. McElfresh, J. D. Thompson: Phys. Rev. Lett. **56**, 185 (1986)

5.24   S. L. Cooper, M. V. Klein, M. B. Maple, M. S. Torikachvili: Phys. Rev. B **36**, 5743 (1987)

5.25   R. Mock, G. Güntherodt: Z. Phys. B **74**, 315 (1989)
5.26a   K. W. Becker, P. Fulde: Europhys. Lett. **1**, 669 (1986)
5.26b   K. W. Becker, P. Fulde: Z. Phys. B **65**, 313 (1987)
5.27   L. G. De Shazer, G. H. Dieke: J. Chem. Phys. **38**, 2190 (1963)
5.28   E. V. Sampathkumaran, L. C. Gupta, R. Vijayaraghavan, K. V. Gopalakrishnan, R. G. Pillay, H. G. Devare: J. Phys. C **14**, L237 (1981)
5.29   G. Schmiester, B. Perscheid, G. Kaindl, J. Zukrowsky: In: *Valence Instabilities*, ed. by P. Wachter, H. Boppart (North-Holland, Amsterdam 1982) p. 219
5.30   E. R. Bauminger, D. Froindlich, J. Nowik, S. Ofer, I. Felner, I. Mayer: Phys. Rev. Lett. **30**, 1053 (1973)
5.31   E. Zirngiebl, S. Blumenröder, G. Güntherodt, A. Jayaraman, B. Batlogg, M. Croft: Phys. Rev. Lett. **54**, 213 (1985)
5.32   G. Güntherodt, A. Jayaraman, E. Anastassakis, E. Bucher, H. Bach: Phys. Rev. Lett. **46**, 855 (1981)
5.33   E. Zirngiebl, S. Blumenröder, G. Güntherodt, E. V. Sampathkumaran: J. Magn. Magn. Mater. **54–57**, 343 (1986)
5.34   G. H. Dieke, H. M. Crosswhite: Appl. Optics **2**, 675 (1963)
5.35   J. Röhler, D. Wohlleben, G. Kaindl: In: *Valence Instabilities* ed. by P. Wachter, H. Boppart (North-Holland, Amsterdam 1982) p. 341
5.36   E. Holland-Moritz, E. Braun, B. Roden, B. Perscheid, E. V. Sampathkumaran, W. Langel: Phys. Rev. B **35**, 3122 (1987)
5.37   I. Mörke, E. Kaldis, P. Wachter: Phys. Rev. B **33**, 3392 (1986)
5.38   E. Zirngiebl, S. Blumenröder, G. Güntherodt, W. Assmus: J. Magn. Magn. Mater. **47 & 48**, 72 (1985)
5.39   I. Mörke: Ph. D. Thesis, ETH Zürich (1985) unpublished
5.40   Y. Baer, H. R. Ott, J. C. Fuggle, L. E. De Long: Phys. Rev. B **24**, 5384 (1981)
5.41   J. C. Fuggle, F. U. Hillebrecht, Z. Zolnierek, R. Läser, Ch. Freiburg, O. Gunnarsson, K. Schönhammer: Phys. Rev. B **27**, 7330 (1983)
5.42   B. Hillebrands, G. Güntherodt, R. Pott, W. König, A. Breitschwert: Solid State Commun. **43**, 891 (1982)
5.43   M. Loewenhaupt, E. Holland-Moritz: J. Appl. Phys. **50**, 7456 (1979)
5.44   P. Fulde, M. Loewenhaupt: In: *Modern Problems in Condensed Matter Sciences*, ed. by A. S. Borovik-Romanov, S. K. Sinha Vol. 22 (Part I), (North-Holland, Amsterdam 1988) and references therein
5.45   K. R. Lea, M. J. M. Leask, W. P. Wolf: J. Phys. Chem. Solids **23**, 1381 (1962)
5.46a   Z. Fisk: Ph.D. Thesis, Univ. of California, San Diego (1969) unpublished
5.46b   S. Horn, F. Steglich, M. Loewenhaupt, H. Scheuer, W. Felsch, K. Winzer: Z. Phys. B **42**, 125 (1981)
5.47   T. Goto, A. Tamaki, S. Kunii, T. Nakajima, T. Fujimura, T. Kasuya, T. Komatsubara, S. B. Woods: J. Magn. Magn. Mat. **31–34**, 419 (1983)
5.48   T. Fujita, M. Suzuki, T. Komatsubara, S. Kunii, T. Kasuya, T. Oktsaka: Solid State Commun. **35**, 569 (1980)
5.49   K. Hanzawa, T. Kasuya: J. Phys. Soc. Japan **53**, 1809 (1984)
5.50   M. Loewenhaupt, J. M. Carpenter: Bull. Am. Phys. Soc. **28**, 286 (1983)
5.51   E. F. Westrum, H. L. Cleaver, J. T. S. Andrews, G. Feick: In: *Rare Earth III*, ed. by L. Eyring (Gordon and Breach, New York 1965) p. 597
5.52   Z. Fisk: Solid State Commun. **18**, 221 (1976)
5.53   M. Loewenhaupt, M. Prager: Z. Phys. B **62**, 195 (1986)
5.54   G. Pofahl, E. Zirngiebl, S. Blumenröder, H. Brenten, G. Güntherodt, K. Winzer: Z. Phys. B **66**, 339 (1986)
5.55   J. D. Verhoeven, D. E. Gibson, M. A. Noack, R. J. Conzemius: J. Cryst. Growth **36**, 115 (1976)
5.56   N. Grewe, F. Steglich: In *Handbook on the Physics and Chemistry of Rare Earths*, Vol. 14, ed. by K. A. Gschneidner Jr., L. Eyring (North-Holland, Amsterdam 1991)
5.57   F. Steglich, C. D. Bredl, M. Loewenhaupt, K. D. Schotte: J. de Phys. **40**, C5-301 (1979)

5.58   H. R. Ott, H. Rudigier, Z. Fisk, J. L. Smith: Phys. Rev. Lett. **50**, 1595 (1983)
5.59   G. R. Stewart, Z. Fisk, J. O. Willis, J. L. Smith: Phys. Rev. Lett. **52**, 679 (1984)
5.60   C. Stassis, B. Batlogg, J. P. Remeika, J. D. Axe, G. Shirane, Y. J. Uemura: Phys. Rev. B **33**, 1680 (1986)
5.61   S. M. Johnson, J. A. C. Bland, P. J. Brown, A. Benoit, H. Capellmann, J. Flouquet, H. Spille, F. Steglich, K. R. A. Ziebeck: Z. Phys. B **59**, 900 (1985)
5.62   S. L. Cooper, M. V. Klein, Z. Fisk, J. L. Smith: Phys. Rev. B **34**, 6235 (1986)
5.63   K. W. Becker, P. Fulde, J. Keller: Z. Phys. B **28**, 9 (1977)
5.64   L. C. Lopes, B. Coqblin: Phys. Rev. B **33**, 1804 (1986)
5.65   G. Güntherodt, R. Merlin: In Ref. [5.2b] p. 243
5.66   J. A. Koningstein: J. Chem. Phys. **46**, 2811 (1967)
5.67   J. A. Koningstein, P. Grünberg: Canad. J. Chem. **44**, 2336 (1971)
5.68   G. Güntherodt, R. Merlin, A. Frey, M. Cardona: Solid State Commun. **27**, 551 (1978)
5.69   M. I. Nathan, F. Holtzberg, J. E. Smith, Jr., J. B. Torrance, J. C. Tsang: Phys. Rev. Lett. **34**, 467 (1975)
5.70   J. E. Smith, Jr., F. Holtzberg, M. I. Nathan, J. C. Tsang: In *Light Scattering in Solids* ed. by M. Balkanski, R. C. C. Leite, S. P. S. Porto (Flammarion, Paris 1976) p. 31
5.71   Y. L. Wang, B. R. Cooper: Phys. Rev. **172**, 539 (1968)
5.72   Y. L. Wang, B. R. Cooper: Phys. Rev. **185**, 696 (1969)
5.73   M. Gronau: Ph. D. thesis, Ruhr-Universität Bochum (1979) unpublished
5.74   J. C. Tsang: Solid State Commun. **18**, 57 (1976)
5.75   H. A. Mook, T. Penney, F. Holtzberg, M. W. Shafer, J. de Phys. **39**, C-6, Suppl. 8, 837 (1978)
5.76   E. Holland-Moritz, E. Zirngiebl, S. Blumenröder: Z. Phys. B **70**, 395 (1988)
5.77   P. H. Frings, J. J. M. Franse, F. R. de Boer, A. Menovsky: J. Magn. Magn. Mater. **31–34**, 240 (1983)
5.78   A. I. Goldman, S. M. Shapiro, G. Shirane, J. L. Smith, Z. Fisk: Phys. Rev. B **33**, 1627 (1986)
5.79   H. A. Mook, B. D. Gaulin, G. Aeppli, Z. Fisk, J. L. Smith: J. Am. Phys. Soc. **32**, 594 (1987)
5.80   G. Aeppli, E. Bucher, G. Shirane, J. L. Smith, Z. Fisk: Phys. Rev. B **32**, 7579 (1985)
5.81   G. Güntherodt, E. Zirngiebl, R. Mock, S. Blumenröder, H. Brenten, J. de Phys. 49, Coll. C8, 681 (1988)
5.82   G. Aeppli, H. Yoshizawa, Y. Endoh, E. Bucher, J. Hufnagl, Y. Onuki, T. Komatsubara: Phys. Rev. Lett. **57**, 122 (1986)
5.83   A. Auerbach, J. H. Kim, K. Levin, M. R. Norman: Phys. Rev. Lett. **60**, 623 (1988)
5.84   A. de Visser, J. J. M. Franse, A. Menovsky, T. T. M. Palstra: Physica B **127**, 442 (1984); A. de Visser: Ph. D. Thesis, University of Amsterdam (1986) unpublished
5.85   M. B. Maple: J. Magn. Magn. Mat. **31–34**, 479 (1983)
5.86   U. Walter, C. K. Loong, M. Loewenhaupt, W. Schlabitz: Phys. Rev. **33**, 7875 (1986)
5.87   G. Aeppli: unpublished
5.88   C. Broholm, J. K. Kjems, W. J. L. Buyers, P. Matthews, T. T. M. Palstra, A. A. Menovsky, J. A. Mydosh: Phys. Rev. Lett. **58**, 1467 (1987)
5.89   A. Mooradian: In *Light Scattering Spectra of Solids*, ed. by G. B. Wright (Springer, Berlin, Heidelberg 1968) p. 285
5.90   M. Chandrasekhar, M. Cardona, E. O. Kane: Phys. Rev. B **16**, 3579 (1977)
5.91   G. Abstreiter, M. Cardona, A. Pinczuk: In Ref. [5.26] p. 5
5.92   D. Pines: *Elementary Excitations in Solids* (Benjamin, New York 1963)
5.93   M. Yoshizawa, B. Lüthi, T. Goto, T. Suzuki, B. Renker, A. de Visser, P. H. Frings, J. J. M. Franse: J. Magn. Magn. Mater. **52**, 413 (1985)
5.94   C. M. Varma: Rev. Mod. Phys. **48**, 219 (1976)

5.95   G. Güntherodt, R. Merlin, A. Frey, F. Holtzberg: In *Lattice Dynamics*, ed. by
         M. Balkanski (Flammarion, Paris 1977) p. 130
5.96   G. Güntherodt, A. Jayaraman, H. Bilz, W. Kress: In: *Valence Fluctuations in Solids*,
         ed. by L. M. Falicov, W. Hanke, M. B. Maple (North-Holland, Amsterdam 1981)
         p. 121
5.97   G. Güntherodt, A. Jayaraman, W. Kress, H. Bilz: Phys. Lett. **82A**, 26 (1981)
5.98   A. Treindl, P. Wachter: Solid State Commun. **32**, 573 (1979)
5.99   A. Treindl, P. Wachter: Solid State Commun. **36**, 901 (1980)
5.100   N. Stüsser, M. Barth, G. Güntherodt, A. Jayaraman: Solid State Commun. **39**, 965
         (1981)
5.101   N. Stüsser, G. Güntherodt, A. Jayaraman, K. Fischer, F. Holtzberg: In *Valence
         Instabilities*, ed. by P. Wachter, H. Boppart (North-Holland, Amsterdam 1982)
         p. 69
5.102   W. Kress, H. Bilz, G. Güntherodt, A. Jayaraman: J. de Phys. **42**, Coll. C6, 3 (1981)
5.103   H. Bilz, G. Güntherodt, W. Kleppmann, W. Kress: Phys. Rev. Lett. **43**, 1988
         (1979)
5.104   S. K. Ghatak, K. H. Bennemann: J. Phys. F **8**, 571 (1978)
5.105   N. Grewe, P. Entel: Z. Phys. B **33**, 331 (1979)
5.106   K. H. Bennemann, M. Avignon: J. Magn. Magn. Mater. **15–18**, 947 (1980)
5.107   P. Entel, N. Grewe, M. Sietz, K. Kowalski: Phys. Rev. Lett. **43**, 2002 (1979);
         P. Entel, M. Sietz: Solid State Commun. **39**, 249 (1981)
5.108   M. Miura, H. Bilz: Solid State Commun. **59**, 143 (1986)
5.109   N. Wakabayashi: Phys. Rev. B **22**, 5833 (1980)
5.110   T. Matsura, R. Kittler, K. H. Bennemann: Phys. Rev. B **21**, 3467 (1980)
5.111   H. A. Mook, D. B. McWhan, F. Holtzberg: Phys. Rev. B **25**, 4321 (1982)
5.112   H. A. Mook, R. M. Nicklow, T. Penney, F. Holtzberg, M. W. Shafer: Phys. Rev.
         B **18**, 2925 (1978)
5.113   B. Hillebrands, G. Güntherodt: Solid State Commun. **47**, 681 (1983)
5.114   H. A. Mook, F. Holtzberg: In *Valence Fluctuations in Solids*, ed. by L. M. Falicov,
         W. Hanke, M. B. Maple (North-Holland, Amsterdam 1981) p. 113
5.115   G. Güntherodt, A. Jayaraman, B. Batlogg, M. Croft: Phys. Rev. Lett. **51**, 2330
         (1983)
5.116   G. Güntherodt, S. Blumenröder, B. Hillebrands, R. Mock, E. Zirngiebl: Z. Phys.
         B **60**, 423 (1985)
5.117   G. Güntherodt, E. Zirngiebl, S. Blumenröder, A. Jayaraman, B. Batlogg, M. Croft:
         J. Magn. Magn. Mater. **47 & 48**, 315 (1985)
5.118   S. Blumenröder, E. Zirngiebl, G. Güntherodt, A. Jayaraman, B. Batlogg, Z. Fisk,
         A. Meyer: J. Magn. Magn. Mater. **47 & 48**, 318 (1985)
5.119   E. Zirngiebl, S. Blumenröder, R. Mock, G. Güntherodt: J. Magn. Magn. Mater.
         54–57, 359 (1986)
5.120   R. Mock, E. Zirngiebl, B. Hillebrands, G. Güntherodt: Phys. Rev. Lett. **57**, 1040
         (1986)
5.121   M. Ishii, M. Aono, S. Muranaka, S. Kawai: Solid State Commun. **20**, 437 (1976)
5.122   E. Holland-Moritz, D. Wohlleben, M. Loewenhaupt: Phys. Rev. B **25**, 7482 (1982)
5.123   A. Menth, E. Buehler, T. H. Geballe: Phys. Rev. Lett. **22**, 295 (1969)
5.124   K. Winzer, W. Felsch: J. de Phys. **39**, C6-832 (1978)
5.125   T. Tanaka, J. Yoshimoto, M. Iskii, E. Bannai, S. Kawai: Solid State Commun. **22**,
         203 (1977)
5.126   M. Ishii, T. Tanaka, E. Bannai, S. Kauni: J. Phys. Soc. Japan **41**, 1075 (1976)
5.127   H. Scholz, W. Bauhofer, K. Ploog: Solid State Commun. **18**, 1539 (1976)
5.128   I. Mörke, V. Dvorak, P. Wachter: Solid State Commun. **40**, 331 (1981)
5.129   H. G. Smith, G. Dolling, S. Kunii, M. Kasaya, B. Liu, K. Takegahara, T. Kasuya,
         T. Goto: Solid State Commun. **53**, 15 (1985)
5.130   H. E. King, Jr., S. J. La Placa, T. Penney, Z. Fisk, in: Valence Fluctuations in
         Solids, ed. by L. M. Falicov, M. B. Maple and W. Hanke (North-Holland, Am-
         sterdam 1981) p. 333

5.131   E. Müller-Hartmann: In *Electron Correlation and Magnetism in Narrow-Band Systems*, ed. by T. Moriya, Springer Ser. Solid-State Sci., Vol. 29 (Springer, Berlin, Heidelberg 1981) p. 178
5.132   M. B. Maple, D. Wohlleben: Phys. Rev. Lett. **27**, 511 (1971)
5.133   E. E. Vainshtein, S. M. Blochin, Yu. B. Paderno: Sov. Phys. Solid State **6**, 2318 (1965)
5.134   W. Reichart, N. Nücker: J. Phys. F **14**, L135 (1983)
5.135   W. Reichart, N. Nücker: unpublished data (1985)
5.136   C. T. Yeh, W. Reichardt, B. Renker, N. Nücker, M. Loewenhaupt: J. de Phys. **42**, C6-371 (1981)
5.137   T. Penney, B. Barbara, R. L. Melcher, T. S. Plaskett, H. E. King, Jr., S. J. LaPlaca: In *Valence Fluctuations in Solids*, ed. by L. M. Falicov, M. B. Maple, W. Hanke (North-Holland, Amsterdam 1981) p. 341
5.138   H. Razafimandimby, P. Fulde, J. Keller: Z. Phys. B **54**, 111 (1984)
5.139   F. Steglich: In *Theory of Heavy Fermions and Valence Fluctuations*, ed. by T. Kasuya, T. Saso Springer Ser. Solid-State Sci., Vol. **62** (Springer, Berlin, Heidelberg 1985) p. 23
5.140   R. Mock, G. Güntherodt: J. Magn. Magn. Mater. **47 & 48**, 312 (1985)
5.141   P. Rossi, R. Marazza, R. Feno: J. Less-Common Met. **66**, 17 (1979)
5.142   E. Holland-Moritz, E. Zirngiebl: Unpublished results (1986)
5.143   R. Mock, B. Hillebrands, H. Schmidt, G. Güntherodt, Z. Fisk, A. Meyer: J. Magn. Magn. Mater. **47 & 48**, 312 (1985)
5.144   D. Wohlleben, J. Röhler: J. Appl. Phys. **55**, 1904 (1984)
5.145   M. J. Besnus, J. P. Kappler, A. Meyer: Solid State Commun. **48**, 835 (1983)
5.146   G. Eynatten, C. F. Wang, L. S. Fritz, S. S. Hanna: Z. Phys. B **51**, 37 (1983)
5.147   D. Lenz, H. Schmidt, S. Ewert, W. Boksch, R. Pott, D. Wohlleben: Solid State Commun. **52**, 759 (1984)
5.148   W. Kaspers: Diploma Thesis, University of Cologne (1983) unpublished
5.149a   R. Merlin, G. Güntherodt, R. Humphreys, M. Cardona, R. Suryanarayanan, F. Holtzberg: Phys. Rev. B **17**, 4951 (1978)
5.149b   G. Güntherodt, P. Grünberg, E. Anastassakis, M. Cardona, H. Hackfort, W. Zinn: Phys. Rev. B **16**, 3504 (1977)
5.150   R. Merlin, R. Zeyher, G. Güntherodt, Phys. Rev. Lett. **39**, 1215 (1977)
5.151   G. Güntherodt, R. Merlin, P. Grünberg: Phys. Rev. B **20**, 2834 (1979)
5.152   R. Zeyher, W. Kress: Phys. Rev. B **20**, 2850 (1979)
5.153   G. Güntherodt, R. Zeyher: In Ref. [5.2b] p. 203
5.154   T. Penney, R. L. Melcher, F. Holtzberg, G. Güntherodt: AIP Conf. Proc. **29**, 392 (1975)
5.155   A. Jayaraman, A. K. Singh, A. Chatterjee, S. Usha Devi: Phys. Rev. B **9**, 2513 (1974)
5.156   J. W. Allen, R. M. Martin, J. B. Boyce, F. Holtzberg: Phys. Rev. Lett. **44**, 1275 (1980)
5.157   W. Weber, E. Holland-Moritz, K. Fischer: Europhys. Lett. **8**, 257 (1989)
5.158   H. J. Schmidt, E. Müller-Hartmann: Z. Phys. B **60**, 363 (1985)
5.159   G. Güntherodt, R. Keller, P. Grünberg, A. Frey, W. Kress, R. Merlin, W. B. Holzapfel, F. Holtzberg: In *Valence Instabilities and Related Narrow Band Phenomena*, ed. by R. D. Parks (Plenum, New York 1977) p. 321
5.160   K. Baba, M. Kobayashi, H. Kaga, I. Yokota: Solid State Commun. **35**, 175 (1980)
5.161   E. Bucher, K. Andres, F. J. Di Salvo, J. P. Maita, A. C. Gossard, A. S. Cooper, G. W. Hull: Phys. Rev. B **11**, 500 (1975)
5.162   M. Campagna, E. Bucher, G. K. Wertheim, D. N. E. Buchanan, L. P. Longinotti: Phys. Rev. Lett. **32**, 885 (1974)
5.163   H. Launois, M. Rewiso, E. Holland-Moritz, R. Pott, D. Wohlleben: Phys. Rev. Lett. **44**, 1271 (1980)
5.163a   H. R. Ott, K. Andres, E. Bucher: AIP Conf. Proc. **24**, 40 (1974)

5.164 B. B. Triplett, N. S. Dixon, P. Boolchand, S. S. Hanna, E. Bucher: J. de Phys. **35**, C6-653 (1974)
5.165 H. Bjerrum Moeller, S. M. Shapiro, R. J. Birgeneau: Phys. Rev. Lett. **39**, 1021 (1977)
5.166 B. Batlogg, H. R. Ott, E. Kaldis, W. Thoni, P. Wachter: Phys. Rev. B **19**, 247 (1979)
5.167 B. H. Grier, S. Shapiro: In *Valence Fluctuations in Solids*, ed. by L. M. Falicov, W. Hanke, M. B. Maple (North-Holland, Amsterdam 1981) p. 325
5.168 E. Holland-Moritz. J. Magn. Magn. Mat. **38**, 253 (1983)
5.169 M. V. Klein: In Ref. [5.2a] p. 121
5.170 H. Boppart, A. Treindl, P. Wachter: In *Valence Fluctuations in Solids*, ed. by L. M. Falicov, M. B. Maple, W. Hanke (North-Holland, Amsterdam 1981) p. 103
5.171 M. Loewenhaupt, E. Holland-Moritz: J. Magn. Magn. Mater. **9**, 50 (1978)
5.172 F. Marabelli, G. Travaglini, P. Wachter, J. J. M. Franse: Solid State Commun. **59**, 381 (1986)
5.173 L. Pintschovius, E. Holland-Moritz, D. K. Wohlleben, S. Stöhr, J. Liebertz: Solid State Commun. **34**, 953 (1980)
5.174 B. Lüthi, C. Lingner: Z. Phys. B **34**, 157 (1979)
5.175 M. Loewenhaupt, B. D. Rainford, F. Steglich: Phys. Rev. Lett. **42**, 1709 (1979)
5.176 P. Thalmeier, P. Fulde: Phys. Rev. Lett. **49**, 1588 (1982)
5.177 P. Thalmeier. J. Phys. C **17**, 4153 (1984)

6. Light Scattering in High-T_c Superconductors

C. Thomsen

With 28 Figures

With the discovery of materials that show superconductivity at temperatures significantly higher than previously known, *Bednorz* and *Müller* [6.1] initiated both a materials race for yet higher transition temperatures (T_c) and an intensive search for an explanation of the mechanism that makes possible this exotic phenomenon. All high-T_c superconductors discovered so far contain copper-oxygen planes and various metallic elements that form perovskite-like structures. One generally divides these materials into classes or families based on the particular metallic ions. Known classes of superconductors are the La_2CuO_4-types, the Y–rare earth, the Bi, and the Tl superconductor families. The highest generally accepted T_c's lie at 125 K for a Tl compound [6.2, 3] and hence nearly 50 K above the boiling point of liquid nitrogen. Reports of even higher critical temperatures exist but have not yet been reproduced by at least one more laboratory. These so-called USO's, unidentified superconducting objects, have been compiled in a survey [6.4]. In spite of the lack of progress since March 1988, there is no compelling evidence that an intrinsic upper limit in T_c has been reached.

While the design of materials with higher transition temperatures has developed quite rapidly and successfully, the understanding of the fundamental mechanisms involved is still far from complete. A number of possible theoretical explanations, some starting from the successful theory of *Bardeen, Cooper* and *Schrieffer* (BCS) [6.5], have been proposed, but so far none of them can consistently explain the observed properties of these materials. It is known from flux-quantization experiments and the inverse ac Josephson effect experiment that the charge of the elementary current-carrying excitation is $2e$ (e being the elementary charge), i.e. a bound state between two carriers is formed just as in conventional superconductors [6.6–8]. Both paired holes and electrons have been observed for the supercurrents in different materials, but most high-T_c superconductors so far designed are hole superconductors. The binding force in conventional superconductors is provided by coupling to phonons, and it may be tempting to suspect a similar pairing mechanism in the high-T_c materials. The very small isotope shift [6.9–12], however, seems to contradict this hypothesis [6.13] and has led many researchers to propose mechanisms other than lattice deformations as the effective binding force. Magnons [6.14], excitons [6.15, 16], virtual electric-quadrupole fluctuations [6.17], polarons [6.18], bipolarons [6.19, 20], antiferromagnetic spin fluctuations [6.21, 22], the spin-bag mechanism [6.23, 24], and the resonating valence bond [6.25] and anyon superconductivity [6.26] have been central themes of explanations for the high-T_c mechanism.

In view of the thorough understanding of conventional superconductors within the BCS theory and its extensions, it appears natural to investigate the lattice vibrational properties of the high-T_c superconductors in spite of the lacking isotope effect. This has been accomplished from early on in quite some detail by light scattering [6.27]. Other, partly complementary, methods for the investigation of lattice properties are far-infrared spectroscopy [6.28, 29], neutron scattering [6.30], and specific heat and thermal conductivity measurements [6.31]. We review here the vibrational properties of the high-T_c superconductors from a light-scattering point-of-view and emphasize effects that relate the phonons to the superconducting electronic system. In Sect. 6.2 we present the group-theoretical analysis of the eigenmodes at $k = 0$ of characteristic copper-oxide superconductors with one, two, and three CuO_2-layers and show typical experiment results. The well-understood eigenfrequencies are listed together with assignments to atomic species and theoretical values in tables.

Light scattering is, of course, not restricted to vibrational phenomena and has in fact provided relevant information on scattering by charge carriers in the superconductors. In Sect. 6.3 we describe light scattering by electrons and present a discussion about the possible observation of a superconductivity-related gap or pseudo-gap. Further we discuss the Fano effect in high-T_c superconductors, an interference effect caused by the coherent scattering of light by electrons and by phonons [6.32, 33]. Light scattering by magnons, typical of the semiconducting variants of the superconductors has been important for the understanding of the magnetic properties of the high-T_c materials and is discussed in Sect. 6.4

In Sect. 6.5 we describe recent important results regarding the question of strong-coupling superconductivity in the high-T_c materials. Light scattering has given rather definite answers on some of the gap-related issues through the observed frequency and linewidth changes of some phonons below T_c as compared to the normal state.

6.1 Introduction to High-T_c Superconductors

The perovskite unit cell is the basic structural building block of all high-T_c superconductors discovered so far. It consists of an oxygen octahedron with a metal ion each at the center and at the corners of a surrounding cube (Fig. 6.1). The perovskites are known to undergo a series of structural phase transitions as a function of temperature, e.g. $SrTiO_3$, which transforms from cubic to tetragonal at 110 K [6.34]. Such a lattice instability is also known for La_2CuO_4, the undoped, semiconducting

Fig. 6.1. A perovskite unit cell ($BaTiO_3$): the basic structural unit from which all high-T_c superconductors can be derived

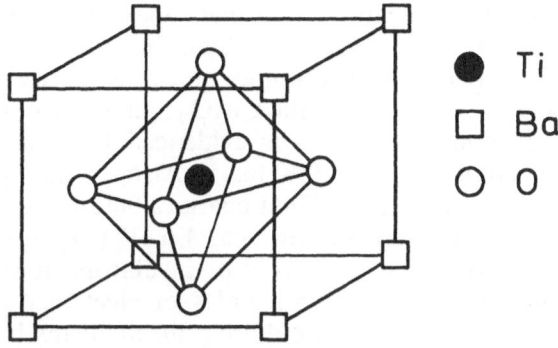

● Ti

☐ Ba

○ O

parent compound of the Bednorz-Müller material. It changes from orthorhombic to tetragonal when heated above the phase-transition temperature of 533 K [6.35]. As far as we know, these structural transformations are not connected with the phenomenon of superconductivity, but they do affect the vibrational properties and, signaling the presence of soft modes, have to be taken into account when analyzing the $k = 0$ eigenmodes. The instabilities presumably stem from the peculiar polarizability properties of the oxygen ions [6.36].

The complete octahedron of the perovskite is not consistently present in all high-T_c superconductors. Some oxygen atoms may be missing, which is why one refers to the superconductors as defect perovskites. Both apex oxygen atoms may be present in a particular structure (La_2CuO_4-family) or just one (e.g. $RBa_2Cu_3O_7$-family); they also may be absent altogether (e.g. Nd_2CuO_4-family), reducing the oxygen octahedra to tetrahedral pyramids or to simple planes. The transition temperature is not affected systematically by the presence or absence of the apex oxygen, making the copper-oxygen planes the important, common-structural element of the high-T_c superconductors. Chemically, the copper-oxygen planes have the formula unit CuO_2 and are hence referred to as CuO_2 planes. $Ba_{1-x}K_xBiO_3$ [6.37], a superconducting perovskite with a T_c of almost 30 K has BiO_2 planes in place of the CuO_2 planes, and is an exception, but it is still being debated whether or not it should be counted a high-T_c material. The authors of [6.38] argued that based on its low carrier density, $Ba_{1-x}K_xBiO_3$ should be. Isotope-shift experiments by these authors and by those in [6.39] ($\alpha \approx 0.30$), however, put this material into the class of conventional BCS superconductors as do the tunneling measurements of [6.40] and we feel that it should be considered as such. Furthermore, theoretical calculations of α and T_c based on strong-coupling theory agree with the experimental values [6.41, 42].

The metal-oxygen planes in perovskites are insulating with large bandgaps – many of the perovskites are transparent to the eye, e.g. $SrTiO_3$ has an optical bandgap of 3.3 eV [6.43] and, by the way, when doped by oxygen vacancies, becomes superconducting with $T_c \approx 0.3$ K [6.44]. The copper-oxygen planes in the defect perovskites are also insulating when they are undoped, although the materials are not transparent in the visible. The La_2CuO_4 and $YBa_2Cu_3O_{7-\delta}$ families have insulating or semiconducting modifications, as is evidenced for example by a strong upturn in resistivity for $T \to 0$ K [6.45]. Also, in the optical properties, a gap may be discerned [6.46, 47] at energies below the visible. $RBa_2Cu_3O_6$, e.g., has an onset of absorption near 1.6 eV (Fig. 6.2) [6.48]. These materials are thus true semiconductors or insulators. It is possible, however, to dope the CuO_2 planes (with holes or electrons) and then they may become superconductors. The doping mechanisms hence play an important role in the high-T_c superconductors. It is a peculiarity of the $RBa_2Cu_3O_{7-\delta}$ family that the compositional phase diagram consists only of a superconducting and an insulating phase. No metallic-non-superconducting member is known, regardless of oxygen concentration. $Y_{1-y}Pr_yBa_2$ $Cu_3O_{7-\delta}$ e.g. has no simply metallic region in the phase diagram over

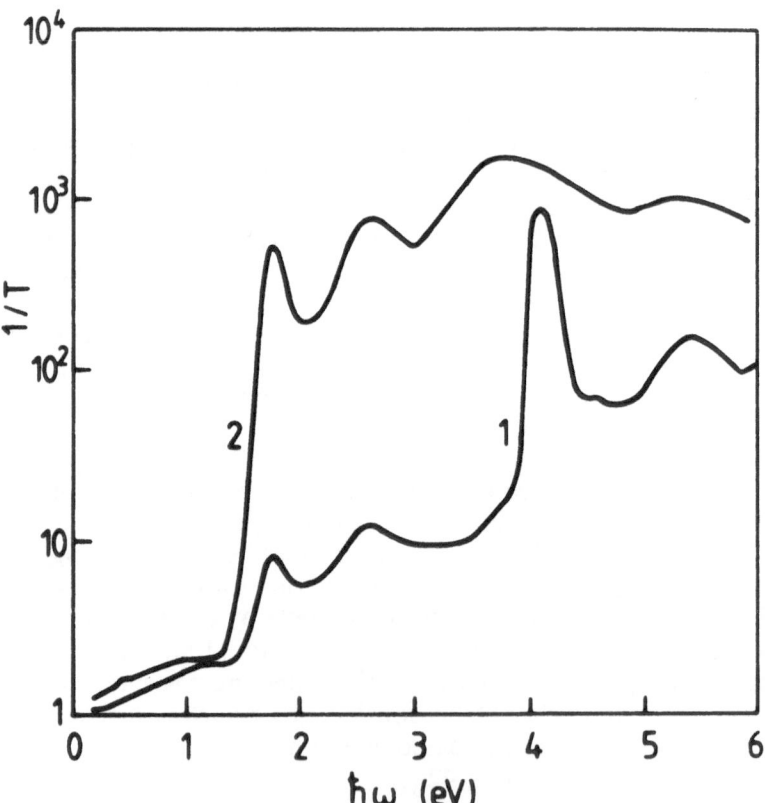

Fig. 6.2. Reciprocal transmittance of thin films of $YBa_2Cu_3O_{7-\delta}$ with $\delta \geq 0.5$. The thickness was (1) $d = 4000$ Å, (2) $d = 7000$ Å. From [6.48] with permission of the authors

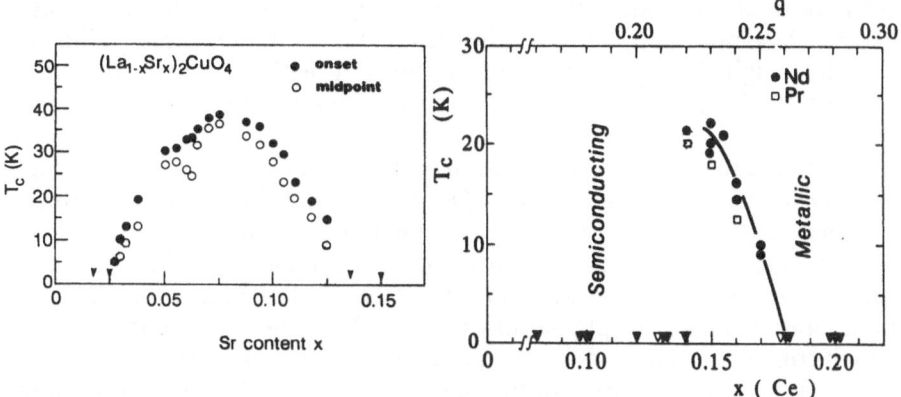

Fig. 6.3. Transition temperature T_c vs. composition x for two high T_c superconductor systems. (a) $La_{2-x}Sr_xCuO_{4+\delta}$ (p-type); from [6.50]. (b) $Ln_{2-x}Ce_xCuO_{4+\delta}$ (n-type) Ln = Nd (circles) Pr (squares). The non-superconducting regions in these compounds may be non-metallic (left) or metallic (right). From [6.51]

the entire y and δ range ($0 \leq y, \delta \leq 1$) [6.49]. This is different, for instance, in $La_{2-x}Sr_xCuO_4$ and $Ln_{2-x}Ce_xCuO_4$ (Ln = Nd, Pr) as is seen in Figs. 6.3a, b [6.50, 51]. For increasing x, both substances are first semiconducting, then enter a region of superconductivity with a maximum T_c for $x \approx 0.15$, and then become simple metals.

In the high-T_c materials, the metal ions other than the copper of the CuO_2 planes have predominantly a structural task. They provide a separation between CuO_2 planes if there is more than one per primitive cell (Ca, rare earth) and they contribute to the doping reservoir of the superconductor. Doping may, for instance, occur through additional oxygen bonded in the reservoir or through the replacement of an atom of a certain valence by a higher or lower-valent one. Examples for oxygen doping are the superconductors $La_2CuO_{4+\delta}$ and $YBa_2Cu_3O_{7-\delta}$, for valence changes $La_{2-x}Ba_xCuO_4$, $Y_{1-x}Ca_xBa_2Cu_3O_6$, $Pb_2Sr_2Y_{1-x}Ca_x$ $Cu_3O_{8+\delta}$, or $Nd_{2-x}Ce_xCuO_{4+\delta}$. When varying the oxygen content from O_6 to O_7 (δ from 1 to 0) $YBa_2Cu_3O_{7-\delta}$ turns superconducting at an oxygen concentration of ~ 6.45. Between ~ 6.6 and 6.8 it forms the so-called 60 K superconductor [6.52] in order to rise to a T_c of 92 K in fully-oxygenated $YBa_2Cu_3O_7$. When doping with different-valent atoms, the oxygen-doping effect is likely to happen simultaneously due to preparational procedures and may not easily be distinguished from the former. This, and general uncertainties in determining oxygen concentrations, often lead to the notation O_x or $O_{n+\delta}$ in the formula for the compounds. Reliable absolute methods for obtaining the oxygen concentration are iodometric titration [6.53], complete calcination [6.54] and neutron-scattering refinements; *relative* changes in oxygen concentration may be determined quite accurately from thermogravimetry. As may be seen by counting valences, $Nd_{2-x}Ce_xCuO_{4+\delta}$ (for $\delta = 0$) should have excess electrons compared to the insulator Nd_2CuO_4 and is, in fact, the first high-T_c superconductor discovered with electrons as charge carriers [6.55]. The other substances mentioned above are hole super-conductors.

The transition temperature within a class of high-T_c materials is thought to increase with the number of CuO_2 planes per primitive cell. Examples are the Bi and the Tl families for which the highest T_c occurs in the three-CuO_2-planes members. However, little concrete evidence, exper-imental or theoretical, of why this should be so has been published. The interaction between the planes may increase T_c, or the carrier concentration which controls superconductivity in these pseudoperovskites may be more easily fine tuned in the multiple-layer structures. Special preparation procedures allow, for example, for the one-layer Tl compound to reach a T_c of near 90 K [6.56] equal to the double-layer value of the $RBa_2Cu_3O_7$-family. Also, the single-layer Bi compound has been reported with T_c between 10 K and 50 K [6.56, 57], the difference probably resulting again from preparation details [6.58].

The high-T_c superconductors are type-II superconductors, i.e. the magnetic penetration depth λ is much larger than the coherence length ξ. Typical values that have been obtained in $YBa_2Cu_3O_7$ are $\lambda \approx 1500$ Å from muon-spin rotation experiments [6.59, 60] while ξ, as determined from measurements of the upper critical field B_{c2} and the relations $\xi^{\parallel} = (\phi_0/2\pi B_{c2}^{\perp})^{1/2}$ and $\xi^{\perp}/\xi^{\parallel} = B_{c2}^{\perp}/B_{c2}^{\parallel}$, are $\xi^{\parallel} \approx 16$ Å and $\xi^{\perp} \approx 3$ Å [6.61]. The symbols $\parallel$ and $\perp$ refer to directions parallel and perpendicular to the CuO_2 planes. The Ginzburg-Landau parameter $\varkappa = \lambda/\xi$ is thus on the order of 100 or more. Values of the critical fields in $YBa_2Cu_3O_7$ have been reported to be $B_{c1}^{\parallel} = 530$ G, $B_{c1}^{\perp} = 189$ G [6.62], and $B_{c2}^{\parallel} = 674$ T, $B_{c2}^{\perp} = 122$ T [6.61]. Thus, even for relatively small magnetic fields, a significant penetration of the superconductor by flux lines occurs. The flux lines have shown to form the characteristic hexagonal Abrikosov lattice at sufficiently low temperatures [6.63, 64] as is expected for a conventional type-II superconductor. New dynamics, such as flux-line entanglement, have been suggested to occur in the extreme type-II superconductors [6.65].

The properties of the flux lines are important for the technical applications of the superconductors. When an electric field is applied, ideally free (unpinned) flux lines cause a flux-flow resistivity due to the viscosity of the flux lines which move due to the Lorentz force acting upon them. The ideal conductivity, on which many technical applications are based, is thus lost in a pinning-free type-II superconductor [6.66]. In a real superconductor there are however many pinning centers that prevent the flow of the flux lines. Oxygen defects, twin boundaries, dislocations or impurities may all act to hold flux lines in place. Then, as long as the Lorentz force is not too large, flux lines do not move and the ideal conductivity is restored. The value where dissipation sets in is the critical current J_c and is often specified for a particular sample or material. At zero temperature J_c depends on the magnetic field in the sample, since the criterion for ideal conductivity is really given by the Lorentz force acting upon the flux lines, i.e. the product BJ_c. Critical currents near zero-applied magnetic fields of $J_c = 3.5 \times 10^7$ A/cm^2 at 4.2 K have been reported for thin films of YBa$_2$Cu$_3$O$_7$ [6.67].

At temperatures significantly away from the $T = 0$ K limit, a mechanism different from the Lorentz force many cause flux motion and thus a decrease in the current-carrying capability of the high-T_c oxides. When the thermal energy becomes comparable to the pinning energy, the flux lines start to move in an activated fashion if a current is applied. This leads to a gradual decrease of the critical current, a phenomenon referred to as flux creep [6.68–71]. Since the operating temperature of devices made from present high-T_c superconductors is aimed to be that of liquid nitrogen, flux creep will be a more important problem than in the conventional (low-T_c) type-II superconductors. Critical currents $J_c > 10^6$ A/cm^2 have been reported at $T = 77$ K and zero external magnetic field [6.72, 73]. At 20 K and a 15 T magnetic field, J_c has been measured to be 5×10^5 A/cm^2, a current density which can be carried in Nb$_3$Sn only up to 8 T at 4.2 K [6.69]. There thus exist regions of B and T where the high-T_c superconductors are superior to conventional superconductors.

The presence of flux lines also determines the details of the levitational properties of the high-T_c materials [6.74]. In spatially inhomogeneous sufficiently-strong magnetic fields the type-II superconductors are held afloat in any position. They appear to offer resistance to a displacement or a rotation. This is due to the dissipation caused by the pinning forces acting on the flux lines which must enter or leave the sample in a changing magnetic field. However, in a magnetic field with higher symmetry, say cylindrical, the type-II superconductor may rotate freely (without resistance) about the high-symmetry axis. Another spectacular phenomenon, the suspension of a type-II superconductor below a magnet, also follows from the presence of pinned flux lines. When an initially applied field B is reduced by moving a superconducting sample some distance away from a magnet, flux pinning will cause the internal field to remain constant, producing a magnitization of the same sign as the applied field (paramagnetic state). There is thus an attractive force on the sample. If the pinning and the external magnetic-field gradient are strong enough, the sample will hang freely below the magnet. Thus, while levitation is a sign of bulk superconductivity in a sample, suspension below a magnet indicates the presence of many effective pinning centers, i.e. possibly a defect-rich specimen. For a detailed discussion of flux-related phenomena, see [6.68].

The Meissner effect, the expulsion of an applied magnetic field smaller than B_{c1} from a sample which turns superconducting, is a frequently-used measure of the bulk-transition temperature and the superconducting volume fraction. Only for low magnetic fields $H < 100$ Oe is a near complete explusion normally found. A Meissner-effect characterization is not as sensitive to small amounts of superconducting phase as is a resistivity measurements, for which the existence of a percolation path suffices to find zero resistance. The magnetic measurement is thus generally a better probe of the bulk quality of a given sample than the determination of the zero-resistivity temperature.

The insulating members of the La_2CuO_4 and $RBa_2Cu_3O_{7-\delta}$ families show an antiferromagnetic ordering of copper spins in the CuO_2 planes. The ordering temperature (Néel temperature T_N) is highest when the CuO_2 planes are undoped ($T_N \approx 300$ K in La_2CuO_4, $T_N \approx 500$ K in YBa_2-Cu_3O_6). With increasing carrier concentration T_N drops and goes to zero near the onset of superconductivity. The antiferromagnetism, which is mediated by the neighboring oxygen via superexchange, is frustrated by the carriers introduced to the CuO_2 planes. In the $RBa_2Cu_3O_{7-\delta}$ family, various experiments have attempted to decide whether or not a region exists with overlapping antiferromagnetism and superconductivity [6.75]. The results have not been conclusive, and we may consider the question for the understanding of superconductivity still unresolved. The region of overlap, if it exists, is small. A fundamental question is whether or not magnetism is involved in the pairing interaction, or whether it is simply not compatible with superconductivity. At very low temperatures the rare-earth atoms are found to order antiferromagnetically also in the superconducting state. The Néel temperature is below 3 K for most elements investigated (exception Pr, see below).

Another fundamental question is what the dimensionality of the superconductivity is. The structure suggests a two-dimensional main effect, and the extremely-short coherence length in the c-direction (~ 3 Å, shorter than the interplanar spacing of $YBa_2Cu_3O_7$) supports this. Because the coherence lengths diverge near T_c there should be a cross-over from three- to two-dimensional behavior. Torque magnetometry has given evidence that such a cross-over does indeed exist [6.76].

Most high-T_c superconductors can be prepared as ceramics (poly-crystalline samples), single crystals or thin films even as superlattices. Thin films are usually oriented with the c-axis perpendicular to the substrate, which is usually $SrTiO_3$, $LaGaO_3$, $LaAlO_3$, MgO or Si, depending on the intended purpose. Light scattering experiments have had the advantage that all forms of the superconductor could be investigated from a very early stage on, even tiny single crystals with the help of a Raman microscope.

6.2 $k = 0$ Vibrational Mode Analysis of the Perovskite-Like Superconductors

From a structural point of view the various high-T_c superconductors are quite similar. They are either exactly tetragonal or slightly distorted into orthorhombic symmetry. The distortion may occur along the edge or along the diagonal of the basal plane of the tetragonal unit cell. The ideally square CuO_2 (but not necessarily flat) planes are then distorted either into a rectangle or into a rhomb. Examples of the former are the $RBa_2Cu_3 O_{7-\delta}$ family, of the latter the Bi family. The crystallographic lattice constants are typically $a \approx b \approx 3.8$ Å when the distortion is into a rectangle and hence similar to those of perovskites (e.g. $SrTiO_3$: $a = b = c = 3.905$ Å in the cubic phase [6.78]), while they are approximately $a' = \sqrt{2}\, a$ and $b' = \sqrt{2}\, b$ for the diagonally-distorted structure. For many purposes the structure of the superconductor can be treated in an approximate tetragonal symmetry. Then, a 45° rotation of the idealized tetragonal unit cell with respect to the crystallographic axes has to be taken into account in the diagonally-distorted structures. A concurrent change in the notation of the various eigenmodes has led to occasional confusion in the literature. For the non-totally symmetric modes it is essential to mention the underlying point group. We will try to keep this notational point as clear as possible and will give conversion tables that should make it easy to switch from one point group to another.

The c-axis parameter of the unit cell of La_2CuO_4 is $c = 13.15$ Å. In the other superconductors the c-axis length may be estimated if the chemical formula is known. A typical distance between CuO_2 planes within a unit cell is 3.4 Å, the distance of the CuO_2 plane to the closest Pb/Bi/Tl/Cu layer ~ 4.1 Å. In addition, there is the distance $d*$ if two of these metal-oxide layers are present in a structure. In $YBa_2Cu_3O_7$ and Tl_1 compounds $d* = 0$, in Bi_2 compounds $d* \approx 1.5$ Å, in Tl_2 compounds $d* \approx 1.1$ [6.79], and in the Pb superconductors $d* = d$ (Cul-O2) $= 1.8$ Å. The approximate c-axis length of the chemical unit cell is thus obtained by adding $2[4.1$ Å $+ (n - 1)\ 1.7$ Å $+ d*]$, with n being the number of CuO_2 planes of the substance considered. To obtain the crystallographic c-parameter, the number thus obtained must be multiplied by another factor of 2 if the structure is body centered (e.g. the Bi_2 and Tl_2 compounds). The existence of a systematic way to calculate the c-parameter reflects the fact that, structurally, the different high-T_c materials are closely related, which will, in turn, facilitate the normal-mode analysis. A detailed structural analysis of $YBa_2Cu_3O_{7-\delta}$ for a series of δ between 0 and 1 has been given by *Jorgensen* et al. [6.80]. It includes the 90 K ($0 < \delta < 0.2$), the 60 K superconductor ($0.3 < \delta < 0.5$) and the semiconductor ($0.6 < \delta < 1$). Recently, the La_2CuO_4 compound has been reported to exist with two CuO_2 layers separated by Ca and is superconducting as well: $La_2CaCu_2O_6$ is the generic chemical formula [6.81, 82].

A number of important deviations exists from the ideal tetragonal unit cell. On top of the orthorhombic distortion discussed above, the metal-oxide layers may have a reduced symmetry. In the Bi_2 family, for instance, the oxygen positions deviate significantly from their NaCl-like, ideal position, where the structure would not be able to maintain reasonable Bi-O bond distances. These displacements will affect the strict $k = 0$ selection rules for Raman scattering: additional modes may appear. See [6.83, 6.84] for details of the structure. A complete eigenmode analysis, based on the space-group assignment *Amaa* of [6.83], has been published for the Bi_2 one-layer and two-layer materials [6.85].

Another deviation from the ideal structure comes from breaking the translational symmetry. This may happen when doping is achieved by adding heterovalent atoms. If, for example, the replacement takes place in the inversion center (e.g. Ca dopes in the Y site of $YBa_2Cu_3O_6$), a formally IR-active-only mode may become Raman active [6.86].

Parallel to the eigenmode analyses, different types of calculations have appeared attempting to estimate theoretically the eigenfrequencies and eigenvectors. These calculations have been useful both to give insight into various structural aspects of these superconductors and to help assign modes of equal symmetry to particular atomic displacements. Also, they predict a certain degree of mixing between modes of equal symmetry. An important example of such a mixing will be discussed later (Sect. 6.5). The different calculations can be classified into three major types. Force-constant methods assume massless springs between the atoms and attempt to reproduce the observed spectra by adjusting spring constants for best agreement with experiment [6.87, 88]. The trouble with this method is the need to adjust a large number of parameters, which implies a certain knowledge or assumptions about available experimental spectra. Furthermore, the same spring constants are usually not applicable to different crystal structures involving similar bonds, as in the cuprate super-conductors. Also, when deviating from $q = 0$, one has little confidence in the reliability of the calculations. A q-dependent screening, e.g., cannot be reproduced well with this method.

A significantly improved and generalized approach to the lattice dynamics of high-T_c materials has been brought forward by shell-model calculations [6.89–91]. Here the polarizability of electronic shells around the ions is taken into account. Spring constants are replaced by so-called Born-Mayer potentials. The potential parameters, while still determined empirically, are taken from other well-understood substances and reduce the arbitrariness in the calculations considerably, giving them a certain amount of predictive power. Also, since knowledge about the entire Brillouin zone is contained in the calculations, they can be used to reproduce the phonon density-of-states as measured with neutron scattering [6.89, 92] and related thermodynamic properties such as the specific heat [6.93]. Furthermore, within one superconductor family the potentials can be set to be the same with merely the atomic sites varying according to the particular structure. Such self-consistent sets of calculated eigenmodes have been reported for the Tl family by *Kress* et al. [6.94].

Entirely without parametrization are *ab initio* total-energy calculations in the local-density approximation which are, however, quite computer intensive [6.95]. Results reported for the phonon frequencies of La_2CuO_4 [6.96] and $YBa_2Cu_3O_7$ [6.97, 98] agree well with experiment, and the calculations are able to reproduce even subtle effects in the phonon spectra to be discussed later.

We now describe how one obtains the symmetry of the eigenmodes in a one, a two, and a three-layer high-T_c superconductor and analyze particular materials.

6.2.1 Superconductors with One CuO_2 Plane

La_2CuO_4 is the substance which first led to a high-temperature super-conductor [6.1]. *Bednorz* and *Müller* obtained a "possible" transition temperature, as determined by the onset of the drop-off in resistivity at 35 K, by doping La_2CuO_4 with Ba. For many experiments, La_2CuO_4 is still used as the prototype of a superconductor in spite of its now relatively low T_c. It has a comparatively simple structure and only seven atoms per formula unit. The vibrational properties of both the undoped and doped substances have been investigated by a number of groups [6.27, 6.99–105].

The K_2NiF_4-like structure of La_2CuO_4 is shown in the idealized tetragonal form in Fig. 6.4. The complete CuO_6-octahedra are easily

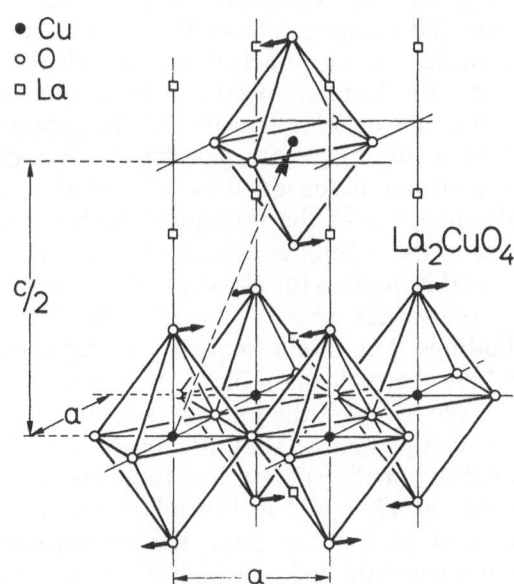

Fig. 6.4. The structure of tetragonal La_2CuO_4. At the structural transition to the orthorhombic phase the copper-oxygen octahedra undergo a tilt as indicated by the arrows. Also shown is the centered character of the crystallographic unit cell (dashed arrow). The atomic-site parameters were taken from [6.111, 112]

recognized. The body-centered character of the crystallographic unit cell is indicated by the dashed arrow. Also indicated by arrows is the distortion which occurs at the transition to the orthorhombic system. It corresponds to staggered rotations of the oxygen octahedra around the $\langle 1, \pm 1, 0 \rangle$ direction of the tetragonal system. Subsequent octahedra along the tetragonal a axes rotate in alternating directions. The Cu-atoms remain undisplaced by this distortion. Since all oxygen atoms within the CuO_2 plane are shared by two octahedra, the rotation of one octahedron determines the change in all of them. The distorted, orthorhombic cell has its a and b axes along the old [110] and 1$\bar{1}$0] directions, i.e. the tetragonal diagonals. The distorted unit cell is thus rotated by 45° and has approximately twice the volume of the old cell. The crystallographic space groups are $I4/mmm$ (D_{4h}^{17}) and $Abma$ (D_{2h}^{18}) for the tetragonal and the orthorhombic structure. The group-theoretical analysis of the eigenmodes yields [6.27, 99–104] $2A_{1g} + 2E_g$ Raman-active and $3A_{2u} + 4E_u$ infrared-active modes and one B_{2u} silent mode for the tetragonal phase. The larger, orthorhombic phase has $5A_g + 3B_{1g} + 6B_{2g} + 4B_{3g}$ Raman-active modes, $6B_{1u} + 4B_{2u} + 7B_{3u}$ infrared-active and $4A_u$ silent modes.

It is easy to see where the Raman-active modes come from in the tetragonal case. Cu lies in an inversion center and thus cannot produce an even mode at $k = 0$. The apex-oxygen site and the lanthanum site are crystallographically equivalent (they differ only by their z-parameter), and these atoms contribute one A_{1g} and one E_g mode each. The $A_{1g}(E_g)$ vibrations correspond to displacements parallel (perpendicular) to the c-axis; both types of vibration have even parity with respect to inversion. In the orthorhombic case, $3A_g$ vibrations describing the structural distortion of the in-plane oxygen, apex oxygen and lanthanum atoms are added to the totally symmetric modes. These additional modes should become soft and disappear from the Raman spectra when the structural phase transition is approached by e.g. raising the temperature. In Fig. 6.5 we show the Raman spectra of tetragonal and orthorhombic La_2CuO_4 as reported by *Weber* et al. [6.110]. The polarizations of incident and scattered field in the figure are parallel to the z-direction. We want to point out the strong mode at 104 cm^{-1} in the orthorhombic phase, which has disappeared in the tetragonal high-temperature phase. It is apparently one of the A_g modes generated by the distortion. *Weber* et al. have shown that this mode is the classical soft mode, whose frequency tends to zero at a second-order phase transition. In Fig. 6.6 we have reproduced these findings. The mode frequency extrapolates indeed to zero near ~ 500 K where the orthorhombic $\rightarrow$ tetragonal transitions occurs.

We mention the known assignments of the other modes of La_2CuO_4 for completeness; they are based on the results of [6.100]. A detailed analysis of the spectra has been given in the references mentioned. The mode at 429 cm^{-1} in Fig. 6.5 is due to A_g motion of the apical oxygens and is present in both the orthorhombic and tetragonal structure. Characteristic for all high-T_c superconductors is the strong Raman

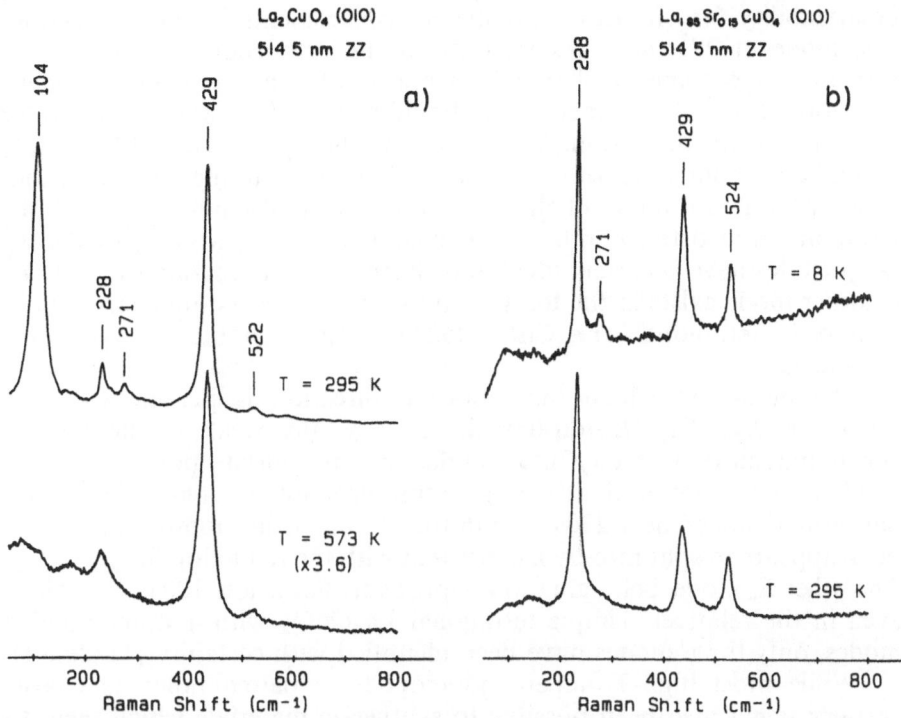

Fig. 6.5. Raman spectra of tetragonal (bottom) and orthorhombic (top) La$_{2-x}$Sr$_x$CuO$_4$. The polarizations are parallel to the tetragonal axis ($\hat{c}$) of the structure (a) undoped ($x = 0$), b) doped ($x = 0.15$)). From [6.100]

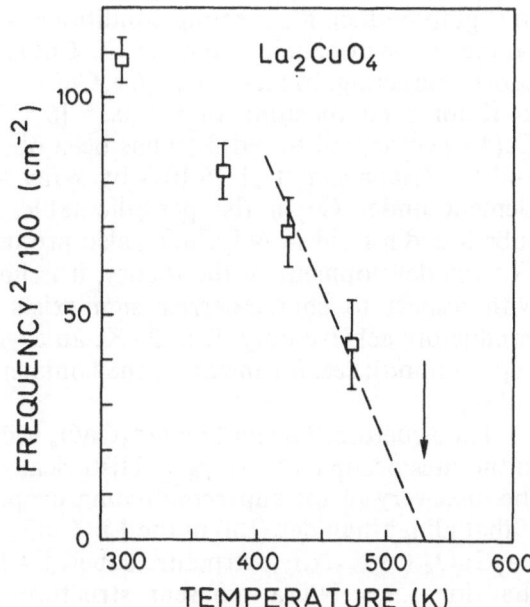

Fig. 6.6. The soft mode of the orthorhombic-tetragonal phase transition in La$_2$CuO$_4$. Plotted is the phonon-frequency squared vs. temperature. The arrow indicates the phase transition. From [6.100]

polarizability for vibrations involving displacements in the z-direction. The mode at 271 cm^{-1} occurs only in the orthorhombic system; the Raman peak becomes weaker and disappears when approaching the phase transition. It has A_g symmetry and should be one of the additional modes generated by the distortion. At 228 cm^{-1} in the A_g spectrum of La$_2$CuO$_4$ we find the second tetragonal A_g mode. It is assigned to the La z-vibration, which produces a mode of this symmetry and is Raman active in both tetragonal and orthorhombic La$_2$CuO$_4$. Thus four (two, all) of the A_g (A_{1g}) modes have been identified in orthorhombic (tetragonal) La$_2$CuO$_4$. Another mode exists in the tetragonal phase. It appears quite strongly in Sr-doped "tetragonal" La$_2$CuO$_4$ (524 cm^{-1}); it should be symmetry forbidden.

The modes in which the vibration direction is perpendicular to the c-axis B_{2g}, B_{3g} (E_g-tetragonal) are generally weak in the high-T_c superconductors. For La$_2$CuO$_4$, under the appropriate polarizations of incident and scattered light (xz), a feature probably related to O motion has been identified near 230 cm^{-1} [6.100]. In the orthorhombic phase, this peak appears to split into two, consistent with the reduction in symmetry. The other E_g mode, belonging to La, probably lies below 100 cm^{-1}. Thus, even in the relatively simple tetragonal La$_2$CuO$_4$ with 4 Raman-active modes, only three modes have been identified with certainty (Table 6.1).

While most high-T_c superconductors have paired holes as charge carriers, it has also been possible to synthesize materials which seem to have electrons in the role of the Cooper pairs. As might be guessed from the general doping concepts in these materials, electron doping can be achieved, at least in principle, by replacing a metal ion of the pseudo-perovskites by an atom with higher valence and by treating the substance in slightly-reducing annealing conditions. The additional electrons, if not bound to the dopant, could enter the CuO$_2$ planes which then might turn superconducting. In fact, Nd$_{2-x}$Ce$_x$CuO$_4$ has a transition temperature of 24 K for a composition of $x \approx 0.15$ [6.55, 106]. The higher valence of Ce(4+) compared to Nd(3+) has been observed e.g. by *Alp* et al. [6.106] and by *Tranquada* et al. [6.107] by x-ray absorption spectroscopy. The element under Ce in the periodic table, Th, is also tetravalent and, substituted for Nd in Nd$_2$CuO$_4$, also produces superconductivity [6.108]. For the development of the theory, it is important that some symmetry with respect to charge-carrier sign exists (although the n-type super-conductors achieve only $T_c \leq 24$ K, an asymmetry which should also be kept in mind); see, for instance, the comments by *Emery* [6.109] and *Rice* [6.110].

The structure of insulating Nd$_2$CuO$_4$ is different from that of La$_2$CuO$_4$ in the sites occupied by oxygen. Historically it was thought – long before the discovery of the superconducting properties of the doped structures – that all lanthanides (Ln) in the Ln$_2$CuO$_4$ composition crystallize in the La$_2$CuO$_4$ (or K$_2$NiF$_4$) structure, labeled T [6.111]. It was then discovered that for Ln = Nd a different structure is formed (T') [6.112]. The

Table 6.1. Observed Raman-active mode-frequencies, their symmetries and the polarizations (pol.) under which they have been seen for La$_2$CuO$_4$ and Nd$_2$CuO$_4$. The symmetries given correspond to tetragonal symmetry except for those modes that *only* occur in orthorhombically distorted La$_2$CuO$_4$. The polarization directions are sequenced in order of decreasing strength of the corresponding peaks. Very-weak peaks have their polarizations given in parentheses. Theoretical values come from total-energy calculations (TEC) and lattice-dynamical calculations (LDC). Note that not all allowed modes have actually been reported and that some symmetry forbidden modes (symm. forb.) are observed. All frequencies are given in cm^{-1}

La$_2$CuO$_4$

atom, symmetry	pol.	experiment[a] undoped	experiment[a] doped	theory TEC[b]	theory LDC[c]
tetragonal					
La, A_{1g}	zz, (xx)	228	228	224	218
La, E_g	xz				105
O, A_{1g}	zz, xx	429	429	415	553
O, E_g	xz	230	230, 250	233	333
orthorhombic (additional modes)					
A_g	zz	104			
B_{1g}	xx	220			
B_{1g}	xx	320			
A_g	zz	271	271		
symm. forb.	zz	(522)	524		

Nd$_2$CuO$_4$

atom, symmetry	pol.	experiment undoped[a]	experiment doped[e]	theory LDC[e]
tetragonal				
Nd, A_{1g}	zz, (xx)	230	228	230
Nd, E_g	xz			161
O, B_{1g}	xx, x'y'	344	328	328
O, E_g	xz	494	480	479
symm. forb.	zz		581	
symm. forb.	xy	589		

a) Ref. [6.100] b) Ref. [6.96] c) Ref. [6.119] d) Ref. [6.114] e) Ref. [6.118]

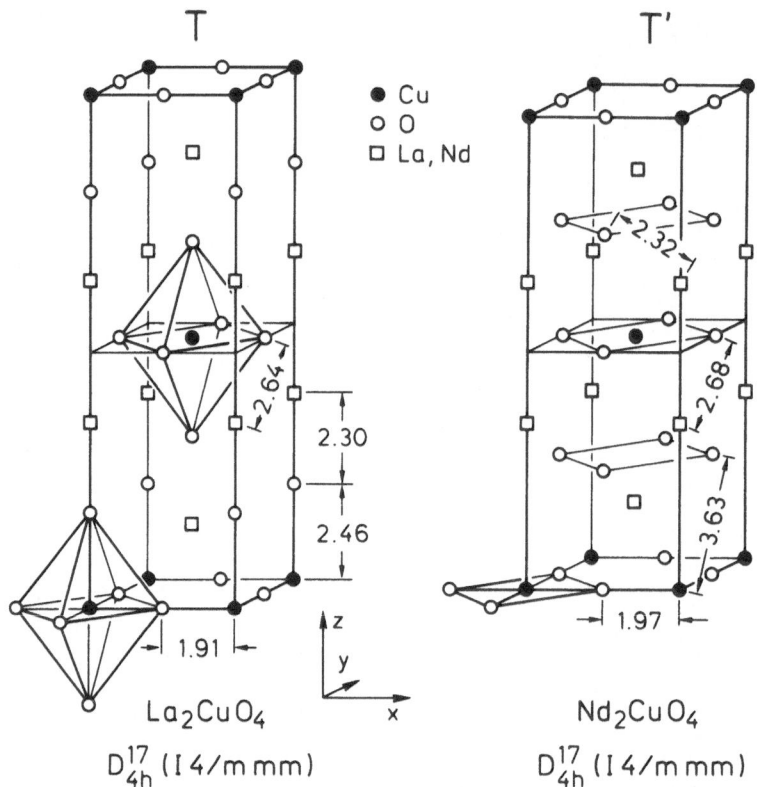

Fig. 6.7. Structure of La_2CuO_4 (T) and Nd_2CuO_4 (T'). The two structures differ in the atomic site of one oxygen atom forming CuO_6 octahedra (left) or simple CuO_4 square planes (right). An intermediate structure between T and T', usually labeled T^*, is also known to exist [6.121] (see text). Distances given in Å. After atomic-site parameters of [6.111, 112]

fundamental difference lies in the quadratic-planar coordination of copper in Nd_2CuO_4. It is now clear that $Ln = La$ is the structural exception and all other lanthanides crystallize in planar-Cu version [6.113]. The chemical unit cells of the two different structures are shown in Fig. 6.7. Both belong to the tetragonal space group $I4/mmm$ (D_{4h}^{17}). (La_2CuO_4 only at high temperature, see discussion above; a low-temperature orthorhombic phase is not known for Nd_2CuO_4.) The essential difference, as far as the lattice vibrations are concerned, is the atomic site of oxygen outside of the CuO_2 planes. In La_2CuO_4 it has site symmetry C_{4v} (Wyckoff position $4e$) and forms the perovskite octahedra, in Nd_2CuO_4 the site symmetry is D_{2d} (Wyckoff position $4d$), a fact which leaves only CuO_2 squares as perovskite elements. The Raman and infrared-active modes change accordingly.

The factor-group analysis of Nd_2CuO_4 has been performed by a number of authors [6.114–116] and yields: $A_{1g} + B_{1g} + 2E_g$ Raman-active and $3A_{2u} + 4E_u$ infrared-active modes and one B_{2u} silent mode (tetragonal symmetry). As far as the light-scattering spectra are concerned, the site-symmetry change in Nd_2CuO_4 produces a new mode symmetry, absent in La_2CuO_4, for the out-of-plane oxygen: its z-direction displacement has B_{1g} symmetry. Spectra of doped [6.116] and undoped [6.114, 115] Nd_2CuO_4 have been published. In Fig. 6.8 we show polarization dependent spectra of $Nd_{2-x}Ce_xCuO_4$ [6.116]. We identify a peak corresponding to the Nd vibration at 228 cm^{-1}. It has A_{1g} symmetry, as can be judged from the absence of the peak in the spectra under crossed polarizations, and has $\alpha_{zz} \gg \alpha_{xx}$. ($\alpha_{ij}$ are the components of the Raman tensor). A B_{1g} mode is clearly identified at 328 cm^{-1}, as it is seen under crossed polarizations along the base diagonals of the crystal $[z(x'y')\bar{z}]$. Since there is only one B_{1g} mode allowed, we can be sure of its eigenvector. It involves out-of-phase vertical vibrations of the oxygen located *outside* of the CuO_2 planes. This B_{1g} mode has the same symmetry and eigenvector as the B_{1g} vibration of oxygen *in* the planes observed in materials with more than one CuO_2 plane and which shows electron-phonon interaction effects in the $RBa_2Cu_3O_{7-\delta}$ system. One E_g mode in Nd_2CuO_4 is identified at 480 cm^{-1}, because of its high frequency it must belong to oxygen, the E_g vibration of Nd has not been detected so far. Due to the large mass of Nd it should be at low frequencies, lattice dynamical calculations predict it near 160 cm^{-1} [6.118].

The mode at 580 cm^{-1} in the $A_{1g}(zz)$ spectrum is not allowed in the ideal structure and is very weak in undoped Nd_2CuO_4 where it is also seen in (xx) and (xy) polarization [6.114]. The systematic dependence of these spectra on doping concentration predominantly shows a variation in intensity of the observed peaks accompanied by a frequency softening (~ 10 cm^{-1}) of the oxygen (E_g) mode. The B_{1g} oxygen and the A_{1g} Nd mode harden slightly with increasing doping. For a detailed study of the T' Raman phonons see [6.117].

Lattice-dynamical calculations (LDC) of the Nd_2CuO_4 structure reproduce well the frequencies observed, in particular the relatively high A_{1g} frequency of the heavy Nd [6.118]. The results of the shell-model [6.91, 119] and total-energy calculations (TEC) are compared with the experimental results in La_2CuO_4 and Nd_2CuO_4 in Table 6.1. Not all of the allowed Raman modes are actually observed. It should also be noted that neither copper nor oxygen of the CuO_2 plane is involved in any Raman-active vibrations in the tetragonal phase. An interaction of the CuO_2-plane related vibrations with superconductivity occurring in the planes can thus only be observed for the modes generated by the ortho-rhombic distortion. For a visual representation of the atomic displacements generated by the orthorhombic-only modes in La_2CuO_4 see [6.120]. These displacements correspond to zone-edge vibrations of the tetragonal phase. Copper is not involved in the distortion and remains Raman-inactive (center of inversion) even in A_g (orthorhombic) symmetry. It is, however, conceivable that the A_g (orthorhombic) oxygen vibration shows anomalies near T_c as they have been observed for instance in $YBa_2Cu_3O_7$ (see

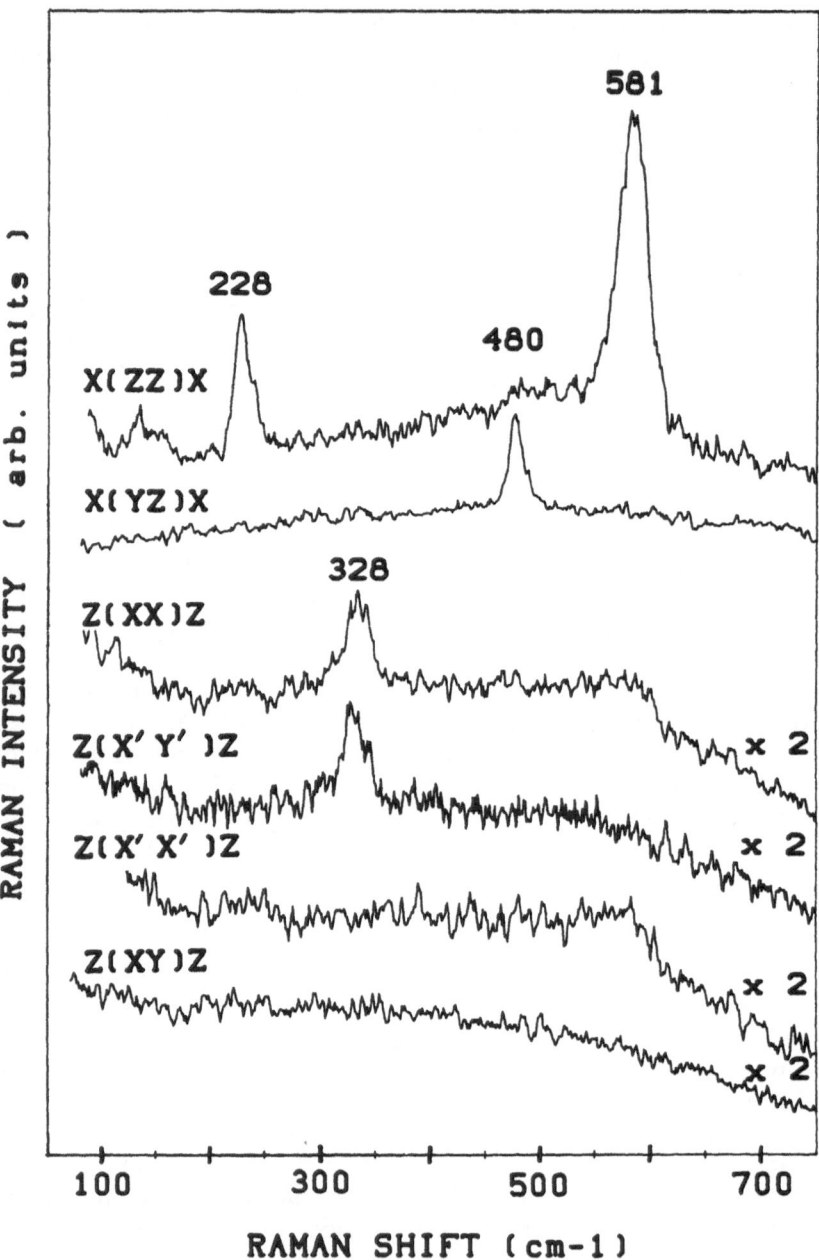

Fig. 6.8. Light-scattering spectra on single crystals of $Nd_{2-x}Ce_xCuO_4$ taken at room temperature. From [6.116]. Note that the mode at $581 \, cm^{-1}$ is not expected for the T' structure (see text)

discussion below). No such anomaly has been reported yet for $La_{2-x}Sr_x$-$CuO_{4+\delta}$ nor for $Nd_{2-x}Ce_xCuO_4$. The displacements of atoms in the four Raman-active modes of Nd_2CuO_4 have been depicted in [6.114] and those of all modes in [6.156].

Also known is a mixture of the two structures shown in Fig. 6.7. *Akimitsu* et al. [6.121] have reported that $(Nd_{0.66}Sr_{0.205}Ce_{0.135})_2CuO_{4-\delta}$ has one apex oxygen missing compared to La_2CuO_4, it is situated on the site of the out-of-plane oxygen in Nd_2CuO_4. This so-called T^* structure (space group $P4/mmm$, D_{4h}^7) has lost the inversion center in the CuO_2 plane; the primitive cell, however, contains two formula units and has a center of inversion halfway between two CuO_2 planes. Its superconducting representative just mentioned has a T_c up to 37 K [6.122]. For a detailed vibrational analysis of the T^* phase, see [6.123].

Finally, we will briefly discuss three other pseudo-perovskites with one CuO_2 plane. Two of them are the members with the lowest T_c in the Bi and Tl families. Their composition is $Bi_2Sr_2CuO_6$ and $Tl_2Ba_2CuO_6$. The structures are approximately body-centered tetragonal (space groups $I4/mmm$ in the idealized description) and have been analyzed by *Torardi* et al. [6.56]. A third, $TlBa_2CuO_5$ is primitive tetragonal ($P4/mmm$). The deviations from the ideal, tetragonal symmetry are along the base diagonal and have been described above. They are larger for the Bi than for the Tl compound. Reports of the transition temperatures vary, $T_c = 50$ K for the Bi compound [6.57], and $T_c = 10-80$ K [6.58], and 90 K for the Tl_2-substance [6.56]. $TlBa_2CuO_5$ has not been reported to be superconducting.

The significance of the one-layer Bi and Tl superconductors lies in that they are part of a larger sequence. Both two and three CuO_2 layers may be built into these compounds, something not possible in the $RBa_2Cu_3O_7$-type superconductors (as many as four layers have been reported for the Tl compound [6.124]). The two-layer compound $La_2CaCu_2O_6$ has recently been synthesized [6.82]. It has been true, so far, that an increase in the number of CuO_2 planes within a family goes along with a higher T_c, although saturation effects or even a slight decrease are reported for the $n = 4$ Tl_2 material [6.124]. The understanding of high-temperature superconductivity has not come far enough, though, to decide whether three layers should be an optimum value for high T_c (for instance due to screening of the central layer by the two adjacent layers), whether more layers could bring higher T_c's (see e.g. USO-reports [6.4, 125]) or whether a refined doping procedure (e.g. high-pressure annealing conditions) will bring up the T_c's in the materials with one or two CuO_2 layers as well.

We summarize here the phonon frequencies obtained from light-scattering spectra of some of these substances. The Bi-structure was analyzed in the low-symmetry space group $A2aa$ (C_{2v}^{13}) together with the corresponding two-layer material by *Cardona* et al. [6.85]. Raman spectra were taken on single crystals and most totally-symmetric modes could be identified. Similar results were obtained on ceramics in [6.57, 99]. The authors of [6.57] included ir spectra showing the corresponding odd-parity

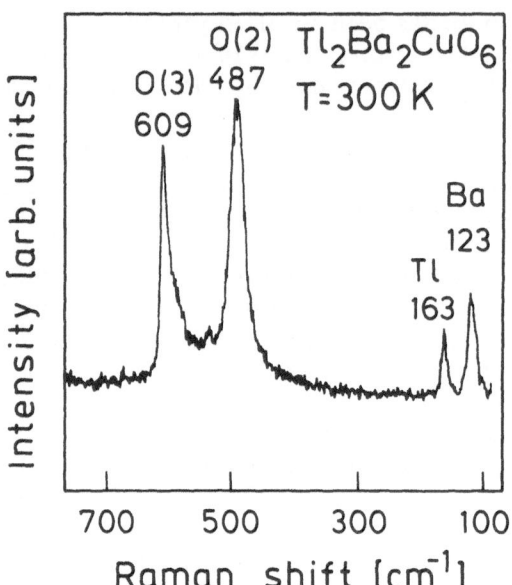

Fig. 6.9. Raman spectrum of $Tl_2Ba_2CuO_6$ taken at room temperature ($\lambda_L = 530.9$ nm). Note the absence of a CuO_2-plane related B_{1g}-mode in the (≈ 300 cm^{-1}) region since there is only one CuO_2 plane per unit cell

modes. The Tl_2 compound has been studied for one, two and three CuO_2 layers by *Krantz* et al. [6.126] and by *Timofeev* et al. [6.127]. In Fig. 6.9 we show our measurement of the Tl_2 ($n = 1$) superconductor which had a T_c of 90 K [6.128]. Clearly identified are two high-frequency oxygen-related peaks and two low-frequency Tl and Ba related peaks. The assignment is based on predictions of the lattice-dynamical calculation by *Kulkarni* et al. [6.94]. In Table 6.2 we compare the experimental frequencies for the one-layer Bi_2 and Tl_2 compounds. Theoretical frequencies are included where available. The calculations should be fairly reliable since the parameters involved are consistent throughout a given family. Total-energy calculations for these compounds have not yet become available.

Table 6.2. Observed Raman-active A_g modes (exp.) in two compounds with one CuO_2 layer per unit cell and theoretical frequencies from a lattice-dynamical calculation (LDC) where available. Weak modes are in parentheses. Frequencies given in cm^{-1}. Note that unit cells with one CuO_2 plane (and inversion symmetry) do not give rise to Raman-active modes involving those planes

$Bi_2Sr_2CuO_6$		$Tl_2Ba_2CuO_6$		
atom	exp.[a]	atom	exp.[b]	LDC[c]
Bi	130	Tl	165	153
Sr	(~ 200)	Ba	125	129
bridging oxygen	462	bridging oxygen	485	445
oxygen	650	oxygen	603	640

a) Ref. [6.57] *b*) Ref. [6.126] *c*) Ref. [6.94]

6.2.2 The Double-CuO$_2$-Plane Superconductors

YBa$_2$Cu$_3$O$_{7-\delta}$, the prototype and most-studied high-T_c superconductor, has two CuO$_2$ planes per unit cell. It was the first compound discovered to have $T_c = 93$ K, higher than the boiling point of liquid nitrogen [6.129]. The electron-phonon interaction and some of its consequences has been studied extensively in this structure. Phonon softening at T_c, linewidth, and lineshapes different from a simple Lorentzian due to interference effects have been reported first for YBa$_2$Cu$_3$O$_{7-\delta}$ [6.27]. The effects are not limited to this particular superconductor but have been studied best in it.

We will discuss here a double-layer structure and the resulting vibrational modes in some more detail. Since a number of reviews exist in the literature for RBa$_2$Cu$_3$O$_7$ [6.27, 101–104], we take Pb$_2$Sr$_2$RCu$_3$O$_8$ here when discussing the group-theoretical aspects particular to the double occurrence of the CuO$_2$ layer. We compare, however, results in both materials whenever applicable, so that this article also serves as a reference for RBa$_2$Cu$_3$O$_7$. Pb$_2$Sr$_2$Y$_{1-x}$Ca$_x$Cu$_3$O$_{8+\delta}$ was discovered to be a super-conductor by *Cava* et al. [6.130] and has a $T_c \approx 70$ K. In many respects it is similar to YBa$_2$Cu$_3$O$_{7-\delta}$ and is well suited for the following discussion. Different from YBa$_2$Cu$_3$O$_{7-\delta}$ is, for instance, the doping mechanism. In the Pb superconductor partial replacement of Y(3+) by Ca(2+) introduces holes into the system. The oxygen annealing conditions in the $x = 0$ system alone are not sufficient to produce superconductivity, a certain amount of "preloading" of the planes is required.

The unit cells of the RBa$_2$Cu$_3$O$_7$ and Pb$_2$Sr$_2$RCu$_3$O$_8$ are shown in Fig. 6.10. The CuO$_2$ planes consist of pyramids, i.e. the octahedron of La$_2$CuO$_4$ has lost one apex oxygen. The two bases of the pyramids in the two CuO$_2$ planes face each other and are separated by a layer consisting of rare-earth atoms. The layer surrounding the CuO$_2$ planes contains BaO or SrO, the oxygen forming the apex atom of the pyramids and the Ba/Sr being located vertically above the R-atom. The next layers are different for the two superconductors; this is what makes the classification into the so-called families. In YBa$_2$Cu$_3$O$_7$ a layer of copper and oxygen follows, both ions being at inversion centers of the cell. This arrangement of copper and oxygen is commonly referred to as *CuO chains* and is only known to occur in the RBa$_2$Cu$_3$O$_7$ family (including the compounds YBa$_2$Cu$_{3.5}$O$_{7.5}$ and YBa$_2$Cu$_4$O$_8$). The chains are indicated in the figure and cause the orthorhombic distortion. Before the discovery that $T_c > 90$ K could be obtained without the chains (i.e. in the Bi and Tl compounds), a great significance was put into them. It is now believed that the chains simply provide a doping reservoir for the CuO$_2$ planes.

Pb$_2$Sr$_2$RCu$_3$O$_8$, on the other hand, has a Pb-O layer and a Cu layer following the SrO layer, making the c-axis parameter somewhat larger than in YBa$_2$Cu$_3$O$_7$. The substance is orthorhombically distorted along the base diagonals of the chemical unit cell. The crystallographic axes, a,

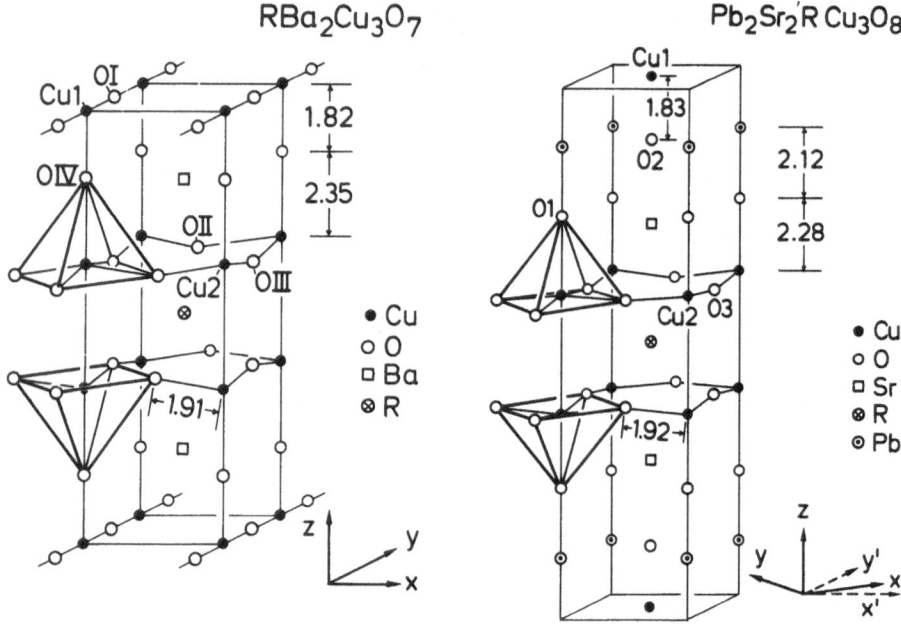

Fig. 6.10. Structure of two double-layer high-T_c superconductors. Both unit cells are nearly tetragonal but slightly distorted into orthorhombic symmetry. $RBa_2Cu_3O_7$ (a) is distorted along the [100] direction of the tetragonal axes and hence has its orthorhombic axes a, b parallel to the ideal tetragonal axes. $Pb_2Sr_2RCu_3O_8$ (b), on the other hand, is distorted along the base diagonals [110] or [1$\bar{1}$0] of the tetragonal cell. The crystallographic axes a, b are thus rotated by 45° compared to the tetragonal a', b' axes (centered structure). Taking $z = 0$ in $YBa_2Cu_3O_7$ in the plane of Cu1 and O_I, the z parameter of Ba is 2.19 Å and that of O_{II}, O_{III} 4.46 Å. The difference in the latter to that of Cu2 describes the buckling of the planes and amounts to 0.29 Å. The lattice parameters are (for R = Y): $a = 3.827$, $b = 3.877$, $c = 11.708$ Å. Structural parameters of $YBa_2Cu_3O_7$ from [6.131], those of $Pb_2Sr_2RCu_3O_8$ are from [6.130]

b point 45° away from the ideal tetragonal axes a', b'. The distortion is believed either to come from Pb-O arrangements due to bond-length considerations or to be due to additional oxygen in the relatively-empty Cu layer [6.132–134].

For $RBa_2Cu_3O_7$ all but Ce, Pr, and Tb (no information is available for unstable Pm) may be placed on the R-site without superconductivity being affected noticeably: all compounds have a $T_c > 87$ K [6.135]. It is particularly interesting that the magnetism of some of the lanthanides appears not to have the pair-breaking effect known for conventional superconductors [6.136]. Instead, the magnetic ions simply order when cooled below a critical temperature. For Gd this Néel temperature is 2.25 K, for Dy 0.9 K, for Sm 0.61 K, for Er 0.60 K, for Nd 0.52 K [6.137], and for Ho 0.17 K [6.138]. Pr in place of Y will allow the structure to be formed, but $PrBa_2Cu_3O_{7-\delta}$ remains semiconducting regardless of the annealing conditions; it has an ordering temperature of 17 K for the Pr ions [6.139]. For R = Ce, Tb the structure is not formed. There have been many attempts to explain why for R = Pr superconductivity is suppressed. They include the valence [6.140] (Pr occurs as 3+ and 4+ while the rare earths for which $RBa_2Cu_3O_7$ is superconducting usually are 3+), Pr^{3+} has the largest ionic radius and hybridization with other electronic bands may occur [6.141, 142], and that the magnetism is of a particular nature [6.143]. Experiments aimed at finding out the significant reason gave partly contradictory results. Raman scattering on ceramics, e.g. indicates through the ionic-radius dependence of the eigenmode frequencies that Pr occurs predominantly as 3+ [6.49] while magnetic-moment measurements indicated the presence [6.144, 145] of both 3+ and 4+. Magnetic effects may be indicated by 1 to 2 orders of magnitude higher Néel temperature of Pr^{3+} in $PrBa_2Cu_3O_7$ ($T_N = 17$ K) [6.139]. Also interesting is the observation of an increasing isotope effect upon admixture or Pr to Y. The mass exponent α was found to be 0.4 for a Pr to Y concentration ratio of 0.4 to 0.6 in [6.146]. We consider the question as not resolved. Interesting is, however, that Pr in place of R in the Pb superconductor still produces a superconductor [6.130]. An intrinsic feature of Pr, such as an unknown magnetic interaction, thus appears to be unlikely.

As is seen in the figure (atomic positions are drawn to scale), the CuO_2 planes of both substances are warped; those of the Pb superconductor somewhat less than those of $YBa_2Cu_3O_7$. The angle of the Cu-O bond with the $a - b$ plane is $\sim 5°$ in the former and $\sim 9°$ in the latter structure. The warping of the planes was also believed to be significant for the explanation of 90 K superconductivity, but the nearly flat CuO_2 planes in the double-layer Bi and Tl [6.79] superconductors with similar or even higher T_c's weakened this hypothesis. Nevertheless, for the Raman spectra the distortion plays an important role which will be discussed below.

We show at the example of $Pb_2Sr_2RCu_3O_8$ how one may obtain the mode symmetries of the vibrations at $k = 0$, i.e. the Raman-active, IR-active and silent modes. The detailed knowledge of these symmetries is a basis for an understanding of experimental spectra, which are rich in features. The lead compound under discussion has $16 \times 3 - 3 = 45$ optical modes, and an assignment without such analysis is impossible. A number of ways exists to obtain the symmetries [6.147–150], and we will give two straightforward ones. A coarse picture of the expected eigenmodes is unveiled when realizing that the structure has an inversion center and that the atoms occur either alone (R-atom, Cu1), in which case they are *necessarily* at an inversion center, or they may be grouped in pairs (all other atoms). Lone atoms, since their displacement is out of a center of inversion, can generate only odd modes: one each for the three space directions. Pairs of atoms, on the other hand, have the possibility to generate vibrations either odd or even with respect to the inversion center. They are referred to as Davidov pairs. Each pair thus gives three even and three odd modes (for the three-space directions). The in-plane oxygen may be considered as two Davidov pairs and gives rise to six even and six odd modes. (This is so for orthorhombic and tetragonal $Pb_2Sr_2RCu_3O_8$, and in tetragonal $RBa_2Cu_3O_6$; in orthorhombic $RBa_2Cu_3O_7$, there are, in fact, two distinct Davidov pairs O_{II} and O_{III}.) In total, there are thus 6×3 Raman-active modes and 8×3 odd modes, of which we substract the three acoustic ones. Whether or not the remaining 21 odd modes are IR-active or (nearly) silent may be estimated from the dipole moment for an anticipated particular atomic displacement. This has to be done with caution, though, since the displacement patterns for odd eigenmodes are generally quite complex due to the need for a center-of-mass conservation (mode mixing). A more detailed mode analysis as shown below will produce the exactly silent modes automatically.

Given the number of modes expected in the Raman and IR spectra, one might want to learn about the polarization selection rules and dipole directions of these modes. For IR-active displacements the case is simple: the vector character of these dipoles requires the incident electric field to be parallel to the atomic displacements (Fig. 6.11a, b).

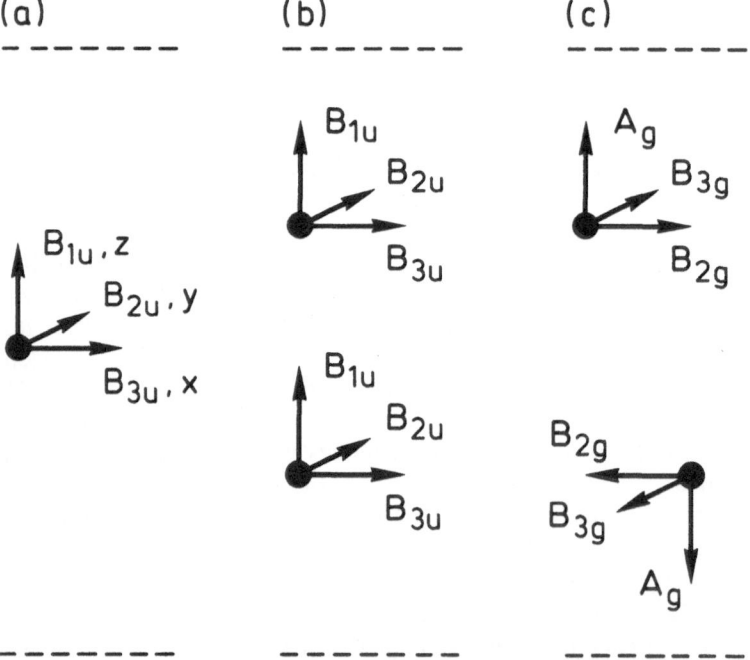

Fig. 6.11. Atomic displacements and their symmetries in a system with inversion center as is typical for the high-T_c superconductors of a) a single equivalent atom (holohedral site symmetry) b), c) pairs of equivalent atoms in an orthorhombic (D_{2h}) system. Odd displacements (b) generate IR-active or silent, even displacements (c) Raman-active modes

The *even* vibrations have displacements as indicated in Fig. 6.11 c. Displacements of individual pairs along the *c*-axis are in a high-symmetry direction, and, in a primitive cell, are totally symmetric (A_g) in the orthorhombic case, i.e. they have the full symmetry of the point group of the crystal. Two degenerate pairs (such as the two pairs of O3 in the CuO$_2$ planes) form one totally symmetric mode (all atoms in a plane moving in the same direction) and one mode of lower symmetry (B_{1g}) when the respective displacements in the plane are out of phase. Vibrations in *x* and *y* directions have lower, *B*-symmetry (or *E* if in tetragonal symmetry) and are denoted B_{2g} and B_{3g}, respectively, in orthorhombic notation (twofold degenerate E_g in tetragonal). The polarizations required for the corresponding Raman tensors to belong to an irreducible representation of the crystal are given in tables [6.148, 6.151] for all point groups. For convenience we reproduce in Table 6.3 all Raman tensors of the D_{2h} and D_{4h} point groups, to which most high-T_c superconductors belong. An exception is, e.g., cubic Ba$_{1-x}$K$_x$BiO$_3$ [6.37]. The selection rules for Raman scattering and IR reflectivity are summarized in Table 6.4.

These results may be obtained systematically by performing a factor-group analysis [6.148, 152]. Given the space group and the atomic sites within the unit cell one may find the symmetries of the eigenmodes by multiplying the representations of the site symmetries in the space group under consideration by the displacement-vector symmetry.

It follows from this rule that atomic sites which have the full symmetry of the space group (even) only produce odd eigenmodes, since the displacement vector has odd symmetry. Taking the space group of Pb$_2$Sr$_2$RCu$_3$O$_8$ to be *Cmmm* (D_{2h}^{19}), as has been suggested by *Cava* et al. [6.130] based on X-ray data, we may read the atomic-site symmetries off the International Crystallographic Tables [6.153]. The point group corresponding to *Cmmm* is *mmm* (D_{2h}) and Cu1 and R thus have the full point-group symmetry if they are chosen to be at the origin or else the point-group symmetry plus a primitive translation (Table 6.5). (The group D_{2h}^{19} is symmorphic, i.e. contains no fractional translation.) The other atoms have lower, *mm2* (C_{2v}) symmetry, except for the oxygen quadruplet, which has only 2 (C_2^z) site symmetry. From the character table of representations of D_{2h} (Table 6.4) [6.148] we may see that the displacement vector has B_{3u}, B_{2u}, B_{1u} representations for its *x*, *y*, and *z* components. This is to say, a displacement in *x* direction changes or does not change sign under the symmetry operations of the point group as indicated for the B_{3u} representation. The atoms with the full point-group symmetry have representation A_g in the table, the product of this representation and the displacement give simply the representations B_{3u}, B_{2u}, B_{1u} for Cu1 and R, i.e. the three odd modes we expected.

The representations of the C_{2v} site symmetry in D_{2h} may be determined by finding rows in Table 6.4 whose characters are left invariant under the

Table 6.3. Irreducible representations (in Mulliken's and Bethe's notation) of the possible Raman tensors in D_{2h} and D_{4h} point groups. Most high-T_c superconductors fall into either of these two groups. To a good first approximation, all orthorhombic structures in these materials may be considered tetragonal. Note that not necessarily all Raman tensors are actually present in a given crystal structure. Except for special conditions (near resonance), $ij = ji$, $i, j = x, y, z$, i.e. the Raman tensors are symmetric

orthorhombic

$$
\begin{pmatrix} xx & & \\ & yy & \\ & & zz \end{pmatrix}
\quad
\begin{pmatrix} & xy & \\ yx & & \\ & & \end{pmatrix}
\quad
\begin{pmatrix} & & xz \\ & & \\ zx & & \end{pmatrix}
\quad
\begin{pmatrix} & & \\ & & yz \\ & zy & \end{pmatrix}
$$

mmm (D_{2h}) $A_g\, \Gamma_1^+$ $B_{1g}\, \Gamma_3^+$ $B_{2g}\, \Gamma_2^+$ $B_{3g}\, \Gamma_4^+$

tetragonal

$$
\begin{pmatrix} xx & & \\ & xx & \\ & & zz \end{pmatrix}
\quad
\begin{pmatrix} & \overline{xy} & \\ -\overline{xy} & & \\ & & \end{pmatrix}
\quad
\begin{pmatrix} xx & & \\ & -xx & \\ & & \end{pmatrix}
\quad
\begin{pmatrix} & xy & \\ xy & & \\ & & \end{pmatrix}
\quad
\begin{pmatrix} & & zx \\ & & \\ zx & & \end{pmatrix}
\begin{pmatrix} & & xz \\ & & \\ xz & & \end{pmatrix}
$$

$4/mmm$ (D_{4h}) $A_{1g}\, \Gamma_1^+$ $A_{2g}\, \Gamma_2^+$ $B_{1g}\, \Gamma_3^+$ $B_{2g}\, \Gamma_4^+$ $E_g\, \Gamma_5^+$

Table 6.4. Character table for the D_{2h} point group, symmetry operations, selection rules for Raman- and IR-active modes and compatibilities (comp.) for tetragonal (D_{4h}) group and from this group to an orthorhombic one rotated by 45° [$D_{2h}(45°)$]. Note that the selection rules refer only to the D_{2h} point group. The symbols have their usual meanings

D_{2h}	E	C_2^z	C_2^y	C_2^x	i	σ^{xy}	σ^{xz}	σ^{yz}	selection rules	comp. D_{4h}	comp. $D_{2h}(45°)$
A_g	1	1	1	1	1	1	1	1	xx, yy, zz	$A_{1g}B_{1g}$	$A_g B_{1g}$
A_u	1	1	1	1	−1	−1	−1	−1	silent	$A_{1u}B_{1u}$	$A_u B_{1u}$
B_{1g}	1	1	−1	−1	1	1	−1	−1	xy	$A_{2g}B_{2g}$	$B_{1g}A_g$
B_{1u}	1	1	−1	−1	−1	−1	1	1	IR(z)	$A_{2u}B_{2u}$	$B_{1u}A_u$
B_{2g}	1	−1	1	−1	1	−1	1	−1	xz	E_g	B_{2g}
B_{2u}	1	−1	1	−1	−1	1	−1	1	IR(y)	E_u	B_{2u}
B_{3g}	1	−1	−1	1	1	−1	−1	1	yz	E_g	B_{3g}
B_{3u}	1	−1	−1	1	−1	1	1	−1	IR(x)	E_u	B_{3u}

symmetry operators of C_{2v}^z. Those are E, C_2^z, σ^{xz}, σ^{yz} and the representations of C_{2v} in D_{2h} are thus A_g and B_{1u}. (Alternatively, they are given in [Ref. 6.148, Table D].) Multiplying these by the displacements we obtain (A_g, B_{1u}) $\otimes$ (B_{3u}, B_{2u}, B_{1u}) $= A_g + B_{2g} + B_{3g} + B_{1u} + B_{2u} + B_{3u}$, i.e., each of the Davidov pairs generates three Raman-active and three IR-active modes. Lastly, the oxygen quadruplet in the CuO$_2$ planes (C_2^z-site symmetry, symmetry operations are E and C_2^z) has representations A_g, A_u, B_{1g}, B_{1u} and generates the following modes (A_g, A_u, B_{1g}, B_{1u}) $\otimes$ (B_{3u}, B_{2u}, B_{1u}) $= A_g + B_{1g} + 2B_{2g} + 2B_{3g} + A_u + B_{1u} + 2B_{2u} + 2B_{3u}$. Table 6.5 summarizes the atomic-site symmetries and the resulting eigenmode symmetries in Pb$_2$Sr$_2$RCu$_3$O$_8$ and RBa$_2$Cu$_3$O$_{7-\delta}$, $\delta = 0, 1$. It should be noted that for the Pb structure, only the modes of a primitive cell are listed.

Following a similar procedure one may obtain the corresponding results for RBa$_2$Cu$_3$O$_{7-\delta}$ in orthorhombic and tetragonal symmetry [6.27]. Some care must be taken when comparing modes in the three structures. The directions indicated for the activities in Table 6.3 and 6.4 are given with respect to the *crystal axes*. Since the CuO$_2$ planes are rotated in the Pb compound with respect to those in RBa$_2$Cu$_3$O$_7$ (Fig. 6.10), the Raman tensors appear rotated as well. For ease of comparison we have included the orthorhombic-orthorhombic (rotated by 45°)-tetragonal compatibilities in Table 6.4. The analyses of the other double-layer superconductors have been performed accordingly for the Bi [6.85] and Tl [6.154, 6.155] families and the structure of the variations YBa$_2$Cu$_{3.5}$O$_{7.5}$ and YBa$_2$Cu$_4$O$_8$ [6.156].

Having discussed the group-theoretical approach, we will now take a look at some experimental results and compare them to available

Table 6.5. Summary of the atomic-site symmetries and the resulting even and odd modes for orthorhombic $RBa_2Cu_3O_7$ and $Pb_2Sr_2RCu_3O_8$. Also given are the corresponding modes for tetragonal $RBa_2Cu_3O_6$

$RBa_2Cu_3O_7$ (mmm D_{2h})

atom	site symmetry	even modes	odd modes
R, Cu1, O$_I$	mmm D_{2h}		B_{1u}, B_{2u}, B_{3u}
O$_{Ib}$, O$_{III}$, O$_{IV}$, Ba, Cu2	$mm2$ C_{2v}	A_g, B_{2g}, B_{3g}	B_{1u}, B_{2u}, B_{3u}

$RBa_2Cu_3O_6$ ($4/mmm$ D_{4h})

atom	site symmetry	even modes	odd modes
R, Cu1	$4/mmm$ D_{4h}		A_{2u}, $2E_u$
O$_{IV}$, Ba, Cu2	$mm4$ C_{4v}	A_{1g}, $2E_g$	A_{2u}, $2E_u$
O$_{II}$	$mm2$ C_{2v}	A_{1g}, B_{1g}, $4E_g$	A_{2u}, B_{2u}, $4E_u$

$Pb_2Sr_2RCu_3O_8$ (mmm D_{2h})

atom	site symmetry	even modes	odd modes
R, Cu1	mmm D_{2h}		B_{1u}, B_{2u}, B_{3u}
Pb, Cu2, O1, O2, Sr,	$mm2$ C_{2v}	A_g, B_{2g}, B_{3g}	B_{1u}, B_{2u}, B_{3u}
O3	2 C_2	A_g, B_{1g} $2B_{2g}$, $2B_{3g}$	A_u, B_{1u} $2B_{2u}$, $2B_{3u}$

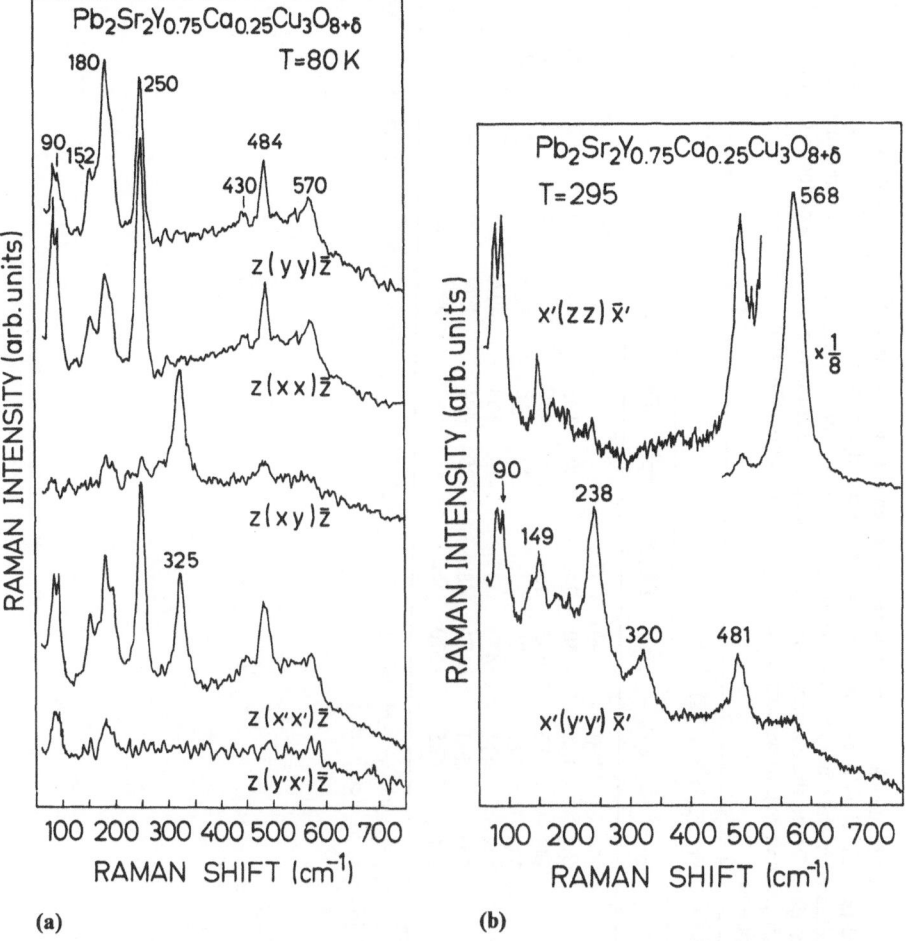

Fig. 6.12. Light-scattering spectra of $Pb_2Sr_2Y_{0.75}Ca_{0.25}Cu_3O_{8+\delta}$ for different polarizations of incident and scattered light. a) $k_L \parallel k_S \parallel \hat{c}$, b) $k_L \parallel k_S \perp \hat{c}$. From [6.158]

lattice-dynamical calculations. In Fig. 6.12 we show the Raman spectra of $Pb_2Sr_2Y_{1-x}Ca_xCu_3O_{8+\delta}$ for various polarizations of incident and scattered light. Comparison of the figure with Tables 6.3 or 6.4 will quickly reveal the symmetries of the observed modes. The uppermost curve, for instance, where incident and scattered fields are parallel to each other and to one crystal axis $[z(yy)\bar{z}]$, yields only A_g modes in orthorhombic D_{2h} symmetry. There is one more A_g mode than we have predicted in Table 6.5. We will show later that a resonance in the excitation spectrum produces the peak at 180 cm^{-1} in Fig. 6.12 with apparent A_g selection rules. However, it does not correspond to a Raman-active eigenmode of the ideal structure. The remaining modes are listed in Table 6.6 in columns Pb together with their assignments to particular atomic displacements. We recognize in the spectra the typical division into low-frequency

Table 6.6. Experimental (exp) and calculated vibrational frequencies in cm^{-1} with displacements in z-direction (A_g, A_{1g}, or B_{1g}) for high-T_c superconductors with two CuO$_2$-planes per unit cell ($n = 2$). The theoretical values refer to lattice-dynamical (LDC) and total-energy calculations (TEC). The asterisk highlights vibrations of atoms situated in the CuO$_2$ planes. The out-of-phase oxygen vibration has B_{1g} (tetragonal) or B_{1g}-like symmetry. The notation in the first three columns Y(ijk) refers to the chemical formula Y$_i$Ba$_j$Cu$_k$O$_{4+k}$ of the compounds. The index on the two Tl superconductors indicates the number of Tl-O planes per unit cell. Rows correspond to the structurally equivalent atomic sites and the assignment is to the atoms appropriate for the particular compounds

assignment	Y(1 2 3)			Y(1 2 3.5)	Y(1 2 4)	Pb		Tl$_1$		Tl$_2$		Bi	
	exp[a]	TEC[b]	LDC[c]	exp[d]	exp[e]	exp[f]	LDC[g]	exp[h]	LDC[i]	exp[k]	LDC[j]	exp[l]	LDC[m]
Cu/Pb/Tl/Bi	–			250	250	90	87	–		130	138	122	87
Ba/Sr	112	116	116	108	104	250	249	120	123	108	119	184	182
*Cu	154	157	157	147	152	152	153	148	137	158	149	156	164
*O out of phase	340	364	355	338	341	325	336	278	296	316[n]	303	282	347
*O in phase	440	411	378	448	438	430	437	475	358	407	370	391	387
bridging oxygen	500	495	508	500	500	484	483	525	494	494	439	469	493
oxygen	–			606	605	570	562	–		599	623	631	517

a) Ref. [6.27]   b) Ref. [6.98]   c) Ref. [6.27, 89]   d) Ref. [6.156]   e) Ref. [6.156]   f) Ref. [6.156, 157]   g) Ref. [6.158]   h) Ref. [6.159]   i) Ref. [6.154]   j) Ref. [6.94]   k) Ref. [6.155, 160]   l) Ref. [6.85]   m) Ref. [6.161]   n) Ref. [6.162]

metallic-ion modes (below $\sim 300 \, \text{cm}^{-1}$) and high-frequency oxygen-related vibrations (above $300 \, \text{cm}^{-1}$). The second curve [$z(xx)\bar{z}$] also shows only A_g modes, but due to the untwinned nature of the specimen we are able to obtain information about the $a - b$ anisotropy of the Raman-scattering amplitude. The Pb-mode at $90 \, \text{cm}^{-1}$ appears to be most strongly affected, suggesting a deviation of the ideal crystal structure in the Pb-O planes. The next curve, for crossed polarizations along the crystal axes of $Pb_2Sr_2Y_{1-x}Ca_xCu_3O_{8+\delta}$, should show only B_{1g} modes according to Table 6.4. In fact, only one such mode is allowed for this particular crystal structure (Table 6.5) and we can thus be sure that the peak at $325 \, \text{cm}^{-1}$ in the spectra corresponds to this eigenmode. Its eigenvector involves the out-of-phase vertical displacement of oxygen in the CuO_2 planes. This mode is of special importance since it exhibits considerable electron-phonon interaction effects in the $RBa_2Cu_3O_7$ system, to be discussed later.

In the next curve, the polarizations of both incident and scattered radiation are along the crystal-base diagonals. By contracting the Raman tensors R of Table 6.3 with the appropriate electric fields [$E_L = 2^{-1/2}(1, 1, 0)$, $E_S = 2^{-1/2}(1, 1, 0)$] one finds that scattering should be allowed for $A_g + B_{1g}$ symmetries. Correspondingly, all seven modes (plus one resonance-induced peak) are seen in the trace. Finally, in the bottom curve with crossed polarizations along the diagonals [$E_L = 2^{-1/2}(1, 1, 0)$, $E_S = 2^{-1/2}(1, -1, 0)$] only differences in the xx and yy tensor components should be seen. The modes at 90 and $180 \, \text{cm}^{-1}$ are thus the only ones with considerable $x - y$ anisotropy. The relative size of the tensor elements has been summarized in [Ref. 6.158, Table I]. The A_g and B_{1g} modes have thus been consistently identified.

An equivalent set of curves for $YBa_2Cu_3O_{7-\delta}$ was first published by *Liu* et al. [6.90]. There, the *approximate B_{1g}* mode (note again that for the Pb superconductor of Fig. 6.12 it is exactly B_{1g}) occurs at $340 \, \text{cm}^{-1}$, a fact that has been also established by using left and right-circularly polarized excitation; see [6.27, 90] for details.

A summary of all known (to date) double-layer high-T_c superconductors and their A_g and B_{1g} Raman frequencies is given in Table 6.6. In addition to the Y-rare earth and the Pb families this includes $Tl_1Ba_2CaCu_2O_7$, $Tl_2Ba_2CaCu_2O_8$, and $Bi_2Sr_2CaCu_2O_8$. The double-layer $La_2CaCu_2O_6$-compound has not been included since vibrational frequencies have not yet become available. Note that the tables have been arranged according to structural aspects and the frequencies are not always increasing when moving down in a particular column. Rather, we emphasize vibrations of a particular layer by listing them in a row. The copper vibration of the CuO_2 plane, for instance, has nearly the same frequency in all materials, the adjacent layer has a higher or lower frequency depending on whether it is occupied by Sr or Ba. The layer(s) furthest removed from the CuO_2 planes have vibrational frequencies that differ according to the mass of the metal ion, Pb being the heaviest atom with the lowest frequency. The oxygen out-of-phase mode ($340 \, \text{cm}^{-1}$ in

$YBa_2Cu_3O_{7-\delta}$) is the lowest in frequency of the oxygen-involving vibrations. It is seen to lie between ~ 280 and $340\,cm^{-1}$ for the different compounds. This is considerably more scatter than found for the Cu mode, but one should keep in mind that even within the $RBa_2Cu_3O_7$ or Pb family this frequency varies by about this amount for the different R's or different oxygen content [6.163–166]. The corresponding in-phase mode lies in most cases near $440\,cm^{-1}$. The bridging-oxygen vibration (or apex oxygen of the CuO_5 pyramids) is found near $500\,cm^{-1}$. Its dependence on oxygen has been taken as an analytic tool for the determination of the oxygen content in $YBa_2Cu_3O_{7-\delta}$. This vibration, in all superconductors, has the strongest transition-matrix element and is predominantly zz polarized. If there are more than one metal-oxide planes between neighboring Ba/Sr oxide planes, another oxygen A_g mode exists. Its frequency lies near $600\,cm^{-1}$, and it is also strongly zz polarized.

Raman spectra of $Bi_2Sr_2CaCu_2O_{8+\delta}$ have also been published in [6.99, 167–172]. While the various experimental frequencies agree mostly, the assigments do not always follow Table 6.6. Additional modes below $100\,cm^{-1}$ have been reported [6.167, 169, 172] and it is quite possible that Bi vibrations (A_g or B_{2g}, B_{3g}) lie so low. Controversial has been the mode at $469\,cm^{-1}$, which is strongest in xx polarization, but weaker than the $630\,cm^{-1}$ mode for zz [6.172]. Based on its strength, the $630\,cm^{-1}$ mode was thought to stem from the bridging oxygen, and based on a temperature anomaly [6.99, 172] the $468\,cm^{-1}$ mode was to come from the CuO_2 planes. Comparison with $Bi_2Sr_2CuO_6$ with no Raman-active modes in the CuO_2 plane (Table 6.2) shows, however, both high-frequency modes. Furthermore, comparing the Tl_1 and Tl_2 compounds (Table 6.6), we see that the mode forbidden in Tl_1 (Tl-O layers) and hence absent in Table 6.6 is indeed the highest-frequency one. Regarding the reported temperature anomaly, we mention that it is not possible, in general, to localize a vibration in the primitive cell based on its electron–phonon interaction, i.e. the CuO_2 planes are not necessarily involved in a vibration showing anomalous temperature effects.

We have included in Table 6.6 theoretical Raman frequencies from total-energy and lattice-dynamical calculations as described above. The agreement with the experiments is generally good to excellent.

In addition to the modes with displacements in z-direction, there are B_{2g} and B_{3g} (E_g in tetragonal) modes with eigenvectors, perpendicular to the high-symmetry axis of the unit cell. They are generally weak, probably due to cancellations in the Raman polarizability. *McCarty* et al. [6.173] have recently found the B_{2g}, B_{3g} modes in $YBa_2Cu_3O_{7-\delta}$, and they are about 100 times smaller in amplitude than the A_g modes. In the double-chain compounds $YBa_2Cu_{3.5}O_{7.5}$ and $YBa_2Cu_4O_8$ the chain-related B_{2g} and B_{3g} modes have been observed as well [6.156]. Experimental and theoretical frequencies (where available) are compared in Table 6.7. For a more detailed analysis of the chain-related modes see [6.156].

Table 6.7. B_{2g}, B_{3g} frequencies (in cm^{-1}) unambiguously identified in the two-layer high-T_c materials. The modes of Y(1 2 3) (for notation, see Table 6.6) are approximately 100 times weaker than those of A_g symmetry while the modes in Y(1 2 3.5) and Y(1 2 4) are comparable in strength to A_g modes. Parentheses indicate weak modes, empty fields that modes have not been observed, and dashes that Raman-active modes do not exist. An asterisk labels atoms in the CuO$_2$ planes

assignment	Y(1 2 3)						Y(1 2 3.5)^d		Y(1 2 4)^d
	exp^a		TEC^b		LDC^c		exp		exp
	B_{2g}	B_{3g}	B_{2g}	B_{3g}	B_{2g}	B_{3g}	B_{2g}	B_{3g}	B_{2g}
Cu	—	—	—	—	—	—	(140)	305	(140) 314
Ba	70	83	57	72	73	92			
* Cu	142	140	133	133	142	137			
bridging oxygen	210		185	257	356	496			
* O in phase		303	365	335	429	412			
* O out of phase	579	526	568	524	564	544			
oxygen	—	—	—	—	—	—	236		228

a) Ref. [6.173]   b) Ref. [6.97]   c) Ref. [6.27, 89]   d) Ref. [6.156]

6.2.3 Superconductors with Three CuO$_2$ Planes

From a group-theoretical point of view the third CuO$_2$ layer in the unit cell does not change the number of Raman-active modes. (It *does* change the number of IR-active modes.) In particular, there is still only one B_{1g} mode, which can thus be identified uniquely in the spectra. Only for a substance with four or more CuO$_2$ planes there are additional B_{1g} modes; one for every two CuO$_2$ planes. Calculations of the Raman frequencies for Tl$_1$Ba$_2$Ca$_3$Cu$_4$O$_x$ exist [6.94], but experimentally the vibrational properties have not yet been reported.

The three-CuO$_2$-layer material of the Tl family is interesting because it has the highest T_c of the currently-known superconductors. On the other hand, because of the health hazards involved handling the substance and because of the increasingly complex lattice, not many experimental results have been reported. The three-layer Bi superconductor has been hard to fabricate in single phase without the addition of e.g. Pb and has thus not yielded very reliable results. We summarize in Table 6.8 available experimental and theoretical phonon frequencies for compounds with three or more CuO$_2$ layers per unit cell.

Table 6.8. A_g, B_{1g} frequencies (in cm^{-1}) of the high-T_c superconductors with more than two CuO$_2$-planes. The vibrational spectra of the Tl$_1$ ($n = 4$) compound have not been reported so far, and we give only theoretical predictions. Empty fields indicate that modes have not been observed yet, dashes that the modes do not exist in a particular compound. An asterisk indicates modes involving atoms in the CuO$_2$ planes. The Bi ($n = 3$) superconductor is difficult to prepare as a single-phase material. The Raman frequencies have not yet been analyzed in detail

assignment	Tl$_1$ ($n = 3$)		Tl$_2$ ($n = 3$)		Bi ($n = 3$)	Tl$_1$ ($n = 4$)	
	exp^a	LDC^b	exp^a	LDC^b	exp	exp	LDC^b
Tl/Bi	—	—	133	126		—	—
Ba/Sr	104	113	99	104			101
* Cu	152	131	159	144			127
* Cu	—	—	—	—	—		142
* O out of phase	238	260	245	256			206
* O out of phase	—	—	—	—	—		298
* O in phase	260	293	270	302			246
* O in phase	—	—	—	—	—		369
O-Ca		449		456			450
bridging oxygen	526	505	498	420			503
oxygen	—	—	601	608		—	—

a) Ref. [6.160]   b) Ref. [6.94]

6.2.4 Resonant Raman Scattering

The technique of resonant Raman scattering, where the light involved in the scattering process is resonant with a strong electronic transition or absorption edge is well-known and rather fruitful in semiconductor research [6.174]. Depending on whether the incident or the scattered photon is in resonance one speaks of incoming or outgoing resonance. Under such conditions, substantial enhancement of the Raman-peak intensity (several orders of magnitude) may be observed. By studying the dependence of the strength of the Raman peaks on the energy of the exciting photons, one may thus extract information about the electronic band structure of a material. If not all peaks are enhanced equally one may localize the origin of the resonant transitions in a subsection of the unit cell, depending on which Raman peaks experience the enhancement. In a more accurate approach, one may calculate the electronic band structure and compare theoretically obtained resonance profiles with the experiment.

Such experiments have been performed on the Pb$_2$Sr$_2$Y$_{1-x}$Ca$_x$Cu$_3$O$_{8+\delta}$ superconductor and the Y-rare-earth system. In Fig. 6.13 we show the resonance profiles of Pb$_2$Sr$_2$Y$_{1-x}$Ca$_x$Cu$_3$O$_{8+\delta}$ and YBa$_2$Cu$_3$O$_{7-\delta}$ measured by *Heyen* et al. [6.175, 176]. We note peaks at ~ 1.7, 2.55 and > 2.7 eV. The two lower-energy peaks are common to the modes at 150, 250, 325, and 435 cm^{-1}, while the near-UV resonance enhancement occurs only for modes involving the Pb-O planes (80 cm^{-1} Pb, 484 and 570 cm^{-1} O). Related peaks are seen in ellipsometric spectra of the material as small bumps in ε_2 [6.175]. Knowing the eigenvectors of the Raman modes thus helps to understand the origin of the electronic transitions.

In the case of Fig. 6.13, the authors of [6.175] conclude that the peaks at 1.7 and 2.55 eV involve predominantly the Cu(2) $d_{x^2-y^2}$-O(3) $p_{x,y}$ complex, i.e. happen within the structural unit of the CuO$_2$-planes. The UV peak is proposed to originate from transitions involving the O(1) 2p $-$ Pb 6p level. This example shows a very-nice separation of electronic transitions made obvious by Resonant Raman Scattering.

Under resonance it is also possible to make active vibrations which are group theoretically not allowed. Figure 6.13b shows a good example.

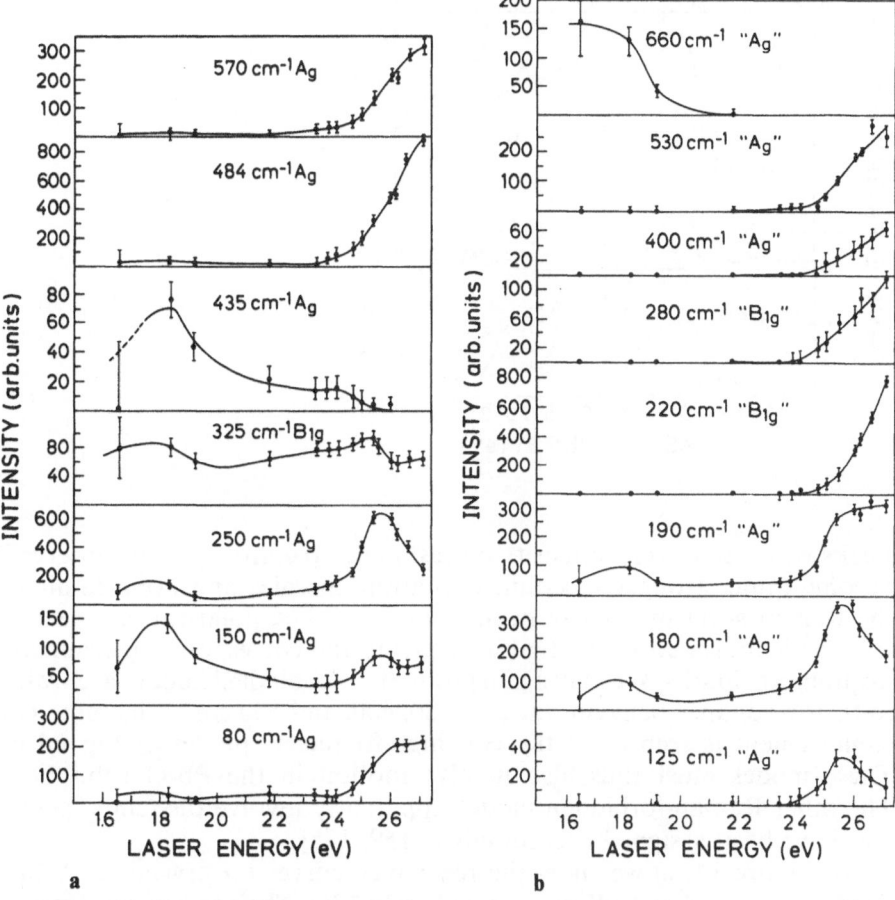

Fig. 6.13a–c. Resonance Raman profiles of Pb$_2$Sr$_2$Y$_{1-x}$Ca$_x$Cu$_3$O$_{8+\delta}$ (a, b) and YBa$_2$-Cu$_3$O$_{7-\delta}$ ($\delta \approx 0$) c). The sample temperature in all spectra is 10 K. (a) Profiles of the true eigenmodes of the crystal structure ($6A_g + 1B_{1g}$) and (b) profiles of peaks with polarization selection rules corresponding to A_g or B_{1g}-type modes. The units on the abcissa in (a, b) are arbitrary but the same for all spectra and the lines are a guide for the eye. From [6.175]. In (c) *absolute* scattering efficiencies are given and comparison is made with theoretical scattering efficiencies obtained from band-structure calculations in the LDA approximation. The full lines correspond to mixed eigenmodes, the dashed lines to pure modes. The dotted lines are a guide to the experimental points (black dots), xx, yy, zz refer to the polarizations of incident and scattered light. From [6.176]

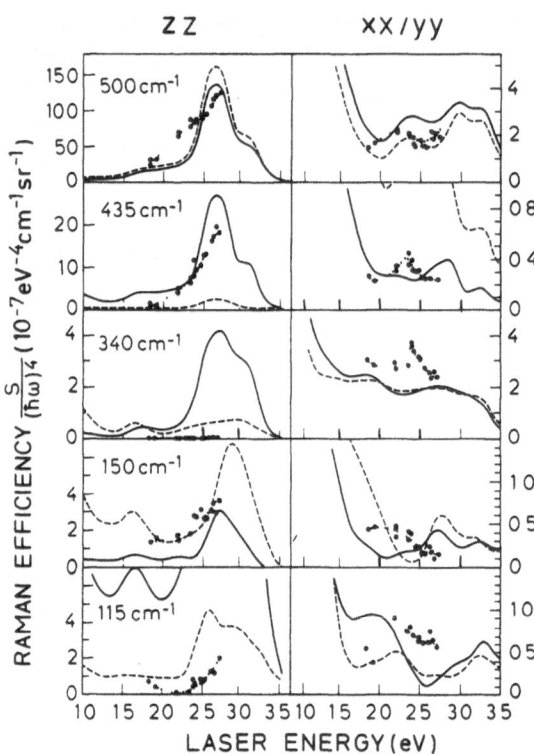

Fig. 6.13c

Peaks which have zero-intensity off resonance (i.e. they are not observed) become quite strong for certain excitation energies and even dominate the Raman spectrum. These peaks may also obey polarization selection rules as is indicated in the figure by giving the corresponding symmetry in quotation marks. We may again group the "forbidden" modes according to their resonance behavior: at 220, 280, 400, and 530 cm^{-1} the same UV enhancement is seen as in the two high-frequency modes of Fig. 6.13a. These modes must thus also involve motion in the Pb-O sublattices. The other Raman-forbidden modes apparently involve the CuO$_2$ planes and have been assigned accordingly [6.159, 175].

In Figure 6.13c) we show the resonance curves for crystals of YBa$_2$-Cu$_3$O$_{7-\delta}$ ($\delta \approx 0$) at 10 K as reported by [6.176]. The experimental points (black dots) are compared to theoretical Raman efficiencies calculated from the electronic band structure obtained in the local-density approximation (LDA). The different calculated curves correspond to different degrees of mixing between the involved eigenmodes (dashed line – no mixing; full line – finite mixing). Note that calculated and experimental efficiencies are given in absolute units and that no parameter has been adjusted in this comparison. We consider the agreement in both magnitude and energy dependence of the scattering efficiencies quite satisfactory.

6.3 Light Scattering by Electronic Excitations

An important contribution to the understanding of superconductivity in the high-T_c material has come from the direct scattering of light by electronic excitations. The phenomenon is known from the research on heavily doped semiconductors where it is understood in some detail [6.33] and also from the conventional superconductors [6.177]. In the high-T_c superconductors such scattering has been clearly identified through its temperature dependence, in spite of its weak strength. Below approximately the transition temperature, the uniform background in the scattering spectra begins to drop for low frequencies, a fact which is generally attributed to the opening of a superconducting gap. If this interpretation is correct, Raman scattering yields direct information about such gap, even though the details of the scattering mechanism are still being discussed. In this section, we present first the experimental evidence for such electronic scattering and then discuss a number of possible explanations.

6.3.1 Electronic Scattering

The presence of a continuous scattering background due to electronic excitations in the high-T_c superconductors was first put forward by *Lyons* et al. [6.178] and by *Ossipyan* et al. [6.179] on ceramics of $YBa_2Cu_3O_{7-\delta}$. Since then the data have been reproduced on single crystals with higher precision and with information about the polarization dependence [6.27, 179–185]. The effect has been found in the two-layer compound of the Bi family [6.185–187] and in the Tl compounds as well [6.188] so that there is little doubt about its existence. In Fig. 6.14 we show spectra taken on single crystals. The main features can be summarized as follows (not all evident from the figure):

1. A broad, nearly linear drop in the background below a certain frequency at low temperatures which, for the lowest temperatures measured, appears to extrapolate to zero scattering at zero frequency.
2. A different roll-off frequency exists for different symmetries.
3. The electronic scattering is absent for some symmetries.
4. The background appears to be constant in amplitude up to very large frequencies ($\omega > 4000$ cm^{-1}) and independent of temperature in the normal state.

These partly peculiar properties have not been fully understood but some of them have given rise to rather fundamental hypotheses about the high-T_c superconductors. See e.g. [6.189–192]. The linear drop to zero scattering at low temperatures, if it is not an effect of oxygen inhomogeneities near the surface, indicates either a significant density of states in an energy gap near the roll-off frequency or even a so-called zero-gap superconductor.

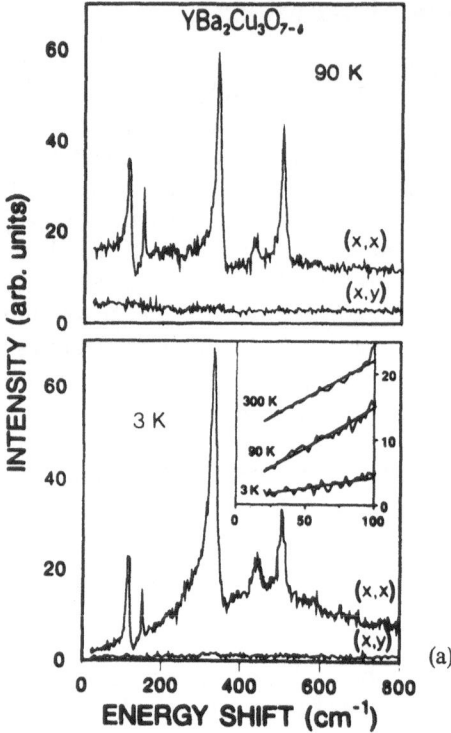

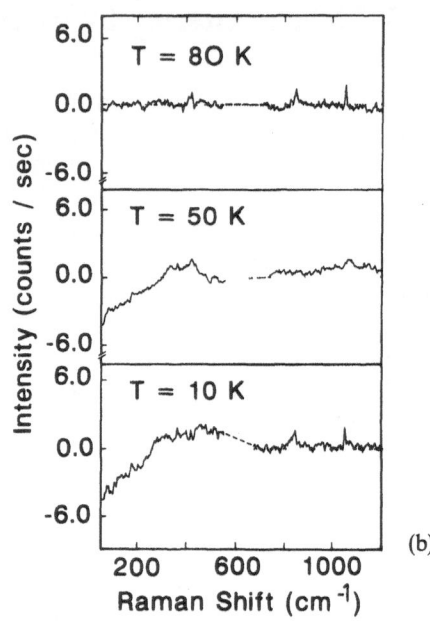

(a)

(b)

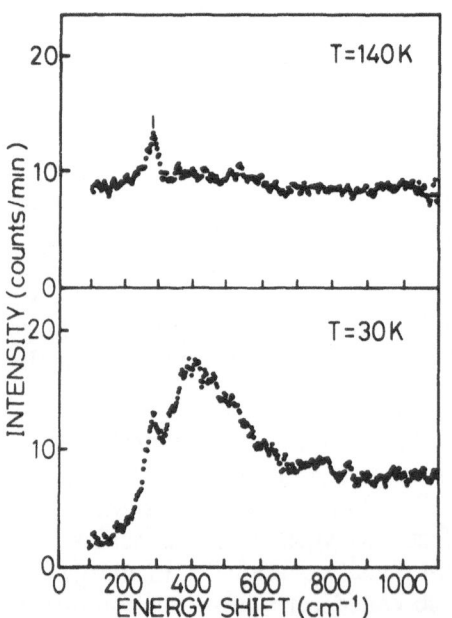

(c)

Fig. 6.14a–c. Electronic Raman scattering in various high-T_c superconductors (a) $YBa_2Cu_3O_{7-\delta}$ [6.181], (b) $Tl_2Ba_2Ca_2Cu_3O_{10}$ [6.188], (c) $Bi_{2.2}$-$Sr_{1.8}CaCu_2O_x$ [6.186]. The temperature-dependent background is interpreted as due to scattering by electronic states which are redistributed when a superconducting gap opens

The latter has nodes at the Fermi surface which yield a non-zero density of states for the particular symmetry studied. *Dierker* et al. [6.177] and *Hackl* et al. [6.193] have shown what such gap scattering for a node-free gap looks like in the conventional superconductors Nb_3Sn and V_3Si. We have reproduced a spectrum of *Hackl* et al. (Fig. 6.15) for comparison with the high-T_c materials. Note the sharp cut-off in scattering near $\omega = 2\Delta$ in contrast to the behavior of high-T_c superconductors. The difference in roll-off frequency for different scattering geometries is not unique to the $YBa_2Cu_3O_7$ system. It has also been observed in the two-layer Bi compound [6.186]. As displayed in Fig. 14 in $YBa_2Cu_3O_{7-\delta}$ the scattering maximum lies at 350 cm^{-1} for A_g symmetry and at ~ 500 cm^{-1} in B_{1g}-like symmetry. For the Bi material the numbers are ~ 300 cm^{-1} (A_g) and ~ 400 cm^{-1} (B_{1g}). The Tl 3-layer compound is seen to have its maximum at 460 cm^{-1} ($A_g + B_{1g}$ symmetry) [6.126]. What precisely the frequencies of the roll-off and maxima in the redistributed spectra mean is still unclear. In particular, without a better understanding of the gap structure in the high-T_c materials, it is hard to assign a single gap to a particular frequency. It is clear though that the redistribution of low-energy states (say below ~ 500 cm^{-1}) is present for a number of high-T_c materials. It thus appears to be a general property of the high-T_c materials and an explanation should not rely on band-structure or structural peculiarities of a particular compound.

A number of theoretical approaches to the understanding of the electronic background in the normal state have been put forward. Note first that a single-particle excitation in a free-electron picture cannot lead to the observed scattering. While in principle inelastic processes exist, they will be screened by the free electron gas introducing a factor $|\varepsilon(q, \omega)|^{-2}$ to the scattering cross section [6.194]. Far below the plasma frequency this factor becomes much too small to explain the observed scattering. In real solids, we may invoke band-structure effects. While scattering of single-particle excitations is still strongly screened when the effective-mass tensor is anisotropic but the same for all carries, anisotropic Fermi surfaces may produce fluctuations of the mass tensor which will lead to excitations unscreened by the free electrons. This may be seen from the scattering Hamiltonian

$$H_{\text{el-rad}} = \frac{e^2}{c^2} A_L \frac{1}{m^*} A_S, \tag{6.1}$$

where $A_{L,S}$ are the vector potentials of the incident and scattered radiation and m^* the effective-mass tensor of the (possibly anisotropic) parabolic bands. The resulting scattering efficiency (Stokes scattering) is proportional to

$$\frac{d^2\sigma}{d\omega\, d\Omega} \sim \left(\hat{e}_L \frac{1}{m^*} \hat{e}_S \right)^2 q^2 [n(\omega) + 1] \, \text{Im} \left[\varepsilon^{-1}(q, \omega) \right], \tag{6.2}$$

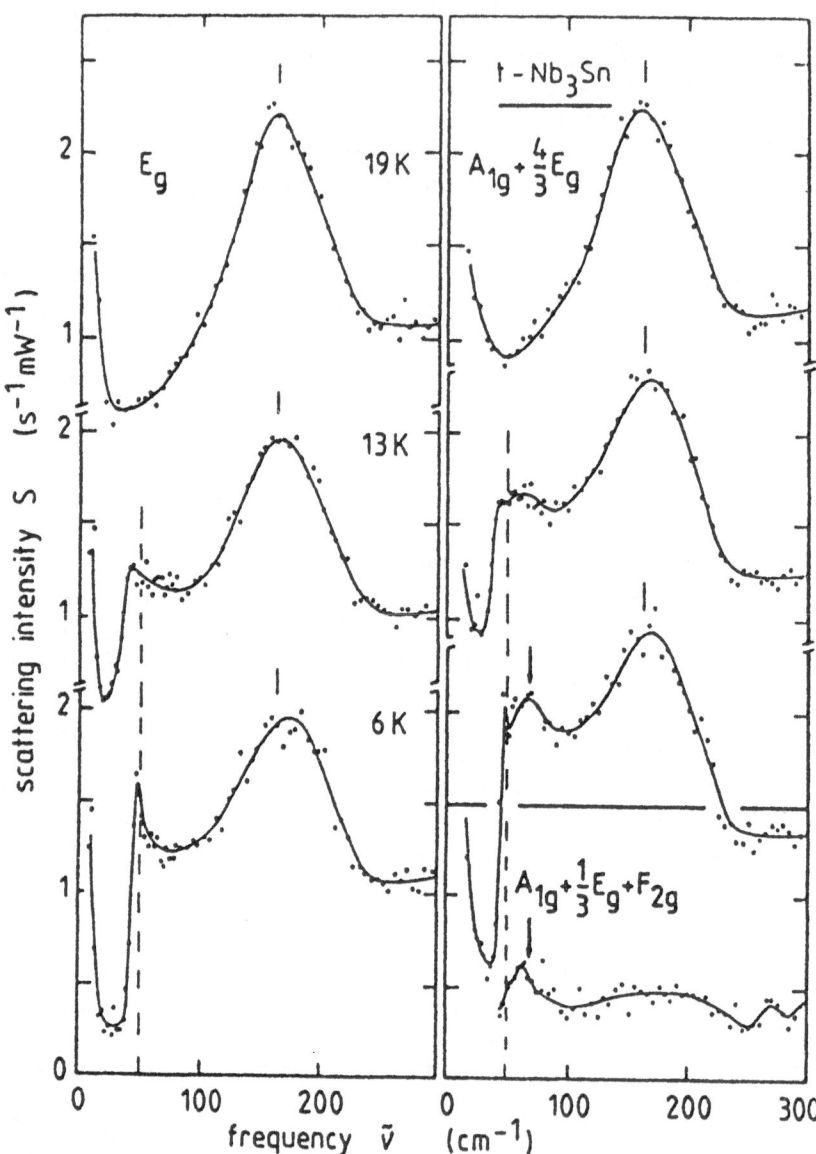

Fig. 6.15. Electronic Raman scattering near the superconducting gap of Nb_3Sn for various symmetries and temperatures. Note the sharp drop in scattering below the gap energy and the appearance of gap modes. The dashed lines indicate the positions of the gap mode with E_g symmetry at 46 cm^{-1}. The structure in A_{1g} symmetry at 68 cm^{-1} (arrow) is presumably of electronic origin, since no allowed phonon modes of A_{1g} symmetry exist. From [6.193] with permission of the authors

For a single, possibly anisotropic valley of carriers of mass m^* the Hamiltonian (6.1) acts like a scalar potential and is screened by the dielectric function of the free electrons. The screening is again very efficient far below the plasma frequency and, for $q \approx 0$ as in light scattering, no observable scattering can come out of this process.

For multivalley systems involving carriers with different $1/m^*$ tensors (referred to common axes) the situation is different (see, for instance, the discussion of *Abstreiter* et al. [6.33]). Physically speaking, carrier-density fluctuations may occur around the Fermi surface with varying effective masses but without fluctuations in the charge density. Free electrons cannot "see" through Coulomb interaction this kind of fluctuation and thus do not screen it. In doped semiconductors, this phenomenon has been observed as scattering from fluctuations between different valleys in [6.195, 196] and other references therein. For Si the transverse and longitudinal masses are different, and deviations of individual effective masses (contracted with $\hat{e}_L$ and $\hat{e}_S$) from the average produce the scattering [6.195]. It is interesting that the relevance of this mechanism can be proven experimentally by removing the anisotropy. In n-type Ge it was shown that upon application of uniaxial pressure (1.5 GPa) the electronic scattering disappears [6.196]. This is so because for the uniaxial pressure applied two of the three valleys move up in energy and become depopulated: the electrons are then all left in a valley with a single mass tensor.

The experimental data obtained on untwinned single crystals show that the amplitude of the electronic background in $RBa_2Cu_3O_7$ is larger in chain (y) direction than in x direction by a factor of ~ 2.5 [6.184, 197]; the corresponding difference in Raman polarizability is even larger, since corrections due to the smaller scattering volume (higher absorption) in y-direction have to be made. It is possible that this anisotropy is related to the difference in band structure, i.e. the chain-related bands appear not to cross the Fermi surface in the $\Gamma - X$ direction, whereas they are strongly dispersive (low effective mass) in the $\Gamma - Y$ direction [6.95, 198]. Without a detailed calculation of the intervalley effective-mass fluctuations for the complete band structure of $YBa_2Cu_3O_7$, a more quantitative statement is not possible, though. Also it is not clear that the observed absence of a significant frequency and temperature dependence would come out of such a calculation.

Another attempt to explain specifically the difference in A_{1g} and B_{1g} (tetragonal) symmetries, involving scattering by both intraband and interband excitations, has been put forward by *Monien* and *Zawadowski* [6.190]. They proposed that the interband scattering responsible for the B_{1g} scattering arises from transitions of a flat band near the Fermi surface to a band crossing the Fermi surface which they suggest is a CuO_2-plane related band. Transitions involve bridging oxygen py orbitals (O_{IV}) and the CuO_2-plane oxygen (O_{III}). This assumption is also capable of explaining the interference phenomenon between the phonon at $340\ cm^{-1}$ and the electronic continuum (Fano effect). The interaction producing the interference with the B_{1g} continuum comes from modulations of these transitions as the O_{III} are displaced in c-direction; the vibrational mode also has B_{1g} symmetry as would be required for such an interference. This explanation of the B_{1g} background scattering, however, relies in a fundamental way on the existence of a flat electronic band near the Fermi surface which does not necessarily exist in all high-T_c superconductors.

Finally, we mention a phenomenological theory of the high-T_c materials which has as a starting point electronic light scattering in the normal state. *Varma* et al. [6.189] made use of the lack of temperature dependence of this scattering over a wide frequency range. From (6.2) the scattering efficiency is proportional to

$$\frac{d^2\sigma}{d\omega \, d\Omega} \sim [n(\omega) + 1] \, \mathrm{Im} \, \{\varepsilon^{-1}(q, \omega)\} , \tag{6.3}$$

$$\sim \begin{cases} T/\omega & \mathrm{Im} \, \{P(q, \omega)\} \quad \text{for} \quad \omega < T; \\ & \mathrm{Im} \, \{P(q, \omega)\} \quad \text{for} \quad \omega > T, \end{cases} \tag{6.4}$$

where $P(q, \omega)$ is an electronic polarizability. A temperature and frequency-independent scattering efficiency is immediately constructed by requiring

$$\mathrm{Im} \, \{P(q, \omega)\} \sim \begin{cases} \omega/T & \text{for} \quad \omega < T; \\ \text{const} & \text{for} \quad \omega > T, \end{cases} \tag{6.5}$$

for small q [6.199]. While (6.5) says nothing about the microscopic reason for such a peculiar polarizability, the authors have pointed out that it may, without any further assumption, describe a number of phenomena typical of the normal state of the superconductor [6.200]. These are the linear dependence of resistivity on temperature [6.45, 201–203] the nuclear-spin relaxation rate, photoemission spectra, and the frequency dependent conductivity [6.204]. *Varma* et al. have coined the term *marginal Fermi Liquid* since the Green's function derived from (6.5) is entirely incoherent directly at the Fermi surface. Another way of saying this is that at low frequencies the electronic excitations are overdamped and the quasiparticle strength goes to zero at the Fermi energy. We consider it still open whether or not the high-T_c superconductors can be described as Fermi liquids [6.205, 206].

6.3.2 The Fano Effect

Fano (also called Breit-Wigner) interference effects in superconductors have been observed and studied most extensively in $YBa_2Cu_3O_7$ [6.180, 181] but they have also been reported for $YBa_2Cu_4O_8$ [6.207]. The significance of the observations for the high-T_c superconductors is twofold. 1) It proves the *existence* of an intrinsic scattering background in the experimental Raman spectra and 2) it proves that an *interaction* exists between the elementary excitation causing the background and those phonons which show the Fano effect. By itself, it *does not* show that the background is due to electronic scattering; this has to be inferred from other data. Furthermore, given that the background is due to scattering by quasi-free carriers in the sample, there is still no reason to believe that the interaction responsible for the Fano effect has a relation to the pairing interaction in the superconductor. Keeping these *caveats* in mind, we shall review the Fano effect and then discuss its implications for high-T_c superconductors.

The Fano effect in Raman scattering is observed as a characteristic change in the usually Lorentzian-lineshape of phonon peaks in the spectra. If two scattering sources in a material, one say a phonon (discrete), and electrons (continuous), do not interact coherently, they will simply produce additive, incoherent contributions with their respective scattering matrix elements T_p and T_e. Such a non-interacting spectrum will look like phonon peaks with a width determined by their incoherent interaction added onto a broad electronic continuum. If, however, a coherent interaction V exists between the two scattering sources, destructive and constructive interference will influence the lineshapes of the spectra. The overall result is that depending on the sign of a parameter q (to be defined below), the light-scattering rate is enhanced on the low-energy side of the phonon peak and depressed on the high-energy side [for sgn $(q) = -1$; for sgn $(q) = +1$ the reverse is true]. This produces an asymmetry of the phonon lineshape, which is referred to as a Fano (Breit-Wigner) effect. It should be clear that the Fano effect is a general phenomenon occurring whenever discrete and continuous excitations may interfere coherently. The effect, in fact, has been proposed by *Fano* [6.208] to explain asymmetries in atomic absorption spectra. The application of his formalism to an approximately flat, energy-independent electronic continuum background interfering with a discrete phonon state in heavily doped Si was first performed by *Cerdeira* et al. [6.33, 209, 210]. Hereby the appearance of the characteristic Fano asymmetry for the optically active phonon at 520 cm^{-1} was found to be correlated with high doping concentrations thus confirming the electronic nature of the scattering continuum.

Assuming constant transition-matrix elements T_e into all continuum states and a phonon transition-matrix element T_p, the light-scattering cross section (Fano profile) is proportional to [6.32]

$$\frac{d^2\sigma}{d\omega \, d\Omega}(E) \sim f(E) = \pi\varrho(\varepsilon) \, T_e^2 \, \frac{(q + \varepsilon)^2}{1 + \varepsilon^2}, \tag{6.6}$$

where

$$\varepsilon = (E - E_p - V^2 R)/\pi\varrho V^2 \tag{6.7}$$

and

$$q = (V T_p/T_e + V^2 R)/\pi\varrho V^2 ; \tag{6.8}$$

$E_p = \hbar\omega_p$ is the phonon energy in the absence of interaction. The functions ϱ and R are the density of electronic excitations (per unit energy) and its Hilbert transform, respectively. While they in general depend on energy, for a broad continuous background as in the normal state of the high-T_c superconductors it is justified to take $\varrho(E) = $ const. We shall assume in the following $R = 0$, in which case the parameters ε and q are simply (in terms of ω)

$$\varepsilon = (\omega - \omega_p)/\Gamma, \qquad q = T_p/\pi\varrho V T_e \quad \text{and} \quad \Gamma = \pi\varrho V^2. \qquad (6.9)$$

The effect of the interaction can thus be represented by a dimensionless (or Fano) lineshape parameter q and a linewidth contribution Γ (HWHM for the limiting Lorentzian) to the total linewidth. Any incoherent form of interaction also adds to the linewidth of the phonon (e.g. decay into two phonons) so that the experimental linewidth should be considered a sum of Γ_{inc} and Γ as given in (6.9). *A priori* it cannot be said how large the interaction-induced, Fano-related broadening is compared to the total linewidth.

In Fig. 6.16 we show a number of lineshapes calculated with (6.6–8). In the limiting case $|q| \to \infty$ a Lorentzian lineshape is recovered. This case corresponds to $\varrho V T_e \to 0$, i.e. no interaction if ϱT_e is not vanishing. For decreasing q the asymmetry of the phonon lineshape becomes more apparent. The analytic expressions for the extrema are

$$\omega_{max} = \omega_p + \Gamma/q, \qquad \omega_{min} = \omega_p - \Gamma q; \qquad (6.10)$$

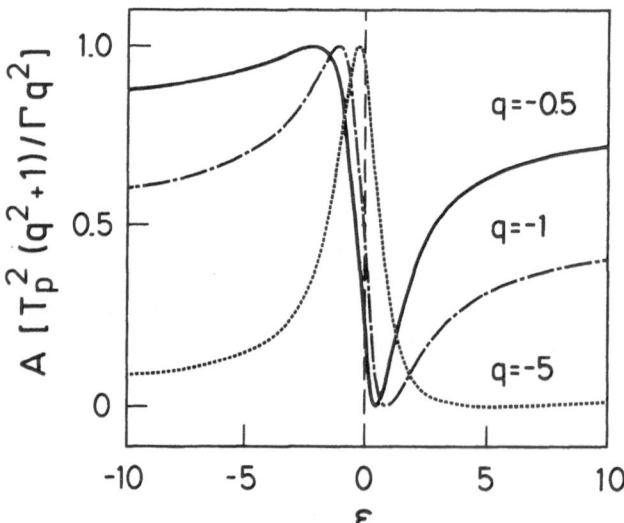

Fig. 6.16. Calculated Fano line shape for various values of the asymmetry parameter q. The curves have been normalized to the amplitude at the maximum. Note that for $q \to \infty$ a Lorentzian line shape is recovered while for $q \to 0$ an inverted Lorenztian appears. For negative q the curves are reflected at $\varepsilon = 0$

ω_{min} is also referred to as *antiresonance*. In the special cases $|q| = 1$ the lineshape is maximally asymmetric, i.e. antisymmetric with respect to ω_p, the bare phonon frequency. In general, the zero crossing of the Fano lineshape (defined with respect to the background) occurs at

$$\omega = \omega_p + \Gamma(1 - q^2)/2q \,. \tag{6.11}$$

A method useful for the geometric extraction of ω_p from the Fano lineshape has recently been given by *Piao* et al. [6.212] who noted that a line connecting the maximum and the minimum crosses the Fano curve at ω_p. The bare phonon frequency may thus be simply evaluated from experimental data.

Care has to be taken when the amplitudes of the Fano lineshape are considered. The amplitudes at the extrema of the Fano profile are according to (6.6)

$$A(\omega_{max}) = T_p^2(q^2 + 1)/\Gamma q^2 = T_p^2/\pi\varrho V^2 + \pi\varrho T_e^2$$

and

$$A(\omega_{min}) = 0 \,. \tag{6.12}$$

A background may exist so that $A(\omega_{min})$ is not necessarily zero experimentally: see discussion below. Equation (6.12) shows that in the maximum the parameters T_p, q, T_e, and Γ are not involved independently. The electronic and phonon transition-matrix elements may be decoupled, though, when considering the limits of the Fano function (6.6) for large $|\omega - \omega_p|/\Gamma$

$$A(\lim |\varepsilon| \to \infty) = \pi\varrho T_e^2 \,. \tag{6.13}$$

Thus, by taking the amplitude in the Fano maximum with respect to the background defined as the scattering level far from ω_p, one obtains a quantity

$$A' = T_p^2/\Gamma \,. \tag{6.14}$$

Equation (6.14) implies that regardless of the asymmetry parameter q, the square of the transition-matrix element $|T_p|$ is related to the amplitude in the maximum of the asymmetric lineshape. It is also possible to extract the ratio $|T_p/T_e|$ without the use of the peak amplitudes through the product

$$\Gamma q^2 = T_p^2/\pi\varrho T_e^2 \,. \tag{6.15}$$

Numerical fits to the experimental data allow then the determination of Γ and q. *Cerdeira* et al. [6.210] have used this approach and found that $q \sim |T_p/T_e|$ in p-type Si scales with $(\omega' - \omega_L)^{-1}$ where ω_L is the laser frequency and ω' a critical point in the band structure of Si ($\omega' = 3.3$ eV).

Another quantity of interest is the integrated strength of the Fano lineshape. For an energy-independent ϱ we find from (6.6)

$$
\begin{aligned}
I(q) &= \pi\varrho T_e^2 \int_{-\infty}^{\infty} [(q + \varepsilon)^2/(1 + \varepsilon^2) - 1] \, d\omega \\
&= \pi^2\varrho T_e^2 \Gamma(q^2 - 1)
\end{aligned}
\tag{6.16}
$$

where we have subtracted the unperturbed electronic background $\pi\varrho T_e^2$ (6.13). The special values $|q| = 1$ in (6.16) correspond to vanishing total area (with respect to the background) due to exact cancellation of integrated scattering by terms proportional to $|T_p|$ and $|T_e|$. Equation (6.16) may also be written as

$$
I(T_p) = \pi(T_p^2 - \pi\varrho T_e^2 \Gamma)
\tag{6.17}
$$

showing the two contributions to I more clearly. The integrated intensity may thus also be used to determine a dependence of $|T_p|$ on photon energy. Equation (6.17) shows furthermore that a Raman-inactive phonon ($T_p = 0$) will show a *negative* Fano intensity as defined by (6.16) if an electron-phonon interaction exists and $\varrho T_e^2 \neq 0$ [6.213]. In the approximation of constant ϱ, the line shape of such an interference-induced phonon is an inverse Lorentzian in the continuous background with a minimum (zero intensity) at ω_p, the bare phonon frequency.

In the experiments, in particular on high-T_c superconductors, it is often observed that, contrary to prediction of (6.12), the amplitude of a Fano resonance is not zero at the minimum. Here we do not mean a background from the dark current of the detector nor stray light. Even after allowance is made for these systematic experimental errors, a finite $A(\omega_{min})$ remains. A number of reasons exist for the light-scattering contribution at the energy where destructive interference between the phonon and the background should cancel the signal. (1) A non-interfering intrinsic, continuous background from another elementary excitation may exist, e.g. two-phonon scattering, two-magnon scattering, or even non-interacting parts of the electronic system. (2) The discrete state (phonon) may have a part, which does not interact with the continuum, as for example a lifetime-limiting decay into two phonons (which could also lead to Fano profiles!). (3) There may be inhomogeneous broadening of both the maximum and minimum, for example through isotopes or varying oxygen concentration (e.g. in the case of $RBa_2Cu_3O_{7-\delta}$). These effects can be described as an incoherent contribution to the Fano form [case (2)] which has been put into analytic form by *Abstreiter* et al. [6.214] by using a complex $\hat{q} = q_r + iq_i$. A practical way of extracting the coherently interacting part of the continuous scattering background from experimental data is found when realizing that, while in principle the electronic background is given by (6.13), it also corresponds to

$$\pi\varrho T_e^2 = A(\omega_{max}) - A(\omega_p),$$ (6.18)

i.e. the difference in Fano amplitudes at the maximum and at the bare phonon frequency.

In the superconductors not only the Raman transition matrix element is of interest. It has been noted that the lineshape, linewidth, and frequency change at the transition to the superconducting state. The lineshape changes are reflected in the Fano parameters and may be interpreted in terms of the quantities ϱ and V. In a sufficiently small temperature range, say 10 K − 100 K, it should be a good assumption that T_p and T_e remain constant. Then, for a Fano resonance the quantities

$$\Gamma q \sim V$$ (6.19)

$$\Gamma q^2 \sim \varrho^{-1}$$ (6.20)

project out V and ϱ independently. If it is assumed that the interaction potential V is also constant in this temperature range, then $\Gamma q = \text{const.}$ should serve as a constraint in the fitting procedure. A change in linewidth is consequently accompanied by a change in q unless it is the *incoherent* linewidth that changes.

A temperature-dependent density of continuum states $\varrho(T)$ is believed to be the origin of the variations in lineshape near T_c in the high-T_c materials. The opening of a gap redistributes the states in the gap to higher energies. To illustrate the constraints of (6.19) and (6.20) we have, under the assumption of an energy-independent ϱ, plotted the Fano function for either ϱ or V fixed while varying the other. The parameters were chosen to resemble a lineshape typical of the 340 cm^{-1} phonon in YBa$_2$Cu$_3$O$_7$. Figure 6.17a shows the scattering cross section (6.6) for three different V's at constant ϱ while (b) demonstrates the effect of varying ϱ at constant V. As can be easily verified, q and Γ, as given in the caption, fulfill the constraints (6.19, 20). Strictly, (6.19, 20) only hold if ϱ is not a strongly varying function of energy, which in the high-T_c materials becomes a poor approximation at low temperature $T \ll T_c$: As shown before (Fig. 6.14) in this case, the continuous background goes to zero at zero frequency. In an accurate numerical analysis q and ε should be used as in (6.7, 8).

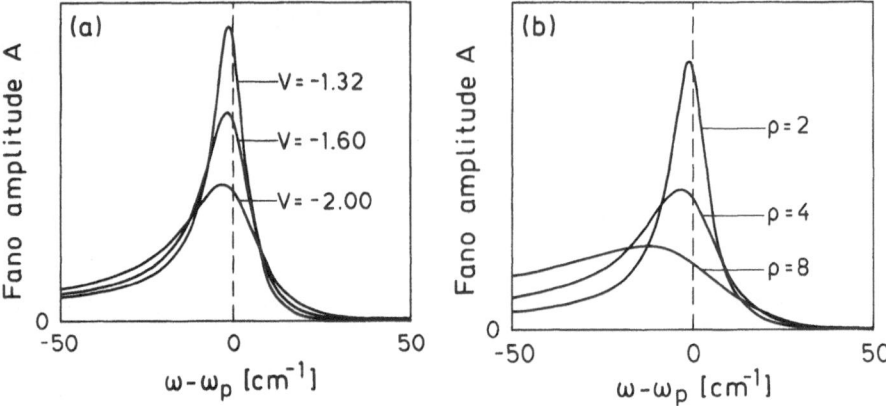

Fig. 6.17a, b. Calculated Fano line shapes with three different interaction potentials V and three different densities of continuum states ϱ and $T_p/T_e = 1$. In (a) we have kept constant ϱ and give V in units of $(\varrho\,\mathrm{cm})^{-1/2}$. The linewidth (HWHM) is $\Gamma = 5.5$, 8.0 and 12.6 cm^{-1} for $V = -1.32$, -1.60, and -2.00. For an assumed $\varrho = 1.6 \times 10^{-3}(1/\mathrm{cm}^{-1})$, the asymmetry parameter is $q = -6$, -4.94, and -3.95, respectively. Note that a change in V affects ω_{max}, the amplitude $A(\omega_{\mathrm{max}})$, q, and Γ. In (b), V remains constant while ϱ (in units of cm$^{-1}/V^2$) is varied. For these curves, $\Gamma = 6.28$, 12.57, and 25.1 cm^{-1} for $\varrho = 2$, 4, 8. For $V = -50.3$ cm^{-1}, $q = -8$, -4, and -2, respectively. Note that in (b) the value for $|\omega| \to \infty$ is different for the three curves, while in (a) the limits are equal

In Fig. 6.18 we show the Fano effect as measured by *Cerdeira* et al. [6.209] in doped Si. The parameter that is different in the curves is the exciting laser energy ω_L. For small q the minimum is nearly an inverted Lorentzian, whereas at large q (far from the resonance) the usual, positive Lorentzian is recovered. Fig. 6.18 serves to demonstrate the Fano effect in a heavily doped semiconductor, which has a carrier concentration not much lower than that of the high-T_c superconductors.

The Fano asymmetry in YBa$_2$Cu$_3$O$_7$ for the phonon at 340 cm^{-1} was first reported by *Thomsen* et al. [6.180] and, independently, by *Cooper* et al. [6.181]. The numerical fit shown in Fig. 6.19 is an excellent representation of the data and demonstrates that the application of the Fano formalism is appropriate. A Lorentzian would certainly not describe the data adequately. The Fano parameters given in the figure caption are typical for this phonon. In particular, ω_{max} is shifted from the bare phonon frequency ω_p by $\Gamma/q = -2$ cm^{-1}, while ω_{min} lies at $\Gamma q = +30$ cm^{-1} (higher than ω_p) and is not very well discernible in the data. The data of Fig. 6.19 were taken on an untwinned single crystal and at an excitation energy of $\omega_L = 2.41$ eV. In this fit a constant background has been subtracted from the data.

In Fig. 6.20 we give a comparison between superconducting ($\delta = 0$) (a) and semiconducting YBa$_2$Cu$_3$O$_{7-\delta}$ ($\delta = 1$) (b) [6.215, 216]. It is clear

Fig. 6.18. Fano line shapes observed for heavily doped p-Si with three different laser wavelengths. The solid lines are theoretical fits to the experimental curves (discrete points). From [6.211]. The fitted q-values are 4.5, 2.6, 0.55 for $\lambda = 4579$, 5145, and 6471 Å

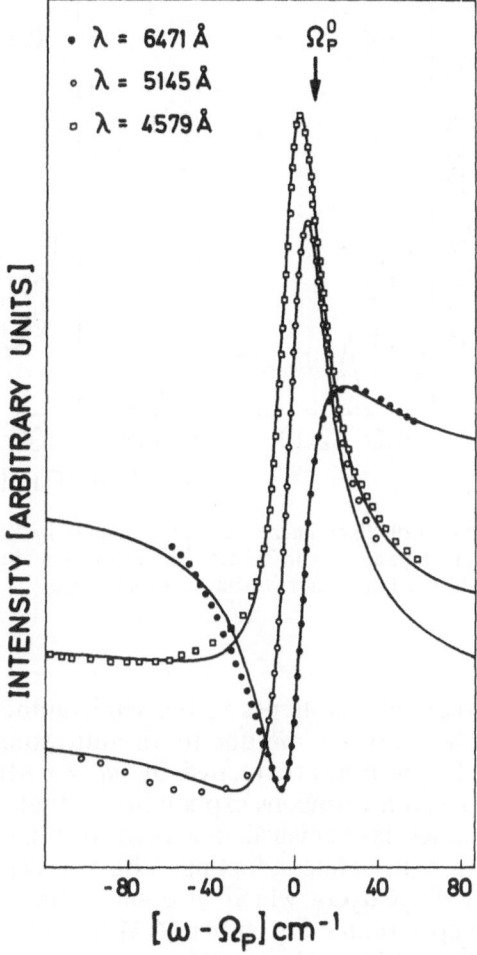

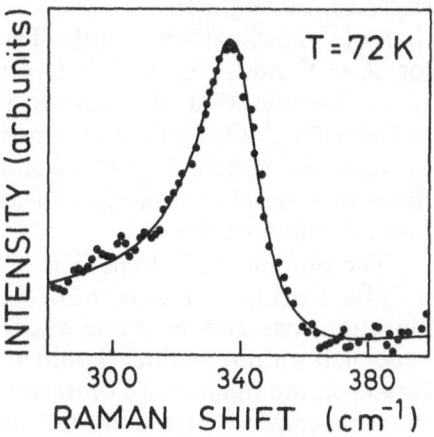

Fig. 6.19. Fano lineshape of the phonon at 340 cm^{-1} in YBa$_2$Cu$_3$O$_{7-\delta}$ (points) and a numerical fit. The parameters of the fits are $\omega_0 = 339.5$ cm^{-1}, $q = -4.3$, and $\Gamma = 9.5$ cm^{-1} (HWHM). From [6.27]

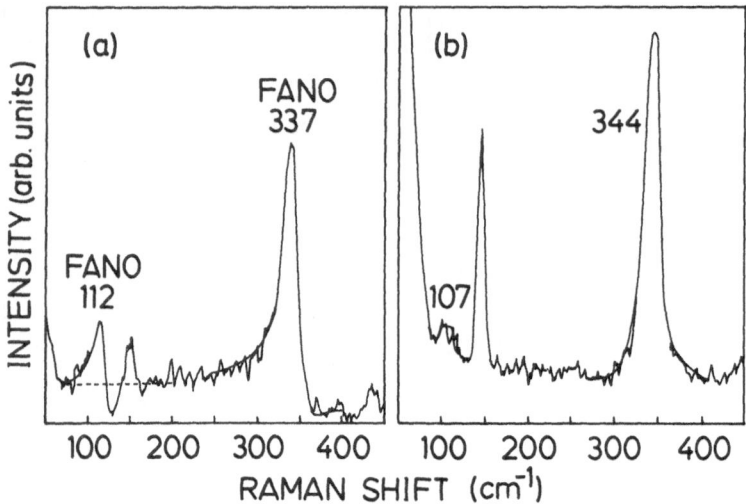

Fig. 6.20. Experimental Fano resonances in $YB_2Cu_3O_{7-\delta}$. The asymmetries are seen for phonons at 112 and 337 cm^{-1} in the superconductor ($\delta \approx 0$) (a). The semiconductor ($\delta = 1$) (b) has Lorentzian lineshapes. From [6.215]

that the lineshapes in the semiconductor are no longer asymmetric and show no minima due to an antiresonance. The asymmetry parameter is thus well approximated by $|q| = \infty$ from which follows that $\varrho V T_e \approx 0$. The most obvious explanation is that the density of electronic continuum states has vanished, in accord with the nearly vanishing carrier concentration in semiconducting $YBa_2Cu_3O_6$. There is also no asymmetry in $PrBa_2Cu_3O_7$ which also has a low carrier concentration and is not superconducting either [6.217]. There is, however, also the possibility that $\varrho V T_e$ is different for another rare earth in place of Y. In Fig. 6.21 we show the Raman B_{1g}-like peak of $ErBa_2Cu_3O_{7-\delta}$ (a), which is superconducting with a substantial Meissner fraction (together with that of $YBa_2Cu_3{}^{18}O_{7-\delta}$ (b)). The asymmetry of the B_{1g}-like phonon for R = Y and $^{16}O \rightarrow {}^{18}O$ is considerably smaller than for R = Er. If we assume that the scattering amplitude $|T_p|$ is similar to that in $ErBa_2Cu_3{}^{16}O_{7-\delta}$, then we must conclude that $\varrho T_e(Y, {}^{18}O) < \varrho T_e(Er)$ or that the coherent part of the interaction potential is different in the two materials. Certainly, this aspect of the Fano interference deserves more detailed studies.

The phonon at 340 cm^{-1} is not the only one with a Fano lineshape in $YBa_2Cu_3O_{7-\delta}$. Figure 6.20 shows that the lowest A_g mode, a Ba-vibration, has also a strong antiresonance for $\delta \approx 0$. It is not obvious why such an interaction should take place for the Ba-vibration [6.218]. However, the significance of the observed asymmetry lies in that it proves the existence of states in the gap in the superconducting state of

Fig. 6.21. The B_{1g}-like phonon in (a) $ErBa_2Cu_3O_{7-\delta}$ and (b) $YBa_2Cu_3{}^{18}O_{7-\delta}$ for temperatures at and below T_c. The line shape in (b) is much more symmetric in spite of the good superconducting quality of both samples. Fits parameters are: (R = Er: ω_v = 380.0 cm^{-1}, Γ = 7.7 cm^{-1}, q = -3.8 at 90 K, ω_v = 340.5 cm^{-1}, Γ = 13.8 cm^{-1}, q = -2.3 at 10 K; R = Y(^{18}O): ω_v = 323.0 cm^{-1}, Γ = 7.2 cm^{-1}, q = -6.2 at 90 K, ω_v = 318.5 cm^{-1}, Γ = 7.7 cm^{-1}, q = -4.9 at 10 K). Note the difference in broadening and softening of this mode for the two materials. From [6.260]

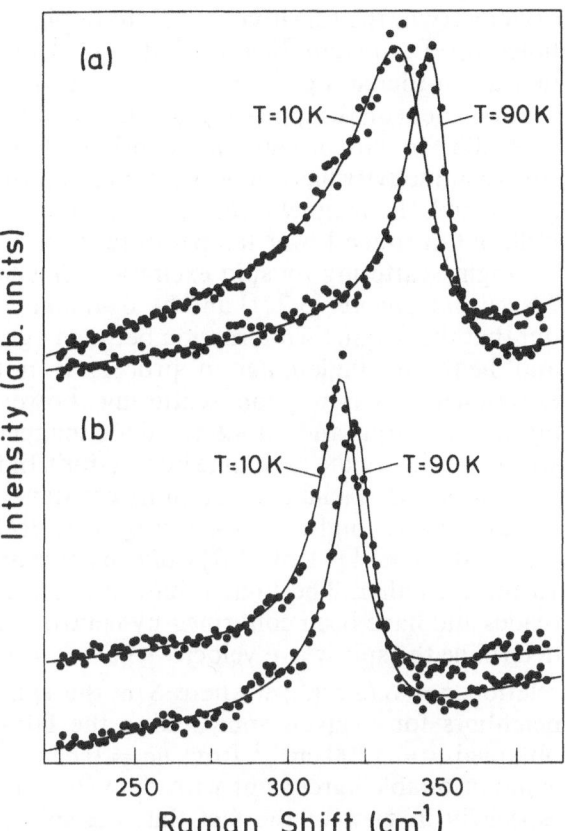

$YBa_2Cu_3O_{7-\delta}$. *Cooper* et al. [6.182] have shown that even at the lowest temperatures (3 K) the antiresonance of the Ba-vibration persists and thus $\varrho \neq 0$ at these temperatures and energies. The low-temperature Fano resonances of this Ba mode may be seen in Fig. 6.14.

6.4 Light Scattering by Spin Excitations

Collective spin excitations (magnons) have been of significant interest in the discussion about possible mechanisms for high-T_c superconductivity. It is observed that when the doping concentration in the cuprate superconductors is reduced and they become semiconductors, the copper spins (Cu2) order antiferromagnetically. The Néel temperature is highest in the undoped compounds, e.g. $T_N \approx 300$ K in La_2CuO_4 [6.219] and $T_N = 500$ K in $YBa_2Cu_3O_6$ [6.220]. With increasing doping T_N decreases and goes to zero before superconductivity appears. There is a notable asymmetry in the sensitivity of T_N to doping between electron-doped and hole-doped superconductors. $Nd_{2-x}Ce_xCuO_{4-\delta}$ remains antiferromagnetically ordered up to $x = 0.13$ while in $La_{2-x}Sr_xCuO_{4-\delta}$ the antiferromagnetism disappears and superconductivity appears already for $x \approx 0.02$. In either case, an overlap of antiferromagnetic order and superconductivity has not been established with certainty [6.75]. As already mentioned, the magnetic rare-earth atoms order antiferromagnetically as well, but at much lower temperature (Sect. 6.2).

Light scattering by spin excitations has been theoretically treated by *Fleury* and *Loudon* [6.221] and by *Cottam* and *Lockwood* [6.222]. It turns out that the magnetic interaction between light and the spins is very weak, and hence a single-magnon process is not observable in a Raman experiment. Two-magnon scattering, however, involving two *electric-dipole* transitions and an intermediate magnetic interaction can be rather strong (excited-state exchange interaction). In the CuO_2 planes of the high-T_c oxides the exchange coupling constant J of the interacting spins is particularly high; in La_2CuO_4, e.g., it has been measured to be $J_{ex} \approx 1000$ cm^{-1} [6.223–227] while in conventional materials J is typically 10 times smaller. The high values of J are characteristic of the high-T_c oxides and have been confirmed by neutron-scattering experiments which determine the spin-wave velocity c near $k = 0$. According to the standard relation $c = SJZa/\sqrt{2}$, where S is the spin, Z the number of nearest neighbors for a given spin, and a the lattice parameter, *Aeppli* et al. obtained $J = 1300$ cm^{-1} from $\hbar c = 0.85$ eVÅ in La_2CuO_4 [6.228]. This is in reasonable agreement with the value obtained from light scattering, especially in view of the fact that the spin-wave velocity is determined near the zone center, while the Raman peak due to two-magnon scattering (in general any two-excitation process) is dominated by the density of states, which is large at the zone boundary. It should be noted that the maximum of the two-magnon peak in the experimental spectra is not at twice the zone-boundary energy of a single magnon as one might naively expect. Rather, due to magnon-magnon interaction, it is lower. The dispersion relation of a two-dimensional magnon neglecting the anisotropy field is [6.229]:

$$E^2(k) = (4JS)^2 [1 - (\cos k_x a + \cos k_y a)^2/4]) \qquad (6.21)$$

which leads to the divergence of the density of states at the zone boundary where $E_{two-mag} = 4J$. Including magnon-magnon interaction leads to an expression derived for the lineshape by *Parkinson* [6.229], with a maximum at $\sim 3J$. A more accurate account of the spin-1/2 nature of the magnetic excitation leads to a peak position at $2.7J$ [6.224]. The latter relation between Raman-peak energy and exchange coupling constant is usually used to determine J.

The symmetry selection rules for scattering of light by magnons have been derived for an isotropic Heisenberg Hamiltonian and nearest neighbor interaction by *Parkinson* [6.229] and have helped to identify the two-magnon nature of the Raman feature. The predominant scattering contribution has B_{1g} symmetry and may be well distinguished in the spectra. Experimentally, however, magnon signals have also been observed with A_{1g} and B_{2g} symmetries (tetragonal notation) (La$_2$CuO$_4$, Nd$_2$CuO$_4$ [6.225]; YBa$_2$Cu$_3$O$_6$ [6.230]). Another apparent discrepancy with the simple theory of scattering by spin fluctuations in the high-T_c materials is the much larger linewidth observed in the spectra leading to contributions at energies even higher than $4J$, the classical cutoff.

Singh et al. [6.231] have proposed that quantum fluctuations in the ground state of the system may become noticeable in the $S = 1/2$ case. In the classical ground state (Néel state) of an antiferromagnet nearest neighbors (*nn*) are aligned antiparallel while next nearest neighbors (*nnn*) are oriented ferromagnetically with respect to each other. The scattering Hamiltonian is [6.229]

$$H \propto \sum_{ij} (E_{\mathrm{inc}}\sigma_{ij}) (E_{\mathrm{sc}}\sigma_{ij}) S_i S_j \tag{6.22}$$

where E_{inc} and E_{sc} are incident and scattered electric fields, and σ_{ij} is a unit vector connecting spins S_i and S_j. In this case the sum over *nn* leads to the B_{1g} classical selection rule. If quantum fluctuations in the ground state are taken into account, *Singh* et al. argue, there is a certain probability that *nnn* spins are antiparallel. Summation in (6.22) over *nnn* leads to scattering contributions in both A_{1g} and B_{2g} geometries [6.231]. The authors of [6.231] have also shown that the observed increase in linewidth in B_{1g} symmetry may come from quantum fluctuations, even if only the *nn* terms are taken into account in (6.22).

Knoll et al. [6.230] have proposed an alternative or additional mechanism for the significant broadening in the B_{1g} spectra of YBa$_2$Cu$_3$O$_6$. They measured linewidth and peak position at temperatures well above the Néel temperature T_N (Fig. 6.22) and found that the linewidth increases with an approximate T^4-law. Such a dependence is characteristic of a decay process into one magnon and one phonon [6.232] and consequently, it appears that spin-lattice interactions should be relevant in YBa$_2$Cu$_3$O$_6$. The reason for why this process is so strong in the high-T_c related materials is that the magnon energy is much higher than in conventional antiferromagnets where primarily acoustic phonons are available for the processes mentioned above. Interaction with many optical branches should reduce the magnon lifetime considerably and may lead to the observed broadening.

There has been some discussion in the literature about a Raman feature observed near 1200 cm^{-1} (near 1550 cm^{-1} in La$_2$CuO$_4$) which has been attributed to either a one-magnon [6.223, 233] or a two-phonon process [6.114, 227]. In the one-magnon interpretation, a zone-boundary excitation would become Raman active e.g. in the vicinity of a vacancy or defect.

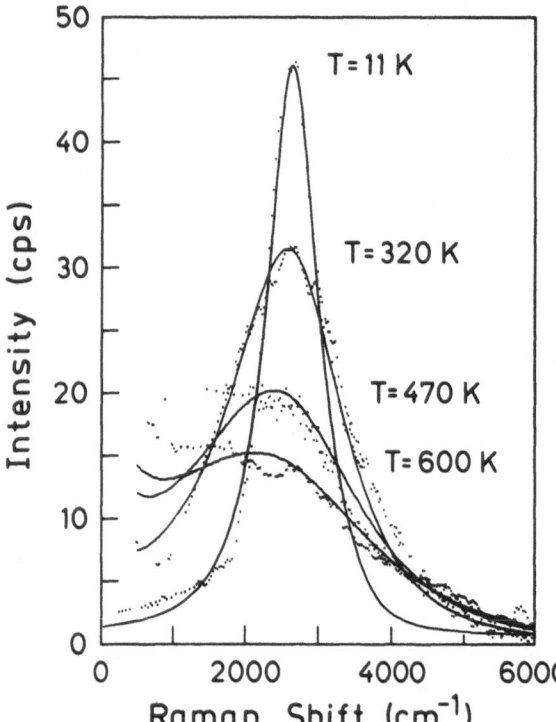

Fig. 6.22. Temperature-dependent spectra of the two-magnon Raman peak in $EuBa_2Cu_3O_6$ for temperatures larger than T_N. The solid lines are fits taking into account spin-lattice interactions. From [6.230]

The two-phonon nature of these peaks, however, has been shown in $YBa_2Cu_3O_6$ by isotopic substitution. The fundamental phonon frequency near 600 cm^{-1} corresponds to IR-active oxygen vibrations, a replacement of ^{16}O by ^{18}O should thus show a corresponding shift in a two-phonon process. The feature near 1200 cm^{-1} has indeed been found at an energy reduced by $\Delta\omega/\omega = (5.1 \pm 0.8)\%$ in the material substituted with the heavy oxygen isotope. The slight reduction from the square-root-of-the-masses value predicted theoretically (6.1%) can be attributed to some admixture of the center-of-mass motion required for odd-parity or IR-active modes [6.234].

6.5 Phonon Self-Energies and the Superconducting Gap

The determination of the superconducting band gap has been one of the most interesting problems in analyzing the high-T_c superconductors. Traditional methods for the observation of the gap 2Δ are far-infrared or microwave absorption [6.235], tunneling [6.236], ultrasonic attenuation [6.237], nuclear magnetic resonance [6.238], or nuclear quadrupole resonance, and thermodynamic methods like specific heat [6.238, 239] and thermal conductivity [6.240]. In most conventional superconductors all these methods yield values $2\Delta/kT_c$ near the BCS prediction of 3.52. They have been applied to the high-temperature superconductors as well, but rather than a single value for a gap, a large range of gap-to-T_c ratios has been found. The spread in values is due partly to the difference in methods, and partly to difficulties in interpreting the experimental results. The jump in electronic specific heat, e.g., is masked by the phonon contribution to the specific heat [6.241, 242], and the far-infrared reflectivity measurements have suffered from the need to assume a single or several different gaps when fitting the data to theoretical expressions [6.243]. Ultrasonic measurements have remained inconclusive so far because large enough single crystals have not yet become available. Pressed ceramics do not give the required consistency in the experimental results to deduce $2\Delta/kT_c$ [6.244]. Magnetic-resonance experiments have yielded two distinct gaps ($2\Delta/kT_c$ = 4.3 and 9.3) in the high-T_c materials which have been associated with the chains and planes in $YBa_2Cu_3O_{7-\delta}$ [6.245–247]. Recent tunneling data, which are very sensitive and accurate to the surface properties because of the very short coherence length in the superconductor (typically $\xi_c \approx$ 2–4 Å, $\xi_{ab} \approx$ 12–30 Å), also suggest two gap values but with $2\Delta/kT_c$ = 1 and 5 for the c-direction and the CuO_2 planes, respectively [6.248]. Summarizing most serious measurements of the gap value in high-T_c materials, a range $0 \leqq 2\Delta/kT_c \leqq 8$ is found using the above methods.

In this section we show how light scattering by phonons provides an extremely sensitive and accurate way of finding a superconducting gap. The method has the advantages that it is probing a significant portion of the bulk (penetration depth is $\approx$ 1000 Å), that the interpretation is unambiguous, and that important qualitative features can be read off the data directly.

6.5.1 Phonon Self-Energies in Superconductors

Anomalous phonon frequency and linewidth dependences on temperature are known from conventional superconductors. From neutron scattering in Nb *Shapiro* et al. [6.249] have determined that both the frequency and linewidth of an acoustical phonon (away from the zone center) change below T_c. The magnitude of the effects is $\Delta\omega/\omega$ = 1.9% and $\Delta\Gamma/\omega$ = 0.8% (Γ = half-width at half maximum). In the high-T_c material $YBa_2Cu_3O_{7-\delta}$ effects of similar magnitude are found for some of the phonons by means of Raman and IR spectroscopy. The mode at 340 cm^{-1}, e.g., shows a softening of $\Delta\omega/\omega$ = 2.0% and a broadening of $\Delta\Gamma/\omega$ = 1.2% (at the zone center). Both anomalies are comparable in strength and have been investigated in quite some detail in the high-T_c materials.

The phonon *softening* was first observed in ceramic $YBa_2Cu_3O_{7-\delta}$ by *Macfarlane* et al. [6.250] and, in single crystals, by *Thomsen* et al. [6.180].

Other groups have confirmed these results [6.181, 251, 252]; there is no doubt about their existence. In Fig. 6.23 from *Thomsen* et al. [6.253] we show the temperature dependence of the B_{1g}-like phonon in $YBa_2Cu_3O_{7-\delta}$ and $TmBa_2Cu_3O_{7-\delta}$. Note the change in eigenfrequency of this mode between 90 K and 10 K, which amounts to $\Delta\omega/\omega = -1.93\%$ (R = Y) and -2.22% (R = Tm). Also shown is the effect of anomalous *hardening* which has been discovered only recently [6.253] for the corresponding A_{1g}-like phonon (O_{II}, O_{III} in phase motion), probably because as a hardening it is more difficult to distinguish from the hardening due to lattice-anharmonicity effects. The anomaly is clearly seen to set in simultaneously with the softening of the B_{1g}-like mode. The other Raman-active A_g modes do not show anomalous frequency shifts of this magnitude.

One of the interesting questions in connection with this phonon softening is whether it bears any relation to the phenomenon of superconductivity. In principle, structural changes may produce such an effect as well. From the structural data available, however, it can be concluded

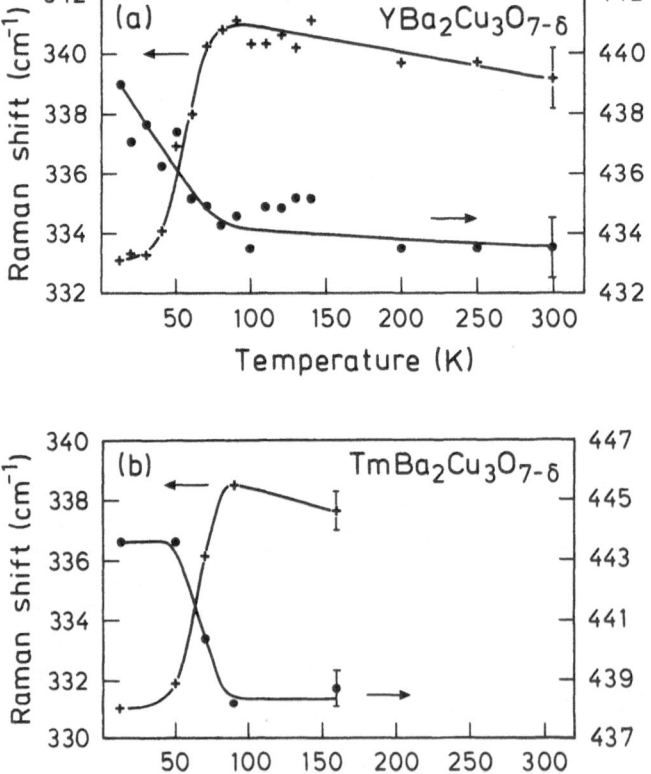

Fig. 6.23a, b. Temperature dependence of two phonon frequencies in ceramic (a) $YBa_2Cu_3O_{7-\delta}$ and (b) $TmBa_2Cu_3O_{7-\delta}$. The phonon at $340\,cm^{-1}$ is seen to soften while the one at $440\,cm^{-1}$ hardens by a similar amount in both materials. Fits to a Fano line shape were used in the evaluation to separate linewidth effects from actual bare frequency changes for the B_{1g}-like phonon. From [6.253]

that a large-enough change is not observed in the internal structural parameters to produce a 2% change in phonon frequency of just one of the observed eigenmodes [6.80, 254]. Experiments on nonsuperconducting substances like $PrBa_2Cu_3O_7$ and $YBa_2Cu_3O_6$ showed no softening [6.217, 255], further suggesting a link to the superconducting transition. The final, convincing proof of this connection was, however, given by an experiment performed on $YBa_2Cu_3O_{7-\delta}$ in a magnetic field. *Ruf* et al. [6.256] were able to show that when a magnetic field is applied, the softening temperature shifts by just the same amount as does the upper critical field. This means that when a superconductor at a given temperature is made normal by a magnetic field, also the anomalous softening reverses. These experiments have been performed with $H \parallel c$, the "easy" direction; the same effect has not yet been observed for $H \parallel a, b$, mainly because the magnetic fields available have not been large enough to quench superconductivity a few degrees away from T_c. It should be noted that the relation $dT_s/dH = dT_c/dH_{c2}$ (T_s is the temperature where the softening occurs) allows a flux-motion-free determination of the temperature dependence of the upper critical field and thus of the coherence length in the superconductor.

Given that electron–phonon interaction is the underlying reason for the softening we shall list some of the properties of this phenomenon:

1. It exists for all rare-earth elements investigated which make $RBa_2Cu_3O_7$ a superconductor;
2. It does not always have the same magnitude [6.253, 257];
3. It depends very sensitively on oxygen content [6.258];
4. IR-active modes near 300 cm^{-1} also show frequency decreases near T_c [6.163, 259];
5. A phenomenon related to the frequency decrease at T_c of a phonon is the observed change in its linewidth.

A number of authors have shown that a broadening of the 340 cm^{-1} Raman mode of $YBa_2Cu_3O_{7-\delta}$ occurs below T_c [6.149, 181, 251, 260]. The sensitivity to oxygen content of the softening has initially been thought to be the reason for variations in $\Delta\omega$. We shall show below, however, that this is not so. Recently, it has become possible to understand qualitatively and quantitatively softening, hardening and broadening of some of the phonons in a unified picture [6.253, 260] in terms of a strong-coupling theory of the superconducting pairing.

6.5.2 Theoretical Evaluation of Phonon Self-Energies

Attempts to explain theoretically the softening of a Raman-active phonon in a superconductor below T_c have been made within the BCS theory by *Klein* and *Dierker* [6.261] and within the Eliashberg theory by *Zeyher* and *Zwicknagl* [6.262, 263]. The derivations are based on an evaluation of the change in the complex electron-phonon self-energy of a phonon when a material becomes superconducting: the real part of the self-energy corresponds to the frequency softening and the imaginary part to the broadening as specified below. The change in this self-energy Σ is [6.253]

$$\Delta\Sigma_\nu/\omega_\nu = \lambda_\nu f(\omega_\nu/2\Delta) \tag{6.23}$$

where ν labels the ν-th mode, λ_ν is its contribution to the electron-phonon coupling constant λ and f a universal function of $\tilde{\omega} \equiv \omega_\nu/2\Delta$. In the BCS limit (weak coupling) f is given by

$$f(\tilde{\omega}) \equiv \begin{cases} -2u/\sin 2u & \text{for} \quad \sin u \equiv \tilde{\omega} < 1 \\ (2v - i\pi)/\sinh 2v & \text{for} \quad \cosh v \equiv \tilde{\omega} > 1 \end{cases} \tag{6.24}$$

The coupling constant for the ν-th mode is defined as

$$\lambda_\nu \equiv 2N(0)\,\langle|g_{\nu kk}|^2\rangle_{\text{FS}}/\omega_\nu \tag{6.25}$$

in terms of the density of states per spin $N(0)$ and the average of the electron-phonon matrix element squared $|g|^2$ over the Fermi surface (FS) in the normal-state. The asymptotic properties for large and small $\tilde{\omega}$ and near the singularity at $\tilde{\omega} = 1$ have been described in [6.253] and may also be seen in the plots of the function $f(\omega)$ in [6.262]. They may also be easily derived from the form for f given in (6.24). The essential feature is the singularity in both real and imaginary parts for a phonon of energy $\omega_\nu = 2\Delta$ and a region of predicted hardening ($\tilde{\omega} > 1$).

In the strong-coupling calculation these singularities are broadened; the divergence is replaced by an extremum which is no longer accurately at but slightly higher than $\tilde{\omega} = 1$. Plots of the strong-coupling curves may be seen in [6.263]. They depend explicitly on temperature and also on a phenomenological impurity scattering rate τ^{-1}. We have reproduced in Fig. 6.24 the curves of *Zeyher* and *Zwicknagl* for $\tau^{-1} = 0$ at different temperatures. The real part of the self-energy (frequency shifts) is given in (a) the imaginary part (broadening) in (b). To obtain experimental values, the relation $\Delta\Sigma = (\omega_\nu\lambda_\nu/2)\,\Pi/N$ is useful.

In order to explain consistently the high transition temperatures on the basis of electron-phonon interaction *Zeyher* and *Zwicknagl* were led to assume a total coupling constant $\lambda_{\text{tot}} = 2.9$, i.e. an average of 0.08 per each of the 36 optical modes. The observed softenings calculated from (6.23) were about one order of magnitude too large without the assumption of impurity scattering, and still about three times too large for reasonable scattering rates. Nevertheless, the theory qualitatively explained the observed softening.

Fig. 6.24a, b. Real (a) and imaginary parts (b) of the polarizability obtained by *Zeyher* and *Zwicknagl* for a strong-coupling superconductor. Phonon-frequency changes at T_c are described by (a) and broadenings by (b). The impurity parameter in these curves was $\tau^{-1} = 0$ (no impurity scattering). From [6.263] with permission of the authors

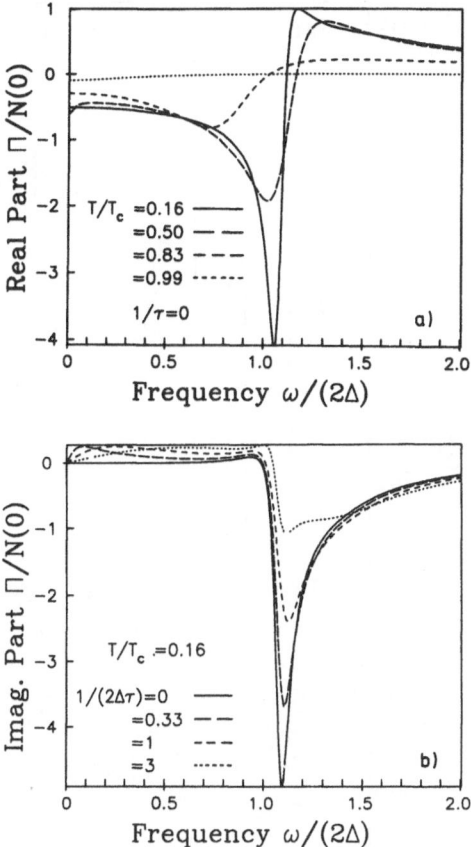

6.5.3 Experimental Evidence for a Sharp Gap

Thomsen et al. and *Friedl* et al. have shown experimentally that the observed phonon frequency shifts [6.253] and broadenings [6.260] near T_c in $RBa_2Cu_3O_7$ fall near the universal curves calculated from strong-coupling theory. They exploited the property that due to changes in the ionic radius (r_I) the B_{1g}-like mode shifts in frequency approximately linearly with r_I [6.163, 165]. We have reproduced data of *Cardona* et al. [6.163] in Fig. 6.25. By replacing systematically the rare-earth atom in the $RBa_2Cu_3O_7$ structure, and under the assumption of a constant gap in the various materials, it was possible to map out the self-energies in the vicinity of such a gap of the superconducting vs. normal state. The authors also used the fact that the phonon frequency is lowered by $\sim 6\%$ upon isotopic replacement of ^{16}O by ^{18}O [6.11].

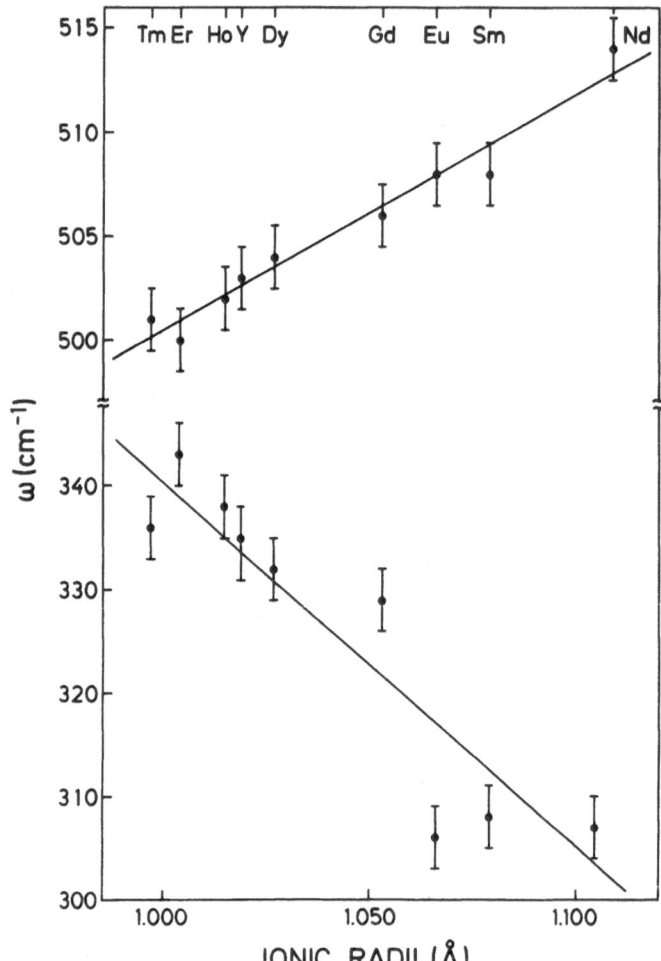

Fig. 6.25. The frequency of the B_{1g}-like mode at 340 cm^{-1} and the A_{1g}-like mode at 500 cm^{-1} for various rare-earth atoms in place of R at 300 K. Note that some deviations from the straight lines are outside the experimental error. From [6.163]

The results of measurements of the frequency changes are shown in Fig. 6.26 and are remarkable in a number of respects.

1. We see that the B_{1g}-type mode in the different rare-earth super-conductors shows a softening which tends to be stronger for higher eigenfrequencies;

2. The quantitative agreement with the theoretical curve calculated from strong-coupling theory and a coupling constant of $\lambda = 0.02$ is excellent, considering that no parameters have been explicitly adjusted; (the impurity scattering rate τ^{-1} was set equal to 2Δ, a typical value [6.29, 264]);

3. At higher frequencies the theory predicts a range of anomalous hardening which is confirmed experimentally for the 440 cm^{-1} mode.

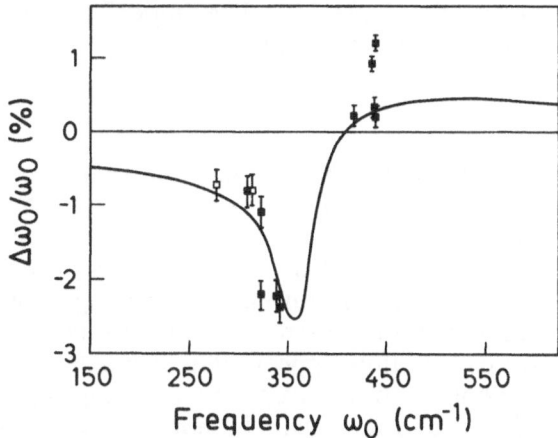

Fig. 6.26. Experimental self-energies (real part) measured for the $RBa_2Cu_3O_{7-\delta}$ super-conductors. The phonon frequency was varied by substituting different rare-earth atoms in place of R. Plotted is the relative shift in per cent vs. the phonon frequency between $T = 90$ K and 10 K for R = Eu, Sm, Tm, $Y^{16}O$, and $Y^{18}O$ (full squares). The open squares represent infrared-active phonons which also show a softening. The parameters of the solid curve, which was taken from a strong-coupling theory, are $2\Delta = 333$ cm^{-1}, impurity scattering rate $\tau^{-1} = 2\Delta$, and $\lambda = 0.02$ per mode. From [6.253]

Two conclusions can be drawn from these data without further analysis:

1. There is one well-defined superconducting gap in the region of 310 to about 420 cm^{-1}, which the phonons couple to independently of the detailed shape of the theoretical curve;
2. The coupling constants of these two modes are obtained reasonably accurately from frozen-phonon one-electron band-structure calculations (to be described below).

One of the consequences of the numerical values of these coupling constants is the broken self-consistency with the average coupling constant required to explain T_c. *Thomsen* et al. [6.253] still used for the superconducting ground state a strong-coupling s-paired Eliashberg state, where the pairing is induced by unspecified *intermediate bosons*, possibly phonons other than the Raman and IR-ones (this would lead, however, to difficulties with the isotope effect [6.13]).

The coupling constants were calculated *ab initio* in the local-density approximation (LDA) for the two modes at 340 cm^{-1} and 440 cm^{-1} in the $YBa_2Cu_3O_7$ system using a full-potential version of the LMTO method. The calculations consisted of 29 different self-consistent band-structure calculations corresponding to different A_g displacement patterns. It is important to note that the density of states was unrenormalized, i.e. a $(1 + \lambda)$ factor was not included in the mass. For more details the reader is referred to [6.98, 253]. The "mode" coupling constants (i.e. per phonon band) thus obtained were

$$\lambda_{340} = 0.02 \quad \text{and} \quad \lambda_{440} = 0.01 \tag{6.26}$$

These mode-specific coupling constants are significantly smaller than the average ones required by the high T_c if phonons are assumed to be the "intermediate bosons". This, again, does not exclude the possibility that other modes or other points in k-space have again a significantly higher coupling constant. Given the coupling constants (6.26), the *quantitative* comparison of the strong-coupling theory with the experiment becomes excellent. It thus appears that through a relatively weak coupling these phonons test or scan a strong-coupling superconducting energy gap. It is important to note that the non-zero values of λ rely on the buckling of the CuO_2 planes. For flat planes, as in the Tl superconductors, λ is calculated to be nearly zero for these modes.

Another important result of the calculation is the degree of mixing of the eigenvectors of the B_{1g}-type phonon (340 cm^{-1}) and A_{1g}-type phonon at 440 cm^{-1}. It is about 80% B_{1g} to 20% A_{1g} for the 340 cm^{-1} mode [6.176, 253]. Furthermore, there is a 20% admixture of the apical oxygen (O_{IV}) to the 440 cm^{-1} vibration [6.198]. At low energies, the calculations predict a significant mixing between the Cu and Ba modes which has, however, been shown not to exist by ^{65}Cu-^{63}Cu isotopic substitution experiments [6.265]. The degrees of mixing may also be inferred from the resonant behavior of the scattering efficiencies (Fig. 6.13c). There the decoupled nature of the Cu and Ba modes and the mixed behavior of the oxygen-related modes becomes apparent in the comparison of the experimental data to predictions based on the band structure [6.176].

The imaginary part of the self-energy shows, even in the strong-coupling calculations, a steep onset of broadening near T_c [6.263]. This reflects the fact that also in the strong-coupling case, a Cooper pair may not be broken up by a phonon with an energy less than the binding energy of the pair (note that in the normal-state phase free electrons do not contribute to the broadening of phonons with $k \simeq 0$ either). *Friedl* et al. [6.260] have investigated the imaginary part of $\Delta\Sigma$ for the series of $RBa_2Cu_3O_7$ superconductors and again found excellent agreement with theory. Similarly as for the real part, they tuned the B_{1g}-like phonon through the gap region by systematically replacing rare-earth atoms for R. In Fig. 6.27 we reproduce their findings for two characteristic phonon energies: one below the gap (open circles) and one above the gap (black dots). Although the phonon frequencies are similar on an absolute scale, they show fundamentally different behavior below T_c. The solid lines are fits to an anharmonic decay into two phonons and represent the usual linewidth narrowing due to lattice anharmonicities. In $EuBa_2Cu_3O_{7-\delta}$ the fit is acceptable for the entire temperature range. For $YBa_2Cu_3O_{7-\delta}$ a significant anomalous broadening occurs below T_c which corresponds

Fig. 6.27. Dependence of linewidth on temperature as determined by a fit to a Fano function for $YBa_2Cu_3O_{7-\delta}$ (closed symbols) and $EuBa_2Cu_3O_{7-\delta}$ (open symbols). While the absolute phonon energies of these modes are similar, the different behavior in linewidth below T_c enables us to determine the superconducting gap energy 2Δ. From [6.260]

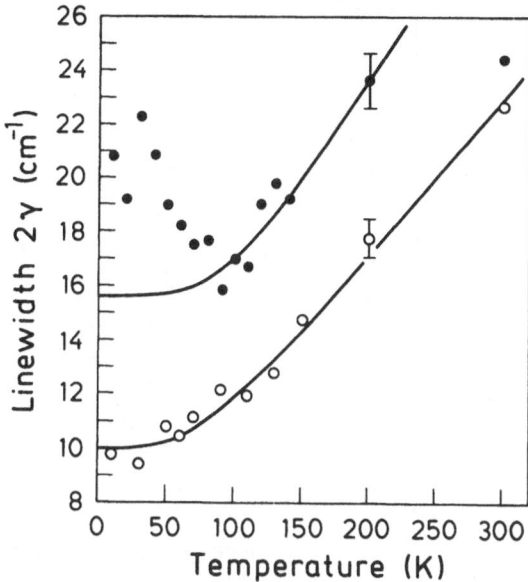

to the self-energy change predicted for a phonon with energy $\omega_v > 2\Delta$. Note that while the B_{1g}-like phonon in $EuBa_2Cu_3O_{7-\delta}$ *does not* broaden, it *does* show a finite softening, in agreement with the theoretical curve of Fig. 6.26.

The results of the linewidth changes below T_c for five different rare-earth superconductors (and one ^{18}O replaced sample) are displayed in Fig. 6.28. The values for the broadening of the phonons between 90 and 10 K are indeed well described by the theoretical curve of *Zeyher* and *Zwicknagl*. The same coupling constants as for the frequency changes were used for the two modes. The superconducting gap is fixed accurately by the onset of broadening and the error in absolute value given by the difference between the point of zero broadening and of smallest detectable finite broadening. Due to the steep onset of broadening near 2Δ (steeper than for the real part) the authors were able to determine

$$2\Delta/kT_c = 4.95 \pm 0.10 \tag{6.27}$$

in the $RBa_2Cu_3O_7$-system [6.260]. We therefore infer that there is only one dominant gap in this energy region which the phonons couple to, since a distribution of gaps would smear out the theoretical dependence of the self-energies considerably and worsen the agreement with experiment. It is of course possible that other gaps exist at energies far from 2Δ as determined here, but the data require that a large stable region with

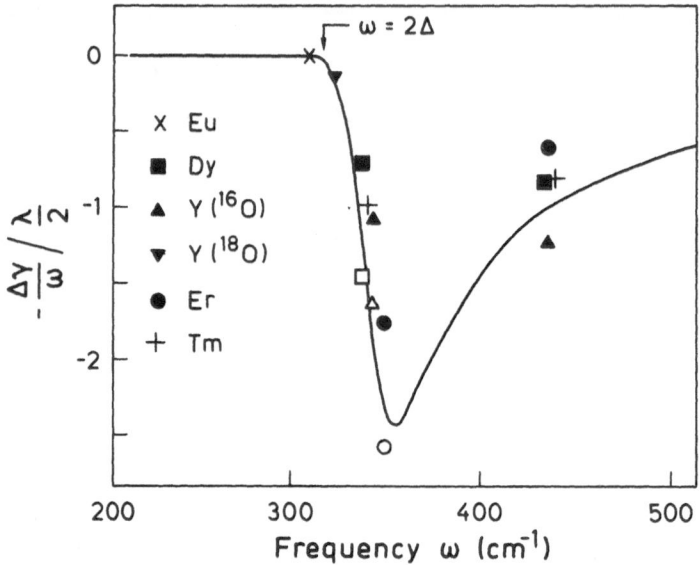

Fig. 6.28. Broadenings of two phonons between 90 K and 10 K in $RBa_2Cu_3O_{7-\delta}$ for various rare-earth atoms R. The data have been normalized to $\lambda_\nu\omega_\nu/2$ which allows quantitative comparison with strong-coupling theory (solid line). The onset of broadening determines the superconducting gap energy (arrow). The open symbols are self-energy corrected shifts as explained in the text.

gap $2\Delta = 4.95kT_c$ exist on the Fermi surface. A couple of remarks are in order concerning Fig. 6.28:

1. The numerical value of the broadening was extracted from Fano fits in the case of the B_{1g}-like mode and from fits to a simple Lorentzian for the 440 cm^{-1} modes;
2. There is a well-known effect in the broadening which does not come from the imaginary part of Σ but rather from changes in the real part of Σ with frequency.

It may be brought into the form [6.266]

$$\Gamma'(\omega) = \frac{\Gamma(\omega)}{1 - [d\Sigma_r/d\omega]_{\omega_\nu}}, \qquad (6.28)$$

were Γ' is the determined linewidth. The correction may be estimated from known data on $d\Sigma_r/d\omega$ and tends to enhance the effects described below 2Δ (note that a reduction of the broadening is, in principle, possible in the case in which $d\Sigma_r/d\omega > 0$). Data points corrected in this fashion are given as open symbols in addition to the uncorrected ones in the figure.

These results constitute independent, convincing evidence for the existence of a sharp gap in the $RBa_2Cu_3O_7$ family of superconductors in this energy region. The conclusion is in quantitative agreement with the tunneling measurements of *Gurvitch* et al. [6.248] mentioned above. There is, however, a problem with the interpretation of the electronic scattering continuum, which suggested the possibility of a distribution of gaps or even a pseudo gap (Sect. 6.3) with nodes at points or lines of the Fermi surface [6.181]. At this time it is not clear how to resolve this controversy, we believe that more work is needed to understand the electronic scattering mechanism in detail.

In this section we have inferred that strong-coupling theory applies to the high-T_c superconductors. The strong-coupling gap is revealed by using as a probe relatively weakly coupling phonons in the adequate energy range. The coupling constants of the phonons, if the same for all phonons as determined experimentally and theoretically in the work cited, are by far not capable of explaining the high transition temperatures in the superconductors. A single, sharp gap value has been determined from the frequency dependence of the phonon softening and broadening at $2\Delta/kT_c = 4.95 \pm 0.10$ in $RBa_2Cu_3O_{7-\delta}$.

6.6 Conclusions

Light scattering has provided a wealth of information on both material properties and physical effects in the high-T_c superconductors. Phonon frequencies, symmetries, assignments to particular atomic displacements, and magnon energies are among the former, while the Fano interferences, the spin-lattice interaction, and the self-energy effects in the $RBa_2Cu_3O_7$-system belong to the latter. Perhaps the most exciting result is the observation of a sharp superconducting gap as described im Sect. 6.5. An important conclusion that can be drawn from the light-scattering experiments is that $RBa_2Cu_3O_7$ is a strong-coupling superconductor. What remains open, though, is the nature of the coupling boson, since the coupling constants of the modes studied here are too small to account for T_c.

A topic not touched upon here is the very high sensitivity of light scattering to the presence of certain foreign phases, in particular such as $BaCuO_2$, R_2BaCuO_5 (the *green phase*), Cu_2O, R_2O_3, $R_2Cu_2O_5$, and BaO. Since most of the foreign phases are insulators, their Raman signals are often larger and may show sharp resonant enhancements. Even a few percent of a foreign phase may be detected easily. The spectra of these phases have been published in the literature [6.57, 267–270]. Together with a Raman microscope with a lateral resolution of a few microns, a spatially resolved, non-destructive sample characterization become possible. It is even feasible to use a Raman instrument in an industrial environment and to employ also as a reference the well-known vibrational

Raman spectra of the substrate material, e.g. Si [6.164]. In addition to the identification of impurity phases one may obtain information about the orientation of, say, a thin film due to the highly anisotropic nature of the vibrational Raman tensors. In the $RBa_2Cu_3O_7$-system in particular, the oxygen content may be determined through its correlation with the Raman frequencies. Another analytic property of vibrational Raman scattering is based on the sensitivity of vibrational energies on the atomic mass of the vibrating species. Through isotopic substitution of ^{16}O by ^{18}O, or natural Cu by ^{65}Cu, related peaks change in energy, or conversely, a change in energy indicates such a substitution. In this way, light scattering has given valuable information for the determination of the isotope shift of T_c [6.9, 11, 234].

Finally we mention a new topic in the optical studies of high-T_c superconductors. Anyon superconductivity, a proposal for the theory of superconductivity, makes the prediction that the superconducting ground state exhibits a violation of parity and time-reversal symmetries. Using, for instance, left and right circularly polarized light *Lyons* et al. [6.271] have found a deviation from a null signal in reflectivity upon cooling from room temperature. For $Bi_2Sr_2CaCu_2O_8$ the deviation occurred near 250 K, for $YBa_2Cu_3O_{7-\delta}$ between 150 and 250 K and increased monotonically to the lowest temperatures measured. *Spielman* et al. [6.272], on the other hand, using a fiber-optic gyroscope measured non-reciprocal phase shifts after transmission through thin films of $YBa_2Cu_3O_{7-\delta}$ and found a null result at all temperatures. In a third experiment of this type *Weber* et al. [6.273] have found again a signal on single crystals of $Bi_2Sr_2CaCu_2O_8$ and $YBa_2Cu_3O_{7-\delta}$; on the Bi compound they were able to measure in transmission, excluding substrate effects as a possible source of error. They also observed that reversing the sign of a magnetic field applied perpendicular to the sample surface could reverse the sign of their signal, which they considered evidence for the time-reversal symmetry breaking in their samples. The onset temperatures of the signals were between 20 and 40 K above the respective T_c's in their samples. In view of the partly disagreeing results and the implications for theoretical work we believe that more work urgently needs to be done in this area.

Acknowledgement. I am grateful to all my colleagues, who have made the collaborative effort on the high-T_c materials possible. I am indebted in particular to M. Cardona for his continuous support in this research, and a critical reading of this manuscript. I have benefitted from many valuable discussions with him on the topic. I thank R. Liu, B. Friedl, T. Ruf, and E. T. Heyen, for their contribution and for obtaining most of the experimental results, O. K. Andersen, W. Kress and R. Zeyher for discussions on theoretical issues, Hj. Mattausch, P. Murugaraj, B. Nick, E. Schönherr, E. Morán, and M. A. Alario-Franco for the preparation of many samples, R. Kremer, L. Viczian, and W. Bauhofer for the sample characterization, H. Hirt, M. Siemers, and P. Wurster for expert technical assistance and S. Birtel for her endurance and expertise in preparing the manuscript. This work was supported financially in part by the Bundesminister für Forschung und Technologie and the European Community.

References

6.1   J.G. Bednorz, K. A. Müller: Z. Phys. B **64**, 189 (1986)
6.2   M.A. Subramanian, C.C. Torardi, J. Gopalakrishnan, P.L. Gai, J.C. Calabrese, T.R. Askew, R.B. Flippen, S.W. Sleight: Science **242**, 249 (1988)
6.3   S.S.P. Parkin, V.Y. Lee, E.M. Engler, A.I. Nazzal, T.C. Huang, G. Gorman, R. Savoy, R. Beyers, S.J. La Placa: Phys. Rev. Lett. **60**, 2539 (1988)
6.4   C.Y. Huang: Int. J. Mod. Phys. B **2**, 355 (1988)
6.5   J. Bardeen, L.N. Cooper, J.R. Schrieffer: Phys. Rev. **106**, 162 (1957) and Phys. Rev. **108**, 1175 (1957)
6.6   R.H. Koch, C.P. Umbach, G.J. Clark, P. Chaudhari, R.B. Laibowitz: Appl. Phys. Lett. **51**, 200 (1987)
6.7   C.E. Gough, M.S. Cochough, E. Morgen, R.G. Jordan, M. Keene, C.M. Muirhead, A.I.M. Rae, N. Thomas, J.S. Abell, S. Sutton: Nature **326**, 855 (1987)
6.8   J. Niemeyer, M.R. Dietrich, C. Politis: Z. Phys. B **67**, 155 (1987)
6.9   B. Batlogg, R.J. Cava, A. Jayaraman, R.B. van Dover, G.A. Kourouklis, S. Sunshine, D.W. Murphy, L.W. Rupp, H.S. Chen, A. White, K.T. Short, A.M. Mujsce, E.A. Rietman: Phys. Rev. Lett. **58**, 2333 (1987)
6.10 K.J. Leary, H.-C. zur Loye, S.W. Keller, T.A. Faltens, W.K. Ham, J.N. Michaels, A.M. Stacy: Phys. Rev. Lett. **59**, 1236 (1987)
6.11 C. Thomsen, Hj. Mattausch, M. Bauer, W. Bauhofer, R. Lui, L. Genzel, M. Cardona: Solid State Commun. **67**, 1069 (1988), and M. Cardona, R. Liu, C. Thomsen, W. Kress, E. Schönherr, M. Bauer, L. Genzel, W. König, Solid State Commun. **67**, 789 (1988)
6.12 H. Katayama-Yoshida, T. Hirooka, A. Oyamada, Y. Okabe, T. Takahashi, T. Sasaki, A. Ochini, T. Suzuki, A.J. Mascarenhas, J.I. Pankove, T.F. Ciszek, S.K. Deb, R.B. Goldfarb, Y. Li: Physica C **156**, 481 (1988)
6.13 P.B. Allen: Nature **339**, 428 (1989)
6.14 R.H. Parmenter: Phys. Rev. Lett. **59**, 923 (1987)
6.15 C.M. Varma, S. Schmitt-Rink, E. Abrahams: Solid State Commun. **62**, 681 (1987)
6.16 K. Kamarás, C.D. Porter, M.G. Doss, S.L. Herr, D.B. Tanner, D.A. Bonn, J.E. Greedan, A.H. O'Reilly, C.V. Stager, T. Timusk: Phys. Rev. Lett. **59**, 919 (1987)
6.17 D.L. Cox, M. Jarrell, C. Jayaprakash, H.R. Krishna-murthy, J. Deisz: Phys. Rev. Lett. **62**, 2188 (1989)
6.18 A.S. Alexandrov: Physica C **158**, 337 (1989)
6.19 P. Prelovsek, T.M. Rice, F.C. Zhang: J. Phys. C **20**, L229 (1987)
6.20 D. Emin: Phys. Rev. Lett. **62**, 337 (1989)
6.21 K.S. Bedell, D. Pines: Phys. Rev. B **37**, 3730 (1988)
6.22 D.J. Scalapino, E. Loh, J. Hirsch: Phys. Rev. B **34**, 8190 (1986)
6.23 J.R. Schrieffer, X.-G. Wen, S.-C. Zhang: Phys. Rev. Lett. **60**, 944 (1988)
6.24 A. Kampf, J.R. Schrieffer: Phys. Rev. B **41**, 6399 (1990)
6.25 P.W. Anderson: Science **235**, 1196 (1987)
6.26 R.B. Laughlin: Science **242**, 525 (1988)
6.27 C. Thomsen, M. Cardona: In Ref. [6.28], p. 409ff
6.28 *High-Temperature Superconductors I and II*, ed. by D.M. Ginsberg (World Scientific, Singapore 1989, 1990)
6.29 T. Timusk, D.B. Tanner: In Ref. [6.28], Vol. I, p. 339
6.30 R.J. Birgeneau, G. Shirane: In Ref. [6.28], Vol. I, P. 151
6.31 M.B. Salamon: In Ref. [6.28], Vol. I, p. 39ff
6.32 M.V. Klein: In *Light Scattering in Solids I*, ed. by M. Cardona, Topics Appl. Phys. **8** (Springer, Berlin, Heidelberg 1983), p. 169ff
6.33 G. Abstreiter, M. Cardona, A. Pinczuk: In *Light Scattering in Solids IV*, Ed. by M. Cardona, G. Güntherodt, Topics Appl. Phys. **54** (Springer, Berlin, Heidelberg 1984), p. 119ff
6.34 W.G. Nilsen, J.G. Skinner: J. Chem. Phys. **48**, 2240 (1968)

6.35 R.J. Birgeneau, C.Y. Chen, D.R. Crabbe, H.P. Jenssen, M.A. Kastner, C.J. Peters, P.J. Picone, T. Thio, T.R. Thurston, H.L. Tuller: Phys. Rev. Lett. **59**, 1329 (1987)

6.36 See, for instance M.E. Lines, A.M. Glass: *Principles and Applications of Ferroelectrics and Related Materials* (Oxford University Press, Oxford 1977)

6.37 R.J. Cava, B. Batlogg, J.J. Krajewski, R. Farrow, L.W. Rupp, Jr., A.E. White, K. Short, W. Peck, T. Kometani: Nature **332**, 814 (1988)

6.38 B. Batlogg, R.W. Rupp, Jr., A.M. Mujsce, J.J. Krajewski, J.P. Rameika, W.F. Reck, Jr., A.S. Cooper, G.P. Espinosa: Phys. Rev. Lett. **61**, 1670 (1988)

6.39 D.G. Hinks, B. Dabrowski, D.R. Richards, J.D. Jorgensen, S. Pei, J.F. Zasadzinski: Physica C **162–164**, 1405 (1989)

6.40 J.F. Zasadzinski, N. Trahlshawala, D.G. Hinks, B. Dabrowski, A.W. Mitchell, D.R. Richards: Physica C **158**, 519 (1989)

6.41 M. Shirai, N. Suzuki, K. Motizuki: In *Phonons 89*, ed. by S. Hunklinger, W. Ludwig, G. Weiss (World Scientific, Singapore 1990) p. 286

6.42 M. Shirai, N. Suzuki, K. Motizuki: J. Phys. C **1**, 2929 (1989)

6.43 Landoldt-Börnstein, Numerical Data and Functional Relationships in Science and Technology, Ed. by K.-H. Hellwege, Vo. III, 16a (Springer, Berlin Heidelberg, 1981) p. 317

6.44 C.S. Koonce, M.L. Cohen, J.F. Schooley, W.R. Hosler, E.R. Pfeiffer: Phys. Rev. **163**, 380 (1967)

6.45 B. Batlogg: In *High Temperature Superconductivity*, The Los Alamos Symposium – 1989, Ed. by K. Bedell, D. Coffey, D. Meltzer, D. Pines, J.R. Schrieffer, (Addison-Wesley, 1990), p. 37

6.46 M. Garriga, J. Humliček, J. Barth, R.L. Johnson, N. Cardona: J. Opt. Soc. Am. B **6**, 470 (1989)

6.47 M.K. Kelly, P. Bardoux, J.-M. Tarascon, D.E. Aspnes, W.A. Bonner, P.A. Morris: Phys. Rev. B **38**, 870 (1988)

6.48 H.P. Geserich, B. Koch, G. Scheiber, J. Geerk, H.C. Li, G. Linker, W. Weber, W. Assmus: Physica C **153–155**, 661 (1988)

6.49 H.B. Radousky, K.F. McCarty, J.L. Peng, R.N. Shelton: Phys. Rev. B **39**, 12383 (1989)

6.50 H. Tagaki, T. Ido, S. Ishibashi, M. Uota, S. Uchida, Y. Tokura: Phys. Rev. B **40**, 2254 (1989)

6.51 H. Tagaki, S. Uchida, Y. Tokura: Phys. Rev. Lett. **62**, 1197 (1989)

6.52 R.J. Cava, B. Batlogg, C.H. Chen, E.A. Rietman, S.M. Zahurek, D. Werder: Nature **329**, 423 (1987)

6.53 J. Maier, P. Murugaraj, C. Lange, A. Rabenau: Angew. Chem. Int. Ed. Engl. **27**, 980 (1988)

6.54 NOA 2003, Leyboldt, Köln

6.55 Y. Tokura, H. Tagaki, S. Uchida: Nature **337**, 345 (1989)

6.56 C.C. Torardi, M.A. Subramanian, J.C. Calabrese, J. Gopalakrishnan, E.M. McCannon, K.J. Morrissey, T.R. Askew, R.B. Flippen, U. Chowdry, A.W. Sleight: Phys. Rev. B **38**, 225 (1988)

6.57 Z.V. Popović, C. Thomsen, M. Cardona, R. Liu, G. Stanisic, R. Kremer, W. König: Solid State Commun. **66**, 965 (1988)

6.58 Y. Kubo, Y. Shimakawa, T. Manako, T. Satoh, S. Iijima, T. Ichihashi, H. Igarashi: Physica C **162–164**, 991 (1989)

6.59 D.R. Harshman, L.F. Schneemeyer, J.V. Waszczak, G. Aeppli, R.J. Cava, B. Batlogg, L.W. Rupp, E.J. Ansaldo, D.L. Williams: Phys. Rev. B **39**, 851 (1989)

6.60 D.W. Cooke, R.L. Hutson, R.S. Kwok, M. Maez, H. Rempp, M.E. Schillaci, J.L. Smith, J.O. Willis, R.L. Lichti, K.C.B. Chan, C. Boekewa, S.P. Weathersby, J. Oostens: Phys. Rev. B **39**, 2748 (1989)

6.61 U. Welp, W.K. Kwok, G.W. Crabtree, K.G. Vandervoort, J.Z. Liu: Phys. Rev. Lett. **62**, 1908 (1989)

6.62 L. Krusin-Elbaum, A.P. Malozemoff, Y. Yeshurun, D.C. Cronemeyer, F. Holtzberg: Phys. Rev. B **39**, 2936 (1989)

6.63 P.L. Gammel, D.J. Bishop, G.J. Dolan, J.R. Kwo, C.A. Murray, L.F. Schneemeyer, J.V. Waszczak: Phys. Rev. Lett. **59**, 2592 (1987)
6.64 G.J. Dolan, F. Holtzberg, C. Feild, T.R. Dinger: Phys. Rev. Lett. **62**, 2184 (1989)
6.65 D.R. Nelson: Physica C **162–164**, 1156 (1989)
6.66 E.H. Brandt: Proceedings of the NASA-Conference, Advances in Materials Science and Applications of High Temperature Superconductors, April 2–6, 1990, Greenbelt, Md., USA, to be published
6.67 J. Lensink, C.F.J. Flipse, J. Roobeek, R. Griessen, B. Dam: Physica C **162–164**, 663 (1989)
6.68 E.H. Brandt: Z. Phys. B **80**, 167 (1990)
6.69 J.D. Hettinger, A.G. Swanson, W.J. Skocpol, J.S. Brooks, J.M. Graybeal, P.M. Mankiewich, R.E. Howard, B.L. Straughn, E.G. Burkhardt: Phys. Rev. Lett. **62**, 2044 (1989)
6.70 I.A. Campbell, L. Fruchter, R. Canabel: Phys. Rev. Lett. **64**, 1561 (1990)
6.71 D.C. Bullock: Physica C **162–164**, 331 (1989)
6.72 P.M. Mankiewich, J.H. Scofield, W.J. Skocpol, R.E. Howard, A.H. Dayem, E. Good: Appl. Phys. Lett. **51**, 1753 (1987)
6.73 C. Neumann, P. Ziemann, J. Geerk, X.X. Xi: Physica C **162–164**, 321 (1989)
6.74 E.H. Brandt: Science **243**, 349 (1989)
6.75 J.M. Tranquada, W.J.L. Buyers, H. Chon, T.E. Mason, M. Sato, S. Shamoto, G. Shirane: Phys. Rev. Lett. **40**, 800 (1990)
6.76 D.E. Farrell, J.P. Rice, D.M. Ginsberg, J.Z. Liu: Phys. Rev. Lett. **64**, 1573 (1990)
6.77 R.M. Hazen: In Ref. [6.28], Vol. II, p. 121
6.78 Ref. [6.43], Vol 7e, (1976) p. 134
6.79 Y. Shimakawa, Y. Kubo, T. Manako, Y. Nakabayashi, H. Igarashi: Physica C **156**, 97 (1988)
6.80 J.D. Jorgensen, B.W. Veal, A.P. Paulikas, L.J. Nowicki, G.W. Crabtree, H. Claus, W.K. Kwok: Phys. Rev. B **41**, 1863 (1990)
6.81 A. Fuertes, X. Obradors, J.M. Navarro, P. Gómez-Romero, N. Casañ-Pastor, F. Pérez, J. Fontcuberta, C. Miravitlles, J. Rodríguez-Carvajal, B. Martínez: Physica C **170**, 153 (1990)
6.82 R.J. Cava; B. Batlogg, R.B. Dover, J.J. Krajewski, J.V. Waszczak, R.M. Fleming, W.F. Peck, Jr., L.W. Rupp, Jr., P. Marsh, A.C.W.P. James, L.F. Schneemeyer, Nature (London) **345**, 602 (1990)
6.83 H.G. von Schnering, L. Walz, M. Schwarz, W. Becker, M. Hartweg, T. Popp, B. Hettich, P. Müller, G. Kämpf: Angew. Chem. Int. Ed. Engl. **27**, 374 (1988)
6.84 P. Bordet, J.J. Capponi, C. Chaillout, J. Chenavas, A.W. Hewat, E.A. Hewat, J.L. Hodeau, M. Marezio, J.L. Tholence, D. Tranquada: Physica C **156**, 189 (1988)
6.85 M. Cardona, C. Thomsen, R. Liu, H.G. von Schnering, M. Hartweg, Y.F. Yan, Z.X. Zhao: Solid State Commun. **66**, 1225 (1988)
6.86 M. Martin et al.: unpublished data
6.87 F.E. Bates, J.E. Eldrigde: Solid State Commun. **64**, 1435 (1987)
6.88 F.E. Bates: Phys. Rev. B **39**, 322 (1989)
6.89 W. Kress, U. Schröder, J. Prade, A.D. Kulkarni, F.W. de Wette: Phys. Rev. B **38**, 2906 (1988)
6.90 R. Liu, C. Thomsen, W. Kress, M. Cardona, B. Gegenheimer, F.W. de Wette, J. Prade, A.D. Kulkarni, U. Schröder: Phys. Rev. B **37**, 7971 (1988)
6.91 M. Mostoller, J. Zhang, A.M. Rao, P.C. Eklund: Phys. Rev. B **41**, 6488 (1990)
6.92 J.J. Rhyne, D.A. Neumann, J.A. Gotaas, F. Beech, L. Toth, S. Lawrence, S. Wolf, M. Osofsky, D.U. Gubser: Phys. Rev. B **36**, 2294 (1987)
6.93 F.W. de Wette, A.D. Kulkarni, J. Prade, U. Schröder, W. Kress: Phys. Rev. B **43**, 5451 (1991)
6.94 A.D. Kulkarni, F.W. de Wette, J. Prade, U. Schröder, W. Kress: Phys. Rev. B **41**, 6409 (1990)
6.95 W.E. Pickett: Rev. Mod. Phys. **61**, 499 (1989)
6.96 R.E. Cohen, W.E. Pickett, H. Krakauer: Phys. Rev. Lett. **62**, 831 (1989)
6.97 R.E. Cohen, W.E. Pickett, H. Krakauer: Phys. Rev. Lett. **64**, 2575 (1990)

6.98   C.O. Rodriguez, A.I. Lichtenstein, I.I. Mazin, O. Jepsen, O.K. Andersen, M. Methfessel: Phys. Rev. B **42**, 2692 (1990)
6.99   G. Burns, F.H. Dacol, M.W. Schafer: Solid State Commun. **62**, 687 (1987)
6.100 W.H. Weber, C.R. Peters, E.M. Logothetis: J. Opt. Soc. Am. **6**, 455 (1989)
6.101 H. Kuzmany, E. Faulques, M. Matus, S. Pekker: In *Status of High Temperature Superconductors, Advances in Research and Application*, Vol. 3, Ed. A. Narlikar (Nova, New York 1989) p. 299
6.102 V.A. Maroni and J.R. Ferraro: In *Practical FT-IR Spectroscopy: Industrial and Laboratory Chemical Analysis*, ed. by J.R. Ferraro, K. Kristinan (Academic, San Diego 1989)
6.103 J.R. Ferraro, V.A. Maroni: Appl. Spectrosc. **44**, 351 (1990)
6.104 R. Feile: Physics C **159**, 1 (1989)
6.105 See features articles in *Optical Studies of High-Temperature Superconductors*, ed. by R.M. Macfarlane, A.J. Sievers: J. Opt. Soc. Am. **6**, 381 (1989)
6.106 E.E. Alp, S.M. Mini, M. Ramanathan, B. Dabrowski, D.R. Richards, D.G. Hinks: Phys. Rev. B **40**, 2617 (1989)
6.107 J. Tranquada, S. Heald, A. Moodenbaugh, G. Liang, M. Croft: Nature (London) **337**, 720 (1989)
6.108 G. Liang, J. Chen, M. Croft, K.V. Ramanujachary, M. Greenblatt, M. Hedge: Phys. Rev. B **40**, 2646 (1989)
6.109 V.J. Emergy: Nature **337**, 306 (1989)
6.110 T.M. Rice: Nature **337**, 689 (1989)
6.111 B. Grande, Hk. Müller-Buschbaum, M. Schweizer: Z. Anorg. All. Chem. **428**, 120 (1977)
6.112 Hk. Müller-Buschbaum, W. Wollschläger: Z. Anorg. Allg. Chem. **414**, 76 (1975)
6.113 Hk. Müller-Buschbaum: Angew. Chem. **101**, 1503 (1989)
6.114 S. Sugai, T. Kobayashi, J. Akimitsu: Phys. Rev. B **40**, 2686 (1989)
6.115 V.G. Hadjiev, I.Z. Kostadinov, L. Bozukov, E. Dinolova, D.M. Mateev: Solid State Commun. **71**, 1093 (1989)
6.116 E.T. Heyen, R. Liu, B. Gegenheimer, C. Thomsen, M. Cardona, S. Piñol, D. McK. Paul: In *Phonons 89*, ed. by S. Hunklinger, W. Ludwig, G. Weiss, (World Scientific, Singapore 1990) p. 346
6.117 E.T. Heyen, R. Liu, M. Cardona, S. Piñol, R.J. Melville, D.McK. Paul, E. Morán, M.A. Alario-Franco: Phys. Rev. B **43**, 2857 (1991); V.M. Orera, M.L. Sanjuan, R. Alcala, J. Fontcuberta, S. Piñol: Physica C **168**, 161 (1990)
6.118 E.T. Heyen, G. Kliche, W. Kress, W. König, M. Cardona, E. Rampf, J. Prade, U. Schröder, A.D. Kulkarni, F.W. de Wette, S. Piñol, D.McK. Paul, E. Morán, M.A. Alario-Franco: Solid State Commun. **74**, 1299 (1990)
6.119 J. Prade, A.D. Kulkarni, F.W. de Wette, W. Kress, M. Cardona, R. Reiger, U. Schröder: Solid State Commun. **64**, 1267 (1987)
6.120 See Fig. 12 on p. 433 of Ref. [6.27]. The three modes of O_{II} in the right most column in the figure of the first edition have been included inadvertently and are not orthorhombic-only modes. For a correction see the errata sheet published in Vol. II
6.121 J. Akimitsu, S. Suzuki, M. Watanabe, H. Sawa: Jap. J. Appl. Phys. **27**, L1859 (1988)
6.122 Z. Fisk, S.-W. Cheong, J.D. Thompson, M.F. Hundley, R.B. Schwarz, G.H. Kwei, J.E. Schirber: Physica C **162–164**, 1681 (1989)
6.123 A.P. Litvinchuk, C. Thomsen, P. Murugaraj, E.T. Heyen, M. Cardona, Phys. Rev. B **43**, 13060 (1991)
6.124 M. Kikuchi, S. Nakajima, Y. Syono, K. Hiraga, T. Oku, D. Shindo, N. Kobayashi, H. Iwasaki, Y. Muto: Physica C **158**, 79 (1989)
6.125 A.I. Akimov, V.I. Gatalskaya, V.C. Gurski, V.V. Dyakin, V.S. Yefanov, V.M. Ogenko, M.A. Tanatar, V.V. Teslenko, A.A. Chniko: Solid State Commun. **73**, 823 (1990)
6.126 M. Krantz, H.J. Rosen, R.M. Macfarlane, V.Y. Lee: Phys. Rev. B **38**, 11962 (1988)
6.127 V.B. Timofeev, A.A. Maksimov, D.V. Misochko, I.I. Tartakovskii: Physica C. **162–164**, 1409 (1989)
6.128 B. Friedl et al.: To be published

6.129   M.K. Wu, J.R. Ashburn, C.J. Torng, P.H. Hor, R.L. Meng, L. Gao, Z.J. Huang, Y.Q. Wang, C.W. Chu: Phys. Rev. Lett. **58**, 908 (1987)

6.130   R.J. Cava, B. Batlogg, J.J. Krajewski, L.W. Rupp, L.F. Schneemeyer, T. Siegrist, R.B. van Dover, P. Marsch, W.F. Peck, P.K. Gallagher, S.H. Glarum, J.M. Marshall, R.C. Farrow, J.V. Waszczak, R. Hull, P. Trevor: Nature **336**, 211 (1988)

6.131   Y. LePage, W.R. McKinnon, J.M. Tarascon, L.H. Greene, G.W. Hull, D.M. Huang: Phys. Rev. B **35**, 7245 (1987)

6.132   J.J. Capponi, P. Bordet, C. Chaillout, J. Chevanas, O. Chmaissem, E.A. Hewat, J.L. Hodeau, W. Korczak, M. Marezio: Physica C **162–164**, 53 (1989)

6.133   H.W. Zandbergen, G. Van Tendeloo, S.S. Amelinckx: Solid State Commun. **72**, 445 (1989)

6.134   R. Hull, J.M. Bonar, L.F. Schneemeyer, R.J. Cava, J.J. Krajewski, J.V. Waszczak: Phys. Rev. B **39**, 9685 (1989)

6.135   J.M. Tarascon, W.R. McKinnon, L.H. Greene, G.W. Hull, E.M. Vogel: Phys. Rev. B **36**, 226 (1987)

6.136   G. Zwicknagl, P. Fulde in: Earlier and Recent Aspects of Superconductivity, eds. J.G. Bednorz and K.A. Müller, Solid State Sciences **90** (Springer, Heidelberg, 1990), p. 326

6.137   M.B. Maple, Y. Dalichaouch, J.M. Ferreira, R.R. Hake, B.W. Lee, J.J. Neumeier, M.S. Torikachvili, K.N. Yang, H. Zhou, R.P. Guerling, M.V. Kuric: Physica B **148**, 155 (1987)

6.138   B.D. Dunlap, M. Slaski, D.G. Hinks, L.S. Soderholm, M. Beno, K. Zhang, C. Segre, G.W. Crabtree, W.K. Kwok, S.K. Malik, I.K. Schulter, J.D. Jorgensen, Z. Sungaila: J. Magn. Mater. **68**, L139 (1987)

6.139   W.H. Li, J.W. Lynn, S. Skanthakumar, T.W. Chinta, A. Kebede, C.S. Lee, J.E. Crow, T. Mihalisin: Phys. Rev. B **40**, 5300 (1989)

6.140   K. Nahm, B.Y. Cha, C.K. Kim: Solid State Commun. **72**, 559 (1989)

6.141   G.Y. Guo, W.M. Temmerman: Phys. Rev. B **41**, 6372 (1990)

6.142   J.J. Neumeier, M.B. Maple, M.S. Torikachvili: Physica C **156**, 574 (1988)

6.143   D.W. Cooke, R.S. Kwok, R.L. Lichti, T.R. Adams, C. Boekema, W.K. Dawson, A. Kebede, J. Schwegler, J.E. Crow, T. Mihalisin: Phys. Rev. B **41**, 4801 (1990)

6.144   E. Morán, U. Amador, M. Barahona, M.A. Alario-Franco, A. Vegas, J. Rodriguez-Carrajal: Solid State Commun. **67**, 369 (1988)

6.145   Y. Dalichaouch, M.S. Torikachrili, E.A. Early, B.W. Lee, C.L. Seaman, K.N. Yang, H. Zhou, M.B. Maple: Solid State Commun. **65**, 1001 (1988)

6.146   J.P. Frank, J. Jung, M.A.-K. Mohamed, S. Gygax, I.G. Sproule: XIX Int. Conf. on Low Temperature Physics, Brighton (1990)

6.147   M. Cardona: In: *Light Scattering in Solids II*, ed. by M. Cardona, G. Güntherodt, Topics Appl. Physics **50**, (Springer, Berlin, Heidelberg 1982), p. 45 ff.

6.148   See, e.g. D.L. Rousseau, R.P. Baumann, S.P.S. Porto: J. Raman Spectr. **10**, 253 (1981)

6.149   R. Liu: Ph.D. thesis, Univ. of Stuttgart, 1990

6.150   R. Evarestov: Lecture Series on Space Group Representations, unpublished

6.151   see Ref. [6.147] Table 2.1, p. 46

6.152   C. Thomsen, M. Cardona, R. Liu, Hj. Mattausch, W. König, F. García-Alvarado, B. Suárez, E. Morán, M. Alario-Franco: Solid State Commun. **69**, 857 (1989)

6.153   *International Tables of Crystallography*: ed. by T. Hahn, revised Ed., Vol. A (Kluwer, Dordrecht 1989)

6.154   K.F. McCarty, D.S. Ginley, D.R. Boehme, R.J. Baughman, E.L. Venturini, B. Morosin: Physica C **156**, 119 (1988)

6.155   K.F. McCarty, D.S. Ginley, D.R. Boehme, R.J. Baughman, B. Morosin: Solid State Commun. **68**, 77 (1988)

6.156   E.T. Heyen, R. Liu, C. Thomsen, R. Kremer, M. Cardona, J. Karpinski, E. Kaldis, S. Rusiecki: Phys. Rev. B **41**, 11058 (1990)

6.157   M.C. Krantz, H.J. Rosen, R.M. Macfarlane, N.G. Asmar, D.E. Morris: Physica C **162–164**, 1089 (1989)

6.158 R. Liu, M. Cardona, B. Gegenheimer, E.T. Heyen, C. Thomsen: Phys. Rev. B **40**, 2654 (1989)
6.159 W. Kress, J. Prade, U. Schröder, A.D. Kulkarni, F.W. de Wette: Physica C **162–164**, 1345 (1989)
6.160 K.F. McCarty, B. Morosin, D.S. Ginley, D.R. Boehme: Physica C **157**, 135 (1989)
6.161 J. Prade, A.D. Kulkarni, F.E. de Wette, U. Schröder, W. Kress: Phys. Rev. B **39**, 2771 (1989)
6.162 B. Friedl: unpublished data
6.163 M. Cardona, R. Liu, C. Thomsen, M. Bauer, L. Genzel, W. König, A. Wittlin, U. Amador, M. Barahona, F. Fernández, C. Otero, R. Sáez: Solid State Commun. **65**, 71 (1988)
6.164 C. Thomsen, M. Cardona, R. Liu: J. Less Common Metals. **150**, 33 (1989)
6.165 H.J. Rosen, R.M. Macfarlane, E.M. Engler, V. Y. Lee, R.D. Jacowitz: Phys. Rev. B **38**, 2460 (1988)
6.166 R.M. Macfarlane, H.J. Rosen, E.M. Engler, R.D. Jacowitz, V.Y. Lee: Phys. Rev. B **38**, 284 (1988)
6.167 J. Sapriel, L. Pierre, D. Morin, J.C. Toledano, J. Schneck, H. Savary, J. Chavignon, J. Primot, C. Dagnet, J. Etrillard, H. Boyer: Phys. Rev. B **39**, 339 (1989)
6.168 P. Knoll, B. Stadlober, M. Pressl, N. Brnicevic: Physica C **162–164**, 1097 (1989)
6.169 L.A. Farrow, L.H. Greene, J.M. Tarascon, P.A. Morris, W.A. Bonner, G.W. Hull: Proc. of the XIth International Conference on Raman Spectroscopy (Wiley, Chichester 1988) p. 417
6.170 M.N. Iliev, V.G. Hadjiev: Physica C **157**, 495 (1989)
6.171 F. Slakey, M.V. Klein, E.D. Bukowski, D.M. Ginsberg: Phys. Rev. B **41**, 2109 (1990)
6.172 M. Boekholt, A. Erle, P.C. Splittgerber-Hünnekes, G. Güntherodt: Solid State Commun. **74**, 1107 (1990)
6.173 K.F. McCarty, J.Z. Liu, R.N. Shelton, H.B. Radousky: Phys. Rev. B **41**, 8792 (1990)
6.174 see Ref. [6.147] p. 19ff
6.175 E.T. Heyen, R. Liu, M. Garriga, B. Gegenheimer, C. Thomsen, M. Cardona: Phys. Rev. B **41**, 830 (1990)
6.176 E.T. Heyen, S.N. Rashkeev, I.I. Mazin, O.K. Andersen, R. Liu, M. Cardona, O. Jepsen: Phys. Rev. Lett. **65**, 3048 (1990)
6.177 S.B. Dierker, M.V. Klein, G. W. Webb, Z. Fisk: Phys. Rev. Lett. **50**, 853 (1983)
6.178 K.B. Lyons, S.H. Liou, M. Hong, H.S. Chen, I. Kwo, T.J. Negran: Phys. Rev. B **36**, 5592 (1987)
6.179 Y.A. Ossipyan, V.B. Timofeev, I.F. Shegolev: Physica C **153–155**, 1133 (1988)
6.180 C. Thomsen, M. Cardona, B. Gegenheimer, R. Liu, A. Simon: Phys. Rev. B **37**, 9860 (1988)
6.181 S.L. Cooper, M.V. Klein, B.G. Pazol, J.P. Rice, D.M. Ginsberg: Phys. Rev. B **37**, 5920 (1988)
6.182 S.L. Cooper, F. Slakey, M.V. Klein, J.P. Rice, E.D. Bukowski, D.M. Ginsberg: Phys. Rev. B **38**, 11934 (1988)
6.183 R. Hackl, W. Gläser, P. Müller, D. Einzel, K. Andres: Phys. Rev. B **38**, 7133 (1988)
6.184 M.N. Iliev, V.G. Hadjiev: J. Phys.: Cond. Matter **2**, 3135 (1990)
6.185 T. Staufer, R. Hackl, P. Müller: Solid State Commun., **75**, 975 (1990)
6.186 A. Yamanaka, T. Kimura, F. Minami, K. Inove, S. Takekawa: Jpn. J. Appl. Phys. **27**, L1902 (1988)
6.187 D. Kirilov, I. Bozovic, T.H. Geballe, A. Kapitulnik, D.B. Mitzi: Phys. Rev. B **38**, 11955 (1988); M. Boegholt, G. Güntherodt: Physica C **169**, 436 (1990)
6.188 M.C. Krantz, H.J. Rosen, H.Y.T. Wei, D.E. Morris: Phys. Rev. B **40**, 2635 (1989); A.A. Maksimov, I.I. Tartakovkii, V.B. Timofeev, L.A. Falkovskii: Sov. Phys. JETP **70**, 588 (1990)
6.189 C.M. Varma, P.B. Littlewood, S. Schmitt-Rink, E. Abrahams, A.E. Ruckenstein: Phys. Rev. Lett. **63**, 1996 (1989)
6.190 H. Monien, A. Zawadowski: Phys. Rev. Lett. **63**, 911 (1989)
6.191 K. Efetov: To be published
6.192 J. Philipps: Phys. Rev. B **40**, 7348 (1989)

6.193   R. Hackl, R. Kaiser, W. Gläser: Physica C **162–164**, 431 (1989)
6.194   See Ref. [6.32] p. 182
6.195   G. Contreras, A.K. Sood, M. Cardona: Phys. Rev. B **32**, 924 (1985)
6.196   G. Contreras, A.K. Sood, M. Cardona: Phys. Rev. B **32**, 930 (1985)
6.197   F. Slakey, S.L. Cooper, M.V. Klein, J.P. Rice, D.M. Ginsberg: Phys. Rev. B **39**, 2781 (1989)
6.198   C. O. Rodriguez, A. Liechtenstein, I. I. Mazin, O. Jepsen, O. K. Andersen: Phys. Rev. B **42**, 2692 (1990)
6.199   M.V. Klein proposed a similar dependence of the polarization in Physica C **162–164**, 1701 (1989)
6.200   C.M. Varma: Int. J. Mod. Phys. B **3**, 2083 (1989)
6.201   R.J. Cava, B. Batlogg, R.B. van Dover, D.W. Murphy, S. Sunshine, T. Siegrist, J.P. Rameika, E.A. Rietman, S. Zahurak, G.P. Espinosa: Phys. Rev. Lett. **58**, 1676 (1987)
6.202   S. Martin, A.T. Fiory, R.M. Fleming, L.F. Schneemeyer, J.V. Waszczak: Phys. Rev. Lett. **60**, 2194 (1988)
6.203   P.B. Allen, Z. Fisk, A. Migliori: In Ref. [6.28] Vol. I, p. 213ff.
6.204   Ref. [6.29] p. 352ff.
6.205   see e.g. H. Kuzmany, M. Mehring, J. Fink (eds.), *Electronic Properties of High-Temperature Superconductors and Related Compounds*, Springer Ser. Solid-State Sci. Vol. 99 (Springer, Berlin, Heidelberg 1990)
6.206   J.H. Kim, K. Levin, A. Auerbach: Phys. Rev. B **39**, 11633 (1989)
6.207   E.T. Heyen, R. Liu, C. Thomsen, R. Kremer, M. Cardona, J. Karpinski, E. Kaldis: Phys. Rev. B **41**, 11058 (1990)
6.208   U. Fano: Phys. Rev. **124**, 1866 (1961)
6.209   F. Cerdeira, T.A. Fjedly, M. Cardona: Solid State Commun. **13**, 325 (1973)
6.210   F. Cerdeira, T.A. Fjedly, M. Cardona: Phys. Rev. B **8**, 4734 (1973)
6.211   F. Cerdeira, T.A. Fjedly, M. Cardona: Phys. Rev. B **9**, 4344 (1974)
6.212   G. Piao, R.A. Lewis, P. Fisher: Solid State Commun. **75**, 835 (1990)
6.213   M. Chandrasekhar, H.R. Chandrasekhar, M. Grimsditch, M. Cardona: Phys. Rev. B **22**, 4852 (1980)
6.214   Ref. [6.33] p. 127ff.
6.215   M. Cardona, C. Thomsen: In *High-T_c Superconductors: Electronic Structure*, ed. by A. Bianconi, A. Marcelli (Pergamon, Oxford 1989) p. 79
6.216   C. Thomsen, R. Liu, M. Bauer, A. Wittlin, L. Genzel, M. Cardona, E. Schönherr, W. Bauhofer, W. König: Solid State Commun. **65**, 55 (1988)
6.217   C. Thomsen, R. Liu, M. Cardona, U. Amador, E. Morán: Solid State Commun. **67**, 271 (1988)
6.218   J.C. Phillips: In *Physics of High-T_c Superconductors* (Academic, Boston 1989)
6.219   D. Vakuin, S.K. Sinha, J.E. Moncton, D.C. Johnston, J.M. Newsam, C.R. Safinya, H.E. King, Jr: Phys. Rev. Lett. **58**, 2802 (1987)
6.220   J.M. Tranquada, D.E. Cox, W. Kunnmann, H. Mondden, G. Shirane, M. Suenaga, P. Zolliker, D. Vaknin, S.K. Sinha, M.S. Alvarez, A.J. Jacobsen, D.C. Johnston: Phys. Rev. Lett. **60**, (1988)
6.221   P.A. Fleury, R. Loudon: Phys. Rev. **166**, 514 (1968)
6.222   M.G. Cottam, D.J. Lockwood: In *Light Scattering in Magnetic Solids* (Wiley, New York 1986)
6.223   K.B. Lyons, P.A. Fleury, L.F. Schneemeyer, J.V. Waszczak: Phys. Rev. Lett. **60**, 732 (1988)
6.224   K.B. Lyons, P.A. Fleury, J.P. Remeika, A.S. Cooper, T. Negran: Phys. Rev. B **37**, 2353 (1988)
6.225   P.E. Sulewski, P.A. Fleury, K.B. Lyons, S.-W. Cheong, Z. Fisk: : Phys. Rev. B **41**, 225 (1990)
6.226   A.A. Maksimov, I.I. Tartakovskii, V.B. Timofeev: Physica C **160**, 249 (1989)
6.227   S. Sugai, S. Shamoto, M. Sato: Phys. Rev. B **38**, 6436 (1988)
6.228   G. Aeppli, S.M. Hayden, H.A. Mook, Z. Fisk, S.-W. Cheong, D. Rytz, J.P. Remeika, G.P. Espinosa, A.S. Cooper: Phys. Rev. Lett. **62**, 2052 (1989)

6.229 J.B. Parkinson: J. Phys. C **2**, 2012 (1969)

6.230 P. Knoll, C. Thomsen, M. Cardona, P. Murugaraj: Phys. Rev. B **42**, 4842 (1990)

6.231 R.R.P. Singh, P.A. Fleury, K.B. Lyons, P.E. Sulewski: Phys. Rev. Lett. **62**, 2736 (1989)

6.232 M.G. Cottam: J. Phys. C **7**, 2901 (1974)

6.233 D.M. Krol, M. Stavola, L.F. Schneemeyer, J.V. Waszczak, H.O'Bryan, S.A. Sunshine: Phys. Rev. B **38**, 11346 (1988)

6.234 C. Thomsen, E. Schönherr, B. Friedl, M. Cardona, Phys. Rev. B **42**, 943 (1990)

6.235 R.E. Glover, III, M. Tinkham: Phys. Rev. **108**, 243 (1957)

6.236 I. Giaever: Phys. Rev. Lett: **5**, 147, 464 (1960)

6.237 R.W. Morse: Progress in Cryogenics, Vol. I, ed. by K. Mendelssohn (Heywood, London 1959) p. 220

6.238 L.C. Hebel, C.P. Slichter: Phys. Rev. **113**, 1504 (1959)

6.239 N.E. Philips: Phys. Rev. **114**, 676 (1959)

6.240 J. Bardeen, G. Rickayzen, L. Tewordt: Phys. Rev. **113**, 982 (1959)

6.241 J. Bardeen, D.M. Ginsberg, M.B. Salamon: In Novel Superconductivity, eds. S.A. Wolf, V.Z. Kresin (Plenum, New York 1987) p. 333

6.242 A. Junod: In Ref. [6.28] Vol. II, p. 13

6.243 Ref. [6.29] p. 379 ff

6.244 for a review see M. Cardona: Ultrasonics International, Proceedings (Butterworths 1, London 1989)

6.245 W.W. Warren, R.E. Walstedt, G.F. Brennert, G.P. Espinosa, J.P. Rameika: Phys. Rev. Lett. **59**, 1860 (1987)

6.246 D. Brinkmann: Physica C **153–155**, 75 (1988)

6.247 C.H. Pennington, C.P. Slichter: In Ref. [6.28] Vol. II, p. 269

6.248 M. Gurvitch, J.M. Valles, Jr., A.M. Cucolo, R.C. Dynes, J.P. Garno, L.F. Schneemeyer, J.V. Waszczak: Phys. Rev. Lett. **63**, 1008 (1989)

6.249 S.M. Shapiro, G. Shirane, J.D. Axe: Phys. Rev. B **12**, 4899 (1975)

6.250 R.M. Macfarlane, H.J. Rosen, H. Seki: Solid State Commun. **63**, 831 (1987)

6.251 R. Feile, U. Schmitt, P. Leiderer, J. Schubert, U. Poppe: Z. Phys. B **72**, 141 and 161 (1988)

6.252 A. Wittlin, R. Liu, M. Cardona, L. Genzel, W. König, W. Bauhofer, H. Mattausch, A. Simon, F. García-Alvarado: Solid State Commun. **64**, 477 (1987)

6.253 C. Thomsen, M. Cardona, B. Friedl, C.O. Rodriguez, I.I. Mazin, O.K. Andersen: Solid State Commun. **75**, 219 (1990)

6.254 P.M. Horn, D.T. Keane, G.A. Held, J.L. Jordon-Sweet, D.L. Kaiser, F. Holtzberg, T.M. Rice: Phys. Rev. Lett. **59**, 2772 (1987)

6.255 C. Thomsen, M. Cardona, W. Kress, R. Liu, L. Genzel, M. Bauer, E. Schönherr, U. Schröder: Solid State Commun. **65**, 1139 (1988)

6.256 T. Ruf, C. Thomsen, R. Liu, M. Cardona: Phys. Rev. B **38**, 11985 (1988)

6.257 R.M. Macfarlane, M.C. Krantz, H.J. Rosen, V.Y. Lee: Physica C **162–164**, 1091 (1989)

6.258 M.C. Krantz, H.J. Rosen, R.M. Macfarlane, V.Y. Lee: Phys. Rev. B **38**, 4992 (1988)

6.259 T. Zetterer, M. Franz, J. Schützmann, W. Ose, H.H. Otto, K.F. Renk: Phys. Rev. B **41**, 9499 (1990)

6.260 B. Friedl, C. Thomsen, M. Cardona: Phys. Rev. Lett. **65**, 915 (1990)

6.261 M.V. Klein and S.B. Dierker: Phys. Rev. B **29**, 4976 (1984)

6.262 R. Zeyher and G. Zwicknagl: Solid State Commun. **66**, 617 (1988)

6.263 R. Zeyher and G. Zwicknagl: Z. Phys. B **78**, 175 (1990)

6.264 Z. Schlesinger, R.T. Collins, F. Holtzberg, C. Feild, S.H. Blonston, U. Welp, G.W. Crabtree, Y. Fang, J.Z. Liu: Phys. Rev. Lett. **65**, 801 (1990)

6.265 A. Mascarenhas, H. Katayama-Yoshida, J. Pankove, S.K. Deb: Phys. Rev. B **39**, 4699 (1989)

6.266   J. Menéndez, M. Cardona: Phys. Rev. B **29**, 2051 (1984)

6.267   H. Rosen, E.M. Engler, T.C. Strand, V.Y. Lee, D. Bethune: Phys. Rev. B **36**, 726 (1987)

6.268   Z.V. Popović, C. Thomsen, M. Cardona, R. Liu, G. Stanisić, W. König: Solid State Commun. **66**, 7971 (1988)

6.269   Z.V. Popović, C. Thomsen, M. Cardona, R. Liu, G. Stanisić, W. König: Z. Phys. B **72**, 13 (1988)

6.270   M. Udagawa, N. Ogita, A. Fukumoto, Y. Utsunomiya, K. Ohbayashi: Jpn. J. Appl. Phys. **26**, L858 (1987)

6.271   K.B. Lyons, J. Kwo, J.F. Dillon, Jr., G.P. Espinonsa, M. McGlashan-Powell, A.P. Ramirez, L.F. Schneemeyer: Phys. Rev. Lett. **64**, 2949 (1990)

6.272   S. Spielmann, K. Fester, C.B. Eom, T.H. Geballe, M.M. Fejer, A. Kapitulnik: Phys. Rev. Lett. **65**, 123 (1990)

6.273   H.J. Weber, D. Weitbrecht, D. Brach, H. Keiter, A.L. Shelankow, W. Weber, T. Wolf, J. Geerk, G. Linker, G. Roth, P.C. Splittgerber-Hunnekes, G. Güntherodt: Solid State Commun., **76**, 511 (1990)

Additional references:

1.) Electron–phonon interactions of Raman active phonon in $YBa_2Cu_3O_{7-y}$; E. Altendorf, J. Chrzanowski, J.C. Irwin, A. O'Reilly, and W.N. Hardy, Physica C **175**, 47 (1991).

2.) Temperature dependence of the linewidth of the Raman-active phonons of $YBa_2Cu_3O_7$: Evidence for a superconducting gap between 440 and 500 cm^{-1}; K.F. McCarty, H.B. Radovsky, J.Z. Liu, and R.N. Shelton, Phys. Rev. B **43**, 13751 (1991)

3.) Effect of substitutional impurities on the superconducting gap of $YBa_2Cu_3O_{7-\delta}$; C. Thomsen, B. Friedl, M. Cieplak, and M. Cardona, Sol. State Commun. **78**, 727 (1991)

Topics in Applied Physics, Vol. 68
© Springer-Verlag Berlin Heidelberg 1991

7. Light Scattering in Silver Halides

Wolf von der Osten

With 29 Figures

The silver halides AgCl and AgBr are ionic crystals with the rocksalt structure like the alkali halides, but despite many similarities with these, some of their properties are very different and even unique. Much of this difference is due to the 4d-electrons present at the silver ion giving rise to non-central "covalent" binding forces, which strongly affect electronic and lattice properties. Regarding the electronic energy bands, the most striking consequence is the indirect character of the lowest bandgap in the Brillouin zone [7.1]. As illustrated in Fig. 7.1 for the example AgBr [7.2], the lowest conduction band has predominantly Ag s-like character. Its minimum at point Γ (symmetry Γ_6^+) in the zone center is isotropic and essentially comparable with the situation typically encountered in alkali halides. Unlike these, however, the valence-band structure is more complex, its uppermost maximum occuring at point L instead of Γ. This pecularity is due to the interaction of the silver 4d and halogen p-states. At point Γ that transforms according to O_h containing inversion, these states with symmetries Γ_8^+, Γ_7^+ and Γ_8^-, Γ_6^-, repectively, are adjacent in energy but do not combine due to different parity. At non-zero wavevector along $\langle 111 \rangle$, however, due to reduced symmetries, like states occur that can mix causing the uppermost valence bands along $\Gamma \rightarrow$ L to rapidly bend upwards and resulting in the highest maximum at L (symmetries L_4^-, L_5^-, with L_6^- split-off by spin-orbit interaction). Derived from this indirect gap, in both AgCl and AgBr the lowest electronic state is an indirect exciton having symmetry $\Gamma_6^+ \otimes L_{4,5}^-$. Being widely separated in energy from the next higher (zone center) direct state it is the origin for the tail of relatively weak (exciton) absorption extending from the ultraviolet into the blue or blue-green region of the visible spectral range.

The presence of the silver d-electrons in the silver halides as compared to alkali halides also gives rise to differences in the phonon dispersion, both quantitatively and qualitatively [7.3]. In a microscopic picture, these differences could be explained as having their origin in the large quadrupolar deformability of the electronic charge distribution that occurs around the silver ion under appropriate displacement of the neighbouring halide ions. Incorporating this additional degree of freedom into the "simple" shell model used for alkali halide crystals, a very good description of the phonon dispersion curves (as well as other lattice properties) could be obtained for the silver halides, as shown for AgBr in Fig. 7.2 (for details and references see [7.4, 5]). One major effect of the silver ion deformability,

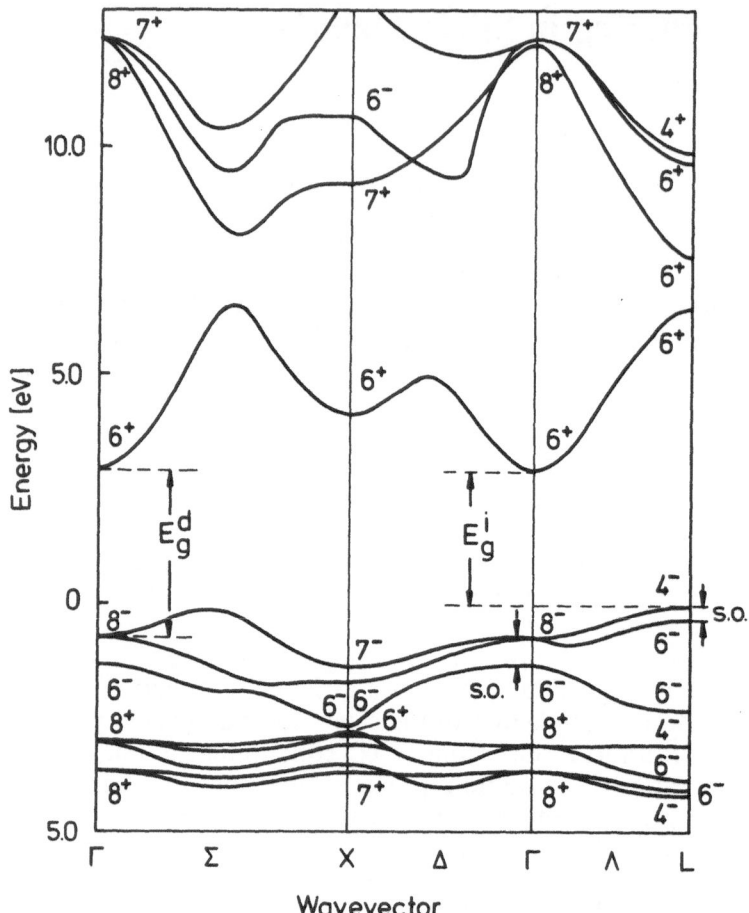

Fig. 7.1. Electronic band stucture including spin-orbit interaction for AgBr along three principal directions of the Brillouin zone. E_g^d and E_g^i denote energies of the lowest direct and indirect gaps, s.o. spin-orbit split valence band states at Γ and L. States Γ_8^+ and L_4^- are degenerate with Γ_7^+ and L_5^-, respectively. From KKR calculations reported in [7.2]

of concern for the discussion throughout this chapter, is the lowering in energy of the TO phonon branch along direction Λ. As a result, in AgCl the TO mode at the L-point comes very close to the TA(L) mode. In AgBr, the lowering is even stronger. It results in a crossing (anticrossing) of the two branches that is hidden in the dispersion curves and leads to an exchange of eigenvectors between these modes so that phonons near L, although belonging to the TO branch, have TA character and vice versa. This effect is important when selection rules for phonon-assisted exciton transitions are considered. It was originally predicted by theory from the shell model with deformable ions [7.3] and later verified by optical luminescence and neutron scattering [7.6–8].

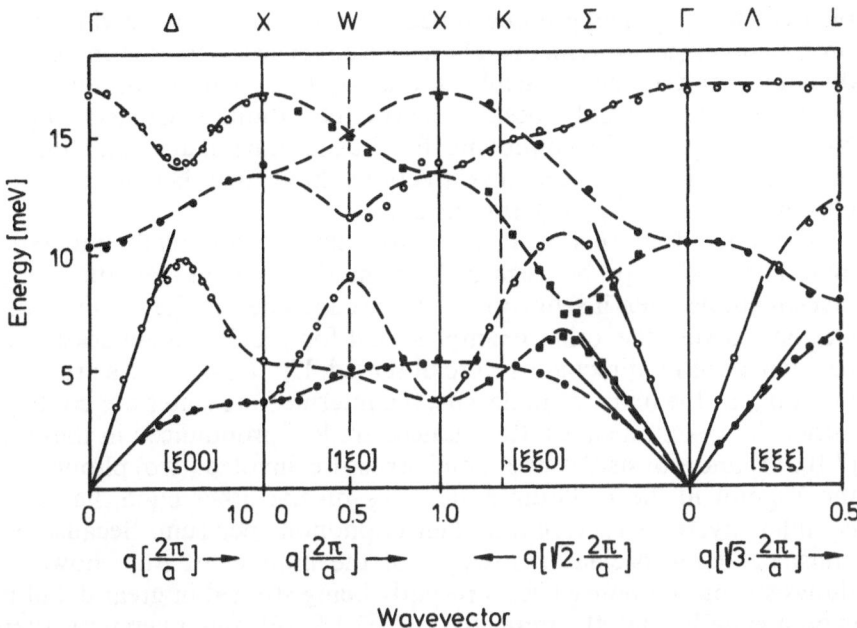

Fig. 7.2. Phonon dispersion along different directions in the Brillouin zone for AgBr. Points are experimental data from inelastic neutron scattering at 85 K. Broken lines are a fit of an extended shell model, from [7.9]

Light scattering in the silver halides has proved to be an extremely powerful method, especially when excited in resonance with the lowest indirect exciton. It naturally allows to extract information on the phonon properties but, in addition, provides even more detailed knowledge on the electronic (or excitonic) states which is not obtainable by means of conventional optical techniques like absorption and photoluminescence measurements. A particular advantage of the indirect exciton being the resonant intermediate state is that with regard to the energy range it is accessible to available tunable lasers. In addition, its relatively weak absorption allows to use convenient scattering geometries and guarantees that volume effects are looked at, a fact which facilitates the investigations and simplifies their interpretation. By employing resonant Raman scattering, thereby varying the excitation photon energy, details of exciton-phonon and exciton-impurity interactions are revealed. These establish a consistent picture of exciton energy relaxation and allow one to quantitatively determine a number of exciton parameters. In contrast to the case of direct excitons, for indirect intermediate states Raman scattering contains spectral information that can be exploited to unambiguously identify the relevant exciton-phonon interactions. In addition and even more important, in silver halides absolute scattering rates for these processes can be measured as a function of exciton energy. To this end, picosecond equipment was developed to perform nearly bandwidth-limited time-resolving measurements. These give direct access to exciton relaxation dynamics and render possible a quantitative analysis of the various processes. Very recently, besides energy relaxation, coherence properties have been explored by employing, for the first time in any solid, quantum beat spectroscopy. These investigations have just begun, the results obtained so far being very promising.

Considered in this way, the silver halides are really outstanding materials and may be regarded as model systems for this type of investigations, even if semiconductors are included in the comparison. In fact only a very few other examples exist for which Raman scattering in resonance with indirect gaps were observed. Up to now, these are Si, GaP and BiI_3. In the first two materials, the intermediate states are continuum rather than exciton states. Resonances are less pronounced in these cases [7.10, 11] and, consistent with band structure, involve (two) phonons near the X-point in the Brillouin zone. BiI_3, on the other hand, has a fairly complex layered structure and, hence, phonon spectrum. Because of the unusually large oscillator strength of the indirect exciton, however, it shows strong resonance effects recently being studied in great detail both experimentally and theoretically; see [7.12–16] and references therein. Finally, one other material has to be mentioned for which resonant light scattering is observed, formally much similar to that in the silver halides: that is Cu_2O [7.17, 18]; see also [7.19, 20] for references and more detailed discussions. Here the resonant intermediate state is the well-known 1s (ortho-) exciton of the yellow series which is a direct but electric-dipole forbidden state. Like in AgCl and AgBr, this exciton can be excited by a phonon-assisted transition giving rise to a resonantly enhanced two-phonon process that, however, involves long wavelength ($\sim \Gamma_{12}^-$) rather than zone-edge (L-point) phonons as in the silver halides. For this material time-resolved measurements were also reported and briefly reviewed previously [7.21, 22].

This chapter presents a comprehensive review of light scattering investigations, both nonresonant and resonant, that have been performed over the years on the two silver halides AgCl and AgBr. Although these substances appear to be very special ones, the results obtained are surely of general interest, in particular with regard to the central question of exciton relaxation which will be elucidated by the light scattering experiments.

Following this introduction, we begin in Sect. 7.1 with a brief presentation of some experimental details specific to the silver halides and will describe special instrumentation developed for picosecond time-resolving measurements reported in later sections. Since the lowest indirect exciton plays a key role as intermediate state in the resonant scattering treated in this article, Sect. 7.2 is devoted to its electronic structure and related optical transitions. In Sect. 7.3, non-resonant Raman scattering including selection rules will briefly be touched upon, the expression for the scattering cross-section then serving as a starting point to treat the resonance case in Sect. 7.4. This section refers to continuous wave (cw) excitation and will describe resonant two-phonon scattering and exciton energy relaxation including measurements under hydrostatic pressure and in high magnetic fields. Section 7.5 is concerned with time-resolved measurements on a subnano- and picosecond time scale providing a quantitative model of exciton relaxation. This part includes most recent observations of quantum beats discovered in the resonant light scattering signal which are important to reveal the coherence properties of the intermediate exciton state. We finally close with a few concluding remarks in Sect. 7.6.

A central question to be addressed at various stages on our way towards a model to describe the observed processes adequately will be the distinction between *resonant Raman scattering* and *hot luminescence*. This problem has been studied theoretically by many authors using various approaches; see e.g. [7.23, 24] and references therein. Both components, resonant Raman scattering and hot luminescence, are in principle contained in the resonant secondary emission. Resonant Raman scattering is considered a one-step coherent process with the coherence retained in the intermediate state between the light absorption-emission sequence. In contrast, hot luminscence clearly is a two-step process. It corresponds to "absorption followed by emission" with the coherence of the intermediate state destroyed by elastic and inelastic scattering processes between the steps. There has been quite some debate in the literature as to whether these processes can be distinguished by any experiment, a question that will be discussed on several occasions below.

7.1 Some Specific Experimental Aspects

Silver halides are the main ingredient of all photographic materials, since they are sensitive to irradiation with visible and ultraviolet light which darkens the crystals. This, however, is true only at temperatures higher than about 70 K where thermally created silver interstitial ions are sufficiently mobile to form larger aggregates of silver atoms stabilized by the photo-created electrons (for details of the mechanism see e.g. [7.4]). In contrast, at very low temperatures where ionic diffusion is suppressed, the processes occuring under optical excitation are purely electronic, leading preferentially to formation and annihilation of free electron–hole pairs or excitons. An inconvenient consequence of this light sensitivity is that not only all optical measurements have to be carried out at low temperatures, but also all handling and preparation of samples, like cutting, grinding, polishing and orienting by X-rays, must be done in the dark or in very low-level dark-red light to make sure that the sample does not "see" room or excitation light unless it is cooled down. An additional complication encountered with silver halides is that, in contrast to alkali halides and many semiconductors, they are soft and not at all cleavable. Therefore orienting crystals is troublesome and special caution is necessary in order to not deform the sample during preparation.

With regard to cw measurements, there is no need to say much about instrumentation since Raman and resonant Raman scattering nowadays are well-established techniques, see e.g. [7.25]. Besides a double monochromator with good stray light rejection and electronic equipment for photon counting, the most important component for resonant Raman experiments is a tunable dye laser. To cover the spectral range of indirect exciton absorption in AgBr and AgCl, we use as dyes coumarin 2, stilbene 3 and polyphenyl 1, all pumped with the UV lines of an Ar^+ laser achieving between 30 and 300 mW output for 3W input power depending on the dye. Because of the small absorption the scattering light can be observed at 90° with respect to the exciting laser beam. The measured intensity is usually corrected for absorption of the *incident* light by properly taking into account the scattering geometry [7.26]. In most cases the *scattered* light intensity need not be corrected since its wavelength lies outside the absorption edges.

Concerning pulsed excitation and the time-resolving experiments, which will be presented in Sect. 7.5, a more detailed discussion of experimental needs and apparatus seems to be appropriate. Depending on energy, typical scattering rates anticipated for acoustic phonons are of the order of 10^{10} s^{-1} and much larger ($> 10^{12}$ s^{-1}) for optical phonons. To study these processes directly in the time-domain, the equipment must be capable of generating and detecting light pulses with picosecond duration. Concomitantly, high spectral resolution is required in order to separate lines in the secondary emission spectrum that originate from different phonon scattering processes.

The set-up illustrated in Fig. 7.3 accounts for these demands [7.27]. Since pump-and-probe type techniques with possible time-resolution in the femtosecond range are not applicable to our problem, we use a "classical" luminescence decay configuration in which the intensity (at scattered photon energy E_s) emitted after pulsed excitation is measured as function of time, but with simultaneous optimum spectral and temporal resolution. The light pulses are produced with a dye laser synchronously pumped by an Ar^+ laser which is mode-locked at the UV lines. Operating at a repetition rate of 76 MHz, this system yields nearly transform-limited pulses of about 8 ps duration, 0.3 meV wide and tunable between about 2.95 and 2.58 eV (420 and 480 nm).

The key feature of the set-up to analyze the emitted light signal is a subtractively mounted double monochromator with 1 m focal length. Due to the symmetric configuration of gratings and rays [7.28, 29], it compensates for the light transit-time spread occuring in each single monochromator and, hence, avoids pulse broadening which, in case of

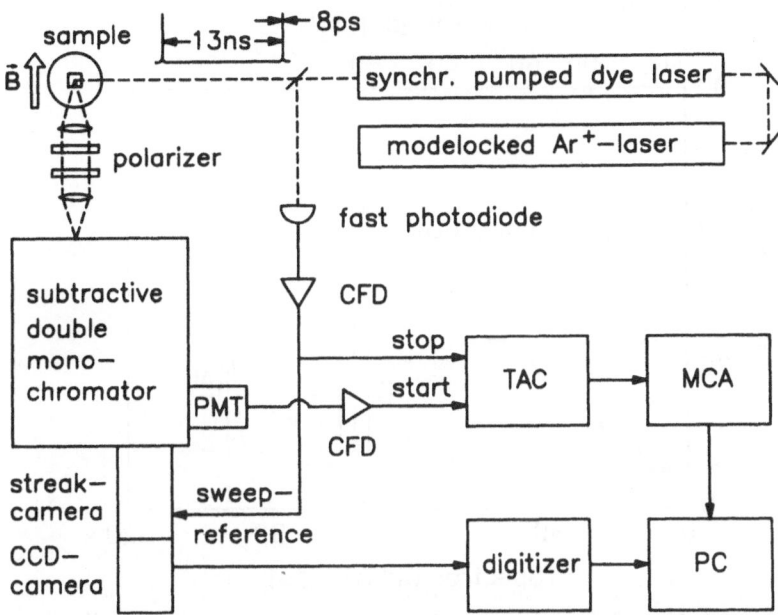

Fig. 7.3. Schematic representation of set-up for picosecond time-resolved measurements described in Sect. 7.5. CFD: constant fraction discriminator; TAC: time-to-amplitude converter; MCA: multichannel analyzer; PC: personal computer; PMT: photomultiplier tube; CCD: charge coupled device. From [7.27]

additive dispersion, would amount to several hundreds of picoseconds. As checked by using a synchroscan streak camera as detector, with this system one obtains transform-limited performance down to about 8 ps. This is demonstrated in Fig. 7.4. It shows the time response Δt of the system versus the inverse bandwidth Δv^{-1} (adjusted by the width of the intermediate spectrometer slit) for an 8 ps light pulse entering the monochromator [7.30]. The full points are experimental data measured at the detector. The dashed line corresponds to $\Delta v \, \Delta t = 0.886$. This relation is expected from Fourier-transforming the rectangular intensity distribution in frequency domain across the intermediate spectrometer slit (inset of Fig. 7.4), the excellent agreement with the measured data confirming the anticipated behavior of the spectrometer.

For reasons of sensitivity, the actual measurements in the silver halides have to be performed with the help of a fast microchannel plate photomultiplier tube as detector. The electronic system comprises especially developed wide-band constant-fraction discriminators, a commercial time-to-amplitude converter and a computer-controlled multichannel analyzer (cf. Fig. 7.3) with inverse single photon counting em-

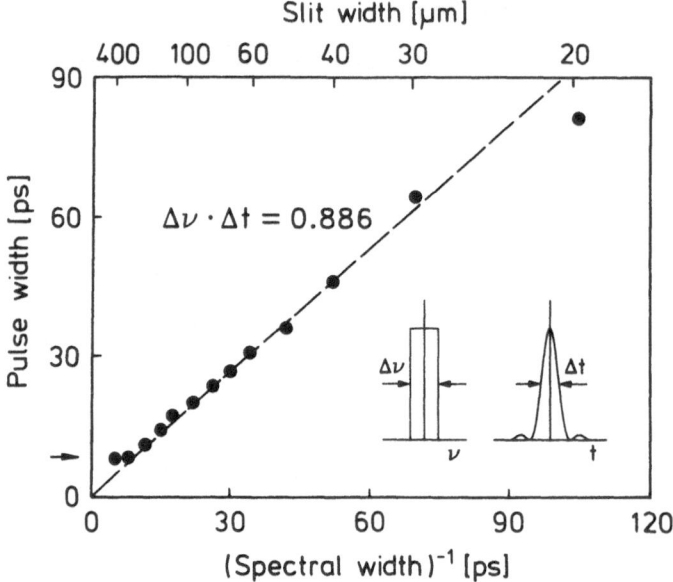

Fig. 7.4. Overall time response of the spectroscopic set-up shown in Fig. 7.3 to an 8 ps laser pulse versus reciprocal spectral width (or slit width) of the spectrometer to demonstrate bandwidth-limited performance down to 8 ps (arrow on ordinate). For further explanation see text. From [7.30]

ployed to register the temporal evolution of the scattered intensity. The overall system response to the laser pulse in this case is better than 40 ps (FWHM) allowing one to determine decay times of about 10 ps after deconvolution. During recent years we have improved the performance of this set-up to reach the present "state-of-the-art" [7.27, 30–33] which is absolutely crucial especially for the quantum beat experiments described in Sect. 7.5.3.

7.2 The $\Gamma_6^+ \otimes L_{4,5}^-$ Indirect Exciton State

According to the electronic band structure of Fig. 7.1, the energetically lowest exciton in both silver halides is composed of electron and hole states with Γ_6^+ and $L_{4,5}^-$ symmetries, respectively. Corresponding to the total wavevector $k_c + k_v = k_L$ (with wavevectors of conduction and valence band states $k_c = 0$ and $k_v = k_L$) this exciton therefore has indirect character with the lowest state at the points L of the Brillouin zone. As originally discussed in the context of studies of magneto-absorption [7.34, 35], see also Sect. 7.4.3, this exciton consists of altogether four states with (zero-field) wavefunctions Φ_j^0 ($j = 1, ..., 4$) that in the missing electron scheme may be represented by [7.33]

$$\Phi_1^0 = -i(X_L + iY_L)\,\alpha_v \otimes S_c\alpha_c\,,$$

$$\Phi_2^0 = -i(X_L + iY_L)\,\alpha_v \otimes S_c\beta_c\,,$$

$$\Phi_3^0 = i(X_L - iY_L)\,\beta_v \otimes S_c\alpha_c\,, \qquad (7.1)$$

$$\Phi_4^0 = i(X_L - iY_L)\,\beta_v \otimes S_v\beta_c\,.$$

In (7.1) the quantization basis is chosen with its z-axis along the direction of one of the four (inequivalent) L-points, i.e. $\|[111]$, while x and y are along $[\bar{1}\bar{1}2]$ and $[1\bar{1}0]$ as usual. X_L, Y_L denote the Bloch functions at point L transforming as x and y under the symmetry operations of D_{3d}, while S_c is a Bloch function at Γ which is totally symmetric under the operations of O_h. $\alpha_{c,v}$ and $\beta_{c,v}$ represent the spin functions with spin-up and -down, each for conduction and valence bands (subscripts c, v), with the hole spin equal to the negative of the missing electron spin. From this it immediately follows that states Φ_2^0, Φ_3^0 have pure triplet character, while the remaining states Φ_1^0, Φ_4^0 are of mixed singlet-triplet type. According to *Matsushita* [7.34], electron–hole exchange interaction removes the fourfold degeneracy of states Φ_j^0. As a result, the singlet-triplet states and the pure triplet states each are (twofold) degenerate with the triplet lowered relative to the singlet states by the exchange energy Δ.

The functions in (7.1) are to be multiplied by appropriate exciton envelope functions $F(r)$ describing the relative motion of the electron and hole (r: position vector). Considering the lowest (1s) exciton only and neglecting excited states, the corresponding energies are given by

$$E^{1s}(k) = E_g^i - E_{bx}^i + \frac{\hbar^2(k - k_L)^2}{2M^*} = E_{gx}^i + \varepsilon(|k - k_L|)\,. \qquad (7.2)$$

In (7.2) a parabolic exciton band is assumed, with its minimum at the L-point with wavevector $k = k_L$, with ε and M^* representing the kinetic part of energy and (isotropic) effective mass of the exciton, respectively. E_g^i and E_{gx}^i denote the energies of the indirect bandgap and the (singlet-triplet) exciton gap, respectively, while E_{bx}^i is the binding energy of the lowest ($n = 1$) exciton state (in AgCl [7.36]: $E_{gx}^i = 3.248$ eV, $E_{bx}^i = 47$ meV; in AgBr [7.2]: $E_{gx}^i = 2.684$ eV, $E_{bx}^i = 28$ meV).

Optical absorption into each of the states Φ_j^0 represented by (7.1) is accomplished by assistance of a momentum-conserving L-point phonon η of energy $\hbar\omega^\eta$ (for comprehensive treatments of exciton states and transitions the reader is referred e.g. to [7.37–39]). For each valley or L-point, time-dependent second-order perturbation theory for this process

[7.40] gives the standard expression for the energy-dependent absorption coefficient α (subscript j omitted)

$$\alpha^\eta(E) = \left(n^\eta + \frac{1}{2} \pm \frac{1}{2}\right)\frac{B}{E}\,|W^\eta|^2\,|F(0)|^2\left(\frac{2M^*}{\hbar^2}\right)^{3/2}$$
$$(E - E^i_{gx} \mp \hbar\omega^\eta)^{1/2}\,, \tag{7.3}$$

where the last factor represents the kinetic energy of the exciton given by

$$\varepsilon = E - E^i_{gx} \mp \hbar\omega^\eta\,. \tag{7.4}$$

While in (7.3) B comprises various constant quantities, we have written the factor $(2M^*/\hbar^2)^{3/2}$ explicitly since together with the square-root dependence of energy it comes from the density of exciton states of the parabolic band, see e.g. [7.41]. $|F(0)|^2$ is the squared envelope wavefunction at $r = 0$, which for the 1s state with exciton Bohr radius a_x is given by $|F(0)|^2 = 1/(\sqrt{\pi}\,a^3_{x,\,1s})$. n^η denotes the occupation number of phonon η. The different signs in the population and energy terms correspond to phonon creation and annihilation, respectively, accounting for the characteristic temperature dependence of the absorption edge [7.1]. Equation (7.4) determines the photon energy E necessary to create (single-triplet) excitons at kinetic energy ε and (for $\varepsilon = 0$) reproduces different absorption threshold energies E^η_{ax} according to the different energies $\hbar\omega^\eta$ of the participating phonons (for triplet excitons replace E^i_{gx} by $E^i_{gx} - \Delta$). Most important in (7.3) is the effective transition matrix-element W^η from which selection rules may be derived. Omitting again the subscript j, it is given by [7.37]

$$W^\eta = \sum_\lambda \frac{\langle c, \Gamma, n^\eta|\,\hat{e}\cdot p\,|\lambda, \Gamma, n^\eta\rangle\,\langle\lambda, \Gamma, n^\eta|\,\mathscr{V}^\eta\,|v, L, n^\eta \pm 1\rangle}{E_v(L) - E_\lambda(\Gamma) \pm \hbar\omega^\eta}$$
$$+ \frac{\langle c, \Gamma, n^\eta|\,\mathscr{V}^\eta\,|\lambda, L, n^\eta \pm 1\rangle\,\langle\lambda, L, n^\eta \pm 1|\,\hat{e}\cdot p\,|v, L, n^\eta \pm 1\rangle}{E_c(\Gamma) - E_\lambda(L) \pm \hbar\omega^\eta}\,. \tag{7.5}$$

Each of the two terms in (7.5) consists of two virtual processes. One is a direct photon transition (at Γ and L, respectively) with the electron-radiation operator $\hat{e}\cdot p$ (p: electron momentum operator, $\hat{e}$: polarization unit vector). The other is phonon scattering (between Γ and L) which is mediated by electron-phonon interaction (operator $\mathscr{V}^\eta$) and may occur in either the valence or the conduction band (v, c). The matrix-elements depend on phonon wavevector, the summation in (7.5) has to be carried out over all available intermediate states denoted by λ.

To discuss the selection rules for absorption into each exciton state, we consider for the moment only assistance of transverse acoustic L-point phonons, i.e. TA(L) in AgCl and TO(L)[1] in AgBr, associated with the most prominent absorption components in the two materials. These phonons have L_3^+ symmetry giving rise to *allowed* valence and conduction band scattering with intermediate states Γ_8^-, Γ_6^- and the pair of $L_{4,5}^+$, L_6^+ states, respectively (cf. Fig. 7.1 for AgBr). Comparison of the denominators in (7.5) with the respective energies in the band structure [7.2] shows that the Γ_8^- and Γ_6^- valence band minima are the energetically favored intermediate states. It implies that valence band scattering represented by the first term in (7.5) is the dominant process, while the second term can be neglected. To discuss the polarization properties further, one can define for each exciton state Φ_j^0 a transition moment M_j related to the transition matrix elements in (7.5) through the polarization vector $\hat{e}_L$ of the incident light by

$$W_j^\eta = \hat{e}_L \cdot M^\eta, \tag{7.6}$$

where η stands for TA(L) in AgCl and TO(L) in AgBr. Performing a complete group-theoretical analysis of the scattering matrix-elements, apart from a common constant factor, the transition moments (index η omitted) are obtained as [7.33]

$$M_1^S = (\hat{e}_x - i\hat{e}_y) - i\sigma \frac{3 - \xi}{3 - 2\xi} \hat{e}_z,$$

$$M_2^T = - \frac{i\xi}{3 - 2\xi} [\sigma(\hat{e}_x + i\hat{e}_y) + 2\hat{e}_z],$$

$$M_3^T = - \frac{i\xi}{3 - 2\xi} [\sigma(\hat{e}_x - i\hat{e}_y) + 2\hat{e}_z],$$

$$M_4^S = - (\hat{e}_x + i\hat{e}_y) + i\sigma \frac{3 - \xi}{3 - 2\xi} \hat{e}_z. \tag{7.7}$$

The superscripts S and T label transition moments of singlet-triplet and pure triplet states, respectively, and $\hat{e}_x$, $\hat{e}_y$, $\hat{e}_z$ are the orthonormal unit vectors of the quantization basis defined above. The parameter σ is given by the ratio of the two independent matrix elements, which, according to group theory, determine the valence band scattering [7.33, 42]. The quantity ξ is determined by the spin-orbit interaction of the valence band states at Γ and the width of the valence band along Γ to L and is given by

$$\xi = \frac{E_v(\Gamma_8^-) - E_v(\Gamma_6^-)}{E_v(L_{4,5}^-) - E_v(\Gamma_6^-)}. \tag{7.8}$$

---

[1] Remember that the eigenvectors for the transverse optical branch at and near L correspond to acoustic phonon displacements (cf. discussion relating to Fig. 7.2).

Inspection of (7.7) shows that with properly polarized light basically all four exciton states can be excited giving rise to (phonon-symmetry-) allowed indirect absorption. But it is also evident that for the triplet states the absorption strength (proportional to $|\hat{e}_L \cdot M_j^\eta|^2$) is smaller by a factor of about $|\xi|^2$ as compared to the singlet-triplet states. Evaluating (7.8) by inserting known band structure data [7.2], one estimates for AgBr $|\xi|^2 \approx 1\%$ and an even smaller value for AgCl. This is consistent with experiment where only singlet-triplet but no pure triplet state absorption is observed.

The (degenerate) single-triplet states Φ_1^0 and Φ_4^0 with transition moments M_1^S and M_4^S can interact with either circularly or plane polarized light. In the case of excitation with plane polarization, e.g. $\parallel y$, the states are coherently superimposed and may be taken as linear combinations

$$\Phi_x = \frac{1}{\sqrt{2}} (\Phi_1^0 + \Phi_4^0) \quad \text{and} \quad \Phi_y = \frac{1}{\sqrt{2}} (\Phi_1^0 - \Phi_4^0). \tag{7.9}$$

The corresponding transition moments following from (7.7) are given by

$$M_x = \hat{e}_x - i\sigma \cdot \frac{3 - \xi}{3 - 2\xi} \hat{e}_z \quad \text{and} \quad M_y = -i\hat{e}_y \tag{7.10}$$

and are oriented parallel or perpendicular to the polarization of the exciting light.

Very similar results can be derived for LA(L) phonon assistance [7.42]. This phonon has L_1^+ symmetry. It also gives rise to allowed scattering and, hence, relatively prominent absorption in AgCl and AgBr, but with transition moments slightly different from those in (7.7) ($\hat{e}_z$-components missing).

The total absorption coefficient for each allowed component is obtained by summing up the contributions from the four L-points. Assuming the matrix elements in (7.5) to be independent of wavevector and, hence, the effective transition matrix-element for absorption W^η to be constant, the shape of the absorption edge involving *allowed* phonons will be given, according to (7.3), by

$$\alpha(\varepsilon) \propto \left(n^\eta + \frac{1}{2} \pm \frac{1}{2} \right) \frac{1}{E} |W_{abs}^\eta|^2 \, \varepsilon^{1/2} \,. \tag{7.11}$$

It is important to note that $|W_{abs}^\eta|^2$ now contains contributions to the absorption from all valleys averaging out all polarization effects and resulting in isotropic absorption as expected in the cubic material.

In case of *forbidden* indirect transition [involving phonons TO(L), LO(L) in AgCl and TA(L), LO(L) in AgBr], which may also occur, the exciton-phonon matrix elements are zero in lowest order. At energies close to the exciton threshold, they may be expanded in powers of exciton wavevector $K = k - k_L$ so that, dropping quadratic and higher terms in $|K|$, we may write (see [7.43] for a detailed treatment)

$$W^\eta_{abs} \approx W^{\eta;0}_{abs} + W^{\eta;1}_{abs} |K| + \ldots \approx W^{\eta;1}_{abs} |K| . \tag{7.12}$$

$W^{\eta;1}_{abs}$ is the first-order exciton-phonon matrix element for coupling to symmetry-forbidden phonons. Assuming again a parabolic exciton band, i.e. $\varepsilon \sim |K|^2$, cf. (7.2), the absorption coefficient obtained from (7.11) is in this case

$$\alpha(\varepsilon) \propto \left(n^\eta + \frac{1}{2} \pm \frac{1}{2} \right) \frac{1}{E} |W^{\eta;1}_{abs}|^2 \, \varepsilon^{2/3} . \tag{7.13}$$

As an example Fig. 7.5 shows a fit to the exciton absorption edge in AgBr including contributions from both forbidden TA(L) and allowed TO(L) and LA(L) processes (labelled 1 to 3) with their respective threshold energies and energy dependences [7.44]. Except for some small deviation between 2.72 eV and 2.76 eV, the overall fit nicely follows the energy dependence predicted by (7.11, 13). At higher energies, one must add an additional contribution (labelled 4 in Fig. 7.5) that exhibits a quadratic energy dependence as expected for absorption into the continuum of free electron-hole pair states [7.40], from which a value of the $1s$ exciton binding-energy $E^i_{bx} = (28 \pm 5)$ meV is derived.

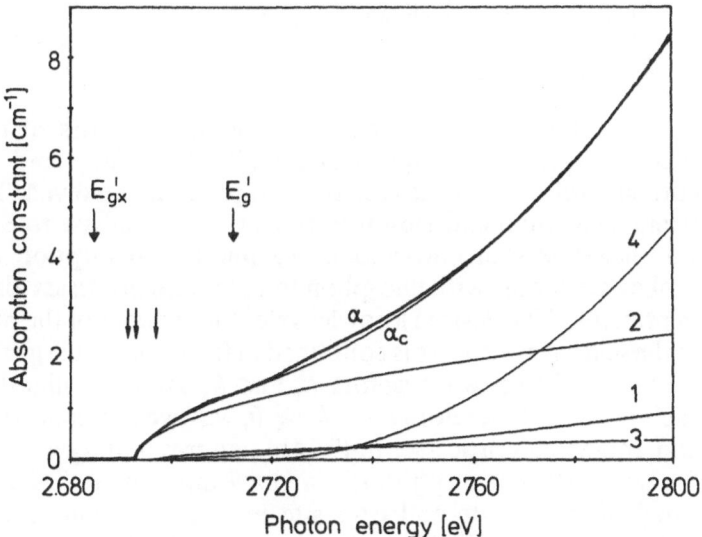

Fig. 7.5. Exciton absorption edge of AgBr at 1.8 K. α: experimental data; α_c: computed from various contributions 1 to 4 according to (7.11, 13). The components are due to the following processes: (1) TA(L), (2) TO(L), (3) LA(L), (4) absorption due to free electron-hole pairs. E^i_{gx}: exciton gap energy. E^i_g: band gap energy obtained from fit. Additional arrows mark thresholds for phonon-assisted processes (1)–(3). From [7.44]

Radiative emission is governed by the same selection rules as absorption. Therefore all formulas are valid for this case, too, except that we have to replace $\hat{e}_L$ in (7.6) by the polarization vector $\hat{e}_S$ of the emitted light and the transition moments M in (7.7) and (7.10) by the complex conjugate expressions because of the reversed order of states in the transition matrix-elements in (7.5). Like absorption, the total luminescence will also be unpolarized and isotropic in the emission process considered.

The indirect $\Gamma_6^+ \otimes L_{4,5}^-$ exciton state described above gives rise to pronounced resonances in Raman scattering described in later sections.

7.3 Non-Resonant Raman Scattering

Before proceeding to resonance effects in light scattering, a very brief description of non-resonant Raman scattering observed in the silver halides AgBr and AgCl is appropriate and will serve as a basis for discussing the resonance case in Sect. 7.4. In silver halide crystals having O_h symmetry each ion on a lattice site sits at a center of inversion. Therefore first-order Raman scattering is forbidden and only second-order processes are observed in these materials, see e.g. [7.45]. The kinematics of these processes is governed by the conservation laws for energy and wavevector (or quasi-momentum)

$$E_i = E_s \pm \hbar\omega_1 \pm \hbar\omega_2 , \tag{7.14a}$$

$$k_i = k_s \pm q_1 \pm q_2 . \tag{7.14b}$$

E_i, E_s and k_i, k_s are the energies and wavevectors of the incident and scattered photons, respectively, while $\hbar\omega_1$, $\hbar\omega_2$ and q_1, q_2 are the corresponding quantities for the two phonons involved. The positive sign refers to phonon emission into the lattice giving rise to Stokes scattering. The negative sign corresponds to phonon absorption and hence anti-Stokes scattering, with two-phonon difference processes also being possible in principle. The desired knowledge of the energies of the phonons involved in the scattering process is contained in the Raman energy shift $E_i - E_s$.

Since photon wavevectors k_i and k_s are only slightly different from each other and therefore $k_i - k_s \approx 0$, wavevector conservation for Stokes and anti-Stokes processes in (7.14b) requires that $q_1 \approx -q_2$, which means the two participating phonons can have any wavevector as long as they are equal in magnitude and opposite in direction. This corresponds to the well-known result [7.45] that in second-order Raman scattering phonons from all over the Brillouin zone, up to the zone boundary, may combine even though the wavevector of the exciting light ($|k_i| \simeq 10^5$ cm^{-1}) is orders of magnitude smaller than the size of the Brillouin zone (of the order of $2\pi/a \approx 10^8$ cm^{-1} for lattice constant a). Since the energy of each of the scattered phonons may range from zero up to the maximum phonon energy, the Raman spectrum is broad. Roughly speaking, it corresponds to the two-phonon density-of-states of the crystal weighted according to the contribution of each phonon pair to the polarizability that gives rise to the scattered intensity.

Early Raman spectra for AgCl and AgBr were taken by *Boettger* and *Damsgard* [7.46] in unoriented crystals and with relatively low spectral resolution. Later, these spectra were measured in oriented samples and could be much improved by exploiting the light polarization [7.47]. As an example, Fig. 7.6 displays spectra for AgBr at 77 K showing both Stokes and anti-Stokes components. The directions of polarizations for incident and scattered light were chosen so as to decompose the spectra into components of irreducible symmetries $\Gamma_1^+(A_{1g})$, $\Gamma_3^+(E_g)$ and $\Gamma_5^+(T_{2g})$ (or simple combinations of these) and to allow tentative assignments of certain structures to Raman-active two-phonon processes (not reproduced here).

The appearance of corresponding Raman spectra in AgCl is quite similar with increased separations between the characteristic singularities consistent with the larger phonon energies in this material as compared

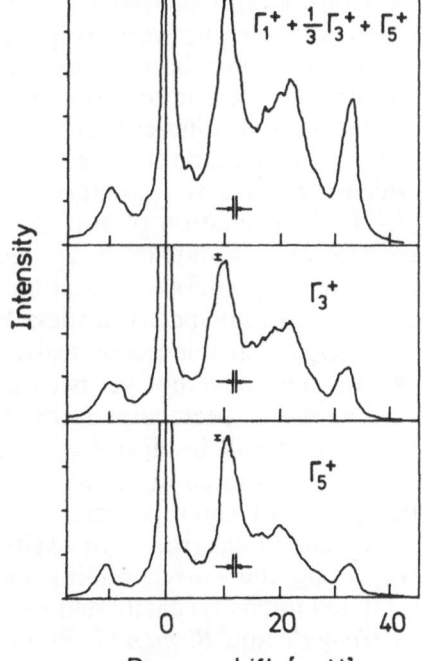

Fig. 7.6. Non-resonant Raman spectra of AgBr at 77 K for excitation at $E_L = 2.541$ eV (488.0 nm). Spectra (from top) are measured with the polarizations of the incident and scattered light along [1$\bar{1}$0] [1$\bar{1}$0], [1$\bar{1}$0] [110], and [001] [1$\bar{1}$0], respectively. From [7.47]

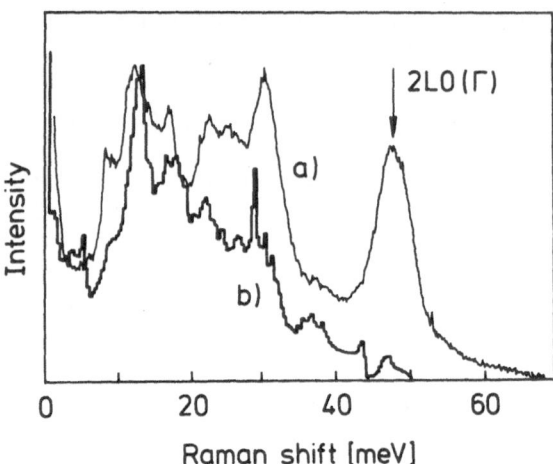

Intensity

Raman shift [meV]

2LO(Γ)

a)

b)

0   20   40   60

Fig. 7.7. Second-order Raman scattering in AgCl. A measured unpolarized spectrum at 77 K (a) is compared with a calculation (b) based on the shell model with deformable ions according to *Fischer* et al. [7.3]. The 2LO(Γ) energy position derived from inelastic neutron scattering is indicated. From [7.47]

to AgBr [7.47]. Figure 7.7 shows a measured unpolarized spectrum together with a calculation based on the modified shell model used to describe the phonon dispersion in silver halides (cf. introductory Section and Fig. 7.2). While the energy positions of all structures are very well reproduced, confirming essentially the validity of the model, with regard to (relative) intensities strong discrepancies exist between experiment and calculation. These are due largely to the fact that in the calculated spectrum the electron–phonon coupling was assumed to be constant, clearly neglecting differences for the various phonons [7.3]. Both for AgCl and AgBr, especially strong discrepancies are found in the 2LO region; they were suggested to be due to resonance effects [7.46, 47]. In AgBr, for instance, excitation with different Ar$^+$ laser lines in the visible (514.5–476.5 nm) reveals a systematic increase of about 20% in the 2LO Raman intensity which is caused by resonance presumably with the lowest *direct* exciton [7.47]. The excitation photon energy in this case is lower by nearly 2 eV than the exciton state (at $\sim$ 290 nm), the occurrence of resonance enhancement at these energies reflecting strong exciton interaction with LO(Γ) phonons as anticipated in these highly polar materials. Investigations of the energy dependence of scattered intensity much closer to the direct exciton state have not yet been performed.

The more spectacular resonance effects, which are observed in silver halides and will be discussed later, occur at the lowest *indirect* exciton state. As illustrated in the diagrams in Fig. 7.8, in principle three types of two-phonon Raman processes are possible; they differ from each other in the details of electron– (or exciton–) phonon interaction, see e.g. [7.48]. Inspecting the corresponding expressions of Raman cross-section (or scattered intensity) calculated in a microscopic theory with exciton effects by *Ganguly* and *Birman* [7.49], process B turns out to be the only one to provide a resonance with an indirect exciton state as encountered in silver

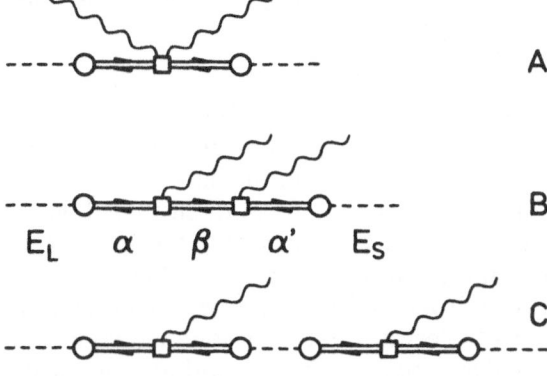

Fig. 7.8. Graphical representation of possible two-phonon Raman processes including exciton effects. The dashed, double and wavy lines respresent photon, exciton and phonon, respectively. Circles and squares are the exciton–phonon and exciton– (one- or two-) phonon interactions. Process B corresponds to resonant Raman scattering described in Sects. 7.4, 5 with intermediate states α, α' and β denoting non-resonant direct and resonant indirect exciton states, respectively. Corresponding expressions for the Raman scattering cross-section are given in [7.49]

halides. Considering only Stokes scattering (appropriate for the experiments at low temperatures described later), in the corresponding process the incident laser photon (with energy E_L and polarization vector $\hat{e}_L$) creates via exciton-photon interaction a direct (virtual) exciton in state α. By successive emission of two phonons ($\hbar\omega_1$, $\hbar\omega_2$), each created through first-order exciton–lattice interaction, this exciton is subsequently scattered into (virtual) states β and α' before it finally is annihilated and the scattered photon (energy E_S, polarization vector $\hat{e}_S$) is emitted. With obvious notations for the states and energies, the scattered two-phonon intensity in this case takes the form [7.49]

$$I(E_L, E_S)$$

$$\propto \sum_q \left| \sum_{\alpha'\beta\alpha} \frac{\langle 0| \hat{e}_S \cdot \boldsymbol{p} |\alpha'\rangle \langle \alpha'| \mathscr{V}^{(1)} |\beta\rangle \langle \beta| \mathscr{V}^{(2)} |\alpha\rangle \langle \alpha| \hat{e}_L \cdot \boldsymbol{p} |0\rangle}{(E_{\alpha'} - E_S)(E_\beta + \hbar\omega_1(\boldsymbol{q}) - E_L)(E_\alpha - E_L)} \right|^2$$

$$\delta(E_L - E_S - \hbar\omega_1 - \hbar\omega_2), \tag{7.15}$$

whereby exciton states are shown appropriate to the case of silver halides. The matrix-elements in (7.15) have already been introduced above, (7.5), and represent the exciton–photon and exciton–phonon interactions in general depending on the intermediate states as well as on photon and phonon wavevectors. The sums have to be carried out over the intermediate exciton states α, β, α' and the final state, i.e. the phonon density of states as represented by phonon wavevector $\boldsymbol{q}$. It is this term (among altogether 24 others representing permutations of the vertices in process B) that can result in a resonance of scattered intensity at an indirect exciton state and will serve as starting point in discussing the resonance effects in Sects. 7.4, 5.

7.4 Resonant Light Scattering at the Indirect Exciton: CW Excitation

Resonant Raman scattering takes advantage of resonances that occur in scattered intensity when the excitation photon energy is tuned close to and across some intermediate electronic state. As originally discovered for AgBr [7.50, 51], in both silver halides narrow line laser excitation in the $\Gamma_6^+ \otimes L_{4,5}^-$ exciton absorption at low temperature leads to strong enhancement of certain well-defined two-phonon processes in the second-order Stokes Raman spectra. From comparison of Raman energy shifts with the crystals lattice phonon energies it is evident that these processes involve L-point phonons and are identical to the momentum-conserving phonons that participate in the indirect exciton absorption and luminescence transitions that were treated in Sect. 7.2. Accordingly, the scattering lines in the two substances involve TA(L) and LA(L) phonons in AgCl and TO(L) as well as LA(L) phonons in AgBr which assist *allowed* exciton transitions. Other L-point phonons [TO(L), LO(L) in AgCl, and TA(L), LO(L) in AgBr] associated with *forbidden* transitions are also selectively enhanced in the second-order resonant Raman spectra; however, they are much weaker in intensity. The situation encountered in case of resonance with the lowest (1s-) indirect L-point exciton state is depicted in Fig. 7.9

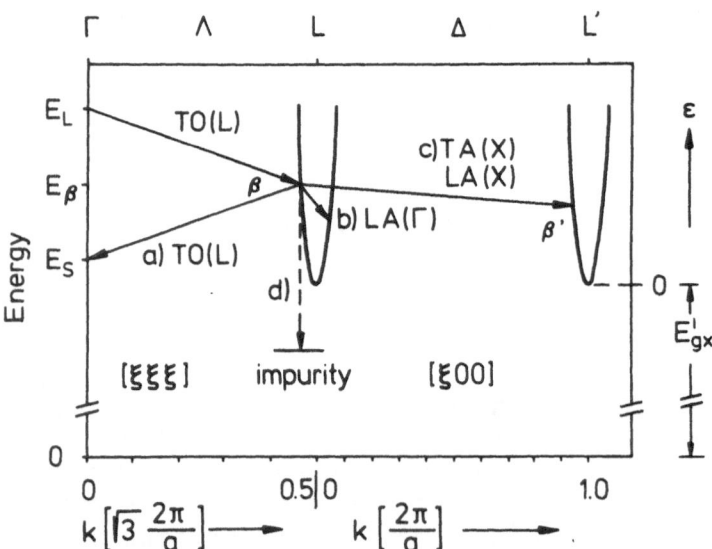

Fig. 7.9. Schematic 1s-exciton dispersion and resonant exciton scattering and trapping processes for AgBr as discussed in the text. L, L′: inequivalent L-points in the Brillouin zone. E_{gx}^i, E_β, E_L, E_S are the energies of the exciton gap, the intermediate state β, the incident (laser) photon and the scattered photon, respectively. ε is the kinetic energy of the exciton. From [7.52]

by process (a) which may be considered a Raman-like two-phonon scattering process [7.52]. At the low temperatures of the experiments, it generates a narrow Stokes line at photon energy E_S shifted with regard to the excitation photon energy E_L by the total energy of the two phonons involved ($E_L - E_S = 2\hbar\omega^{TO(L)}$ in the example shown). This is analogous to the non-resonant case discussed in the previous section, but with the general lattice phonons [$\hbar\omega_1$, $\hbar\omega_2$ in (7.15)] now replaced by specific L-point phonons. As before, the two phonons have equal but opposite wavevectors. They take care of wavevector and energy conservation as clearly illustrated in the figure, the corresponding scattering processes being selectively enhanced by resonance with the intermediate (indirect) exciton state (β in Fig. 7.8, 9). As will be demonstrated below, the intensity of the resonantly enhanced processes depends on exciton energy and reflects the exciton lifetime determined by various relaxation processes. In addition, the resonant scattering spectra exhibit lines due to higher-order processes, e.g. (b) and (c) in Fig. 7.9. They directly reflect the relaxation of the resonant intermediate state to exciton states with lower energy from which recombination to the crystal ground state may occur. The analysis of these processes allow a fairly complete and consistent interpretation of the exciton relaxation behavior.

In trying a first crude distinction of processes with regard to their physical nature, as made in the introductory section, those involving two L-point phonons may be considered *Raman-like*, although this statement later will need further refined considerations. In contrast, the processes that involve additional phonons resulting in exciton relaxation, are *hot luminescence-like*, since scattering in this case can be decribed as a multiple step process with a loss of phase memory between the steps. Unless absolutely necessary, however, a strict distinction between these types of processes will not be made in the following and the term "resonant Raman scattering" will sometimes be used synonymously with "hot luminescence".

7.4.1 Energy Dependence of Two-Phonon Scattering Intensities

In order to derive expressions for the two-phonon scattering intensity as a function of energy, we proceed from the explicitly written term in (7.15) which contains the resonance with the indirect exciton state (state β). This term may be further simplified by making the following reasonable assumptions [7.51]:

1. The intermediate direct exciton states (α, α', not shown in Fig. 7.9) are higher in energy than the indirect state (β) by 1.9 eV and 1.6 eV in AgCl and AgBr, respectively [7.2]. Therefore under resonance conditions with the indirect edge we are far off resonance with the direct states i.e. $E_\alpha, E_{\alpha'} \gg E_\beta \cong E_L, E_S$, so that the terms ($E_{\alpha'} - E_S$) and ($E_\alpha - E_L$) in (7.15) may be regarded constant and the sum over α, α' may be dropped.

2. Since higher exciton states contribute to the total absorption strength
   an amount proportional to $1/n^2$ (n = quantum number), they may be
   neglected leaving as intermediate state only the lowest (1s) exciton
   energy $E_\beta = E^{1s}(\mathbf{k})$ according to (7.2).

With these assumptions the two-phonon scattered intensity can be
calculated. Considering first of all only one L-point and one exciton state
j (thereby completely neglecting the weak triplet states), according to (7.15)
this intensity may be written as

$$I^{\eta_1 + \eta_2} \propto \sum_q \left| \frac{W^{\eta_2}_{abs} \, W^{\eta_1}_{em}}{E_\beta + \hbar\omega^{\eta_1}(\mathbf{q}) - E_L} \right|^2 . \tag{7.16}$$

Here we have introduced and correspondingly marked the effective
transition matrix-elements W^η for absorption and emission according to
(7.5). As in Sect. 7.2, η denotes the momentum-conserving (allowed and/or
forbidden) L-point phonons assisting the indirect transitions. Restricting
ourselves to allowed phonons, the matrix-elements are independent of
phonon wavevector $\mathbf{q}$ and constant. Replacing $E_\beta = E^{1s}(\mathbf{k})$, with
$\mathbf{k} = \mathbf{k}_L - \mathbf{q}$, by the kinetic energy ε according to (7.2) and noting that
the phonons at and around the L-point are nearly dispersionless, the sum
can be changed into an integral

$$\sum_q \rightarrow \int_0^{\varepsilon_0} \frac{\varepsilon^{1/2} \, d\varepsilon}{|E^i_{gx} + \varepsilon + \hbar\omega^{\eta_1} - E_L|^2} , \tag{7.17}$$

where the integration has to be performed over the entire exciton band.
Since the band is continuous, resonance always occurs for $\varepsilon \geq 0$, i.e. as
long as the incident photon energy $E_L \geq E^i_{gx} + \hbar\omega^{\eta_1}$ (cf. Fig. 7.9); the
integral diverges unless we take into account exciton damping. This can
be phenomenologically introduced in (7.17) by replacing ε in the denomina-
tor by the complex energy $\varepsilon - (i/2) \, \gamma_{tot}(\varepsilon)$. Here $\gamma_{tot}(\varepsilon)$ is the total lifetime
broadening of the exciton state related to the corresponding total scattering
rate $\Gamma_{tot}(\varepsilon)$ and population lifetime $T_1(\varepsilon)$, respectively, by

$$\gamma_{tot}(\varepsilon) = \hbar\Gamma_{tot}(\varepsilon) = \hbar/T_1(\varepsilon) . \tag{7.18}$$

Assuming a parabolic exciton band as in (7.2), one can integrate (7.17)
finally obtaining for the scattered intensity

$$I^{\eta_1 + \eta_2}(\varepsilon) \propto |W^{\eta_2}_{abs}|^2 \cdot |W^{\eta_1}_{em}|^2 \, \frac{\varepsilon^{1/2}}{\Gamma_{tot}(\varepsilon)}$$

$$\propto |\hat{e}_L \cdot M^{\eta_1}_{abs}|^2 \cdot |\hat{e}_S \cdot M^{\eta_2}_{em}|^2 \, \frac{\varepsilon^{1/2}}{\Gamma_{tot}(\varepsilon)} , \tag{7.19}$$

whereby we have used (7.6) to explicitly show the transition moments M^η for absorption and emission. The kinetic energy ε of the intermediate state is determined by the excitation photon energy (cf. (7.4) and Fig. 7.9). In deriving (7.19), the total lifetime broadening had to be small compared to the kinetic energy. This assumption is justified for all actually measured lifetimes, which can be as short as 500 ps, corresponding to a maximum value of $\gamma_{tot} \approx 10^{-3}\,\mathrm{meV}$ (Sect. 7.5).

While the intensity in (7.19) refers only to one valley, the total measured intensity consists of contributions due to all L-points. Since intervalley mixing in case of free exciton states in forbidden [7.53], the total intensity can be obtained by calculating the intensities for each L-point according to (7.19) and simply summing them up. An essential point to be stressed in this context is the fact that the scattered intensity will in general be polarized [7.33]. Even though the silver halides are cubic and, hence, optically isotropic with regard to both absorption and emission (Sect. 7.2), in resonant scattering according to (7.19) polarized absorption may lead to preferential excitation of certain exciton states in the various valleys that can be probed by polarized emission. This fact is expressed by the effective transition moments $|\hat{e}_L \cdot M_{abs}^{\eta_1}|^2$ and $|\hat{e}_S \cdot M_{em}^{\eta_2}|^2$ in (7.19), which give rise to polarization effects not observed by conventional absorption or emission spectroscopy. This phenomenon is completely analogous to the case of anisotropic (non-cubic) local centers where polarized excitation results in polarized luminescence or anisotropic bleaching effects utilized to determine center symmetries (for references see e.g. [7.54]). In resonant Raman scattering from silver halides, polarization is indeed observed and exploited in the time-resolved measurements presented in later sections (7.5.2, 3).

Equation (7.19) formally corresponds to the intensity (or scattering probability) that can be expected for a process of "absorption followed by emission" and follows, by definition, from the product of absorption probability times the quantum efficiency for a radiative exciton transition to the crystal ground state. According to (7.11), in the case of an indirect (allowed) exciton the absorption probability is given by the absorption coefficient $\alpha \propto |W_{abs}^\eta|^2 \varepsilon^{1/2}$. The quantum efficiency, on the other hand, is defined as ratio of total and radiative lifetimes, i.e. $Q = T_{tot}/T_{rad}$. Since $T_{rad} \propto 1/|W_{em}^\eta|^2$ and the matrix element is energy independent and constant (see above), multiplication of the two quantities results immediately in an intensity expression identical to (7.19). An important implication of these considerations is that resonant Raman scattering in the present case of an indirect exciton as intermediate (resonant) state appears to be equivalent to absorption followed by emission or hot luminescence probing the non-thermal distribution of population in the exciton state.

The question of whether resonant Raman scattering, in certain cases, is equivalent to hot luminescence was originally raised [7.55, 56] in the context of experiments in Cu_2O and has subsequently led to considerable controversy in the literature [7.57–60]. As already pointed out above, the difference between resonant Raman scattering and hot luminescence is connected in essence with the dephasing of the resonant intermediate state between light absorption and emission. Formally this problem was treated by *Shen* [7.57] who considered the effect of random fields that scatter the intermediate state. Emission of light originating from such randomized states gives rise to the dephased component, i.e. to hot luminescence, while the non-dephased component corresponds to resonant Raman scattering.

In deriving (7.19) from the expression for the two-phonon Raman intensity in (7.15) we restricted ourselves to a single intermediate state [state β in (7.15, 16)]. Since interference effects as expected for degenerate or nearly degenerate states cannot occur in this case, the phase of the state has no significance and the distinction between hot luminescence and resonant Raman scattering is meaningless. We have also assumed, by introducing the population lifetime T_1 to account for the damping of the state, (7.18), that only energy relaxation affects the intermediate state, thereby neglecting the elastic dephasing that will certainly contribute to the homogeneous linewidth (or coherence time). Due to these assumptions, it seems evident at this point that Raman processes cannot be distinguished from hot luminescence. This is formally expressed by (7.19) which accounts for the intensities of both kinds of processes under the conditions described above.

From the discussion in Sect. 7.2 we know that the intermediate exciton state is actually twofold degenerate. In this case elastic dephasing may lead to a concomitant change in polarization, which can be used to probe the coherence of the intermediate state. These polarization effects even may show up in cw experiments which therefore in principle allow us to differentiate between resonant Raman scattering and hot luminescence. Regarding the cw measurements discussed in this section, polarization properties were not exploited but will be discussed in the context of time-resolved measurements in Sects. 7.5.2, 3.

While the intensity expression in (7.19) is valid for scattering involving symmetry-allowed phonons, forbidden scattering will result in a quite different energy dependence of intensity. Naturally scattering associated with *two* forbidden phonons is extremely weak making it hard to detect this process at all in the spectrum. However, two-phonon scattering involving one forbidden and one allowed phonon [in AgBr: TA(L) and TO(L), respectively] is clearly found and can be analyzed [7.44]. The expansion of matrix elements necessary in this case, cf. (7.12), now has to be performed for either absorption or emission depending on which of the processes is accomplished by the forbidden phonon. One also has to be aware that, according to (7.4), different intermediate states with energies ε_1, ε_2 are reached for a given incident photon energy depending on the order in which the phonons participate. Writing the contributions of the two possible processes separately, the intensity finally becomes

$$I^{\eta_1 + \eta_2} + I^{\eta_2 + \eta_1} \propto |W^{\eta_1, 1}|^2 |W^{\eta_2, 0}|^2 \left(\frac{\varepsilon_1^{3/2}}{\Gamma_{\text{tot}}(\varepsilon_1)} + \frac{\varepsilon_2^{3/2}}{\Gamma_{\text{tot}}(\varepsilon_2)} \right). \tag{7.20}$$

As introduced in (7.12), $|W^{\eta_1,1}|$ is the first-order matrix element for coupling to forbidden phonons (η_1) and $|W^{\eta_2,0}|$ the zero-order matrix element involving allowed phonons (η_2). Since $|W^{\eta_1,1}_{abs}|^2 = |W^{\eta_1,1}_{em}|^2$ and $|W^{\eta_2,0}_{abs}|^2 = |W^{\eta_2,0}_{em}|^2$, we have dropped the subscripts referring to absorption and emission, assuming that one of the two matrix elements in (7.20) accounts for absorption and the other for emission. Recalling the concept of "absorption followed by emission" to describe resonant Raman scattering, the $\varepsilon^{3/2}$-dependence of scattered intensity in (7.20) could also have been obtained by replacing in (7.19) the absorption probability for allowed processes (7.11) by that for forbidden processes represented by (7.13).

Equations (7.19, 20) are crucial in studying the exciton by means of resonant Raman scattering. They suggest that $\Gamma_{tot}(\varepsilon)$ and, hence, the total exciton lifetime according to (7.18) can be investigated by measuring the scattered two-phonon intensity as a function of kinetic energy ε. The energy dependence directly reveals the effects of the various relaxation processes undergone by the exciton (Sect. 7.4.2). According to (7.4) variation of ε can easily be achieved by varying the incident photon energy of a tunable laser.

7.4.2 Exciton Energy Relaxation

Resonant Raman scattering in AgCl and AgBr under cw laser excitation has allowed the exploration of relevant relaxation processes in quite some detail and has provided a consistent picture of exciton energy relaxation. Generally, two distinctly different types of process, by definition inelastic in nature, are anticipated to affect the exciton lifetime. One is *trapping of free excitons* at defects and impurities, a process that reduces the number of excitons in the exciton band. The other is *exciton scattering by phonons* involving various exciton–phonon interaction mechanisms. In this case, the total number of excitons remains unchanged but population is transferred into exciton states of different energies. These processes are both found in the silver halides and show up very clearly in the scattering spectra. They contribute to the total scattering rate $\Gamma_{tot}(\varepsilon)$ (or lifetime broadening $\gamma_{tot}(\varepsilon)$) resulting, according to (7.19, 20) in characteristic energy dependences of two-phonon scattered intensities. Even more important, scattering of excitons with certain phonons shows up in the scattering spectra where they produce well-defined phonon sidebands to the leading two-phonon scattering lines. These reflect the relaxation processes undergone by the exciton and are crucial in the analysis and for assignments.

As an example, Fig. 7.10 illustrates typical low temperature resonant Raman scattering spectra of AgBr [7.52], each excited at slightly different exciton kinetic energies ε [related to excitation photon energy by (7.4)]. In each spectrum the scattered intensity is plotted versus the Stokes Raman shift $E_L - E_S$. The baselines are arranged along the ordinate in proportion to exciton energy ε. This kind of plot allows one to discriminate Raman

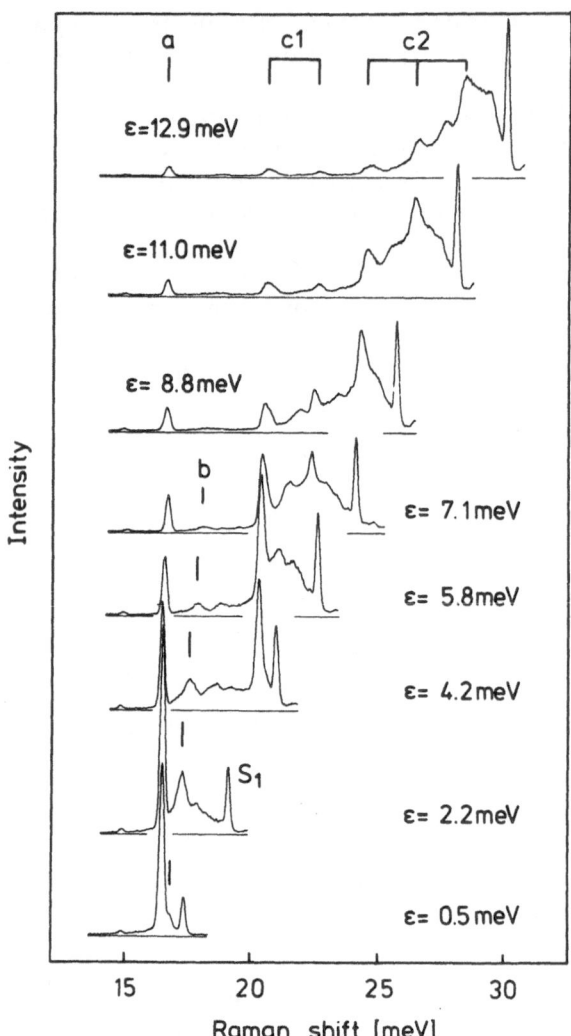

Fig. 7.10. Resonant light scattering spectra of AgBr at 1.8 K for excitation at different kinetic exciton energies ε. Notation: (a) 2 TO(L) scattering; (b) intravalley scattering by dispersive LA(Γ) phonons; (c1, c2) intervalley scattering by TA(X), LA(X) and 2TA(X), TA(X) + LA(X), 2LA(X) phonons, respectively. Forbidden TO(L) + TA(L) scattering at 14.8 meV. S_1 marks a TO(L) assisted recombination line of a weakly localized exciton state (localization energy $E_1 = 0.36$ meV). Its approximate energy position at $E_{gx}^i - \hbar\omega^{TO(L)}$ may be utilized as indicator on the energy scale in each spectrum. From [7.52]

lines (occurring at constant positions like lines a, c1, c2) from luminescence lines (shifting linearly with ε or E_L like line S_1). Scattering lines involving dispersive phonons also occur and are identified by the nonlinear dependence of their Raman shift on ε (line b).

At low ε, the dominant process in AgBr (line a) involves two (allowed) momentum-conserving TO(L) phonons. Two-phonon lines involving the allowed LA(L) phonon [TO(L) + LA(L) and 2 LA(L)] are much weaker and in most spectra stronger processes are superimposed. The other lines emerging in the spectra for higher excitation (or exciton) energies originate from intra- and intervalley scattering of excitons by means of various well-defined phonons which, besides exciton trapping, was found to be

Fig. 7.11. Resonant light scattering spectra of AgCl at 1.8 K. The lowest spectrum is excited slightly below the indirect exciton absorption edge and corresponds to second-order non-resonant Raman scattering (cf. Fig. 7.7). (D^0, X) denotes TA(L) and LA(L) assisted localized exciton transitions. From [7.62]

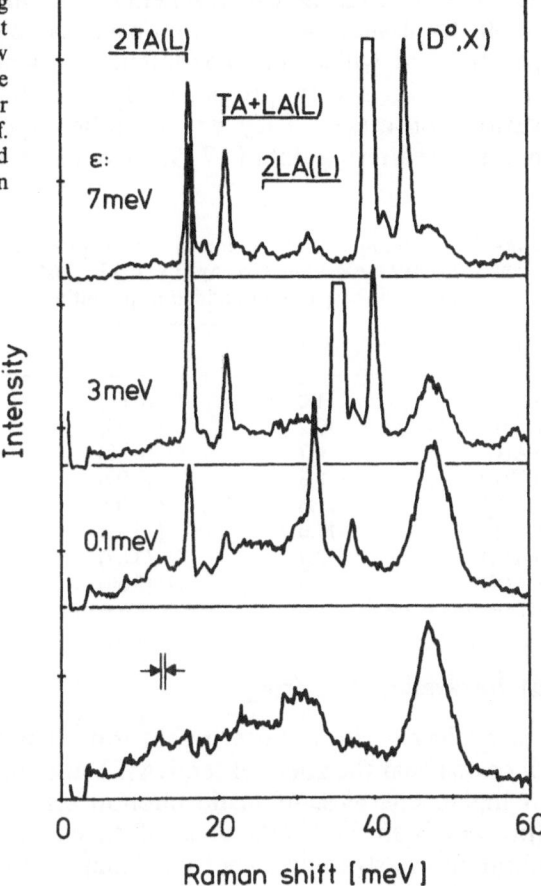

essential in exciton energy relaxation. In AgCl, the appearance of corresponding spectra, reproduced in Fig. 7.11, is slightly different [7.61, 62]. The resonantly scattered overall intensities are considerably smaller than in AgBr and comparable with the intensity of the underlying non-resonant second-order Raman scattering. As in AgBr, allowed two-phonon scattering is observed, this time with all three possible phonon combinations [processes 2 TA(L), TA(L) + LA(L) and 2 LA(L) in Fig. 7.11], while intra- and intervalley scattering is weak. These differences, however, are of quantitative rather than of qualitative origin, the relaxation processes effectively being the same in the two substances (Sect. 7.4.2d).

A detailed interpretation of the resonant scattering spectra in Figs. 7.10, 11 in terms of exciton relaxation relies on the Raman energy shifts and the resonance behavior. These will be described in the following and particular attention will be paid to the relative contribution of each process to the total lifetime broadening of the exciton. To this end, we also refer to Fig. 7.9, which schematically illustrates the various processes implicitly assuming only phonon emission as expected at the low temperatures at which the experiments are generally performed. Of course, the scattering spectra provide energies of the phonons associated with the various processes. They are identified by comparison with measured neutron scattering data [7.7–9, 63, 64] and are listed in Table 7.1.

Table 7.1. Phonon energies $\hbar\omega$ [meV] obtained from resonant light scattering at $T = 1.8$ K. Accuracy $\pm$ 0.2 meV. Values for AgBr at 4.4 K [7.64] and AgCl at 78 K [63] determined from inelastic neutron scattering are added in parentheses.

Phonon	AgBr [7.52]		AgCl [7.61, 62]	
TO(Γ)	11.0	(10.69)	–	
LO(Γ)	17.2	(17.24)	24.0	(23.9)
TA(L)	6.7	(6.68)	8.2	(8.9)
TO(L)	8.3	(7.99)	–	
LA(L)	11.8	(12.00)	12.9	(12.4)
LO(L)	17.0	(16.51)	–	
TA(X)	3.9	(3.87)	4.3	(4.1)
LA(X)	5.8	(5.66)	–	

a) Intravalley Scattering

Intravalley exciton scattering, at low exciton energies, involves acoustic phonons near the zone center, LA(Γ), and optical phonons, mainly LO(Γ), at higher energies. Acoustic phonon scattering as was studied in AgBr (process b in Figs. 7.9, 10) is of special interest since for $q \to 0$ these phonons are strongly dispersive [7.52]. As illustrated in detail in Fig. 7.12, the exciton is scattered from the initial into the final state (wavevectors k_i, k_f) by emission of a phonon with wavevector q. This process generates a sideband to the 2 TO(L) line having a total (maximum) Raman energy

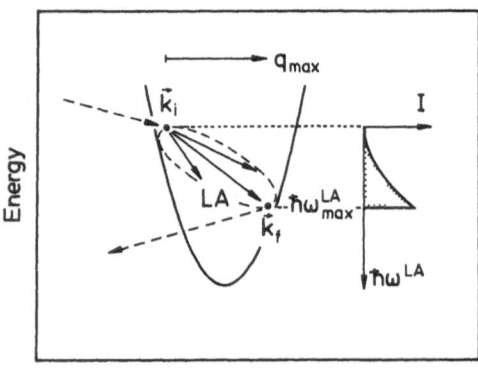

Fig. 7.12. Schematic diagram of exciton dispersion and intravalley scattering by long wavelength LA phonons. The dashed arrows correspond to TO(L) phonons. The final states k_f are along the dashed conic section. Also shown is the lineshape of the scattering spectrum expected from this scattering process

shift $E_L - E_S = 2\hbar\omega^{TO(L)} + \hbar\omega^{LA(\Gamma)}_{max}(\varepsilon)$ with the LA(Γ) phonon energy depending on the exciton (or excitation photon) energy. The phonon wavevector q is determined by conservation of wavevector and energy:

$$k_i = k_f + q,\tag{7.21}$$

$$\frac{\hbar^2 k_i^2}{2M^*} = \frac{\hbar^2 k_f^2}{2M^*} + \hbar v\,|q|,\tag{7.22}$$

where a linear phonon dispersion with sound velocity v (Fig. 7.2) and an effective (isotropic) exciton mass M^* are assumed.

The contribution $\Gamma_{ac}(\varepsilon)$ of intravalley scattering to the total scattering rate $\Gamma_{tot}(\varepsilon)$ may be calculated by Fermi's Golden Rule giving

$$\Gamma_{ac}(\varepsilon) \propto \sum_q |\langle M \rangle|^2\, \delta\left(\frac{\hbar^2 k_i^2}{2M^*} - \frac{\hbar^2 k_f^2}{2M^*} - \hbar v\,|q|\right),\tag{7.23}$$

where the sum runs over all possible phonon wavevectors. $\langle M \rangle$ is the matrix element for exciton–phonon interaction via the deformation potential which, according to *Toyozawa* [7.65], is given by

$$\langle M \rangle \equiv \langle 1s(k_f), q |\, \mathscr{V}^{LA(\Gamma)}\,|1s(k_i), 0 \rangle \propto |q|^{1/2}.\tag{7.24}$$

Carrying out the summation in (7.23) running over all possible phonon wavevectors and noting that in the wavevector region of interest $M^*v \ll |\hbar k_i|$, one finally obtains the scattering rate

$$\Gamma_{ac}(\varepsilon) = C_{ac}\varepsilon.\tag{7.25}$$

A very striking feature of scattering involving LA(Γ) phonons is the energy-dependent Raman shift (Fig. 7.13) which offers the possibility of deriving a value for the effective exciton mass [7.52, 66]. Using the requirements of wavevector and energy conservation in (7.21, 22) and introducing the kinetic energy ε of the exciton, it is straightforward to derive the energy dependence of the (peak) LA-phonon energy as

$$\hbar\omega^{LA(\Gamma)}_{max} = 2M^*v^2\left(\sqrt{\frac{2\varepsilon}{M^*v^2}} - 1\right).\tag{7.26}$$

As shown by the full line in Fig. 7.13, very good agreement of this dependence is obtained with the experimental data (points) for scattering involving one LA(Γ) phonon, with an effective mass $M^* = 1.5m_e$ as fitting parameter (v known from independent data; see also [7.52]). Using a similar analysis, a good fit is also achieved with the same mass value for the observed 2 TO(L) + 2 LA(Γ) process. In spite of this consistent result, the derived magnitude of M^* has to be treated with caution since the

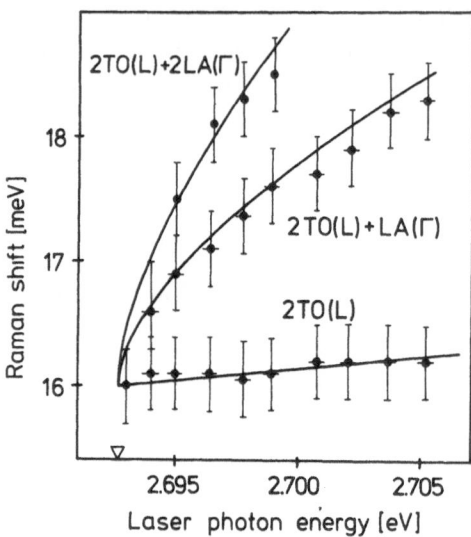

Fig. 7.13. Raman energy shift vs excitation photon energy for dispersive intravalley scattering in AgBr at 1.8 K involving one and two LA(Γ) phonons. The points are experimental data, the solid lines fits using $M^* = 1.5m_e$. The linear shift of the 2 TO(L) line is due to small dispersion of this phonon around L. The mark at the abscissa denotes the energy of the TO(L) assisted absorption edge. From [7.66]

anisotropy of the hole mass is completely neglected in the analysis so that the M^* obtained must correspond to an average value only (for further considerations concerning the exciton effective mass see [7.41]).

In principle, the effective mass can also be determined by analyzing the lineshape of the LA(Γ) scattering process. It is obvious from Fig. 7.12 that the energies of scattered phonons in this process range continuously from zero to a maximum value $\hbar\omega_{max}^{TA(L)}$, corresponding to the emitted phonon with maximum wavevector q_{max}. Accordingly, the lineshape of the phonon sideband, determined by the density of phonon states, increases up to its maximum at this cut-off energy. Taking in (7.23) the sum only over the direction of q, with $|q|$ determined by $\hbar\omega^{LA(\Gamma)} = \hbar v\,|q|$, one obtains the transition probability (or scattered intensity) in the LA(Γ) phonon sideband as

$$I(E_s) \propto \frac{q^2}{|k_i|} \propto (\hbar\omega^{LA(\Gamma)})^2 , \qquad (7.27)$$

whereby the excitation photon energy E_L was held fixed and hence $|k_i|$ was constant, see (7.2, 4). Compared to the prediction of (7.27), the experimentally observed lineshape (cf. line b in Fig. 7.9 for $\varepsilon = 2.2$ or 4.2 meV) appears more symmetric, reflecting the fact that the anisotropy of M^* must indeed be taken into account in a quantitative analysis. Essentially based on (7.27), lineshapes have also been calculated for multiple LA(Γ) scattering occurring at higher excitation energies [7.52, 67]. These calculations give overall good agreement with the experimental results although they do not include the effective mass anisotropy either.

b) Intervalley Scattering

Especially in AgBr, scattering of excitons by intervalley phonons represents a very efficient relaxation mechanism [7.68]. As shown in Fig. 7.9 (process c), exciton scattering across the Brillouin zone takes place between inequivalent L-points (having wavevectors k_L and $k_{L'}$) and, at low temperature, is accompanied by emission of an X-point phonon required by wavevector conservation ($q_X = k_L + k_{L'}$). The total Raman shift is $E_L - E_S = 2\hbar\omega^{TO(L)} + \Sigma \hbar\omega_i^X$ where $\hbar\omega_i^X$ stands for the energy of the participating phonons TA(X) or LA(X) which are observed. Fig. 7.14 displays spectra in AgBr most clearly representing the processes associated with emission of one or two X-point phonons.

From Fig. 7.9 it is evident that these processes should have an activation threshold at initial exciton state energies $\varepsilon = \Sigma \hbar\omega_i^X$. This is entirely confirmed and obvious from inspecting Fig. 7.15 in which the Raman shift of various intervalley scattering lines in AgBr is plotted versus excitation photon energy. As compared with the 2 TO(L) process [activated at the TO(L) assisted absorption threshold], scattering with additional X-

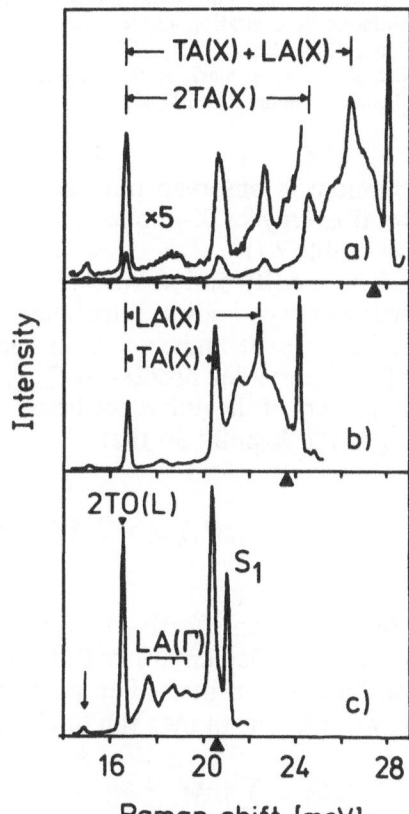

Fig. 7.14. Intervalley scattering by X-point phonons in AgBr at 1.8 K for different excitation photon energies. E_L: (a) 2.7037 eV, (b) 2.6998 eV, (c) 2.6969 eV. LA(Γ) denotes multiple intravalley scattering, S_1 is a luminescence line. The arrow on the absicssa marks the energy $E_{gx}^i - \hbar\omega^{TO(L)}$. From [7.68]

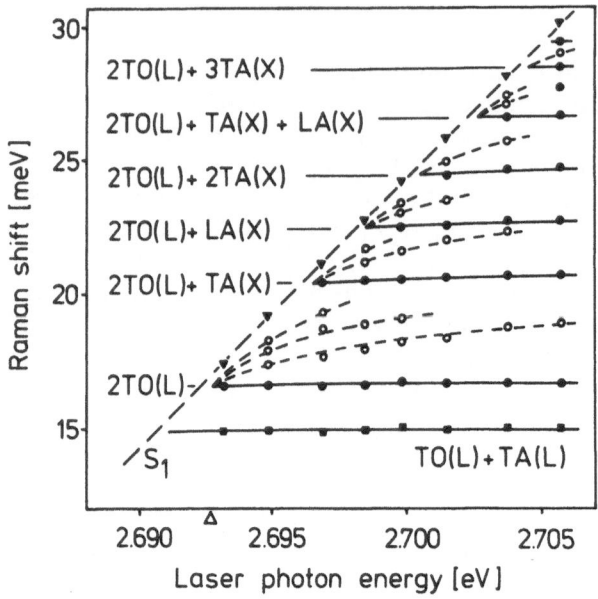

Fig. 7.15. Raman shift of various scattering processes in AgBr involving TA(X) and LA(X) phonons vs excitation photon energy. The diagonal marks the energy position of the luminescence line S_l. Dispersive LA(Γ) and forbidden TO(L) + TA(L) phonon scattering is also shown. The mark at the abscissa denotes the energy of the TO(L) assisted exciton absorption edge. From [7.68]

phonons is observed only at excitation energies higher than this by the total energy of X-phonons that participate (for corresponding energies see Table 7.1).

As TA(X) and LA(X) phonons have odd parity, exciton scattering between two L-points is forbidden [7.51] becoming gradually allowed away from L, i.e. at higher exciton energies. To derive the contribution $\Gamma_{iv}(\varepsilon)$ of the intervalley process to $\Gamma_{tot}(\varepsilon)$, the matrix element has to be expanded in powers of the difference between phonon wavevector $\boldsymbol{q}$ and wavevector $\boldsymbol{k}_X$ at the X-point so that

$$\langle M \rangle \equiv \langle 1s(\boldsymbol{k}_f), \boldsymbol{q}| \, \mathcal{V}^X \, |1s(\boldsymbol{k}_i), 0 \rangle = \left(\frac{\hbar}{2Nm\omega^X} \right)^{1/2} D_{iv} |\boldsymbol{q} - \boldsymbol{k}_X|. \quad (7.28)$$

Here D_{iv} is an effective first-order deformation potential [7.81], m is the mass of an elementary cell of the crystal and N the number of cells. The probability for intervalley scattering of an exciton from the initial (energy ε, wavevector $\boldsymbol{k}_i$) into the final exciton state (ε', $\boldsymbol{k}_f$) will then be given by

$$\Gamma_{iv}(\varepsilon) \propto \sum_{\boldsymbol{k}_f} |\langle M \rangle|^2 \, \delta(\varepsilon - \varepsilon' - \hbar\omega^X). \quad (7.29)$$

Neglecting dispersion of the participating X-phonon, the summation can be carried out [7.68] giving

$$\Gamma_{iv}(\varepsilon) = C_{iv}(\varepsilon - \hbar\omega^X)^{1/2} (2\varepsilon - \hbar\omega^X), \tag{7.30}$$

where $\Gamma_{iv}(\varepsilon) = 0$ for $\varepsilon \lesssim \hbar\omega^X$ as anticipated from the considerations above.

c) Exciton Trapping at Defects and Impurities

Excitation of the two silver halides in the free exciton absorption results not only in scattering but also in luminescence occurring at energies below the exciton gaps. This luminescence is due to recombination of excitons at bound states of a quite different nature (for a recent review see [7.36]). It occurs with appreciable intensity even in very pure material containing only residual impurities, indicating that trapping of excitons can be a very efficient relaxation channel. Of course, the trapped excitons no longer contribute to the resonant scattering spectrum reducing its overall intensity. Since resonant trapping of free excitons [7.69] is found experimentally to have no effect on the two-phonon scattered intensity, we assume the exciton trapping probability to be independent of exciton energy ε resulting in a constant contribution to $\Gamma_{tot}(\varepsilon)$ given by

$$\Gamma_{trap} = C_{trap}. \tag{7.31}$$

Clearly we expect Γ_{trap} to be sample dependent due to different types and concentrations of defect and impurity states.

d) Comparison of the Model with Experimental Results

The relaxation processes considered above all contribute to the total lifetime broadening of the exciton state and determine its energy dependence. In terms of scattering rates, $\Gamma_{tot}(\varepsilon)$ is therefore given by

$$\begin{aligned} \Gamma_{tot}(\varepsilon) &= \Gamma_{trap} + \Gamma_{ac}(\varepsilon) + \Gamma_{iv}(\varepsilon) \\ &= C_{trap} + C_{ac}\varepsilon + C_{iv}(\varepsilon - \hbar\omega^{TA(X)})^{1/2} (2\varepsilon - \hbar\omega^{TA(X)}), \end{aligned} \tag{7.32}$$

with $\Gamma_{iv}(\varepsilon) = 0$ for $\varepsilon \lesssim \hbar\omega^{TA(X)}$. To obtain the explicit ε-dependence in (7.32), we have used (7.25, 30, 31), but omitted a term describing LA(X) intervalley scattering, since it is weaker than TA(X) scattering and plays a role only at somewhat higher ε. Also neglected is the effect of radiative recombination, since for indirect exciton transitions, being second-order processes, the radiative lifetime is expected to be long and, hence, $\Gamma_{rad} \propto 1/T_{rad}$ is small [7.70]. Based on the absorption strength, the radiative lifetime of the indirect exciton in AgBr is estimated as $T_{rad} \simeq 10^{-4}$ s [7.71]. This value has to be compared with the (total) lifetimes actually measured in AgBr. These are sample dependent and in all cases shorter by orders of magnitude (between about 60 ns and 500 ps; Sect. 7.5) showing that the radiative lifetime broadening is always negligible compared with contributions from other relaxation processes.

Table 7.2. Relative fitting parameters (normalized to C_{ac}) describing exciton trapping, acoustic LA(Γ) intervalley and TA(X) intravalley scattering, and relative matrix elements. The data are obtained from cw resonant Raman scattering except the ratio of matrix elements in AgBr which follows from analyzing the exciton absorption edge. Notation for phonons: η_1: TA(L) for AgCl and TO(L) for AgBr; η_2: LA(L) in AgCl and AgBr

C_{trap} [s^{-1}]	C_{ac} [s^{-1}(meV)$^{-1}$]	C_{iv} [s^{-1}(meV)$^{-3/2}$]	$|W^{\eta_2}|^2/|W^{\eta_1}|^2$
AgCl 47.6[a]	1[a]	0.48[a]	0.34[a]
AgBr 0.77[b]	1[b]	0.80[b]	0.16[c]

[a] From Ref. [7.62]
[b] From Ref. [7.52]
[c] From Ref. [7.44]

By means of (7.32), the expressions for the two-phonon scattered intensities in (7.19, 20) can be written explicitly as a function of exciton kinetic energy and used to fit the measured data by determining the relative contribution of each relaxation process. For the observed allowed scattering [in AgCl: 2 TA(L), TA(L) + LA(L), 2 LA(L); in AgBr: 2 TO(L)] the model in both silver halides nicely reproduces the measured intensities with the parameters (normalized to C_{ac}, see below) listed in Table 7.2. In case of AgBr, which is illustrated as an example in Fig. 7.16, some

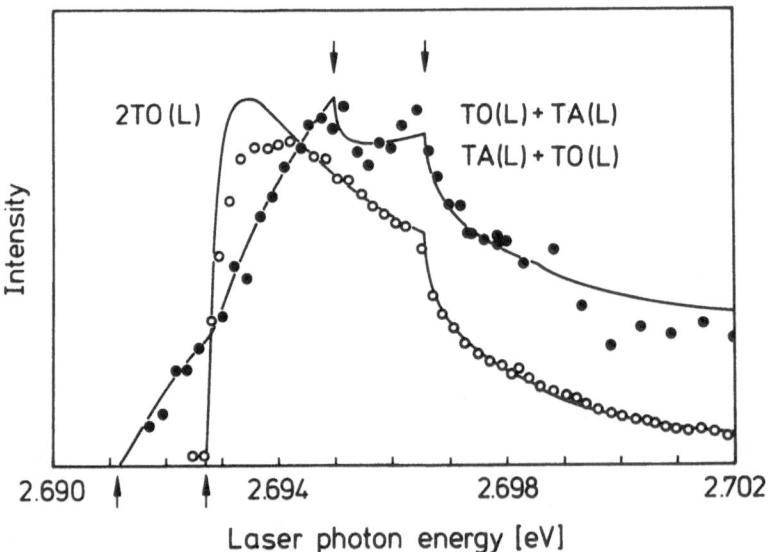

Fig. 7.16. Integrated scattered intensity as function of excitation photon energy for allowed 2 TO(L) and partially forbidden TO(L) + TA(L) and TA(L) + TO(L) resonant Raman scattering in AgBr at 1.8 K. The open and full circles are experimental data, the solid lines fits according to (7.19, 20), respectively. The intensities are normalized with the 2 TO(L) curve reduced by a factor of about 50. Upward arrows mark TA(L) and TO(L) assisted absorption thresholds, downward arrows corresponding onsets of intervalley TA(X) scattering. Data taken from [7.44]

discrepancy between theory and experiment occurs near the TO(L) absorption edge. This is presumably related to polarization effects or additional exciton trapping at weakly bound states neglected in the analysis. In AgCl (not shown) qualitatively very similar results are obtained, the larger trapping probability in this material (Table 7.2) resulting in a different appearance of the corresponding dependence [Ref. 7.62, Fig. 3]. Since here scattering is observed involving both possible allowed L-point phonons, from the energy-dependent line intensities in this case one can additionally determine the ratio of corresponding matrix elements.

Included in Fig. 7.16 for AgBr are the data for the partially forbidden process involving one TA(L) and one TO(L) phonon (line at 14.8 meV in Figs. 7.10, 15), together with a fit according to (7.20) [7.44]. The set of fitting parameters is very close to that in Table 7.2 except for C_{trap}, which is reasonable because the measurements were performed in different samples. Even though the measured intensities in the case of forbidden scattering are relatively small and the data scatter quite a bit, the general trend is well reproduced by the model. In particular very good agreement is found at the thresholds where specific scattering processes are initiated and the intensities exhibit pronounced discontinuities. In contrast to allowed 2 TO(L) scattering, forbidden scattering is activated at slightly lower energy, namely at the onset of forbidden TA(L) absorption at energy $E_L = E_{gx}^i + \hbar\omega^{TA(L)}$ ($\varepsilon_1 = 0$ in (7.20)). Considering the process as "absorption followed by emission" (Sect. 7.4.1), here absorption is aided by the forbidden TA(L) phonon while emission is associated with the allowed TO(L) phonon. The reverse process, i.e. TO(L)-assisted absorption followed by TA(L)-assisted emission, like 2 TO(L) scattering, is activated at the TO(L) edge [$\varepsilon_2 = 0$ in (7.20)] and in Fig. 7.16 gives rise to a clearly visible steeper increase in (total) scattering intensity. Corresponding to the two thresholds for TA(L) + TO(L) and TO(L) + TA(L) scattering, consistent with expectation, intervalley scattering by means of TA(X) phonons leads to reductions in line intensity at energies $\varepsilon_1 = \hbar\omega^{TA(X)}$ and $\varepsilon_2 = \hbar\omega^{TA(X)}$, which are quite obvious in the figure. In addition, from the fit the ratio of first to zero-order matrix elements can be deduced and is found to be $|W^{TA(L), 1}|^2/|W^{TO(L), 0}|^2 = 5.5 \times 10^{-3}$ (meV)$^{-1}$ [44].

It is worth mentioning that forbidden exciton transitions in silver halides had not been observed previously, either in absorption or luminescence. Since they are weak, the forbidden components for most exciton energies are hidden behind the much stronger allowed components and therefore totally obscured (cf. e.g. Fig. 7.5). The advantage of resonant Raman scattering here is evident: the (partially) forbidden and the allowed components in the scattering spectra give rise to two well-separated distinct narrow lines at different Raman energy shifts (in the example discussed at $\hbar\omega^{TA(L)} + \hbar\omega^{TO(L)}$ and $2\hbar\omega^{TO(L)}$). These lines can be investigated independently, the desired information on matrix elements and effective relaxation processes being contained in the energy-dependent intensities.

Although the scattering processes in AgCl and AgBr are basically identical, the fitting parameters in Table 7.2 show important quantitative differences between the two substances. While for AgBr C_{trap}, C_{ac} and C_{iv} are all similar in magnitude, in AgCl trapping is predominant. Actually, compared to C_{ac} and C_{iv}, for this material the parameter C_{trap} is larger by nearly two orders of magnitude. This reflects the well-known fact that, contrary to the case of AgBr, free excitons in AgCl are effectively trapped during relaxation to form deeply localized states known to be additionally stabilized by local lattice distortions; see e.g. [7.36].

To compare the effect of the various relaxation processes at different exciton energies and also their relative importance in AgCl and AgBr, it is reasonable to assume that the deformation potentials for intravalley LA(Γ) phonon scattering are not very different in the two materials. In Table 7.2 we have therefore normalized the fitting parameters to C_{ac} making $\Gamma_{ac}(\varepsilon)$ identical in both materials. With this assumption, all contributions to the total scattering rate $\Gamma_{tot}(\varepsilon)$ are plotted in Fig. 7.17 as a function of exciton kinetic energy [7.72]. The figure demonstrates that in AgCl exciton trapping is the dominant relaxation mechanism in the entire energy range investigated. In contrast, for AgBr trapping is much less important and plays a role only very close to the exciton absorption edge (at $\varepsilon = 0$) while at only slightly higher energies intra- and intervalley exciton scattering prevail and become increasingly stronger.

As a result of the efficient exciton trapping in AgCl, luminescence due to free excitons does not occur at all. Resonance Raman scattering in this material is thus really unique since it provides information on the

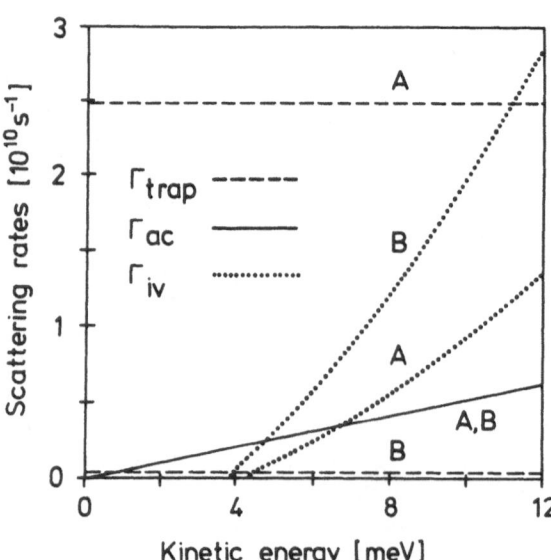

Fig. 7.17. Relative rates for exciton trapping and intra- and intervalley scattering as function of exciton kinetic energy ε in AgCl (A) and AgBr (B). $\Gamma_{ac}(\varepsilon)$ is assumed to be equal in the two materials. The energy relaxation time $T_1 = 2.5$ ns measured in the AgBr sample close to $\varepsilon = 0$ was used to obtain the scale of the ordinate. From [7.72]

(unrelaxed) free exciton state not available by any other means. In particular it allows some estimate of the free exciton lifetime [7.62], which, due to the lack of free exciton recombination, is otherwise unknown. Since very close to the absorption edge in the two silver halides the total scattering rate is merely determined by exciton trapping, the free exciton lifetime in AgCl can be deduced from the measured lifetime in AgBr and the ratio of trapping probabilities according to

$$T_1(\text{AgCl}) = T_1(\text{AgBr}) \frac{\Gamma_{\text{trap}}(\text{AgBr})}{\Gamma_{\text{trap}}(\text{AgCl})}. \tag{7.33}$$

With the value $T_1(\text{AgBr}) = 2.5$ ns measured in the sample used for the comparison, the free exciton lifetime in AgCl becomes $T_1(\text{AgCl}) = 65$ ps. This magnitude suggests that in AgCl, like in AgBr (cf. Sect. 7.5), picosecond time-resolved methods may be applicable to resonant Raman scattering to directly study the dynamics of exciton relaxation.

e) Scattering at Higher Exciton Energies

To explore the effect of excited ($n = 2, 3 \ldots$) exciton states and in an attempt to derive a value for the exciton binding energy, the resonance Raman measurements in AgBr were extended to exciton kinetic energies as high as $\varepsilon \approx 40$ meV. As pointed out before (Sect. 7.2), due to the $1/n^2$-dependence of absorption strength, scattering that could be unambiguously attributed to these states was not revealed [7.41]. Although at these energies some structure is observed in the 2 TO(L) scattering cross-section, the intensities are much too weak to allow definite assignments. Instead, the main effect found in this energy range is intravalley scattering by zone-center optical phonons $\text{LO}(\Gamma)$ [7.73]. These have energy $\hbar\omega^{\text{LO}(\Gamma)} = 17.0$ meV and lead to especially drastic changes in the scattering spectrum if excitons are excited at multiples of this value above the bottom of the exciton band, i.e. at $\varepsilon = m\hbar\omega^{\text{LO}(\Gamma)}$ ($m = 1, 2 \ldots$) or corresponding excitation energies $E_L = E_{\text{gx}}^i + \hbar\omega^\eta + m\hbar\omega^{\text{LO}(\Gamma)}$ [η = momentum-conserving L-point phonon; see (7.4)]. Figure 7.18 reproduces resonant scattering spectra excited around $\varepsilon = 1\hbar\omega^{\text{LO}(\Gamma)}$ making the effect of LO scattering very obvious. All spectra are recorded in the energy range around $E_{\text{gx}}^i - \hbar\omega^{\text{TO}(L)}$ as approximately marked by the position of the luminescence line S_1 (cf. caption of Fig. 7.10). In spectrum (a) the kinetic energy of the exciton is slightly smaller than the $\text{LO}(\Gamma)$ phonon energy. Relaxation to lower energies here can proceed exclusively by multiple scattering of intravalley $\text{LA}(\Gamma)$ and intervalley $\text{TA}(X)$ and $\text{LA}(X)$ phonons. The superposition of all these processes smears out any detailed structure and gives rise to the relatively broad spectral feature. Spectrum (b) corresponds to excitons created at energies just slightly above $\varepsilon = 1\hbar\omega^{\text{LO}(\Gamma)}$, the narrow peak at Raman shift $E_L - E_S = 2\,\text{TO(L)} + \text{LO}(\Gamma) = 33.8$ meV reflecting the fact that intravalley LO scattering prevails over all other relaxation

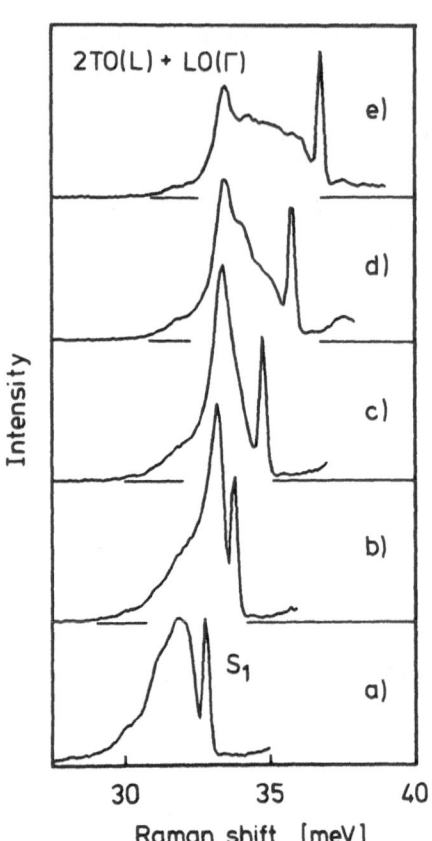

Fig. 7.18. Resonant light scattering in AgBr at 1.8 K at high excitation photon energies. E_L: (a) 2.7090 eV, (b) 2.7100 eV, (c) 2.7115 eV, (d) 2.7125 eV, (e) 2.7135 eV. The resonantly enhanced strong peak denoted 2 TO(L) + LO(Γ) corresponds to intravalley scattering of excitons by LO(Γ) phonons. S_1 labels the luminescence line (cf. caption of Fig. 7.10). From [7.73]

channels. At still higher ε [spectra (c–e)], all scattering lines [2 TO(L) + LA(Γ), 2 TO(L) + TA(X), etc.] of Fig. 7.10 practically repeat, shifted however in Raman energy by an additional $\hbar\omega^{LO(\Gamma)}$ and considerably broadened due to the LO(Γ) participation. Because of this broadening, a quantitative analysis is quite complex and has not been performed so far. Analogous spectra, increasingly broadened, are observed for exciton energies allowing multiple LO(Γ) scattering. In these processes only very weak intensity is scattered [e.g. by TO(L) emission] from the intermediate states reached when the exciton cascades down the dispersion curve by multiple phonon emission. This implies that the LO(Γ) relaxation is fast and determines the lifetime in these states [7.73].

Exciton relaxation by multiple phonon scattering as discussed above is crucial to explain the origin of the recombination luminescence lineshape in AgBr found under (monochromatic) excitation at energies much higher than those considered before. As anticipated for indirect free excitons (see e.g. [7.41]), the luminescence band at low temperature is Maxwellian in shape, suggesting, at first glance, thermalized excitons. The detailed analysis, however, shows that the exciton temperature derived from a lineshape fit is considerably larger than the lattice (or sample) temperature [7.6]. Like for excitation at low ε (see above), this implies luminescence from unrelaxed states. It means that even in case of multiple exciton-phonon scattering thermal equilibrium between the exciton and the phonon system cannot be completely accomplished within the exciton lifetime, and the lineshape is determined by these scattering processes. At least part of the total emission, however, originates from thermalized excitons and may be considered ordinary luminescence. This can be seen in the series of spectra in Figs. 7.10, 18 where a background of intensity is seen to shift in a "luminescence-like" fashion parallel to the S_1 line.

7.4.3 Resonant Raman Scattering in External Fields

A problem encountered in analyzing the effect of external perturbations on the indirect exciton state in silver halides with conventional spectroscopic techniques is related to the continuous absorption and the relatively broad luminescence spectra. Unless special modulation techniques are applied, external fields here give rise only to small changes in lineshape [7.34, 35, 74] rather than to the line splittings and clear intensity variations that often occur in the case of narrow direct excitons. As already discussed in connection with forbidden scattering processes (Sect. 7.4.2), in resonant Raman scattering clearly discernible narrow lines occur due to well-defined scattering processes that can be analyzed in a straightforward way. It is the line intensity, in principle easily detectable, that contains information on the resonant intermediate exciton state. Mixing of states, or energy shifts and splittings, if caused by the external perturbation, will therefore lead to typical changes of intensity either through effects on matrix elements or resonance behavior.

Exploiting these advantages, resonant Raman scattering has been successfully used to investigate the $\Gamma_6^+ \otimes L_{4,5}^-$ indirect exciton in AgBr under hydrostatic pressure up to 0.7 GPa and in magnetic fields up to 20 T. These investigations allowed a detailed analysis of states and the determination of important quantities like deformation potentials and various magnetic parameters.

a) Effects of Hydrostatic Pressure

To investigate the effect of hydrostatic pressure on the indirect exciton state, resonant light scattering as described above is an ideal method and was employed in a study in AgBr [7.75]. Hydrostatic pressure, through the change in unit cell volume, affects not only the electronic (or excitonic) but also the phonon energies. The resonant scattering spectra reflect both

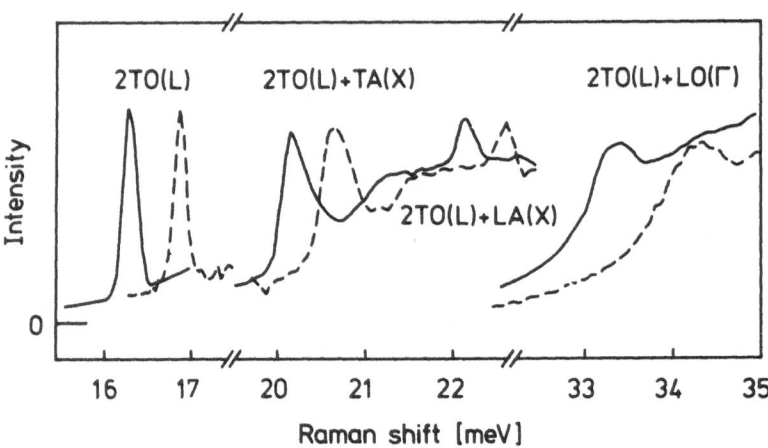

Fig. 7.19. Effect of hydrostatic pressure on various resonant scattering lines in AgBr at 7 K. Solid and dashed lines correspond to zero pressure and $P = 0.573$ GPa, respectively. The excitons were excited with kinetic energies $\varepsilon = 3.1$ meV for 2TO(L), $\varepsilon = 7.7$ meV for 2 TO(L) + TA(X) and 2 TO(L) + LA(X), $\varepsilon = 20.6$ meV for 2 TO(L) + LO(Γ). From [7.75]

these changes simultaneously and allow one to separate the corresponding contributions to the pressure shift of the absorption edge. If we consider for example the prominent TO(L) phonon-aided exciton absorption edge, the pressure change of its energy according to (7.4) (setting $\varepsilon = 0$) is given by

$$dE_{ax}^{TO(L)}/dP = dE_{gx}^{i}/dP + d(\hbar\omega^{TO(L)})/dP \qquad (7.34)$$

with the (pressure-dependent) indirect exciton gap energy E_{gx}^{i} and phonon energy $\hbar\omega^{TO(L)}$. Experimentally $dE_{ax}^{TO(L)}/dP$ can be directly deduced from the threshold energy at with 2 TO(L) scattering is activated (e.g. Fig. 7.16 for zero pressure) and which shifts as function of pressure. The pressure dependence of the phonon energy, on the other hand, is taken from the observed 2 TO(L) Raman shift (Fig. 7.19) allowing us to determine dE_{gx}^{i}/dP. The experiments are in principle straightforward. The cryostat in the spectroscopic setup (cf. Sect. 7.1) was replaced by one equipped with an optical high pressure cell that allowed measurements up to 0.7 GPa at temperatures as low as 7 K. In AgBr, the data obtained from a complete analysis of all results include the indirect exciton deformation potential and mode Grüneisen parameters for several phonons, in particular the off-center phonons TA(X) and LA(X) associated with exciton relaxation. All quantities derived are summarized in Table 7.3; the electronic band structure and lattice properties of silver halides [7.75] will not be considered further here.

Table 7.3. Quantities deduced from resonant light scattering in AgBr at 7 K under hydrostatic pressure [7.75]

a. Phonon energies $\hbar\omega_0$ at normal pressure, pressure shift of phonon energies $d(\hbar\omega)/dP$ and mode Grüneisen parameter γ_G

Phonon	$\hbar\omega_0$	$d(\hbar\omega)/dP$	γ_G
LO(Γ)	17.1	+0.66	$+1.7 \pm 0.8$
TO(L)	8.2	+0.51	$+2.8 \pm 0.4$
TA(X)	3.8	−0.27	-3.2 ± 0.8
LA(X)	5.8	−0.21	-1.6 ± 0.8

b. Pressure coefficients and deformation potential of indirect exciton

absorption edge energy $dE_{ax}^{TO(L)}/dP$: $-(14.6 \pm 0.1)$ meV/GPa
exciton gap energy dE_{gx}^i/dP: $-(15.1 \pm 0.2)$ meV/GPa
deformation potential D_{gx}^i: (0.68 ± 0.02) eV

b) Effects of High Magnetic Fields

A beautiful example of the use of resonant Raman scattering in the silver halides is the investigation of magnetic properties of the indirect $\Gamma_6^+ \otimes L_{4,5}^-$ exciton carried out in AgBr [7.76, 77]. Here again this technique is superior to the usual magneto-optical methods and enables one to obtain reliable values of the electron–hole exchange energy Δ and g-values of states. As already discussed in Sect. 7.2, at zero field the exciton consists of two pairs of singlet-triplet and pure triplet states, each twofold degenerate and separated in energy by Δ with the triplet state absorption being negligibly small. Basically the application of a magnetic field removes the spin-degeneracies resulting in shifts and mixing of states. Using, in the missing electron scheme, effective hole and electron spins $\sigma_z = \pm\frac{1}{2}$ ($L_{4,5}^-$) and $s_z = \pm\frac{1}{2}$ (Γ_6^+), the effective Hamiltonian describing the shift and mixing of the states in a magnetic field $\boldsymbol{B}$ is given by [7.34, 77]

$$\mathscr{H}_{ex} = 2\Delta\left(\sigma_z s_z - \frac{1}{4}\right) + g_c\mu_B \boldsymbol{s} \cdot \boldsymbol{B} - g_v^\parallel \mu_B \sigma_z B_z$$

$$+ \frac{1}{2E_{bx}^i \mu^{*2}} \mu_B^2 B^2 + C_0(2B_z^2 - B_x^2 - B_y^2). \tag{7.35}$$

In (7.35) the first term describes the exchange interaction, whereby the energy scale without magnetic field is chosen to be zero at the position of the degenerate singlet-triplet states. The next two terms represent the spin-Zeeman interaction of the conduction and valence bands with effective electron and (longitudinal) hole g-factors (g_c, $g_v^\parallel$) and Bohr magneton μ_B.

The remaining terms account for the diamagnetic shift that consists of an isotropic part (depending on exciton binding energy E_{bx}^i and reduced effective exciton mass μ^*) and an anisotropic part due to the effective mass anisotropy [7.78]. The components of the magnetic field are referred to the [111] direction as z-axis. The $L_{4,5}^-$ hole symmetry results in the peculiar form of the exchange term and in the transverse hole g-factor $g_v^\perp = 0$ [7.34].

In the actual experiments, the magnetic field direction was chosen as $\boldsymbol{B} \parallel [100]$. This highly symmetric geometry greatly facilitates the analysis since the four L-point excitons "feel" equal field components and the anisotropy term vanishes. The energies of the four exciton states $(j = 1, ..., 4)$ are determined by the eigenvalues of $\mathscr{H}_{ex}$ and, relative to the singlet-triplet exciton gap (at $E_{1,4} = 0$ for $B = 0$), in this case take the form

$$\Delta E_{1,2} = -\frac{1}{2}(\Delta + g_v^\parallel \mu_B B_z) \pm \frac{1}{2}[(\Delta + g_c \mu_B B_z)^2$$

$$+ g_c^2 \mu_B^2 (\boldsymbol{B}^2 - B_z)]^{1/2} + \frac{\mu_B^2 \boldsymbol{B}^2}{2 E_{bx}^i \mu^{*2}},$$

$$\Delta E_{3,4} = \frac{1}{2}(-\Delta + g_v^\parallel \mu_B B_z) \mp \frac{1}{2}[(\Delta - g_c \mu_B B_z)^2$$

$$+ g_c^2 \mu_B^2 (\boldsymbol{B}^2 - B_z)]^{1/2} + \frac{\mu_B^2 \boldsymbol{B}^2}{2 E_{bx}^i \mu^{*2}}. \tag{7.36}$$

The corresponding wavefunctions $\Phi_j (j = 1, ..., 4)$ are obtained from linear combinations of the zero-field functions represented by (7.1). They are given by [7.77]

$$\Phi_{1,2} = \alpha_{1,2}^S \Phi_1^0 + \alpha_{1,2}^T \Phi_2^0, \qquad \Phi_{3,4} = \alpha_{3,4}^T \Phi_3^0 + \alpha_{3,4}^S \Phi_4^0, \tag{7.37}$$

where α_j^S, α_j^T denote the field-dependent singlet (S) and triplet (T) amplitudes, which can be shown to have the form

$$\frac{1}{|\alpha_1^S|^2} = 1 + 4 \frac{[\Delta E_1 - \frac{1}{2}(g_c - g_v^\parallel)\mu_B B_z]^2}{g_c^2 \mu_B^2 (\boldsymbol{B}^2 - B_z^2)} \tag{7.38}$$

and similarly for the other coefficients.

Experimentally the effect of the magnetic field on the exciton states was investigated by means of the 2 TO(L) scattering line in AgBr. Assuming, as before, simple parabolic exciton bands with equal effective masses for all exciton states j and bearing in mind that each exciton

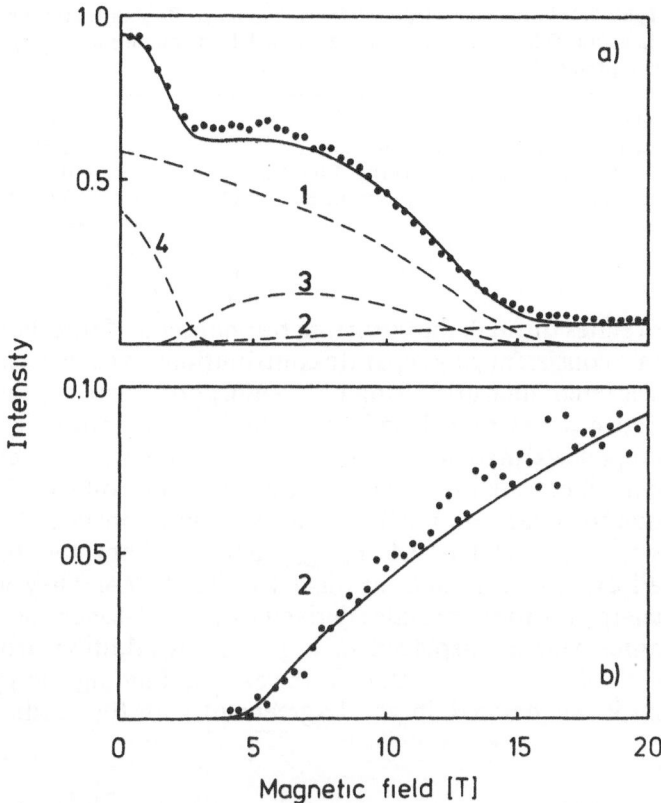

Fig. 7.20. 2 TO(L) scattered intensity as function of magnetic field. (a) and (b) correspond to excitation above ($E_L = 2.6928_3$ eV) and below ($E_L = 2.6921_4$ eV) the zero-field TO(L) absorption edge which was determined here at $E_{ax}^{TO(L)} = 2.6927_1$ eV. Points: experimental data. Solid lines: computed from (7.38–40) as described in the text. Dashed lines correspond to individual contributions from exciton states $j = 1, ..., 4$. From [7.77]

contributes to the total intensity only via its singlet component α_j^S, the 2 TO(L) intensity according to (7.19) may be written as

$$I^{2\,TO(L)} \propto \sum_{j=1}^{4} |\alpha_j^S|^4 \cdot \frac{\varepsilon_j^{1/2}}{\Gamma_{tot,j}(\varepsilon_j)}, \qquad (7.39)$$

where the kinetic energy of each state now is given as [cf. (7.4)]

$$\varepsilon_j = E_L - E_{gx}^i - \Delta E_j - \hbar\omega^{TO(L)}. \qquad (7.40)$$

Figure 7.20 represents typical experimental data from a series of measurements taken with various excitation photon energies E_L. In each of these measurements E_L was kept constant and the 2 TO(L) scattered intensity was detected as function of magnetic field. This scheme of measurement enables one to identify the (field-dependent) energy positions

Table 7.4. Parameters deduced from resonance Raman scattering in AgBr at 2 K in high magnetic fields [7.77]. Values determined from quantum beat spectroscopy are added in parentheses [7.27]

g-factor of electron state g_c:	1.46 ± 0.05	(1.41)
g-factor of (longitudinal) hole state $g_v^{\parallel}$:	2.61 ± 0.05	(2.64)
electron–hole exchange energy Δ (meV):	0.17 ± 0.01	(0.13)
binding energy of indirect exciton E_{bx}^i (meV):	32 ± 5	–

of states through the occurring resonances and thresholds and to determine in a consistent way separate contributions of each exciton state to the total scattered intensity. Using (7.38–40), good agreement is generally achieved between the model and the experimental results, except for some discrepancies in relative intensities between states 1 and 4 which should contribute equally to the total intensity in zero-field. These are presumably due to polarization effects that were not considered in the analysis. The set of parameters (Δ, g_c, $g_v^{\parallel}$) derived from the best fit of (7.39) to all experimental data are listed in Table 7.4. They were used to fit the energy positions of states derived from the measurements, also giving good agreement as displayed in Fig. 7.21. In addition, from the diamagnetic shift of the states, a value for the exciton binding energy of $E_{bx}^i = (32 \pm 5)$ meV was derived, in good agreement with that deduced in Sect. 7.2.

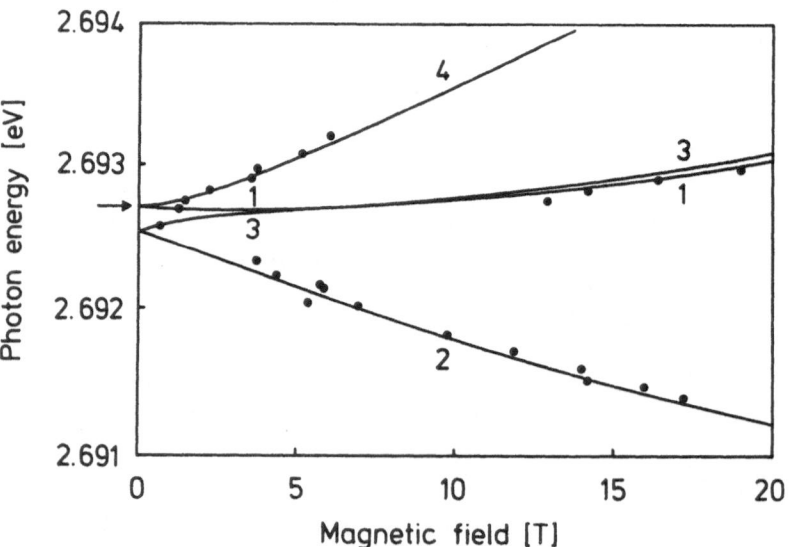

Fig. 7.21. Energies of singlet-triplet (1, 4) and pure triplet (2, 3) exciton states as function of magnetic field. The experimental data (points) are obtained from the field dependence of the 2 TO(L) intensity (Fig. 7.20), the solid lines are fits according to (7.36). The arrow marks the TO(L) assisted indirect absorption edge at zero-field. From [7.77]

7.5 Picosecond Time-Resolved Light Scattering

From the discussion in the preceding section, it is quite obvious that time-resolved measurements of resonant Raman scattering will provide additional and more direct knowledge of exciton relaxation, in particular of its dynamics. As we have shown, the steady-state scattering spectrum in AgCl and AgBr can be considered as due to hot luminescence directly imaging the non-thermal energy distribution of the excitons. Quite generally, excitons created at some high energy with a narrow energy distribution will thermalize by scattering with lattice phonons. In order to finally attain thermal equilibrium with the lattice, the exciton lifetime prior to annihilation either by radiative and non-radiative recombination or by exciton trapping has to be sufficiently long for a large number of scattering events to take place. This is not the case in silver halides. Here the lifetime is instead determined by trapping, which, for low exciton energies, occurs with rates comparable to exciton–phonon scattering. Furthermore, in this energy range, due to the restrictions of energy and wavevector conservation, only a few selected phonon processes are possible, whereby at low temperature phonon absorption is nearly absent, which could support thermalization of the exciton–lattice system.

The advantage of time-resolved spectroscopy is that the various times (or rates) involved in exciton relaxation can be directly measured. In particular, they can be followed up as function of exciton energy to test the model developed in the previous sections and make it quantitative. The absolute exciton–phonon scattering rates obtained from the measurements enable one to deduce the corresponding interaction energies so that additional important information becomes available beyond that of the cw experiments.

It is has been suggested that, although resonant Raman scattering and hot luminescence under certain circumstances are indistinguishable, time-resolved measurements may allow their separation [7.22, 57]. In attempting to tackle this question, besides usual time-resolving techniques, we have recently employed quantum beat spectroscopy to reveal the coherence properties of the exciton and to include these into a complete analysis of exciton relaxation. Correspondingly, it became necessary to progressively refine the model used to analyze the measurements and finally to include dephasing processes in addition to mere population effects. The sections below follow essentially these developments treating successively time-resolved energy relaxation, time-resolved depolarization and quantum beat experiments. All these investigations refer to AgBr and exclusively to the (allowed) 2 TO(L) scattering process as will be tacitly presumed in the following. With regard to the special experimental instrumentation used in these studies, the reader is referred to Sect. 7.1 where it is briefly described.

7.5.1 Dynamics of Energy Relaxation

If resonant Raman scattering is interpreted as absorption followed by emission, the 2 TO(L) process may be described in a simple two-level scheme representing the crystal ground state and the intermediate resonant exciton state, respectively. Considering again initially only one exciton state (therefore omitting subscript j) and introducing the population density n in the excited state, its change with time is determined by a rate equation

$$dn/dt = G(t) - \Gamma_{\text{tot}}(\varepsilon)\, n \,. \tag{7.41}$$

$G(t) = \alpha I_L(t)$ is the rate of generation given by the product of absorption coefficient $\alpha \propto |\hat{e}_L \cdot M^{\text{TO(L)}}|^2$ and incident laser intensity I_L (in units of photons/sm^2). $\Gamma_{\text{tot}}(\varepsilon)$ is the total scattering probability containing the contributions of all inelastic scattering processes of an exciton at energy ε, (7.32) and related to the energy relaxation time of exciton population by $\Gamma_{\text{tot}}(\varepsilon) = 1/T_1(\varepsilon)$ according to (7.18). Assuming the initial population of the state to be $n(t = 0) = 0$, the general solution for arbitrary $I_L(t)$ is

$$n(t) = \alpha \exp\left[-\Gamma_{\text{tot}}(\varepsilon)\, t\right] \int_0^t I_L(t') \exp\left[\Gamma_{\text{tot}}(\varepsilon)\, t'\right] dt' \,. \tag{7.42}$$

While it is straightforward to show from (7.42) that cw excitation [i.e. $I_L(t)$ = const.] yields the steady-state emission intensity as given by (7.19), the case of interest to us here is short light pulse excitation. Approximation of the pulse by a δ-function $I_L(t) = I^*\delta(t)$ and integration of (7.42) yields

$$n(t) = \alpha I^* \exp\left[-\Gamma_{\text{tot}}(\varepsilon)\, t\right] \,. \tag{7.43}$$

Finally, expressing the emitted intensity in terms of radiative transition probability $\Gamma_{\text{rad}} = |\hat{e}_S \cdot M_{\text{em}}^{\text{TO(L)}}|^2$ and population density n, the transient 2 TO(L) intensity takes the form

$$I^{2\,\text{TO(L)}}(\varepsilon, t) \propto \Gamma_{\text{rad}} n \propto |\hat{e}_L \cdot M_{\text{abs}}^{\text{TO(L)}}|^2\, |\hat{e}_S \cdot M_{\text{em}}^{\text{TO(L)}}|^2\, \varepsilon^{1/2} \exp\left[-\Gamma_{\text{tot}}(\varepsilon)\, t\right] \,. \tag{7.44}$$

In contrast to the cw intensity in (7.19), in (7.44) $\Gamma_{\text{tot}}(\varepsilon)$ is contained in an exponential function and can be determined as an absolute quantity by measuring the time-dependence of the 2 TO(L) intensity following pulsed excitation at different energies. As before, the contributions of all L-point excitons must be properly taken into account to obtain the total scattering intensity.

Representative experimental results of time-resolved measurements are shown in Fig. 7.22 with spectra excited slightly above the indirect absorption edge at two different exciton kinetic energies [7.33, 79]. In addition to the spectral information, which is comparable to that in

Fig. 7.22. Time-resolved resonant scattering spectra in AgBr at 3.8 K excited at $E_L = 2.700$ eV (above) and $E_L = 2.6968$ eV (below) with ε_L denoting the corresponding exciton kinetic energies. Spectral and temporal resolutions are 0.2 meV and 50 ps, respectively. Inset: schematic representation of exciton states and scattering processes. After [7.32, 33]

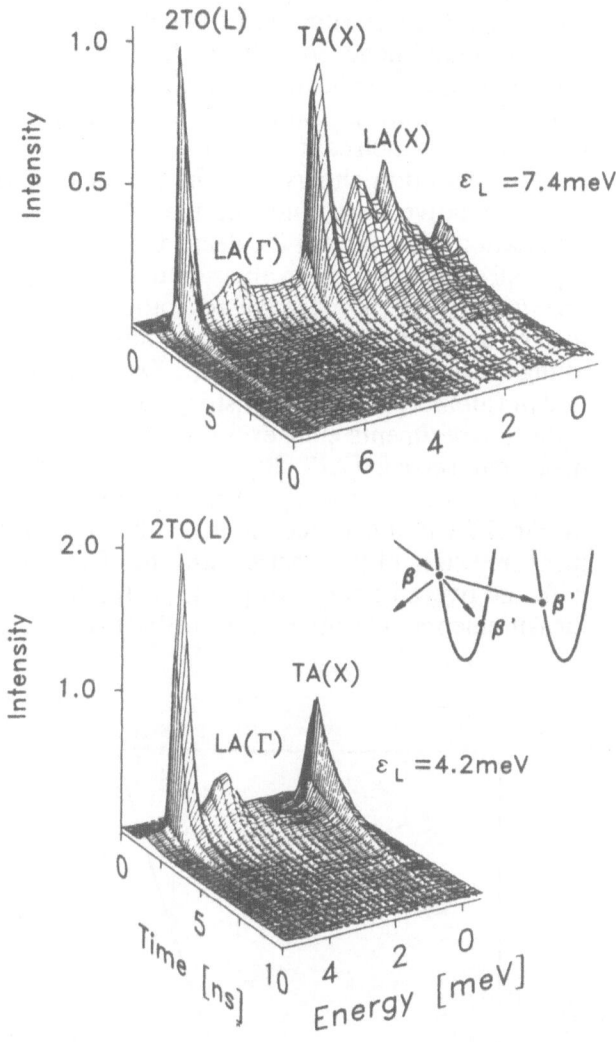

Fig. 7.10, the intensities of the scattering lines are shown now as they evolve in time after excitation with a 15 ps pulse. In accordance with (7.44), the 2 TO(L) decay directly reflects the total lifetime T_1 of the exciton in the intermediate state (state β; see inset Fig. 7.22). As anticipated, because of stronger relaxation, T_1 is seen to decrease for excitation at higher exciton energies with simultaneous reduction in scattered intensity (note different intensity scales). The time-dependent intensities of the remaining lines [LA(Γ), TA(X), LA(X), related to additional relaxation] reflect the population of final states β'. They are more complex and not only determined by the exciton lifetime but also by relaxation time since population in states β' is built up from population decay at higher states

(β in Fig. 7.22). Assuming, as is reasonable, an exponential function for the decay [and hence for $G(t) \propto I_L(t)$ which generates population in state β'], the temporal behavior of the line intensities can be readily calculated from (7.42) and is found to be represented by a difference of two exponentials which contain the information on relaxation time and lifetime. Closer inspection shows that the actual time dependence is determined by the relative magnitude of the two times. Only in the case of fast population ($\beta \rightarrow \beta'$) will the decay be exponential with the lifetime of state β'; in all other cases the evaluation of times requires a more comprehensive analysis [7.79]. The effect of relaxation becomes directly obvious in the lower part of Fig. 7.22 from the shift of the higher-order LA(Γ) peaks along the time axis to longer times. Depending on sample impurity content and presumably also on prehistory and treatment (extensive use in previous optical experiments etc.), exciton lifetimes ranging from 0.4 to 60 ns were measured, see e.g. [7.27, 80]

The scattering rate $\Gamma_{tot}(\varepsilon)$ extracted from the measured 2 TO(L) decay in Fig. 7.22 is illustrated in Fig. 7.23. The data are obtained from the measured signal by a least-squares fit of an exponential decay convoluted with the overall system response to the laser excitation function (in the measurements presented: 120 ps FWHM for a 15 ps laser pulse). Making

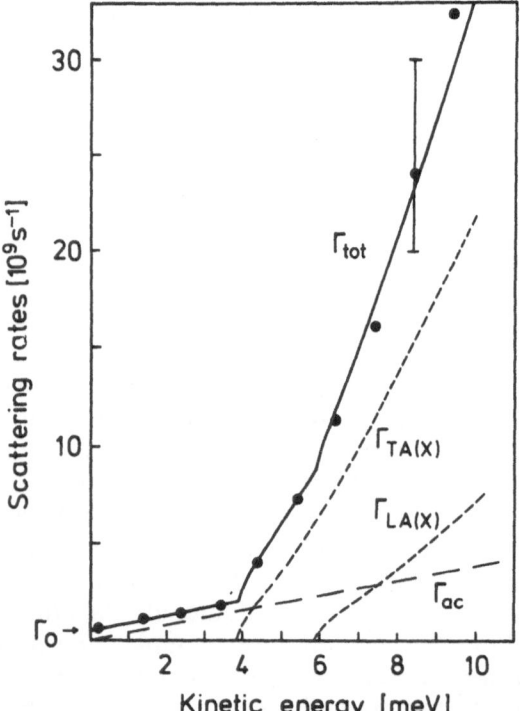

Fig. 7.23. Scattering probability of the 2 TO(L) process in AgBr at 1.8 K as a function of exciton kinetic energy ε. The total probability is shown together with contributions due to various relaxation processes as indicated. Γ_0 corresponds to trapping at impurities and defects assumed to be independent on energy. The points represent experimental data, the solid and dashed lines calculations based on the model described in the text. From [7.32, 33]

use of the explicit energy dependence according to (7.32), it is decomposed into the different contributions so as to give the best fit to the experimental data. The fitting parameters obtained in this case are [7.31, 32]

$$C_{\text{trap}} = (5.2 \pm 0.1)\,10^8\,\text{s}^{-1}; \quad C_{\text{ac}} = (3.8 \pm 0.1)\,10^8\,\text{s}^{-1}\,(\text{meV})^{-1};$$

$$C_{\text{TA(X)}} = (5.5 \pm 0.5)\,10^8\,\text{s}^{-1}\,(\text{meV})^{-3/2};$$

$$C_{\text{LA(X)}} = (2.5 \pm 0.5)\,10^8\,\text{s}^{-1}\,(\text{meV})^{-3/2}.$$

Normalized to C_{ac}, these values are consistent with the parameters deduced from stationary light scattering [7.52] except for C_{trap}, which, as pointed out before, is extrinsic in origin and hence sample dependent (note that the quantitative analysis of the cw experiments in Sect. 7.4.2 did not include LA(X) scattering and therefore resulted in somewhat different values (Table 7.2). The parameters that describe exciton–phonon scattering are directly related to the corresponding interaction energies. For long wavelength acoustic phonons, according to *Toyozawa* [7.65] C_{ac} is given by

$$C_{\text{ac}} = \frac{16}{3\pi}\frac{(D_{\text{c}} - D_{\text{v}})^2}{\varrho v^5 \hbar^4}\,\varepsilon_0^2,\tag{7.45}$$

where D_{c} and D_{v} are the deformation potentials for conduction and valence bands, ϱ is the density of the material and $\varepsilon_0 = \frac{1}{2}M^*v^2$ is the kinetic energy of excitons travelling with sound velocity v. Similarly C_{iv} is related to the effective first-order deformation potential D_{iv} [7.81] for each intervalley phonon [TA(X), LA(X)] as defined in (7.28) and may be written as

$$C_{\text{iv}} = \frac{8}{\pi}\frac{D_{\text{iv}}^2}{\varrho v^5 \hbar^4 \hbar\omega_{\text{iv}}}\,\varepsilon_0^{5/2},\tag{7.46}$$

Using the values for C_{ac} and C_{iv}, the deformation potentials are $|D_{\text{c}} - D_{\text{v}}| = 0.90\,\text{eV}$, $D_{\text{TA(X)}} = 4.0\,\text{eV}$ and $D_{\text{LA(X)}} = 3.4\,\text{eV}$, respectively. The magnitude of $|D_{\text{c}} - D_{\text{v}}|$ for coupling to long wavelength acoustic phonons has to be compared with the deformation potential $D_{\text{gx}}^{\text{i}} = 0.68\,\text{eV}$ derived from the hydrostatic pressure experiments described in Sect. 7.4.3 (Table 7.3). Since the latter is very directly obtained from the energy shift of the exciton edge we believe it is the more reliable value, the discrepancy with that from the time-resolved measurements being most probably due to the neglect of the effective-mass anisotropy in (7.45, 46). For the same reason, the values for $D_{\text{TA(X)}}$ and $D_{\text{LA(X)}}$ may bear some uncertainty, but we emphasize that there is no other means to obtain these deformation potentials for off-center phonons.

7.5.2 Polarized 2 TO (L) Scattering

As an explicit criterion to distinguish resonant Raman scattering from hot luminescence, the polarization properties of the scattered light may be considered [7.19]. The distinction rests upon the fact that exciton states in highly symmetric crystals are often degenerate as is the case in the silver halides (Sect.7.2). Excitation with polarized light prepares the degenerate state with well-defined polarization which, assuming that the excitons are sufficiently long-lived, may subsequently decay by elastic scattering into another degenerate state having different polarization. Although under these circumstances the excitons do not suffer inelastic scattering between absorption and emission, the polarization of the emitted light loses correlation with that of the incident light implying hot luminescence as origin. On the other hand, if no depolarization (meaning here a random loss of polarization) is observed, the emitted light would be due to resonance Raman scattering occurring without loss of phase memory in the intermediate state.

The depolarization effects expected in the case of hot luminescence are quite analogous to spin orientation of excitons by optical pumping ([7.82], for a recent review see e.g. [7.83] and references therein). In cw experiments they may show up as polarization of the steady-state emission reflecting that the exciton–lattice system is not thermalized. Depending on the relative magnitude of scattering rates involved, these effects are sometimes small and experimentally not easy to discriminate from unintentional polarization produced by the spectroscopic apparatus. However, depolarization manifests itself very clearly in time-resolved measurements in which polarized light is used for excitation and emission. The observation made in an oriented sample of AgBr is that pulsed excitation in the indirect exciton e.g. with plane-polarized light gives rise to a distinctly different time behavior of the 2 TO(L) intensity dependent on the emitted light polarization. Despite the small polarization found under cw excitation, this directly demonstrates that several time constants are involved and, concurrent with energy relaxation, depolarization processes do in fact occur.

To examine in detail what we should expect, the two-level model of Sect. 7.5.1 to describe energy relaxation is extended to account for the twofold degeneracy of the singlet-triplet exciton states as shown by the inset in Fig. 7.24 [7.33]. According to (7.10), for linear light polarization the degenerate exciton states Φ_x, Φ_y interact with radiation as described by absorption coefficients $\alpha_x \propto |\hat{e}_L \cdot M_x|^2$ and $\alpha_y \propto |\hat{e}_L \cdot M_y|^2$ and corresponding radiative emission probabilities $\Gamma_{rad,x} \propto |\hat{e}_S \cdot M_x^*|^2$ and $\Gamma_{rad,y} \propto |\hat{e}_S \cdot M_y^*|^2$. Considering exciton states only at one L-point and excluding elastic scattering between different L-point valleys [7.53], the population transfer between states Φ_x and Φ_y (with population densities n_x and n_y) is phenomenologically taken into account by introducing cross-relaxation with scattering rate w which tends to equalize excess population created by polarized absorption in one of the states. In contrast to the case of the nondegenerate two-level system described by (7.41), the rate equations now contain appropriate coupling terms and are given by

$$dn_x/dt = \alpha_x I_L(t) - (1/T_1) n_x - w(n_x - n_y),$$

$$dn_y/dt = \alpha_y I_L(t) - (1/T_1) n_y + w(n_x - n_y). \tag{7.47}$$

Again assuming δ-pulse excitation and zero initial populations n_x and n_y, the solution of (7.47) is straightforward. Choosing the directions of polarization as in the actual experiments ($\hat{e}_L \parallel [110]$; $\hat{e}_s^{\parallel} \equiv \hat{e}_s \parallel [110]$; $\hat{e}_s^{\perp} \equiv \hat{e}_s \parallel [1\bar{1}0]$) and taking into account the contributions from all L-points appropriate for this geometry, one obtains as result the polarized 2 TO(L) intensities

$$I^{\parallel}(t) = \left(1 + \frac{a^2}{9}\right)\left[\exp\left(-\frac{t}{T_1}\right) + \exp\left(-\frac{t}{T_p}\right)\right],$$

$$I^{\perp}(t) = \frac{2a}{3}\left[\exp\left(-\frac{t}{T_1}\right) - \exp\left(-\frac{t}{T_p}\right)\right]. \tag{7.48}$$

The parameter $a = 1 + 2[\sigma(3 - \xi)/(3 - 2\xi)]^2$ must be determined experimentally and involves quantities already introduced in Sect.7.2 [eqs. (7.7, 8)]. T_p is the effective depolarization time for the degenerate two-level system being given by

$$\frac{1}{T_p} = \frac{1}{T_1} + 2w. \tag{7.49}$$

As seen from (7.48, 49), in the case of degeneracy, in addition to energy relaxation, cross-relaxation also affects the decay of the 2 TO(L) intensity and, consistent with experiment, gives rise to the polarization-dependent time behavior.

Results obtained from measuring the 2 TO(L) decay using the polarization directions quoted above are reproduced in Fig. 7.24. They are represented so as to most clearly demonstrate the effects of energy relaxation and depolarization. As implied by (7.48), the two processes each contribute to $I^{\parallel}(t)$ and to $I^{\perp}(t)$ but can be separated by computing $I^{\pm} = I^{\parallel} \pm f I^{\perp}$ where f is an appropriate factor to be obtained from (7.48), see [7.33] for details. Accordingly, the experimental data were processed so as to give $I^+(t)$ and $I^-(t)$ determined by T_1 and T_p, respectively (Fig. 7.24). As expected, the decay is exponential, giving straight lines in the semilog plot of the figure. This nicely confirms the underlying model and allows us to fit the data and extract T_1 and T_p. Their values are indicated in the figure for the examples presented there. They are used to

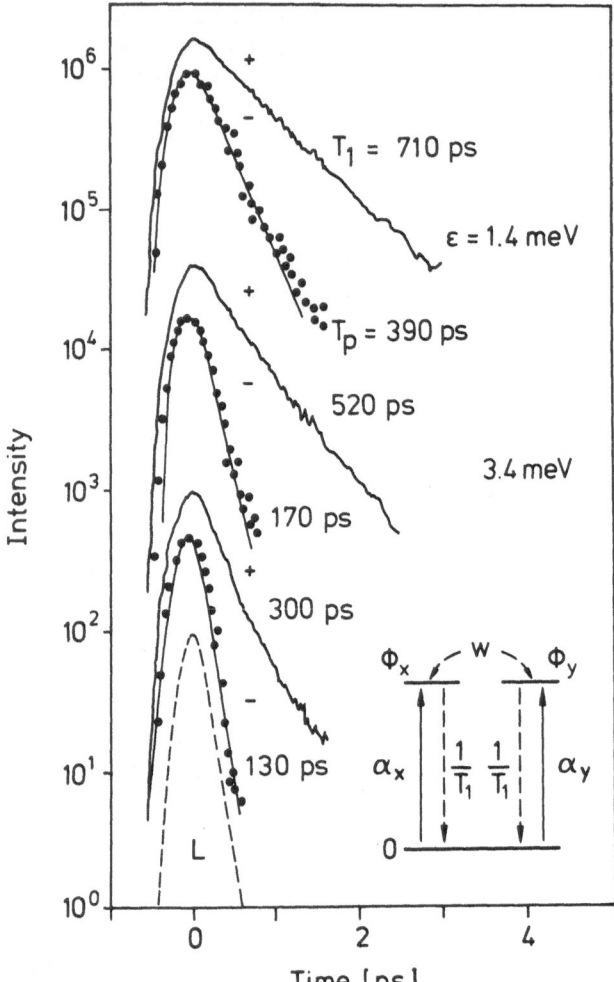

Fig. 7.24. Time-dependence of polarized 2 TO(L) intensity. Excitons are excited by a 15 ps laser pulse at different kinetic energies ε (lowest curves excited at $\varepsilon = 4.4$ meV). Denoted by + are the values of $I^{\parallel} + fI^{\perp}$ while those of $I^{\parallel} - fI^{\perp}$ are denoted by −, the measured data being represented by points and the fits by solid lines as described in the text. *L* is the system response to the laser pulse. The inset shows the degenerate two-level system used for the analysis. The effective depolarization time T_{p} is defined by (7.49). After [7.33, 79]

construct Fig. 7.25 showing the energy dependence of rates for elastic (w) and inelastic ($1/T_{\mathrm{l}}$) scattering according to the model.

The appearance of depolarization effects in the time-dependent polarized 2 TO(L) intensity demonstrates that, in a comprehensive picture of exciton relaxation dynamics, the degeneracy of exciton states must be taken into account. Besides energy relaxation, elastic scattering between the degenerate states occurs, and is manifest in the resulting depolarization. As to the physical origin of the elastic scattering and exciton depolarization, one has to exclude the longitudinal–transverse splitting, which for direct excitons is the dominant mechanism [7.83], but does not exist in our case of indirect exciton states. Also, at the relatively low exciton densities reached experimentally (typically on the order of $10^{11}–10^{12}$ cm^{-3}), we may rule out exciton–exciton interaction leaving scattering at impurities and

Fig. 7.25. Energy relaxation rate $1/T_1$ and cross relaxation rate w determined from the polarized $2\,TO(L)$ decay as function of exciton kinetic energy ε. The full and open circles represent experimental data. The full line corresponds to a fit of $\Gamma_{tot}(\varepsilon) = 1/T_1(\varepsilon)$ according to (7.32), the dashed line is a guide to the eye. After [7.79]

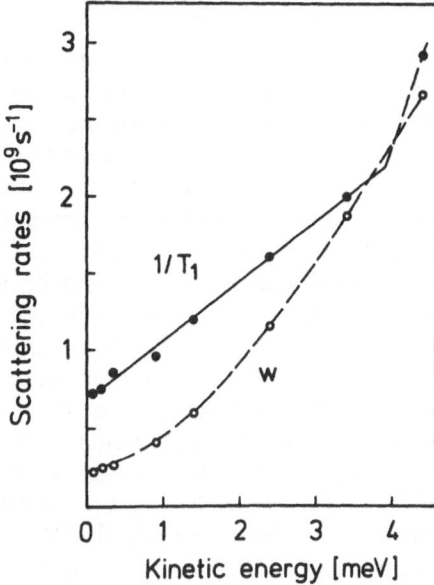

lattice defects as the effective mechanism. The tendency found for the rate w to increase with kinetic energy of the excitons (Fig. 7.25) would be consistent with this suggestion although the data are not sufficient to establish a model that can be tested quantitatively.

It was emphasized in Sect. 7.2 that by excitation with linearly polarized light, (7.9, 10), excitons are created in a coherent state with well-defined wavevector. In principle, the effect of elastic scattering is to change both the wavevector and the coherence properties. Basically, any change in wavevector should be observable in the Raman spectrum, via the associated change in phonon energy, as a shift or at least a broadening of lines. However, since momentum-conserving L-point phonons with negligible dispersion participate in the $2\,TO(L)$ scattering process, any change in exciton wavevector will have no effect on the spectrum. Coherence, however, manifests itself in the polarization which is seen to be time dependent reflecting the dephasing of the intermediate state following polarized (pulsed) excitation. Since, due to the selection rules, resonant Raman scattering will only give rise to intensity in parallel polarization [i.e. $I^\perp = 0$ in (7.48)], the depolarized component $I^\perp$ may be regarded as entirely due to hot luminescence. Correspondingly, $I^\parallel - I^\perp$ monitors the Raman contribution, its time dependence (roughly given by $I^-(t)$ in Fig. 7.24) representing the loss of coherence due to elastic scattering. The model used for the analysis is based on the two-step process of absorption followed by emission as represented by the rate equations for population (7.47). The interpretation of the dephasing as population transfer between appropriate states nevertheless allows a clear distinction between resonant Raman scattering and hot luminescence [7.33].

7.5.3 Recent Advances: Quantum Beats in Resonant Light Scattering

To reveal directly contributions of resonant Raman scattering to the secondary emission signal and to distinguish them from hot luminescence, one must employ a type of spectroscopy capable of probing the coherence of the resonantly excited intermediate state. In the case of extended excited states, such as excitons in solids, coherence properties have been investigated until now by means of nonlinear techniques like transient four-wave-mixing or others; see [7.27] and references therein. In these experiments an excite-and-probe scheme is usually employed and the *optical coherence*, i.e. the coherent coupling of the excitation relative to the external light field, is measured. As is well known from atomic spectroscopy [7.84], coherence can also manifest itself in form of *quantum coherence* appearing as a beat signal that originates from the coherent superposition of nearly degenerate excited states.

Suppose that two states with wavefunctions Φ_a, Φ_b and (slightly different) energies E_a, E_b are simultaneously excited by a light pulse of duration t_p short compared with the reciprocal of the splitting frequency of the states $[t_p \lesssim \hbar/(E_a - E_b) = 1/(2\pi\nu_{ab})]$, a quantum mechanical state is prepared which at $t = 0$ is represented as a superposition of the two eigenstates

$$|\Phi(0)\rangle = \alpha_a |\Phi_a(0)\rangle + \alpha_b |\Phi_b(0)\rangle \,, \tag{7.50}$$

where α_a, α_b are probability amplitudes. The temporal evolution of this state gives, at some later time,

$$\begin{aligned} |\Phi(t)\rangle = &\alpha_a \exp\left[-(\mathrm{i}/\hbar) E_a t - \mathrm{i}\varphi_a - t/2\tau_a\right] |\Phi_a(0)\rangle \\ &+ \alpha_b \exp\left[-(\mathrm{i}/\hbar) E_b t - \mathrm{i}\varphi_b - t/2\tau_b\right] |\Phi_b(0)\rangle \,, \end{aligned} \tag{7.51}$$

where for each state we have written the phase factor explicitly (i.e., α_a and α_b are taken to be real) and introduced a phenomenological damping term to account for the finite lifetimes (τ_a, τ_b) of the states. The emitted time-dependent intensity in the case of an electric-dipole allowed transition can be calculated from the matrix element between ground and excited states according to

$$\begin{aligned} I(t) \propto &|\langle\Phi_0| \boldsymbol{D} |\Phi(t)\rangle|^2 \\ \propto &\alpha_a^2 \exp\left(-t/\tau_a\right) |\langle\Phi_0| \boldsymbol{D} |\Phi_a\rangle|^2 + \alpha_b^2 \exp\left(-t/\tau_b\right) |\langle\Phi_0| \boldsymbol{D} |\Phi_b\rangle|^2 \\ &+ 2 \,\mathrm{Re}\{\alpha_a\alpha_b \exp\left[-(t/2\tau_a) - (t/2\tau_b)\right] \langle\Phi_0| \boldsymbol{D} |\Phi_a\rangle \langle\Phi_0| \boldsymbol{D} |\Phi_b\rangle \\ &\exp\left[-\mathrm{i}((E_a - E_b) t/\hbar + (\varphi_a - \varphi_b))\right]\} \,, \end{aligned} \tag{7.52}$$

with Φ_0 denoting the ground state wavefunction and $\boldsymbol{D}$ the electric dipole operator. The intensity expression (7.52) consists of altogether three terms. The two quadratic terms decay with the corresponding lifetimes of states determined by energy relaxation. Since the two states are very close in energy and thus affected by almost the same relaxation processes, one may write $\tau_a \approx \tau_b \approx T_1$, these terms leading then to an exponentially decaying background intensity. Most important is the interference term predicting intensity oscillations with a beat frequency $\nu_{ab} = (E_a - E_b)/h$ superimposed on the more slowly decaying signals. The intensity contributed by this term is time-dependent due to damping from two different sources. As before, one corresponds to energy relaxation with time constant $(1/2\tau_a) + (1/2\tau_b) \approx 1/T_1$, the other to the time variation of the phases of the two states, which originates from elastic scattering processes. These processes are taken into account by a contribution $2w$ to the inverse coherence time that accordingly may be written as

$$\frac{1}{\tau_{\text{coh}}} = \frac{1}{T_1} + 2w \qquad (7.53)$$

and is related to the homogeneous linewidth by $\gamma_{\text{hom}} = \hbar/\tau_{\text{coh}}$. It is the interference term which reflects the coherence as a function of time and offers the possibility to investigate, in addition to energy relaxation, phase-perturbing processes.

Investigations employing this coherent type of spectroscopy have just started, but have allowed us, for the first time in any solid, to discover quantum oscillations for the $\Gamma_6^+ \otimes L_{4,5}^-$ indirect exciton state in AgBr [7.27]. In these experiments, the degenerate singlet-triplet mixed exciton states (Φ_1^0, Φ_4^0, cf. Sects. 7.2 and 7.4.3) were slightly split in a magnetic field and the 2 TO(L) scattering process investigated after short-pulse ($t_p = 8$ ps) excitation in the indirect exciton absorption. The applied magnetic field was varied up to 1 T to give an energy splitting of the two sublevels appropriate to the spectral width of the laser pulse (~ 0.3 meV). Also in this low-field regime the singlet-triplet exciton states largely retain their character since mixing with the lower lying triplet states (Φ_2^0, Φ_3^0) is negligible (cf. Sect. 7.4.3, Figs. 7.20, 21). The coherent superposition of states Φ_1^0 and Φ_4^0 accomplished under these conditions of excitation leads to the anticipated oscillations, which occur as a modulation of the 2 TO(L) intensity decay.

Representative experimental results to demonstrate the effect of the magnetic field in comparison to the zero-field case are illustrated in Fig. 7.26. In the example shown, the magnetic field is directed along [001]. The exciting light is polarized along $\hat{e}_L \parallel$ [110] while the scattered light, observed in Faraday geometry, is left and right circularly polarized. The decay for zero magnetic field is fairly exponential. With field applied, i.e. with the states Φ_1 and Φ_4 slightly split (frequency difference ν_{41}), clear oscillations are detected that are unambiguously ascribed to quantum beats. For $B = 0.5$ T the oscillation period is about 200 ps with the signals for σ^+ and σ^- polarizations phase-shifted by 180°. Qualitatively similar results are obtained at different magnetic field strengths and for other directions of polarization.

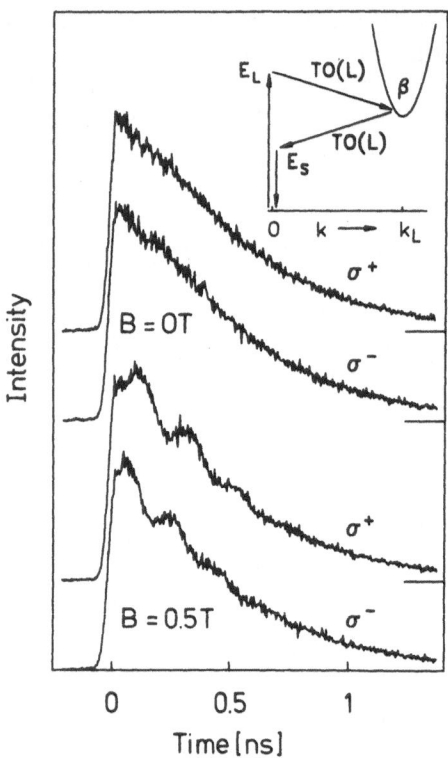

Fig. 7.26. Quantum beats observed in the decay of the 2 TO(L) intensity in AgBr at 2 K. The excitation is accomplished by an 8 ps laser pulse at $E_L = 2.6931$ eV ($\varepsilon = 0.3$ meV) polarized along [110]. The scattered light (at $E_S = 2.6765$ eV) is observed along [001] with right (σ^-) and left (σ^+) circular polarization. The upper traces give the decay without field while the lower ones represent the decay with a field $B = 0.5$ T applied along [001]. The system response to the laser pulse (not shown) is faster than 40 ps (FWHM). The inset represents schematically the 2 TO(L) scattering process. From [7.27]

The data were analyzed by applying the density-matrix formalism [7.85] by which both energy relaxation and dephasing can be phenomenologically taken into account. The temporal evolution of the density-matrix ϱ_{ij} follows from solving the Liouville equation, which is of the form

$$\partial\varrho/\partial t = -(i/\hbar)[\mathscr{H}_{\mathrm{ex}}, \varrho] - R\varrho + G. \tag{7.54}$$

$\mathscr{H}_{\mathrm{ex}}$ is the Hamiltonian in (7.35), through which the magnetic-field-induced energy splitting enters. The explicit time dependence is described by the relaxation operator R, whereby the diagonal and off-diagonal density-matrix elements are assumed to decay with energy relaxation time T_1 and dephasing (or coherence) time τ_{coh} of the excited sublevels, respectively. The generation matrix G contains the interaction of the exciton system with the exciting laser light. Taking into account the selection rules for TO(L) absorption and emission (Sect. 7.2), one obtains $\varrho_{ij}(t)$ and from this finally the polarized 2 TO(L) intensities as function of time. As an example, (7.55) reproduces the expressions derived for the conditions of polarization and field direction in Fig. 7.26. The intensities in this case are given by [7.27]

$$I(\sigma^\pm, t) \propto \exp\left(-\frac{t}{T_1}\right) \pm \sqrt{\frac{2}{27}} \exp\left(-\frac{t}{\tau_{\mathrm{coh}}}\right) \sin(2\pi\nu_{41}t). \tag{7.55}$$

In order to derive (7.55), the intensity contributions from excitons at different L-points in the Brillouin zone had to be taken into account and the effect of the pure triplet states was neglected (see above). Also, the parameter introduced in Sect. 7.2 in connection with the transition moments in (7.7) was assumed to be $\sigma = 1$ as suggested by the experimental results.

Equations (7.55) are entirely consistent with the experimental observation in Fig. 7.26. They imply that, without magnetic field (i.e. splitting $v_{41} = 0$), the 2 TO(L) intensity for polarisation σ^+ and σ^- exhibits identical decay determined essentially by the energy relaxation time T_1 (effects of depolarization as described in Sect. 7.5.2 cancel in this case). For $B \neq 0$ and, hence, $v_{41} \neq 0$, the expressions also reproduce the oscillatory structure now seen to decay with τ_{coh} and having the experimentally observed phase shift. With the help of (7.55), the experimental data can be fitted and values for the two decay times and the field-induced frequency splitting v_{41} are deduced. The latter can be obtained with improved accuracy from the intensity difference $I(\sigma^+) - I(\sigma^-)$, for which, according to (7.55), the first terms cancel and only the beat signal is retained. Experimental data processed in this way are displayed in Fig. 7.27 for two magnetic field strengths. They nicely demonstrate the anticipated behavior, in particular the decay of beat amplitude with τ_{coh} and the expected increase in oscillation frequency at higher field.

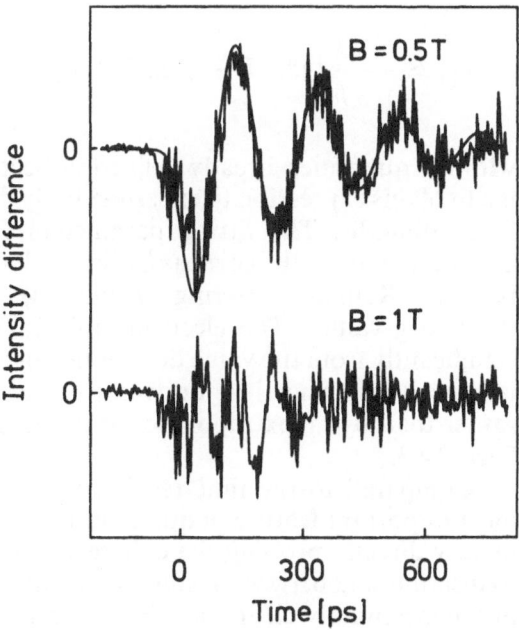

Fig. 7.27. Intensity difference between left and right circularly polarized 2 TO(L) decay in AgBr at two magnetic fields as indicated. For $B = 0.5\,T$ a fit according to (7.55) is shown by the smooth full line together with the experimental data. From [7.27]

Fitting all data for the investigated crystal, e.g. at $B = 0.5$ T, one obtains $T_1 = 450$ ps and $\tau_{coh} = 400$ ps. In the light of (7.53), these numbers imply that quantum coherence in this specific sample is predominantly destroyed by inelastic processes, i.e. by energy relaxation depopulating the exciton state (Sects. 7.4.2, 7.5.1), while elastic scattering (at rate $2w$) is found to be of minor importance. Somewhat shorter times are obtained at higher fields, although these values are less accurate due to the worse signal-to-noise ratio.

From the intensity difference in Fig. 7.27 an oscillation frequency $\nu_{41} = (5.12 \pm 0.25)$ GHz is found to describe the experimental data excellently, as recognized from the quality of the fit shown for this case in the figure. This value corresponds to an energy splitting between states Φ_1 and Φ_4 of only 21 μeV and demonstrates that much higher spectral resolution can be achieved by quantum beat spectroscopy than by conventional resonant Raman scattering (Sect. 7.4.2).

The reasons for this are quite obvious: First, quantum beat spectroscopy is a coherent technique and hence spectral resolution is not limited by inhomogeneous broadening of lines (see below). Secondly, the measurements, being performed in the time domain, are determined by the temporal resolution of the spectroscopic equipment (cf. Sect. 7.1), which means that small splittings of states are easily accessible, since they correspond to larger period oscillations. Of course, the measurements are limited by τ_{coh}, which has to be sufficiently long to allow the observation of the quantum beat signal over at least one period.

For the chosen field direction $\boldsymbol{B} \parallel [001]$, in the low field limit the beating frequency between states Φ_1 and Φ_4 from (7.36) is given by

$$\nu_{41} \simeq \frac{1}{\sqrt{3}\,h} \{g_v^{\parallel}\mu_B B - g_c\mu_B B + [(g_c\mu_B B)^3/3\Delta^2]\}\,, \tag{7.56}$$

with the quantities already defined in Sect. 7.4.3. The full line in Fig. 7.28 is a fit of this expression to experimentally deduced points at three magnetic field strengths. The fitting parameters (Δ, g_c, $g_v^{\parallel}$; see figure) have to be compared with corresponding values derived from conventional resonant Raman scattering (Table 7.4). Good agreement is found for the two g-values. The electron–hole exchange energy, however, differs significantly from the value determined in Sect.7.4.3 ($\Delta = 0.17$ meV) which is conceivable since it is derived from experimental data near zero-field rather than by an extrapolation from data taken at very high fields as in Sect. 7.4.3.

Compared to the time-resolving measurements in Sects. 7.5.1, 2, the most important feature of quantum beat spectroscopy is that it is capable of very directly probing the coherence of states and therefore allows one to discriminate between resonant Raman scattering and hot luminescence in a conceptually clear way. Provided the same initial, intermediate and

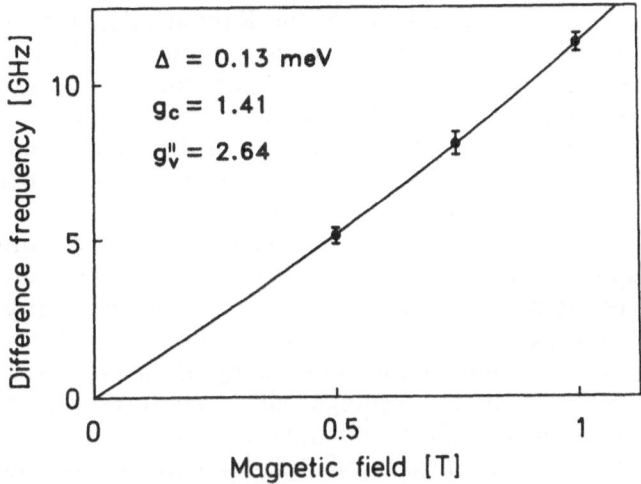

Fig. 7.28. Frequency difference ν_{41} between exciton sublevels Φ_1 and Φ_4 as a function of magnetic field. The full points represent experimental data deduced from the beating frequency while the solid line describes the fit obtained with (7.56) using the set of fitting parameters shown in the figure. From [7.27]

final states are involved, the difference between these two processes (both second order) rests on whether or not quantum coherence is conserved during the light absorption–emission sequence, see e.g. [7.86]. As schematically illustrated in Fig. 7.29, a light pulse having finite spectral width ΔE will, because of exciton dispersion, create excitons with a distribution of wavevectors giving rise to an inherent inhomogeneous broadening of the exciton line. Of these, only excitons with same wavevector will interfere and contribute to the quantum beat signal while recombination of all others results in an incoherent background intensity. One may therefore take, at fixed time, the ratio of the beating amplitude to the background

Fig. 7.29. Schematic representation of exciton states (Φ_1, Φ_4) split in a magnetic field to explain coherent Raman-like and incoherent luminescence-like contributions to the scattered intensity after excitation with a transform-limited light pulse of spectral width ΔE. States having equal wavevector are coherently excited and superimposed within wavevector range Δk

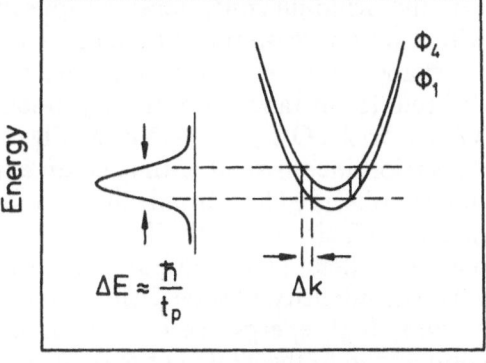

intensity as a measure of the Raman relative to the luminescence-type contributions to the total emission intensity. As is evident from Fig. 7.29, this ratio depends on exciton dispersion as well as on energy position and spectral width of the exciting light. Up to now, no calculations exist that enable a quantitative estimation of these dependences, but it is probable that they explain at least part of the discrepancy between experiment and the prediction (7.55) for this ratio.

In the light of these considerations, the quantum beat signal corresponds to the coherent portion of the resonant secondary emission spectrum and unambiguously probes the resonant Raman component. As implied by its time-dependence, coherence is destroyed in about 400 ps, mainly by depopulation but also by elastic processes. It is to be emphasized that τ_{coh} deduced for energies close to the exciton gap is comparable in magnitude to the effective depolarization time T_p determined in Sect. 7.5.2, based on the degenerate two-level model. It confirms our earlier assumption that the depolarization describes the loss of coherence expressed formally by the correspondence of (7.49) and (7.53). Both effects are caused by elastic scattering of the excitons at crystal defects and impurities, which destroys coherence.

Before closing this section, a brief comment is appropriate concerning the two-phonon resonant Raman linewidths in comparison to the exciton lifetime data extracted from the investigations above. In cw measurements with high spectral resolution [7.87], the width of the 2 TO(L) scattering line close to the exciton absorption edge (exciton kinetic energies $\varepsilon = 0\text{--}2.5$ meV) is found to be $\delta E \simeq 0.15$ meV (FWHM) increasing linearly with ε for $\varepsilon > 2.5$ meV (e.g. $\delta E \simeq 0.23$ meV at $\varepsilon = 8$ meV). Since after termination of the 2 TO(L) scattering process the final state contains one photon plus one TO(L) phonon (but no exciton), the 2 TO(L) linewidth should be determined by the phonon lifetime. This seems to indeed be the case at low ε, where the phonon lifetime (~ 30 ps $\Rightarrow \delta E = 0.12$ meV [7.88]) results in an energy width of about the right magnitude. The broadening observed at larger ε might, at first glance, be associated with the effect of dispersion of the TO phonon branch near and around point L in the Brillouin zone, since phonons at these wavevectors are associated with exciton scattering at higher energies. Detailed inspection of the lineshape, however, shows that the broadening is symmetric [distinctly different for instance from the asymmetric broadening of lines 2 TO(L) + TA(X) or 2 TO(L) + LA(X); cf. Fig. 7.14] and in view of the phonon dispersion inconsistent with such an explanation. We therefore speculate that the observed line broadening is (indirectly) correlated with the exciton lifetime which is drastically reduced in this energy range as directly measured (Sect. 7.5.2) and also confirmed by the decrease in integrated 2 TO(L) intensity observed simultaneously (Fig. 7.16). It is evident that at some high energy the exciton becomes considerably shorter lived in comparison to the TO(L) phonon leaving behind in the final state, besides the photon, *two* TO(L) phonons. The linewidth in this case, as in a two-phonon Raman process, would be given by $\sqrt{2(\delta E)^2} \simeq 0.21$ meV which would agree nicely with the actually measured value at higher exciton energy. Detailed investigations of linewidths as function of exciton energy need to be performed to really substantiate this intriguing picture.

7.6 Concluding Remarks

The main processes of light scattering in AgCl and AgBr described in this article are quite well understood. In our discussions we have considered only these two materials since they provide clear results that can be interpreted in a straightforward and consistent way. We have deliberately omitted further investigations carried out in mixed crystals of type $AgBr_{1-x}Cl_x$. Depending on x, due to compositional disorder, these systems are characterized by different kinds of localized exciton states and exhibit an interesting but complex phonon mode behavior [7.89]. By employing resonant light scattering (first order in this case) both these aspects could be investigated and have given informative results, some of which were recently reviewed [7.36].

Most promising for future work are the new possibilities provided by quantum beat spectroscopy, which we have shown to be applicable to excitonic systems. The investigations of intermediate state coherence that become possible by this method are in their infancy and more systematic work needs to be done. One unanswered question concerns the coherence times. These are found to be surprisingly long in AgBr, for instance as compared to CuCl or GaAs in which (optical) dephasing times for excitonic states of only a few picoseconds have been determined; see [7.27] for references. Moreover, a refined quantitative analysis of our results seems to reveal a fundamental deficiency in the model used up to now. This manifests itself in the clearly nonexponential intensity decay found in the picosecond experiments with bandwidth-limited resolution (Fig. 7.26 for $B = 0$). It might indicate that in order to properly account for the time-resolved light scattering, one needs a theory which goes beyond the presently used phenomenological description. It may also be necessary to consider the temporal properties of the excitation light and the photodetection process [7.90, 91].

Acknowledgment. Among many others I would like to thank especially my coworkers Dr. H. Stolz, Dr. E. Schreiber and Mr. V. Langer who, in recent years, have participated in this work and contributed significantly to it. I also appreciate the support of the Deutsche Forschungsgemeinschaft and the Ministerium für Wissenschaft und Forschung des Landes Nordrhein-Westfalen.

References

7.1   B.L. Joesten, F.C. Brown: Phys. Rev. **148**, 919 (1966)
7.2   "I–VII Compounds", in *Numerical Data and Functional Relationships in Science and Technology*, ed. by O. Madelung, Landolt-Börnstein, New Series III – Semiconductors, (Springer, Berlin, Heidelberg), Vol. 17b (1982); Vol. 22a (1987)
7.3   K. Fischer, H. Bilz, R. Haberkorn, W. Weber: Phys. Status Solidi B **54**, 285 (1972)
7.4   H. Bilz, W. Weber: "Lattice Dynamics of Silver Halides and Dynamical Aspects of the Formation of the Latent Image in the Photographic Process", in *The Physics of Latent Image Formation in Silver Halides*, ed. by A. Baldereschi, W. Czaja, E. Tosatti, M. Tosi (World Scientific, Singapore 1984) p. 25
7.5   H. Bilz: Cryst. Latt. Def. Amorph. Mat. **12**, 31 (1985)
7.6   W. von der Osten, J. Weber: Solid State Commun. **14**, 1133 (1974)
7.7   H. Kazaki, S. Sakuragi, S. Hoshino, G. Shirane: Solid State Commun. **15**, 1547 (1974)
7.8   W. von der Osten, B. Dorner: Solid State Commun. **16**, 431 (1975)
7.9   B. Dorner, W. von der Osten, W. Bührer: J. Phys. C **9**, 723 (1976)
7.10   P.B. Klein, H. Masui, J.J. Song, R.K. Chang: Solid State Commun. **14**, 1163 (1974)
7.11   J. Wagner, M. Cardona: Solid State Commun. **48**, 301 (1983)
7.12   Y. Petroff, P.Y. Yu, Y.R. Shen: Phys. Status Solidi B **61**, 419 (1974)
7.13   T. Komatsu, T. Karasawa, T. Iida, K. Miyata, Y. Kaifu: J. Lumin. **24/25**, 679 (1981)
7.14   T. Karasawa, K. Miyata, T. Komatsu, Y. Kaifu: J. Phys. Soc. Jpn. **52**, 2592 (1983)
7.15   T. Iida, M. Sakai, T. Karasawa, T. Komatsu, Y. Kaifu: J. Phys. C **16**, 4719 (1983)
7.16   T. Karasawa, T. Iida, M. Sakai, T. Komatsu, Y. Kaifu: J. Phys. C **18**, 4043 (1985)
7.17   P.Y. Yu, Y.R. Shen: Phys. Rev. B **12**, 1377 (1975)
7.18   A.Z. Genack, H.Z. Cummins, M.A. Washington, A. Compaan: Phys. Rev. B **12**, 2478 (1975)
7.19   R.M. Martin, L.M. Falicov: "Resonant Raman Scattering", in *Light Scattering in Solids*, ed. by M. Cardona, Topics Appl. Phys. **8** (Springer, Berlin, Heidelberg 1975) p. 80
7.20   M. Cardona: "Resonance Phenomena", in *Light Scattering in Solids II*, ed. by M. Cardona, G. Güntherodt, Topics Appl. Phys. **50** (Springer, Berlin, Heidelberg 1982) p. 19
7.21   J.S. Weiner, P.Y. Yu: Solid State Commun. **50**, 493 (1984)
7.22   P.Y. Yu: Comments Solid State Phys. **12**, 33 (1985)
7.23   Y. Toyozawa: "Theory of Excitons in Phonon Fields" in *Excitonic Processes in Solids*, ed. by M. Ueta, H. Kanzaki, K. Kobayashi, Y. Toyozawa, E. Hanamura, Springer Ser. Solid State Sci., Vol. 60 (Springer, Berlin, Heidelberg 1986) p. 203
7.24   T. Kushida, S. Kinoshita: "Excitation Profiles of Raman Scattering and Luminescence" in *Raman Spectroscopy: Sixty Years On*, ed. by H.D. Bist, J.R. Durig, J.F. Sullivan, Vibrational Spectra and Structure, Vol. 17B (Elsevier, Amsterdam 1989) p. 495
7.25   M. Cardona, G. Güntherodt: "Instrumentation, Techniques" in *Light Scattering in Solids II*, ed. by M. Cardona, G. Güntherodt, Topics Appl. Phys. **50** (Springer, Berlin, Heidelberg 1982) Sect. 1.5.1, p. 6
7.26   W. Richter: "Resonant Raman Scattering in Semiconductors", in Springer Tracts Mod. Phys., Vol. 78 (Springer, Berlin, Heidelberg 1976) p. 121
7.27   V. Langer, H. Stolz, W. von der Osten: Phys. Rev. Lett. **64**, 854 (1990); J. Lumin. **45**, 406 (1990)
7.28   P. Saari, J. Aaviksoo, A. Freiberg, K. Timpmann: Opt. Commun. **39**, 94 (1981)
7.29   A. Freiberg, P. Saari: IEEE J. **QE-19**, 622 (1983)
7.30   E. Schreiber: "Picosekunden-Spektroskopie lokalisierter Elektronenzustände und Evidenz von Silberclustern in Silberhalogeniden"; Ph.D. Thesis, Universität Paderborn (1989)
7.31   H. Stolz, W. von der Osten: Solid State Commun. **49**, 1035 (1984)

7.32   H. Stolz, E. Schreiber, W. von der Osten: In Proc. 17th Int. Conf. Phys. Semiconductors, San Francisco, USA, 1984, ed. by J.D. Chadi, W.A. Harrison (Springer, New York 1985) p. 1271

7.33   H. Stolz, W. von der Osten: Cryst. Latt. Def. Amorph. Mat. 12, 293 (1985)

7.34   M. Matsushita: J. Phys. Soc. Jpn. 35, 1688 (1973)

7.35   S. Kurita, K. Kobayashi: J. Phys. Soc. Jpn. 44, 1583 (1978)

7.36   W. von der Osten, H. Stolz: J. Phys. Chem. Solids 51, 765 (1990)

7.37   F. Bassani, G. Pastori Parrevicini: in Electronic States and Optical Transitions in Solids, ed. by R.A. Ballinger, The Science of the Solid State Int. Series, Vol. 8 (Pergamon, Oxford 1975)

7.38   R.S. Knox: "Theory of Excitons" in Solid State Physics, ed. by H. Ehrenreich, F. Seitz, D. Turnbull, Suppl. 5 (Academic, New York 1963)

7.39   R.J. Elliott: "Introduction to the Theory of Excitons" in Polarons and Excitons in Polar Semiconductors and Ionic Crystals in NATO ASI Series B: Physics, ed. by J.T. Devreese, F. Peeters, Vol. 108 (Plenum, New York 1984) p. 271

7.40   R.J. Elliott: Phys. Rev. 108, 1384 (1957)

7.41   W. von der Osten: "Excitons and Exciton Relaxation in Silver Halides", in Polarons and Excitons in Polar Semiconductors and Ionic Crystals, NATO ASI Series B: Physics, ed. by J.T: Devreese, F. Peeters, Vol. 108 (Plenum, New York 1984) p. 293

7.42   Calculation communicated by H. Stolz, Habilitationsschrift, Universität Paderborn (in preparation)

7.43   J. Nakahara, K. Kobayashi, A. Fujii: J. Phys. Soc. Jpn. 37, 1312 (1974)

7.44   U. Sliwczuk, H. Stolz, W. von der Osten: Phys. Status Solidi B 122, 203 (1984)

7.45   W. Hayes, R. Loudon (eds.): Scattering of Light by Crystals (Wiley, New York 1978)

7.46   G.L. Bottger, C.V. Damsgard: Solid State Commun. 9, 1277 (1971)

7.47   W. von der Osten: Phys. Rev. B 9, 789 (1974)

7.48   R. Loudon: Adv. Phys. 13, 423 (1964)

7.49   A.K. Ganguly, J.L. Birman: Phys. Rev. 162, 806 (1967)

7.50   W. von der Osten, J. Weber, G. Schaack: Solid State Commun. 15, 1561 (1974)

7.51   J. Weber, W. von der Osten: Z. Phys. B 24, 343 (1976)

7.52   J. Windscheif, W. von der Osten: J. Phys. C 13, 6299 (1980)

7.53   K.L. Shaklee, R.E. Nahory: Phys. Rev. Lett. 24, 942 (1970)

7.54   W. von der Osten: "Vacancy Aggregate Centers in Ionic Crystals" in Defects and their Structure in Nonmetallic Solids, ed. by B. Henderson, A.E. Hughes, NATO ASI Series B : Physics, Vol. 19 (Plenum, New York 1976) p. 237

7.55   M.V. Klein: Phys. Rev. B 8, 919 (1973)

7.56   P.Y. Yu, Y.R. Shen, Y. Petroff, L. M. Falicov: Phys. Rev. Lett. 30, 283 (1973)

7.57   Y.R. Shen: Phys. Rev. B 9, 622 (1974)

7.58   J.R. Solin, H. Merkelo: Phys. Rev. B 12, 624 (1975); Phys. Rev. B 14, 1775 (1976)

7.59   L.K. Aminov: Phys. Rev. B 12, 3490 (1975)

7.60   V.V. Hizhnyakov, A.V. Sherman: Phys. Status Solidi B 85, 51 (1978)

7.61   K. Nakamura, J. Windscheif, W. von der Osten: Solid State Commun. 39, 381 (1981)

7.62   K. Nakamura, W. von der Osten: J. Phys. C 16, 6669 (1983)

7.63   P.R. Vijayaraghavan, R.M. Nicklow, H.G. Smith, M.K. Wilkinson: Phys. Rev. B 1, 4819 (1970)

7.64   Y. Fujii, S. Hoshino, S. Sakuragi, H. Kanzaki, J.W. Lynn, G. Shirane: Phys. Rev. B 15, 358 (1977)

7.65   Y. Toyozawa: Progr. Theor. Phys. 20, 53 (1958)

7.66   J. Windscheif, H. Stolz, W. von der Osten: Solid State Commun. 24, 607 (1977)

7.67   T. Iida, M. Sakai: J. Phys. C 20, 4953 (1987)

7.68   J. Windscheif, H. Stolz, W. von der Osten: Solid State Commun. 28, 911 (1978)

7.69   U. Sliwczuk, W. von der Osten: J. Imag. Science 32, 106 (1988)

7.70   D.L. Dexter: "Theory of the Optical Properties of Imperfections in Nonmetals" in Solid State Phys., Vol. 6, ed. by F. Seitz, D. Turnbull (Academic, New York 1958) p. 353

7.71   J. Windscheif: "Untersuchung der Exzitonenrelaxation in Silberbromid mittels Resonanz-Raman-Streuung", Ph.D. Thesis, Universität Paderborn (1979)

7.72   W. von der Osten: "Electronic Properties of Silver Halides", in *The Physics of Latent Image Formation in Silver Halides*, ed. by A. Baldereschi, W. Czaja, E. Tosatti, M. Tosi (World Scientific, Singapore 1984) p. 1

7.73   U. Sliwczuk: "Resonanz-Raman- und Anregungsspektroskopie an Exzitonen in Silberbromid"; Diplomarbeit, Universität Paderborn (1981)

7.74   A.D. Brothers, D.W. Lynch: Phys. Rev. **180**, 911 (1969)

7.75   W. Waßmuth, H. Stolz, W. von der Osten: J. Phys. Condens. Matter **2**, 919 (1990)

7.76   H. Stolz, W. Waßmuth, W. von der Osten, Ch. Uihlein: Physica **117B & 118B**, 383 (1983)

7.77   H. Stolz, W. Waßmuth, W. von der Osten, Ch. Uihlein: J. Phys. C **16**, 955 (1983)

7.78   K. Cho, S. Suga, W. Dreybrodt, F. Willmann: Phys. Rev. B **11**, 1512 (1975)

7.79   E. Schreiber: "Zeitaufgelöste Spektroskopie an Exzitonen in Silberbromid"; Diplomarbeit, Universität Paderborn (1984)

7.80   H. Stolz, W. von der Osten, J. Weber: In Proc. 13th Int. Conf. Physics of Semiconductors, Rome 1976, ed. by F.G. Fumi, (Tipografia Marves, Rome 1976) p. 865

7.81   B.R. Nag: In *Theory of Electrical Transport in Semiconductors*, ed. by B.R. Pamplin, The Science of the Solid State Int. Series, Vol. 3 (Pergamon, Oxford 1972) p. 61

7.82   A. Bonnot, R. Planel, C. Benoit à la Guillaume: Phys. Rev. B **9**, 690 (1974)

7.83   G.E. Picus, E. L. Ivchenko: "Optical Orientation and Polarized Luminescence" in *Excitons*, ed. by E.I. Rashba, M.D. Sturge, Mod. Probl. Cond. Matter Sc., Vol. 2 (North Holland, Amsterdam 1982) p. 205

7.84   S. Haroche: "Quantum Beats and Time-Resolved Fluorescence Spectroscopy" in *High Resolution Laser Spectroscopy*, ed. by K. Shimoda, Topics Appl. Phys. **13** (Springer, Berlin, Heidelberg 1976) p. 253

7.85   K. Blum: *Density Matrix Theory and Applications* (Plenum, New York 1981)

7.86   T. Kushida: Solid State Commun. **32**, 33 (1979)

7.87   Unpublished measurements by Th. Weber (1989), Universität Paderborn

7.88   J. Weber: Phys. Status Solidi B **78**, 699 (1976)

7.89   A. Fujii, H. Stolz, W. von der Osten: J. Phys. C **16**, 1713 (1983)

7.90   J.H. Eberly, K. Wódkiewicz: J. Opt. Soc. Am. **67**, 1252 (1977)

7.91   T. Takagahara: "Resonant Raman Scattering and Luminescence", in Relaxation of Elementary Excitations, ed. by R. Kubo, E. Hanamura, Springer Ser. Solid State Sci., Vol. 18 (Springer, Berlin, Heidelberg 1980) p. 45

8. Light Scattering and Other Secondary Emission Studies of Dynamic Processes in Semiconductors

J. A. Kash and J. C. Tsang

With 49 Figures

Almost all of the results discussed in this series of volumes on light scattering in solids have been obtained on samples that are at or close to thermal equilibrium. In this chapter, we show that inelastic light scattering and hot luminescence are powerful tools for the study of solids that are far from thermal equilibrium and review our current understanding of how nonequilibrium distributions of carriers and phonons approach thermal equilibrium.

In a semiconductor like GaAs, energetic electrons and holes that are far from thermal equilibrium will approach equilibrium via a number of different interactions. They can relax by giving their excess energy to the vibrational modes of the lattice and other excitations of the system to which they are coupled. They can also scatter off each other and randomize their energy and momentum in these collisions. Since the characteristic interaction times of these different processes can be in the subpicoseond range, quantitative understanding of these relaxation processes requires experimental probes that can measure the evolution of the nonequilibrium carrier and phonon distributions with time resolution on the subpicosecond scale. In addition to subpicosecond time resolution, spectral resolution on the scale of the energies of the relevant elementary excitations of the system is also required for a detailed understanding of these processes. We show in this chapter that the requirements of the temporal and spectroscopic resolution of the individual electron, hole, and vibrational distribution functions can be met in group IV and III–V semiconductors by a combination of Raman scattering and hot luminescence spectroscopy.

The problems to be discussed in this article include both the dynamics of optical phonons and highly energetic electrons and holes. Understanding this dynamics has applications to the behavior of devices that have dimensions comparable to typical carrier scattering lengths or contain electric fields strong enough to significantly perturb the equilibrium distribution functions for the quasiparticles. We review the physical mechanisms coupling carriers and phonons which are responsible for their relaxation. A theoretical basis for our experimental requirements with regards to the time scales of interest is provided. This discussion also touches on the limits of the theoretical approaches to the problem. It shows why spectroscopic information about the energy relaxation is required and how, for example, processes that are important in one region of the hot carrier phase space may be unimportant in another region. The use of inelastic light scattering and secondary emission (hot luminescence) spectroscopy for the characterization of this type of dynamic behavior will be described. Then, specific applications of these techniques to the problems of phonon dynamics and carrier kinetics in group IV and III–V bulk semiconductors will be reviewed. Experimentally generated values for various parameters describing carrier and phonon relaxation in semiconductors are given. Recent extensions of this work to two-dimensional quantum wells and superlattice structures will be described briefly. A summary including a tabulation and discussion of existing experimental results of phonon dynamics and hot carrier kinetics in GaAs and of questions which remain to be addressed, close this article.

Although most of this chapter will be devoted to a very specific subject, that of energy relaxation in III–V semiconductors on the picosecond time scale, our subject in fact has very general conceptual significance. In 1876, James Clerk Maxwell [8.1] declared that "All the Physical Sciences relate to the passage of energy, under its various forms, from one body to another ..." He classified the various parts of any experiment in terms of (1) a source of energy, (2) channels which carry the energy to where it does its work, (3) restraints which limit its work, (4) reservoirs which store it, (5) apparatus which allow it to escape, (6) regulators for equalizing the work done, (7) indicators that are acted on by the forces involved in the experiment and (8) a means of reading off the indicators. The first five of Maxwell's categories can be used to describe the processes by which energy in the form of monochromatic photons is injected into a solid and turns into heat. In our case, we always begin with photons in the visible which are absorbed with the creation of electron–hole pairs in a semiconductor. The energy and momentum of the excited electron–hole pairs can be randomized by collisions between different carriers. The energy can leave the electronic system through interaction of the carriers with phonons. The phonons will interact with one another before the initially deposited energy turns into heat. It is these distinct steps, the thermalization of the hot carriers, their relaxation by phonon emission and the relaxation of the hot phonon population, that form the subject matter of this chapter.

In the previous chapters in these volumes, the populations of the different phonon modes have always been defined by the Bose-Einstein distribution with only two parameters, the phonon energy $\hbar\omega_b(q)$ where b defines the branch of the dispersion curve and q the phonon wavevector, and the temperature of the sample, T. Similarly, the electron and hole distribution functions have been described by Fermi-Dirac statistics with kinetic energies E_e and E_h with respect to the individual band edges, well defined Fermi levels, and again a single temperature. These distributions represent the behavior of a solid at thermal equilibrium and their derivations depend only on very general statistical mechanical arguments. On the other hand, as Maxwell pointed out, our experimental understanding of physical phenomena relies on our ability to measure the consequences of the input of well defined energy into a real physical system.

This input of energy can drive any physical system away from thermal equilibrium raising the question of how the nonequilibrium system is to be described. Experiments involving steady state sources generally produce a weak perturbation of the thermal equilibrium distributions, usually only changing the temperature of the system so that it is no longer equal to the ambient temperature. In the limit of high excitation levels, steady state experiments can produce circumstances where the different distributions of the system are described by different temperatures. The temporal evolution of these temperatures towards thermal equilibrium where all temperatures are equal usually occurs on the time scale of tens to hundreds of picoseconds and will not be discussed here. On the other hand, if energy is added to a system very rapidly, it is possible to create conditions where the populations of elementary excitations in the system cannot be described by thermal distributions. The creation of this type of non-thermal distribution provides one of the definitions of the condition "very rapidly" used in the previous sentence. The number of excitations of each type are then not determined simply by their energy, a single temperature parameter and in the case of fermions, by a chemical potential. For example, irradiation of GaAs by a 100 fs pulse of 2.0 eV light which injects 10^{15} cm^{-3} carriers will result in three narrow groups of electrons occupying states about 420, 290 and 90 meV above the bottom of the conduction band immediately after the light is absorbed [8.2]. This structure in the excited electron distribution corresponds to excitations from the three different valence bands present in GaAs. The spread in energy of each of these groups of electrons immediately after the excitation of the carriers (i.e., before the carriers have scattered) will be determined only by the spectral width of the optical pulse and details of the band structure.

The time evolution of these nonequilibrium distributions of the optically excited carriers into the well-known equilibrium thermal distributions is governed by the interaction of the different elementary excitations of these systems. These relaxation processes can proceed in several different ways. The optically excited carriers will give up energy to the lattice via the emission of phonons in a sequential cascade when the interaction between carriers can be neglected [8.3, 4]. Under these circumstances, a thermal distribution of carriers occurs only after almost all of the excess energy initially deposited in the carrier system is lost to the lattice. Alternatively, in the limit where the collisions between the excited carriers occur frequently compared to the excitation of phonons by the hot carriers, it is possible for the carriers to achieve a thermalized distribution before significant loss of energy to the lattice [8.4, 5]. Similar considerations apply to the relaxation of the phonons which are excited by the carriers. If the phonon lifetimes are long compared to the rate at which the excited carriers generate phonons, then large nonequilibrium phonon populations can be generated by the hot carriers. The reabsorption of these phonons by the cooling carriers can reduce the net rate at which energy can be transferred from the electronic system to the lattice system [8.6]. The parameters that determine these various scattering rates are the focus of this article. While information about these processses can be inferred from steady state measurements, the most complete description of the physical processes involved in the energy relaxation is obtained when the relaxation processes are both temporally and spectrally resolved. We shall see that the scattering times of interest are typically of order 200 fs. Note that an electron in GaAs with 0.25 eV kinetic energy travels about 600 Å in 200 fs. For hot electrons in submicron devices, then, the usual models of diffusive transport may not be valid [8.7, 8]. Thus, the quantitative determination of the different scattering rates that occur in a solid provides the foundation for understanding energetic carrier transport in small devices.

Traditional cw spectroscopic tools such as photoemission and Raman scattering provide information on the energy levels of electrons, holes, and phonons in semiconductors. The energy levels by themselves do not describe how they interact with each other or relax to their ground state when excited. The study of the dynamics of these excitations can provide a direct measure of the interactions and tests our ability to calculate them. The physical processes described in this chapter include (1) the decay of optical phonon modes (Sect. 8.3.1) which in its simplest form involves the anharmonic corrections to the lattice dynamics of a solid and also the interaction with other excitations such as electronic transitions, (2) the generation of small wavevector and optical phonons by energetic charged carriers (Sects. 8.3.2, 3 and 8.4.1a) which proceeds mainly through the Fröhlich term in the electron–phonon interaction, (3) the intervalley scattering of electrons between the different minima of the conduction bands of III–V compounds (Sect. 8.4.1b), which is mediated by the deformation potential for the scattering of charged carriers by large wavevector phonons, (4) carrier–carrier scattering including electron–electron, hole–hole and electron–hole processes (Sect. 8.4.2), and (5) the effects of dimensionality on these processes (Sect. 8.5). From an experimental point of view, these phenomena raise interesting questions about the character of Raman scattering at short times (Sect. 8.2.1). They revive questions about the difference between Raman scattering and hot luminescence (Sect. 8.2.3). Finally, from a practical standpoint, these experiments often test our ability to detect small signals against large backgrounds (Sect. 8.2.5).

The focus here will be on light scattering and secondary emission probes of phenomena occurring during the first 10 ps after carrier injection by photons. In particular, we will not consider processes involving longer time scales (e.g. laser annealing) where there is substantial evidence that the observed behavior is associated with the rapid heating and cooling of the sample [8.9]. While the vibrational and electronic systems of the annealed sample are highly excited on a nanosecond time scale, both systems can still be defined by temperatures in most annealing experiments. We will also not review ultra-fast behavior in semiconductors using first order optical probes such as modulated reflectivity and absorption [8.10], nor non-linear optical probes such as CARS. Finally, although they have been of considerable recent interest, we will not discuss the dynamic properties of II–VI compounds and their magnetic alloys.

8.1 Dynamic Processes in Solids

It is customary to begin most treatises on the physics of solids with the following Hamiltonian for the complete system:

$$H_T = \sum_l \frac{P_l^2}{2M_l} + \sum_{l,m} U(R_l - R_m) + \sum_i \frac{p_i^2}{2m_i} + \sum_{i,l} V(r_i - R_l)$$
$$+ \sum_{i,j} \frac{e^2}{4\pi\varepsilon_0(r_i - r_j)}. \tag{8.1}$$

All the symbols have their normal identities as either ionic (uppercase coordinates and the subscripts l and m) or electronic (lower case coordinates or the subscripts i and j) terms. If the system wavefunction is written as the product of electronic and lattice wavefunctions, then H_T separates into a lattice term H_L, an electronic term H_e, and a collection of other terms H_ε which are the reason for this chapter. H_L is usually solved in the harmonic approximation to generate the phonon spectrum. The phonon dispersion curves for GaAs, which will provide many of the examples in the following sections are shown in Fig. 8.1 [8.11]. For this chapter, the most significant features of the phonon dispersion curves in Fig. 8.1 are the splitting of the LO and TO branches at $q = 0$ (i.e. Γ) into $\omega_{LO}(0)$ and $\omega_{TO}(0)$, the presence of states in the acoustic phonon branches for energies near $\hbar\omega_{LO}/2$, and the phonon frequencies near the L and X points. The LO-TO splitting reflects the ionic character of the Ga–As bond, and provides the basis for the strong Fröhlich interaction between the long

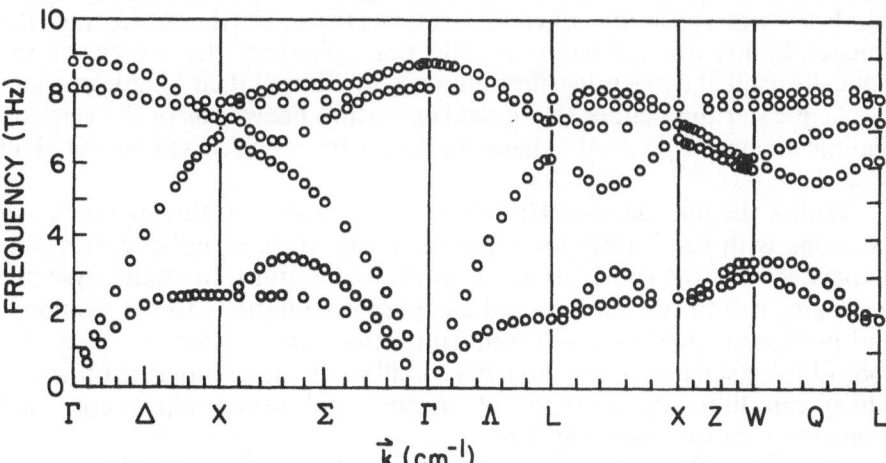

Fig. 8.1. The phonon dispersion curves for GaAs as obtained by neutron scattering [8.11]

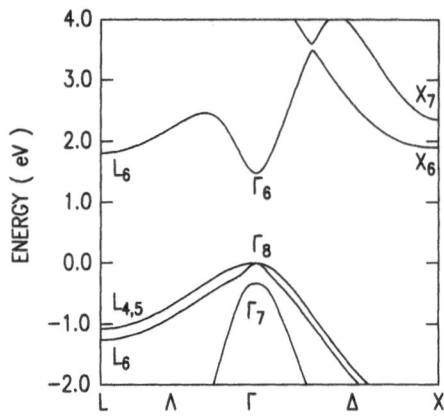

Fig. 8.2. The band structure of GaAs in the vicinity of the direct band gap for the (100) and (111) direction [8.12]

wavelength LO phonons and free carriers. The acoustic phonons near $\hbar\omega_{LO}/2$ provide a channel for the relaxation of the small wavevector optical phonons. The phonons near the X and L points are important in intervalley scattering processes.

The electronic term H_e is usually treated in the one-electron approximation to calculate the band structure. The electronic band structure in the vicinity of the direct energy gap of our prototypical material, GaAs, is shown in Fig. 8.2 [8.12]. The most relevant features of this band structure for our purposes are the Γ valley, the subsidiary conduction band minima at the L and X points, and the splitting of the valence bands into heavy and light hole bands and a spin orbit split valence band. This splitting of the valence band into three separate bands means that the use of a monochromatic optical source to excite GaAs will not generate a monochromatic distribution of electrons in the conduction band except for excitation energies right at the direct gap. Each of the three valence bands can act as a source of conduction electrons. Furthermore, since the valence bands are neither parabolic nor spherical, the widths of the optically excited carrier distributions as a function of their kinetic energies can be greater than either the optical excitation linewidths or the electron lifetime linewidth [8.2, 4]. These matters will be discussed in detail in Sect. 8.4.1.

Within the one-electron Hamiltonian, the details of the interaction of electrons with each other are ignored. There are a series of corrections associated with the consideration of these interactions. In atomic systems, these give rise to the Hartree and Fock approximations to the Coulomb and exchange correlations. In semiconductors, these interactions are responsible for screening, plasmon-type oscillations, and scattering processes which can limit the lifetimes of carriers and randomize energy and momentum in the carrier system.

The lattice Hamiltonian is solved in the harmonic approximation to obtain the phonon dispersion curves. The first correction to the harmonic approximation involves cubic terms in the displacements of the atoms away from their equilibrium positions. The harmonic approximation always assumes infinitesimal excursions from the equilibrium positions. The cubic terms account for how the harmonic Hamiltonian is changed when the displacements have finite amplitudes. Since the eigenfunctions of the harmonic lattice are orthogonal and do not mix, the cubic terms provide the means by which the phonons can interact with each other. The anharmonic lattice terms in the Hamiltonian allow phonons to scatter from other phonons and are responsible for such effects as the expansion of the lattice with increasing temperature and the finite lifetimes of most phonons [8.13].

8.1.1 Electron–Phonon Interactions

The above corrections to the single-electron Hamiltonian for H_e and the harmonic approximation for H_L do not couple the electronic and lattice solutions which come out of the Born-Oppenheimer approximation. The "collection of other terms" which comes directly out of the Born-Oppenheimer approximation also includes electron–phonon interactions, which mix these wavefunctions. While the electronic structure is calculated for the case where the atoms of the lattice are at rest, expansion of $V(r_i - R_l)$ about the equilibrium positions of the ions results in a contribution to the Hamiltonian describing the perturbation of the electronic eigenfunctions and eigenvalues by the phonons. This series of terms in the Hamiltonian has been discussed in many of the other chapters in this series and describes how a charged carrier can be scattered from one state to another by the absorption or emission of a phonon. Our discussion of these terms is based on their treatment by *Ridley* in his recent text and the full derivations can be found there [8.14]. Beginning with Fermi's golden rule for the rate of transitions between an initial state i and a final state f generated by a perturbation H_g, it is possible to derive a general expression for the transition rate, $\Gamma(k)$, describing the scattering of a carrier in a state defined by the wavevector k to the state with wavevector k' by a phonon with wavevector q from the b branch of the phonon dispersion curves.

$$\Gamma(k) = \frac{V_0}{8\pi^2 M}$$
$$\times \int \frac{C_{q,b}^2 I^2(k, k')}{\omega_b(q)} \left(n_{q,b} + \frac{1}{2} \mp \frac{1}{2} \right)$$
$$\times \delta_{k-k'\pm q, K} \delta(E_k - E_{k'} \pm \hbar\omega_b(q)) \, dk' \, . \tag{8.2}$$

Here $\omega_b(q)$ is the frequency of the phonon mode with mode occupation $n_{q,b}$, V_0 is the volume of a unit cell, K is any reciprocal lattice vector, and M is the appropriate mass for the particular vibrational oscillator (i.e., the mass per unit cell for acoustic modes, the reduced mass for optical modes). The upper signs describe absorption of the phonon while the lower describe its emission. $I^2(k, k')$ is the overlap integral between the initial and final state electronic periodic Bloch functions, equal to one for intraband transitions near the band edges. In the presence of nonparabolicity (i.e., small gap or large $|k - k'|$), $I^2(k, k')$, can be less than one. $C_{q,b}$ is a coupling parameter which can be derived from the matrix element for the electron-coordinate dependent part of H_ε for a particular phonon. Most of the experimental work to be described is aimed at obtaining quantitative values for $C_{q,b}$. Of special concern in this review is how different physical mechanisms that contribute to $C_{q,b}$ can alter the details of the hot carrier relaxation. The expression inside the integral just describes scattering between the electronic states k and k' through the absorption or emission of a phonon where $k = k' \mp q$ for normal (non-umklapp) processes. The integral in (8.2) sums over all final states to obtain the scattering rate for the state k.

In the simplest approximation, a long wavelength acoustic phonon produces a change in the energy of each electron similar to that produced in a solid by a uniform external strain. This description leads to the name *deformation potential* for the perturbation. For long wavelength acoustic phonons, Ξ_{ij} is defined as the deformation potential where $H_\varepsilon = \sum \Xi_{ij} S_{ij}$, and S_{ij} are the strain components associated with the phonon eigenvector. Acoustic phonon scattering in a Γ valley is due only to the dilation of the lattice so $\Xi_{ij} = \Xi_d$ and $C_q^2 = \Xi_d^2 q^2$. Substituting this into (8.2) produces a high-temperature ($k_B T \gg \hbar\omega_{\text{phonon}}$) scattering rate for electrons in a parabolic band of

$$\Gamma(k) = \frac{2\pi \Xi_d^2 k_B T N(E_e)}{\hbar \varrho v_L^2}, \tag{8.3}$$

where $N(E_e)$ is the electronic density of states, ϱ is the density of the sample and v_L is the velocity of the longitudinal acoustic waves. Acoustic phonon scattering via the deformation potential interaction in the Γ conduction band valley of a material like GaAs results in electron scattering times that range between 2 and 20 ps, depending on the electron kinetic energy and assuming a deformation potential of 6 eV as in [8.15][1]. We will see that the dominant scattering times are 200 fs or shorter in GaAs, so that long wavelength acoustic phonon scattering can usually be neglected. For a review of deformation potentials of GaAs see [8.15a].

---

[1] The largest estimate [8.16] places Ξ_d at 16 eV, which leads to a scattering time of about 0.7 ps for 0.3 eV electrons. This is still much longer than the 200 fs polar scattering time discussed in the next section.

a) Fröhlich Interaction

The lowest order deformation potential interaction for intravalley scattering of electrons in the Γ valley by optical phonons is symmetry forbidden [8.14]. However, the ionicity of the polar optical phonon modes introduces an additional contribution to the electron–phonon interaction in ionic materials; this polar or Fröhlich interaction dominates the intravalley scattering of electrons in the Γ valley and so is treated in detail here.

The energy relaxation in the single particle limit of nonequilibrium electrons in polar semiconductors is dominated by their interaction with long wavelength LO phonons. As seen in Fig. 8.1, the threefold degeneracy of these modes at $k = 0$ in the group IV semiconductors is reduced by their splitting into a non-degenerate LO mode and a two fold degenerate TO mode in the III–V compounds. For the long wavelength LO phonons, the anion and cation in each unit cell move in opposite directions, generating a long range oscillating dipole field in the direction of propagation of the phonon. The amplitude of this electric field is given by [8.15]

$$eE_{LO} = \frac{m^* e \omega_{LO}}{\hbar} \left(\frac{1}{\varepsilon_\infty} - \frac{1}{\varepsilon_0} \right), \tag{8.4}$$

where m^* is the electron effective mass, ω_{LO} is the long wavelength LO phonon frequency and ε_∞ and ε_0 are the high frequency and static dielectric constants. The interaction of the ionic charges with this field results in the splitting of the long wavelength LO and TO phonons which is described by the Lyddane-Sachs-Teller (LST) relation [8.17]

$$\omega_{LO} \left(\frac{1}{\varepsilon_\infty} - \frac{1}{\varepsilon_0} \right) = \frac{\omega_{LO}^2 - \omega_{TO}^2}{\varepsilon_\infty \omega_{LO}}. \tag{8.5}$$

The LST relation allows the determination of E_{LO} in terms of the experimentally measured spectroscopic frequencies of the $q = 0$ LO and TO phonons (accurate to better than 1 cm^{-1} out of 300) and only the high frequency dielectric response. The LST relation eliminates errors associated with the determination of the static dielectric constant ε_0, which can be substantial in conducting systems. The interaction of E_{LO} with a charged carrier is described by the Fröhlich Hamiltonian. In terms of the coupling constant $C_{q,b}$ in (8.2) for scattering an electron in a state k into a state $k + q$ by an LO phonon of wavevector q,

$$C_{q,LO}^2 = \frac{4\pi e^2 M \omega_{LO}^2}{V_0} \left(\frac{1}{\varepsilon_\infty} - \frac{1}{\varepsilon_0} \right) \frac{q^2}{(q^2 + q_D^2)^2}. \tag{8.6}$$

Here q_D is the inverse Debye screening length which for undoped semiconductors is zero at 0 K. The q^{-2} dependence of the coupling coefficient for the Fröhlich interaction arises from the long range nature of the dipole field associated with the anions and cations. It produces a strong interaction between small q phonons and a charged carrier.

Evaluating (8.2) using $C_{q,LO}^2$ gives the total scattering rate by LO phonons via the Fröhlich interaction of an electron in a state with a wavevector k and energy E_e in a spherical, parabolic band as

$$
\Gamma_{e-LO}(E_e) = \sqrt{\frac{2m^*}{E_e}} \frac{V_0 e^2 \omega_{LO}}{\hbar} \left(\frac{1}{\varepsilon_\infty} - \frac{1}{\varepsilon_0} \right)
$$
$$
\times \left[n_{LO} \sinh^{-1}\left(\sqrt{\frac{E_e}{\hbar\omega_{LO}}} \right) \right.
$$
$$
\left. + (n_{LO} + 1) \sinh^{-1}\left(\sqrt{\frac{E_e}{\hbar\omega_{LO}} - 1} \right) \right]. \tag{8.7}
$$

It is assumed that the LO phonon occupation n_{LO} is independent of phonon wavevector, such as for a thermal phonon population. The dispersion of the small wavevector LO phonons is also neglected. The first term in the bracket describes phonon absorption. The second term describes phonon emission, and is zero if $E_e < \hbar\omega_{LO}$ since no final states are available. Electron scattering rates obtained from (8.7) are shown in Fig. 8.3 for electrons in the Γ valley of GaAs (effective mass = 0.067)

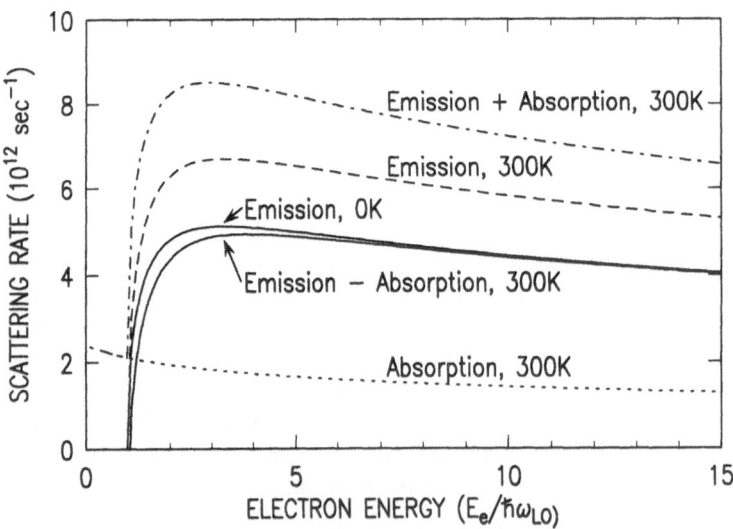

Fig. 8.3. The dependence on the electron kinetic energy of the rate of emission and absorption of LO phonons in GaAs by energetic electrons in the Γ valley of the conduction band at low temperatures and 300 K, from (8.7)

[8.18]. At 0 K, $n_{LO} = 0$ and only phonon emission is present. At 300 K, $n = 0.31$, so both emission and absorption are present. Note, however, that the *net* phonon emission rate (i.e. emission minus absorption) is almost independent of temperature for $E_e > \hbar\omega_{LO}$. For $E_e > 3\hbar\omega_{LO}$, the scattering rate of hot electrons due to the Fröhlich interaction is relatively constant[2] and corresponds to a scattering time of between 150 and 200 fs. From Fig. 8.3, we find that for a single electron injected 300 meV above the bottom of the conduction band of GaAs, it will take about 2 ps for it to reach an energy less than $\hbar\omega_{LO}$ by LO phonon emission. This time provides one of the basic scales for the relaxation of a hot carrier in a material like GaAs.

Equation (8.7) gives the total scattering rate at energy E_e due to the Fröhlich interaction. The distribution of LO phonons in wavevector space generated when electrons scatter from E_e to $E_e - \hbar\omega_{LO}$ can also be calculated from $C_{q,LO}^2$ [8.19]. Figure 8.4 shows this calculated distribution in momentum space for the LO phonons emitted by electrons in GaAs with energy $E_e = 350$ meV $\simeq 10\hbar\omega_{LO}$. In Fig. 8.4, we again consider a simple parabolic band using the conduction band effective mass 0.067 of GaAs. Nonparabolicity makes only a minor change in this distribution. Figures 8.3, 4 are closely related since Fig. 8.3 shows the rate at which hot electrons emit and absorb LO phonons of any allowed wavevector

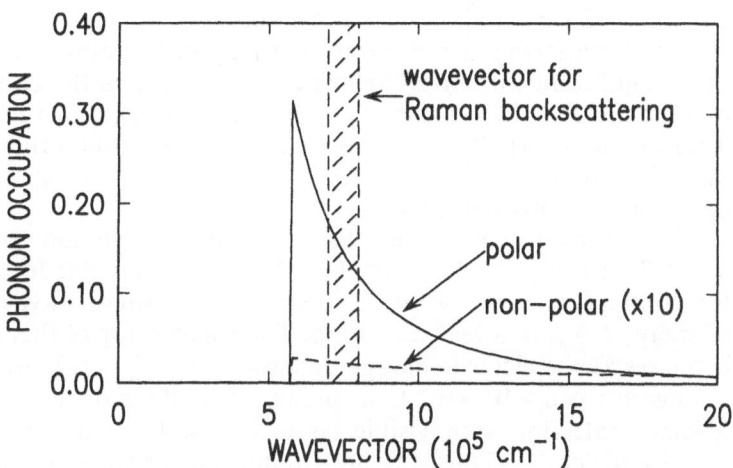

Fig. 8.4. The dependence on the scattered phonon wavevector of the nonequilibrium LO phonon population generated by 350 meV electrons in the Γ valley of the conduction band of GaAs. The vertical axis corresponds to the excitation of LO phonons by 10^{16} cm^{-3} electrons. The hatched region indicates the range of phonon wavevectors that can be excited in a Raman scattering experiment using the backscattering geometry [8.19]

---

[2] *Fawcett* et al. [8.18] have shown that inclusion of nonparabolicity increases the net emission rate slightly at high energies, making the net emission rate even more constant.

while Fig. 8.4 shows the distribution in momentum space of the LO phonons emitted at $E_e \simeq 10\hbar\omega_{LO}$. The absolute nonequilibrium mode occupations shown in Fig. 8.4 were calculated assuming an electron density of $10^{16}\,\text{cm}^{-3}$. At the wavevectors typical for Raman back-scattering with visible lasers, as indicated in the figure, $n \simeq 0.1$. Such mode occupations are readily observable. Measurements discussed in Sect. 8.3.2 show that the calculated nonequilibrium populations are quite close to what is experimentally achieved with the optical pumping of GaAs.

The decrease in the mode occupation of phonons created at larger wavevectors as seen in Fig. 8.4 reflects the fact that the Fröhlich Hamiltonian itself depends on the phonon wavevectors as $1/q$. Figure 8.4 also shows the nonequilibrium phonon mode occupation that would be generated by 350 meV electrons in GaAs through a wavevector independent coupling of the electrons to the optical phonons (dashed line) [8.20]. It is interesting to note that because the band structure limits the final available states for the electrons, there is a wavevector dependence to the phonon distribution even when generated by a wavevector independent coupling. The peak mode occupation is however much smaller ($\simeq 1\%$ here) than for the distribution generated by the Fröhlich interaction. In any case, as mentioned earlier, deformation potential scattering by optical phonons for electrons in the Γ valley is symmetry forbidden [8.14]. Because of their larger masses and lower symmetries, the situation is more complicated for conduction band minima at the L point. Deformation potential scattering of hot electrons by optical phonons has been shown to be significant for carrier relaxation in Ge where the conduction band minima are at L. This has been treated by *Conwell* and *Vassell* [8.15] and *Young* et al. [8.21]. *Pötz* and *Kocevar* [8.22] have noted that TO phonons may contribute to the relaxation of hot holes in GaAs through the deformation potential interaction.

The abrupt cut-off in Fig. 8.4 in the mode occupation at $q = 5.7 \times 10^5\,\text{cm}^{-1}$ reflects the fact that it is impossible for an electron in the conduction band of GaAs to emit a very small wavevector phonon of energy $\hbar\omega_{LO}$ in a real transition. The wavevector of this abrupt cut-off increases as the electron energy decreases. For $E_e = 120$ meV, the cut-off wavevector is $8 \times 10^5\,\text{cm}^{-1}$. As indicated by the hatched area in Fig. 8.4, Raman scattering with visible lasers is unable to detect phonons with $q > 8 \times 10^5\,\text{cm}^{-1}$. Thus, the nonequilibrium phonons detected with Raman scattering originate from electrons with kinetic energy greater than 120 meV.

The dominant role of the small wavevector LO phonon scattering channel in the relaxation of energetic carriers in III–V compounds is one of the foci of this chapter. This dominance has been shown previously in a variety of measurements, including conventional cw Raman scattering [8.23], photoluminescence [8.24], and photoconductivity [8.25, 26] studies. For example, in cw Raman scattering, when resonant effects in the

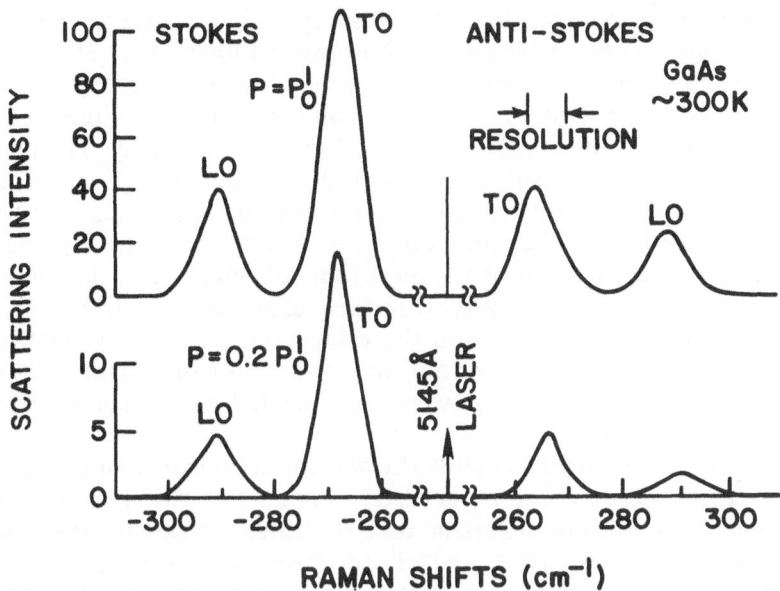

Fig. 8.5. The Raman spectrum of (111) oriented GaAs excited at two different power levels [8.23]

cross-section can be ignored[3], the ratio of Stokes to anti-Stokes intensities for a phonon mode occupation n is $(n + 1)/n$. In an early cw measurement, *Shah* et al. [8.23] determined the mode occupations of Raman active LO and TO phonons in GaAs as a function of laser power. As shown in Fig. 8.5, the TO mode occupation of the sample increased only slightly with laser power. At the highest power, the TO mode occupation was consistent with a heating of the sample to 420 K, above the room temperature ambient. At the same time, the LO phonon mode occupation corresponded to an equivalent LO phonon population temperature of above 800 K. This showed there was a significant nonequilibrium LO phonon population as opposed to the thermal TO phonon population.

Another example of the dominant role of the emission of LO phonons in the relaxation of energetic electrons in GaAs was observed by *Shaw* [8.26], *Nahory* [8.25], and *Weisbuch* et al. [8.24]. They measured the dependence of the magnitude of the photoconductivity or the intensity of the band edge photoluminescence on the excitation photon energy. In both cases, the excitation spectra showed a periodic modulation of as much as 40%. The modulation period is equal to $\hbar\omega_{LO}(1 + m^*/m_{hh})$. The multi-

---

[3] The use of the Stokes to anti-Stokes ratio to obtain the mode occupations has been shown to be subject to significant errors in systems where the electronic structure and scattering matrix elements can also depend on the effective temperature as in the case of laser annealed Si [8.9].

plying factor in parenthesis for this expression is due to the division of the kinetic energy between the photoexcited heavy hole at mass m_{hh} and electron of mass m^* when an electron–hole pair is excited by light. As a result, for an electron in the conduction band, the photoluminescence and photoconductivity is periodic in the LO phonon energy. The dependence of the modulation period on the LO phonon frequency for these two effects showed that the relaxation of the hot electrons in GaAs is dominated by the emission of LO phonons. It was estimated that the LO phonon emission rate was at least an order of magnitude larger than the TO phonon emission rate. *Weisbuch* also found that the strength of the modulation depended on the excitation intensity. At high densities, the carrier–carrier scattering processes which can broaden the optically injected carrier distribution were responsible for a reduction in the depth of the modulation.

While these measurements showed the significance of the LO phonon Fröhlich interaction for energetic carrier relaxation, none were able to determine the characteristic time or scattering rate for this interaction. Transport measurements, which produce average scattering times integrated over a range of states are consistent with the theoretical calculations shown in Fig. 8.3 [8.18, 27, 28]. The transport measurements, however, are not a direct measure of the scattering time. In this paper, we review how time resolved Raman spectroscopy can be used to directly obtain a value for the total time required for the cascade emission of LO phonons by hot electrons. From this measurement, one can obtain the LO phonon scattering time itself. In addition, we will discuss studies of the hot luminescence spectra due to the recombination of free electrons at neutral acceptors which have provided alternative methods of measuring the electron–LO phonon scattering time as well as other important scattering mechanisms for the electrons.

b) Deformation Potentials and Intervalley Scattering

The Fröhlich interaction is present only in polar semiconductors and is significant only for scattering by long wavelength LO phonons. In contrast, the deformation potential coupling for scattering of carriers by phonons is present in all semiconductors and does not single out small wavevector phonons. The strength of the deformation potential interaction as measured in resonant Raman experiments shows only a weak dependence on the phonon wavevector in contrast to the strong $1/q$ dependence of the Fröhlich interaction [8.29]. The deformation potential scattering of carriers by optical phonons is symmetry forbidden at $q = 0$ in the Γ valley. However, it has nonzero strength for intervalley scattering processes such as those which take an energetic carrier from the Γ valley to the L and X valleys shown in Fig. 8.2 [8.14]. Since the phonon branches are dispersionless at the zone edge, it is natural to write the electron–phonon coupling for the optical phonon branch b as a constant

$$C_{q,b}^2 = \left(\frac{M}{M_1 + M_2}\right) D_{ij,b}^2,\qquad(8.8)$$

where $D_{ij,b}$ is the deformation potential for intervalley scattering from valley i to valley j[4]. Within this approximation, and considering only a single phonon branch so that the subscript b may be dropped, the rate for scattering an electron from energy E_e in valley i to a state in a parabolic valley j characterized by a mass m_j, by a near zone edge phonon of energy ω_{ze} and mode occupation n_{ze} is [8.15],

$$\frac{1}{\tau_{i-j}} = \frac{D_{ij}^2 m_j^{3/2}}{\sqrt{2}\pi\hbar^3 \varrho\omega_{ze}}\left[(n_{ze} + 1)\sqrt{E_e - \hbar\omega_{ze} - E_{ij}}\right.$$
$$\left. + n_{ze}\sqrt{E_e + \hbar\omega_{ze} - E_{ij}}\right].\qquad(8.9)$$

Again the first term in the brackets describes the phonon emission process while the second term describes the phonon absorption process. E_{ij} is the energy difference between the bottoms of valleys j and i. Note that the energy dependence of τ_{i-j} results solely from the density of final states, which is proportional to $m_j^{3/2}\sqrt{E_e \mp \hbar\omega_{ze} - E_{ij}}$. Equation (8.9) represents scattering to a single valley. In the case of multiple equivalent valleys, such as the 6 X or 4 L valleys [8.28], the net scattering rate is increased proportionately. Also, (8.9) includes contributions from only a single phonon branch. If more than a single branch is involved, a sum over the different branches must be performed. Note that the density of the crystal ϱ appears in the denominator of (8.9) for historical reasons related to scattering by long wavelength acoustic phonons (chap. 7).

Intervalley scattering plays a dominant role in the high electric field transport of GaAs and is responsible for the Gunn effect [8.30]. The rates at which electrons can make transitions from the Γ valley to the L and X valleys (Fig. 8.1) under various types of excitations including the presence of high electric fields, have been the the center of considerable interest over many years due to the technological significance of the Gunn effect and other high field transport phenomena [8.28]. Because the different conduction band minima are separated by large wavevectors, electron scattering between the valleys must be mediated by the emission or absorption of large wavevector phonons. Therefore, the deformation potential interaction determines the magnitude of the intervalley scattering

---

[4] The displacement pattern of zone edge or optical phonons cannot be obtained by a mechanical deformation of the crystal. Therefore, the deformation potential for zone edge or optical phonons is defined somewhat differently than for small q acoustic phonons. The units are also different: eV/cm for $D_{ij,b}$ and eV for Ξ_{ij}.

processes. As pointed out earlier, the intravalley deformation potential scattering for energetic electrons where $E_e > \hbar\omega_{LO}$ in the Γ valley of GaAs is small compared to the Fröhlich scattering. The heavier masses of the L $(0.22m_e)$ and X $(0.407m_e)$ valleys [8.19] compared to the Γ valley $(0.067m_e)$ and the presence of 6 X and 4 L valleys results in an order of magnitude larger density of states for the satellite valleys. As a result, it is possible for the intervalley scattering rate to be as fast as the polar (intravalley) scattering rate. Similarly, the smaller density of electronic states in the Γ valley compared to the L and X valleys means that the scattering of carriers from the Γ valley to the L and X valleys will be much faster than the scattering in the reverse direction. $D_{\Gamma L}$ and $D_{\Gamma X}$ are 10^8 to 10^9 eV/cm, which results in $\tau_{\Gamma-L}$ and $\tau_{\Gamma-X}$ at low temperatures in the range of ~ 30 fs to 3 ps for E_e roughly 100 meV above the threshold for intervalley scattering [8.15]. Since the LO phonon scattering time is seen in Fig. 8.3 to be about 200 fs for energetic electrons near the threshold for intervalley scattering, intervalley scattering has the possibility of dominating the intravalley scattering processes for highly energetic electrons in GaAs. Because of intervalley scattering, the kinetics of energetic electrons in a semiconductor like GaAs can be qualitatively different for electron energies above and below the threshold for intervalley scattering. Because the scattering times depend quadratically on D_{ij}, a factor of 3 error in the deformation potential can result in a qualitatively incorrect picture of the relaxation of a nonequilibrium electron. Also, because the mass of the satellite valley must be assumed to connect the deformation potential to the scattering rate, the density of states mass employed should be given (but not always is) in a discussion of the deformation potential and intervalley scattering.

The measurement of negative differential mobility in the velocity-field characteristics of *n* type GaAs clearly shows the importance of intervalley scattering. Such transport measurements are not spectroscopic, and extensive modeling is required to obtain the actual intervalley scattering rates. Unfortunately, the measurement is not sensitive to the scattering rates [8.31]. In addition, transport measurements have difficulty in separating the contributions of the L and X valleys. In Sect. 8.4.1b, we show how hot luminescence spectroscopy of the recombination of energetic electrons at neutral acceptors provides several spectroscopic means of directly determining the intervalley scattering rates as a function of hot electron energy.

8.1.2 Phonon–Phonon Interactions and Phonon Lifetimes

If the electron–LO phonon scattering time is short compared to the LO phonon lifetime, large nonequilibrium LO phonon populations can be excited by the rapid relaxation of highly energetic carriers. The magnitude of the nonequilibrium phonon distribution is significant for the carrier dynamics because the carriers can re-absorb the phonons in addition to exciting them. Quantitative knowledge of the phonon lifetimes is necessary to model this effect. The lifetimes of the LO phonons which are excited by relaxing hot carriers also help to set the time scales of importance to our subject. Within the simple relaxation time approximation, the mode occupation $n_{qb}(t)$ of nonequilibrium phonons with momentum q from the b branch of the phonon dispersion curves, can be described by [8.23]

$$\frac{dn_{qb}(t)}{dt} = G_{qb} - \frac{n_{qb}(t) - n_{qb}^0}{\tau_{qb}}, \tag{8.10}$$

where G_{qb} is the generation rate, τ_{qb} is the mode lifetime, and n_{qb}^0 is the equilibrium mode occupation. cw measurements of the Raman spectra of solids such as GaAs, Ge, ZnSe, and GaP show that the optical phonons have linewidths between 0.25 and 2 cm^{-1} at low temperatures [8.32, 33]. These widths broaden by a factor of 2–3 at room temperature. In a well-ordered solid where the lineshape of a Raman active phonon is broadened only because of the lifetime, the Raman lineshape is a Lorentzian with full width at half maximum of Δv_{qb} where [8.33]

$$\tau_{qb} = \frac{1}{2\pi \, \Delta v_{qb}}. \tag{8.11}$$

In practice, the experimentally measured spectral widths may not rigorously define the phonon lifetimes since there can be inhomogeneous contributions to the linewidths. Inhomogeneous contributions can be especially troublesome in disordered solids such as semiconductor alloys where the picture of a single Raman active mode breaks down [8.34, 35]. However, since the different contributions to phonon linewidths are independent of each other and combine cumulatively, the spectroscopic linewidth provides a lower bound to the lifetime of the excitation. The lifetime of the excitation can be longer than $(2\pi \, \Delta v_{qb})^{-1}$ but will not be shorter. Given this caveat, the phonon lifetimes obtained from their linewidths dictate interest in dynamic phenomena occurring on the 2–10 ps time scale. When GaAs is excited by a 2 eV laser, nonequilibrium LO phonons will be emitted by relaxing electrons over a time period equal to about 200 fs × $(E_e/\hbar\omega_{LO})$. This period of time over which the hot phonons are generated is roughly equal to the phonon lifetime, so a quantitative understanding of both hot phonon and hot electron dynamics will require an accurate measurement of the phonon lifetime.

Aside from cw Raman and infrared measurements, phonon lifetimes can be calculated theoretically [8.13, 33, 36]. For phonon populations that are not too far from equilibrium and where the mode occupations are relatively modest, the lifetimes of optical phonons are determined by anharmonic processes involving the excitation of multiple, low energy phonons. These can be described by the cubic anharmonic terms mentioned earlier. These contributions have been treated theoretically and the requirements of the conservation of momentum and energy mean that the relaxation of small wavevector optical phonons is dominated by the emission of two large wavevector phonons with equal and oppositely directed values of q and whose energies sum to the optical phonon energy. The temperature dependence of the lifetime $\tau_{0,b}$ of a $q = 0$ phonon in branch b decaying anharmonically into a pair of lower energy phonons on branches b' and b'' is

$$\frac{1}{\tau_{0,b}(T)} = \frac{1}{\tau_{0,b}(0)} \left(1 + n_{q,b'} + n_{-q,b''}\right). \tag{8.12}$$

For the decay of optical phonons in semiconductors such as GaAs, there are a number of possible pairs of phonons including pairs of acoustic phonons and sums of one acoustic and one optical phonon. The identifica-

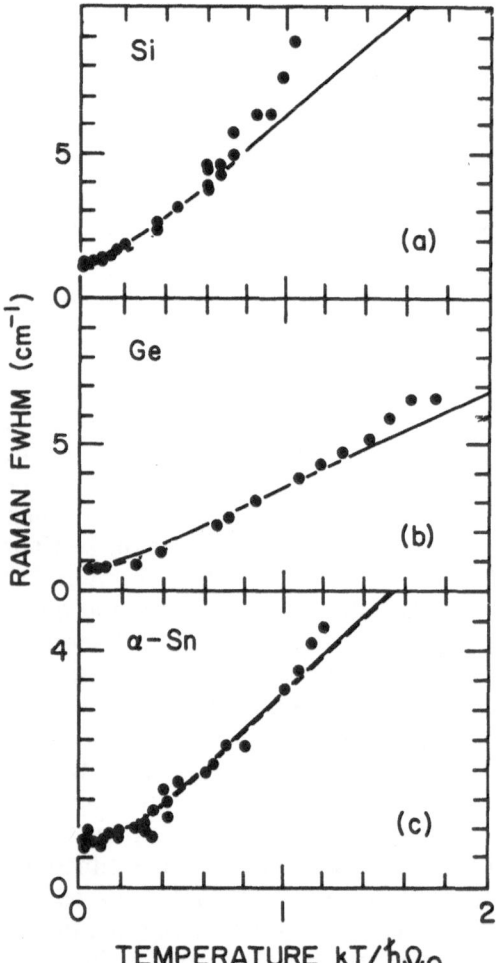

Fig. 8.6. The temperature dependence of the linewidths of the Raman active, long wavelength optical phonons of Si, Ge and Sn. The solid lines are theoretical fits to the linewidths based on relaxation by two phonon decay [8.33]

tion of the particular phonon pair responsible for the relaxation is facilitated by the distinctive temperature dependence. For the group IV semiconductors Si, Ge and Sn, the temperature dependences of the phonon linewidths, illustrated in Fig. 8.6, [8.33] have been used to show that the phonon lifetime is determined by relaxation processes involving the emission of an optical phonon whose energy is about $0.65\omega_{LO}$ and an acoustic phonon whose energy is about $0.35\omega_{LO}$.

The low temperature lifetime $\tau_{0,b}(0)$ depends on a quantitative knowledge of coefficients for the anharmonic terms in the lattice Hamiltonian. Theoretical calculations of the $\tau_{0,b}(0)$ are usually only accurate to within a factor of 5, so that experimental results must be relied on for accurate values. For alloys and other disordered materials such as polycrystalline materials, the Raman lineshape will reflect the presence of disorder in addition to the lifetime broadening, so that extraction of the lifetime from the linewidth is difficult [8.35]. Furthermore, for the problem at hand, that of the energy relaxation of excited states, the phonon lifetime extracted from the linewidth may include scattering processes that are not relevant to this problem. The phonon linewidth might include contributions from dephasing events and from processes which scatter the optical phonons within a particular branch of the dispersion curve [8.37]. It is the scattering of the phonons out of the LO phonon branch which is the determining process in the relaxation of hot carriers in a compound like GaAs and we review in this paper how time dependent Raman scattering provides a direct means of measuring the population lifetime of the LO phonons.

8.1.3 Carrier–Carrier Scattering Processes

Carrier–carrier scattering processes are an efficient means of randomizing both momentum and energy in the carrier system and can rapidly turn a monochromatic or quasi-monochromatic energy distribution into a thermalized distribution. Such processes must be considered if the relaxation of hot carriers at intermediate to high carrier densities is to be understood.

The problem of the scattering of hot carriers from other carriers can be viewed as the rate of loss of energy by a fast electron with energy E_e to a cold sea of other electrons with density n_e. Within the Born approximation and using a host of simplifying assumptions, an expression can be derived [8.14, 15] where

$$\frac{dE_e}{dt} \simeq -\frac{4\pi n_e\, e^{*4}}{\sqrt{2m^* E_e}}. \tag{8.13}$$

Here e^* is an effective charge which can be approximated as $e/\sqrt{\varepsilon(\omega)}$. In GaAs, the average value of $\varepsilon(\omega)$ is about 12. The rigorous derivation of this scattering rate for the case at hand remains an open question and it is known that (8.13) suffers from a variety of possible errors related to

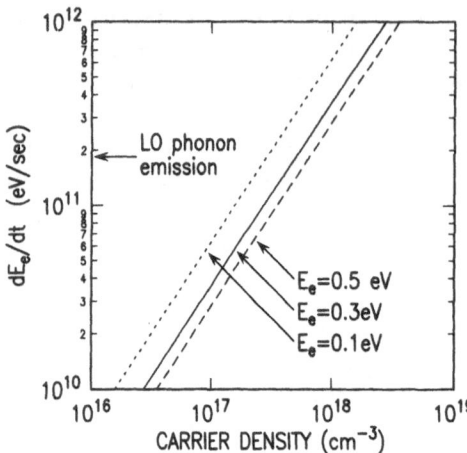

Fig. 8.7. The dependence on the density of carriers of the rate of loss of energy of a fast electron to a sea of electrons in GaAs for three different values of the carrier kinetic energy from (8.13). Also shown is the value of the energy loss rate due to LO phonon emission

the fact that an unscreened Coulomb potential has an infinite scattering cross section. It does show the expected linear dependence on the carrier density, 4th power dependence on the effective electrical charge and inverse dependence on momentum. Typical values for dE_e/dt using this simple expression for carrier–carrier scattering are shown in Fig. 8.7. Also shown in this figure is the position of dE_e/dt for the electron–LO phonon scattering process when $dE_e/dt = \hbar\omega_{LO}/0.190$ ps^{-1} for an energetic electron. The average energy loss per collision for the hot electrons can be much larger than the LO phonon energy for relatively energetic electrons colliding with cold electrons. It is clear from Fig. 8.7 that for carrier concentrations similar to those often found in bulk semiconductors, i.e. between 10^{14}–10^{18} cm^{-3}, carrier–carrier scattering will continuously vary from an insignificant correction to the LO phonon emission rate to the dominant scattering rate. As a result, if $n_e < 10^{14}$ cm^{-3} and $E_e < 400$ meV, the electron–LO phonon scattering time is short enough that statistically less than one collision with another carrier will occur before the energetic electron loses over 90% of its energy to the lattice by LO phonon emission. Under these conditions, the relaxation of hot electrons in a semiconductor like GaAs is described by a cascade process which is schematically shown in Fig. 8.8. Relaxing carriers are found only at energies that are an integer number of LO phonons below the energy at which the electrons were initially created. In contrast, if the carrier–carrier scattering time is short compared to the LO phonon emission time, then the carrier distribution will come into thermal equilibrium with itself before it can lose any energy to the lattice. If the carriers are all initially injected into the conduction band with kinetic energy E_e and the carrier–carrier scattering rate is fast compared to the LO phonon emission rate, then before a single LO phonon is emitted, the carrier distribution will be Maxwellian with an initial temperature proportional to E_e. This model of carrier relaxation is illustrated in Fig. 8.9. The cooling of the electron

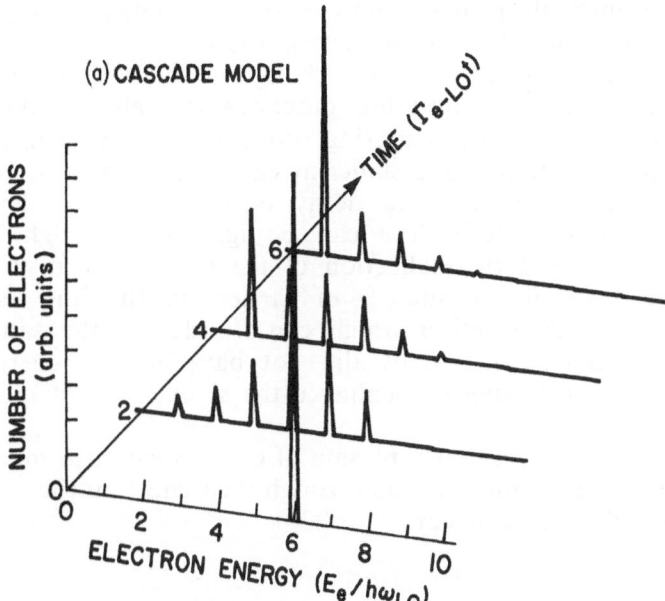

Fig. 8.8. A schematic depiction of the relaxation of a monochromatic distribution of energetic carriers in the limit where the electron–phonon scattering time is much faster than the carrier–carrier scattering time

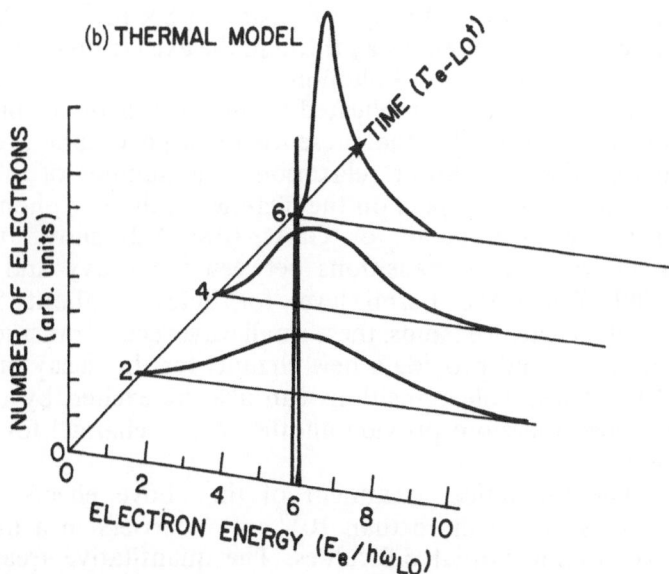

Fig. 8.9. A schematic depiction of the relaxation of an initially monochromatic distribution of hot carriers in the limit where the carrier–carrier scattering time is much faster than the electron–phonon scattering time

system by the emission of LO phonons will then correspond to a reduction in the temperature of the distribution.

Aside from the drastic change in the carrier distribution function, carrier–carrier scattering processes can also modify the significance of the intervalley scattering processes. At low densities, if the carriers do not initially have sufficient energy to scatter to one of the satellite valleys, then they will decay only in the Γ valley and the higher energy satellite valleys can be ignored. At high densities, because of the rapid thermalization of the hot carrier distribution, there can be a significant number of carriers in the high energy tail of the thermal distribution which can transfer to the satellite valleys even if each carrier initially did not have enough energy. Thus carrier–carrier scattering can enhance the significance of intervalley scattering processes.

The presence of a plasma of carriers can also modify the relaxation of the individual carriers through their effect on the phonon spectrum. If the plasma frequency,

$$\omega_p = (4\pi n e^2/\varepsilon_0 m^*)^{1/2} \tag{8.14}$$

is comparable to ω_{LO}, then the elementary excitation spectrum of the system will be composed of the mixed excitations of the plasmon–phonon system [8.38]. In GaAs, $\omega_p = \omega_{LO}$ at $n = 8 \times 10^{17}$ cm^{-3}. At such densities, hot carriers will then decay through the excitation of these coupled modes instead of the bare LO phonon.

Because of the complicated valence bands of zincblende semiconductors such as GaAs, the presence of large numbers of holes can also modify the hot carrier relaxation in a number of different ways. One is through their impact on the lifetimes of the LO phonons. Examination of the valence bands of GaAs (Fig. 8.2) shows the possibility of small wavevector transitions between the heavy and light hole bands [8.39]. When the Fermi level for holes is about 5 meV below the top of the valence bands, these small wavevector transitions are degenerate with $\hbar\omega_{LO}$ and provide a new channel for the decay of the LO phonons [8.40]. These hole transitions can also be excited by energetic electrons and also therefore provide another decay channel for the hot electrons [8.41].

The theoretical treatment of the above effects which arise from the presence of more than 10^{17} cm^{-3} carriers in a material like GaAs involves substantial difficulties. The quantitative treatment of carrier–carrier scattering by itself is a formidable problem. In all cases, direct experimental measurements of the scattering rates represent the simplest approach to understanding these effects.

8.1.4 Homopolar Semiconductors

While the above discussion has taken place in the context of hot carrier relaxation in polar semiconductors, the same general questions can be asked of homopolar semiconductors such as Si and Ge. By and large, with one critical exception, most of the considerations discussed above apply to the homopolar semiconductors. The critical exception is the leading role of the Fröhlich interaction in guiding the transfer of energy from the electronic system to the lattice. In the homopolar materials, the electron–phonon interaction must proceed through the deformation potential interaction. Since this interaction is only weakly dependent on the wavevector of the scattered phonon, the well defined series of LO phonon cascade steps which is characteristic of hot carrier relaxation at low densities in ionic materials will not occur. All optical and acoustic phonons can contribute to the relaxation [8.21]. For excitation energies in the visible, the indirect nature of these materials can also have an effect on the carrier relaxation. While the hot electrons and holes in GaAs are all initially created in the Γ valleys by most methods of pumping including light and electrical injection, the optical response of materials such as Ge and Si is dominated by the creation of electrons and holes in the bands running out towards the zone edges. For example, in Ge, under optical excitation between about 2 and 2.5 eV, the hot electrons and holes are created in the bands responsible for the $E_1 - E_1 + \Delta_1$ transitions which run parallel to each other in the (111) direction [8.42, 43]. The relaxation of the electrons is then towards the L minimum in the conduction band while the holes relax towards Γ. The complicated structure of the valence band means that phonon relaxation processes can also be supplemented by relaxation processes involving transitions between the spin–orbit split valence bands.

8.2 Experimental Considerations

In the previous section, we showed that the effects which govern the dynamics of highly energetic carriers in zincblende structure semiconductors operate on a time scale from about 10 fs to 10 ps. The characteristic times can be strong functions of the carrier kinetic energies. Given such time scales and the need to obtain energy resolved data, it is clear that many conventional measurement techniques will have difficulty in adequately studying these time dependent phenomena. Most electrical techniques will have difficulties in surmounting the hurdles associated with stray capacitances and inductances in any circuit which will tend to stretch the response time of the measuring system to beyond 10 ps. The electrical measurements also generally involve averages over the carrier distribution functions, complicating the process of obtaining the energy dependences of the parameters of interest. Optical techniques such as band edge photoluminescence will be hampered by the fact that the recombination times for band edge carriers in even direct gap semi-conductors such as GaAs tend towards hundreds of picoseconds. Aside from these somewhat long recombination times relative to the relaxation times discussed in the previous section, the band edge emission is usually several hundred meV below the energies of interest so that the hot carrier relaxation must be reconstructed from events that occur after many scattering steps. Finally, most first order optical processes such as absorption or luminescence involve the creation or annihilation of electron–hole pairs. The derivation of the hot electron or hot hole distribution functions from such a measurement requires the ability to separate the hot electron and hot hole dynamics which can be difficult. Two techniques that have good time and spectroscopic resolution and can measure the hole and electron distribution functions individually are inelastic light scattering and hot luminescence involving impurity levels. We shall show now that both of these techniques are capable of subpicosecond time and milli-electron volt energy resolution, making them suitable for the problem of the quantitative description of nonequilibrium carrier relaxation in semiconductors.

8.2.1 Time-Resolved Raman Spectroscopy

In 1980, *von der Linde* et al. [8.32] showed that Raman spectroscopy could be used to directly measure the temporal evolution of an optically generated population of nonequilibrium LO phonons. They were able to obtain a value of 7 ps for the population lifetime of the LO phonons in GaAs at 77 K. The temporal resolution of this experiment was limited by the 3 ps width of the exciting laser so that faster processes could have been measured with a shorter pulse laser. This experiment showed that Raman spectroscopy had a useful future as a probe of subpicosecond dynamics in semiconductors.

The application of Raman spectroscopy to picosecond and sub-picosecond phenomena raises a number of questions. In these experiments, we are interested both in the spectral information and the magnitudes of the scattered signals. If the time varying intensity of the Raman scattering is used to monitor the temporal variation of a nonequilibrium phonon population, it is necessary to establish that the normal linear dependence of the scattering cross section on the mode occupation remains valid for picosecond excitations.

If a Raman scattering experiment is performed using a pulsed laser, by causality, only physical properties that are present in the sample when the laser pulse excites the sample can produce a signal. Raman scattering involves a transition between the ground state and a low lying excited state via the virtual excitation of intermediate, more energetic excited states [8.29, 44]. The time scale for the Raman scattering is set by the detuning of the probing pulse from the virtual intermediate states of the Raman transition and the lifetimes of the intermediate and final states. If the detuning of the optical field from the excited electronic states of the solid is greater than 100 meV, then the characteristic time of the Raman transitions will be below 10 fs using the uncertainty relationship connecting energy and time to describe the virtual transition. Changes in the populations of the real states involved in the Raman transition will be detectable on this time scale. On the other hand, in resonant Raman scattering, the detuning of the transition can go to zero. Then the temporal resolution of the Raman scattering will depend on the lifetime of the resonant intermediate state. If the phonon population has a lifetime short compared to that of the resonant intermediate state, the time dependence measured by phonon Raman scattering will be characteristic of the intermediate electronic state instead of the phonon wavefunction or population. This effect was demonstrated by *Weiner* and *Yu* [8.45], who studied the temporal evolution of the phonon Raman scattering in Cu_2O under resonant excitation. Raman scattering was resonantly excited via the zone center ortho-exciton transition by 10 ps laser pulses. The intensity of the Raman scattered light was measured as a function of time with a temporal resolution of 100 ps. The intensity of the phonon Raman scattering in the symmetry allowed scattering geometry was found to decay exponentially after the excitation with two time constants, 1.5 ns and 0.7 ns. The vibrational Raman scattering therefore persisted long after the exciting pulse was turned off. The first of these lifetimes was the population lifetime for the exciton transition being resonantly excited. The second lifetime arises from the complicated structure of the exciton transition in Cu_2O. The domination of the Raman scattering cross section by these long-lived intermediate states limits the rapidity with which an observed Raman signal can change. The observation by Raman scattering of picosecond and subpicosecond processes using light in resonance with these excitons would be difficult. The exciton lifetime effectively broadens the temporal response function of the system.

For the Raman scattering experiments discussed in this chapter, the laser excitation is above the band gap and therefore in resonance with the single particle continuum. The time resolution can then be limited by the lifetime of the continuum states. For a material like GaAs, this will be of the order of 200 fs, i.e., the electron–phonon polar coupling time [8.15]. In fact, the strength of the Raman scattering for above band gap continuum excitation has strong contributions from the off-resonant neighboring states and need not be dominated by the single strongly resonant transition [8.29]. As a result, the temporal resolution can be considerably faster than the continuum state lifetime. In fact, for materials like Ge and GaAs, the dielectric response in the visible is dominated by the higher energy critical points of the band structure such as E_1 or E_2. Most of the strength of the Raman scattering will come from these resonances which have widths of the order of 40 meV [8.42]. The question of the time resolution in such an experiment when these gaps are resonantly excited has yet to be studied in detail.

A related question is how the Raman scattering process itself can be modified by the use of short pulses when the pulse length becomes comparable to the lifetime or vibrational period of the excitation being probed. *Jha* et al. [8.46] have treated this problem quantitatively and find that Raman scattering using subpicosecond pulses differs only trivially from cw Raman scattering. For a system governed by the time dependent Schrödinger equation, the electron–photon interaction was treated as a time dependent perturbation which turned on at a time $t = 0$ and was nonzero over an interval corresponding to the pulse length t_p. They found that the nonresonant Raman scattering efficiency between an initial state i and a final state f differs from the steady state expression only by the presence of a temporal evolution function $g_{fi}(t)$. For a laser pulse of constant intensity switched on at $t = 0$ and off at $t = t_p$,

$$g_{fi}(t_p) \cong \langle |b_{if}^{(0)}|^2 \rangle_{t_p} \left| \int_0^{t_p} dt \, e^{-\Gamma_{fi} t} e^{-i(\omega_{fi} - \omega)t} \right|^2 . \tag{8.15}$$

Here, $\langle |b_{if}^{(0)}|^2 \rangle_{t_p}$ is the average value of the initial state occupation, which is assumed to vary slowly. Γ_{fi} is the inverse of the lifetime of the final state, ω_{fi} is the frequency difference between final and initial states, and ω is the frequency difference between ingoing and outgoing photons.

Equation (8.15) has two limiting cases. The first case is that of a pulse long compared to the damping time, $t_p \gg 1/\Gamma_{fi}$, where

$$g_{fi}(t_p) \cong \langle |b_{if}^{(0)}|^2 \rangle_{t_p} \left[\frac{\pi}{\Gamma_{fi}} \right] \frac{\Gamma_{fi}/\pi}{[(\omega_{fi} - \omega)^2 + \Gamma_{fi}^2]} . \tag{8.16}$$

Equation (8.16) is the usual Lorentzian form for the Raman lineshape and is illustrated by $t_p = 10$ ps in Fig. 8.10. The other limiting case of (8.15) is $t_p \ll 1/\Gamma_{fi}$, i.e. a very short pulse. Here,

$$g_{fi}(t_p) \cong \langle |b_{if}^{(0)}|^2 \rangle_{t_p} \frac{4 \sin^2 (\omega_{fi} - \omega) t_p}{(\omega_{fi} - \omega)^2} , \tag{8.17}$$

which shows that the Raman lineshape is now the Fourier transform of the laser pulse. This case is shown in Fig. 8.10 with $t_p = 0.1$ ps and $t_p = 0.5$ ps. In all cases, the integrated area of the Raman peak remains constant as the spectral lineshape changes.

Jha et al. [8.46] showed that Raman spectroscopy could be used with subpicosecond pulse sources to obtain quantitative information about the dynamics of nonequilibrium phonon populations. In the subpicosecond regime, the intensities of the Stokes and anti-Stokes Raman scattering still vary as $\langle |b_{if}^{(0)}|^2 \rangle_{t_p}$, i.e. $n + 1$ and n, where n is the average mode occupation of the phonons during the pulse. As a result, the intensity of the Raman

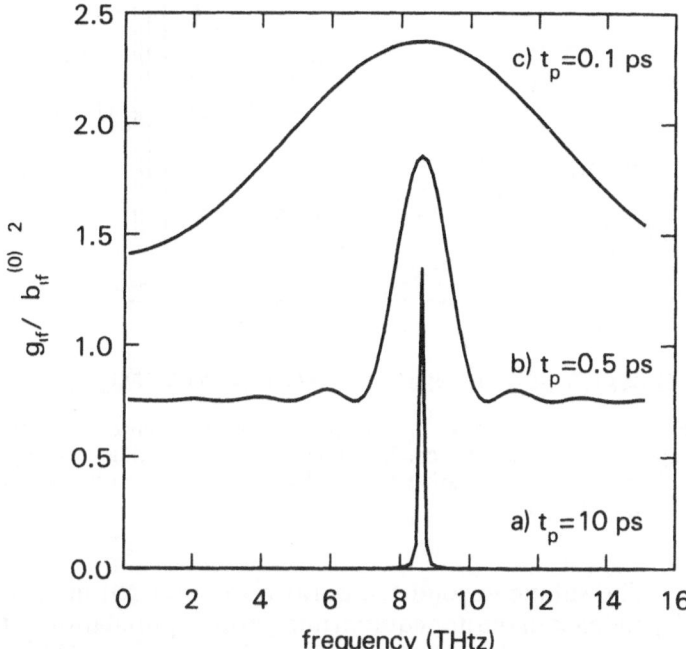

Fig. 8.10. The Raman scattering response function $g_{if}(t_p)$ (8.15) for a mode with a lifetime of 4 ps, a frequency of 8.6 THz, (the LO phonon frequency of GaAs) and several different values of the probe pulse duration t_p

scattering excited by a short pulse monitors the population of the LO phonons during the pulse. This conclusion, when combined with the range of phonon wavevectors that can be studied by Raman scattering in an opaque semiconductor like GaAs and the distribution in momentum space shown in Fig. 8.4 of the nonequilibrium phonons generated by highly energetic carriers in GaAs, makes Raman scattering a good tool for monitoring the behavior of the nonequilibrium phonon population as a function of time after the injection of hot carriers.

For an opaque semiconductor with a high dielectric constant ($\varepsilon \simeq 12$), Raman scattering is practical only in the forward scattering geometry (where phonons with wavevectors less than about 1.5×10^5 cm^{-1} can be observed) or the backscattering geometry (where phonons with wavevectors of about 6×10^5 cm^{-1} can be observed) [8.40]. These scattering geometries are shown schematically in Fig. 8.11. These two particular phonon wavevectors stem from the requirement of momentum conservation in the Raman scattering process. As seen in Fig. 8.4, the backscattering wavevector is close to the peak in the distribution of nonequilibrium phonons while the forward scattering geometry samples a range of wavevectors where there should be no nonequilibrium phonons generated by relaxing hot carriers.

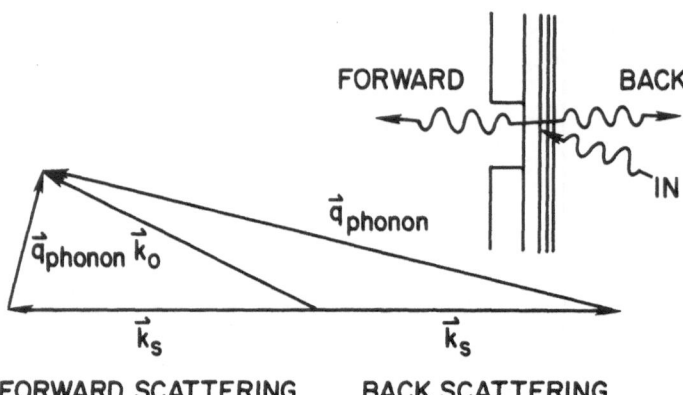

Fig. 8.11. The kinematics of Raman scattering from an opaque semiconductor. The two experimentally accessible configurations are backscattering from any surface of the sample and forward scattering using a thinned window structure with the substrate for support as sketched in the insert

The subpicosecond temporal resolution required to obtain the time dependence of the nonequilibrium phonon population at these wavevectors can be obtained using the pump-probe technique. The pump beam creates hot electron–hole pairs. As the electrons relax they generate the nonequilibrium phonons. The probe beam detects these phonons by Raman scattering. The excitation of the electron–hole pairs by the light occurs on the time scale of the inverse frequency of the light ($\simeq 2$ fs), which is much shorter than the 200 fs LO phonon emission time discussed earlier. Given the momentum distribution of Fig. 8.4, many of the LO phonons emitted by the relaxing hot electrons can be detected by Raman back-scattering. Since for about 2.0 eV excitation, there are approximately 10 Raman active LO phonons created sequentially by each hot carrier, the nonequilibrium Raman signal will increase over about 2 ps.

8.2.2 Electronic Raman Scattering from Nonequilibrium Carriers and Coupled LO Phonon-Plasmons

cw Raman scattering has been used to monitor vibrational, electronic and magnetic excitations [8.47]. *Jha* et al. have also considered the use of time resolved Raman spectroscopy to directly measure the temporal evolution of a hot electron distribution [8.46]. For the nonequilibrium phonons discussed above, it is the amplitude of the Raman signal which changes with time. In contrast, Raman studies of the time dependent changes in the hot carrier distribution function show changes in both the shape and amplitude of the Raman spectra. Both the Raman scattering from single particle excitations due to intraband transitions of a highly excited distribution of carriers and from the coupled LO phonon-plasmon modes should show characteristic changes in frequency as the carrier distribution cools.

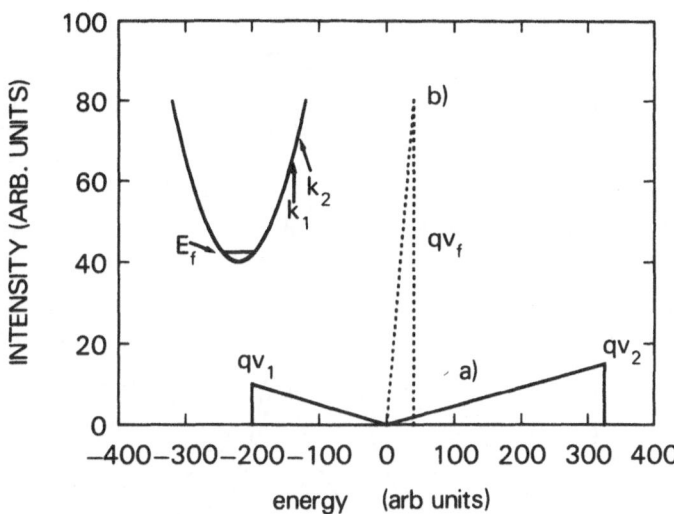

Fig. 8.12. A schematic description of electronic Raman scattering from intraband scattering for a highly excited distribution of carriers fully occupying a set of states with wavevectors between k_1 and k_2 (solid line) and a $T = 0$ K equilibrium distribution of carriers (dotted curve). v_i and k_i are related by the expression $\hbar k_i = m_e v_i$. The inset shows these distributions schematically [8.46]

Figure 8.12 [8.46] contrasts the single particle intraband scattering from a collection of hot electrons occupying states with wavevectors between k_1 and k_2 as shown in the insert with the single particle scattering expected from a cold distribution of carriers at the bottom of the conduction band. For the cold carriers, the Pauli exclusion principle forbids anti-Stokes scattering where an electron would make a transition to a filled lower energy state. Such transitions are possible for the nonequilibrium electron distribution, resulting in the presence of Stokes and anti-Stokes scattering. The high energy cut-offs in Fig. 8.12 where $v_i = \hbar k_i / m_e$ are determined by the minimum and maximum wavevectors of the band of occupied electronic states. These cutoffs provide a means of determining the energy distribution of the hot carriers.

Figure 8.13 [8.46] shows how the coupled mode spectrum of GaAs can be modified by the injection of hot carriers. The effect of these carriers can differ from the presence of cold carriers. These differences are due to the dependence of the plasma frequency at nonzero q on both the density of carriers n_e and their average velocity squared $\langle v^2(t) \rangle$ at a time t:

$$\omega_p^2(q) = \frac{4\pi n_e e^2}{m^* \varepsilon_\infty} + q^2 \langle v^2(t) \rangle . \tag{8.18}$$

In Fig. 8.13, the Raman spectrum was calculated assuming that the observed width of the phonon scattering was determined by the spectral width of the laser. The phonon spectra were characterized by a Lorentzian

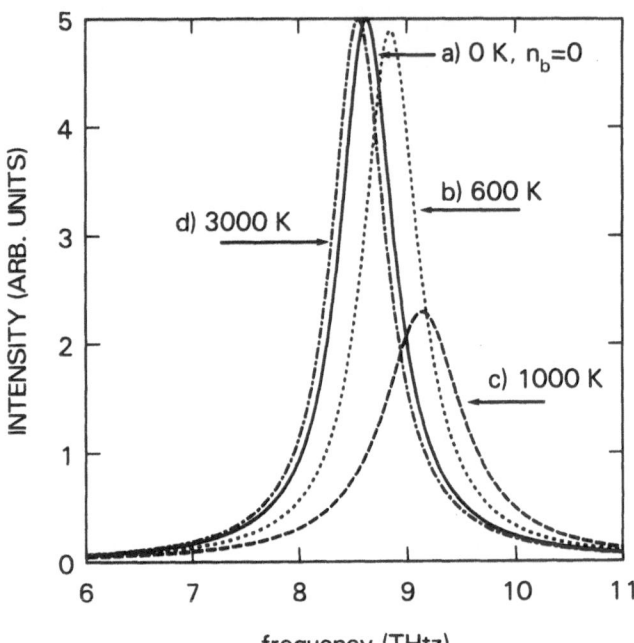

Fig. 8.13. The spectral density of the Raman active LO phonon of GaAs under the injection of hot electrons with a density of 10^{17} cm^{-3} for three carrier temperatures and the excitation spectral width of a transform limited 0.5 ps pulse. The solid curve is the response in the absence of carriers [8.46]

shape with a nominal spectral width of 0.3 THz, corresponding to a transform limited 0.5 ps laser pulse. The magnitude of q was taken as 5×10^5 cm^{-1}, typical for backscattering from GaAs in the visible. The plasma frequency at $q = 0$ was set to $\omega_p(0) = 3$ THz corresponding to an injected carrier density of 10^{17} cm^{-3} and well below the LO phonon energy. The curves in Fig. 8.13 show that the changes in the effective temperature of the excited carriers, defined as $kT_{\text{eff}} = 3/2\, m^* \langle v^2 \rangle$, can have a strong effect on the Raman spectra. When $\omega_p(q)$ is large compared to ω_{LO} and $\omega_p(0)$, the plasmon contribution to the phonon energy and lifetime is small. As the carriers cool and $\omega_p(q)$ decreases to where $\omega_p(0) \ll \omega_p(q) < \omega_{\text{LO}}$, the coupled mode shifts to higher energies. Finally, as $\omega_p(q)$ approaches $\omega_p(0)$, the coupled mode frequency again decreases.

8.2.3 Hot Photoluminescence Involving Neutral Acceptors

While Raman spectroscopy can readily be used to observe the temporal evolution of hot phonon distributions excited by the relaxation of energetic carriers, it can be more difficult to use in monitoring the time evolution of the electron distribution function. Practical difficulties include the smaller magnitudes of the Raman scattering cross sections for electronic transitions and difficulties in separating electronic Raman scattering and hot luminescence excited by the pump beam from the Raman scattering excited by the probe beam [8.48].

However, there are a number of techniques relying on the measurement of the recombination spectra of hot carriers which can directly observe the hot electron distribution function in compounds such as GaAs. In 1981, *Mirlin* et al. [8.49] reported the observation of photoluminescence associated with the recombination of nonequilibrium electrons high in the conduction band with neutral acceptors. This luminescence shared a common feature with the excitation wavelength dependence of the photoconductivity and band edge photoluminescence discussed in Sect. 8.1.1 a. All three show oscillations in intensity with a periodicity comparable to the LO phonon energy. In the present hot luminescence case, the period is just the LO phonon energy. In contrast, for the excitation spectra, the period is slightly larger than $\hbar\omega_{LO}$ because of the kinetic energy given to the hole. This hot luminescence and its relationship to the electronic structure of GaAs is shown in Fig. 8.14 [8.50]. The spectra can be understood in the context of Fig. 8.8 which showed that at low injected carrier densities, the time integrated electron distribution function of the hot carriers consists of a series of peaks below the energy of the initially injected electrons. These peaks are spaced by the LO phonon energy. Fig. 8.14 shows that for incident photon energies below 1.8 eV, there are two distinct groups of optically injected hot electrons corresponding to excitations from the heavy and light hole valence bands of GaAs. Combining this with Fig. 8.8 produces two series of LO-phonon-spaced peaks. The highest energy peak in each series is shifted from the laser photon excitation energy by an amount equal to the energy between the acceptor ground state and the valence band state from which the photoexcited hot electron originates. The positions of the first or zero phonon member of each series is shown in Fig. 8.14 and labeled as light hole (LH) or heavy hole (HH) honoring its origins. For photon energies above 1.86 eV, transitions from the spin-orbit split valence band also become possible [8.2]. These transitions cannot be easily resolved due to the small fraction of electrons generated from this band. These spin-orbit split transitions will not be further discussed here. For photon energies greater than 1.88 eV, intervalley scattering of electrons from the Γ valley, where they are initially injected to the L and X conduction band valleys, can occur. Because of the large density of states in the satellite valleys as compared to the Γ valley, the electrons in the X and L valleys relax to the bottom of the L valley before they scatter back to Γ. The spectra in Fig. 8.14 show this return of the carriers from the L valley as the series of peaks labeled R (for reentrant).

The peaks in the carrier distributions are shown schematically in Fig. 8.8 as much sharper than the LO phonon spacing. The measured hot luminescence peaks in Fig. 8.14 are broadened by the short lifetimes of the hot electrons and the warping of the valence bands [8.2, 50, 51]. Because of this warping, monochromatic photons can excite electrons and holes with different kinetic energies at different positions in k space. Nonetheless, the various peaks in Fig. 8.14 clearly provide a means of directly observing

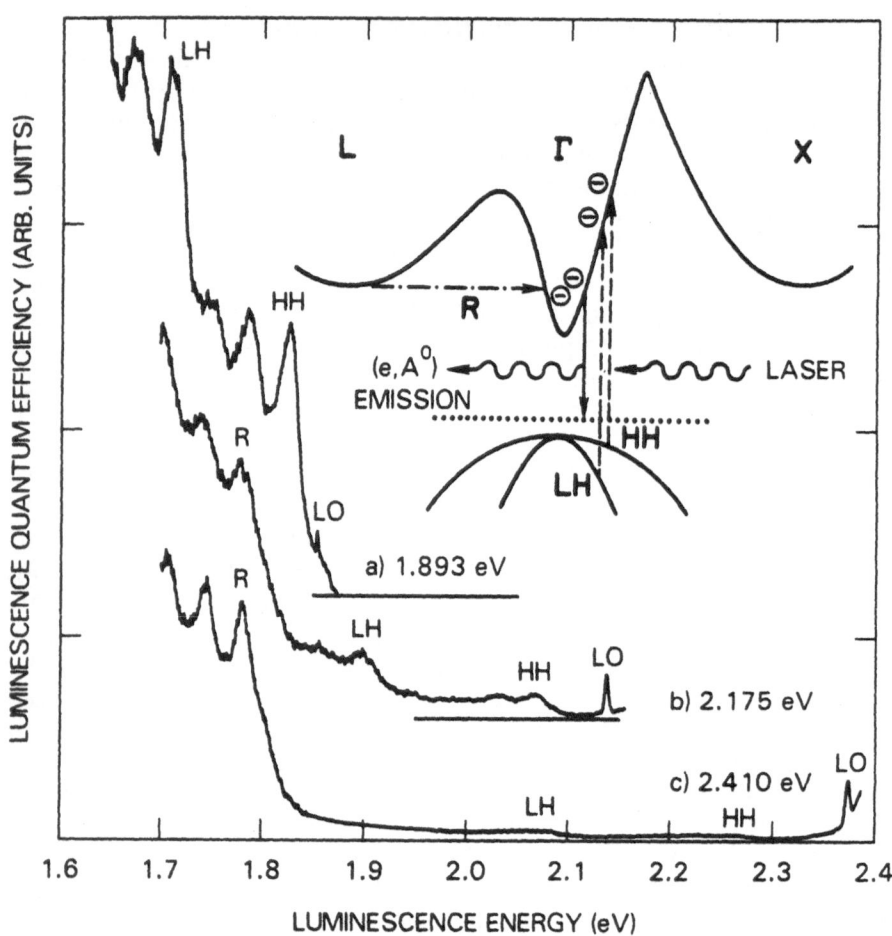

Fig. 8.14. The high energy photoluminescence spectra of GaAs doped with the neutral acceptor Mg and excited at three different photon energies. The spectra were measured at 2 K. A schematic of the band structure of GaAs is shown in the figure to indicate the origins of the three different emission bands [8.50]

the time-integrated hot electron distributions in p-doped materials. The highest energy peak comes from electrons generated from the heavy hole valence band before these electrons lose energy. If LO phonon emission is the only important energy loss mechanism, this peak originates during the first 200 fs after the excitation of the electron. The lower energy peaks can occur only after the emission of a suitable number of LO phonons. The use of the acceptor-bound states to provide the holes in the recombination process instead of the photogenerated holes in the valence bands means that the hole energy distribution function is completely determined. All of the kinetics observed must be due to the evolution of the electron distribution function. It is important to note that the fraction of nonequilibrium electrons which contribute to the hot luminescence is extremely small, so that the kinetics of the electrons are not appreciably affected by the presence of the acceptors. This hot emission can be contrasted with the much stronger emission involving band edge electrons and neutral acceptors. The band edge emission is 5–7 orders of magnitude stronger than the hot luminescence [8.51]. About 4 orders of magnitude of this increase in intensity results from the longer lifetime of the band edge electrons compared to the hot electrons (~ 1 ns vs 200 fs). The remaining 1 to 3 orders of magnitude come from the rapid decrease in the recombination matrix element $M(k)$ as E_e is increased.

The optical transitions responsible for the emission lines in Fig. 8.14 have been treated by *Dumke* [8.52], who showed that the matrix element $M(k)$ for the recombination of an electron with wavevector k at a neutral acceptor is

$$M(k) \propto \left[1 + \left(\frac{m^* E_e}{m_a E_a} \right) \right]^{-2}, \tag{8.19}$$

where m_a and E_a are the mass and energy of the bound hole, assumed to be effective mass-like. In the spherical approximation, *Baldereschi* and *Lipari* [8.53] have determined that $m_a = 0.310$ and $E_a = 27$ meV for GaAs. The latter is in good agreement for many acceptors in GaAs such as Be (28 meV), C (26 meV), Mg (28 meV), and Zn (31 meV). $M(k)$ is proportional to the Fourier component of the wavefunction of a hole bound to an acceptor at the wavevector k of the recombining electron. While *Dumke* described a photoluminescence process with a real excited state, Fig. 8.14 shows that this luminescence will have some "Raman-like" features. In particular, its high energy threshold will track the laser line in a manner similar to that of electronic Raman scattering involving the excitation of the bound hole into the valence band at the wavevector necessary for the resonant excitation of the electron–hole pair. We shall show in Sect. 8.4 how this hot electron neutral acceptor emission provides a very useful tool for the characterization of the hot electron distribution on the time scale of the LO phonon scattering time for the energetic electrons.

8.2.4 Hot Bimolecular Recombination in Photoexcited Semiconductors

At low carrier densities, Fig. 8.8 showed that the electron relaxation proceeds via the sequential emission of LO phonons. In the absence of neutral acceptors to provide holes with which the photogenerated energetic electrons can recombine, electron–hole luminescence at energies well above the band gap is very weak. This is because the emission of light by an electron-hole pair requires that the momentum of the pair be equal to the momentum of the emitted photon. Since the photon momentum is small compared to the hot electron or hot hole momentum, the electron and hole must be at nearly the same wavevector. The optically injected electrons and holes are highly correlated in k space when they are first created due to momentum conservation, but they spend only a negligible fraction of their lifetimes in this state[5] and do not overlap in momentum space in a direct gap semiconductor like GaAs while they relax until they reach the band edges. In the absence of this momentum overlap, the above band gap luminescence can occur only from phonon and/or defect mediated radiative recombination which is usually extremely weak.

Comparison of Figs. 8.8 and 8.9 shows how the hot carrier distribution function changes drastically as the density of carriers is increased. At high densities the relaxation involves the cooling of a thermalized distribution. For a thermalized distribution, there is a uniform occupation of all degenerate energy states. There can be a significant overlap between thermalized electrons and holes in k space so hot luminescence well above the band edge is possible even without neutral acceptors. The intensity of this intrinsic hot luminescence is proportional to the densities of both hot electrons and hot holes (i.e. bimolecular) and is therefore proportional to the square of the laser intensity [8.54, 55]. In contrast, the emission involving neutral acceptors is linear in laser intensity. The observed dependence on laser intensity of this intrinsic hot luminescence is shown in Fig. 8.15 [8.55]. The intensity of the luminescence quadruples when the

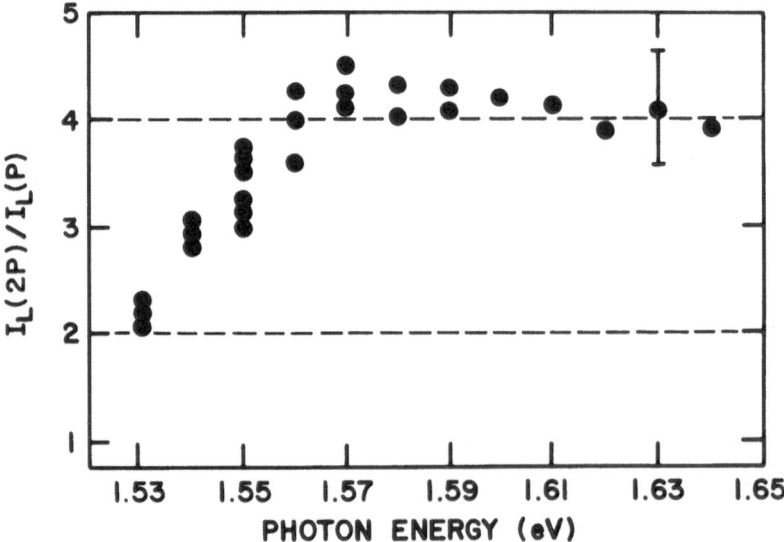

Fig. 8.15. The bimolecular character of above band edge emission in GaAs excited by 2.5 ps pulses at 2.16 eV. $I(P)$ is the luminescence intensity excited using a laser beam of power P. $I(2P)$ is the luminescence intensity excited using $2P$ power [8.55]

---

[5] As a practical matter, the photons emitted in this state are at the same energy as the incident photons, and hence contribute only to the Rayleigh (elastic) scattering.

excitation level is doubled. In order to avoid the sample heating that would occur if cw lasers were used to excite hot carrier densities in the range of 10^{17} cm^{-3}, picosecond or femtosecond laser pulses are used to excite the bimolecular luminescence. The quadratic dependence on excitation intensity is a distinctive signature of the bimolecular emission since the bound acceptor photoluminescence, the Raman scattering from thermal phonons and the band edge emission will all vary linearly with excitation intensity. Note in Fig. 8.15 that at emission energies near the band edge, where the carrier lifetimes reflect the radiative and nonradiative recombination rates, the intrinsic electron-hole recombination is linear in excitation intensity. The nonlinear dependence of the GaAs anti-Stokes bimolecular luminescence intensity on the incident power is shown in Fig. 8.16 [8.56]. At injected carrier densities below those shown in this figure, the anti-Stokes spectrum is dominated by the Raman scattering from the GaAs LO phonon with no luminescence background. As the injected carrier density increases, the spectrum comes to be dominated by the hot luminescence background which grows much more rapidly than the LO phonon scattering.

The anti-Stokes component (i.e. at energies higher than the laser photon energy) of this bimolecular recombination provides an interesting complement to the free-electron to neutral-acceptor recombination described in the previous section [8.56]. In order to have significant amounts of this anti-Stokes emission, the carrier thermalization must occur before the

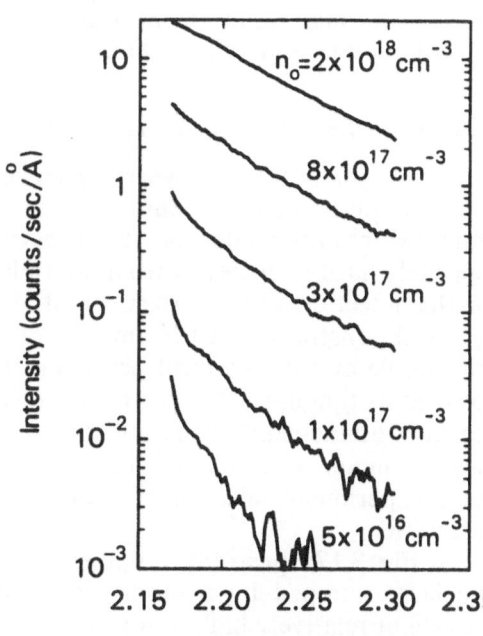

Fig. 8.16. The anti-Stokes secondary emission spectra of GaAs as a function of increasing excitation power. The spectra were excited by 2.13 eV photons with a pulse width of 0.5 ps [8.56]

electrons can lose energy to the LO phonons. Such rapid thermalization of the injected carrier distribution will produce a smoothing out of the peaked structure associated with the acceptor luminescence. Hence one expects the anti-Stokes bimolecular emission to appear just as the structure disappears in the acceptor luminescence.

Because of its quadratic dependence on excitation intensity, the bimolecular recombination of free holes and electrons is well suited to pump-probe type experiments where the temporal resolution is obained by varying the time delay between the subpicosecond duration pump and probe pulses. For carrier injection with two short pulses, there is bimolecular emission from each pulse separately. In addition, hot electrons in the first pulse can recombine with hot holes from the second pulse and vice-versa, generating a "luminescence correlation" signal which depends on the delay between the pulses [8.55, 56]. While quantitative analysis of the correlation signal will be difficult, the qualitative behavior in a number of different limits can be understood. If the injected carrier densities are small so that the bimolecular processes are weak, there will be little bimolecular luminescence. If the delay between the two pulses is large compared to the time required for the carriers to cool, then the correlation signal will again be small. Only when (1) the injected carrier density is high enough that thermalization is fast and (2) the delay between pump and probe is small compared to the cooling rate will one expect to see a significant anti-Stokes correlation signal. Measurement of the luminescence correlation signal at a given luminescence energy, especially in the anti-Stokes regime, is a qualitative indication of how long there are hot, thermalized electrons and holes at the corresponding energies in the conduction and valence bands.

8.2.5 The Measurement of Picosecond Dynamics in Semiconductors

A typical system for the measurement of time resolved Raman scattering using the pump-probe technique is sketched in Fig. 8.17 [8.57]. While the temporal requirements for the experiments of interest here are not especially hard to achieve with modern laser technology, the requirements for the detection system are substantial. This arises from the presence of practical constraints on the amount of power that can be used to excite the sample and the spectral properties of the exciting light. Some of the constraints that apply to time resolved Raman scattering are modified for the hot luminescence experiments. Yet, because of the requirement of measurements over much broader spectral regions, we shall find that these experiments also have similar requirements for high sensitivity detection.

In Fig. 8.17, spectrally tunable picosecond pulses are generated by a dye laser pumped by a mode-locked Ar laser. Such systems typically operate at relatively high frequencies near 100 MHz so that the energy in each pulse and the number of carriers generated by a single pulse can be

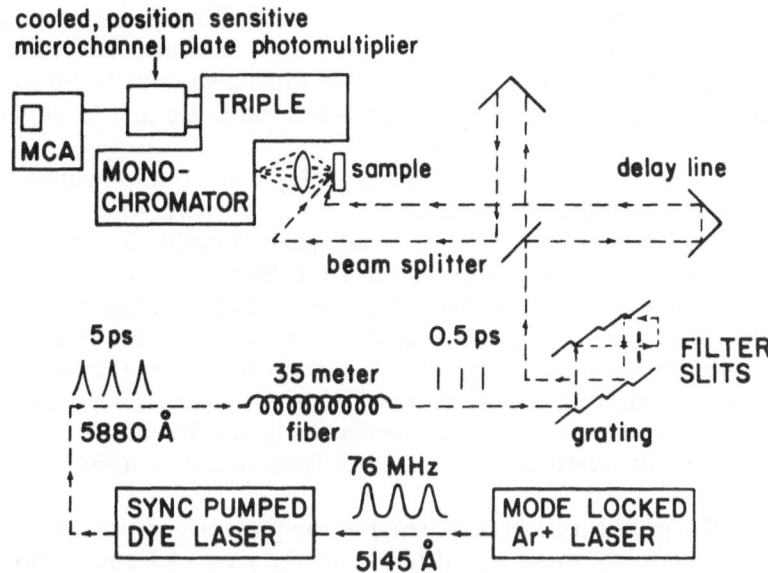

Fig. 8.17. A schematic of a typical experimental system for the measurement of time dependent Raman scattering using the pump-probe technique

modest. The dye laser emission has a temporal width of the order of 5 ps when it is pumped by a mode-locked Ar laser as shown in Fig. 8.17 or by some other source such as a frequency-doubled cw mode-locked Nd: YAG laser. In order to temporally resolve dynamic processes on the 1 ps time scale, the 5 ps pulses from the dye laser can be compressed to below a picosecond using a fiber optic pulse compressor. The compressed pulse spectral width is greater than 30 cm^{-1}, while the transform-limited pulse has a spectral width of about 10 cm^{-1}. In both cases, the width of the pulse is much larger than the spectral width of the GaAs Raman line which at room temperature is about 3 cm^{-1} [8.58]. Similar subpicosecond pulses can be obtained directly from the dye laser by adding a second saturable absorber jet to the cavity or by pumping the dye laser with shorter pulses. In some cases, these alternate approaches result in pulses which are closer to the transform limit, which is desirable for the Raman measurements. Considerably shorter transform-limited pulses can be obtained using colliding pulse mode-locked (CPM) lasers. However, experiments using such lasers are restricted to a single energy near 2.0 eV which is very restrictive for Raman scattering in general and rather inconvenient for GaAs in particular. This is due to the difficulty of separating the different scattering mechanisms including intervalley processes that become important near 2 eV [8.15]. For any dye laser, a means must be provided to filter out the luminescence from the dye that inevitably copropagates with the laser beam. Just as with cw Raman scattering, this luminescence will otherwise obscure the inelastic Raman signal. In

Fig. 8.17, this filter function is provided for by a set of slits at the appropriate place in the pulse compressor. Similar solutions are available for other laser systems, although care must always be taken to ensure that the filter does not temporally broaden the pulse due to either a narrow bandpass or spectral dispersion.

Figure 8.17 shows how the time delay in the pump-probe experiment is achieved using a beam splitter and an optical delay. (A 1 ps delay corresponds to an extra 0.3 mm path length). The two beams are recombined on the sample. One beam is used to excite hot carriers in the sample and is called the pump beam. The second beam is used to excite the Raman scattering from the phonons or electronic excitations created by the pump beam and is called the probe beam. The scattered light is polarization analyzed to improve its signal to noise as described below, spectrally analyzed to reject the elastically scattered light and identify the inelastically scattered signals, and then detected using a multichannel detector.

The physical effects discussed in the previous section are all associated with relatively small signal levels. In the case of Raman scattering from LO phonons, the scattering efficiency of a material like GaAs for excitation energies near 2 eV is about 1.4×10^{-3}/steradian/meter [8.59]. In general, the intensities of the electronic scattering processes are considerably smaller. For time dependent Raman experiments attempting to measure the time dependent changes in the LO phonon population after the injection of hot carriers, there are a number of constraints which limit the maximum magnitude of the scattered signals. With present laser technology, the requirement of temporal resolution on the subpicosecond scale means that most experiments have been done with pump and probe beams derived from a single laser beam. Since both the pump and the probe are the same wavelength, the pump pulse must be more intense than the probe pulse. The analysis of the experiment will also be simplified if the pump beam irradiates a larger spot on the sample than the probe beam so that the probe beam sees a reasonably homogeneous spot on the sample. Therefore, of the total power available for a pump-probe, time dependent Raman scattering experiment, only about 5–10% can be used to generate the actual Raman signal with the bulk of the power being used to excite the hot carriers.

The limited power available for the probe beam is further restricted by the maximum amount of power that can be used in the pump beam. We have indicated in Section 8.1.3 how carrier–carrier scattering processes compete with the carrier–LO phonon emission process. This competition can greatly complicate the analysis of any experimental results. In addition, the possibility of coupled mode behavior where the LO phonon can mix with the optically generated plasmons must be avoided since this can change the scattering cross-section, making it difficult to obtain quantitative results on the time dependent changes in the phonon population. These considerations limit the pump beam to power levels where the density of optically injected carriers will be well below 10^{17} cm^{-3}. Since the probe illuminated spot on the sample must be efficiently imaged through the entrance slits of a monochromator, which typically are less than 300 μm wide, the illuminated spot must be less than 100 μm wide. Even if a line focus with a height of 3 mm on the sample is used, since the penetration depth of 2 eV light in GaAs is about 3000 Å, for a mode locked picosecond laser system operated at 80 MHz, the maximum pump beam power must be less than 50 mW. As a result, the maximum probe beam power must be an order of magnitude smaller. Such power levels are an order of magnitude lower than normally used for Raman scattering from materials such as GaAs.

A further difficulty in obtaining reasonable single to noise ratios in picosecond Raman scattering experiments arises from the spectral widths of the exciting laser pulses when subpicosecond laser sources are used. As mentioned previously, the spectral widths of 0.5 ps pulses are considerably broader than the natural widths of the Raman lines themselves. Therefore, the peak intensities of the Raman lines under subpicosecond excitation can be an order of magnitude smaller than under cw excitation since the scattering efficiency of the transition will be spread over 10–30 cm^{-1} instead of the 1–3 cm^{-1} of the natural linewidths. The signal to noise ratio of the Raman lines will also be degraded by background signals generated by the pump pulses such as the bimolecular luminescence discussed in the previous section. In contrast, the dominant background contribution to above bandgap cw Raman scattering from Group IV and III-V semiconductors is the second order Raman scattering which is between 20 and 100 times weaker than the first order scattering, and is linear in laser intensity [8.60]. Taken together, subpicosecond pump-probe studies of materials like GaAs will require the detection of peak signal levels between 10 and 100 times smaller than those detected in conventional cw Raman experiments on these materials. In addition, these signals can be superimposed on background contributions that are comparable in magnitude or greater. While the time resolved experiments can be done using conventional detection capabilities, the degradation of the signal to noise ratio shows why multichannel detection systems using either imaging, microchannel plate photomultipliers or charge coupled device (CCD) imagers have been widely used in these studies. The performance of these detectors has been reviewed in Vol. V of this series by one of the present authors [8.61], and the experiments described in this article make good use of the ability of these detectors to readily observe signals at the level of 0.1 to 1 detected photon per second.

It is worth pointing out that the total number of Raman scattered photons is much larger than the number of detected photons. This results from the optical properties of opaque semiconductors such as GaAs which limit the number of Raman scattered photons which escape the sample and are captured by any reasonably fast collection optics to a tiny fraction of the total number of scattered photons [8.61–63]. An additional reduction results from the low efficiency of multiple grating monochromators whose throughputs are generally in the range of 10%.

The system shown in Fig. 8.17 can also be used to measure both hot luminescence associated with the recombination of free electrons with bound holes on neutral acceptors and the anti-Stokes bimolecular recombination of free holes and free electrons. As shown in Fig. 8.14, the intensity of the acceptor related luminescence for excitation energies above 2 eV is comparable to the intensity of the symmetry allowed Raman scattering from the LO phonon. However, in contrast to the Raman scattering where the spectral region of interest is only $300-400$ cm^{-1}, in the hot-luminescence case, the spectral range of interest can be 3000 to 4000 cm^{-1}. As in the case of the Raman scattering, the requirement that the optically injected carrier densities be below 10^{16} cm^{-3} restricts the maximum power that can be used to excite the sample so that the signals cannot be arbitrarily increased by increasing the excitation level. In these situations, the ability of the multichannel detectors to simultaneously collect the complete spectrum or a substantial fraction of it greatly facilitates the experimental effort.

We have emphasized the limitations of the experimental system shown in Fig 8.17 with respect to the detection of small signals. Another limitation, which is especially severe for the Raman scattering experiments, arises from the intimate connection between the temporal and the spectral resolution. Since the Raman experiments depend on the ability to spectrally distinguish the Raman scattered light from the exciting light, the spectral width of the exciting light must be much smaller than the frequency shifts associated with the Raman transitions. For example, since a 100 fs pulse will have a spectral full width at half maximum of at least 70 cm^{-1}, any Raman scattering excited by this pulse will also have a spectral width of > 70 cm^{-1}. Given the finite rejection of any multi-stage monochromator and reasonable estimates of the amount of elastically scattered light collected along with the inelastically scattered light, such spectral widths represent the limit at which a Raman shifted line about 300 cm^{-1} away from the exciting light can be resolved [8.46].

The poor signal-to-noise ratio arising from the intensity of the background Raman scattering excited by the pump beam when both the pump and the probe have the same wavelength can be easily improved in the case of phonon Raman scattering from zincblende structure semiconductors. This solution draws on the symmetry selection rules which govern Raman scattering in these materials [8.32]. Depending on the orientation of the polarization of the exciting light with respect to either the [100] or [110] axes of the (100) face of a zincblende structure crystal, the Raman scattered light can be either polarized parallel to or perpendicular to the incident polarization. If the pump and probe beams are oriented perpendicular to each other, then the polarization properties of the scattered light can be used to reject the Raman scattering excited by the pump beam. Rejection ratios of over 100 are common. Since the pump beam is often an order of magnitude stronger than the probe beam, the polarization properties of the scattered light can be used to reduce the intensity of the pump induced scattering to where it is less than 10% of the intensity of the probe excited scattering. This is a much more favorable situation from the signal-to-noise point of view than a situation where the pump induced background Raman scattering is an order of magnitude stronger than the probe excited signal. If, as in cases such as Raman scattering from the (111) face of GaAs, the Raman scattered light occurs in both polarizations, this advantage is not possible.

In the discussion of Fig. 8.17, the temporal resolution has been achieved using the pump-probe technique. This assumes that there is some nonlinear coupling of the pump and probe beams in the sample. If we consider the hot luminescence associated with the recombination of free electrons to bound holes in the low density limit where carrier–carrier scattering is unimportant, there is no obvious way to use the pump-probe technique for example, to time resolve the series of peaks arising from the sequential LO phonon emission in Fig. 8.14. *Shah* et al. [8.64] have used an up-conversion technique to obtain subpicosecond temporal resolution for linear optical processes. A schematic drawing of their system is shown in Fig. 8.18. In this scheme, a signal is generated in a sample and the time resolution obtained via the pump-probe method by mixing the signal in

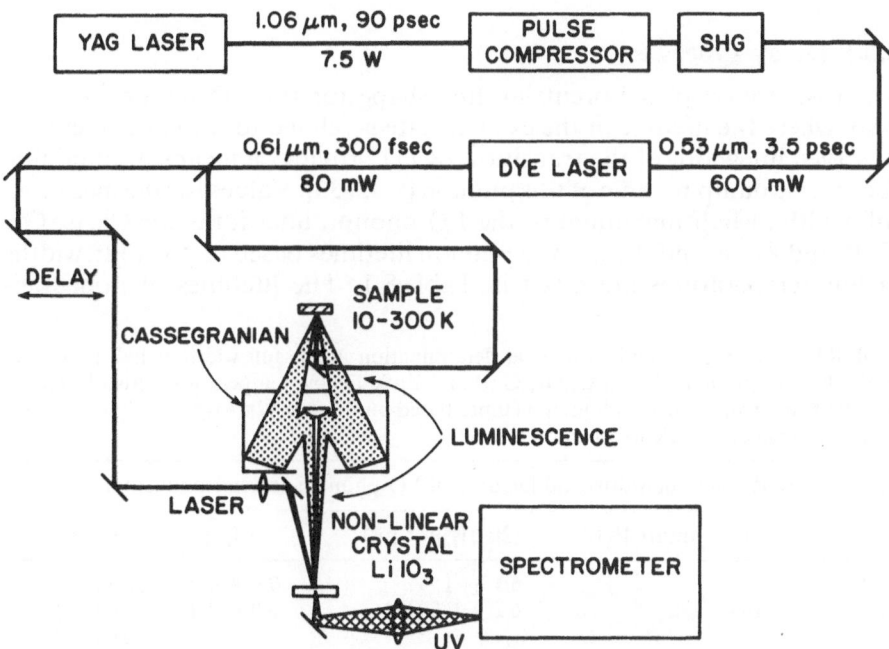

Fig. 8.18. A schematic drawing of a time resolved optical spectroscopy system based on the up-conversion of the experimetally significant light in an nonlinear optical crystal [8.64]

an external nonlinear $LiIO_3$ crystal with a pump beam to generate light at the sum frequency. The up-conversion also makes it possible to study emission at infrared wevelengths where conventional photodetectors lose sensitivity. When applied to study photoluminescence near the band edge, *Shah* et al. have been able to achieve temporal resolution better than 1 ps. A major difficulty of this technique with respect to the signal levels of interest in this review is the inefficiency of the up-conversion process. At the present level of technology, such processes generally produce less than one up-converted photon for every 1000 photons collected from the sample. As a result, signal levels for the up-converted photons will be below 0.001 counts per second (cps) for Raman experiments and below 0.1 cps even for the hot electron–neutral acceptor emission. These count rates are at or below the present limits of detectability.

8.3 Phonon Dynamics

In this section, we describe results obtained by Raman scattering and hot electron photoluminescence on the dynamics of nonequilibrium phonons in III–V and IV semiconductors. We will first consider experimental studies of the optical phonon lifetimes and then describe measurements of the rate at which energetic electrons can generate LO phonons in these materials and their alloys.

8.3.1 Decay Processes

The assumption of a Lorentzian line shape for the LO phonon Raman peak where the lifetime of the excited state is related to the experimentally observed linewidth of the transition (8.11) provides a means of obtaining the inverse damping time of this phonon [8.32, 33]. Values for the measured full width at half maximum of the LO phonon linewidths for CaAs, Ge, GaP, and ZnSe, and the derived phonon lifetimes based on their linewidths at low temperatures are given in Table 8.1. The lifetimes of nonequili-

Table 8.1. Experimental and theoretical determination of the full width at half maximum of the Raman phonons in GaAs, Ge, GaP and ZnSe at low temperatures. Also listed are the calculated values for the phonon lifetimes based on the $\tau = (2\pi \Delta \nu_{LO})^{-1}$ relationship for a homogeneously broadened phonon

Table 8.1 cw Raman linewidths and lifetimes of LO phonons in semiconductors

Material	Temperature [K]	$(2\pi \Delta \nu_{LO})^{-1}$ [ps]	τ_{LO} [ps]	References
GaAs	77	6.3 ± 1	7.0 ± 1	[8.32]
Ge	105	6.2 ± 0.6 (1)	8.0 ± 1 (2)	(1) [8.33]
				(2) [8.37]
GaP	4.2	23 ± 0.4	26 ± 3	[8.77]
ZnSe	4.2	4.4 ± 0.1	4.8 ± 0.2	[8.112]

brium populations of these phonons for GaAs and Ge measured by light scattering are also listed as are the dephasing times for the LO modes in GaP and ZnSe as measured by coherent anti-Stokes Raman scattering. The experimental values show excellent agreement. Theoretical estimates of these quantities show a much larger spread of values [8.33]. The measurements of the phonon linewidths suffer from a number of problems including the finite resolution of the spectrometers used in these measurements. This necessitates the deconvolution of the response function of the spectrometer which can be as much as 30% of the measured linewidth. In addition, the finite penetration depth of light into the sample means that for a material like Ge, where this depth in the visible is below 500 Å, the results can be very sensitive to surface preparation. Strain induced by polishing etc. can significantly broaden the measured phonon linewidth [8.65]. The measured linewidth reflects the coherent lifetime of an individual phonon and also inhomogeneous broadening effects (e.g. those due to strains). Thus, the inverse linewidth need not be the same as the lifetime of the nonequilibrium phonon population [8.37].

In 1970, *Shah* et al. [8.23] used their observation of highly nonequilibrium LO phonon populations in GaAs under cw pumping (Fig. 8.5) to estimate the nonequilibrium LO phonon population lifetime. This was done by assuming that the nonequilibrium LO phonon population could be described in the linear relaxation time approximation of (8.10). The absorption of light by electron–hole pairs and the generation of Raman active optical phonons by the hot electrons was modeled in a simple manner. It was also assumed the hot electron relaxation was dominated by emission of phonons from carriers in the Γ valley. In fact, since the signals were excited by 2.54 eV light, the possibility also existed of significant contributions by relaxation processes involving the L and X conduction band minima. In any event, for a steady state population excited by a cw source, $G(k)\,\tau = n(k) - n_0(k)$ as shown in Sect. 8.1.1 a; (8.10). By mesuring the power dependence of the nonequilibrium phonon population and separating out the increase in the phonon population due to thermal heating of the sample by the optical excitation, *Shah* et al. were able to estimate an LO phonon population lifetime of approximately 5 ps at room temperature. This was consistent with the lifetime obtained from the linewidth studies.

The temporal evolution of a nonequilibrium phonon population was first directly measured by *von der Linde* et al. [8.32] in 1980 and subsequently by *Kash* et al. [8.66] using experimental systems similar to that shown in Fig. 8.17. These measurements assumed that in the pump-probe arrangement, the detected changes in the Raman scattering intensities arose solely from changes in the phonon populations. By concentrating on changes occurring between 2 and 20 ps after the pump pulses excite the sample, it was not expected that the matrix elements or energy denominator terms in the Raman cross section would show any significant time-dependent changes. For these times, the optically injected hot carrier distributions were thought to be largely thermalized and cooling at a relatively slow rate while the carrier concentration stayed constant. *Von der Linde*'s measurements used 2–3 ps laser pulses in a pump-probe experiment where the pump injected hot carrier densities were about 10^{17} cm^{-3}. The measurements were made at an excitation energy of 2.16 eV, above the thresholds for both Γ to L and Γ to X intervally scattering processes. The Raman active LO phonon population at 77 K was found to decay with a lifetime of 7 ± 1 ps. Measurements were made on both (100) and (111) faces. The experiments on the (111) face showed Raman scattering from both the 291 cm^{-1} LO and 268 cm^{-1} TO phonons with only a nonequilibrium population associated with the LO modes, consistent with the earlier results of *Shah* et al. [8.23]. *Von der Linde* et al. also measured the LO phonon Raman linewidths of their samples. At 77 K, the LO phonon linewidth was observed to be 0.85 ± 0.1 cm^{-1}. As shown in Table 8.1, this value of the linewidth was in good agreement with the measured lifetime. From this agreement, it was concluded that the LO phonon linewidth was not inhomogeneously broadened and that the LO phonon lifetime was limited by its anharmonic decay into two acoustic phonons. The lifetime of the phonon was set by energy relaxation rather than phase relaxation processes so that the dephasing time and the phonon population lifetime were identical.

Kash et al. [8.40, 66, 67] extended the time dependent Raman measurements of the LO phonon lifetime to lower carrier densities and considered how the measured lifetimes would be affected by the ability of the photogenerated carriers to scatter to the L and X conduction band mimima. Their experiments were performed for injected carrier densities of about 2×10^{16} cm^{-3} as estimated from the incident power level, the area of the illuminated sample spot and the published optical constants of GaAs [8.68]. The time dependences of the nonequilibrium LO phonon population at three different temperatures are shown in Fig. 8.19. The lifetime measured at 80 K is 6.4 ± 1 ps in agreement with *von der Linde* et al. [8.32]. The observed decrease in the lifetime with temperature is consistent with the decay of the LO phonon into two acoustic phonons, each of energy near $\hbar\omega_{LO}/2$. Careful measurements for

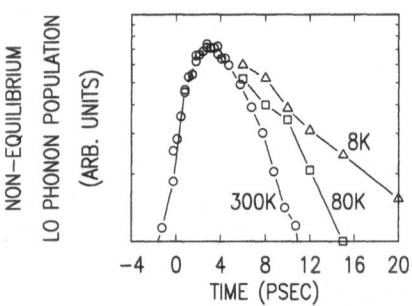

Fig. 8.19. The time dependence of the magnitude of the hot electron generated nonequilibrium LO phonon population in GaAs injected with about 2×10^{16} cm^{-3} electron-hole pairs at 2.16 eV. The results are shown for three different sample temperatures [8.40, 66, 67]

Fig. 8.20. The dependence on the optically injected carrier density of the magnitude of the nonequilibrium LO phonon population and the nonequilibrium LO phonon lifetime in GaAs [8.40, 69]

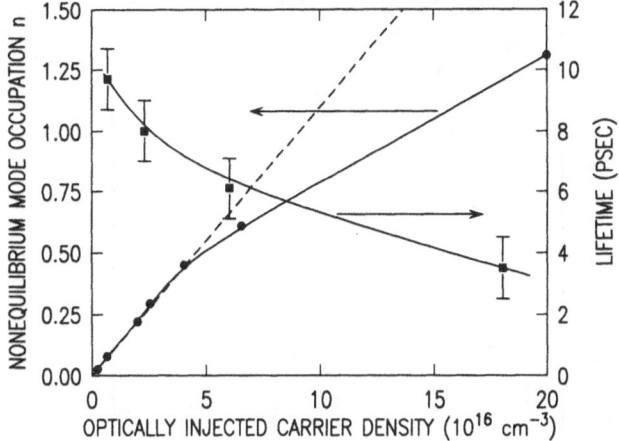

pump/probe delays of 2–10 ps showed no effects in the probe excited Raman scattering due to the presence of nonequilibrium phonons generated by the relaxation of hot carriers after returning to the Γ valley from X and L.

The LO phonon lifetime was also measured as a function of the pump injected carrier density. Figure 8.20 shows both the lifetime and the peak magnitude of the nonequilibrium LO phonon mode occupation as a function of injected carrier density under 2.07 eV excitation [8.40, 69]. For densities below 5×10^{16} cm^{-3} the phonon population lifetime is constant and the mode occupation increases linearly with the injected density, as expected. Above 5×10^{16} cm^{-3} the lifetime decreases and the increase in the mode occupation becomes sub-linear in the injected density. The density of 5×10^{16} cm^{-3} is too low to attribute these effects to the coupling of the LO phonon to the plasmon since the plasma frequency is only 9 meV. *Kash* et al. have suggested that the LO phonons couple to the intervalence band transitions associated with the photoinjected holes [8.40]. As pointed out in Sect. 8.1.3, a heavy hole with about 5 meV kinetic energy can make an LO phonon assisted transition to the light hole band with small momentum change. A hole density of 3×10^{17} cm^{-3} corresponds to a quasi Fermi energy of about 4 meV. Further studies of the importance of intervalence band transitions are necessary to fully understand these effects which should be included for accurate modeling of hot carrier dynamics.

Kash et al. [8.70, 71] and *Tsen* et al. [8.72, 73] have used time dependent Raman scattering to measure the LO phonon population lifetimes in the two-mode system Al$_x$Ga$_{1-x}$As. This system is described as a two-mode system because its Raman spectrum shows a pair of optical phonons, one at an energy close to that of the GaAs optical phonon ("GaAs-like") and one close to that of the AlAs optical phonon ("AlAs-like") [8.74, 75].

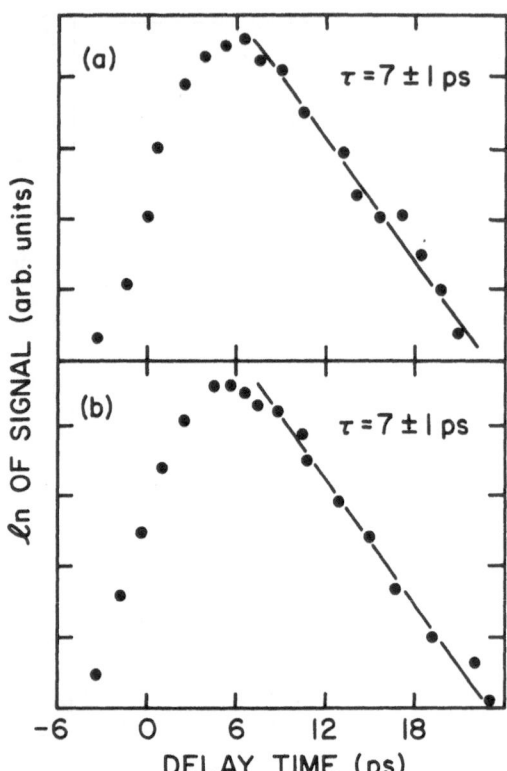

Fig. 8.21. The time dependence of the intensity of the anti-Stokes Raman signal from the (a) GaAs- and (b) AlAs-like modes of $Al_{0.3}Ga_{0.7}As$ [8.73]

The lattice dynamics of this system has been described in the random element isodisplacement model where the GaAs and AlAs sublattices vibrate independently. The small wavevector GaAs-like phonon mode corresponds to Ga atoms vibrating against the As atoms as in pure GaAs and the AlAs-like mode corresponds to the Al atoms vibrating against the As atoms. The amplitudes of the GaAs-like mode on the AlAs sublattice sites and vice-versa are very small. The time-resolved Raman scattering measurements indicated that the nonequilibrium phonon populations of both the GaAs-like and the AlAs-like vibrational modes decay with a lifetime of about 3.5 ps at 300 K and about 7 ps at 10 K as shown in Fig. 8.21. These results are, within the error of the measurements, identical to the results in pure GaAs.

The $Al_xGa_{1-x}As$ phonon lifetime results are interesting for two reasons. The first is the relationship between the population lifetime and the Raman linewidth. The Raman spectra of the alloy systems shaw that these modes are asymmetrically broadened to lower energies [8.35, 76]. The measured linewidths of the phonons in the alloys are much greater than the widths of the lines in the pure compounds. The disagreement between the Raman linewidths and the excess phonon population lifetimes shows that the Raman lines in the alloy are inhomogeneously broadened so that (8.11) is no longer applicable. This excess broadening has been discussed by several authors and attributed to a number of different effects including the localization of these optical phonons in the disordered alloy or the activation by disorder of Raman scattering from large wavevector optical modes [8.35, 70, 76]. This point will be discussed further in Sect. 8.3.4.

The second interesting point involves the lifetime of the AlAs-like mode. The similarities in the decay times of the two different LO phonons is surprising. Energy conservation requires that the AlAs-like mode decays into different phonons than those for the GaAs-like mode. To first order one might guess that the AlAs-like mode would decay in a time comparable to the LO phonon in pure AlAs. Although the LO phonon lifetime has not been measured in pure AlAs, it has been measured in GaP [8.77]. GaP and AlAs have similar phonon dispersion relations since the bonding is almost identical as are the masses of the As and Ga atoms and the Al and P atoms. *Menendez* and *Cardona* [8.33] showed that the LO phonon lifetime depends strongly on the details of the phonon dispersion curves. This suggests that since AlAs more closely resembles GaP than GaAs, the phonon decay processes in AlAs might resemble those in GaP more than in GaAs. In Fig. 8.22, we show the results of *Kuhl* et al. [8.77], who found that the LO phonon linewidth in GaP at low temperatures is about 0.25 cm^{-1}, corresponding to a lifetime of about 25 ps, more than three times narrower and longer than for the LO phonon of GaAs. Thus the identical lifetimes of the two LO modes in $Al_xGa_{1-x}As$ are unexpected.

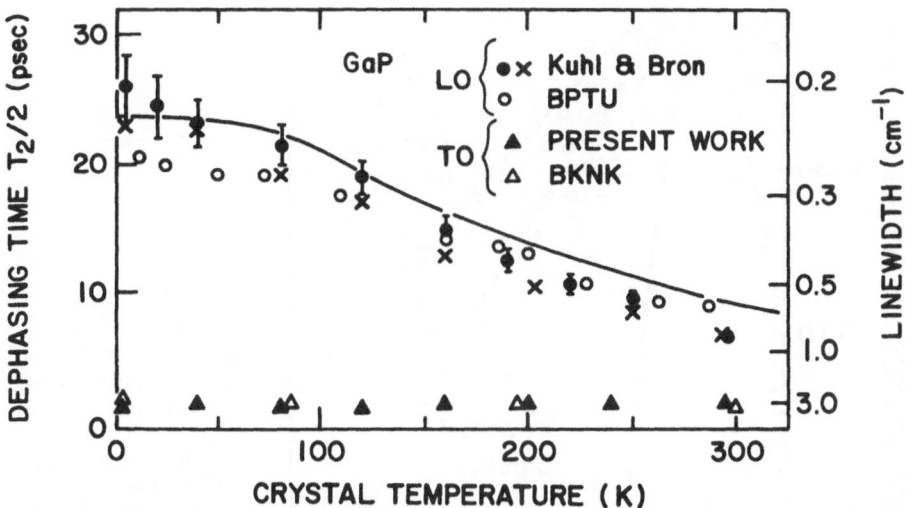

Fig. 8.22. The temperature dependence of the LO phonon linewidth (dots) and dephasing time (crosses) for GaP as measured by spontaneous Raman scattering and coherent anti-Stokes Raman scattering [8.77]. Open symbols refer to other work

The dynamics of optically generated nonequilibrium phonon populations have been studied also in semiconducting Ge and the alloy system Ge_xSi_{1-x} for $0.75 < x < 1$. *Young* et al. [8.20, 21] and *Genack* et al. [8.37] have shown that it is possible to optically induce large changes in the intensity of the Raman scattering from Ge and Ge_xSi_{1-x} alloys using the pump-probe technique. Since there is no Fröhlich interaction in homopolar Ge, one may ask about the physical origins of the observed changes in the Raman scattering. *Young* et al. [8.21] have studied this problem and concluded that kinematic restrictions associated with the band structure of Ge can result in a measurable nonequilibrium, Raman active phonon population generated by the cooling of hot holes. The dependence of the intensity of the anti-Stokes Raman scattering on the delay between pump and probe beams for Ge is shown in Fig. 8.23 [8.37]. While the low temperature results obtained from both the phonon linewidths and the population decay times appear to converge towards a lifetime of 8–9 ps, *Genack*, et al. [8.37] reported apparent deviations between the two measurements at temperature near 300 K. The decay rate derived from the phonon linewidth was faster than the decay rate measured from the lifetime of the nonequilibrium population. The similarities of the low temperature results suggest that the differences observed near 300 K by *Genack* et al. cannot arise from experimental artifacts such as the static broadening of the cw phonon Raman scattering by disorder or surface strains. *Genack* et al. have explained their results by considering the effects

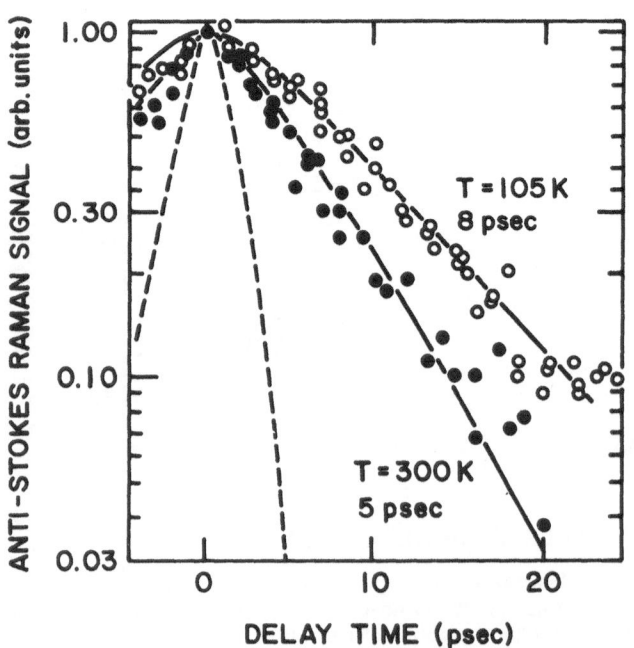

Fig. 8.23. The time dependence of the intensity of the optically excited 1st order phonon Raman scattering in Ge. The dashed line is the laser pulse autocorrelation [8.37]

of dephasing transitions within the optical phonon band. The rich range of different, naturally occurring isotopes of Ge can modify the kinetics of the Ge phonons. There are six different isotopes of Ge with atomic masses between 70 and 76 and relative abundances between 7.8% and 36.5%. The disorder associated with isotopic fluctuations can scatter optical phonons into states with different k vectors but at the same energy, i.e. dephasing transitions responsible for the T_2 derived from lineshape experiments. Such dephasing will not affect a direct measurement of the phonon population lifetime T_1. The phonon linewidth and lifetime in this situation would be related by the expression [8.37]

$$2\Gamma = \frac{1}{(2\pi\tau_{LO})} + A(2\Gamma)^{0.5}(n_{LO} + 1). \tag{8.20}$$

Genack et al. derived a value for the parameter A of $0.37 \text{ cm}^{-0.5}$ to explain their results. They suggested that the failure to observe similar effects in GaAs was due to the fact that the isotope effects in GaAs should be much smaller since there is only one isotope of As and two isotopes of Ga with masses 69 and 71 and relative abundances of 60% and 40% respectively. It should be noted that *Young* et al. obtained a somewhat shorter lifetime then *Genack* et al. at 300 K, 4 ± 1 ps and did not consider the differences between the linewidth and direct lifetime measurements to be experimentally significant. Current experiments with isotopically pure Ge crystals may resolve this issue [8.78] (preliminary data for ^{70}Ge indicate no difference in Raman linewidth with respect to natural Ge to within 0.03 cm^{-1} [8.78a]).

8.3.2 The Generation of Nonequilibrium Phonon Distributions in Bulk GaAs

In the previous section, we considered the behavior of nonequilibrium LO phonon populations and described their lifetimes and how these lifetimes were related to the linewidths of the Raman active LO modes. We now show how time-dependent Raman scattering and secondary emission measurements have been used to quantitatively describe the scattering process by which such a nonequilibrium LO phonon distribution is created. We describe how the scattering rates for the Fröhlich interaction between hot electrons and small wavevector LO phonons can be measured spectroscopically.

The experimental results in Fig. 8.19 show both a temporally resolvable increase in the nonequilibrium phonon population as well as the decrease which was analyzed in the previous section. Figure 8.24 presents in detail, the initial stages of the time evolution of the anti-Stokes Raman scattering intensity for a pump-probe experiment on GaAs using subpicosecond laser pulses. Included in this figure are data obtained using two different excitation energies at 1.91 and 2.09 eV and also the autocorrelation trace

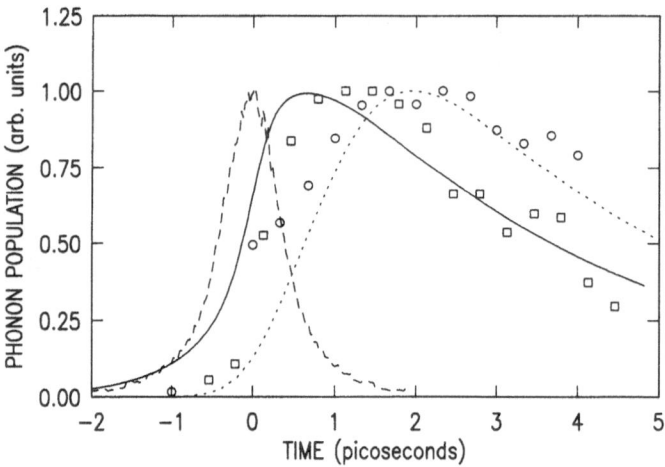

Fig. 8.24. The time dependence of the nonequilibrium phonon population of GaAs at 300 K generated by using either 1.91 (squares) or 2.09 eV (circles) light to inject about 2×10^{16} cm^{-3} electron–hole pairs. The symmetric peak at $t = 0$ is the autocorrelation of the 1.91 eV laser pulse and the solid line is the system response for the instantaneous generation of a full nonequilibrium population of Raman active phonons with a 3.5 ps population lifetime. The dotted curve for electron relaxation by an LO phonon cascade is described in the text [8.40, 66, 67]

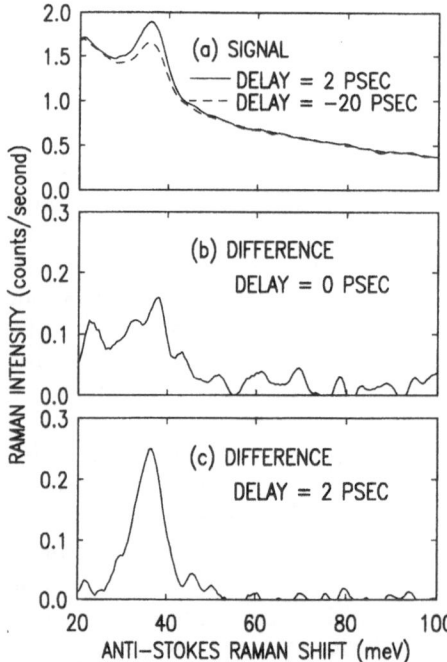

Fig. 8.25. Raman spectra of GaAs generated using the pump-probe technique at several different pump/probe delays. (a) The measured spectrum for a delay of −20 ps where the probe beam precedes the pump. This is used to generate the background signal. (b, c) The difference spectra due to the optically excited LO phonon population obtained by subtracting the background spectrum from the spectrum obtained with the specified pump/probe delay [8.66]

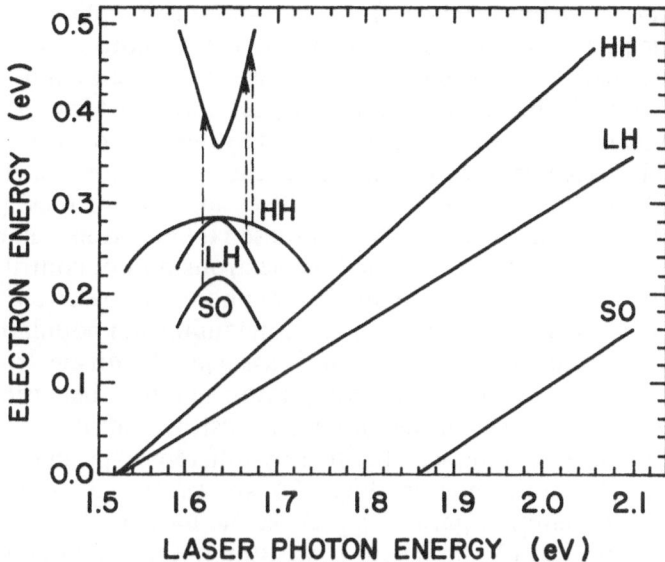

Fig. 8.26. The dependence on the optical photon energy of the kinetic energies of hot electrons excited in GaAs by these photons [8.2]

for the 1.91 eV laser pulse [8.40, 66, 67]. The rise time of the nonequilibrium Raman signal is clearly resolvable in Fig. 8.24. Examples of the actual Raman spectra used to derive the results shown in Fig. 8.24 are displayed in Fig. 8.25 [8.66]. These spectra were excited at 300 K using 2.09 eV light to inject about 10^{16} cm^{-3} carriers into the GaAs.

These measured risetimes can be directly related to the electron LO phonon scattering time. Figure 8.26 shows how the kinetic energy of electrons excited in GaAs varies with the incident photon energy [8.2]. As discussed in connection with Fig. 8.14, optically excited hot electrons can be derived from three possible types of holes. These are associated with the heavy hole (HH), light hole (LH), and spin orbit split (SO) valence bands. For excitation energies between 1.91 and 2.09 eV, electrons can be generated in room temperature GaAs at three different energies in the Γ valley of the conduction band. The band structure parameters are such that less than 20% of the hot electrons are excited from the spin orbit split valence band over this energy range. Because the heavy and light hole valence bands are parallel at these energies, approximately equal numbers of photoexcited electrons come from each of these two valence bands. The contribution to the Raman active nonequilibrium phonon population from electrons excited from the SO valence band is small because of the small number of SO electrons and because their kinetic energies are close to the bottom of the conduction band where the wavevectors of the nonequilibrium phonons created by the hot electrons exceed what is detectable with Raman scattering. For these two reasons, the electrons excited from the spin-orbit split valence bands can be neglected in most experiments at these photon engergies. The electrons excited from the light hole valence band are at most slightly above the 324 meV threshold for intervally scattering. At low enough densities their kinetics will therefore be described by the cascade process shown in Fig. 8.8. The electrons originating from the heavy hole valence bands are well above the threshold for intervalley scattering for these photon energies. As discussed in Sect. 8.1, most of these HH electrons will undergo intervalley scattering. As a result, the HH electrons do not contribute significantly to the Raman-active nonequilibrium phonon population. The dominant contribution to the generation of a nonequilibrium population of Raman active LO phonons by hot electrons is associated with electrons excited from the light hole valence band for photon energies between 1.91 and 2.09 eV. This is a significant detail which must be considered when results on the hot electron kinetics obtained by different experiments are compared.

The solid line in Fig. 8.24 shows the temporal response of the experimental apparatus based on the measured autocorrelation function of the 1.91 eV laser pulse, assuming the instantaneous generation of nonequilibrium phonons and a phonon population lifetime of 3.5 ps. The phonon population excited by the 1.91 eV pump photon in Fig. 8.24 shows that the peak of the population is delayed by 0.7 ps with respect to the system response and by about 1.3 ps for the 2.09 eV excited emission. Figure 8.26 indicates that the hot LH electrons have an initial kinetic energy of 0.38 eV (2.09 eV laser) or 0.29 eV (1.91 eV laser). Raman active LO phonons are emitted only for $E_e > 0.12$ eV (Sect. 8.1.1). Therefore there are 8 steps (2.09 eV laser) or 5 steps (1.91 eV laser) in the cascade during which Raman active LO phonons are emitted. During these steps in the cascade, the net LO phonon emission rate is essentially constant (Fig. 8.3). In the simplest approximation, the difference of the emission of three LO phonons for incident pump photon energies between 1.91 and 2.09 eV produces an additional delay in reaching the peak of the LO phonon population of 600 fs so that each of the three steps can be said to involve a delay of about 200 fs. A more sophisticated model, which weights each step of the cascade of the LH electrons by the number of phonons created at the Raman active wavevector in the backscattering geometry and also includes the 3.5 ps phonon population lifetime at 300 K, leads to a phonon emission time of 190 fs. This time is in excellent agreement with the predicted value of 200 fs from Fig. 8.3 for the net emission rate (emission-absorption) at 300 K. Experiments at 2 K [8.40, 67] give the same risetime, again in accord with Fig. 8.3.

The absolute mode occupation of the nonequilibrium LO phonons can be readily determined by normalizing against the Raman scattering from thermal phonons excited by the same pulses[6]. For this reason,

---

[6] Of course, at low temperature this is not possible, and one must use the Stokes to anti-Stokes ratio.

Fig. 8.25 shows Raman spectra obtained using an experimental system similar to that presented in Fig. 8.17 for two different values of the delay between the pump and the probe. For the 76 MHz repetition rate of the laser pulses, there is more than 13 ns between each pulse. Therefore, before each pulse excites the sample, the optically injected carriers and nonequilibrium phonon population from the previous pulse will have had ample time to decay to zero. This situation can be verified from the time decay of the bandedge photoluminescence. The -20 ps curve in Fig. 8.25 corresponds to an experimental situation where the probe beam precedes the pump by 20 ps, when there are no carriers or nonequilibrium phonons in the sample. Since pump-excited Raman scattering is rejected using polarization selection rules, this curve corresponds to Raman scattering from the thermal phonon population with known mode occupation $n_{thermal} = 0.31$ at 300 K. As an example, in Fig. 8.25, at 2 ps the integrated intensity of the nonequilibrium LO phonon scattering is 60% of the thermal peak so that the nonequilibrium LO phonon mode occupation number is about 0.2. Since only about 40% of the electrons are from the light hole valence band, the effective density of electrons generating Raman active nonequilibrium phonons is only about 2.5×10^{15} cm^{-3}, and there are 8 steps in the cascade for generating the phonons. Figure 8.4 shows that the injection of 10^{16} cm^{-3} hot electrons should generate a Raman active mode occupation of 0.1, which is in reasonably good agreement with the measured mode occupation.

8.3.3 Wavevector Distribution for Nonequilibrium Phonons in GaAs

The Fröhlich interaction, when combined with the bandstructure of the Γ valley in GaAs, was shown in Fig. 8.4 to generate a distribution of nonequilibrium LO phonons strongly peaked for wavevetors between 6 and 8×10^5 cm^{-1}. On the other hand, the excitation of LO phonons with wavevectors smaller than 4×10^5 cm^{-1} by hot electrons created by the absorption of photons around 2 eV is forbidden. In Fig. 8.11, we showed the constraints placed on Raman scattering in the backscattering and the forward scattering configurations by the conservation of crystal momentum in a solid. For 2.0 eV excitation, in the backscattering geometry, Raman scattering couples to phonons with wavevectors near 7×10^5 cm^{-1} for scattering wavevector normal to the sample surface. In the forward scattering geometry, phonons with a wavevector of less than 1.5×10^4 cm^{-1} (wavevector parallel to the surface of the sample) are observed by Raman scattering. Comparing these wavevectors with Fig. 8.4 leads to the expectation that a nonequilibrium phonon population will only be observed in Raman experiments using the backscattering geometry for bulk GaAs. In the forward scattering geometry, no nonequilibrium phonons should be detected.

These expectations have been verified by *Kash* et al. [8.40, 58] using thin epitaxial layers of GaAs grown on 1 μm Al$_{0.7}$Ga$_{0.3}$As support layers on GaAs substrates. The 1 μm $Al_{0.7}Ga_{0.3}As$ support layer was used as an etch stop to permit the removal of the thick GaAs substrate [8.79]. Forward scattering experiments involving the GaAs epitaxial layer were then possible since 2.0 eV light could be transmitted through the sample. The $Al_{0.7}Ga_{0.3}As$ layer was also used to provide mechanical support for the GaAs layer whose thickness was 3300 Å. The thin GaAs could be used to study the Raman active phonon population in both back and forward scattering. The support layer was transparent to the laser beam so that no nonequilibrium phonons were excited in it by the pump beam. An antireflection coating was evaporated on top of the GaAs layer to eliminate crosstalk between the forward and backscattering light. For a GaAs layer of thickness L, the phonon wavevector in the growth direction is uncertain by an amount $\Delta q \simeq \pi/L$. Therefore, the 3300 Å film is thick enough to be essentially bulk, since $\Delta q \simeq 10^5 \, cm^{-1}$.

Figure 8.27 [8.67] shows both the thermal and the nonequilibrium LO phonon scattering excited in the 3300 Å GaAs layer in the forward and backscattering geometries where the pump and probe beams propagated in the same direction. In the backscattering geometry, a large nonequilibrium LO phonon population was observed by the probe beam consistent with the results shown previously for bulk samples. In the forward scattering geometry, the detected nonequilibrium phonon mode occupation was at least a factor of 10 smaller than in backscattering and essentially zero to within the signal to noise of the experiment. The thermal spectra in Fig. 8.27 show contributions from the $Al_{0.7}Ga_{0.3}As$ support layer. The thermal Raman spectrum of GaAs in the forward scattering geometry also includes contributions from TO phonon scattering, which is symmetry allowed in (100) GaAs for this experimental configuration. The thermal

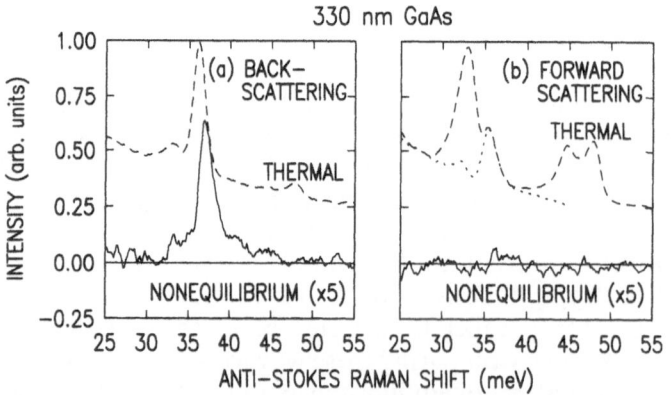

Fig. 8.27. The nonequilibrium Raman spectra of GaAs at room temperature obtained from a 3300 Å (100)-oriented epitaxal layer of GaAs in the (a) backscattering and (b) forward scattering geometry. The spectra were excited using 5 ps pulses of 2.1 eV light with a pump/probe delay of 2 ps. In (b), the thermal phonon peaks of the $Al_{0.7}Ga_{0.3}As$ supporting layer are more pronounced because of the experimental geometry and the subtraction of these peaks results in the dotted curve which is just the thermal GaAs LO phonon population [8.67]

LO phonon signal from just the GaAs layer is indicated as the dotted curve. The results in Fig. 8.27 provide confirmation of the predicted distribution in momentum space of nonequilibrium LO phonons generated by hot electrons in GaAs as shown in Fig. 8.4.

An interesting question is what happens when the GaAs layer becomes thin enough that $\Delta q \simeq 5 \times 10^5$ cm^{-1}. The results of such experiments are described as part of the next section.

8.3.4 Nonequilibrium Phonon Generation in Semiconductor Alloys

In sect. 8.3.1, the two mode character of $Al_xGa_{1-x}As$ where there are two LO phonon modes which are called AlAs-like and GaAs-like was discussed. The population lifetimes of each were shown to be the same as the LO phonon in GaAs. The existence of two LO phonon modes in the alloy system raises a question of how the Fröhlich coupling is modified from that of a single mode system as described in (8.4–7). In addition, although the lifetimes of the LO phonons in $Al_xGa_{1-x}As$ are the same as in GaAs, the linewidth of the phonons in the alloy is significantly broader than in the binary, showing that inhomogeneous broadening effects must be considered [8.35, 76].

The behavior of the Fröhlich interaction in the two mode $Al_xGa_{1-x}As$ system is closely related to the variation in the energy differences between the LO and TO phonons with alloy composition x. The composition dependences of these energy splittings are shown in Fig. 8.28 [8.76]. The important feature for the two-mode Fröhlich interaction problem is the increase in the LO-TO splitting for the AlAs-like mode with increasing Al concentration, while the LO-TO splitting for the GaAs-like mode decreases. It was shown in Sect. 8.1 that the LO-TO splitting in a one mode system can be used to define the ionicity of the system through the LST relation. The ionicity is a measure of the electric field of the LO phonon, which determines the strength of the Fröhlich interaction. The increase in the AlAs-like LO-TO splitting with x shows that as the Al concentration increases, the strength of the Fröhlich interaction for the AlAs-like vibrations also increases. The strength of the interaction for the GaAs-like vibrations decreases. The Fröhlich Hamiltonian for the two mode system can be expressed as [8.71]:

$$
H_F = \left(\frac{4\pi Ne^2}{\varepsilon_\infty}\right)^{1/2} \sum_q \frac{i}{q} \Bigg[\Delta_1 \left(1 + \frac{\Delta_2^2}{(\omega_{1LO}^2 - \omega_{2LO}^2)}\right)^{1/2} \left(\frac{\hbar}{2\omega_{1LO}}\right)^{1/2}
$$

$$
\times \left(\frac{\omega_{1LO}^2 - \langle v^2 \rangle q^2}{\omega_{1LO}^2 - \omega_{pq}^2}\right)^{1/2} (a_{1q} e^{iq \cdot r} - a_{1q}^\dagger e^{-iq \cdot r})
$$

$$
+ \Delta_2 \left(1 + \frac{\Delta_1^2}{(\omega_{2LO}^2 - \omega_{1LO}^2)}\right)^{1/2} \left(\frac{\hbar}{2\omega_{2LO}}\right)^{1/2}
$$

$$
\times \left(\frac{\omega_{2LO}^2 - \langle v^2 \rangle q^2}{\omega_{2LO}^2 - \omega_{pq}^2}\right)^{1/2} (a_{2q} e^{iq \cdot r} - a_{2q}^\dagger e^{-iq \cdot r}) \Bigg]. \tag{8.21}
$$

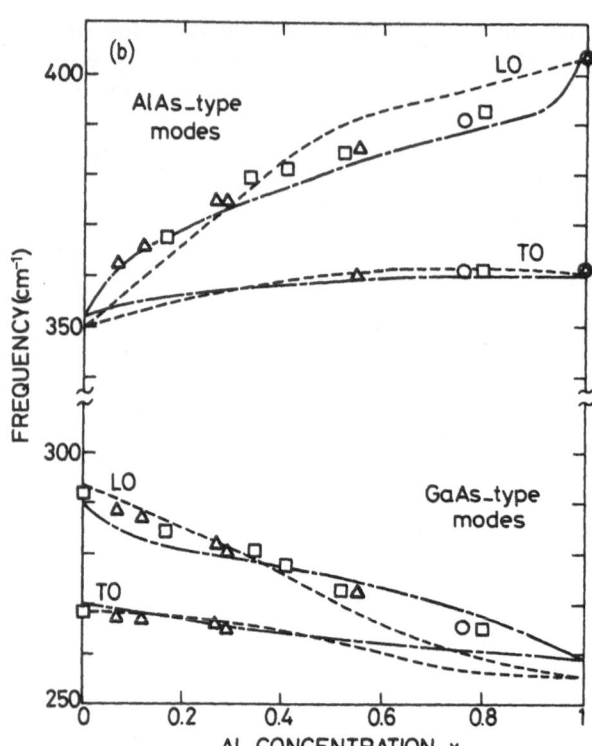

Fig. 8.28. Composition dependence of the optical phonon energies in the alloy system $Al_xGa_{1-x}As$ [8.76]

Here $\Delta_i^2 = \omega_{iLO}^2 - \omega_{iTO}^2$, where $i = 1$ for the AlAs-like mode and $i = 2$ for the GaAs-like mode. The TO phonon frequencies are ω_{iTO}, while ω_{iLO} are the LO phonon frequencies. The boson creation and annihilation operators are a_{iq} and $a_{iq}^\dagger$, while $\langle v^2 \rangle$ is the average carrier velocity defined earlier and ω_{pq} is the coupled mode plasmon frequency [8.38]. The complexity of this expression derives from the fact that the effective charges for the Al and Ga atoms have been expressed in terms of the LO-TO phonon splittings Δ_i, and also because the coupling between the LO phonons and the plasmon is included. From this expression, *Kash* et al. [8.70, 71, 80] calculated the scattering rates for the AlAs- and GaAs-like LO phonons. The scattering rate for the AlAs-like phonon is proportional to

$$\Gamma_{1,e-LO} \propto \frac{\Delta_1^2}{\varepsilon_\infty \omega_{1LO} q^2} \left[1 + \left(\frac{\Delta_2^2}{\omega_{1LO}^2 - \omega_{2LO}^2} \right) \left(\frac{\omega_{1LO}^2 - \langle v^2 \rangle q^2}{\omega_{1LO}^2 - \omega_{pq}^2} \right) \right].$$

$$(8.22)$$

A similar expression applies for the GaAs-like mode. These expressions simplify in the limit of low carrier densities ($\omega_{pq} \ll \omega_{iLO}$) and first order in Δ_i^2 to

$$\Gamma_{i,e-LO} \propto \frac{(\Delta_i^2)}{(\varepsilon_\infty \omega_{iLO} q^2)} \tag{8.23}$$

and can be related to (8.7) describing the electron-LO phonon scattering in the one-mode system. The linear dependence of $\Gamma_{i,e-LO}$ on Δ_i^2 shows how the Fröhlich interaction reflects the composition dependence of the LO-TO splittings shown in Fig. 8.28. Also, since $(\Delta_1/\omega_{1LO}) + (\Delta_2/\omega_{2LO})$ is approximately independent of the alloy composition, the total polar scattering rate $\Gamma_{1,e-LO} + \Gamma_{2,e-LO}$ at any composition will not be very different from the rate in pure GaAs or AlAs. This is in contrast to the behavior of the short range alloy scattering in systems such as Ge_xSi_{1-x} where the scattering shows a pronounced maximum at intermediate compositions [8.81]. When the plasma frequency is small compared to the LO phonon frequencies, a hot electron in $Al_xGa_{1-x}As$ will lose energy by emitting either a GaAs-like or an AlAs-like LO phonon. From (8.23) the fraction of nonequilibrium LO phonons that are AlAs-like is [8.71, 80]

$$f_1 = \frac{\Delta_1^2}{\Delta_1^2 + \Delta_2^2}\left(1 + \frac{\Delta_2^2}{\omega_{1LO}^2 - \omega_{2LO}^2}\right), \tag{8.24}$$

where we have kept terms to $O(\Delta_i^4)$. In deriving this expression, *Kash* et al. [8.70] ignored the small difference in the density of final electronic states for an electron which emits an AlAs-like LO phonon as opposed to a GaAs-like LO phonon.

Figure 8.29 [8.70] shows pump-probe Raman spectra for $Al_xGa_{1-x}As$. The spectra were excited using 5 ps pulses and a 2 ps delay between the pump and the probe. These spectra demonstrate the presence of nonequilibrium LO phonons in both the GaAs-like and AlAs-like modes. Since the lifetimes of both nonequilibrium phonon populations are identical (Fig. 8.21) [8.70, 73], the nonequilibrium mode occupations measured in this data reflect the strength of each LO mode. From spectra like those of Fig. 8.29 obtained from samples of different alloy concentration, the fraction of the total nonequilibrium phonon population that is AlAs-like, i.e. $f_1 = n_1/(n_1 + n_2)$, is shown in Fig. 8.30 [8.80] by the open circles. The crosses are the theoretically derived values for this fraction based on the experimentally measured phonon frequencies for these samples and (8.24). The smooth curve connecting the crosses is derived from the phonon frequencies shown in Fig. 8.28. Because of the weak strength of the AlAs-like Fröhlich coupling for low aluminum concentration, the injected carrier density to generate the experimental open circle point in Fig. 8.30 for $x = 0.7$ was about 10^{17} cm^{-3}. At this density, one must begin to consider the influence of the plasmons on (8.24), which will

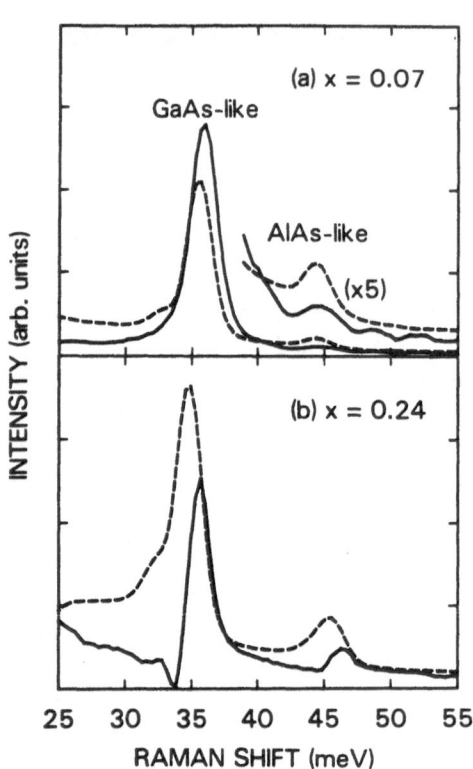

Fig. 8.29. Thermal (dashed lines) and nonequilibrium (solid curves) Raman spectra for two compositions of $Al_xGa_{1-x}As$ excited by 5 ps optical pulses at 2.19 eV using the pump-probe technique with a 2 ps delay [8.70]

reduce the predicted value of f_1 by about 10%. Nevertheless, Fig. 8.30 shows that the simple long wavelength dielectric response picture for the relaxation of hot carriers by the Fröhlich interaction used to derive (8.21–24) provides a good picture of the LO phonon coupling in two mode systems.

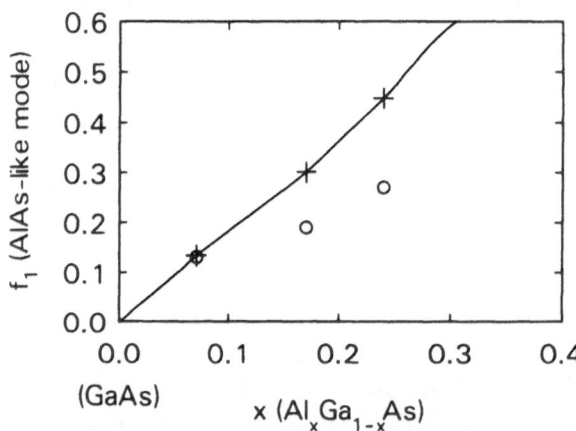

Fig. 8.30. The fraction f_1 of nonequilibrium phonons in photoexcited $Al_xGa_{1-x}As$ which are AlAs-like versus alloy composition. The open circles are experimental results while the crosses and solid line are derived from (8.24) [8.71, 80]

Fig. 8.31. The dependence on the phonon wavevector of the magnitude of the nonequilibrium phonon population generated by the relaxation of hot electrons in the Γ conduction band of GaAs for several different localization conditions. The solid line describes bulk like behavior in GaAs and also localization in a 3300 Å slab which is essentially bulk like. The dashed line describes localization by a 500 Å slab. The dash-dotted line and the dotted lines describe the behavior of the nonequilibrium population when the phonons are localized three-dimensionally on the scales of 700 and 150 Å respectively

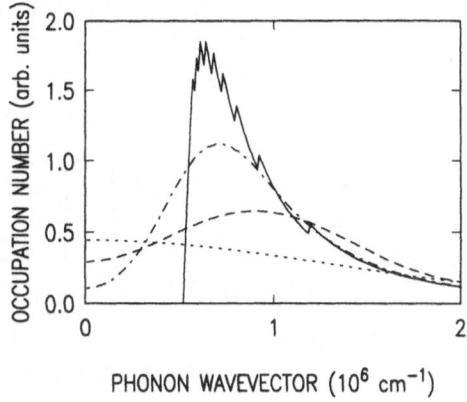

PHONON WAVEVECTOR (10^6 cm^{-1})

Figure 8.27 showed how *Kash* et al. [8.40, 58] used Raman scattering to verify that the distribution of hot phonons in momentum space generated by energetic electrons in the zone center conduction band of GaAs was in agreement with the predictions of Fig. 8.4. A large nonequilibrium population was observed in backscattering while none was observed for the small wavevector phonons excited in forward scattering. The LO phonon lineshape measured with cw Raman scattering in $Al_xGa_{1-x}As$ is broadened, especially to lower energy, as compared to pure GaAs [8.76]. This asymmetric broadening has been interpreted in terms of the existence of disorder localized phonons with a correlation length which varies with x, typically having values of the order of $100-200$ Å for $0.1 < x < 0.3$ for the GaAs-like mode [8.35]. Figure 8.31 [8.40] shows how the distribution of hot electron generated GaAs-like LO phonons in momentum space would be modified if the phonons were spatially localized with a three-dimensional correlation length of either 700 Å (dash-dotted curve) or 150 Å (dotted curve). For a 150 Å correlation length (the correlation length obtained by *Parayanthal* and *Pollak* [8.35] for $Al_{0.11}Ga_{0.89}As$), there should be little difference between forward scattering (phonon wavevector $< 1.5 \times 10^5$ cm^{-1}) and backscattering (phonon wavevector $> 8 \times 10^5$ cm^{-1}). Figure 8.32 [8.58] shows the results of *Kash* et al. for the strength of the nonequilibrium LO phonon scattering excited by a 2.10 eV pump in the forward and backscattering geometries for a 3300 Å layer of $Al_{0.11}Ga_{0.89}As$. All of the details of the experiment were identical to those for Fig. 8.27 on a 3300 Å epitaxial layer of GaAs. The results of the experiment are also identical with a large nonequilibrium LO phonon population observed in the backscattering geometry and, within the signal to noise, none observed in the forward scattering geometry. This was in contradiction to the expectation shown in Fig. 8.31 (dotted curve) for phonons localized on a 150 Å scale. The experimental result on the 3300 Å alloy layer that the forward scattering from the nonequilibrium LO phonon population was at least an order of magnitude weaker than in backscattering showed that any localization length for the GaAs-like LO phonons was greater than 700 Å.

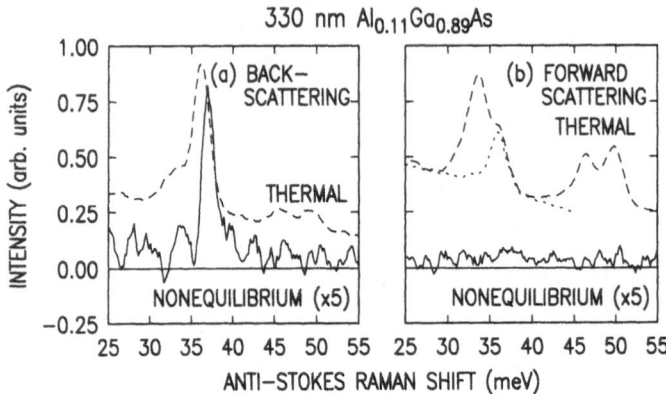

330 nm Al$_{0.11}$Ga$_{0.89}$As

Fig. 8.32. Same as Fig. 8.27 but for a 3300 Å layer of Al$_{0.11}$Ga$_{0.89}$As in (a) the backscattering geometry and (b) the forward scattering geometry [8.58]

To verify the predictions of the effects of spatial localization on the wavevector distribution of nonequilibrium phonons, a comparison of forward and backscattering was also performed in a GaAs layer 500 Å thick by *Kash* [8.67]. Here, the one-dimensional localization to a slab of thickness L means that the nonequilibrium phonon distribution of the bulk semiconductor should be convolved with the function $(\sin (x)/x)^2$, where $x = qL/2$. This convolution is shown as the dashed curve in Fig. 8.31, which predicts similar mode occupations in forward and backscattering. Experimental results are shown in Fig. 8.33 where significant and similar nonequilibrium populations are observed in the two scattering geometries. The observation therefore of good wavevector selection rules in the 3300 Å thick layer of Al$_{0.11}$Ga$_{0.89}$As shows that the phonon wavefunctions are not localized on a length scale of less than 700 Å in the alloy. As a result, the asymmetric lineshapes of the GaAs-like

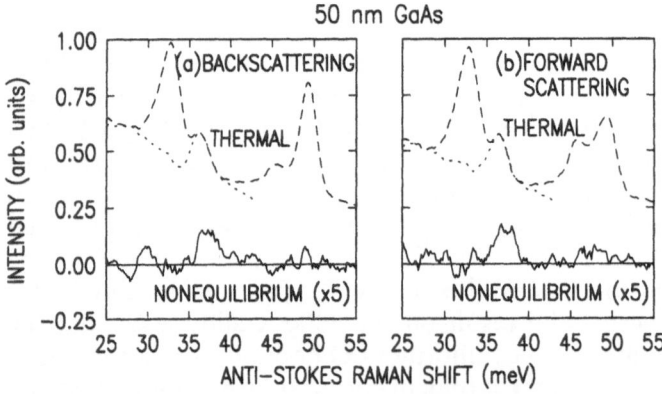

50 nm GaAs

Fig. 8.33. Same as Fig. 8.27 except that the active layer is a 500 Å GaAs layer [8.58]

and AlAs-like optical phonons cannot be explained by localization. *Kash* et al. suggested that the disorder in the alloy lattice activated normally Raman inactive large wavevector modes, but did not affect the smaller wavevector modes which are Raman active in the perfect lattice [8.40].

8.3.5 Second-Order Raman Scattering from Nonequilibrium Phonons

As mentioned in the introductory section, the phonon-assisted relaxation of hot carriers in GaAs also involves the emission of large wavevector phonons when the hot electron kinetic energy is high enough to permit intervalley scattering. These phonons are in principle observable by Raman scattering in second order, although their wavevectors are too large for their observation by first order Raman scattering. Experimental efforts to observe these phonons have not been successful. Given the shape of the $n(k)$ curve in Fig. 8.4, second order Raman scattering has also been used in an attempt to detect nonequilibrium Raman scattering from LO phonons with wavevector above 10^6 cm^{-1}. *Olego* et al. have shown tht resonant Raman scattering near the $E_0 + \Delta_0$ direct gap allows the observation of overtone scattering at twice the optical phonon frequency for wavevectors near 10^6 cm^{-1}, depending on the detuning of the excitation frequency from the gap [8.82]. However, efforts to observe nonequilibrium effects in the second order scattering near resonance have been hampered by luminescence associated with this resonance. *Bray* and coworkers have also attempted to observe second order Raman scattering from zone edge and near zone edge nonequilibrium acoustic phonons [8.83]. Here, electronic Raman scattering associated with trace impurities of zinc in the GaAs masked any such second order scattering.

8.4 Light Scattering and Hot Luminescence Probes of Electron Kinetics in Bulk III–V Compounds

In the previous section, we showed how time resolved Raman scattering has been used to measure the strength of the *Fröhlich* interaction between small wavevector LO phonons and hot electrons in semiconductors such as GaAs. In order to understand ballistic and quasi-ballistic transport in semiconductors and the behavior of carriers at high fields in small structures, it is also necessary to measure the distribution functions characterizing the energies and momenta of the hot carriers. In this section we show how the time varying, nonequilibrium distribution functions describing the populations of energetic electrons in III–V semiconductors can be measured by time dependent Raman scattering and hot photo-luminescence. Because of the central role of the *Fröhlich* interaction in the relaxation of hot electrons through the emission of LO phonons, there will be considerable overlap in this section with some of the results discussed in Sect. 8.3. We shall show how measurements of the electron-LO phonon interaction studied in Sect. 8.3 by time-dependent Raman scattering from the nonequilibrium LO phonon population, can also be derived from the behavior of the hot electron luminescence.

In this section, we shall review the experimental studies of the hot carrier kinetics in several different regimes. We begin with low density electron-hole gases (Sect. 8.4.1) which correspond to the cascade model of Fig. 8.8 and show how hot luminescence spectroscopy provides a means of measuring the hot electron scattering rates. These scattering rates include the LO phonon emission process and intervally scattering processes. Efforts to use Raman scattering to obtain the intervally scattering rates will also be discussed. We then discuss the regime of high carrier densities (Sect. 8.4.2) where interactions between carriers become significant and the carrier kinetics are qualitatively described in terms of Fig. 8.9. The use of hot luminescence from free electrons recombining at neutral acceptors and anti-Stokes bimolecular recombination to provide information about what happens when carrier–carrier scattering rates become comparable to or faster than the carrier–LO phonon scattering rates will be reviewed. The use of Raman spectroscopy to describe the cooling of a very hot thermalized carrier distribution, the screening effects from an optically generated electron–hole plasma, and the spatial diffusion of hot carriers will also be discussed in this section.

8.4.1 Electron Kinetics in the Low Density Regime

a) LO Phonon Emission

In Sect. 8.2, we described the luminescence emitted by an isolated, energetic electron in the conduction band of GaAs when it recombines at a neutral acceptor (Fig. 8.14). The richly structured spectrum associated with this luminescence process provides a number of different ways of measuring the hot electron lifetime as a function of the electron kinetic energy. All of these experimental approaches make certain assumptions about the band structure of GaAs or require a detailed knowledge of the band structure parameters both near the top of the valence band and well above the bottom of the conduction band in order to obtain quantitative values for the hot electron lifetimes. Uncertainties in our knowledge of these parameters introduce uncertainties in the lifetimes derived from the experimental data. This means that a variety of different experimental approaches should be used to obtain reliable accuracy.

In Fig. 8.34 [8.51], we show in detail the most energetic peaks of the hot luminescence spectrum that was shown in general in Fig. 8.14. The photon energy used to excite this spectrum is below the threshold for intervalley scattering so that in the low density limit, no carriers are ever scattered to the X and L conduction band minima. All of the electrons

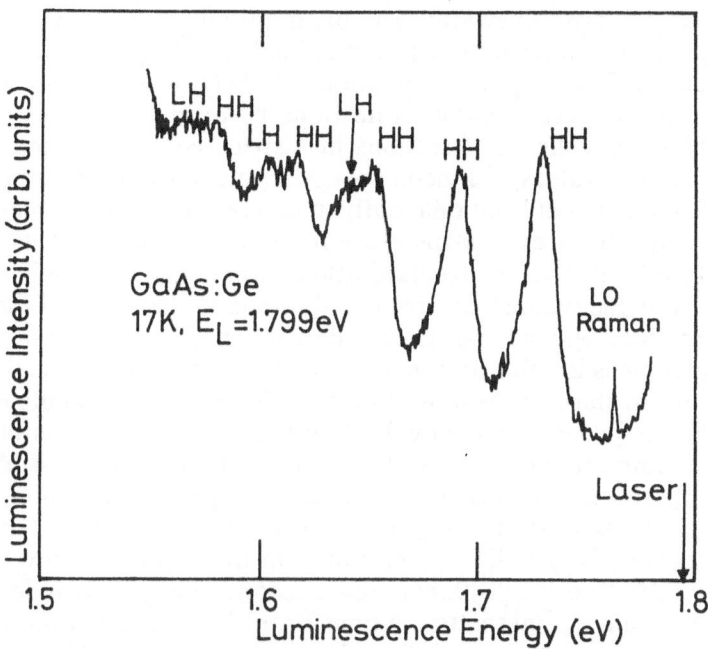

Fig. 8.34. The highest energy peaks in the hot luminescence spectrum of acceptor doped GaAs under 1.79 eV excitation [8.51]

excited into the conduction band can participate in the emission shown in Fig. 8.34. The full widths at half maximum of the emission peaks have been interpreted by *Ulbrich* et al. [8.50] and *Fasol* et al. [8.51] as originating in part from electron lifetime effects. They assumed that these peaks arise from a luminescence process involving a real excited state which is uncorrelated with the incident radiation field (i.e. hot luminescence as opposed to resonant Raman scattering). This is an important assumption since it means that there is no contribution to the width of the luminescence due to the lifetime of the free hole state remaining after the hot electron and bound hole recombine. If the width of the acceptor energy level is narrow, then the linewidth of the luminescence transition will be related to the lifetime of the hot electron in the same way that the linewidths of the Raman phonons in Sect. 8.3 reflect the total scattering times for these phonons. The 12 meV linewidth of the highest energy peak in Fig. 8.34 imposes the constraint that the excited state lifetime for the nonequilibrium electron is more than 50 fs due to the uncertainty relation (8.11). This limit for the electron scattering time is much shorter than the 190 fs electron-LO phonon emission time obtained in Sect. 8.3 which is expected to dominate the hot electron relaxation. Fasol et al. [8.51] have considered in detail the observed lineshapes for the hot luminescence. They find that band structure considerations are responsible for a significant fraction of the observed linewidth. The band structure effects include spin splitting and, more importantly, the directional anisotropy or warping of the valence bands. The warping, as mentioned in Section. 8.2, means that for photons of fixed energy, optical transitions between the valence and conduction bands in different directions in k-space create electrons and holes with different values of kinetic energy. The total kinetic energy of the pair is fixed, but is divided differently in different directions because the constant energy surfaces in k-space are not perfectly spherical. *Fasol* et al. used a 16×16 $\mathbf{k} \cdot \mathbf{p}$ band structure calculation of the electronic energy bands to derive the band structure contributions to the shape of the highest energy hot luminescence peak. This derived shape, which assumes the electron lifetime is infinite, and that the neutral acceptor state makes no contribution to the linewidth, is shown in Fig. 8.35. The infinite electron lifetime lineshape was then convolved with a Lorentzian to simulate the effect of lifetime broadening. *Fasol* et al. found that it was possible to obtain the experimentally measured luminescence peak by assuming a hot electron scattering time of 132 ± 10 fs. In their error analysis, *Fasol* et al. did not account for possible uncertainties in the band structure calculations.

Mirlin et al. [8.49, 84] used a second, completely independent method to obtain a total scattering time for the hot carriers from the higher energy luminescence peaks of Fig. 8.34. Under excitation by a linearly polarized laser pump beam, they found that the hot luminescence is partially polarized. The excess linear polarization $\varrho_1 = (I_{\parallel} - I_{\perp})/(I_{\parallel} + I_{\perp})$. $I_{\parallel}$ is the intensity of emission polarized parallel to the excitation while $I_{\perp}$ is the emission intensity polarized perpendicular to the excitation E field. For the case of band-to-band emission, they reported that $\varrho > 0.25$ at low temperatures for emission energies close to the excitation energy.

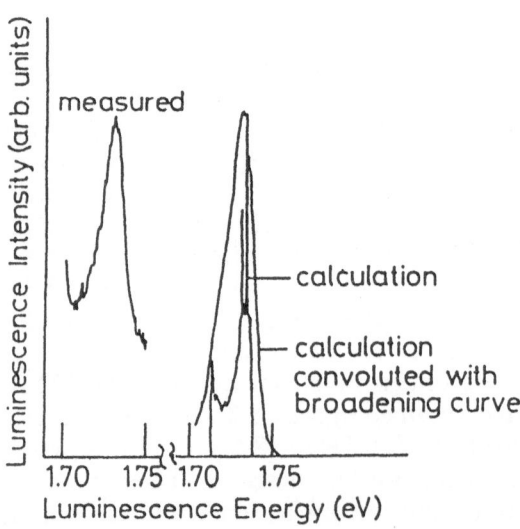

Fig. 8.35. The highest energy peak of the hot electron spectrum of acceptor doped GaAs under 1.79 eV excitation. The theoretically calculated curves show the lineshape of the emission due to band structure effects without lifetime broadening and the effects of lifetime broadening of 132 fs on this lineshape [8.51]

There are at least two distinct mechanisms that can contribute to the partial linear polarization of the hot luminescence in GaAs under linearly polarized excitation [8.84]. The momenta of the optically excited electrons can be aligned in the plane normal to the incident oscillating electric field. If this alignment is preserved, then recombination with thermalized holes which are unaligned will show $\varrho_1 = 0.14$. The warping of the valence bands can also modify the bound acceptor wavefunctions. *Zakharchenya* et al. [8.84] suggested that this warping results in a distribution of the hole momenta at the acceptors that is strongly peaked in the [111] direction, i.e., there is an effective alignment of the bound hole wavefunctions. Given this additional alignment which is tied to the crystal axes, the hot electron–neutral acceptor luminescence is dominated by electrons occupying states associated with the [111] direction of the band structure[7]. The linear polarization of the luminescence due to the optical alignment of the electron momentum and the selection of the electronic states in the (111) direction means that ϱ_1 can be used to measure the scattering of the electrons. The application of a magnetic field in the Faraday geometry will modify the time evolution of the optically excited electronic states. In the quasi-classical approximation, the momenta of the optically generated electrons will precess about the magnetic field H at the cyclotron frequency $\omega_c = eH/(m_c c)$ where m_c is the cyclotron mass. For example, for a field of 10 T and $m_c = 0.067$, $\hbar\omega_c = 17$ meV, corresponding to a cyclotron period of 240 fs. If the optically excited electrons have a lifetime of τ_e, then in a field H, the optically aligned momenta will rotate by an angle $\omega_c\tau_e$ away from their original value. The observed polarization of the time-integrated luminescence will decrease in the magnetic field as compared to the zero field case, with the decrease depending on $\omega_c\tau_e$ as (Hanle effect) [8.49]

$$\frac{\varrho_1(H)}{\varrho_1(0)} = \frac{1}{1 + 4\omega_c^2\tau_e^2} . \tag{8.25}$$

The external magnetic field means that ω_c can be used as a "clock" against which the electron scattering rate can be timed. *Mirlin* et al. measured the magnetic field dependence of the degree of linear polarization of the highest energy luminescence peaks at several different laser photon energies and obtained the results shown in Fig. 8.36 [8.49]. If (1) the cyclotron frequency of GaAs is known at a kinetic energy of 300 meV, (2) the magnetic energies are small compared to the energies of the significant scattering mechanisms, and (3) the relaxation time approximation is indeed applicable to this situation, then the value of τ_e can be

---

[7] Note that this explanation contradicts one of the assumptions of *Fasol* et al. above. The lineshape analysis of *Fasol* et al. assumes that the acceptor wavefunctions are spherically symmetric so that electrons in all directions of k-space contribute equally to the hot luminescence.

obtained from the data of Fig. 8.36 with (8.25). The curves in Fig. 8.36 were fitted using (8.25) with a cyclotron mass described by the Kane expression [8.84]:

$$m_c(E_e) = m_c^0 \left(1 + \frac{2E_e}{E_g} \right), \tag{8.26}$$

where $m_c^0 = 0.067m_e$ is the cyclotron mass at the bottom of the conduction band. At 1.83 eV (just below the threshold for intervally scattering) *Mirlin* et al. obtained an electron scattering time of about 100 fs, 30% smaller than the value obtained by *Fasol* and about a factor of two smaller than expected from Fig. 8.3 and the time resolved Raman scattering experiments discussed in Sect. 8.3. Because of the uncertainty in the cyclotron mass for high energy electrons and the fact that the cyclotron energy at 4 T is already 20% of the LO phonon energy, this value should perhaps be taken as only an estimate of the polar scattering rate. *Zakharchenya* et al. [8.84] have also pointed out that while their model describes the magnitude of the observed low temperature polarization, the 300 K linear polarization is much stronger than expected suggesting that the luminescence polarization may include contributions from factors other than the above two

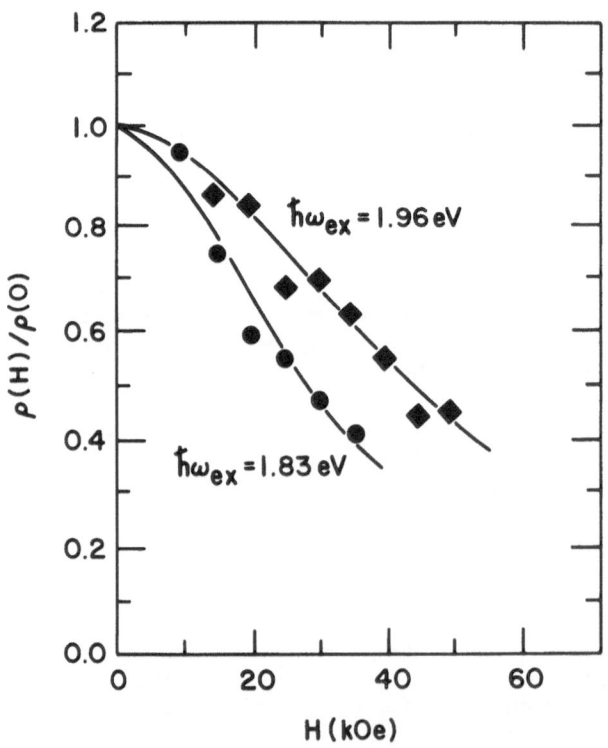

Fig. 8.36. The magnetic field dependence of the depolarization of the highest energy hot electron-bound acceptor recombination peak in GaAs under 1.83 and 1.96 eV excitation. The fit to the 1.83 eV excited emission corresponds to a $\tau_e = 100$ fs while the fit to the higher energy spectrum yields $\tau_e = 70$ fs [8.49]

mechanisms. We will return to the variations between these various measurements and the theory in Sect. 8.6.2.

b) Intervalley Scattering

The discussion in the previous section considered the electron lifetime for electron kinetic energies where only polar phonon emission is important. When the electron energy is increased above the threshold for intervalley scattering to the satellite X and L valleys, additional scattering channels open up for an energetic electron, reducing its lifetime to

$$\tau_e^{-1} = \tau_{po}^{-1} + \tau_{\Gamma-L}^{-1} + \tau_{\Gamma-X}^{-1} .$$ (8.27)

In this case each of the scattering times except τ_{po} depends strongly on the electron kinetic energy E_e as seen in Sect. 8.1.1. (Taking τ_{po} constant simply restricts application of (8.27) to $E_e > 3\hbar\omega_{LO}$, which is always the case here.) The relationship of the intervalley scattering time to the band structure parameters was given in (8.9). Assuming the deformation potential is independent of wavevector and intervalley scattering is dominated by a zone edge optical phonon of energy $\hbar\omega_{ze}$, then intervalley scattering from the Γ valley to an L valley depends on E_e only as the density of final states in the L valleys. Hence, at low temperatures where the phonon mode occupation numbers are small, we can parameterize $\tau_{\Gamma-L}$ as[8]

$$\tau_{\Gamma-L} = \tau_{\Gamma-L}^0 \left(\frac{E_e - E_L - \hbar\omega_{ze}}{\hbar\omega_{ze}}\right)^{-0.5} ,$$ (8.28)

where $E_L = 296$ meV is the energy of the L valley minimum with respect to the Γ minimum. $\tau_{\Gamma-L}^0$ is the intervalley scattering time for energy $\hbar\omega_{ze}$ above the threshold. For GaAs, a typical zone edge optical phonon energy for the (111) direction is $\hbar\omega_{ze} = 28$ meV.[9] A similar equation can be written

---

[8] The results in this section are expressed as scattering times rather than as a deformation potential. This choice is dictated by the fact that the experiments measure scattering times and not the deformation potential. The relation between the deformation potential $D_{\Gamma-L}$ and $\tau_{\Gamma-L}^0$ is shown in (8.9, 28) and discussed in Sect. 8.6.3.

[9] The form of $\tau_{\Gamma-L}$ is not changed significantly by a different choice of $\hbar\omega_{ze}$. The essential part of this equation is just the density of states dependence of the numerator. *Zollner* et al. [8.85] have pointed out that because the phonons involved in intervalley scattering have a wavevector that decreases somewhat as E_e increases, (8.9) is not quite correct as the phonon density of states varies with wavevector. They conclude however that the approximation is valid, at least for low temperatures.

for $\tau_{\Gamma-X}$, where $E_X = 460$ meV. The temperature dependence of the intervalley scattering rates can be seen from (8.9) to depend on the identities of the phonons involved in the scattering process. Because the different zone edge modes in GaAs have energies between 10 and 28 meV (Fig. 8.1) their populations can change significantly between 2 K and 300 K. In particular, *Zollner* et al. [8.85] have shown that the relative contributions to the intervalley scattering by the 10 meV TA phonons increase rapidly with temperature near 300 K because of the strong temperature dependence of their mode occupations as compared to that of the higher energy phonons. Therefore, the important phonon modes for $\Gamma - L$ scattering will differ for samples at 2 K and at 300 K. Equation (8.28) is valid only at low temperatures. The full expression in (8.9) must be summed over all the contributing phonons to obtain the correct temperature dependence of the intervalley scattering process.

The techniques discussed in the previous section can be applied to obtain the change in the electron lifetime due to intervalley scattering. Both *Fasol* et al. [8.51] and *Mirlin* et al. [8.49] observed a decrease in the electron lifetime for hot electron energies greater than 310 meV, which is the threshold for scattering to the L valleys. The extraction of the intervalley scattering time from their experimental results requires the subtraction of the inverse LO phonon scattering time τ_{po}^{-1} from the measured inverse electronic scattering time τ_e^{-1}. This means that uncertainties in the measurement of τ_{po} will also effect the derived intervally scattering rates. *Fasol* et al., using $\tau_{po} = 132$ fs, obtain $\tau_{\Gamma-L} = 150 \pm 75$ fs for electrons with kinetic energies between 340 and 440 meV, i.e. above the threshold for phonon-assisted transitions to L and below the threshold for transitions to X. Interestingly, their results do not indicate any dependence of $\tau_{\Gamma-L}$ on the electron energy, in contradiction to (8.28). This is surprising given the expected dependence of the scattering rate on the density of final states in the L valleys.

Mirlin et al. have measured the degree of polarization of hot luminescence in a magnetic field at different laser photon energies above and below the threshold for intervalley scattering from Γ to L and from Γ to X [8.49, 86]. Results are shown in Fig. 8.36 corresponding to initial electron energies of 0.27 eV (i.e. no intervalley scattering, $\hbar\omega_{ex} = 1.83$ eV) and 0.385 eV (Γ to L scattering allowed, $\hbar\omega_{ex} = 1.96$ eV). For an electron with energy 380 meV, this corresponds to $\tau_{\Gamma-L} = 250$ fs, about 100 fs longer than the results of *Fasol* et al. When the excitation energy was increased to generate hot electrons where $E_e = 570$ meV, above the threshold for $\Gamma - X$ scattering, *Mirlin* et al. derived a value for $\tau_{\Gamma-X}$ of 30 ± 10 fs.

Ulbrich et al. [8.50] have used the hot electron–neutral bound acceptor emission to derive the intervally scattering rates in GaAs in two additional ways. In contrast to the work of *Mirlin* et al. who imposed an external magnetic field on the sample in order to use the cyclotron frequency as a clock to measure the electron dephasing time, *Ulbrich* et al. used the electron-LO phonon scattering time as an "internal" clock to measure the intervalley scattering rates. Equation (8.19) in Sect. 8.2 showed the matrix element $M(k)$ for the recombination of a electron of wavevector k and energy E_e at a neutral acceptor. For any single peak in the hot luminescence, the intensity under the peak per electron is, from Fermi's golden rule,

$$I \propto |M(k)|^2 \, \tau_e , \qquad (8.29)$$

where τ_e is the electron lifetime at the energy corresponding to that peak. Measurements of the intensity of the highest energy peak (labelled HH in Fig. 8.14) as a function of the photon excitation energy is shown in Fig. 8.37a [8.50]. Here the intensity is defined as the number of photons per second under the integrated area under the peak, divided by the incident photon flux [10]. With the assumption that τ_e is constant over this energy range, I varies just as $|M(k)|^2$, which is shown as the dashed curve in Fig. 8.37 (overlapping the solid curve for photon energies below 1.9 eV).

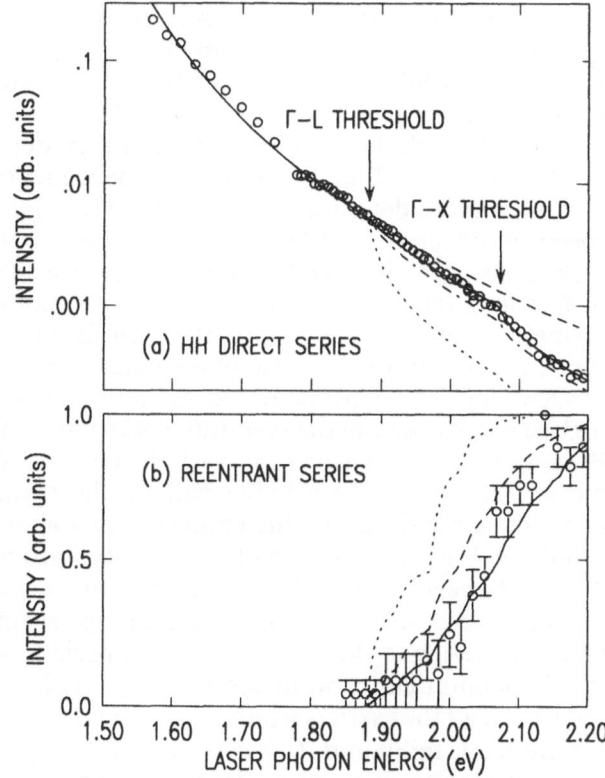

Fig. 8.37. The dependence on excitation energy of the efficiency of the hot electron–neutral acceptor luminescence in GaAs. The open circles in (a) are the intensities of the highest energy heavy hole series peak while the open circles with error bars in (b) give the intensities of the highest energy line in the reentrant spectra. The solid and broken lines are calculations described in the text [8.50]

---

[10] *Ulbrich* et al. assume that the fraction of electrons generated from the heavy hole band is constant. If the fraction is decreasing, the scattering times would be slightly longer than those which they derived.

Ulbrich et al. used the effective mass approximation of *Baldereschi* and *Lipari* [8.53] based on a spherical approximation for the valence band structure of GaAs to calculate $M(k)$. In doing so, they neglected the cubic corrections to the acceptor wavefunctions which *Zakharchenya* et al. [8.84] used to explain the polarization of the hot luminescence. For electron energies below the threshold for intervalley scattering, this curve describes the data very well, providing support for the effective mass approximation made in evaluating (8.29)[11]. Above the threshold for intervalley scattering to the L point conduction band minima, the data falls below the dashed curve. This result shows that there is a shorter lifetime for the hot electrons above the threshold. Since the data in Fig. 8.37 for photon energies below the intervalley scattering threshold do not give the value of τ_{po}, *Ulbrich* et al. could only determine the ratios $\tau_{\Gamma-L}^0/\tau_{po}$ and $\tau_{\Gamma-X}^0/\tau_{po}$ in using (8.27) for the lifetime τ_e. The other curves shown in Fig. 8.37a are for $\tau_{\Gamma-L}^0/\tau_{po} = 7.14$, 3.175, and 0.595 for the solid, dot-dashed, and dotted curves respectively, with $\tau_{\Gamma-X}^0/\tau_{po} = 1.79$ for all three curves. The solid curve gives the best fit to the data, and corresponds to a Γ to L scattering rate for a 380 meV electron of $5.0\,\tau_{po}$ and a Γ to X scattering rate for a 580 meV electron of τ_{po}. The ratio $\tau_{\Gamma-L}^0/\tau_{po}$ is a factor of 1.9 greater than that obtained by *Mirlin* et al., while $\tau_{\Gamma-X}^0/\tau_{po}$ is a factor of 3.5 greater. These differences are physically quite significant.

Ulbrich et al. used the excitation energy dependence of the reentrant spectra (labelled R in Fig. 8.14) to independently verify the intervalley scattering times determined from Fig. 8.37a. The intensity of the reentrant spectrum depends on how many electrons are scattered to the satellite valleys. As noted in Sect. 8.1, almost every electron that scatters to the X or L valleys returns to the Γ valley from the bottom of the L valley. This assumption is consistent with the well-defined peaks of the reentrant spectra and with the absence of a similar series of peaks associated with the X valley. The dependence on excitation energy of the intensity of the highest energy peak of the reentrant spectrum is shown in Fig. 8.37b [8.50]. Because this peak is superimposed on the emission due to electrons that did not scatter to the satellite valleys, the relative uncertainties in the measured intensities for this emission are somewhat larger than for the results in Fig. 8.37a, especially for excitation energies between 1.8 and 2.0 eV. However, the gradual increase in the intensity of the reentrant hot luminescence over a 300 meV energy range strongly suggests that intervalley scattering to the L valleys is not especially fast compared to the rate of LO phonon emission in the Γ valley. With the assumption discussed above that all the carriers which scattered into the X and L valleys return to Γ from the L minimum, *Ulbrich* et al. made a quantitative estimate of the intervalley scattering rate using the cascade model and (8.27). Since the reentrant spectrum is fixed in energy and the kinetics of electrons in the Γ

---

[11] In a later publication, *Baldereschi* and *Lipari* suggested that the cubic corrections to the acceptor energy levels were small [8.87].

valley upon their return from an L valley are independent of laser photon energy, it was not necessary to use (8.29), nor to calibrate the quantum efficiency of the detection system with respect to the emission wavelength to obtain the results in Fig. 8.37b. Thus this measurements is essentially independent of the assumptions used for the analysis of Fig. 8.37a, in particular, the behavior of the acceptor matrix element. *Ulbrich* et al. found that their best fit for intervalley scattering times derived from the HH emission in Fig. 8.37a also provided a good fit to the excitation wavelength dependence of the reentrant emission as shown by the solid line in Fig. 8.37b. The data in Fig. 8.37b are clearly inconsistent with any assumption that $\tau_{\Gamma-L}$ is much smaller than τ_{po}, as demonstrated by the dotted line which corresponds to $\tau_{\Gamma-L}^0/\tau_{po} = 0.595$, i.e. a scattering time at 380 meV of about $0.42\tau_{po}$ [8.88]. The dashed curve in Fig. 8.37b was derived for a ratio of the intervalley scattering parameter to the optical phonon scattering time of 3.175, or a scattering time at 380 meV of about $2.2\tau_{po}$, which clearly represents the shortest scattering time for any reasonable fit to the data.

Although the experimental uncertainties are larger in the data shown in Fig. 8.37b than for the results in Fig. 8.37a, the analysis of the excitation wavelength dependence of the reentrant series intensity requires a less detailed understanding of the band structure of GaAs than the other three approaches discussed above, since the emission energy remains fixed and no external fields are applied. The chief assumptions that are made in the analysis of the reetrant emission are that all electrons which are scattered to the satellite valleys return from the L minimum and that the electrons do not diffuse into the bulk of the sample beyond the optical penetration/escape depth of GaAs (about 5000 Å) during the 5–10 ps before returning from an L valley. Scattering to the X valleys does not occur until the laser photon energy exceeds 2.07 eV, where the reentrant spectrum has already nearly reached its maximum amplitude. Therefore, analysis of the reentrant spectra gives a good measure of $\tau_{\Gamma-L}^0/\tau_{po}$, but may not give an accurate value of $\tau_{\Gamma-X}^0/\tau_{po}$ given the small size of the intensity changes for excitation energies between 2.0 and 2.2 eV.

Raman scattering has also been used to obtain values for the intervalley scattering rates in GaAs. *Collins* and *Yu* [8.19] used the laser photon energy dependence of Raman scattering from nonequilibrium LO phonons to obtain values for these parameters. If the photon flux used to pump the sample is held constant as the excitation photon energy increases, the increasing kinetic energy of each electron will allow the generation of more LO phonons. Many of these phonons are Raman active, so this produces an increase in the nonequilibrium phonon mode occupation measured by the ratio of Stokes to anti-Stokes scattering. On the other hand, if the carriers are lost from the Γ valley by intervalley scattering, then they will not be able to excite Raman active LO phonons given the large masses of the zone edge conduction band minima. This will reduce the expected increase with increasing excitation photon energy of the nonequilibrium phonon mode occupation. *Collins* and *Yu* used a single 10 ps pulse to both generate the nonequilibrium electrons and sample the LO phonon mode occupation by Raman scattering. The measured mode occupation as a function of laser photon energy is shown in Fig. 8.38 [8.19]. The initial increase in the nonequilibrium phonon population with increasing photon energy is in agreement with the expectation that increasing the energy of each hot electron increases the number of nonequilibrium LO phonons it can generate. At the 1.88 eV threshold for hot electrons generated from the heavy hole band to scatter to the L valleys, *Collins* and *Yu* observed a relatively small change in the nonequilibrium mode occupation. A much larger effect was seen at the onset of Γ to X transition near 2.1 eV. The small influence of the Γ to L transitions on their data led *Collins* and *Yu* to conclude that the scattering rate for this intervalley process was very much smaller than the polar LO phonon scattering rate. The Γ to L intervalley scattering rate they derived was about 7 times slower than the results of *Ulbrich* et al. The discrepancy with the results of *Mirlin* et al. and *Fasol* et al. is even larger. However, in modeling their experiment, *Collins* and *Yu* only considered the generation of hot phonons by the electrons excited from the heavy hole band and did not include the contribution due to electrons from the light hole bands. As shown in Fig. 8.26, there is a significant and in some cases dominant contribution to the hot electron distribution from the light hole valence band. These electrons do not undergo intervalley scattering until the laser photon energy exceeds 2 eV, so the sensitivity of the measurements of *Collins* and *Yu* to intervalley scattering is reduced. Their results provide therefore a lower limit to the intervalley scattering rates.

There have been many other attempts to measure the intervalley scattering rates in semiconductors such as GaAs and InP using other techniques such as transport measurements [8.31, 89], pump-probe absorp-

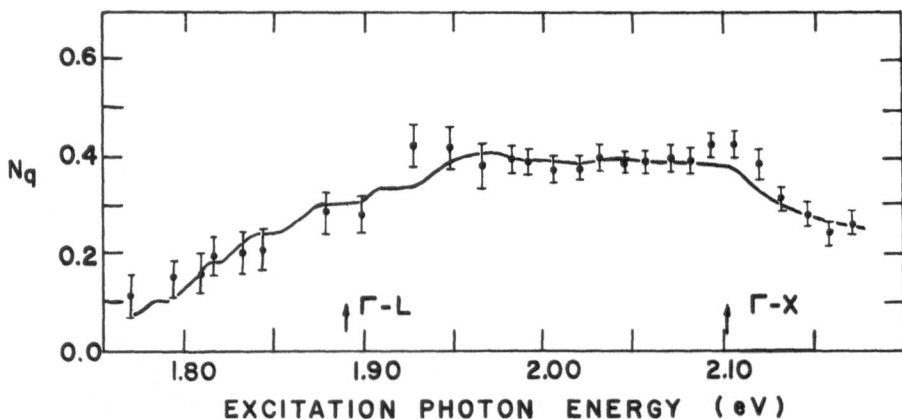

Fig. 8.38. The excitation energy dependence of the nonequilibrium LO phonon mode occupation generated by the photoexcition of electron hole pairs in GaAs as measured by picosecond Raman scattering [8.19]

tion and reflectivity [8.88], and time-resolved band edge photoluminescence [8.90]. These experiments will be briefly summarized in Sect. 8.6.3, as
will recent theoretical calculations of *Grinyaev* et al. [8.91, 92] and *Zollner*
et al. [8.85, 93]. The wide range of intervalley scattering times obtained
in these experiments will also be analyzed.

8.4.2 Electron Dynamics in the Strongly Interacting Carrier Regime

The qualitative resemblance of the schematic drawing in Fig. 8.8 of the
time-integrated distribution function for the energy relaxation of a single
electron in a conduction band to the hot luminescence spectra in Fig. 8.14
is due to the fact that the data in Fig. 8.14 were obtained for optically
injected carrier densities below $3 \times 10^{15} \mathrm{cm}^{-1}$. For these densities, comparing Fig. 8.3 and 8.7, carrier–carrier scattering is insignificant compared
to the Fröhlich interaction. The hot luminescence spectra in Fig. 8.14 in
fact provide direct confirmation that hot carriers in GaAs relax by a LO
phonon cascade process at low injected densities.

Kash [8.94] has quantitatively studied how the cascade process will be
modified by the presence of free carriers. A two color pump-probe
experiment was done where one laser at 1.64 eV was used to inject a high
density of free carriers near the band edge of GaAs. A second laser at
1.88 eV was used to generate a small number of nonequilibrium electrons
which generate the hot electron-neutral acceptor luminescence. A schematic diagram of this experiment is displayed in Fig. 8.39 [8.94], which
shows the two different injected distributions and the different scattering
processes that will occur. The probe-injected carrier density was held
below $2 \times 10^{15} \mathrm{cm}^{-1}$ where no carrier–carrier scattering effects were
observed in the absence of the pump. The pump injected density was as
high as $8 \times 10^{16} \mathrm{cm}^{-3}$. The upper bound was set to avoid pump induced
modifications in the Fröhlich interaction due to the screening of the LO
phonon. The probe beam was delayed by 30 ps with respect to the pump
beam to allow the pump-injected carriers to thermalize and cool to about
100 K. Therefore, the experiment measured the interaction of a single
nonequilibrium electron with a cool electron–hole plasma. To prevent
the pump-injected carriers from diffusing to the surface or into the bulk
during the 30 ps delay, the samples were clad with $Al_{0.7}Ga_{0.3}As$ layers.
Figure 8.40a [8.94] shows the spectrum when both pump and probe are
present, with a pump injected density of $2.2 \times 10^{16} \mathrm{cm}^{-3}$. For the dashed
curve, the probe precedes the pump by 30 ps; for the solid curve the probe
follows the pump by 30 ps. The injection of carriers produces a decrease
in the intensity of the hot luminescence peaks given by the difference
between the curves in Fig. 8.40a. This difference is shown in Fig. 8.40b,
and the difference for a higher pump-injected density is shown in Fig. 8.40c.
From (8.27), the decrease in the integrated intensity of the highest energy
hot luminescence peak corresponds to the decrease in the lifetime τ_e of
the hot electrons from Γ_{po}^{-1} to $(\Gamma_{po} + \Gamma_{e-p})^{-1}$ where Γ_{e-p} is the electron–

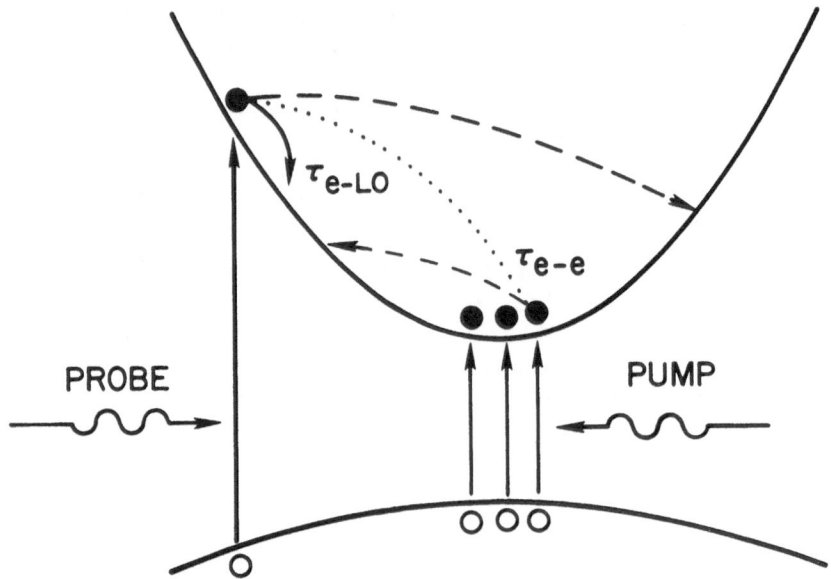

Fig. 8.39. A schematic of a pump-probe experiment where the pump is used to inject relatively cold carriers and the probe injects a relatively small number of hot carriers which can relax either by LO phonon emission (τ_{e-LO}) or by scattering from the cold carriers (τ_{e-e})

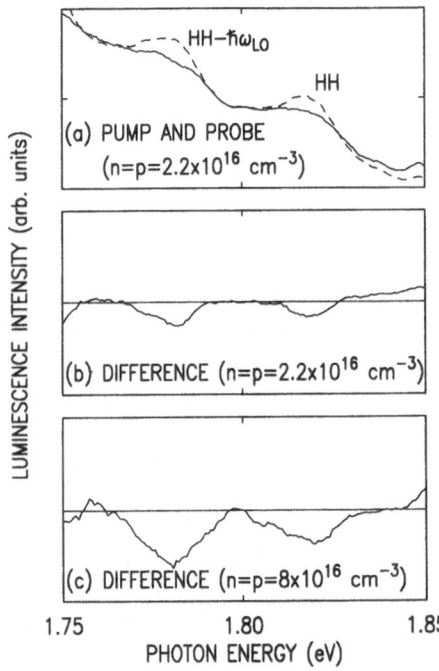

Fig. 8.40. (a) The luminescence spectra of LO phonon cascade electron–neutral acceptor luminescence with the 1.88 eV probe pulse preceding the pump pulse by 30 ps (dashed curve) and with the probe following the pump by 30 ps (solid curve). (b) The difference between the two curves in (a). (c) The difference spectrum for a higher pump injected carrier density of 8×10^{16} cm^{-3}. The probe injected carrier density is below 2×10^{15} cm^{-3}. All spectra are to the same scale [8.94]

plasma scattering rate[12] and $\Gamma_{po} = 1/\tau_{po}$. From the measured decrease, *Kash* concluded that $\Gamma_{e-p} = \Gamma_{po}$ when the carrier density is about 8×10^{16} cm^{-3}. This density is consistent with recent theoretical estimates [8.41]. This result is also consistent with other experimental data such as the hot bimolecular luminescence spectra shown in Fig. 8.16.

It is also interesting to compare the results in Fig. 8.40 with the measurements of the dependence of the nonequilibrium LO phonon population on optically injected carrier density in Fig. 8.20. Figure 8.40 shows that there are significant carrier–carrier scattering effects on the hot carrier distribution functions at injected carrier densities below 2×10^{16} cm^{-3}. However, no deviations from linearity in the dependence of the nonequilibrium phonon population on injected carrier density are observed until the injected densities are at least a factor of two higher. These results are consistent with the fact that the carrier–carrier scattering processes do not remove energy from the carrier system but merely redistribute it among the carriers so that the same amount of energy still has to go into the lattice.

In Fig. 8.16 of Sect. 8.2.4 it was shown that the subpicosecond injection of highly energetic electron–hole pairs at densities above 5×10^{16} cm^{-3} produces observable bimolecular recombination involving free holes and electrons at energies above the excitation energy. The dependence of this anti-Stokes luminescence emission on incident power was seen in Fig. 8.16 to be quadratic in laser intensity for injected densities greater than 3×10^{17} cm^{-3}. Because of its nonlinear dependence on excitation power level, *Tsang* and *Kash* [8.56] were able to make luminescence correlation measurements (Sect. 8.2.4) using pump and probe beams each injecting about 5×10^{17} cm^{-3} carriers. These results are shown in Fig. 8.41. The anti-Stokes emission in the two narrowest solid curves is characterized by correlation times as short as 1 ps. While quantitative interpretation of such results requires substantial modeling, the subpicosecond correlation times observed for the anti-Stokes emission show that a high density photoinjected carrier distribution can thermalize on a subpicosecond time scale, before it loses much energy to the lattice.

The results of *Kash* in Fig. 8.40 were restricted to carrier densities below 10^{17} cm^{-3} to avoid the possibility of optically injected carrier induced changes in the Fröhlich interaction, i.e. screening. Screening by optically injected carriers has been observed in Raman scattering from GaAs and InP by *Nather* et al. [8.95] and by *Young* et al. [8.96, 97]. Figure 8.42 shows the results of *Young* et al. [8.96, 97] for the Raman spectrum of InP after the optical injection of 4×10^{17} cm^{-3} electron–hole pairs. The top curve shows the spectrum of the unexcited sample with the single symmetry allowed LO phonon line while the bottom curve shows the

---

[12] Γ_{e-p} is the rate for inelastic collisions in which the nonequilibrium electrons lose more than about $\hbar\omega_{LO}$ of energy to the cold carriers.

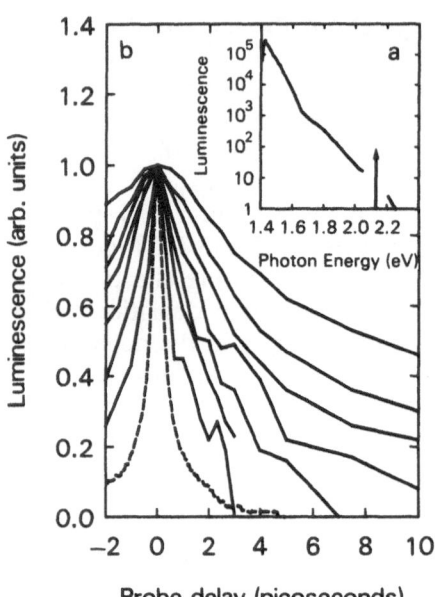

Fig. 8.41 The variation with delay between the pump and the probe of the intensity of the time correlated photoluminescence from GaAs under 0.5 ps excitation at 2.13 eV using injected carrier densities for both the pump and the probe of 5×10^{17} cm^{-3} per pulse. The emission energy for each correlation curve is, going from the narrowest to the broadest, 2.45, 2.22, 2.03, 1.99, 1.94, 1.80 and 1.65 eV. The broken line is the autocorrelation of the exciting pulse. The insert shows the actual luminescence spectrum [8.56]

spectrum probed 62 ps after the injection of the carriers. Here the characteristic coupled LO phonon–plasmon mode at 360 cm^{-1} is seen. Since the coupled mode frequency is related to the electron density, *Young* et al. could measure the surface plasma density as a function of the pump probe delay. They obtained a value for the ambipolar diffusion constant and surface recombination rate of InP. The diffusion constant was found to be 5 cm^2 s^{-1} while the surface recombination velocity was 10^5 cm s^{-1} in good agreement with results obtained from transport studies. Extensions of these experiments to bulk GaAs showed that the plasmon effects were reduced by strong surface recombination effects which produced highly inhomogeneous electron–hole distributions. *Nather* et al. [8.95] and *Young* et al. [8.20] were able to observe coupled mode effects in GaAs by capping the GaAs layer with a thin layer of Al$_x$Ga$_{1-x}$As.

The experiments of *Young* et al. [8.20] and *Nather* et al. [8.95] which have a timescale $\gtrsim 50$ ps, show that the equilibrium phonon spectrum is modified due to screening by photoinjected carriers. The nonequilibrium phonon spectrum is also modified by the phonon–plasmon coupling. In Figs. 8.27, 29, 32, the Raman peak due to nonequilibrium phonons is shifted to higher energies than the peak due to the thermal phonons. The spectrum of the thermal phonons is taken before photoinjection by the pump pulse. The nonequlibrium phonons are generated in the presence of the photoinjected carriers, and hence are coupled to the plasmon where $\omega_p < \omega_{LO}$.

In addition to observing the screening, *Nather* et al. [8.95] detected single-particle electronic Raman scattering from the photoinjected plasma.

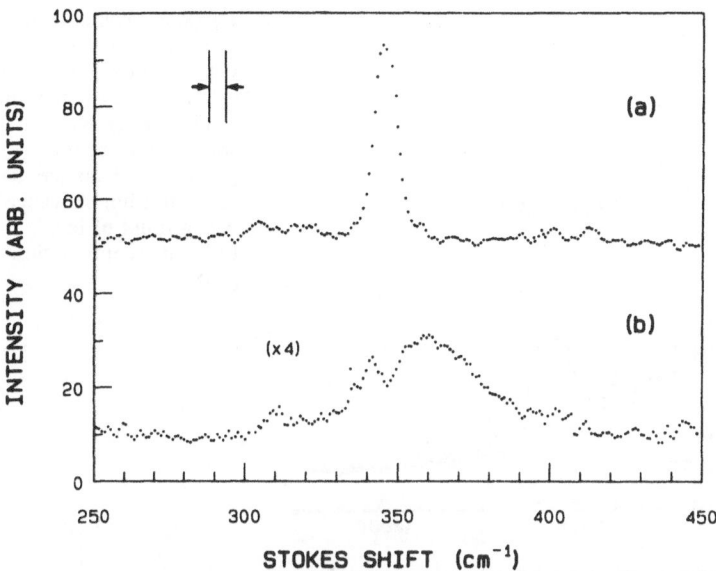

Fig. 8.42. (a) The Raman spectrum of InP showing the unscreened LO phonon. (b) The Raman spectrum of this sample 62 ps after it was excited by 2.19 eV light injecting 4×10^{17} cm^{-3} electron hole pairs [8.96]

Because of the nanosecond time scale of these experiments, the photo-injected plasma temperature was typically 100 K. The optical injection of hot carriers for photon energies near 2 eV produces an average carrier kinetic energy of about 250 meV. If the hot electrons and holes both thermalize on a time scale short compared to the electron–LO phonon interaction time, the initial carrier temperature will exceed 2500 K. In order to have this rapid thermalization, the injected carrier density should exceed 5×10^{17} cm^{-3} as can be seen in Fig. 8.41. Observation of the single particle Raman scattering under such conditions poses some substantial experimental difficulties since the Raman spectrum will be superimposed on a strong luminescence background as shown in Fig. 8.43 [8.48]. In this figure, *Huang* and *Yu* have indicated the relationship of the single-particle Raman scattering in the vicinity of the laser line and the luminescence background present in a picosecond pump-probe Raman scattering experiment. In Fig. 8.43, 5 ps excitation pulses at 2.13 eV were used to inject a carrier density of about 5×10^{18} cm^{-3}. By subtracting away the luminescence background, *Huang* and *Yu* [8.48] were able to obtain the spectra in Fig. 8.44 which arose from Raman scattering from a cooling electron gas. The 0 ps data (actually an average over the 5 ps pulses) were fitted with a temperature of 800 K and a 20 fs damping time. The 20 ps data were fitted with a temperature of 400 K and 70 fs damping time. The carrier density was found to decay by a factor of 6 during this time by measuring the frequency of the coupled LO phonon-plasmon. The

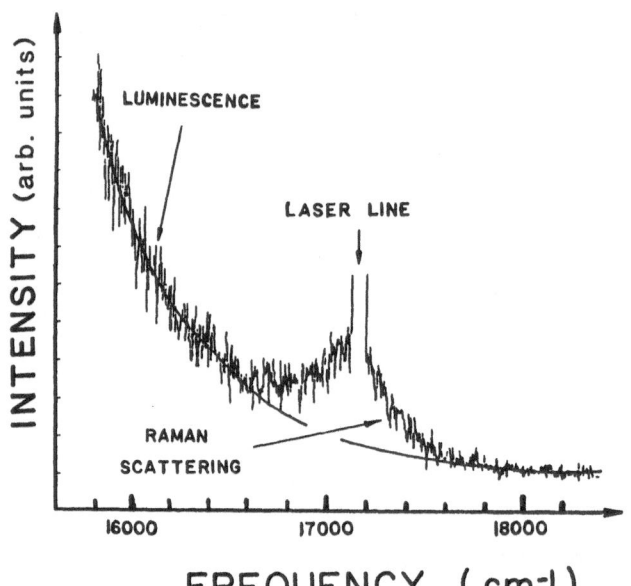

Fig. 8.43. The optical emission spectrum of GaAs using 4.5 ps pulses at 2.13 eV to excite both single particle Raman scattering from the photoexcited carriers and a hot luminescence background of band-to-band recombination [8.48]

short scattering time was attributed to electron–hole scattering while the change in carrier concentration was associated with the expansion of the plasma into the sample. Similar times have also been measured in pump probe studies of various optical properties such as the reflectivity which have yielded comparable values for the carrier distribution temperatures, scattering times and cooling rates.

Raman scattering from single particle excitations have also been used by *Tsen* et al. [8.98] to measure the spatial expansion of photoexcited plasmas in bulk Si. In these experiments, they showed how the high temporal resolution of Raman spectroscopy using picosecond lasers and the good spatial resolution of optical spectroscopy where spot sizes of less than 30 µm are achieved easily, can be combined. This allowed them to measure the diffusion coefficients for carriers in Si.

Most of the discussion on carrier relaxation in semiconductors has focused on relaxation in energy and momentum. The results of *Tsen* and *Sankey* [8.98], *Collins* and *Yu* [8.99] and *Tsang* et al. [8.57] all show that Raman scattering can be used to study the relaxation of a spatially localized carrier distribution. In the first case, this can be done by spatially resolving the lateral motion of the injected carriers. In the last two cases, the diffusion of carriers away from the surface region in highly absorbing materials like GaAs (penetration depth at 2 eV of about 3000 Å) and InAs (penetration depth at 2 eV of about 600 Å) reduces both the electronic and vibrational Raman signals generated by the carriers within an absorption length of the surface.

Fig. 8.44. The Raman spectra due to single particle scattering by a photoexcited plasma at three different time delays. The luminescence background shown in Fig. 8.43 has been subtracted from each spectrum. The broken lines are a theoretical fit used to generate the carrier temperatures and scattering times discussed in the text [8.48]

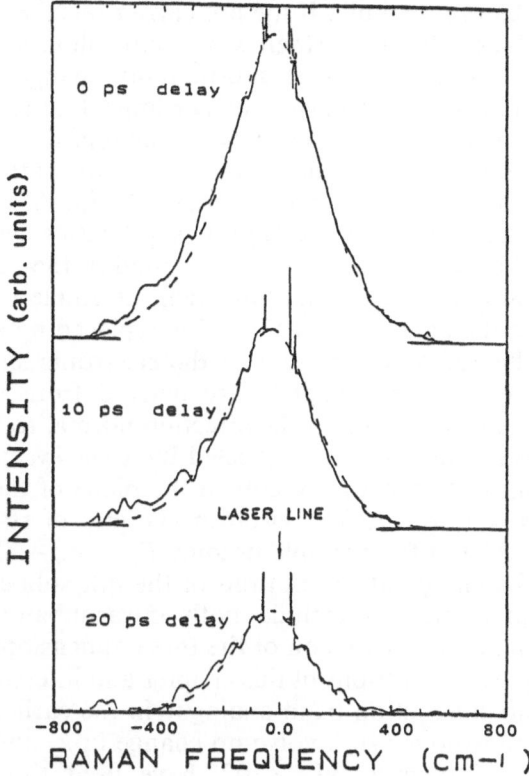

8.5 Picosecond Dynamics in Quantum Wells and Superlattices

The questions posed in previous sections about the relaxation mechanisms and rates for electrons and phonons in bulk III–V and group IV semiconductors are also of interest for quantum wells and superlattices of these materials. In this section we briefly review some of the recent work applying Raman scattering and hot luminescence to this area. It has been shown in previous contributions to this series that the electronic [8.100] and vibrational [8.101] eigenfunctions and eigenvalues of bulk materials can be substantially altered in superlattices and quantum wells. This raises the possibility that their dynamics will also be changed. The reader is referred to the articles by *Jusserand* and *Cardona* [8.101] and *Pinczuk* and *Abstreiter* [8.100] in volume V of this series for detailed descriptions of the properties of the lattice vibrations and the electronic structure of semiconductor superlattices and quantum wells.

Substantial changes in the vibrational and electronic eigenvalues of an ordered semiconductor structure as compared to the bulk behavior have been demonstrated. Because of the zone folding in the superlattice direction, there are many more small wavevector modes in a superlattice than in the bulk. If the hot carrier relaxation proceeds primarily through the excitation of small wavevector phonons, then the carrier dynamics in superlattices can be substantially changed from the bulk. The LO-TO splitting which played such an important role in the bulk carrier relaxation has been shown to depend on superlattice periodicity, actually decreasing from the bulk value by almost a factor of three for the GaAs-like phonons and increasing by more than 50% for AlAs-like phonons in the extreme case of the $(AlAs)_1(GaAs)_1$ superlattice system [8.101]. These and other changes can be expected to modify the carrier and phonon relaxation processes. In a quantum well, the single family of transitions beginning at the band edge in Fig. 8.2 is replaced by a series of transitions between the different subbands of the electronic states localized in the quantum well. These subbands are derived from the confinement of the bulk electronic states in the direction normal to the layer planes. The confined states are no longer labeled by a wavevector in the three directions but instead by a wavevector in the plane of the layers and a subband index $n = 1, 2, 3...$ The electron energies of states near the bottom of each subband for example become $E_e = E_n + (\hbar^2/2m^*)(k_x^2 + k_y^2)$ where E_n is the energy at the bottom of the nth subband. Carrier relaxation in this situation will include both intrasubband transitions which are two-dimensional version of the three-dimensional transitions described in the previous sections of this chapter and intersubband transitions which have no conduction band analogue in the earlier sections. In particular, these transitions can involve no change in k_x and k_y.

Tatham et al. [8.102] have used time resolved electronic Raman scattering to measure the relaxation time for intersubband transitions in 14.6 nm wide GaAs quantum wells with 15.7 nm $Al_{0.36}Ga_{0.64}As$ barriers. The normally low intensity of the Raman scattering from the electronic transitions was enhanced by resonant Raman scattering with care taken to guarantee that the excitation energies were below the threshold for intervalley scattering. The injected carrier densities were low so that the hot carrier relaxation was dominated by the electron–LO phonon interaction and the effects of carrier–carrier scattering processes could be neglected on the subpicosecond time scale. Figure 8.45 [8.102] shows time resolved anti-Stokes Raman spectra obtained by *Tatham* et al. at $T = 30$ K. Because of the low lattice temperature, all anti-Stokes Raman scattering results from nonthermal excitations. The nonequilibrium GaAs LO phonon scattering displays a strong time dependence similar to that described in Sect. 8.3 for pump-probe experiments on the LO phonons in bulk GaAs. There is an additional feature at 51 meV in the spectra which is due to electronic Raman scattering from an electron making a transition from the second subband (C_2) above the bottom of the conduction band to the lowest lying subband (C_1). The time dependences of the nonequilibrium Raman intensities of the LO phonon and the intersubband transitions are shown in Fig. 8.46. The LO phonon shows a decay time of 5 ps, somewhat shorter than the bulk value of 7 ps. The

Fig. 8.45. Time-resolved, anti-Stokes resonant Raman scattering from a 14.6 nm GaAs multiple quantum well excited at the $C_4 - HH_4$ resonance. The 36 meV peak is the GaAs LO phonon while the 51 meV peak is the C_2 to C_1 intersubband scattering. The dotted lines are fits to the phonon and luminescence parts of the spectra [8.102]

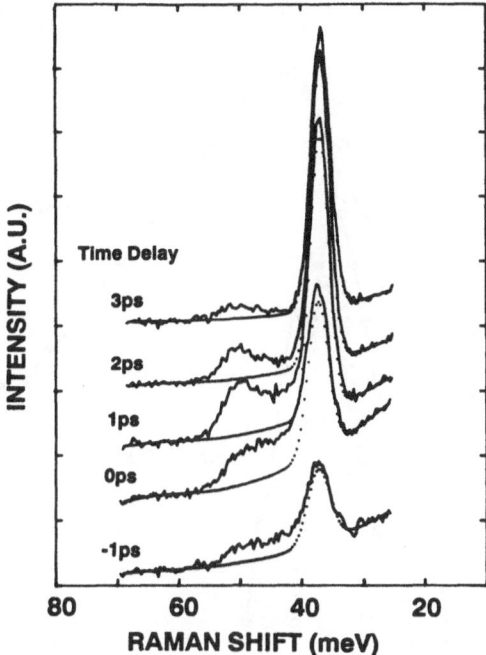

time dependence of the intersubband Raman scattering intensity indicates that the lifetimes of electrons in the $n = 2$ subband is less than 1 ps. *Tatham* et al. showed that this value was consistent with theoretical expectations based on several different descriptions of the phonon wavefunctions involved in the relaxation processes. The small delay in the rise time of the intersubband signal with respect to the excitation pulse auto-correlation function was also shown to be experimentally significant. This delay reflects the $\simeq 500$ fs intersubband relaxation time since the carriers were optically injected into a different subband (C_4) from that of the excited state (C_2) of the electronic Raman transition. Similar Raman studies by *Oberli* et al. [8.103] and others have shown that such experiments provide a sensitive probe of the relaxation of carriers in these structures as a function of excitation energy and injected carrier density.

The appearance of nonequilibrium LO phonons in the experiments of *Tatham* et al. and in the earlier experiments of *Tsen* et al. [8.104] in narrower 50 Å quantum wells raises interesting questions about carrier relaxation and Raman scattering in quantum wells. An important question is the origin of the nonequilibrium optical phonon population. These phonons are presumably generated as the electrons in the quantum well relax. Since the electrons are quasi-two-dimensional, such LO phonons are kinematically required to have in-plane wavevectors greater than

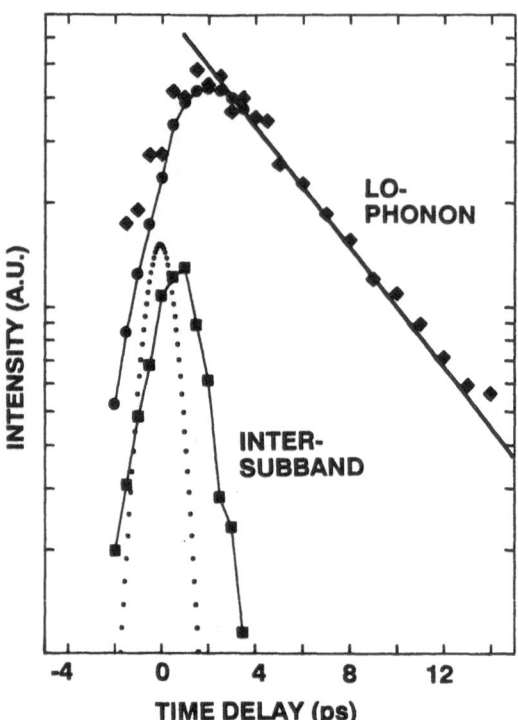

Fig. 8.46. The time dependence of the intensities of the anti-Stokes Raman scattering in a 14.6 nm GaAs quantum well from the LO phonon and the $C_2 - C_1$ intersubband transitions under resonant excitation at the $C_4 - HH_4$ transition. The dotted line is the autocorrelation of the laser pulse [8.102]

4×10^5 cm^{-1} if they result from the intrasubband relaxation[13] of the electrons. For a backscattering geometry, as noted in Sect. 8.1, the in-plane wavevector for Raman active phonons is less than 1.5×10^5 cm^{-1}. Therefore, these nonequilibrium phonons should not be detectable by Raman scattering. To explain this anomaly, *Tsen* et al. and *Tatham* et al. suggested that wavevector conservation breaks down because they are performing their Raman scattering experiments with excitation energies in resonance with electronic transitions in the superlattices. Imperfections in the GaAs/Al$_x$Ga$_{1-x}$As interfaces then cause break down of wavevector conservation [8.105, 106]. The nonequilibrium phonons observed by Raman scattering in 50 Å quantum wells in fact occur at 287 cm^{-1}, an energy below the LO phonon (296 cm^{-1}) of the well layer (Fig. 8.47) [8.104]. These modes have been identified as propagating in the plane of the layers at finite wavevectors [8.106]. Accepting this explanation means that the wavevector of the detected nonequilibrium phonons is not known, and therefore it is difficult to quantitatively interpret the measured kinetics, especially the rise time of the nonequilibrium population.

---

[13] For intersubband relaxation in a 150 Å quantum well, electrons which have ~ 300 meV of kinetic energy in the second subband can relax to the lowest subband with emission of a 36.5 meV LO phonon with in-plane wavevector $\simeq 1.5 \times 10^5$ cm^{-1}. This might be barely detectable if wavevector conservation is required. For a 50 Å well, in-plane wavevector is always greater than 4×10^5 cm^{-1}, even including intersubband scattering.

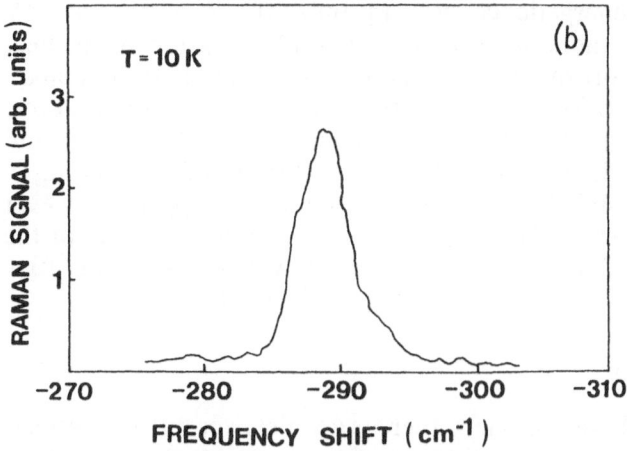

Fig. 8.47. The resonant anti-Stokes spectrum at 10 K of a 50 Å GaAs quantum well under pump–probe excitation with a delay of 0 ps. The GaAs LO phonon is at 296 cm^{-1} and, in the backscattering configuration used, the non-equilibrium Raman scattering is a normally symmetry forbidden phonon excitation [8.104]

Zakharchenya et al. [8.107] have studied hot electron — neutral acceptor luminescence in C-doped GaAs quantum wells of 70 Å width with 100 Å barriers of $Al_{0.3}Ga_{0.7}As$. Under a linearly polarized excitation at 1.65 eV, the initial kinetic energy of the photoexcited electrons was 60 meV. They found that the degree of linear polarization of the highest energy hot luminescence was 0.06. This was close to the theoretical value of 0.075 based on conventional band structure parameters. They measured the depolarization of this emission as a function of magnetic field as described in Sect. 8.4, fitted their results to (8.25) and obtained a value for the hot electron lifetime of 150 ± 10 fs. The appearance of oscillatory structure in the hot luminescence along with the 150 fs lifetime indicates that, as in the case of the bulk studies, the electron–LO phonon scattering process is the major contributor to the hot electron lifetimes.

Because of the great interest in both the scientific problems raised by these materials and their possible technological applications, it is likely that Raman and hot luminescence spectroscopy will be widely applied in the future to these quantum structures. A question that needs investigation is the time resolution of Raman scattering in the resonant Raman experiments. As noted in Sect. 8.2.1, the resonant Raman process need not be instantaneous on the subpicosecond time scale [8.45].

8.6 Discussion

We have shown how Raman scattering and hot luminescence spectroscopy can be used to quantitatively describe the processes responsible for the relaxation of highly energetic carriers in bulk III–V and group IV semiconductors. While there has been considerable progress in this field over the last decade, many questions remain to be answered. In particular, for all of the cases considered here, the dual goals of quantitative consistency in experimental data and agreement between experiment and theory have yet to be fully achieved. A good qualitative understanding of the mechanisms responsible for the hot carrier relaxation appears however to exist at present. In this section, we attempt to highlight some of the questions that have still to be answered before we have a quantitative understanding of hot carrier relaxation in semiconductors.

8.6.1. Phonon Lifetimes

From the experimental point of view, there is a widely accepted consensus concerning the low temperature lifetimes of the long wavelength optical phonons which participate in the initial stages of the hot carrier relaxation in materials such as GaAs and Ge. The good quantitative agreement between the experimentally measured phonon lifetimes in these materials and the linewidths of these phonons as measured by cw Raman scattering produces considerable confidence in these results. In Table 8.1, we have tabulated the optical phonon lifetimes in GaAs, Ge, GaP, and ZnSe as measured directly by time resolved spectroscopy and/or inferred from the linewidths of these Raman active modes. The excellent agreement between the results obtained using the different techniques is obvious. Theoretical treatments of these lifetimes have been able to explain the experimentally observed temperature dependences showing that the measured lifetimes mainly arise from anharmonic lattice effects. However, as discussed in Sect. 8.2, the direct calculation of the low temperature phonon lifetime $\tau_p(0)$ remains a significant challenge. Past treatments have produced phonon lifetimes that have ranged from an order of magnitude below to an order of magnitude above the measured values. At the present time, *Menendez* et al. [8.33] have shown that we cannot predict the lifetime of the Raman active optical phonons in a particular semiconductor even if we know its phonon dispersion curves and the lifetime and phonon dispersion curves of a closely related, isostructural material. We do not know how to "scale" the phonon lifetimes as we can the phonon energies between two similar materials. Therefore, the lifetime of the phonons in each material and for the individual modes in the different alloys of these materials must be directly measured for a quantitative understanding of the hot carrier relaxation. The magnitude of this task suggests that the establishment of the theoretical underpinning for a model that would at least allow the derivation of the phonon lifetimes in semiconductor alloys given their measured lifetimes in the binary constituents would be very useful.

8.6.2 Fröhlich Interaction

The high degree of experimental consistency and the inability of theory
to accurately describe the experimental results in the above discussion of
the phonon lifetimes does not characterize the situation with regard to
our understanding of the electron–phonon interaction in materials such
as GaAs. At the present time, experimental measurements of the various
scattering rates involved in the relaxation of energetic electrons in GaAs
show significant disagreements. In contrast to the phonon lifetimes at
0 K, the theoretical values for these rates can be readily reconciled with
at least some of the experimental points. Our understanding of the hot
electron–LO phonon relaxation processes discussed in this chapter is
summarized in Fig. 8.48 and Table 8.2. The solid curve in Fig. 8.48
represent a theoretical treatment of the electron–LO phonon scattering
time at $T = 0$ K. Scattering rates increase with temperature as discussed in
Sects. 8.1.1 and 8.4. This curve duplicates the $T = 0$ K results for the
Fröhlich coupling shown in Fig. 8.3. The derivation of this theoretical
curve in Sect. 8.1.1 is quite straightforward, and we noted that the
parameters needed to determine the absolute magnitude of the polar
scattering rate can be spectroscopically measured with high accuracy. The
major sources of error, the determination of ε_∞ and the effective mass for
energetic electrons, lead to at most about a 15% error in the absolute
magnitude of the Fröhlich coupling. The analysis of the temperature dependence for 100 K $< T <$ 400 K of the electron mobility in n-type GaAs
by various workers [8.18, 27] gives excellent agreement with the theory
for electron scattering by LO phonons which produced the solid curve in
Fig. 8.48. The time resolved Raman data of *Kash* et al. discussed in Sect. 8.3
(solid diamonds and dotted line in Fig. 8.48, indicating that the measurement is an average over electron energies below the injection energy) also
give good agreement with the theory. Further, the measurements of the

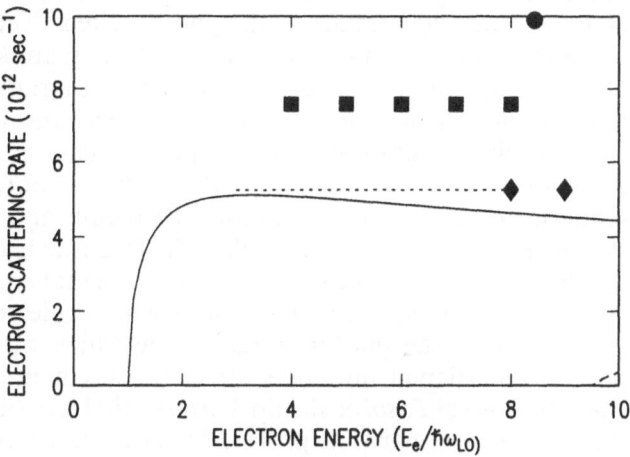

Fig. 8.48. The magnitude
of the electron
scattering rate arising
from the polar LO
phonon interaction at
0 K as calculated from
(8.7) (solid line) with
several experimental
values for this scattering
time as described in the
text from [8.18, 27]. The
dashed line to the right
of the figure indicates
the onset of
Γ-L intervalley
scattering

Table 8.2. Fröhlich (polar-optical) scattering rates for GaAs as measured and calculated

Table 8.2 Fröhlich
Scattering in GaAs

Reference	Technique	τ_{e-LO} at 0 K
Theory [8.18]	–	see Fig. 8.3 and (8.7)
[8.27]	Low field mobility vs. temperature	Equation (8.7) explains mobility for $100 < T < 400$ K
[8.49] Sect. 8.4.1 a	Magnetic depolarization of hot electron neutral acceptor luminescence	100 ± 5fs $E_e = 0.26$ eV
[8.44, 66] Sect. 8.3.2	Time-resolved Raman scattering	190 ± 20 fs $100 < E_e < 350$ meV
[8.51] Sect. 8.4.1 a	Lineshape of hot electron neutral acceptor luminescence	132 ± 10 fs $100 < E_e < 300$ meV

Note: Net LO phonon emission time is independent of temperature, see Fig. 8.3

strength of the polar coupling in $Al_xGa_{1-x}As$ (Sect. 8.3) demonstrate clearly the importance of the energy splitting of the LO and TO phonons for the Fröhlich coupling strength. These various direct measures of the polar coupling provide strong evidence that the theory is correct both in functional form and absolute magnitude. The range of experimental values for the magnitude of the Fröhlich interaction is summarized in Table 8.2.

The results of *Fasol* et al. [8.51] (solid squares in Fig. 8.48) and *Mirlin* et al. [8.49] (solid circle) for the hot electron lifetime at kinetic energies below the threshold for intervalley transitions are both significantly faster than the theoretical polar scattering rate. There are two possible reason for the discrepancy between the electron lifetime measurements of *Fasol* et al. and *Mirlin* et al. and the measurements of the polar scattering rate. The first is possible errors in the analysis of the lifetime measurements as discussed in Sect. 8.4.1a. A more intriguing possibility is that the difference is experimentally significant. Note that *Fasol* et al. and *Mirlin* et al. measure the lifetime of the photoexcited electron in its initial state, while the other measurements and the theory are specifically for the LO phonon emission time, usually averaged over a range of electron energies. Perhaps there is an additional dephasing mechanism for a photogenerated electron which is not related to LO phonon emission. One such possibility is acoustic phonon scattering although, as noted in Sect. 8.1.1, the scattering rate for acoustic phonons is generally agreed to be at least 5 times slower than for polar scattering [8.85]. This rate is far too slow to explain the difference between the various experimental results. Another possible dephasing mechanism for the photogenerated electron is related to its decoupling from the photogenerated hole which was created by the same photon. Such dephasing would affect measurements of the electron lifetime (such as those of *Fasol* et al. and *Mirlin* et al.) but would not be important in the LO phonon emission process for electron energy relaxation.

8.6.3 Intervalley Scattering in GaAs

The intervalley scattering rate measurements described in Sect. 8.4 show considerable scatter. Other measurements not discussed here yield still different rates. For example, time-resolved band edge photoluminescence measurements of *Shah* et al. [8.90] have been analyzed by Monte Carlo simulation to yield $\tau_{\Gamma-L} = 100$ fs for an electron energy of 500 meV and a lattice temperature of 300 K. Early measurements of electron velocity vs electric field in n-type GaAs [8.31] showed a maximum velocity at 3 kV/cm, while more recent results [8.31a] find the peak at 4.5 kV/cm, implying a substantially slower scattering rate [8.31]. Measurements of transient reflectivity of transmission using laser pulses below 100 fs, have claimed extremely fast $\Gamma-L$ scattering times, as fast as 30 fs [8.88]. These measurements are done at carrier densities exceeding 10^{17} cm^{-3} and are dominated by carrier–carrier scattering. The rapid rate of carrier–carrier scattering in these experiments will tend to mask or possibly even change the intervalley scattering rates.

None of the measurements are convincing enough to be considered final. Most have significant experimental or analytic difficulties. The continuing interest in intervalley scattering rates stems from their importance in device modeling of GaAs-based devices. The important parameter for device modeling is not the absolute scattering rates as much as the relative scattering rates compared to the LO phonon emission rate. This "branching ratio" is what actually determines the fraction of carriers which scatter to the low mobility satellite valleys. We will see in the discussion below that comparing this branching ratio [i. e. $\tau_{\Gamma-L}^0/\tau_{po}$ — see (8.27, 28)] significantly reduces the experimental uncertainties since the experiments that have produced the fastest intervalley scattering rates often produce fast polar scattering rates as well.

Various experimental and theoretical results for intervalley scattering are summarized in Fig. 8.49 and Tables 8.3, 4. The solid curve in Fig. 8.49 is based on (8.9) and shows the dependence of the intervalley scattering from the Γ to the L valleys on the hot electron kinetic energy at low sample temperatures. Non-parabolic effects are neglected here which means the calculated curve will lose accuracy at energies substantially above the bottom of the L minimum. If this scattering is dominated by the emission of 28 meV zone edge optical phonons [8.85, 91, 92], then the threshold energy $E_{\Gamma-L} + \hbar\omega_{ze}$ in (8.9) is 324 meV. The density of states mass for one of the four equivalent L valleys is 0.22 m$_e$ [14] [8.19]. The $\Gamma-L$ deformation

---

[14] The size of the uncertainties associated with any effort to connect scattering rates and deformation potentials for transitions to the L valleys of GaAs must be greater than the uncertainty about the correct values for these masses. For example, *Blakemore* [8.28] has argued that the density of states mass for *all four* ellipsoids at L combined (i.e., including degeneracy factor) is only $0.56m_e$. This difference in the correct value of the density of states mass can produce an 80% difference in the scattering time derived from the deformation potential.

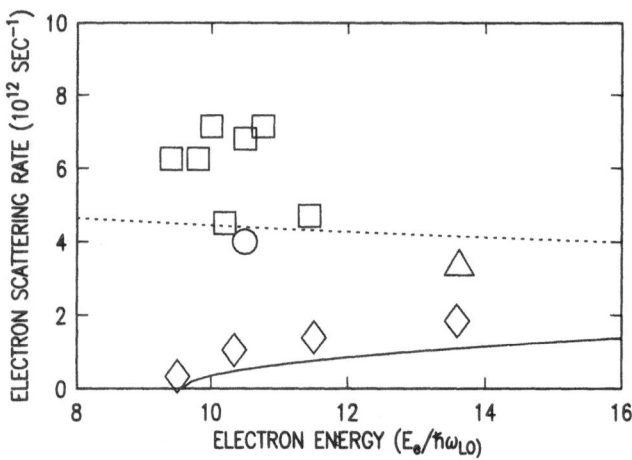

Fig. 8.49. The magnitude of the electron scattering rate arising from Γ to L valley transitions as calculated from (8.9) for a deformation potential of 3×10^8 eV cm^{-1} and a L valley density of states mass of $0.22m_e$ (solid line). The effects of nonparabolicity are ignored. The dotted line in the middle of the figure indicates the magnitude of the e–LO phonon emission rate from (8.7). Also shown are a number of experimentally measured scattering rates as described in the text. In particular, the 300 K results of *Shah* et al. [8.90] (triangles) have been scaled by a factor three to produce a low temperature deformation potential [8.85]. The other experimental results are all for low temperatures. Sources of other data: diamonds [8.50], squares [8.90], circles [8.51]

potential is 3.0×10^8 eV cm^{-1} using the recent calculation of *Zollner* et al. [8.85]. *Grinyaev* et al. [8.91, 92] obtained a value of 3.6×10^8 eV cm^{-1} for this deformation potential which would increase the calculated scattering rate in Fig. 8.49 by 44%.

In Fig. 8.49 the Γ–L intervalley scattering rates measured by *Ulbrich* et al. [8.50], *Shah* et al. [8.90], *Fasol* et al. [8.51] and *Mirlin* et al. [8.49] are shown by the open diamonds, triangle, squares and circle, respectively. These scattering rates, as well as those determined in other experiments are tabulated in Tables 8.3, 4. While the experimentally measured polar scattering rates show a factor of two variation, the Γ–L scattering rates vary by more than an order of magnitude. Although the calculations of *Zollner* et al. and *Grinyaev* et al. and the experiments of *Ulbrich* et al. are in reasonable agreement, this agreement should not be taken as definitive since the calculation of intervalley scattering rates requires many assumptions and approximations.

In spite of the wide range of measured intervalley scattering times, the kinetics of nonequilibrium electrons in GaAs are better understood than might appear by simple inspection of the published deformation potentials. A basic problem is that what is measured are scattering times or the ratio of intervalley to polar scattering rates. In most cases, the authors use different

Table 8.3. Experimental determinations of the Γ to L valley scattering rates in GaAs. The deformation potentials have been derived using the results of *Zollner* et al. [8.85] for the temperature dependence

Table 8.3. $\Gamma-L$ Scattering in GaAs

Reference	Technique	Remarks	Measured Scattering Time	Temperature [K]	$D_{\Gamma-L}$ [eV/cm], No Transverse Phonons	$\tau^0_{\Gamma-L}/\tau_{po}$ at $T = 0$ K Including TA Phonons*	$D_{\Gamma-L}$ [eV/cm], Including TA Phonons*
[8.31]	Velocity-field measurements	Monte Carlo simulations.	—	$\simeq 300$	10^9	3.7	8.0×10^8
[8.19]	Nonequilibrium phonon population	Ignores light holes. (Sect. 8.4.1b)	—	10	$<1.5 \times 10^8$	>50	$<1.5 \times 10^8$
[8.49]	Magnetic depolarization of hot electron–neutral acceptor luminescence	Magnetic field 5–10 T Assumes cyclotron mass. Also obtains $\tau_{po} = 100$ fs (Sect. 8.4.1b).	250 fs at $E_e = 0.385$ eV	2	$8.0 \pm 1.5 \times 10^8$	3.7 ± 1.5	$8.9 \pm 1.5 \times 10^8$
[8.90]	Time-resolved band edge photoluminescence	Monte Carlo simulation.	100 fs at $E_e = 0.50$ eV	300	$6.5 \pm 1.5 \times 10^8$	4.0 ± 1.8	$5.3 \pm 1.3 \times 10^8$
[8.50]	Intensity of hot electron–neutral acceptor luminescence	Assumes acceptor wavefunction and $\tau_{po} = 190$ fs. (Sect. 8.4.1b)	$\tau_{\Gamma-L} = 5\tau_{po}$ at $E_e = 0.38$ eV	2	$4.0 \pm 0.5 \times 10^8$	7.1 ± 1.5	$4.0 \pm 0.5 \times 10^8$
[8.88]	Time-resolved transmission	High carrier densities. Monte Carlo simulation for $D_{\Gamma-L}$. $Al_{0.3}Ga_{0.7}As$.	34 fs at $E_e = 0.13$ eV	300	9.0×10^8	—	—
[8.51]	Lineshape of hot electron–neutral acceptor luminescence	$16 \times 16\ k \cdot p$ calculation assumed exact. Also obtains $\tau_{po} = 132$ fs. (Sect. 8.4.1b)	$\tau_{\Gamma-L} = 150 \pm 75$ fs at $0.34 < E_e < 0.44$ eV	10	$9.5 \pm 1.5 \times 10^8$	1.9 ± 0.6	$9.5 \pm 1.5 \times 10^8$

Note: *Assumes $\tau_{\Gamma-L}$ at 0 K is three times that as 300 K. See [8.84]

Table 8.4. Scattering rates describing the Γ to X transitions in GaAs. The deformation potential for the results of *Ulbrich* et al. [8.50] is scaled from the results of *Mirlin* et al. [8.86]

Table 8.4. Γ–X Scattering in GaAs

Refer-ence	Technique	Measured Scattering Time	Temper-ature [K]	$D_{\Gamma-X}$[eV/cm]	$\tau^0_{\Gamma-X}/\tau_{po}$ at $T = 0$ K*	$\tau^0_{\Gamma-L}/\tau^0_{\Gamma-X}$ at $T = 0$ K*
[8.19]	Nonequilibrium phonon popu-lation	–	10	$1.0 \pm 0.1 \times 10^9$	1.1 ± 0.2	>45
[8.86]	Magnetic depolarization of hot electron– neutral acceptor luminescence	30 ± 10 fs at $E_e = 0.57$ eV	2	$1.5 \pm 0.3 \times 10^9$	0.5 ± 0.2	7.2 ± 4.0
[8.50]	Intensity of hot electron–neutral acceptor luminescence	$\tau_{\Gamma-X} = \tau_{po}$ at $E_e = 0.58$ eV	2	$6.0 \pm 1.0 \times 10^8$	1.8 ± 0.5	4.0 ± 1.5

Note: *See (8.28)

prescriptions and parameters to obtain deformation potentials $D_{\Gamma L}$ from their data. In Table 8.3, we present the measured scattering time and the stated $D_{\Gamma L}$ for some of the experiments. To convert to a meaningful $D_{\Gamma L}$, it is necessary to use a consistent band structure. Also, if the measurement is not at low temperatures, it is necessary to specify which phonons are assumed to be important because the scattering rate depends on the phonon mode occupation[15]. Examination of the various experimental methods reveals that there are two experiments which make the least demands in analyzing the data with respect to knowledge of detailed band structure parameters, external perturbations (high carrier densities, high temperatures, or large fields) and at the same time are sensitive to intervalley scattering. These are the experiments of *Shah* et al. [8.90] and *Ulbrich* et al. [8.50]. The experiments of *Shah* et al. were at 300 K and were analyzed including only scattering by zone edge longitudinal phonons to obtain a deformation potential of $D_{\Gamma L} = 6.5 \pm 1.5 \times 10^8$ eV/cm. Including

---

[15] For scattering from $k = 0$ to the L or X point, only longitudinal phonons can contribute due to symmetry considerations. Zone edge LA and LO phonons have energies of 24 and 30 meV, respectively. Including only these phonons leads to a factor of two increase in the intervalley scattering rate on raising the temperature to 300 K. In fact, intervalley scattering occurs between points not exactly at these high symmetry points, so TA phonons with energy about 9 meV can also contribute. At wavevectors near the intervalley scattering threshold, *Zollner* et al. [8.85] found that the deformation potential for TA phonons is about one third that of the longitudinal phonons. Because of the low energy of the TA phonons, they have a mode occupation of about 2.3 at 300 K, and make a large contribution to room temperature intervalley scattering. *Zollner* et al. [8.85] found that including the TA phonons means that the scattering rate changes by a factor of three on raising the temperature from 0 K to 300 K.

contributions from the TA phonons, as in the recent analysis of *Zollner* et al. [8.85], means that extrapolation to 0 K of the 1987 results of *Shah* will reduce $D_{\Gamma L}$ to 5.3 $\pm$ 1.3 eV/cm, which is in reasonable agreement with the 0 K results of *Ulbrich* et al., $D_{\Gamma L}$ = 4.0 $\pm$ 0.5 eV/cm. In addition, the Monte Carlo analysis of *Shah* et al. found a negligible contribution from the X valleys, even though the electrons generated from the heavy hole valence band had enough energy to scatter to an X valley. Including scattering to the X valley should decrease the derived $D_{\Gamma L}$ slightly, improving the agreement between these two experiments.

The above analysis suggests that $D_{\Gamma L} = 4 - 5 \times 10^8$ eV/cm. If this analysis is correct, then for an electron at energy E_e = 480 meV, $\tau_{\Gamma - L} = 2 - 3\tau_{po}$ at low temperatures. E_e = 480 meV is just below the threshold for scattering to the X valleys, so this relatively slow scattering time suggests that the contribution of the X valleys to intervalley scattering dynamics will be important since at excitation energies near 2.1 eV, Fig. 8.37 showed that almost all the excited carriers are transferred from the Γ valley to the zone boundary minima. There are fewer experiments which measure $\tau_{\Gamma - X}$ in the limit of relatively low carrier densities where carrier–carrier scattering effects can clearly be ignored. All of these experiments, as summarized in Table 8.4, agree that scattering to the X valleys is significantly faster than scattering to the L valleys, although there are substantial disagreements as to how much faster. If one compares intervalley scattering rates as derived by the same experimental technique, then the disagreement in $\tau_{\Gamma - L}^0 / \tau_{\Gamma - X}^0$ is almost a factor of two between *Mirlin* et al. [8.86] and *Ulbrich* et al. [8.50]. (Again the scattering times observed by *Mirlin* et al. are much faster than those measured by *Ulbrich* et al., but taking ratios with respect to the individual LO phonon or Γ–L scattering times substantially reduce the differences.) Recent work on hot electron devices by *Berthold* et al. [8.108] suggests that $\tau_{\Gamma - L}/\tau_{\Gamma - X} \sim 3$ at E_e = 0.58 eV. At this kinetic energy, *Mirlin* et al. obtain a ratio of 4, while *Ulbrich* et al. find a value of 2.4. Thus, although there is disagreement between these measurements as to the absolute magnitude, all the experiments agree that scattering to the X valleys dominates scattering to the L valleys.

The large discrepancies between the different measurements of the Γ–X scattering rates are enhanced when efforts are made to normalize the experimental results and discuss them in terms of the deformation potentials for these scattering processes. Again, a major contributor to this problem is the uncertainty about the parameters describing the band structure near the X point. These uncertainties include the degeneracy of the valley (6 if it is a camel-back structure and 3 if it is at the zone boundary), its effective mass, and the size of the effects of nonparabolicity. As in the case of the L valleys, published values for the total density of states mass for the X valleys differ by at least 50% as demonstrated by *Blakemore* [8.28]. As mentioned in the previous section, this uncertainty means that the $m_j^{3/2}$ dependence in (8.9) can produce an 80% uncertainty in the relation between the scattering rate and the deformation potential. In Table 8.4, we use the results of *Mirlin* et al. [8.86] to provide a connection between $\tau_{\Gamma-X}$ and $D_{\Gamma-X}$. The value of the deformation potential listed for the experimental results of *Ulbrich* et al. [8.50] for the scattering time is then scaled from the values of *Mirlin* et al.

8.6.4. Hole Relaxation, Carrier–Carrier Scattering and Screening

Our understanding of optical phonon and hot electron relaxation in the single electron limit, still exceeds our understanding of carrier–carrier scattering and hot hole relaxation processes. cw and time–resolved photoluminescence measurements have observed interesting differences between the relaxation of electrons and holes in GaAs [8.109, 110]. However, at present, the temporal evolution of the hole distribution functions has not yet been directly measured. While the hot holes should also relax through the emission of LO phonons until they are close to the top of the valence bands, these LO phonons have yet to be directly observed because the wavevectors are too large to be Raman active. The complicated valence bands of a material like GaAs mean that the hole relaxation can involve interband transitions as well as intraband processes. Theoretical estimates suggest that these are in fact quite important in the hot hole relaxation, but their particular contributions to these processes have yet to be isolated [8.84].

While it is clear from our discussion that carrier–carrier scattering processes play only an insignificant role in the initial relaxation of hot carriers for carrier densities below 10^{16} cm^{-1}, work remains to be done on how such processes come to dominate the hot carrier relaxation and on the nature of hot carrier relaxation in the presence of a high density of either hot or cold carriers. The experiments of *Tsang* and *Kash* [8.56] and *Kash* [8.94] have provided insight into carrier–carrier scattering rates, but have not been able to separate the different mechanisms, such as electron–electron, hole–hole, and electron–hole scattering. Here again the complications of the valence bands make both theory [8.41] and experimental interpretation difficult. A related questions is how carrier–carrier scattering is modified for nonequilibrium carriers or a hot plasma, as opposed to the relatively cool plasma in the experiments of *Kash*. Questions such as the effects on the carrier relaxation of the screening of the Fröhlich interaction in heavily doped materials also remain to be answered [8.111].

8.6.5 Concluding Remarks

We have seen that time-resolved Raman scattering is a useful probe of small wavevector nonequilibrium LO phonons. The argument that the generation of large populations of nonequilibrium phonons by relaxing carriers can in turn modify the hot carrier relaxation through the absorption of energy by the carriers from the hot phonons [8.6] raises an interesting challenge for phonon spectroscopy and Raman scattering in the direct observation of these phonons. The relevant LO phonon modes here have larger wavevectors (typically 10^7 cm^{-1}) than those which have up to now been observed with Raman scattering. Because of their wavevectors, they are Raman inactive in first order Raman scattering from bulk materials. The advent of microstructures where the Raman selection rules can be modified and the use of higher order Raman processes which see the full phonon density of states might provide a new means of studying these excitations and their populations in time resolved, pump-probe experiments.

In addition to the old questions which remain to be answered, the capabilities of Raman spectroscopy and hot luminescence studies combined with recent advances in the fabrication of heterostructure quantum wells and superlattices raise a variety of new questions. Our discussion in Sect. 8.5 provides a hint about the character of these new questions. The dynamics of hot carriers in a system where the electron wavefunctions are Bloch-like in only two dimensions and the phonons are localized in a single layer or at an interface can be significantly different from the dynamics in the bulk. Given the interest in using these structures in fast devices, the detailed characterization of these relaxation processes is a major challenge. In addition to the dependence of the carrier relaxation processes on simple structures, questions can also be raised on how external perturbations such as applied magnetic fields can modify the carrier relaxation. The fact that fields in the 10–20 T range can produce Landau level spacings comparable to the LO phonon splitting suggests that the carrier relaxation will be sensitive to such external perturbations [8.113].

The understanding of the relaxation of energetic carriers in materials such as GaAs requires experimental probes with subpicosecond time resolution, energy resolution at the meV level and the ability to clearly distinguish between different types of excitations such as electrons, holes, phonons and plasmons, etc. Raman spectroscopy and secondary emission processes such as the radiative recombination of hot electrons at neutral acceptors have been and will remain useful tools in the study of these problems.

Acknowledgements. We have benefited greatly from the knowledge, energy and wisdom of our collaborators over the last several years including Professors J. M. Hvam, S. S. Jha and R. G. Ulbrich. Dr. T. F. Kuech has provided many useful insights into the materials considerations associated with the studies described in this chapter. We are grateful to Professors E. Burstein and M. Cardona and Dr. M. Fischetti who supplied many valuable insights and corrected many misconceptions.

Note added in proof: Recent advances in Ti:Sapphire laser technology [8.114] may substantialy expand the range of laser frequencies and times available for time-resolved Raman spectroscopy. The extremely high peak powers available from these lasers will also improve the sensitivity of luminescence up-conversion experiments.

References

8.1   J.C.M. Maxwell: *Scientific Papers of James Clark Maxwell*, ed. by D. Nivan, (Dover, New York 1965) p. 505
8.2   G. Fasol, H.P. Hughes: Phys. Rev. B **33**, 2953 (1986)
8.3   R.G. Ulbrich: Solid State Electron. **21**, 51 (1978)
8.4   S.A. Lyon: J. Luminesc. **35**, 121 (1986)
8.5   J. Shah: Solid State Electron. **21**, 43 (1978)
8.6   P. Price: Physica **134** B, 164 (1985)
8.7   M. Heiblum, D.C. Thomas, C.M. Knoedler, M.I. Nathan: Appl. Phys. Lett. **47**, 1105 (1985)
8.8   J.R. Hayes, A.F.J. Levi, W. Wiegmann: Phys. Rev. Lett. **54**, 1570 (1985)
8.9   G. Wartmann, M. Kemmlar, D. von der Linde: Phys. Rev. B **30**, 4850 (1984)
8.10   C.V. Shank, B.P. Zakharchenya (eds.): *Spectroscopy of Non-Equilibrium Electrons and Phonons* (North Holland, Amsterdam 1992)
8.11   D. Strauch, B. Dorner: J. Phys. Condens. Matter **2**, 1457 (1990)
8.12   M. Fischetti: private communication (1990); M.L. Cohen, J.R. Chelikowski: *Electronic Structure and Optical Properties of semiconductors*, 2nd ed., Springer Ser. Solid-state Sci. Vol. 75 (Springer, Berlin, Heidelberg 1989) p. 103
8.13   R.A. Cowley: Adv. Phys. **12**, 421 (1963)
8.14   B.K. Ridley: In *Quantum Processes in Semiconductors*, 2nd ed. (Clarendon, Oxford 1988)
8.15   E.M. Conwell, M.O. Vassell: IEEE Trans. Electron. Devices ED **13**, 22 (1966)
8.15a S. Zollner, M. Cardona: *Properties of Gallium Arsenide* (IEE, London 1990) p. 126
8.16   H.J. Lee, J. Basinski, L.Y. Juravel, J.C. Wooley: Can. J. Phys. C **11**, 233 (1978)
8.17   N.M. Ashcroft, N.D. Mermin: *Solid State Physics* (Holt, Rinehart and Winston, New York 1976) pp. 548–559
8.18   W. Fawcett, A.D. Boardman, S. Swain: J. Phys. Chem. Solids **131**, 1963 (1970)
8.19   C.L. Collins, P.Y. Yu: Phys. Rev. B **30**, 4501 (1984)
8.20   J.F. Young, K. Wan, H.M. van Driel: Solid-State Electron. **31**, 455 (1988)
8.21   J.F. Young, K. Wan, D.J. Lockwood, J.-M. Baribeau, O. Othonos, H.M. van Driel: in SPIE Proc. Ultrafast Laser Probe Phenomena in Bulk and Microstructure Semiconductors II, ed. by R. Alfano, (SPIE, Bellingham 1988) Vol. 942, p. 124
8.22   W. Pötz, P. Kocevar: Phys. Rev. B **28**, 7040 (1983)
8.23   J. Shah, R.C.C. Leite, J.F. Scott: Solid State Commun. **8**, 1089 (1970)
8.24   C. Weisbuch: Solid-State Electron. **21**, 179 (1978)
8.25   R.E. Nahory: Phys. Rev. **178**, 1293 (1969)
8.26   R.W. Shaw: Phys. Rev. B **3**, 3283 (1971)
8.27   D.L. Rode, S. Knight: Phys. Rev. B **3**, 2534 (1971)
8.28   J.S. Blakemore: J. Appl. Phys. **53**, R 123 (1982)
8.29   M. Cardona: Basic Concepts and Instrumentation, in *Light Scattering in Solids II*, ed. by M. Cardona, G. Güntherodt; Topics Appl. Phys. **50** (Springer, Berlin, Heidelberg 1982) p. 49
8.30   B.K. Ridley, G.B. Watkins: Proc. Phys. Soc. **78**, 293 (1961)
8.31   M. Littlejohn, J. Hauser, T. Glisson: J. Appl. Phys. **48**, 4587 (1987)
8.31a W.T. Masselink, N. Braslau, W.I. Wang, S.L. Wright: Appl. Phys. Lett. **51**, 1533 (1987)
8.32   D. von der Linde, J. Kuhl, H. Klingenberg: Phys. Rev. Lett. **44**, 1505 (1980)
8.33   J. Menéndez, M. Cardona: Phys. Rev. B **29**, 2051 (1984)
8.34   M.H. Brodsky: In *Light Scattering in Solids I, Introductory Concepts*, 2nd Ed., ed. by M. Cardona, Topics Appl. Phys. **8**, (Springer, Berlin, Heidelberg 1983) p. 205

8.35 P. Parayanthal, F.H. Pollak: Phys. Rev. Lett. **52**, 1822 (1984)
8.36 P.G. Klemens: Phys. Rev. **148**, 845 (1966)
8.37 A.Z. Genack, L. Ye, C.B. Roxlo: In SPIE Proc. Ultrafast Laser Phenomena in Bulk and Microstructure Semiconductors II, ed. by R. Alfano, (SPIE, Bellingham 1988) Vol. 942, p. 130
8.38 A. Mooradian, G.L. Wright: Phys. Rev. Lett. **16**, 999 (1966)
9.39 D. Olego, M. Cardona: Phys. Rev. B **23**, 6592 (1981)
8.40 J.A. Kash, J.C. Tsang: Solid-State Electron. **31**, 419 (1988)
8.41 J.F. Young, N.L. Henry, P.J. Kelly: Solid-State Electron. **32**, 1567 (1989)
8.42 S. Adachi: Phys. Rev. B **38**, 12966 (1988); P. Lautenschlager, M. Garriga, S. Logothetidis, M. Cardona: Phys. Rev. B **35**, 9174 (1987)
8.43 M. Chandrasekhar and F.H. Pollak, Phys. Rev. B **15**, 2127 (1977)
8.44 A. Pinczuk, E. Burstein: In *Light Scattering in Solids I*, 2nd ed., ed. by M. Cardona, Topics Appl. Phys. **8** (Springer, Berlin, Heidelberg 1983), p. 23
8.45 J.S. Weiner, P.Y. Yu: Solid State Commun. **50**, 493 (1984)
8.46 S.S. Jha, J.A. Kash, J.C. Tsang: Phys. Rev. B **34**, 5498 (1986)
8.47 M. Cardona: In *Light Scattering in Solids V*, ed. by M. Cardona, G. Güntherodt, Topics Appl. Phys. **66** (Springer, Berlin, Heidelberg 1989)
8.48 Y. Huang, P.Y. Yu: Solid State Commun. **63**, 109 (1987)
8.49 D.N. Mirlin, L.P. Nikitin, I.I. Reshina, V.F. Sapega: Solid State Commun. **37**, 757 (1981)
8.50 R.G. Ulbrich, J.A. Kash, J.C. Tsang: Phys. Rev. Lett. **62**, 949 (1989)
8.51 G. Fasol, W. Hackenberg, H.P. Hughes, K. Ploog, E. Bauser, M. Kano: Phys. Rev. B **41**, 1461 (1990)
8.52 W.P. Dumke: Phys. Rev. **132**, 1998 (1963)
8.53 A. Baldereschi, N.O. Lipari: Phys. Rev. B **8**, 2697 (1973)
8.54 A. Mooradian, H.Y. Fan: Phys. Rev. **148**, 873 (1966)
8.55 D. von der Linde, J. Kuhl, E. Rosengart: J. Lumines. **24/25**, 675 (1981)
8.56 J.C. Tsang, J.A. Kash: Phys. Rev. B **34**, 6003 (1986)
8.57 J.C. Tsang, J.A. Kash, J.M. Hvam: J. de Phys. C **7**, 235 (1985)
8.58 J.A. Kash, J.M. Hvam, J.C. Tsang, T.F. Kuech: Phys. Rev. B **38**, 5776 (1988)
8.59 M. Cardona, M.H. Grimsditch, D. Olego: In *Light Scattering in Solids*, ed. by J.L. Birman, H.Z. Cummins, K.K. Rebane (Plenum, New York 1979) p. 249
8.60 P.A. Temple, C.E. Hathaway: Phys. Rev. B **7**, 3685 (1973)
8.61 J.C. Tsang: In *Light Scattering in Solids V*, ed. by M. Cardona, G. Günterodt, Topics Appl. Phys. **66** (Springer, Berlin, Heidelberg 1989) pp. 233–282
8.62 E. Anastassakis, Y.S. Raptis: J. Appl. Phys. **57**, 920 (1985)
8.63 A. Anastassiado, Y.S. Raptis, E. Anastassakis: J. Appl. Phys. **59**, 627 (1986)
8.64 J. Shah, T.C. Damen, B. Deveaud, D. Block: Appl. Phys. Lett. **50**, 1307 (1987)
8.65 H. Shen, F.H. Pollak: Appl. Phys. Lett. **45**, 692 (1984)
8.66 J.A. Kash, J.C. Tsang, J.M. Hvam: Phys. Rev. Lett. **54**, 2151 (1985)
8.67 J.A. Kash: In *Ultrafast Laser Probe Phenomena in Bulk and Microstructure Semiconductors*, ed. by R. Alfano (SPIE, Bellingham 1988) Vol. 942, p. 143
8.68 D.E. Aspnes, A.A. Studna: Phys. Rev. B **27**, 985 (1983)
8.69 J.C. Tsang, J.A. Kash, S.S. Jha: Physica **134** B, 184 (1985)
8.70 J.A. Kash, S.S. Jha, J.C. Tsang: Phys. Rev. Lett. **58**, 1869 (1987)
8.71 J.A. Kash, S.S. Jha, J.C. Tsang: Phys. Rev. Lett. **60**, 864 (1988)
8.72 K.T. Tsen, H. Morkoç: Phys. Rev. B **38**, 5615 (1988)
8.73 K.T. Tsen, H. Morkoç: Phys. Rev. B **37**, 7137 (1988)
8.74 M. Ilegems, G.L. Pearson: Phys. Rev. B **1**, 1576 (1970)
8.75 I.F. Chang, S.S. Mitra: Adv. Phys. **20**, 539 (1971)
8.76 B. Jusserand, J. Sapriel: Phys. Rev. B **24**, 7194 (1981)
8.77 J. Kuhl, W.E. Bron: Solid State Commun. **49**, 935 (1984)
8.78 M. Cardona: Private communication
8.78a H.D. Fuchs, C.H. Grein, R.I. Devlen, J. Kuhl, M. Cardona, Phys. Rev. in press
8.79 J.J. LePore, J. Appl. Phys. **51**, 6441 (1980)

8.80   K.J. Nash, M.S. Skolnick: Phys. Rev. Lett. **60**, 863 (1988)
8.81   A. Amith: Phys. Rev. **139**, A1624 (1964)
8.82   D. Olego, M. Cardona: Solid State Commun. **39**, 1071 (1981)
8.83   R. Bray, K. Wan: J. Luminescence **30**, 375 (1985)
8.84   B.P. Zakharchenya, D.N. Mirlin, V.I. Perel, I.I. Reshina: Sov. Phys. Usp. **25**, 143 (1982)
8.85   S. Zollner, M. Cardona, S. Gopalan: To be published
8.86   D.N. Mirlin, I. Karlick, V.F. Sapega: Solid State Commun. **65**, 171 (1988)
8.87   A. Baldereschi, N.O. Lipari: Phys. Rev. B **8**, 1525 (1974)
8.88   M.J. Rosker, F.W. Wise, C.L. Tang: Solid-State Electron. **31**, 439 (1988)
8.89   J. Pozelas, A. Reklaitis: Solid-State Electron. **23**, 927 (1980)
8.90   J. Shah, B. Deveaud, T.C. Damen, W.T. Tsang. A.C. Gossard, P. Lugli: Phys. Rev. Lett. **59**, 2222 (1987)
8.91   S.N. Grinyaev, G.F. Karavaev, V.G. Tyuterev, V.A. Chaldeyshev: Sov. Phys. Solid State **30**, 1586 (1988)
8.92   S.N. Grinyaev, G.F. Karavaev, V.G. Tyuterev: Sov. Phys. Semicond. **23**, 905 (1989)
8.93   S. Zollner, S. Gopalan, M. Cardona: Appl. Phys. Lett. **54**, 614 (1989)
8.94   J.A. Kash: Phys. Rev. B **40**, 3455 (1989)
8.95   H. Nather, L.G. Quagliano: J. Luminesc. **30**, 50 (1985)
8.96   J.F. Young, K. Wan: Phys. Rev. B**35**, 2544 (1987)
8.97   J.F. Young, K. Wan, A.J. Springthorpe, P. Mandeville: Phys. Rev. B **36**, 1316 (1987)
8.98   K.T. Tsen, O.F. Sankey: Phys. Rev. B **37**, 4321 (1988)
8.99   C.L. Collins, P.Y. Yu: Solid State Commun. **51**, 123 (1984)
8.100 A. Pinczuk, G. Abstreiter: In *Light Scattering in Solids V*, ed. by M. Cardona, G. Güntherodt, Topics Appl. Phys. **66**, (Springer, Berlin, Heidelberg 1989) p. 153
8.101 B. Jusserand, M. Cardona: In *Light Scattering in Solids V*, ed. by M. Cardona, G. Güntherodt, Topics Appl. Phys. **66** (Springer, Berlin, Heidelberg 1989) p. 49
8.102 M. Tatham, J.F. Ryan, C.T. Foxon: Phys. Rev. Lett. **63**, 1637 (1989)
8.103 D.Y. Oberli, D.P. Wake, M.V. Klein, J. Clem, T. Henderson, H. Morkoç: Phys. Rev. Lett. **59**, 696 (1987)
8.104 K.T. Tsen, R.P. Joshi, D.K. Ferry, H. Morkoc: Phys. Rev. B **39**, 1446 (1989)
8.105 P.J. Colwell, M.V. Klein: Solid State Commun. **8**, 2095 (1970)
8.106 A.K. Sood, J. Menendez, M. Cardona, K. Ploog: Phys. Rev. Lett. **59**, 2114 (1985)
8.107 B.P. Zakharchenya, P.S. Kopev, D.N. Mirlin, D.G. Polakov, I.I. Reshina, V.F. Sapega, A.A. Sirenko: Solid State Commun. **69**, 203 (1989)
8.108 K. Berthold, A.F.J. Levi, J. Walker, R.J. Malik: Appl. Phys. Lett. **54**, 813 (1989)
8.109 J. Shah, A. Pinczuk, A.C. Gossard, W. Wiegmann: Phys. Rev. Lett. **54**, 2045 (1985)
8.110 H.J. Polland, W.W. Rühle, J. Kuhl, K. Ploog, K. Fujiwara, T. Nakayama: Proceedings of the 18th International Conference of the Physics of Semiconductors, ed. by O. Engström, (World Scientific, Singapore 1986) p. 1315
8.111 S. Das Sarma, J.K. Jain, R. Jalabert: Solid-State Electron. **31**, 695 (1988)
8.112 W.E. Bron, J. Kuhl, B.K. Rhee: Phys. Rev. B **34**, 6961 (1986)
8.113 T. Ruf, C. Trallero, R.T. Phillips, M. Cardona: Solid State Commun. **72**, 67 (1989)
8.114 J. Goodberlet, J. Wang, J.G. Fujimoto, P.A. Schulz: Opt. Lett. **14**, 1125 (1989); D.E. Spence, P.N. Kean, W. Sibbett: Opt. Lett. **16**, 42 (1991)

Subject Index

Contents of Brevious Volumes

Light Scattering in Solids I

Light Scattering in Solids II

Light Scattering in Solids III

Light Scattering in Solids IV

Light Scattering in Solids V